Practical Civil Engineering

Practical Civil Engineering

P.K. Jayasree, K Balan and V Rani

CRC Press
Taylor & Francis Group
Boca Raton London New York

CRC Press is an imprint of the
Taylor & Francis Group, an **informa** business

First edition published 2021
by CRC Press
6000 Broken Sound Parkway NW, Suite 300, Boca Raton, FL 33487-2742

and by CRC Press
2 Park Square, Milton Park, Abingdon, Oxon, OX14 4RN

© 2021 Taylor & Francis Group, LLC

CRC Press is an imprint of Taylor & Francis Group, LLC

ISBN: 978-1-138-03313-9 (hbk)
ISBN: 978-0-429-09481-1 (ebk)

Typeset in Times
by codeMantra

Contents

22.5   Renewable Energy Assets..595
    22.5.1 Wind Power 595
    22.5.2 Hydropower 596
    22.5.3 Solar Energy 596
    22.5.4 Geothermal Energy 596
    22.5.5 Bioenergy 596
22.6   Introduction to Green Buildings..597
    22.6.1 Goals of Green Buildings 597
22.7   Green Building Foundations 597
    22.7.1 Green Foundation Transition 598
22.8   Ecological Design...598
    22.8.1 Applications in Design 598
22.9   Assessing High-Performance Green Buildings...599
22.10 Assessment of Green Buildings...599
22.11 Green Building Rating Systems 599
    22.11.1 Indian Green Building Council 599
    22.11.2 Green Rating for Integrated Habitat Assessment 600
    22.11.3 Bureau of Energy Efficiency..600
22.12 The Green Building Design Process 600
22.13 The Sustainable Site and Landscape 601
22.14 Energy and Carbon Footprint Reduction 601
    22.14.1 Ways to Lessen Carbon Impression..602
22.15 Built Environment Hydrologic Cycle 602
22.16 LCA of Building Materials and Products...603
22.17 Indoor Environmental Quality 603
22.18 Green Building Economics 603
    22.18.1 Economic Benefits of Green Buildings 604
22.19 Sustainable Construction...604
    22.19.1 Sustainable Materials in Construction...604

Bibliography 607
Index..613

Authors

P K Jayasree is a Professor of Civil Engineering at College of Engineering Trivandrum, Thiruvananthapuram, Kerala. She has 20 years of experience in her academic and research career. She is a doctorate holder and also has postdoctoral experience from IIT Madras. She has a number of laurels to her credit like Dr. T. S. Ramanatha Ayyar Endowment in 1995, M. Tech. degree first rank holder in 1997 from University of Kerala, Researcher of the Year Award in 2012 instituted by the Centre for Engineering Research and Development, Government of Kerala, and Prof. V K M John National Award for the Best Engineering College Teacher from Kerala in 2015. She is a reviewer of a number of reputed international journals like ASCE International Journal of Geomechanics, ASCE Journal of Materials in Civil Engineering, Journal of natural fibers, Case Studies in Construction Materials, Construction and Building Materials and Ground Improvement. She is a life member of a number of professional bodies in India like Indian Society for Technical Education, Indian Geotechnical Society, and Institution of Engineers, India. She has undertaken several sponsored research projects in her institute by receiving funds from various funding agencies and the results from these works are published in various technical journals all over the world. She has authored over 100 journal and conference papers and two books – (i) Gabion retaining walls and (ii) Design and Engineering.

K. Balan Former Professor of Civil Engineering, College of Engineering Trivandrum, Thiruvananthapuram, Kerala, obtained doctoral degree holder from Indian Institute of Technology, Delhi in 1996. Along with teaching and research, he has served efficiently in various administrative posts like First Director or Governing Body Member in National Coir Research and Management Institute, Government of Kerala, Trivandrum; Director of Centre for Engineering Research and Development (CERD); Professor in Charge of Trivandrum Engineering Science and Technology Research Park (TrEST Research Park); R&D Manager in Alyaf Industrial Co. Ltd., Dammam, KSA; Director of Centre for Development of Coir Technology, Trivandrum, Kerala; Managing Director of The Kerala State Coir Corporation Ltd., Alleppey, Kerala; and Assistant Town Planner, Greater Cochin Development Authority, Kerala. He is a recipient of awards like Prof. V.K.M. John National Award of Indian Society for Technical Education for the best engineering college teacher of Kerala State, 2012, and the Lifetime Achievement Award for contributions made in the R&D of Geosynthetics (natural fiber geotextiles) in India by Central Board of Irrigation and Power New Delhi and International Geosynthetics Society. Apart from authoring over 100 technical publications in journals and conference proceedings, he has authored books entitled Coir Geotextiles-Emerging Trends and Coir Geotextiles for Sustainable Infrastructure. He is the Editorial Board Member, Indian Journal on Geosynthetics and Ground Improvement, published by International Geosynthetics Society (India). He is also the Founder Chairman of Indian Geotechnical Society Trivandrum Chapter, and also the Executive Committee member of Indian Geotechnical Society, International Geosynthetics Society (Indian Chapter), and Indian Society for Training Development, Trivandrum Chapter.

V. Rani is an Associate Professor in Civil Engineering, Marian College of Engineering, Trivandrum. She has a teaching experience of over 16 years and an industrial experience 6 years. She has proven her ability as a teacher, handling more than 15 courses over these years. She has guided over more than 20 project works over these years. She is a life member in various professional bodies like Indian Society for Technical Education, Institution of Engineers India, and Indian Geotechnical Society. She is the joint secretary of Indian Geotechnical Society Trivandrum Chapter in Kerala, India. She is a very active researcher carrying out a number of sponsored projects by obtaining funds from various research funding agencies. She has proved her ability in research by publishing the outcomes of her research in various international journals and also presenting them on various international platforms. She has authored more than twenty technical research papers.

1 Introduction to Civil Engineering

1.1 SCOPE OF CIVIL ENGINEERING

Engineering discipline is a vast area that offers unlimited specialization. It can be defined as the systematic presentation of scientific principles employed for the advantage of the public. It is the career in which the understanding of mathematics and natural sciences, achieved by learning, understanding, and rehearsal, is supported by wisdom to acquire methods to consume the material and forces of the environment for the utility of society in a cautious way. The conventional divisions of engineering are civil, mechanical, and electrical engineering. Other than these basic divisions, there are numerous other branches. To name a few, computer engineering, aeronautical engineering, industrial engineering, chemical engineering, marine engineering, automobile engineering, polymer engineering, textile engineering, ceramic engineering, and so on.

As engineers with specific skills established their logical knowledge and professional organizations, the opportunity of civil engineering was confined to construction, which was the first to utilize scientific principles. For example, the theory of cantilever was first explored by Galileo Galilei. Civil engineering is an engineering specialty and is considered to be the original one among all branches of engineering. The Institution of Civil Engineers (ICE), UK, in its royal charter has defined Civil Engineering as "the art of directing the great sources of power in nature for use and convenience of the man." It includes planning, design, erection as well as repairs of the built structures like bridges, buildings, roads, dams, canals, railways, reservoirs, towers, spillways, and chimneys. However, Civil Engineering alone cannot sustain independently. It has to join hands with other branches of Engineering. In fact, different branches of engineering should balance and also complement each other, though each one including Civil Engineering has its own distinct and precise roles in the development and progress of the society.

An engineer means the one who converts the concepts into realism using the developed scientific information and accessible resources for the profit of the community. Civil engineers have one of the world's most significant jobs. They work with structures. Civil engineers are constantly engaged in developing approaches and amenities to survive with most of the earth's severe worries. In the present condition of polluted air, rotten cities, roadways, bridges, congested airports and highways, contaminated rivers, lakes and streams, the civil engineer is being called on to propose solutions that are feasible, sustainable, and economical. He will be responsible for refining the class of living in these areas and much more. Civil engineers are the principal manipulators of present day's cutting-edge technology from all branches of engineering. They utilize the up-to-date perceptions in the computer-aided design during planning, designing, and construction.

Various types of software are being used in project scheduling and cost constraints. It can be justifiably believed that we cannot have current advancements without civil engineers. Thus the key scope of civil engineering is planning, designing, estimating, contracting, and management of diverse construction events for the benefit of the society.

1.2 RESPONSIBILITIES AND ROLE OF A CIVIL ENGINEER

Civil engineers are considered to be the engineers of progress. A building is designed by civil engineers by attributing a group of structural elements that sustain the architectural 3-D scattering. A civil engineer plans a highway by forming a plane that can sufficiently withstand the stress and strains of the moving vehicles and of the centrifugal forces at the curves. Civil engineers design water resource system with provision for the water intake at a suitable source, a water transmission arrangement to carry the water to a water treatment plant, and a water distribution network to supply the water to the end user. Waste water disposal and waste water treatment plant that empties the depleted water away from the user, treats it to neutralize the pathogen carriers and harmful organic materials, and carefully empties it into a suitable carrier without harmfully impairing its natural environment are also designed by a civil engineer.

Civil engineers are the leading performers of development during serene situations. They also design and construct the essential infrastructure on which fiscal progress of a nation relies.

The key responsibilities of a civil engineer can be attributed as given:

- To gather information through exploration and assessment of property and resources.
- To choose construction procedures which result in the accomplishment of the project with optimal period and rate.
- To design the project in such a manner that it allows utmost benefits by scientific and engineering codes.
- To confirm the usage of materials, manpower, and equipment present at the location of a project, which cuts the rates down.
- To do surveying and leveling with survey instruments.
- To make maps, plans, and all other appropriate sketches.
- To do soil investigation.
- To carry out scheduling and overseeing the project.
- To formulate estimates, cost analysis, and tenders with specifications and conditions of contract for the project.

- To design a structure by means of recent approaches to structural analysis.
- To collaborate with the architect for attaining the needed exterior and overall aesthetics of the project.
- To do the estimation of land and building.
- To take up all duties and at times risks associated with the implementation of a project.

The role of the civil engineer with respect to the level he is working can be enlisted as follows:

i. **When he designs a building**: Civil engineer plans the building, according to different architectural principles and to meet the basic needs of the proposed building like residential, commercial, or industrial. He should consider the orientation, aspect, prospect, roominess, privacy, ventilation and lighting, flexibility and circulation of the building and also keeps in mind the basic services like plumbing, electrical fitting, water supply, and drainage system.

ii. **At a sector or colony level**: At a sector or colony level, a greater emphasis is laid on social infrastructure like a community hall, dispensary, primary school, temple, planning of open spaces.

iii. **At a city or town level**: At a city or town level, a civil engineer works on the city/town development plan. He thinks and plans the circulation pattern, because, on the basis of the circulation pattern, the form of the city is decided. This is followed by zoning and thus the city is divided into various zones like residential zone, commercial zone, agricultural zone, industrial zone, vegetation zone, etc.

iv. **At a regional level**: At a regional level, the civil engineers work on accessibility among different centers. For achieving this they plan out various modes of transportation. They also plan out the various complexes for the government and government employees.

v. **At a state level**: At a state level, the civil engineers are working for the connectivity of various regions, basically the state highways. They also plan out various hydraulic structures and also go for watershed development and management.

vi. **At a national level**: The responsibilities of civil engineer increase manifolds at the national level. Interconnectivity of various states takes his prime concern and he works for national highways, railways and also develops airports, docks, and harbors. Apart from this, he also has to plan for several constructions which are important for a country from the strategic point of view.

1.3 HISTORY OF CIVIL ENGINEERING

Earlier, engineering was divided into just two heads as shown in Figure 1.1.

Military engineering: During the former half of the 18th century, military engineers were mainly responsible for

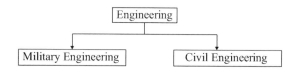

FIGURE 1.1 Engineering was divided into just two heads.

large-scale construction work. Military engineering involved works like design as well as construction of bridges and roads, topographical map preparation, the construction of docks and ports, etc., which are needed for the easy movement of military regiments.

However, presently, military engineering deals with employing engineering sciences for military purposes. It has grown into a specialized field of engineering and has resulted in separate engineering subdisciplines such as electrical engineering to provide solutions to all problems of radio, telephone, telegraph, and various other types of communications, chemical engineering for the research, analysis, and the development of propellants and the Signal Corps ordnance, also includes mechanical engineering for the development of guns.

Civil engineering: The term civil engineering was conceived in the 18th century. It describes the engineering effort that was done by ordinary civilians for nonmilitary uses. In other words, the various applications of scientific knowledge done for refining the standard and value of living of the civilians are categorized under civil engineering. Hence, it can be said that "Civilization is through Civil Engineering."

Civil engineering has always played a vital role in human life. This oldest branch of engineering, civil engineering, had begun to commence during the period 4000 and 2000 BC. It is believed to have developed in ancient Egypt and Mesopotamia. When man began to put an end to his nomadic life, there developed a necessity for the construction of a home for his protection and shelter. Along with this, transportation became inevitable, since it was necessary to carry the construction materials. The Pyramids of Egypt (2700–2500 BC) are still regarded as the first large construction made. Some similar large prehistoric civil engineering structures are Apian Way of Rome (312 BC), the Great Wall of China (220 BC), and the Parthenon in Greece (447–438 BC). Historical evidence of construction of civil engineering structures like aqueduct, bridges, dams, reservoirs, and roads by the Romans has also been traced. Archimedes principle is considered as one of the ancient instances of a scientific methodology applied to mathematical and physical problems. It was applied to civil engineering in the third century BC and gives the explanation of buoyancy and some other practical problems like an Archimedes' screw.

Civil engineering and architecture were unable to be distinguished until modern times. The terms "engineer" and "architect" were used to denote both the individuals. John Smeaton, who constructed the Eddystone Lighthouse, was the first civil engineer. In 1771, Smeaton and his colleagues constituted the first society for the civil engineering profession called the Smeatonian Society of Civil Engineers. The ICE was later established in London in 1818. The distinguished engineer Thomas Telford was elected as its first president in the year 1820. The institution obtained a Royal Charter in the year 1828 and thereafter civil engineering was formally identified as a profession. The ICE, London, recognizes civil engineering as "the art of directing the great sources of power in nature for the convenience of man, as the means of production of traffic in states, both for external and internal trade." This consists of road construction, bridges, canals, aqueducts, docks, and aquatic transportation for internal communications and exchange. Also, additional construction of harbors, ports, lighthouses and breakwaters, drainage systems in cities and town has become essential. Norwich University, which was founded in the year 1819 by Captain Alden Partridge, became the first private college in the United States to teach Civil Engineering. Rensselaer Polytechnic Institute was the first to award in Civil Engineering in the United States in 1835. Nora Stanton Blatch was the first woman to obtain a degree in Civil Engineering in 1905 from Cornell University, founded by Ezra Cornell in 1865 Ithaca, New York. Civil Engineering is an extensive profession, and it includes many different specialized subfields like structures, geology, hydrology, soils, geography, environment, material science, mechanics, and others.

1.4 BRANCHES OF CIVIL ENGINEERING

Civil Engineering is considered to be ancient and looked upon as mostly traditional and unchanging. However, challenges of people's life and works demand that Civil Engineering should go in tune with these changes and modern needs. It basically deals with various types of construction works like buildings, foundations, roads, hydraulic structures, etc., which involves the operation of planning, analysis and design, execution, and finally maintenance. Civil Engineering extends through many disciplines which are dependent and interact with each other. The branches of Civil Engineering are illustrated in Figure 1.2 and are explained in detail as follows.

1.4.1 STRUCTURAL ENGINEERING

Structural engineering is the branch of civil engineering which involves structural design using perfect structural analysis by calculating stresses in structural components, considering the behavior of concrete and loads according to the type of structure.

This branch of engineering includes the study of the following:

a. Design of Reinforced Cement Concrete (RCC) structures like multistoried buildings, retaining walls, water tanks, bridges, tunnels, public buildings, residential buildings, etc.
b. Design of steel structures like railway platform, factory sheds, steel bridges, etc.
c. Design of earthquake-resistant structures.

1.4.2 BUILDING CONSTRUCTION

This discipline includes the construction of different structures, the different types of buildings, various building components like foundations, brickwork, doors and windows, RCC work for beams, slabs, columns, lintels, floors, roofs, etc. It also involves the study of various engineering materials like cement, steel, bricks, stones, timber, paints, glass, etc.

1.4.3 GEOTECHNICAL AND FOUNDATION ENGINEERING

Geotechnical and foundation engineering is concerned mainly with soil mechanics and foundation of structures. The behavior of soils under the combined effect of forces and soil water interactions is studied under this discipline of civil engineering. This principle is applied to the foundation designs, earth dams, retaining walls, clay liners, and geosynthetics for containment of waste materials. The fundamental goals of a geotechnical engineer are the design of foundations, the

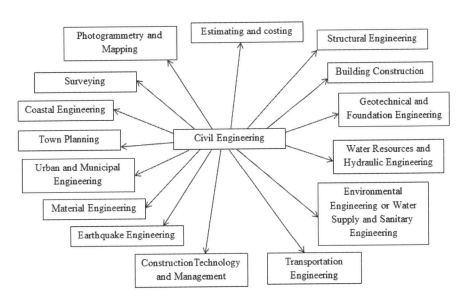

FIGURE 1.2 Branches of Civil Engineering.

design of temporary excavation supports, route selection for highways and railways, site selection and design of landfill, prevention of groundwater contamination and waste disposal. A geotechnical engineer is therefore required to monitor the field as well as the laboratory examinations. This aids him to explore the engineering properties of the soils and rocks at the site and in the further analysis. The stability of slope of soil, testing of soils for its bearing capacity, pile and other types of foundation and footing construction and design are also studied in this discipline. The foundation has become a vital part of modern high-rise building that is being constructed in cities due to the restriction of space. The role of structural and geotechnical engineers is a necessity for the design of foundation of tall buildings.

1.4.4 WATER RESOURCES AND HYDRAULIC ENGINEERING

Water is vital for the existence of mankind, animals, and plants. Water resources engineering relates to the gathering, management, and exploitation of the natural water from rivers, streams, lakes, reservoirs, and from the underground. It combines open channel flow, hydraulics, hydrology, irrigation and water power engineering, flood control and erosion, geology, meteorology, conservation of soil, and management of resources. This branch of civil engineering predicts and manages the quantitative and qualitative aspect of water available in both surface and subsurface sources. Its applications include the design of urban storm-sewer systems, flood forecasting, and the management of the urban water supply.

Hydraulic engineering deals with the flow of water and its distribution. It includes the application of fluid mechanics to analyze the flow of water through a closed medium like a pipe or in an open channel. The primary concern of civil engineers is open channel flow. The applications of hydraulic engineering include the design of hydraulic structures like dams, breakwaters, and sewage conduits and the management of waterways like flood protection and erosion protection. It also comprises environmental management likes forecasting the details regarding mixing or transport of pollutants in the flow of water. Water supply, irrigation, navigation, and hydro-electric power development are some of the applications of water resources engineering which involves the use of water for constructive purposes. Recently, the sustainability concept of concern for the preservation of our natural surroundings has amplified the significance of water resources engineering.

The following subbranches are covered under this discipline:

i. Fluid mechanics, which deals with the behavior of all liquids and gases which can be considered more or less incompressible, when subjected to pressure changes, frictional resistance, flow through various outlets, impacts of jets, etc.
ii. Hydrology, which deals with the study of sources of water, measurement, and study of rainfall, runoff, flood and flood control.
iii. Irrigation engineering, which deals with designing of hydraulic structures like dams, canals, barrages, reservoirs, etc., and water power engineering.

1.4.5 ENVIRONMENTAL ENGINEERING

Environmental engineering involves with the treatment and purification of drinking water, purification of waste from water chemically and biologically, air purification, and the identification of remedial or decontamination measures for sites contaminated due to waste disposal. Air pollution, transport of pollutants, purification of water, treatment of waste water, management of solid state and its treatment, and management of hazardous waste are some of the topics covered by environmental engineering.

Environmental engineers must actively participate in industrial ecology, reduction of pollution, and green engineering. Environmental engineering is related to the collection of information and assessing the impact due to some proposed civil engineering constructions like hydro-projects, chemical projects, thermal projects, etc., on the environment.

Sanitary engineering is a part of environmental engineering nowadays. However, conventionally sanitary engineering did not include management of hazardous waste and works involving environmental remediation which are presently capped under Environmental Engineering. Some alternate terms used for sanitary engineering are environmental health engineering and public health engineering.

1.4.6 TRANSPORTATION ENGINEERING

Transportation engineering is related to the movement of people, goods, and vehicles safely and efficiently without consuming any extra time with the help of different ways of transportation. This includes design, construction, and maintenance of transportation infrastructures like roads, traffic islands, railways, highways, canals, ports, airports, and heavy mass transit. It mainly includes transportation planning, transportation design, traffic engineering, and some concepts of pavement engineering, urban engineering, queuing theory, infrastructure management, and Intelligent Transportation System.

The following subbranches are covered under this discipline:

i. **Highway engineering**: Highway engineering deals with highways, covering the topics of planning, designing, and construction. It also covers questions pertaining to geometrics, materials of the structural design of highways.
ii. **Railway engineering**: Railway engineering involves planning and construction of a surface railway, setting an alignment for the tunnels, devising the signaling systems of meeting the traffic objectives, construction of station buildings and yards.
iii. **Waterways engineering**: Waterways engineering deals with the transportation of people and goods in

vehicles that float upon water. It deals with the construction and development of docks and harbors.

iv. **Airport engineering**: Airport engineering is related to constructing, developing, and maintaining various elements of airports like runways, taxiways, airport pavements, etc.

v. **Traffic engineering**: Traffic engineering is that branch of engineering which deals with the initial planning, the design of geometry, and the operations of traffic related to streets, roadway and highway networks, terminals, etc., for the accomplishment of effective and convenient movement of people, as well as goods in a safe manner. It uses principles of engineering to analyze the problems of transportation by bearing in mind the psychological habits of the commuters and obtain the most optimal solution.

1.4.7 CONSTRUCTION TECHNOLOGY AND MANAGEMENT

It is very much related to environmental and structural engineering. A suitable environment is built by the construction of buildings and other structures. Codes of building bye-laws make sure of the fact that construction is good and sound. Construction engineering thus consists of various techniques of construction for different materials adopted for various site conditions. Construction machinery, management of labor, materials, methods of construction adopted relating to the site, the fund available, all are under the purview of this discipline. The management division of this subject takes into account the managerial aspect of the construction side. It deals with the various project scheduling techniques like PERT, CPM, etc.

1.4.8 EARTHQUAKE ENGINEERING

Earthquake engineering deals with the design of prominent structures to withstand the forces induced to them at the site of construction. It is actually a secondary branch of structural engineering. The primary aims of earthquake engineering can be categorized as follows:

a. Force and acceleration on a building due to an earthquake, i.e., to know how the structures interact with the vibrating ground.

b. Building response characteristics of different shaped structures.

c. The drift of the building and other high structures under the action of earthquake lateral force.

d. Design, construction, maintenance of a building and other structures to resist fully the hazardous earthquake force.

Earthquake engineering has great importance for further study and research due to the drainage of weak structures, tall building, failure of dams, etc., in earthquake-prone areas in different parts of the world.

1.4.9 MATERIALS ENGINEERING

Materials engineering also called as material science is a prominent discipline in civil engineering. Materials engineering comprises understanding the characteristic behavior of construction materials such as cement, sands, concrete, metals, lime, bricks, stones, ferrous and nonferrous materials, timber, paints, plastics, etc. It also concentrates on the behavior of materials with higher strength like steel, light-weight materials like aluminum and alloys, and other specialized materials like polyvinyl chloride, damp proofing materials, termite proofing materials, glass and carbon fibres, etc. Most of the materials of this branch are building materials.

1.4.10 URBAN AND MUNICIPAL ENGINEERING

Urban and municipal engineering is a branch of civil engineering which is mainly dealing with the municipal infrastructure. This branch of civil engineering includes specifications, design, construction, and maintenance of water supply network systems, sewage systems, streets, lighting of streets, sidewalks, bicycle paths, public parks, storage depots for bulk materials which are mainly used for maintenance and public works, and management of municipal solid waste and its disposal. It also comprises the civil engineering components of the local distribution networks for telecommunication and electrical services, especially in the case of underground utility networks like conduits and access chambers. It also deals with planning and implementation of garbage collection and service networks in a locality or a city. Municipal engineering, however, mainly throws light on the management of these infrastructure services and networks, because they are mostly constructed concurrently, and controlled by the same authority over time.

1.4.11 TOWN PLANNING

Town planning includes planning and arranging different units of a town in such a way that the town accomplishes the importance of a living organism. It comprises identifying and suggesting various ways and means to be implemented for the improvements of existing towns or those needed for the extension of towns. It also aids in identifying the most efficient ways of the situation of town considering the use of its land and surrounding environments. In town planning, areas of towns are divided into various zones such as residential zone, commercial zone, industrial zone, etc., town should be planned using certain principles of town planning like providing green belts on the fringes of the town, development of building of various categories following the zoning and building bye-laws, sufficient public buildings and recreation centers, proper roads and transport facilities, etc.

1.4.12 COASTAL ENGINEERING

Coastal engineering mainly deals with the management of the coastal areas. In certain jurisdictions, the terms coastal protection and sea management mean guarding the land against flooding and erosion, respectively. Coastal management has developed into a prevalent area nowadays, as the field has itself expanded to comprise methods and systems which allow erosion to claim land.

1.4.13 SURVEYING

This is another branch of civil engineering, which is essential in different fields of civil engineering. It is defined as the art of determination of relative positions of points above or below the surface of the earth by means of linear and angular measurements, distance, elevation, and direction. Surveying further may be defined as the science and technique for accurately measuring or calculating the three-dimensional position of points on earth as well as the distances and angles between them. The relative altitude of points is ascertained by leveling which is another branch of surveying. The maps of the different points of an area are drawn on a plane or in a vertical section with different scales which are essential for different engineering projects like roads, highways, railroad, water supply, irrigation, transmission lines, bridges, buildings, dams, and reservoirs. It is required for the location of different geological structures, showing the geographical boundaries of districts, states, and countries.

1.4.14 PHOTOGRAMMETRY AND MAPPING

Photogrammetry and mapping are those branches of civil engineering which deals with the accurate measurement of points on the earth's surface to get dependable information for the location and design of engineering projects. For this purpose, the current practices are the use of aerial and terrestrial photogrammetry, satellites, and computer processing of the photographic images obtained from the former two sources. Radio and TV signals received from the satellites, scans by laser, and sonic beams are converted to maps. This gives highly accurate measurements for construction of highways and dams, boring tunnels, locating flood control and irrigation projects, etc.

1.4.15 ESTIMATING AND COSTING

Estimating and costing deals with the financial aspect of construction. Before execution of any construction project, a civil engineer must know about the total cost of the project. It involves in the calculation of the various quantities of materials required and subsequently the total material cost, the number, and type of laborers required and subsequently the total labor cost, the cost of machineries and equipment (if bought), or the rent of machineries and equipment (if hired) along with their operating cost.

1.5 CIVIL ENGINEERING FOR INFRASTRUCTURE DEVELOPMENT OF A COUNTRY

The importance of infrastructure for continued economic development is well acknowledged. Insufficient and incompetent infrastructure always results in high transaction cost.

This inhibits the economy from attaining full growth potential irrespective of the advancement in other areas.

Physical infrastructure covers power, communication, transportation, etc. It facilitates growth through its to-and-fro linkage. Social infrastructure includes the supply of water, disposal of sewage, sanitation, health, and education. All these are in the nature of primary services and make an impression on the life quality. The observable symbol of deficits in capacity and inadequacies includes power failures, increasing congested roads, shortage of drinking water, and long waiting lists for installation of telephones. This exemplifies the broadening distance between supply and demand of infrastructure. It also raises queries regarding the sustainability of future economic growth.

The effectiveness of private sector involvement in the infrastructure growth depends on their ability to commercialize the development projects. The retrieval of their investments can be made through a scheme of implementation of user charges. There is also a prospect for public-private partnerships to contribute more and bridge the infrastructure gap in a country.

1.6 CIVIL ENGINEERING STRUCTURES

An engineering structure roughly means something that is built or constructed. It is also defined as a system of connected parts used to support a load. The major structures of interest to civil engineers are buildings, bridges, dams, walls, cable structures, towers, shells, etc. These structures are made of one or more solid elements organized in such a way that their components, as well as the whole structure, are able to hold themselves during loading and unloading without considerable changes in geometry.

The four major objectives that must be satisfied by the design of a structure are as follows:

Utility: The structure should be able to satisfy the requirements for performance.
Safety: The structure should be able to carry the loads coming on it safely.
Economy: The structure should be cost-effective in construction, material, and cost.
Aesthetics: The structure should have a pleasing appearance.

1.6.1 BUILDING

A building is an artificial permanent or temporary stationary structure consisting of roof and walls built for the shelter of mankind or animals or goods from climatic agencies. Examples of a building are house, factory, sty, cow shed, chicken coop, farmhouse, skyscraper, warehouse, etc. Buildings are built in diverse sizes and shapes for various functions. The type of a building depends on a wide range of factors starting from the availability of building materials to weather conditions along with other factors like ground conditions, land prices, specific uses, and aesthetic reasons. The biggest building in the world, the New Century Global

FIGURE 1.3 The New Century Global Centre, Chengdu, China, which is the biggest building in the world.

FIGURE 1.5 The longest bridge in the world is located in China and is the world famous Danyang Kunshan Grand Bridge.

FIGURE 1.4 The 828 m tall Burj Khalifa in Dubai which has been the tallest building in the world since 2008.

FIGURE 1.6 The Three Gorges Dam which is a hydroelectric dam that spans the Yangtze River in China.

Centre, Chengdu, China, is shown in Figure 1.3. Figure 1.4 shows the tallest building in the world, Burj Khalifa in Dubai, which is 828 m high and was built in 2008.

1.6.2 Bridges

Bridges are structures constructed to span physical spaces like a water body, road, or valley. They are mainly built for the function of providing a route over any hindrance. The design of bridges depends on various factors like the type of the ground where the bridge is constructed and anchored, the function of the bridge, the materials used for construction, and the funds available for construction. The longest bridge in the world is located in China and is the world famous Danyang Kunshan Grand Bridge shown in Figure 1.5. Spanning a length of 164,800 m, this high-speed rail bridge connects Beijing and Shanghai in China. This bridge was opened for public in 2010. Construction of bridges is a great challenge to the civil engineer. Bridges are of different types like arch bridges, cable-stayed bridges, box girder bridges, cantilever bridges, suspension bridges, truss bridges, and so on.

1.6.3 Dams

Dams are enormous obstructions which are constructed across flowing water bodies like streams and rivers. They are built to restrain and direct the flow of water for human use like the generation of electricity and irrigation. This detention of water creates reservoirs behind the dams. Figure 1.6 shows the Three Gorges Dam, which is a hydroelectric dam spanning the Yangtze River in China. The Three Gorges Dam is the world's largest power station in terms of its installed capacity (22,500 MW).

The following are the reasons to build a dam:

- **Power generation**: For the generation of electricity.
- **Irrigation**: Diversion dams are built for this purpose, which will obstruct the natural due course of the river ultimately diverting the water away to a different place with water demand.
- **Control flooding**: Detention dams are generally constructed to either retard or terminate the water flow in a river.

There are different types of dams classified based on function and construction material. Based on the material of construction, dams are categorized as earthen dam, masonry dam, and concrete dam. Based on the function of dams, they are categorized as storage dam, flood control dam, diversion dam, and coffer dam. A storage dam is constructed for water storage.

The stored water may be used for drinking, irrigation, or hydroelectric power generation. A flood control dam is constructed for the temporary storage of flood water and then to discharge it gently so that the downstream side is shielded against the destructive effects of floods. A diversion dam is constructed to divert water from a river to a channel for irrigation. A cofferdam is a temporary structure constructed to divert water so that a safe construction area can be prepared during the construction of bridges and other underwater constructions.

1.6.4 ROADS

Roads are used for the transportation of men and material from one part of a country to another part. Roads may be classified as freeways, arterials, collectors, and local roads. For commuting through a road, a pavement is constructed. The components of a road pavement are—wearing coat, base course, subbase course, and subgrade. Before the commencement of a new road construction or for widening an existing road, a traffic survey has to be carried out. This is done to gather the details like the direction of movement of vehicles, traffic density, type of soil, type of vehicles, origin, and destination of vehicles and so on.

Figure 1.7 shows Australia's Highway 1 which is basically a network of highways that is circumnavigating the Australian continent, which joins all state capitals of the mainland. It is the longest of the national highways in the world which has a total length of almost 14,500 km. It is longer than the Trans-Siberian Highway (over 11,000 km) and the Trans-Canada Highway (8030 km) which are considered as longest highways in the Asian and American continents, respectively.

1.6.5 RUNWAYS

A runway is constructed in an airport. It is a rectangular strip of land which is constructed for the takeoff and landing of an aircraft. It can be a natural surface or man-made one (concrete, asphalt, or a mixture including both). It is generally paved with shoulders provided on its either side. The shoulders act as safety zones in case an aircraft drifts out of the runway during takeoff or landing.

Runways are generally named using numbers from 01 to 36. This number represents the magnetic azimuth of the runway's heading measured in deca degrees. For example, a runway numbered as 09 points toward east (90°). While runway 18 points toward south (180°), runway 27 points west (270°). A runway pointing toward north (360°) is always designated as runway 36. It is never measured as 0° nor designated as runway 0. Denver International Airport's 16R/34L runway is designated as the longest commercial runway (16,000 ft) in North America.

1.6.6 RAILWAYS

Railways play an important role in the development of a nation. Railway network is very essential for the transportation of men and material. Figure 1.8 shows the world's longest high-speed rail line in China which creates a link between Beijing and the southern metropolis of Guangzhou.

A train runs on a railway track. A track consists of two rails is placed at a certain distance apart on a transverse member called as sleeper. Sleepers are usually made of good quality timber. Nowadays, concrete sleepers are being used. The lateral distance between two rails is called a gauge. If the distance between two rails is 1676 mm, the track is called as Broad gauge track. If the width is 1000 mm, it is known as the meter gauge track. A layer of broken stones (ballast) is provided about 20 cm below the rails which is called as ballast. Its function is to absorb shocks from the fast moving train.

1.6.7 WATER TANKS

A water tank is a container to store water. The stored water may be later distributed for drinking purpose, irrigation, or for putting out fire. A water tank can be placed either at the

FIGURE 1.7 Australia's Highway 1 which is a network of highways that circumnavigate the Australian continent, which joins all mainland state capitals.

FIGURE 1.8 The world's longest high-speed rail line in China which creates a link between Beijing and the southern metropolis of Guangzhou.

FIGURE 1.9 A typical example of water tank.

ground level or on an elevated stage depending on the head available for flow. It can be made of steel, concrete, or plastics. It can be of rectangular or circular in shape. Tanks are also used for rainwater harvesting. Rainwater runoff from roof tops can be collected in a rainwater tank (also known as a rain barrel) via rain gutters. The stored rainwater may later be used for agriculture, in washing machines, watering gardens, washing cars, flushing toilets, and so on. A typical example of elevated drinking water tank is shown in Figure 1.9.

1.6.8 Retaining Walls

The walls constructed to retain soil are called retaining walls. They made of concrete or masonry. They are usually built to protect soil at a certain elevation from erosion, for making gardens in sloping grounds, to construct approach roads for bridges, etc. The main criterion checked while designing a retaining wall is that the wall should be strong enough to resist the lateral pressure, which is exerted by the soil by its self-weight or friction. The lateral earth pressure is determined using Rankine's theory or Coulomb's theory.

The lateral earth pressure depends on soil parameters and height of the wall. The soil parameters required for the computation of lateral earth pressure are unit weight, cohesion, and angle of internal friction. Based on the mode of resisting the lateral earth pressure, the retaining walls are classified as gravity retaining walls, cantilevered retaining walls, reinforced soil retaining walls, counterfort retaining walls, crib walls, buttress walls, etc. Depending on the material used for the construction of retaining walls, they are categorized as dry rubble walls, masonry walls, concrete walls, gabion retaining walls, etc.

1.6.9 Towers

A very tall structure which is not intended for living is called as a tower. There are different types of towers such as transmission tower, clock tower, bell tower, communication tower, radio tower, etc. A tower is built for definite purposes as follows:

Bell tower: for hanging bells in churches
Communication tower: for transmission of communication signals
Radio tower: for transmission of radio signals
Tourist tower: as a tourist attraction
Transmission tower: for distribution of electricity

Steel sections like angle section, I section, channel section, square section, etc., are used for the construction of towers. They are connected at junctions by riveting, welding, or using nuts and bolts. Figure 1.10 shows transmission towers.

1.6.10 Chimneys

A chimney is a tall structure constructed for the purpose of discharging smoke or hot gases which emanate from a furnace or a fireplace to the outside atmosphere. It is usually kept in a vertical position so that gases can pass smoothly. Chimneys can be found in buildings, factories, ships, steam locomotives, brick kilns, etc. The gases which are discharged

FIGURE 1.10 Transmission towers.

FIGURE 1.11   A gigantic twin chimney, once a part of old Croydon Power station.

FIGURE 1.12   Pipeline in Alaska.

from factories are usually pollutant gases. Since they are emitted out through very tall towers to the higher portion of the atmosphere, the surrounding area is not polluted. Figure 1.11 shows a gigantic twin chimney, once a part of old Croydon Power station, London.

A chimney may be built of circular, square, or rectangular cross sections having smooth finish inside. In ordinary buildings, the top of the chimney stack is to be kept at least one meter above the roof level. By safety and stability considerations, it is also stipulated that the total height of the chimney shaft should not be greater than 12 times external diameter at the base (for chimneys having circular cross section) or 10 times the least lateral dimension at the base (for chimneys of rectangular cross section). While designing a chimney, wind pressure should also be taken into account.

1.6.11 PIPELINES

Pipelines are constructed for the transportation of flowing goods like oil or water from one place to another through a pipe. Considering nearly 120 countries in the world, there is a total of about 3.5 million km of pipeline altogether.

Pipelines are commonly used for the transportation of fuels including oil, natural gas, and biofuels, crude and refined petroleum, as well as other fluids like sewage, slurry, and water. It is also used for carrying water for irrigation or drinking over long distances. Solid capsules can be transported using pneumatic tubes that use compressed air. Figure 1.12 shows pipeline in Alaska.

1.6.12 CANALS

It is an artificial channel or waterway constructed for drainage or navigation purpose or for land irrigation on the ground to carry water over long distances from the source of water to the field. Figure 1.13 shows a canal at Westbury court garden.

The main sources of water are tanks, reservoirs, or rivers. Canals are usually trapezoidal in section. They are used

FIGURE 1.13   A canal at Westbury court garden.

to carry water to tanks for water supply, to power house for power generation, to field for irrigation, etc.

1.7 CODES AND SPECIFICATIONS

Civil engineering plays a vital role in our day-to-day life. In various fields like agriculture, housing, industry, transport, power, irrigation, health, or education, construction programs are interlinked in large measures in all sectors. Since a large amount of finance is involved, it is very necessary that wastage in construction has to be avoided and thus optimum results are assured. National codes are formulated in various countries to give recommendations for planning, design, and construction of structures. They represent wisdom and knowledge of engineers gained over the years. They are also intermittently reviewed and revised to include the output of current research. It should be remembered that the codes are not meant to replace basic engineering knowledge.

There are several agencies which formulate and prepare the codes for Civil Engineers in various countries. The American Standards for Testing Materials (ASTM) in US, the British Standards (BS) in UK, the Eurocode in Europe generally, the Australian Standards in Australia, the Japanese Society of Civil Engineers guidelines in Japan, the Bureau of Indian Standards in India are some of them.

1.7.1 FUNCTIONS OF CODES

1. Codes safeguard sufficient structural safety by stipulating definite vital and minimum requirements for a safe design.
2. Codes simplify the job of designing by making available simple formulae for complicated design.
3. Codes ensure consistency among different designers.
4. Codes have legal validity.

1.7.2 ASTM INTERNATIONAL

ASTM International is an international standards organization which mainly involves in developing and publishing deliberate agreement of technical standards which are used for a wide range of systems, raw materials, services, and products. The standards were known by the name the American Society for Testing and Materials (ASTM) until 2001. The total number of ASTM voluntary consensus standards counts up to 12,575. The headquarters of ASTM is located in Philadelphia. ASTM is considered to be the largest developer of standards in the world. Thousands of technical committees are grouped under ASTM, which are formed by members from around the world. More than 12,000 standards are developed and maintained by the members of these committees collectively. The ASTM Standards, which is the annual book of ASTM Standards, is published by ASTM International every year in print, online versions, and compact disc.

1.7.3 BRITISH STANDARDS

The standards developed by BSI (British Standard Institute) Group are called as BS. The BSI is incorporated under a Royal Charter (which was formally designated as the National Standards Body for the UK). The standards developed by BSI are always named in a particular format as: British Standard AAAA[-B]:CCCC where AAAA indicates the number specifying the standard, B indicates the number of the part of the standard (usually the standards are divided into several parts), and CCCC represents the year in which the standard was developed. BSI Group has more than 27,000 active standards at present. The "Kitemark" is used to indicate certification by BSI.

1.7.4 EUROCODES

The Eurocodes consist of a set of coordinated technical rules which were formulated by the European Committee for Standardization. They are developed especially for the countries grouped under European Union for the structural design of construction works. There are ten Eurocodes for Civil Engineering subdivided into 58 parts. They cover the basics of design, action of structural forces, elemental structural design in steel, concrete, composite concrete and steel, masonry, aluminum and timber, along with seismic and geotechnical design.

Eurocodes are considered as a means to substantiate acquiescence with the requirement for safety, stability, and strength established by European Union law. They form a fundamental basis for engineering and construction contract specification. They form an outline for forming synchronized technical specification for building-related products.

Eurocodes became compulsory for the civil works of Europe by March 2010. They are envisioned to turn out to be the de facto civil sector standards. They therefore substitute the prevailing national building codes circulated by national standard bodies. In addition, each country should develop a National Annex to Eurocodes which will require referencing for a specific country (for example, the UK National Annex). However, the civil engineering projects in the private sector do not follow Eurocodes. Instead they still widely use the existing national codes.

The ten sections developed and published by the Eurocodes are (1) EN 1990: Basis of structural design, (2) EN 1991: (Eurocode 1) Actions on structures, (3) EN 1992: (Eurocode 2) Design of concrete structures, (4) EN 1993: (Eurocode 3) Design of steel structures, (5) EN 1994: (Eurocode 4) Design of composite steel and concrete structures, (6) EN 1995: (Eurocode 5) Design of timber structures, (7) EN 1996: (Eurocode 6) Design of masonry structures, (8) EN 1997: (Eurocode 7) Geotechnical design, (9) EN 1998: (Eurocode 8) Design of structures for earthquake resistance, and (10) EN 1999: (Eurocode 9) Design of aluminum structures.

1.7.5 AUSTRALIAN STANDARDS

Standards Australia issues the Australian standards. The Civil Engineering Group develops Standards which deals with (1) building materials stipulations, (2) methods to test the construction materials properties, (3) construction and design rules, (4) key or glossaries of terms, and (5) other matters which are related to civil engineering like standard fire tests and various other rules which prevails while carrying out demolition of old buildings. Every few years, the Standards are reviewed and restructured. Standards typically stipulate minimum necessities needed for the functioning and security of specific products related to construction.

2 Units, Measurements, and Symbols

2.1 SYSTEMS OF MEASUREMENTS

The most commonly used and universally recognized four systems of measurements are as follows:
1. C.G.S. units; 2. F.P.S. units; 3. M.K.S. units; 4. S.I. units

2.1.1 THE C.G.S. SYSTEM

The C.G.S. system of units, also known as the common scientific system, derives its name from the units for length, mass, and time in the system: the centimeter, the gram, and the second, respectively.

2.1.2 THE M.K.S. SYSTEM

The M.K.S system of units, also derives its name from the units for length, mass, and time used in the system: the meter, the kilogram, and the second, respectively.

2.1.3 THE F.P.S. SYSTEM

The F.P.S. system or American Engineering system of units, used commonly in the United States, again derives its name from the units for length, mass, and time in the system: the foot, the pound, and the second, respectively.

2.2 THE S.I. SYSTEM

To facilitate the exchange of scientific information, it was necessary to establish a single system of units of measurement that would be acceptable internationally. A metric system is now generally employed around the world. This was first devised in France, 1960, and it is known (from French: Système international d'unités) as the S.I. system. In this system, units are divided into fundamental units or base units and derived units, which together form what is called "the coherent system of SI units." This system of units is now being used in many countries.

2.2.1 FUNDAMENTAL UNITS

One of the most important operations in engineering is the measurement of physical quantities. Every quantity is measured in terms of some arbitrary, but internationally accepted units, called fundamental units. All the physical quantities, met with in Applied Mechanics and Strength of Materials, are expressed in terms of three fundamental quantities, i.e., Length, Mass, Time, and the corresponding fundamental units are meter (m), kilogram (kg), and second (s), respectively.

2.2.1.1 Meter

The international meter may be defined as the shortest distance (at $0°C$) between two parallel lines engraved upon the polished surfaces of the Platinum-Iridium bar, kept at the International Bureau of Weights and Measures at Sevres near Paris.

2.2.1.2 Kilogram

The International kilogram may be defined as the mass of the Platinum-Iridium cylinder, which is also kept at the International Bureau of Weights and Measures at Sevres near Paris.

2.2.1.3 Second

The fundamental unit of time for all the four systems is second, which is $1/(24 \times 60 \times 60) = 1/86,400$th of the mean solar day. A solar day may be defined as the time interval between the instants at which the sun crosses the meridian on two consecutive days. This value varies throughout the year. The average of all the solar days, of one year, is called the mean solar day.

2.2.2 DERIVED UNITS

Sometimes, physical quantities are expressed in terms of other units, which are derived from fundamental units and they are known as derived units, e.g., units of area, velocity, acceleration, pressure, etc.

The base units of the SI system are those from which all other units are derived. For completeness, all seven base units are given in Table 2.1, derived units are given in Table 2.2, supplementary units are given in Table 2.3, and the prefixes that are commonly added to the names of base units given in Table 2.4.

TABLE 2.1
Base Units in SI

Quantity	Name of Unit	Symbol
Length	meter	m
Mass	kilogram	kg
Time	second	s
Electric current	Ampere	A
Temperature	Kelvin	K
Luminous intensity	candela	cd
Amount of substance	mole	mol

TABLE 2.2
Derived Units Commonly Used

Quantity	Unit	Symbol
Acceleration	m/s^2	a
Angle	radian	-
Angular acceleration	rad/s^2	-
Angular velocity	rad/s	-
Area	m^2	-
Mass density	kg/m^3	ρ
Energy	Nm	J
Force	Newton	N
Impulse	Ns	-
Moment	Nm	-
Velocity	m/s	-
Volume	m^3	-
Work	Joule	J
Power	Watt	W

TABLE 2.3
Supplementary Units in SI

Quantity	Unit	Symbol
Plane angle	Radian	Rad
Solid angle	Steridian	Sr

TABLE 2.4
SI prefixes and symbols

Multiplication Factor	Prefix	Symbol
10^{18}	Exa	E
10^{15}	peta	P
10^{12}	tera	T
10^9	giga	G
10^6	mega	M
10^3	kilo	K
10^2	hecta	H
10^1	decca	da
10^{-1}	deci	d
10^{-2}	centi	c
10^{-3}	milli	m
10^{-6}	micro	μ
10^{-9}	nano	n
10^{-12}	pico	p
10^{-15}	femto	f
10^{-18}	atto	a

2.3 RULES FOR SI UNITS

The 11th General Conference on Weights and Measures (Paris, 1960) recommended only the fundamental and derived units of S.I. units. But it did not elaborate the rules for the usage of these units. Later on, many scientists and engineers held a number of meetings for the style and usage of S.I. units. Some of the rules to follow are:

Abbreviations not approved for SI units are not allowed; only standard unit, symbols, quantities, and names are used.

Correct: s, cm^3, m/s
Incorrect: sec, cc, mps

Symbols are unaltered when expressing plural quantities.

Correct: l = 75 cm
Incorrect: l = 75 cms

In order to represent multiplication of units a space or centered dot is used; a slash, horizontal line, or negative exponent is used for division. If a slash is used, it must not appear more than once in the same expression unless parentheses are used, to avoid ambiguities.

Correct: The speed of sound is about 344 m s^{-1} (meters per second) or m/s or m.s^{-1}.
Incorrect: The speed of sound is about 344 ms^{-1} (reciprocal milliseconds).

There is a space between the unit symbol and the numerical value, except for superscripts to represent plane angles (degrees, arc minutes, arc seconds).

Correct: (a) A 25 kg sphere.
                (b) An angle of 2°3′4″
Incorrect: (a) A 25-kg sphere
                (b) A 25kg sphere
                (c) An angle of 2 ° 3 ′ 4 ″

For numbers having 5 or more digits, the digits should be placed in groups of three separated by spaces (instead of commas) counting both to the left and right of the decimal point.

2.4 MEASURES OF LENGTH

Among the various measures of length, the meter is used for ordinary measurements, the centimeter or millimeter for reckoning very small distances or measurements, and the kilometer for roads or great distances. The commonly used measures of length are given below. Table 2.5 gives the conversion factors related to the different measures of length.

1 megameter (Mm) = 1,000,000 meters
1 hecto kilometer = 100,000 meters
1 myriameter (mym) = 10,000 meters
1 kilometer (km) = 1000 meters
1 hectometer (hm) = 100 meters
1 dekameter (dkm or dam) = 10 meters
1 meter (m) = 10 dm
1 decimeter (dm) = 1/10 meters = 10 cm
1 centimeter (cm) = 1 / 100 meters = 10 millimeters
1 millimeter (mm) = 1 / 1000 meters

TABLE 2.5
Length Conversion Factors

Quantity	To Convert From	To	Multiply by
Length	Feet	Centimeters	30.48
		Inches	12
		Kilometers	3.048×10^{-4}
		Meters	0.3048
	Inches	Centimeters	2.540
		Feet	0.08333
		Kilometers	2.54×10^{-5}
		Meters	0.0254
	Kilometers	Feet	3.2808×10^{3}
		Meters	1000
		Yards	1.0936×10^{3}
	Meters	Feet	3.2808
		Inches	39.370
	Micrometers	Centimeters	1.0×10^{-3}
		Feet	3.2808×10^{-6}
		Inches	3.9370×10^{-5}
		Meters	1.0×10^{-6}
		Millimeters	0.001
		Nanometers	1000
	Millimeters	Centimeters	0.1
		Inches	0.03937
		Meters	0.001
		Micrometers	1000
	Nanometers	Angstrom units	10
		Centimeters	1.0×10^{-7}
		Inches	3.937×10^{-8}
		Micrometers	0.001
		Millimeters	1.0×10^{-6}
	Yards	Centimeters	91.44
		Meters	0.9144

1 micron or micrometer (μm) = 1 / 1000 millimeter
1 millimicron = one-millionth of a millimeter

2.5 MEASURES OF WEIGHTS

The various measures of weight are listed below. The gram is defined as the weight of one cubic centimeter or one milliliter of water; the kilogram is defined as the weight of one liter of water; and the ton is the weight of one cubic meter of water. The gram is used in weighing silver, gold, letters, and small quantities of things. The kilogram is used by grocers and is the primary unit of mass. The ton is used for heavy articles. The conversion factors for the various measures of weight are given in Table 2.6.

1 ton (t) = 1,000,000 grams = 1000 kilograms
1 quintal (q) = 100,000 grams = 100 kilograms
1 myriagram (myg) = 10,000 grams = 10 kilograms
1 kilogram (kg) = 1000 grams
1 hectogram (hg) = 100 grams = 1 / 10 kilograms
1 dekagram (dkg or dag) = 10 grams = 1 / 100 kilograms
1 gram (g) = 1 gram = 1 / 1000 kilograms = 1000 mg

TABLE 2.6
Weight Conversion Factors

Quantity	To Convert From	To	Multiply by
Mass	Grains	Grams	0.064799
		Milligrams	64.799
		Tons (metric)	6.4799×10^{-8}
	Grams	Grains	15.432
		Kilograms	0.001
		Micrograms	1×10^{6}
		Tons	1×10^{-6}
	Kilograms	Grains	1.5432×10^{4}
		Pounds	70.932
		Tons (metric)	0.001
	Megagrams	Tons (metric)	1.0
	Milligrams	Grains	0.01543
		Grams	1.0×10^{-3}

1 decigram (dg) = 1 / 10 gram = 10 cg
1 centigram (cg) = 1 / 100 gram = 10 mg
1 milligram (mg) = 1 / 1000 gram
1 microgram (μg) = 1 / 100,000 gram

2.6 MEASURES OF SURFACE AREA

The Square meter is the primary unit of ordinary surfaces or small areas. The are or Square decameter, hectare, or Square hectometer are the units of land measures. The commonly used measures of surface area are given below. The conversion factors related to surface area are listed in Table 2.7.

1 Square kilometer (km²) = 1,000,000 Square meters = 100 hectares (ha)
1 Square hectometer (hm²) = 10,000 Square meters = 100 ares = 1 ha
1 Square decameter (dkm² or dam²) = 100 Square meters = 1 are
1 Square meter (m²) = 100 Square dm
1 Square decimeter (dm²) = 1 / 100 Square meters = 100 Square centimeters
1 Square centimeter (cm²) = 1 / 10,000 Square meters = 100 Square millimeters
centare or centiare = Square meter
hectare = Square hectometer
are = Square decameter

2.7 CUBIC MEASURES

The commonly used measures of volume are given below. The conversion factors related to volume are listed in Table 2.8.

1000 cubic millimeter = 1 cubic cm
1000 cubic centimeters = 1 cubic dm
1000 cubic decimeters = 1 cubic meter

The conversion factors for the derived units like velocity, pressure, and density are given in Table 2.9.

TABLE 2.7
Area Conversion Factors

Quantity	To Convert From	To	Multiply by
Area	Acres	Square feet	4.356×10^4
		Square kilometers	4.0469×10^{-3}
		Square meters	4.0469×10^3
	Square feet	Acres	2.2957×10^{-5}
		Square centimeters	929.03
		Square inches	144.0
		Square meters	0.092903
		Square miles	3.587×10^{-8}
		Square yards	0.111111
	Square inches	Square feet	6.9444×10^{-3}
		Square meters	6.4516×10^{-4}
		Square millimeters	645.16
	Square kilometers	Acres	247.1
		Square feet	1.0764×10^7
		Square meters	1.0×10^6
		Square miles	0.386102
		Square yards	1.196×10^6
	Square meters	Square centimeters	1.0×10^4
		Square feet	10.764
		Square inches	1.55×10^3
		Square kilometers	1.0×10^{-6}
		Square miles	3.861×10^{-7}
		Square millimeters	1.0×10^6
		Square yards	1.196
	Square miles	Acres	640
		Square feet	2.7878×10^7
		Square kilometers	2.590
		Square meters	2.59×10^6
		Square yards	3.0976×10^6
	Square yards	Acres	2.0661×10^{-4}
		Square centimeters	8.3613×10^3
		Square feet	9.0
		Square inches	1.296×10^3
		Square meters	0.83613
		Square miles	3.2283×10^{-7}

TABLE 2.8
Volume Conversion Factors

Quantity	To Convert From	To	Multiply by
Volume	Cubic centimeters	Cubic feet	3.5315×10^{-5}
		Cubic inches	0.06102
		Cubic meters	1.0×10^{-6}
		Cubic yards	1.308×10^{-6}
	Cubic feet	Cubic centimeters	2.8317×10^4
		Cubic meters	0.028317
		Liters	28.317
	Cubic inches	Cubic centimeters	16.387
		Cubic feet	5.787×10^{-4}
		Cubic meters	1.6387×10^{-5}
		Cubic yards`	2.1433×10^{-5}
		Liters	0.01639
	Cubic meters	Cubic centimeters	1.0×10^6
		Cubic feet	35.315
		Cubic inches	6.1024×10^4
		Cubic yards	1.308
		Liters	1000
	Cubic yards	Cubic centimeters	7.6455×10^5
		Cubic feet	27
		Cubic inches	4.6656×10^4
		Cubic meters	0.76455
		Gallons	201.97
		Liters	764.55
		Quarts	807.90
	Liters	Cubic centimeters	1000
		Cubic feet	0.035315
		Cubic inches	61.024
		Cubic meters	0.001

temperature of one gram of water to 1°C. The kilogram-calorie or kilo-calorie (kg-cal or kilo-cal) which is equal to 1000 gram calories is in more frequent practical use, which is the heat required to raise the temperature of one kilogram of water by 1°C.

1 kilo-calorie/kilogram = calorie/gram = centigrade heat unit/second.

A practical unit of energy, usually applied to heat, is the Joule.

2.8 POWER, WORK, ENERGY, AND HEAT

Power is defined as the work done in unit time. The unit of power in the technical system of measurement is the horsepower. The fundamental metric unit of power is the Watt.

1 Watt = 1 Joule/second

The fundamental metric unit of work is Joule.

2.8.1 UNIT OF HEAT

In the metric system, the unit of heat is the gram-calorie or calorie, which is the quantity of heat required to raise the

2.8.2 UNIT OF FORCE

The Newton (symbol: N) is the International System of Units (SI) derived unit of force. It is named after Sir Isaac Newton in recognition of his work on classical mechanics, specifically Newton's second law of motion. One Newton is the force needed to accelerate one kilogram of mass at the rate of one meter per second squared.

In SI base units: 1 N = 1 kg·m/s²

TABLE 2.9
Miscellaneous Conversion Factors

Quantity	To Convert From	To	Multiply by
Pressure	Atmospheres	Centimeters of H_2O (4°C)	1.033×10^3
		Inches of Hg (32°F)	29.9213
		Kilogram / square centimeter	1.033
		Millimeters of Hg (0°C)	760
		Pounds / square inch	14.696
	Inches of Hg (60°F)	Atmospheres	0.03333
		Grams / square centimeter	34.434
		Millimeters of Hg (60°F)	25.4
	Inches of H_2O (4°C)	Atmospheres	2.458×10^{-3}
		Inches of Hg (32°F)	0.07355
		Kilogram/square meter	25.399
		Pounds/square feet	5.2022
		Pounds/square inch	0.036126
	Kilograms/square centimeter	Atmospheres	0.96784
		Centimeters of Hg (0°C)	73.556
		Feet of H_2O (39.2°F)	32.809
		Inches of Hg (32°F)	28.959
		Pounds/square inch	14.223
	Millimeters of Hg (0°C)	Atmospheres	1.3158×10^{-3}
		Grams/square centimeters	1.3595
		Pounds/square inch	0.019337
Velocity	Centimeters/second	Feet/minute	1.9685
		Feet/second	0.0328
		Kilometers/hour	0.036
		Meters/minute	0.6
		Miles/hour	0.02237
	Feet/minute	Centimeters/second	0.508
		Kilometers/hour	0.01829
		Meters/minute	0.3048
		Meters/second	5.08×10^{-3}
	Feet/second	Centimeters/second	30.48
		Kilometers/hour	1.0973
		Meters/minute	18.288
		Miles/hour	0.6818
	Kilometers/hour	Centimeters/second	27.778
		Feet/hour	3.2808×10^3
		Feet/minute	54.681
		Meters/second	0.27778
		Miles (statute)/hour	0.62137
	Meters/minute	Centimeters/second	1.6667
		Feet/minute	3.2808
		Feet/second	0.05468
		Kilometers/hour	0.06
Density	Dynes/cubic centimeter	Grams/cubic centimeter	1.0197×10^{-3}
	Grains/cubic foot	Grams/cubic meter	2.28835
	Grams/cubic centimeter	Dynes/cubic centimeter	980.665
		Grains/milliliter	15.433
		Grams/milliliter	1.0
		Pounds/cubic inch	1.162
		Pounds/cubic foot	62.428

3 Preliminary Mathematics

3.1 MATHEMATICAL SIGNS, SYMBOLS, AND ABBREVIATIONS

A list of symbols used within all branches of mathematics to express a formula or to represent a constant is detailed in this section. It is arranged according to the type of symbol. It is also planned to enable the identification of an unfamiliar symbol by its graphical form. Only those notations that are used frequently in mathematics are incorporated in this list, since it is practically difficult to catalog all the notations used in mathematics. Supplementary information on the notations and their meanings, if any, needed, is given in the respective sections at the point of their occurrence. Table 3.1 shows a set of symbols, abbreviations, and their meanings commonly employed in Mathematics.

3.2 MENSURATION

Mensuration is the branch of mathematics which deals with the study of geometric shapes, their area, volume, and related parameters. i.e., it deals with measurement of various parameters of geometric figures. It is all about the method of quantifying. It is done using geometric computations and algebraic equations to deliver data related to depth, width, length, area, or volume of a given entity. However, the measurement results got using mensuration are mere approximations. Hence actual physical measurements are always considered to be accurate.

There are two types of geometric shapes: (1) 2D and (2) 3D. 2D regular shapes have a surface area and are categorized as circle, triangle, square, rectangle, parallelogram, rhombus, and trapezium. 3D shapes have surface area as well as volume. They are cube, rectangular prism (cuboid), cylinder, cone, sphere, hemisphere, prism, and pyramid.

3.2.1 MENSURATION OF AREAS

3.2.1.1 Circle

For a circle of diameter d as shown in Figure 3.1 having circumference C,

$$\text{Area}, A = \frac{1}{4}\pi d^2 \tag{3.1a}$$

$$= \pi r^2 \tag{3.1b}$$

$$= 0.07958\, C^2 \tag{3.1c}$$

$$= \frac{1}{4} C \times d \tag{3.1d}$$

$$\text{Circumference}, C = \pi d \tag{3.2a}$$

$$= 3.5449\sqrt{\text{area}} \tag{3.2b}$$

TABLE 3.1
Symbols and Abbreviations in Mathematics

Notations		Meanings	
+ Plus		Positive	
− Minus		Negative	
±		Plus or minus	
=		Equal to	
≠		Not equal to	
>		Greater than	
<		Less than	
≥		Greater than or equal to	
≤		Less than or equal to	
×		Multiplied by	
÷		Divided by	
∟		Right angle	
∠		Any angle	
⊥		Perpendicular to	
∵		Since; because	
∴		Therefore; hence	
( ), [ ]		parenthesis, brackets	
n°		n degree	
n′		n minutes	
n″		n seconds	
Δ	delta	difference	
∞		infinity	
√		square root	
$\sqrt[3]{}$		cube root	
Σ		sum of finite quantities	
Φ		phi	Any angle
θ		theta	
π		Pi	
∝		varies as; proportional to	
a²		a squared	
a³		a cubed	
aⁿ		a raised to the power of n	
:		is to	
: :		so is	

$$\text{Side of a square with the same area}, A = 0.8862d \tag{3.3a}$$

$$= 0.285C \tag{3.3b}$$

$$\text{Side of an inscribed square} = 0.707d \tag{3.4a}$$

$$= 0.225C \tag{3.4b}$$

$$\text{Side of an inscribed equilateral triangle} = 0.86d \tag{3.5}$$

$$\text{Side of a square of equal periphery as a circle} = 0.785d \tag{3.6}$$

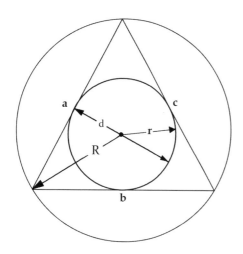

FIGURE 3.1 A triangle with a circumscribed and inscribed circle.

3.2.1.2 Square

$$\text{Area} = \text{side}^2 = 1.2732 \times \text{area of inscribed circle} \quad (3.7)$$

$$\text{Diagonal} = \sqrt{2} \times \text{side} \quad (3.8)$$

Circumference of a circle circumscribing a square

$$= 4.443 \times \text{side of square} \quad (3.9)$$

Diameter of a circle circumscribed about square

$$= 1.414 \times \text{side of square} \quad (3.10)$$

Diameter of a circle equal in area to square

$$= 1.129 \times \text{side of square} \quad (3.11)$$

Diameter of a circle of equal periphery as square

$$= 1.273 \times \text{side of square} \quad (3.12)$$

3.2.1.3 Triangle

Consider the triangle having side lengths a, b, and c as shown in Figure 3.1. R is the radius of circumscribed circle and r is the radius of inscribed circle.

$$\text{Semi perimeter}, s = \frac{a+b+c}{2} \quad (3.13)$$

For an equilateral triangle, if h is the height and a is the side

$$h = \frac{\sqrt{3}}{2}a \quad (3.14a)$$

$$r = \frac{a}{2\sqrt{3}} = \frac{\sqrt{3}}{6}a \quad (3.14b)$$

$$R = \frac{a}{\sqrt{3}} = 2r \quad (3.14c)$$

$$\text{Area} = \frac{\sqrt{3}}{4}a^2 \quad (3.14d)$$

For an isosceles triangle with base length, c, and length of equal sides, a,

$$\text{Area} = \frac{c\sqrt{\left(4a^2 - c^2\right)}}{4} \quad (3.15)$$

3.2.1.4 Semicircle

Consider a semicircle of radius r as shown in Figure 3.2. Its center of gravity lies at a distance of $\frac{4r}{3\pi}$ from the base.

3.2.1.5 Arc of a Circle

An arc is any part of a circumference BCD as shown in Figure 3.3. If r is the radius of the arc and ϕ is the central angle of the arc in degrees,

$$\text{Length of an arc} = \frac{2\pi r\phi}{360} = \frac{r\phi}{57.3} = 0.01745\,r\phi \quad (3.16)$$

If b is the length of the chord of half arc and a is half the length of the chord of arc as shown in Figure 3.3,

$$\text{Length of an arc} = \frac{2 \times (\text{area of sector})}{r} \quad \text{or} \quad \frac{8b - 2a}{3} \text{ (approx.)}$$

$$(3.17)$$

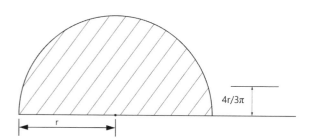

FIGURE 3.2 A semicircle of radius "r."

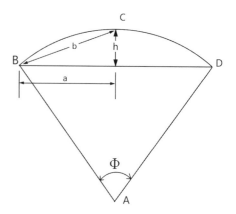

FIGURE 3.3 An arc of a circle.

3.2.1.6 Sector of a Circle

A sector is the space included between an arc and two radii drawn to the center. ABCD in Figure 3.3 is a sector. If r is the radius of the sector and ϕ is the central angle of the sector in degrees,

$$\text{Length of arc BCD,} \quad L = \frac{\pi r \phi}{180} \qquad (3.18)$$

$$\text{Area of the sector} = \frac{\phi}{360} \times \pi r^2 = \frac{1}{2} \times r \times L \qquad (3.19)$$

The position of center of gravity of a sector of a circle (ABCD)

$$= 2/3\, r \times \frac{(2a)}{L} = \frac{4ar}{3L} \text{ from A on the line AC} \qquad (3.20a)$$

$$= 2/3\, r \times \frac{\sin \phi}{\phi}$$

$$= \frac{2r \sin \phi}{3\phi} \text{ from A on the line AC where} \phi \text{is in radians} \qquad (3.20b)$$

3.2.1.7 Segment of a Circle

A segment is that part of the circle contained between the arc (BCD) and its chord (BD) as in Figure 3.3. If h is the rise, a is half the length of the chord (BD), and L is the length of the chord,

$$\text{Area} = (4/3)\, h \sqrt{\left(a^2 + (2/5)h^2\right)} \qquad (3.21a)$$

$$= \frac{(L \times r) - 2a \times (r - h)}{2} \qquad (3.21b)$$

$$= (2/3)\, Lh \qquad (3.21c)$$

$$= (4/3)\, ah \text{ (approx.)} \qquad (3.21d)$$

The position of center of gravity of a segment of a circle (BCD)

$$= \frac{(2a)^3}{12 \times \text{Area}} \quad \text{from A on the line AC} \qquad (3.22)$$

Other relations are established as follows:

$$\text{Radius of the segment,} \ r = \frac{h^2 + a^2}{2h} \qquad (3.23a)$$

$$= \frac{b^2}{2h} \qquad (3.23b)$$

$$\text{Rise of the segment,} \ h = r - \sqrt{\left(r^2 - a^2\right)} \qquad (3.24a)$$

$$= \frac{b^2}{2r} \qquad (3.24b)$$

$$\text{Half the length of the chord,} \ a = \sqrt{h(2r - h)} \qquad (3.25a)$$

$$= \sqrt{r^2 - (r - h)^2} \qquad (3.25b)$$

$$= \frac{8b - 3L}{2} \qquad (3.25c)$$

$$\text{Length of the chord of half arc,} \ b = \sqrt{2\,rh} \qquad (3.26a)$$

$$= \frac{3L + 2a}{8} \qquad (3.26b)$$

3.2.1.8 Polygons

A polygon is a plane figure which is surrounded by a fixed number of straight line segments closing in a loop. There are two types of polygons—regular and irregular polygons. A regular polygon has equal sides and equal angles. A polygon with unequal sides and unequal angles is called an irregular polygon.

Area of any regular polygon = ½ × radius of inscribed circle

(perpendicular drawn from the center of the figure to

the center of side, r shown in Figure 3.1)

× sum of all sides $\qquad (3.27)$

Sum of interior angles of any polygon(regular or irregular)

$= 180° \times (\text{number of sides} - 2) \qquad (3.28)$

Table 3.2 below gives the values to find the area, radii of inscribed (r in Figure 3.1) and circumscribed (R in Figure 3.1) circles, included angle between two adjacent sides (angle) and the length of side (S) of regular polygons in an easy way.

Area of trapezium = $(1/2) \times$ sum of lengths of parallel sides

× height $\qquad (3.29)$

For irregular figures, Simpson's rule is used to calculate the area. Divide the area or figure into an even number (n) of parallel strips by means of ($n + 1$) ordinates, spaced at equal distances, d.

Area = 1/3 d

$$\begin{bmatrix} \text{first ordinate + last ordinate + 2} \\ \times (\text{sum of all intermediate odd numbered ordinates}) \\ + 4 \times (\text{sum of all intermediate even numbered ordinates}) \end{bmatrix}$$

$$(3.30)$$

TABLE 3.2

Data Related to Polygons

Name of Polygon	No. of Sides	Angle Deg.	Length of Sides, $S =$		Area $= S^2 \times$	Radius of Circumscribed Circle, $R = S \times$	Radius of Inscribed Circle, $r = S \times$
			$R \times$	$r \times$			
Triangle	3	60°	1.732	3.464	0.433	0.577	0.289
Tetragon	4	90°	1.414	1.000	1.000	0.707	0.500
Pentagon	5	108°	1.176	1.454	1.721	0.851	0.688
Hexagon	6	120°	1.000	1.155	2.598	1.000	0.866
Octagon	8	135°	0.765	0.828	4.828	1.307	1.207
Decagon	10	144°	0.618	0.650	7.694	1.618	1.538
Dodecagon	12	150°	0.517	0.543	11.196	1.932	1.866

3.2.2 MENSURATION OF VOLUMES

3.2.2.1 Cube

A cube is a 3D solid object bounded by 6 square faces and 6 edges of equal lengths.

$$\text{Diagonal of a cube} = \sqrt{3} \times \text{edge length of the cube} \quad (3.31)$$

3.2.2.2 Cuboid

A cuboid is a 3D rectangular solid bounded by 6 rectangular faces and 6 edges. Opposite faces are rectangles of equal dimensions and opposite edges have equal lengths.

$$\text{Diagonal of a cuboid} = \sqrt{\left(\text{length}^2 + \text{breadth}^2 + \text{depth}^2\right)} \quad (3.32)$$

3.2.2.3 Pyramid

A pyramid is a solid whose base is a polygon and whose sides are triangles uniting at a common point called the vertex.

For a right rectangular pyramid as shown in Figure 3.4 having l, w, and h as the base length, base width, and altitude, respectively,

$$\text{Volume}, V = l \times w \times h/3 \quad (3.33)$$

For a regular tetrahedron, where all edges have equal lengths, say a,

$$\text{Volume}, V = \frac{2\sqrt{2}a^3}{3} \quad (3.34)$$

$$\text{Altitude}, h = (2a) \times \sqrt{\frac{2}{3}} \quad (3.35)$$

$$\text{Surface area}, A = (2a)^2 \times \sqrt{3} \quad (3.36)$$

3.2.2.4 Circular Cone

A cone shown in Figure 3.5 is a solid whose base is a circle and whose convex surface tapers uniformly to a point called the vertex. If r is the radius of the base and h is the height of the cone,

$$\text{Volume} = 1/3 \times \text{area of base} \times \text{vertical height} = \pi r^2 h/3 \quad (3.37)$$

$$\text{Convex area} = \frac{1}{2} \times \text{perimeter of base} \times \text{slant height}$$

$$= \pi r \sqrt{\left(r^2 + h^2\right)} \quad (3.38)$$

$$\text{Position of centre of gravity} = h/4 \text{ above the base of the cone} \quad (3.39)$$

3.2.2.5 Wedge

A wedge is 3D solid with two principal faces meeting in a sharply acute angle making a sharp edge in one direction.

For a wedge on a rectangular base as shown in Figure 3.6, a is the base length, b is the base width, h is the height, and e is the width of the sharp edge.

$$\text{Volume}, V = (bh/6)(2a + e) \quad (3.40a)$$

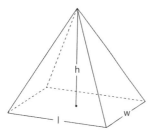

FIGURE 3.4 A right rectangular pyramid.

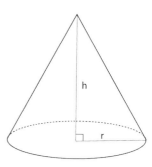

FIGURE 3.5 A circular cone.

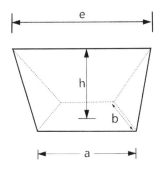

FIGURE 3.6 A eedge on a rectangular base.

$$= (A/3)(2a + e) \tag{3.40b}$$

$$= \text{Area of base} \times (h/2)(\text{approx.}) \tag{3.40c}$$

If the side opposite to "a" is of different size than "a," take "a + the other side" instead of "$2a$."

3.2.2.6 Sphere

A sphere is a 3D object where all the points on the surface of the object are equidistant from a common center. If d is the diameter, r is the radius, and C is the circumference of the sphere,

$$\text{Surface area}, A = \pi d^2 \tag{3.41a}$$

$$= 4\pi r^2 \tag{3.41b}$$

$$= Cd \tag{3.41c}$$

$$= 0.3183C^2 \tag{3.41d}$$

$$\text{Volume}, V = (A/6)d \tag{3.42d}$$

$$= \pi d^3/6 \tag{3.42b}$$

$$= 0.5236d^3 \tag{3.42c}$$

$$= 4/3\pi r^3 \tag{3.42d}$$

3.2.2.7 Hemisphere
Hemisphere is a half sphere.

$$\text{Total surface area}, A = 3\pi r^2 \tag{3.43}$$

$$\text{Volume}, V = 2/3\pi r^3 \tag{3.44}$$

3.2.2.8 Spherical Sector
A sector of a sphere is a portion of the sphere defined by a conical boundary with apex at the center of the sphere as shown in Figure 3.7. It consists of a cylindrical cone with a

spherical cap. As shown in the figure, r is the radius of the sphere, c is the diameter of the circle which forms the base of the spherical cap, and h is the height of the topmost point of the spherical cap measured from the center of the circle which forms the base of the spherical cap.

$$\text{Volume}, V = 2/3\pi r^2 h \tag{3.45a}$$

$$= 2.094r^2 h \tag{3.45b}$$

Total area of the conical and the spherical surface, S

$$= \pi r \left(2h + \frac{1}{2}c \right) \tag{3.46}$$

Position of center of gravity = 3/8 r above the spherical center
$$\tag{3.47}$$

3.2.2.9 Spherical Zone
A spherical zone is the outer portion of the surface of the sphere included between two parallel planes. The altitude of the zone is the perpendicular distance between the two parallel planes.

The bases of the zone are the circumference of the sections made by these two parallel planes. Since it is bounded by two parallel planes, it is called as a spherical zone with two bases. The entire 3D portion between the two parallel planes is called as a spherical segment or a spherical frustum. Figure 3.8 shows a spherical segment with altitude h, radius r, and c_1 and c_2 as the circumference of the two bases.

$$\text{Volume of the spherical segment} = \frac{\pi h}{6} \left(\frac{3c_1^2}{4} + \frac{3c_2^2}{4} + h^2 \right) \tag{3.48}$$

Area of the spherical zone with two bases

$$= \text{Area of spherical surface only} = 2\pi rh \tag{3.49}$$

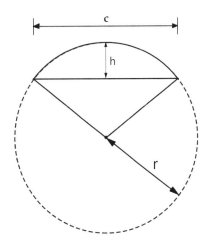

FIGURE 3.7 A dector of sphere.

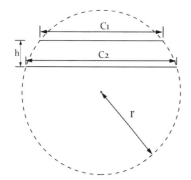

FIGURE 3.8 A spherical segment with altitude "h," radius "r," C_1 and C_2 as circumference of two bases.

If one of the bounding parallel planes is tangent to the sphere, the surface bounded is called as a spherical zone of one base. The entire 3D portion is called as a spherical cap. A spherical cap is the region of a sphere which lies above (or below) a given plane. If the plane passes through the center of the sphere, the cap is a called a hemisphere. Figure 3.9 shows a spherical cap with altitude h, radius r, and circumference of base, c.

$$\text{Volume of the spherical cap} = \pi h^2 (r - h/3) \qquad (3.50a)$$

$$= \pi h \left(c^2/8 + h^2/6 \right) \qquad (3.50b)$$

Area of the spherical zone with one base, A

$$= \text{Area of spherical surface only}$$

$$= 2\pi r h \qquad (3.51a)$$

$$= \pi \left(c^2/4 + h^2 \right) \qquad (3.51b)$$

3.2.2.10 Spherical Wedge

A spherical wedge is a 3D solid formed by revolving a semicircle about its diameter by less than 360°. Figure 3.10 shows a spherical wedge with radius r and included angle θ.

$$\text{Volume} = \frac{\theta°}{360°} \times \frac{4\pi r^3}{3} \qquad (3.52)$$

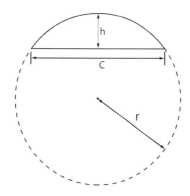

FIGURE 3.9 A spherical cap of altitude "h," radius "r," circumference of base "c."

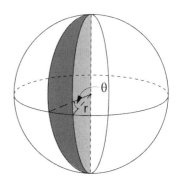

FIGURE 3.10 A spherical wedge with radius "r" and included angle θ.

$$\text{Area of spherical surface} = \frac{\theta°}{360°} \times 4\pi r^2 \qquad (3.53)$$

3.2.2.11 Hollow Sphere or Spherical Shell

A spherical shell is the region between two concentric spheres of differing radii. Figure 3.11 shows a hollow sphere with R as the outer diameter and r as the inner diameter.

$$\text{Volume} = \frac{4}{3}\pi \left(R^3 - r^3 \right) \qquad (3.54a)$$

When the thickness (h) of the shell is very small compared with its outer diameter (D),

$$\text{Volume} = \pi D^2 h \ (\text{approx.}) \qquad (3.54b)$$

3.2.2.12 Cylindrical Ring or Torus

A torus is a surface of revolution generated by revolving a circle in 3D space about an axis coplanar with the circle. If D is the mean diameter of the annular ring, d is its thickness, and R and r are the corresponding radii as shown in Figure 3.12.

$$\text{Volume}, V = 2\pi^2 R r^2 \qquad (3.55a)$$

$$= \frac{1}{4}\pi^2 D d^2 \qquad (3.55b)$$

$$= 2.47 D d^2 \qquad (3.55c)$$

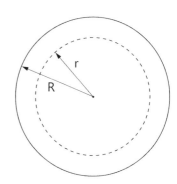

FIGURE 3.11 A hollow sphere with "R" as outer diameter and "r" as inner diameter.

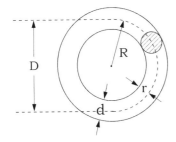

FIGURE 3.12 A cylindrical ring with mean diameter "*D*," thickness as "*d*."

Area of spherical surface, $A = 4\pi^2 Rr$ (3.56a)

$$= \pi^2 Dd \qquad (3.56b)$$

$$= 9.8696\, Dd \qquad (3.56c)$$

3.3 ALGEBRAIC FORMULAE

Some commonly used algebraic formulae are listed in this section. The list, although incomplete, covers all the essential formulae generally employed in Civil Engineering practice. The derivations of the same are beyond the scope of this book and can be found elsewhere. A set of Mathematics books are given in the Bibliography. The scholar may make use of them for further reading.

3.3.1 EXPONENT RULES

The base a raised to the power of *n* is equal to the multiplication of a, *n* times:

$a^n = a \times a \times \ldots \times a$, *n* times where *a* is the base and *n* is the exponent.

$$a^n = \text{a raised to the } n\text{th power}$$

$$a^{-1} = \frac{1}{a} \qquad (3.57a)$$

$$a^{-2} = \frac{1}{a^2} \qquad (3.57b)$$

$$a^{-m} = \frac{1}{a^m} \qquad (3.57c)$$

$$a^m = \left(\frac{1}{a}\right)^{-m} \qquad (3.57d)$$

$$a^{1/2} = \sqrt{a} \qquad (3.58a)$$

$$a^{3/2} = \sqrt{a^3} \qquad (3.58b)$$

$$a^m \times a^n = a^{m+n} \qquad (3.59a)$$

$$a^m \times b^m = (ab)^m \qquad (3.59b)$$

$$\left(a^m\right)^n = a^{mn} \qquad (3.59c)$$

$$a^m \times a^n \times a^x = a^{m+n+x} \qquad (3.59d)$$

$$a^m \div b^m = \left(\frac{a}{b}\right)^m \qquad (3.59e)$$

$$a^m \div a^n = a^{m-n} \qquad (3.59f)$$

3.3.2 POWERS OF TEN

$$10^3 = 1000 \qquad (3.60a)$$

$$10^2 = 100 \qquad (3.60b)$$

$$10^1 = 10 \qquad (3.60c)$$

$$10^0 = 1 \qquad (3.60d)$$

$$10^{-1} = \frac{1}{10} = 0.1 \qquad (3.60e)$$

$$10^{-2} = \frac{1}{100} = 0.01 \qquad (3.60f)$$

$$10^{-3} = \frac{1}{1000} = 0.001 \qquad (3.60g)$$

3.3.3 SIMPLE ALGEBRA

$$n + (-m) = n - m = -(m - n) \qquad (3.61a)$$

$$n - (+m) = n + (-m) \qquad (3.61b)$$

$$n - (-m) = n + (+m) = n + m \qquad (3.61c)$$

$$(-n) \times (-m) = +nm \qquad (3.61d)$$

$$(+n) \times (-m) = -nm \qquad (3.61e)$$

$$(-m) \div (-n) = m \div n = +\frac{m}{n} \qquad (3.61f)$$

$$(-m) \div (+n) = (+m) \div (-n) = -\frac{m}{n} \qquad (3.61g)$$

3.3.4 FORMULAE FOR ABRIDGED MULTIPLICATION

$$(a + b)^2 = a^2 + 2ab + b^2 \qquad (3.62a)$$

$$(a - b)^2 = a^2 - 2ab + b^2 \qquad (3.62b)$$

$$(a + b)(a - b) = a^2 - b^2 \qquad (3.62c)$$

$$(a + b + c)^2 = a^2 + b^2 + c^2 + 2ab + 2bc + 2ac \qquad (3.62d)$$

$$(a + b - c)^2 = a^2 + b^2 + c^2 + 2ab - 2bc - 2ac \qquad (3.62e)$$

$$(a + b)^3 = a^3 + 3a^2b + 3ab^2 + b^3 = a^3 + b^3 + 3ab(a + b) \qquad (3.62f)$$

$$(a-b)^3 = a^3 - 3a^2b + 3ab^2 - b^3 = a^3 - b^3 - 3ab(a-b) \quad \text{(3.62g)}$$

$$(a+b)\left(a^2 - ab + b^2\right) = a^3 + b^3 = (a+b)^3 - 3ab(a+b) \quad \text{(3.62h)}$$

$$(a-b)\left(a^2 + ab + b^2\right) = a^3 - b^3 = (a-b)^3 + 3ab(a+b) \quad \text{(3.62i)}$$

$$a^2 + b^2 = (a+b)^2 - 2ab \quad \text{(3.62j)}$$

$$(a+b)^2 \times (a-b)^2 = 4ab \quad \text{(3.62k)}$$

$$\left(\frac{a+b}{2}\right)^2 - \left(\frac{a-b}{2}\right)^2 = ab \quad \text{(3.62l)}$$

$$(x+a) \times (x+b) = x^2 + (a+b)x + ab \quad \text{(3.62m)}$$

$$(x-a) \times (x-b) = x^2 - (a+b)x + ab \quad \text{(3.62n)}$$

$$(x+a) \times (x-b) = x^2 + (a-b)x - ab \quad \text{(3.62o)}$$

$$(x+a) \times (x+b) \times (x+c) = x^3 + (a+b+c)x^2$$
$$+ (bc + ac + ab)x + abc \quad \text{(3.62p)}$$

3.3.5 Quadratic Equations

If $ax^2 + bx + c = 0$ then $x = \dfrac{-b \pm \sqrt{b^2 - 4ac}}{2a}$ (3.63a)

If $x^2 + ax = b$, then $x = -\dfrac{a}{2} \pm \sqrt{b + \left(\dfrac{a}{2}\right)^2}$ (3.63b)

If $x^{2n} + ax^n = b$ then $x = \sqrt[n]{-\dfrac{a}{2} \pm \sqrt{b + \left(\dfrac{a}{2}\right)^2}}$ (3.63c)

3.3.6 Cubic Equations

If $x^3 + ax + b = 0$, then $x = \left(-\dfrac{b}{2} + \sqrt{\dfrac{a^3}{27} + \dfrac{b^2}{4}}\right)^{1/3}$

$$+ \left(-\dfrac{b}{2} - \sqrt{\dfrac{a^3}{27} + \dfrac{b^2}{a}}\right)^{1/3} \quad \text{(3.64a)}$$

If $x + y = s$, and $xy = p$, then $x = \dfrac{s + \sqrt{s^2 - 4p}}{2}$, y

$$= \dfrac{s - \sqrt{s^2 - 4p}}{2} \quad \text{(3.64b)}$$

3.4 TRIGONOMETRY

Trigonometry is a branch of mathematics which treats the properties of angular functions and their application to the solution of triangles.

A radian is an angle subtended at the center of a circle by an arc whose length is equal to the radius of the circle and is a constant angular measurement.

$$1 \text{ radian} = 57°17'44'' = 180/\pi \text{ degrees} = 57.296 \text{ degrees} \quad \text{(3.65a)}$$

$$1 \text{ degree} = 0.0175 \text{ radian} \quad \text{(3.65b)}$$

$$\pi \text{ radians} = 180 \text{ degrees} \quad \text{(3.65c)}$$

3.4.1 Trigonometric Functions

Sine = sin
    Cosine = cos
    Tangent = tan
    Secant = sec
    Cosecant = cosec
    Cotangent = cot

In Figure 3.13, a, b, c are the lengths of sides of a right-angled triangle.

$$\frac{b}{c} = \sin\theta \quad \text{(3.66a)}$$

$$\frac{a}{c} = \cos\theta \quad \text{(3.66b)}$$

$$\frac{b}{a} = \frac{\sin\theta}{\cos\theta} = \tan\theta \quad \text{(3.66c)}$$

$$\frac{c}{a} = \frac{1}{\sin\theta} = \mathrm{cosec}\,\theta \quad \text{(3.66d)}$$

$$\frac{a}{b} = \frac{\cos\theta}{\sin\theta} = \frac{1}{\tan\theta} = \cot\theta \quad \text{(3.66e)}$$

$$\frac{c}{a} = \frac{1}{\cos\theta} = \sec\theta \quad \text{(3.66f)}$$

$$\mathrm{ver\,sine}\,\theta = 1 - \cos\theta \quad \text{(3.66g)}$$

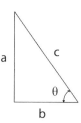

FIGURE 3.13 A right-angled triangle with sides a, b, and c.

$$\operatorname{cover}\sin\theta = 1 - \sin\theta \qquad (3.66\text{h})$$

$$(\sin\theta)^{-1} = \frac{1}{\sin\theta} \qquad (3.66\text{i})$$

Table 3.3 shows the values of trigonometrical functions for some typical angles.

In the first quadrant (i.e., 0°–90°), all the trigonometrical ratios are positive.

In the second quadrant (i.e., 90°–180°), only $\sin\theta$ and $\operatorname{cosec}\theta$ are positive.

In the third quadrant (i.e., 180°–270°), only $\tan\theta$ and $\cot\theta$ are positive.

In the fourth quadrant (i.e., 270°–360°), only $\cos\theta$ and $\sec\theta$ are positive.

3.4.2 Trigonometric Relations

$$\sin^2\theta + \cos^2\theta = 1 \qquad (3.67\text{a})$$

$$1 + \tan^2\theta = \sec^2\theta \qquad (3.67\text{b})$$

$$1 + \cot^2\theta = \operatorname{cosec}^2\theta \qquad (3.67\text{c})$$

$$1 + \cos\theta = 2\cos^2\frac{\theta}{2} \qquad (3.67\text{d})$$

$$1 - \cos\theta = 2\sin^2\frac{\theta}{2} \qquad (3.67\text{e})$$

$$2\cos A\sin B = \sin(A+B) - \sin(A-B) \qquad (3.67\text{f})$$

3.4.3 Functions of the Sum and Differences of Two Angles

$$\sin(A+B) = \sin A\cos B + \cos A\sin B \qquad (3.68\text{a})$$

$$\sin(A-B) = \sin A\cos B - \cos A\sin B \qquad (3.68\text{b})$$

$$\cos(A+B) = \cos A\cos B - \sin A\sin B \qquad (3.68\text{c})$$

$$\cos(A-B) = \cos A\cos B + \sin A\sin B \qquad (3.68\text{d})$$

$$\tan(A+B) = \frac{\tan A + \tan B}{1 - \tan A\tan B} \qquad (3.68\text{e})$$

$$\tan(A-B) = \frac{\tan A - \tan B}{1 + \tan A\tan B} \qquad (3.68\text{f})$$

3.4.4 Functions of ½ A

$$\sin\frac{1}{2}A = \sqrt{\frac{1}{2}(1-\cos A)} = \frac{1}{2}\sqrt{(1+\sin A)} - \frac{1}{2}\sqrt{(1-\sin A)} \qquad (3.69\text{a})$$

$$\cos\frac{1}{2}A = \sqrt{\frac{1}{2}(1+\cos A)} = \frac{1}{2}\sqrt{(1+\sin A)} + \frac{1}{2}\sqrt{(1-\sin A)} \qquad (3.69\text{b})$$

$$\tan\frac{1}{2}A = \sqrt{\frac{1-\cos A}{1+\cos A}} = \frac{1-\cos A}{\sin A} = \frac{\sin A}{1+\cos A}$$

$$= \frac{\tan A}{1+\sec A} = \operatorname{cosec}A - \cot A \qquad (3.69\text{c})$$

$$\cot\frac{1}{2}A = \sqrt{\frac{1+\cos A}{1-\cos A}} = \frac{1+\cos A}{\sin A}$$

$$= \frac{\sin A}{1-\cos A} = \frac{\tan A}{1+\sec A} = \frac{1}{\operatorname{cosec}A - \cot A} \qquad (3.69\text{d})$$

3.4.5 Functions of 2A

$$\sin 2A = 2\sin A\cos A = \frac{2\tan A}{1+\tan^2 A} \qquad (3.70\text{a})$$

$$\cos 2A = \cos^2 A - \sin^2 A = 2\cos^2 A - 1 = 1 - 2\sin^2 A \qquad (3.70\text{b})$$

$$\tan 2A = \frac{2\tan A}{1-\tan^2 A} \qquad (3.70\text{c})$$

$$\cot 2A = \frac{\cot^2 A - 1}{2\cot A} \qquad (3.70\text{d})$$

3.4.6 Products and Powers of Functions

$$\sin A\cos B = \frac{1}{2}\left[\sin(A+B) + \sin(A-B)\right] \qquad (3.71\text{a})$$

$$\cos A\sin B = \frac{1}{2}\left[\sin(A+B) - \sin(A-B)\right] \qquad (3.71\text{b})$$

$$\cos A\cos B = \frac{1}{2}\left[\cos(A+B) + \cos(A-B)\right] \qquad (3.71\text{c})$$

$$\sin A\sin B = \frac{1}{2}\left[\cos(A-B) - \cos(A+B)\right] \qquad (3.71\text{d})$$

$$\sin^2 A = \frac{1}{2}(1-\cos 2A) \qquad (3.71\text{e})$$

TABLE 3.3
Values of Trigonometrical Functions for Some Typical Angles

Angle	0°	30°	45°	60°	90°
Sin	0	1/2	1/√2	√3/2	1
Cos	1	√3/2	1/√2	1/2	0
Tan	0	1/√3	1	√3	∞

$$\cos^2 A = \frac{1}{2}(1 + \cos 2A) \qquad (3.71f)$$

3.4.7 Sums and Differences of Functions

$$\sin A + \sin B = 2\sin\frac{1}{2}(A+B)\cos\frac{1}{2}(A-B) \qquad (3.72a)$$

$$\sin A - \sin B = 2\cos\frac{1}{2}(A+B)\sin\frac{1}{2}(A-B) \qquad (3.72b)$$

$$\cos A + \cos B = 2\cos\frac{1}{2}(A+B)\cos\frac{1}{2}(A-B) \qquad (3.72c)$$

$$\cos A - \cos B = 2\sin\frac{1}{2}(A+B)\sin\frac{1}{2}(B-A) \qquad (3.72d)$$

$$\tan A + \tan B = \frac{\sin(A+B)}{\cos A \cos B} \qquad (3.72e)$$

$$\tan A - \tan B = \frac{\sin(A-B)}{\cos A \cos B} \qquad (3.72f)$$

$$\cot A + \cot B = \frac{\sin(A+B)}{\sin A \sin B} \qquad (3.72g)$$

$$\cot A - \cot B = \frac{\sin(A-B)}{\sin A \sin B} \qquad (3.72h)$$

$$\sin^2 A - \sin^2 B = \sin(A+B)\sin(A-B) = \cos^2 B - \cos^2 A \,(3.72i)$$

$$\cos^2 A - \cos^2 B = \sin(A+B)\sin(B-A) \qquad (3.72j)$$

$$\cos^2 A - \sin^2 B = \cos(A+B)\cos(A-B) \qquad (3.72k)$$

3.4.8 Rule for the Change of Trigonometrical Ratios

$$(A)\begin{cases} \sin(-\theta) = -\sin\theta \\ \cos(-\theta) = \cos\theta \\ \tan(-\theta) = -\tan\theta \\ \cot(-\theta) = -\cot\theta \\ \sec(-\theta) = \sec\theta \\ \operatorname{cosec}(-\theta) = -\operatorname{cosec}\theta \end{cases} \qquad (3.73a \text{ i to vi})$$

$$(B)\begin{cases} \sin(90°-\theta) = \cos\theta \\ \cos(90°-\theta) = \sin\theta \\ \tan(90°-\theta) = \cot\theta \\ \cot(90°-\theta) = \tan\theta \\ \sec(90°-\theta) = \operatorname{cosec}\theta \\ \operatorname{cosec}(90°-\theta) = \sec\theta \end{cases} \qquad (3.73b \text{ i to vi})$$

$$(C)\begin{cases} \sin(90°+\theta) = \cos\theta \\ \cos(90°+\theta) = -\sin\theta \\ \tan(90°+\theta) = -\cot\theta \\ \cot(90°+\theta) = -\tan\theta \\ \sec(90°+\theta) = -\operatorname{cosec}\theta \\ \operatorname{cosec}(90°+\theta) = \sec\theta \end{cases} \qquad (3.73c \text{ i to vi})$$

$$(D)\begin{cases} \sin(180°-\theta) = \sin\theta \\ \sin(180°-\theta) = -\cos\theta \\ \sin(180°-\theta) = -\tan\theta \\ \sin(180°-\theta) = -\cot\theta \\ \sin(180°-\theta) = -\sec\theta \\ \sin(180°-\theta) = \operatorname{cosec}\theta \end{cases} \qquad (3.73d \text{ i to vi})$$

$$(E)\begin{cases} \sin(180°+\theta) = -\sin\theta \\ \cos(180°+\theta) = -\cos\theta \\ \tan(180^0+\theta) = \tan\theta \\ \cot(180°+\theta) = \cot\theta \\ \sec(180°+\theta) = -\sec\theta \\ \operatorname{cosec}(180°+\theta) = -\operatorname{cosec}\theta \end{cases} \qquad (3.73e \text{ i to vi})$$

Following are the rules to remember the above 30 formulae.

Rule 1: Trigonometrical ratio changes only when the angle is $(90° - \theta)$ or $(90° + \theta)$. In all other cases, trigonometrical ratio remains the same.

Following is the law of change:

Sin changes into cos; cos changes into sin,

Tan changes into cot; cot changes into tan,

Sec changes into cosec, cosec changes into sec.

Rule 2: The angle θ is always between 0° and 90° and put the proper sign as per the above formulae.

3.5 SOLUTION OF TRIANGLES

3.5.1 Relation between Angles and Sides of Plane Triangles

In any triangle ABC, if a, b, and c are the lengths of the three sides of the triangle and A, B, and C are included angles opposite to the sides a, b, and c, respectively.

$$A + B + C = 180° \qquad (3.74)$$

3.5.1.1 Law of Sines

$$\frac{a}{\sin A} = \frac{b}{\sin B} = \frac{c}{\sin C} = \frac{abc}{2\,\text{Area}} \qquad (3.75)$$

If R = radius of scribed or circumscribing circle and r is the radius of inscribed circle as shown in Figure 3.1,

$$R = \frac{a}{2\sin A} = \frac{b}{2\sin B} = \frac{c}{2\sin C} \qquad (3.76)$$

$$r = \frac{\text{Area}}{s} = (s-a)\tan\frac{1}{2}A = (s-b)\tan\frac{1}{2}B = (s-c)\tan\frac{1}{2}C \qquad (3.77)$$

$$\text{Area of the triangle} = \frac{1}{2}(ab\sin C) = \frac{1}{2}(bc\sin A)$$

$$= \frac{1}{2}(ac\sin B) = \sqrt{s(s-a)(s-b)(s-c)} \qquad (3.78)$$

$$\text{where } s = \frac{1}{2}(a+b+c) \qquad (3.79)$$

If two angles and a side opposite to one angle are known, the side opposite to the other angle may be computed as:

$$a = b\frac{\sin A}{\sin B} \qquad (3.80a)$$

$$c = b\frac{\sin C}{\sin B} \qquad (3.80b)$$

3.5.1.2 Law of Cosines

If two sides and the included angle are known, the unknown side may be computed as:

$$a = \sqrt{b^2 + c^2 - 2bc\cos A} \qquad (3.81a)$$

$$b = \sqrt{c^2 + a^2 - 2ca\cos B} \qquad (3.81b)$$

$$c = \sqrt{a^2 + b^2 - 2ab\cos C} \qquad (3.81c)$$

If all the three sides are known, the included angles may be determined as:

$$\cos A = \frac{b^2 + c^2 - a^2}{2bc} \qquad (3.82a)$$

$$\cos B = \frac{c^2 + a^2 - b^2}{2ca} \qquad (3.82b)$$

$$\cos C = \frac{a^2 + b^2 - c^2}{2ab} \qquad (3.82c)$$

If two sides and the included angles opposite to them are known, the third side can be determined as:

$$a = b\cos C + c\cos B \qquad (3.83a)$$

$$b = c\cos A + a\cos C \qquad (3.83b)$$

$$c = a\cos B + b\cos A \qquad (3.83c)$$

3.6 MATRIX

A matrix is a rectangular array of numbers, symbols, or expressions, arranged in rows and columns. The individual items in a matrix are called its elements or entries.

$$\text{e.g.} \begin{bmatrix} 1 & 2 \\ 3 & 4 \end{bmatrix}, \begin{bmatrix} 1 & 2 & 3 \\ 4 & 5 & 6 \\ 7 & 8 & 9 \end{bmatrix}, \begin{bmatrix} 1 & 2 & 3 \\ 4 & 5 & 6 \end{bmatrix},$$

$$\begin{bmatrix} 1 & 2 \\ 3 & 4 \\ 5 & 6 \end{bmatrix}$$

$$\text{In general, } [A] = \begin{bmatrix} a_{11} & a_{12} & a_{13} & \cdots & a_{1n} \\ a_{21} & a_{22} & a_{23} & \cdots & a_{2n} \\ \cdots & \cdots & \cdots & \cdots & \cdots \\ \cdots & \cdots & \cdots & \cdots & \cdots \\ a_{m1} & a_{m2} & a_{m3} & \cdots & a_{mn} \end{bmatrix} \text{ where}$$

m and n are the number of rows and columns, respectively.

3.6.1 TRANSPOSE OF A MATRIX

The transpose of a matrix is a new matrix whose rows are the columns of the original matrix. (This makes the columns of the new matrix the rows of the original matrix.)

The transpose of a matrix is denoted by a superscript "T."

The transpose of a matrix is formed by placing the element at row r column c in the original at row c column r of the new matrix. Thus, the element a_{rc} of the original matrix becomes element a_{cr} in the transposed matrix.

$$\text{e.g. } [A] = \begin{bmatrix} a_{11} & a_{12} & a_{13} & \cdots & a_{1n} \\ a_{21} & a_{22} & a_{23} & \cdots & a_{2n} \\ \cdots & \cdots & \cdots & \cdots & \cdots \\ \cdots & \cdots & \cdots & \cdots & \cdots \\ a_{m1} & a_{m2} & a_{m3} & \cdots & a_{mn} \end{bmatrix},$$

$$[A]^{\mathrm{T}} = \begin{bmatrix} a_{11} & a_{21} & \cdots & \cdots & a_{m1} \\ a_{12} & a_{22} & \cdots & \cdots & a_{m2} \\ a_{13} & a_{23} & \cdots & \cdots & a_{m3} \\ \cdots & \cdots & \cdots & \cdots & \cdots \\ a_{1n} & a_{2n} & \cdots & \cdots & a_{mn} \end{bmatrix}$$

3.6.2 TYPES OF MATRIX

3.6.2.1 Square Matrix

In this matrix, the number of rows is equal to the number of columns.

i.e., $A = \lfloor a_{ij} \rfloor_{m \times n}$ is a square matrix if $m = n$

$$\text{e.g.} \begin{bmatrix} 1 & 2 \\ 3 & 4 \end{bmatrix}, \begin{bmatrix} 1 & 2 & 3 \\ 4 & 5 & 6 \\ 7 & 8 & 9 \end{bmatrix}$$

3.6.2.2 Rectangular Matrix

A matrix in which the number of rows is not equal to the number of columns.

i.e., $A = \lfloor a_{ij} \rfloor_{m \times n}$ is a rectangular matrix if $m \neq n$

$$\text{e.g.} \begin{bmatrix} 1 & 2 & 3 \\ 4 & 5 & 6 \end{bmatrix}, \begin{bmatrix} 1 & 2 \\ 3 & 4 \\ 5 & 6 \end{bmatrix}$$

3.6.2.3 Null Matrix or Zero Matrix

A matrix whose all elements are zeros is called a null matrix or zero matrix and is denoted by 0.

$$\text{e.g.} \begin{bmatrix} 0 & 0 \\ 0 & 0 \end{bmatrix}$$

3.6.2.4 Triangular Matrix

There are two types of triangular matrices—upper and lower triangular matrices.

a. **Upper triangular matrix**: A square matrix in which all the elements below the principal diagonal are zeros is called an upper triangular matrix.
   i.e., $A = \lfloor a_{ij} \rfloor_{n \times n}$ is a upper triangular if $a_{ij} = 0$ for $i > j$

$$\text{e.g.} \begin{bmatrix} 1 & 2 & 3 \\ 0 & 4 & 5 \\ 0 & 0 & 6 \end{bmatrix}$$

b. **Lower triangular matrix**: A square matrix in which all the elements above principal diagonal are zeros.
   i.e., $A = \lfloor a_{ij} \rfloor_{n \times n}$ is a lower triangular if $a_{ij} = 0$ for $i < j$

$$\text{e.g.,} \begin{bmatrix} 1 & 0 & 0 \\ 2 & 3 & 0 \\ 4 & 5 & 6 \end{bmatrix}$$

A matrix which is either upper triangular or lower triangular is called a triangular matrix.

3.6.2.5 Diagonal Matrix

A square matrix having nonzero entries only on the main diagonal is called a diagonal matrix.

i.e., $A = \lfloor a_{ij} \rfloor_{n \times n}$ is a diagonal matrix if $a_{ij} = 0$ for $i \neq j$

$$\text{e.g.,} \begin{bmatrix} 1 & 0 & 0 \\ 0 & 2 & 0 \\ 0 & 0 & 3 \end{bmatrix} \text{ is a diagonal matrix of order 3}$$

3.6.2.6 Scalar Matrix

Scalar matrix is a diagonal matrix with all its diagonal elements equal.

$$\text{e.g.} \begin{bmatrix} 9 & 0 & 0 \\ 0 & 9 & 0 \\ 0 & 0 & 9 \end{bmatrix}$$

3.6.2.7 Unit Matrix or Identity Matrix

A unit matrix is a square matrix in which every nondiagonal element is zero and all diagonal elements are unity. $a_{ij} = 0$ when $i = j$ and $a_{ij} = 0$ when $i \neq j$.

$$\text{e.g.,} \begin{bmatrix} 1 & 0 & 0 \\ 0 & 1 & 0 \\ 0 & 0 & 1 \end{bmatrix}$$

3.6.2.8 Row Matrix

A matrix which has only one row and any number of columns is called a row matrix.

$$\text{e.g.,} \begin{bmatrix} 1 & 2 & 3 \end{bmatrix}$$

3.6.2.9 Column Matrix

A matrix which has only one column and any number of rows is called column matrix.

$$\text{e.g.} \begin{bmatrix} 1 \\ 2 \\ 3 \end{bmatrix}$$

3.6.2.10 Nilpotent Matrix

A square matrix A is said to be nilpotent, if $A^m = 0$, where m is any integer and 0 is a null matrix of same order as of A.

$$\text{e.g. } A = \begin{bmatrix} 2 & -2 \\ 2 & -2 \end{bmatrix}$$

$$A^2 = \begin{bmatrix} 2 & -2 \\ 2 & -2 \end{bmatrix} \times \begin{bmatrix} 2 & -2 \\ 2 & -2 \end{bmatrix} = \begin{bmatrix} 4-4 & -4+4 \\ 4-4 & -4+4 \end{bmatrix}$$

$$= \begin{bmatrix} 0 & 0 \\ 0 & 0 \end{bmatrix}$$

Since $[A]^2 = 0$, hence, the matrix $[A]$ is nilpotent.

3.6.2.11 Symmetrical Matrix

A symmetrical matrix is a square matrix in which $a_{ij} = a_{ji}$ for all i, j. If A is symmetric, then $[A]^T = [A]$.

$$\text{e.g.} \begin{bmatrix} 1 & 2 & 3 \\ 2 & 5 & 6 \\ 3 & 6 & 9 \end{bmatrix}$$

3.6.2.12 Skew Symmetrical Matrix

It is a square matrix in which $a_{ij} = -a_{ji}$ for all i and j. All elements along the diagonal are zero in a skew symmetrical matrix.

$$\text{e.g.,} \begin{bmatrix} 0 & 2 & 3 \\ -2 & 0 & 6 \\ -3 & -6 & 0 \end{bmatrix}$$

3.6.2.13 Orthogonal Matrix

A square matrix A can be called an orthogonal matrix if, $[A][A]^T = [A]^T[A] = 1$

$$[A] = \begin{bmatrix} 0 & -0.8 & -0.6 \\ 0.8 & -0.36 & 0.48 \\ 0.6 & 0.8 & -0.64 \end{bmatrix}; [A]^T$$

$$= \begin{bmatrix} 0 & 0.8 & 0.6 \\ -0.8 & -0.36 & 0.8 \\ -0.6 & 0.48 & -0.64 \end{bmatrix}$$

3.6.3 Submatrix

Any matrix obtained by omitting some rows and columns from a given $(m \times n)$ matrix A is called submatrix of A.

The matrix A itself is a submatrix of A as it can be obtained from A by omitting no rows or columns.

$$\text{e.g. } [A] = \begin{bmatrix} 1 & 2 & 3 \\ 4 & 5 & 6 \\ 7 & 8 & 9 \end{bmatrix}$$

Submatrix of $[A]$ obtained by removing the second row and second column is a 2×2 matrix $= \begin{bmatrix} 1 & 3 \\ 7 & 9 \end{bmatrix}$

3.6.4 Determinant of a Matrix

A determinant is a real number associated with every square matrix. The determinant of a matrix A is denoted as $\det A$ or ΔA or $|A|$. Only square matrix can have determinants.

Determinant of matrix $[A] = \Delta A$

$$= \begin{vmatrix} a_{11} & a_{12} & a_{13} & \dots & a_{1n} \\ a_{21} & a_{22} & a_{23} & \dots & a_{2n} \\ \dots & \dots & \dots & \dots & \dots \\ \dots & \dots & \dots & \dots & \dots \\ a_{n1} & a_{n2} & a_{n3} & \dots & a_{nn} \end{vmatrix}$$

3.6.5 Minor of a Matrix

If a_{ij} is an element which is in the ith row and jth column of a square matrix A, then the determinant of the matrix obtained by deleting ith row and jth column of the matrix A is called as minor of a_{ij} and is denoted as M_{ij}.

If $[A] = \begin{bmatrix} a_{11} & a_{12} & a_{13} \\ a_{21} & a_{22} & a_{23} \\ a_{31} & a_{32} & a_{33} \end{bmatrix}$, then

$$M_{11} = \min \text{or of } a_{11} = \begin{vmatrix} a_{22} & a_{23} \\ a_{32} & a_{33} \end{vmatrix} = a_{22}a_{33} - a_{32}a_{23}$$

$$M_{12} = \min \text{or of } a_{12} = \begin{vmatrix} a_{21} & a_{23} \\ a_{31} & a_{33} \end{vmatrix} = a_{21}a_{33} - a_{31}a_{23} \text{ and so on}$$

3.6.6 Cofactor of a Matrix

The signed minor is called the cofactor. The cofactor of the element in the ith row and jth column is denoted by C_{ij}.

$$C_{ij} = (-1)^{i+j} M_{ij} \tag{3.84}$$

If $[A] = \begin{bmatrix} a_{11} & a_{12} & a_{13} \\ a_{21} & a_{22} & a_{23} \\ a_{31} & a_{32} & a_{33} \end{bmatrix}$, then

$$\text{Cofactor of } a_{11} = +M_{11} = + \begin{vmatrix} a_{22} & a_{23} \\ a_{32} & a_{33} \end{vmatrix}$$

$$\text{Cofactor of } a_{12} = -M_{12} = - \begin{vmatrix} a_{21} & a_{23} \\ a_{31} & a_{33} \end{vmatrix} \text{ and so on}$$

3.6.7 Adjoint of a Matrix

Let A be the square matrix given by,

$$A = \begin{bmatrix} a_{11} & a_{12} & a_{13} & \dots & a_{n1} \\ a_{21} & a_{22} & a_{23} & \dots & a_{n2} \\ \dots & \dots & \dots & \dots & \dots \\ \dots & \dots & \dots & \dots & \dots \\ a_{n1} & a_{n2} & a_{n3} & \dots & a_{nn} \end{bmatrix}$$

Let A_{ij} be the cofactor of a_{ij}. Then transpose of the matrix $[A_{ij}]$ is called adjoint of the matrix A.

3.6.8 Equality of Two Matrices

Two matrices $[A]$ and $[B]$ are said to be equal, if

i. both are the same size, and
ii. $a_{ij} = b_{ij}$ for each pair of subscripts i and j

If two matrices A and B are equal, then we write, $[A] = [B]$

3.6.9 Matrix Operations

3.6.9.1 Matrix Addition

Let $[A]$ and $[B]$ be two matrices of the same order $m \times n$.

If $[A] = \left[a_{ij}\right]_{m \times n}$, $[B] = \left[b_{ij}\right]_{m \times n}$, then $[A] + [B] = \left[a_{ij} + b_{ij}\right]_{m \times n}$

$$\tag{3.85}$$

3.6.9.2 Matrix Multiplication

a. By a scalar

Let C be any real number called scalar and $[B]$ be any $m \times n$ matrix, then

$$C[B] = [B]C = \left[cb_{ij} \right]_{m \times n} \text{ where, } \left[b_{ij} \right]_{m \times n} = [B] \quad (3.86)$$

b. By a matrix

Let $\left[A \right] = \left[a_{ij} \right]_{m \times n}$ and $\left[B \right] = \left[b_{jk} \right]_{n \times p}$ be two matrices such that number of columns in $[A]$ is equal to number of rows in $[B]$. Then $[A] \cdot [B] = [C]$ which is a $m \times p$ matrix $[C] = \left[C_{ik} \right]_{m \times p}$ such that $C_{ik} = \sum_{j=1}^{n} a_{ij} b_{jk}$ is called product of the matrices A and B.

3.6.9.3 Inverse of a Matrix

Let $[A]$ and $[B]$ be two square matrices of the same order, such that $[A][B] = [B][A] = 1$, then $[B]$ is called the inverse of $[A]$ and is denoted as $[A]^{-1}$. It is computed as:

$$A^{-1} = \frac{(\text{adj } A)}{|A|} \quad (3.87)$$

3.7 CALCULUS

3.7.1 Limits

3.7.1.1 Functions of Single-Variable Limits

Using limits, we are able to find out behavior of a function $f(x)$ near a point c, when distance between x and c is very small, i.e., $|x - c|$ is small.

Specifically, we should like to know whether there is some real number k such that $f(x)$ approaches k as x approaches c.

$$\text{We write as } \lim_{x \to c} f(x) = k \quad (3.88a)$$

3.7.1.2 Right Hand and Left Hand Limits

$$\text{Left hand limit} : \lim_{x \to c^-} f(x) = k \quad (3.88b)$$

$$\text{Right hand limit} : \lim_{x \to c^+} f(x) = k \quad (3.88c)$$

3.7.1.3 Theorems on Limits

If $\lim_{x \to c} f(x)$ and $\lim_{x \to c} g(x)$ exist, then

1. $\lim_{x \to c} (f(x) \pm g(x)) = \lim_{x \to c} f(x) \pm \lim_{x \to c} g(x) \quad (3.89)$

2. $\lim_{x \to c} (f(x) \cdot g(x)) = \lim_{x \to c} f(x) \cdot \lim_{x \to c} g(x) \quad (3.90)$

3. $\lim_{x \to c} \left(\frac{f(x)}{g(x)} \right) = \frac{\lim_{x \to c} f(x)}{\lim_{x \to c} g(x)} \text{ where } \left(\lim_{x \to c} g(x) \neq 0 \right) \quad (3.91)$

3.7.1.4 Some Useful Limits

1. $\lim_{x \to 0} \frac{\sin x}{x} = 1 = \lim_{x \to 0} \cos x = \lim_{x \to 0} \frac{\tan x}{x} \quad (3.92)$

2. $\lim_{x \to 0} (1 + x)^{\frac{1}{x}} = e = \lim_{x \to 0} \left(1 + \frac{1}{x} \right)^x \quad (3.93)$

3. $\lim_{x \to 0} \frac{\log_a(1 + x)}{x} = \log_a e \quad (a > 0, a \neq 0) \quad (3.94)$

4. $\lim_{x \to 0} \frac{\log (1 + x)}{x} = 1 \quad (3.95)$

5. $\lim_{x \to 0} \frac{a^x - 1}{x} = \log a \quad (a > 0) \quad (3.96)$

6. $\lim_{x \to 0} \frac{(1 + x)^m - 1}{x} = m \quad (3.97)$

7. $\lim_{x \to \infty} \frac{\log x}{x^m} = 0 \quad (m > 0) \quad (3.98)$

8. $\lim_{x \to 0} \frac{x^n - a^n}{x - a} = n a^{n-1} \quad (m > 0) \quad (3.99)$

3.7.1.5 L'Hospital's Rule

The rule states that for functions f and g, if $\lim_{x \to c} f(x) = \lim_{x \to c} g(x) = 0$ or $\lim_{x \to c} g(x) = \pm \infty$ (condition that $\lim_{x \to c} f(x) = \pm \infty$ is not necessary) and if $\lim_{x \to c} \frac{f'(x)}{g'(x)}$ exists then

$$\lim_{x \to c} \frac{f(x)}{g(x)} = \lim_{x \to c} \frac{f'(x)}{g'(x)} \quad (3.100)$$

3.7.2 Differentiability

A function $f(x)$ is said to be differentiable at $x = a$, if both $\lim_{h \to 0} \frac{f(a + h) - f(a)}{h}$, $h > 0$ and $\lim_{h \to 0} \frac{f(a - h) - f(a)}{-h}$, $h > 0$ exist and have common value (finite). This common value is called derivative of $f(x)$ at the point $x = a$ and is denoted by $f(a)$.

3.7.2.1 Rules of Differentiation

$$(f + g)' = f' + g' \rightarrow \text{Sum Rule} \quad (3.101)$$

$$(f - g)' = f' - g' \rightarrow \text{Difference Rule} \quad (3.102)$$

$$(f \cdot g)' = fg' + gf' \rightarrow \text{Product Rule} \quad (3.103)$$

$$\left(\frac{f}{g} \right)' = \frac{gf' - fg'}{g^2} \rightarrow \text{Quotient Rule} \quad (3.104)$$

$$\frac{1}{dx}\big(f(g(x))\big) = \frac{df}{dg}\cdot\frac{dg}{dx} \rightarrow \text{Chain Rule} \qquad (3.105)$$

3.7.2.2 Some Standard Derivatives

Table 3.4 shows some functions in x and their standard derivatives.

3.7.2.3 Differentiation by Substitution

If a function contains an expression of the form

1. $a^2 - x^2$; put $x = a\sin l$ or $x = a\cos l$ $\qquad (3.106)$

2. $a^2 + x^2$; put $x = a\tan l$ or $x = a\cos l$ $\qquad (3.107)$

3. $x^2 - a^2$; put $x = a\sec l$ or $x = a\csc l$ $\qquad (3.108)$

4. $\sqrt{\dfrac{a-x}{a+x}}$ or $\sqrt{\dfrac{a+x}{a-x}}$; put $x = a\cos l$ $\qquad (3.109)$

5. $a\cos x \pm b\sin x$; put $a = r\cos\theta$, and $b = r\sin\theta$ $r > 0$ $\qquad (3.110)$

3.7.2.4 Mean Value Theorems

3.7.2.4.1 Rolle's Theorem

If a function $f(x)$ is continuous in closed interval $[a, b]$, derivable in the open interval (a, b), and $f(a) = f(b)$, then there exists at least one real c in (a, b) such that $f'(c) = 0$.

3.7.2.4.2 Lagrange's Mean Value Theorem

If $f(x)$ is a function such that $f(x)$ is said to be continuous in $[a, b]$, and $f(x)$ exists in (a, b), then there exists at least one c, such that

$$f'(c) = \frac{f(b) - f(a)}{b - a} \qquad (3.111)$$

3.7.2.4.3 Cauchy's Mean Value Theorem

If $f(x)$ and $g(x)$ be two functions such that $f(x)$ and $g(x)$ both are continuous in $[a, b]$
$f'(c)$ and $g'(x)$ both exist in (a, b)

$$g'(x) \neq 0 \forall x \in (a, b) \qquad (3.112a)$$

Then there exists at least one point c in the closed interval (a, b) such that

$$\frac{f'(c)}{g'(c)} = \frac{f(b) - f(a)}{g(b) - g(a)} \qquad (3.112b)$$

3.7.2.4.4 Taylor's Theorem

If $f(x)$ is a continuous function such that $f'(x), f''(x), f'''(x)\dots f^{n-1}(x)$ are all continuous in $[a, a+h]$, $f^n(x)$ exists in $(a, a+h)$, where $h = b - a$, then

$$f(a+h) = f(a) + hf'(a) + \frac{h^2}{2!}f''(a) + \cdots \frac{h^{n-1}}{n-1!}f^{n-1}(a) + \frac{h^n}{n!}f^n(a) \qquad (3.113a)$$

If we put $a = 0$, $h = x$, then the series becomes

$$f(x) = f(0) + xf'(0) + \frac{x^2}{2!}f''(0) + \cdots \qquad (3.113b)$$

3.7.2.5 Partial Derivatives

The partial differential coefficient of $z = f(x, y)$ with respect to x is defined as

$$\mathop{Lt}_{\delta x \to 0} \frac{f(x + \delta x, y) - f(x, y)}{\delta x} \qquad (3.114a)$$

TABLE 3.4
Standard Derivatives

$f(x)$	$f'(x)$
x^n	$n\,x^{n-1}$
$\ln x$	$1/x$
$\log_a x$	$\log_a e \cdot 1/x$
e^x	e^x
a^x	$a^x \log_a e$
$\sin x$	$\cos x$
$\cos x$	$\sin x$
$\tan x$	$\sec^2 x$
$\sec x$	$\sec x \tan x$
$\csc x$	$-\csc x \tan x$
$\cot x$	$-\csc^2 x$
$\sin hx$	$\cos hx$
$\cos hx$	$\sin hx$
$\sin^{-1} x$	$\dfrac{1}{\sqrt{1-x^2}}$
$\cos^{-1} x$	$\dfrac{-1}{\sqrt{1-x^2}}$
$\tan^{-1} x$	$\dfrac{1}{1+x^2}$
$\csc^{-1} x$	$\dfrac{-1}{x^2\sqrt{x^2-1}}$
$\sec^{-1} x$	$\dfrac{1}{x\sqrt{x^2-1}}$
$\cot^{-1} x$	$\dfrac{1}{1+x^2}$

Provided this limit exists and is denoted by $\dfrac{\partial f}{\partial x}$ or $\dfrac{\partial z}{\partial x}$ or f_x or z_x

Similarly, partial derivatives of $f(x, y)$ with respect to y are defined as

$$\underset{\delta y \to 0}{\mathrm{Lt}} \frac{f(x, y + \delta y) - f(x, y)}{\delta y} \tag{3.114b}$$

Provided this limit exists and is denoted by $\dfrac{\partial f}{\partial y}$ or $\dfrac{\partial z}{\partial y}$ or f_y or z_y.

$\dfrac{\partial z}{\partial x}$ and $\dfrac{\partial z}{\partial y}$ are called first-order partial derivatives of z.

3.7.3 INTEGRATION

3.7.3.1 Some Standard Integrations

$$\int x^n = \frac{x^{n+1}}{n+1}, (n \neq -1) \tag{3.115}$$

$$\int \frac{1}{x} dx = \log x \tag{3.116}$$

$$\int e^x \, dx = e^x \tag{3.117}$$

$$\int a^x \, dx = \frac{a^x}{\log a} \tag{3.118}$$

$$\int \sin x \, dx = -\cos x \tag{3.119}$$

$$\int \cos x \, dx = \sin x \tag{3.120}$$

$$\int \sec^2 x \, dx = \tan x \tag{3.121}$$

$$\int \operatorname{cosec}^2 x \, dx = -\cot x \tag{3.122}$$

$$\int \sec x \tan x \, dx = \sec x \tag{3.123}$$

$$\int \operatorname{cosec} x \cot x \, dx = -\operatorname{cosec} x \tag{3.124}$$

$$\int \tan x \, dx = \log \sec x = -\log \cos x \tag{3.125}$$

$$\int \cot x \, dx = \log \sin x \tag{3.126}$$

$$\int \sec x \, dx = \log (\sec x + \tan x) \tag{3.127}$$

$$\int \operatorname{cosec} x \, dx = -\log(\operatorname{cosec} x + \cot x) = \log \tan \frac{x}{2} \tag{3.128}$$

$$\int \frac{1}{\sqrt{1 - x^2}} dx = \sin^{-1} x \text{ or } -\cos^{-1} x \tag{3.129}$$

$$\int \frac{1}{\sqrt{1 + x^2}} dx = \sinh^{-1} x \tag{3.130}$$

$$\int \frac{1}{a^2 + x^2} dx = \frac{1}{a} \tan^{-1}\left(\frac{x}{a}\right) \tag{3.131}$$

$$\int \frac{1}{x\sqrt{x^2 - 1}} dx = \sec^{-1} x = \cos^{-1} \frac{1}{x} \tag{3.132}$$

$$\int \sin hx \, dx = \cos hx \tag{3.133}$$

$$\int \cos hx \, dx = \sin hx \tag{3.134}$$

$$\int \frac{1}{\sqrt{x^2 - a^2}} dx = \log\left(x + \sqrt{x^2 - a^2}\right) = \cosh^{-1}\left(\frac{x}{a}\right) \tag{3.135}$$

$$\int \frac{1}{\sqrt{a^2 + x^2}} dx = \log\left(x + \sqrt{a^2 + x^2}\right) = \sinh^{-1}\left(\frac{x}{a}\right) \tag{3.136}$$

$$\int \frac{dx}{x\sqrt{x^2 - a^2}} = \frac{1}{a} \sec^{-1}\left(\frac{x}{a}\right) \tag{3.137}$$

$$\int \frac{dx}{x^2 + a^2} = \frac{1}{a} \tan^{-1}\left(\frac{x}{a}\right) \tag{3.138}$$

$$\int \frac{dx}{x^2 - a^2} = \frac{1}{2a} \log\left(\frac{x - a}{x + a}\right) \quad \text{if } x > a \tag{3.139}$$

$$\int \frac{dx}{a^2 - x^2} = \frac{1}{2a} \log\left(\frac{a + x}{a - x}\right) \quad \text{if } x < a \tag{3.140}$$

$$\int u \cdot v \, dx = u \int v \, dx - \int \left[\left(\frac{du}{dx}\right)\left(\int v \, dx\right)\right] dx \tag{3.141}$$

where u and v are the functions of x, u is a first function, and v as a second function.

$$\int e^x \left[f(x) + f'(x) \right] dx = e^x f(x) + c \tag{3.142}$$

$$\int \frac{f'(x)}{f(x)} dx = \log f(x) + c \tag{3.143}$$

3.7.3.2 Definite Integral

If a function is defined in the interval $[a, b]$, then the definite integral of $f(x)$ is

$$\int_a^b f(x) \, dx = \left[F(x) \right]_a^b = F(b) - F(a) \tag{3.144}$$

where $F(x)$ is an integral of $f(x)$, $f(b)$ is the value of $f(x)$ at $x = b$, and $f(a)$ is the value of $f(x)$ at $x = a$. a and b are called lower and upper limits, respectively. Geometrically, definite integral represents the area bounded by curve $y = f(x)$, x-axis, and the lines $x = a$ and $x = b$ as shown in Figure 3.14.

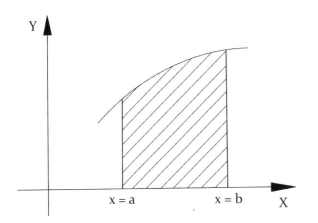

FIGURE 3.14 An area bound by curve $y = f(x)$, x axis, and lines $x = a$ and $x = b$.

3.7.3.2.1 Properties of Definite Integral

Let $\int f(x)\,dx = F(x)$. Then

$$\int_a^b f(x)\,dx = \int_a^b f(y)\,dy \qquad (3.145)$$

$$\int_a^b f(x)\,dx = -\int_b^a f(x)\,dx \qquad (3.146)$$

$$\int_a^b f(x)\,dx = \int_a^c f(x)\,dx + \int_c^b f(x)\,dx \qquad (3.147)$$

where c is a point lies between the interval $[a, b]$, and $f(x)$ integrable over $[a, c]$ and $[c, b]$

$$\int_0^a f(x)\,dx = \int_0^a f(a - x)\,dx \qquad (3.148)$$

$$\int_{-a}^a f(x)\,dx = \int_0^a f(x)\,dx + \int_0^a f(-x)\,dx \qquad (3.149a)$$

If the function is even, i.e., $f(-x) = f(x)$, then

$$\int_{-a}^a f(x)\,dx = 2\int_0^a f(x)\,dx \qquad (3.149b)$$

If the function is odd, i.e., $f(-x) = -f(x)$, then

$$\int_{-a}^a f(x)\,dx = 0 \qquad (3.149c)$$

$$\int_0^{2a} f(x)\,dx = \int_0^a f(x)\,dx + \int_0^{2a} f(2a - x)\,dx \qquad (3.150a)$$

If $f(x) = f(2a - x)$, then $\int_0^{2a} f(x)\,dx = 2\int_0^a f(x)\,dx$ (3.150b)

and

If $f(x) = -f(2a - x)$, then $\int_0^{2a} f(x)\,dx = 0$ (3.150c)

3.7.3.2.2 Definite Integral as Limit of Sum

If a function $f(x)$ is either increasing or decreasing on $[a, b]$, then

$$\int_a^b f(x)\,dx = \underset{h \to 0}{\mathrm{Lt}}\, h\big[f(a) + f(a + h) + \cdots + f(a + (n-1)h) \big]$$

$$= \underset{h \to 0}{\mathrm{Lt}}\, h\big[f(a + h) + f(a + 2h) + \cdots + f(a + nh) \big]$$

(3.151a)

$$\text{where } h = \frac{b - a}{n} \qquad (3.151b)$$

3.7.3.3 Improper Integral

Definite integral $\int_0^a f(x)\,dx$ is called improper integral, if:

Range of integration is infinite and integrand is bounded.
Range of integration is definite and integrand is unbounded.
Neither range of integration neither is finite nor is integrand bounded over it.

3.7.3.4 Multiple Integrals

Let a single valued and bounded function $f(x, y)$ of two independent variables x, y be defined in a closed region R of the xy-plane. Divide the region R into subregions by drawing lines parallel to coordinate axes. Number the rectangles which lie entirely inside the region from 1 to n.

Let (x_r, y_r) be any point inside the rth rectangle whose area is δA_r

$$f(x_1, y_1)\delta A_1 + f(x_2, y_2)\delta A_2 + \cdots + f(x_n, y_n)\delta A_n$$

$$= \sum_{r=1}^n f(x_r, y_r)\delta A_r \qquad (3.152)$$

Let the number of these subregions increases indefinitely such that the largest linear dimension (i.e., diagonal) of δA_r approaches zero. The limit of the sum (i), if it exists, irrespective of the mode of subdivision, is called double integral of $f(x, y)$ over the region R and is denoted by $\iint_R f(x, y)\,dA$.

In other words, $\underset{\substack{n \to \infty \\ \delta A \to 0}}{\mathrm{Lt}} \sum_{r=1}^n f(x_r, y_r)\delta A_r$

$$= \iint_R f(x, y)\,dA \text{ or } \iint_R f(x, y)\,dx\,dy \text{ or } \iint_R f(x, y)\,dy\,dx \qquad (3.153)$$

3.7.3.5 Change of Order of Integration

In a double integral, if limits of integration are constant, then order of integration is immaterial, provided limits of integration are changed accordingly. Thus

$$\int_c^d \int_a^b f(x,y)\,dx\,dy = \int_a^b \int_c^d f(x,y)\,dy\,dx \qquad (3.154)$$

But if limits of integration are variable, a change in the order of integration changes in the limits of integration.

3.7.3.6 Triple Integrals

Consider a function $f(x,y,z)$ which is continuous at every point of the 3 dimensional finite region V. Divide V into n elementary volumes $\delta V_1, \delta V_2, \ldots \delta V_n$. Let (x_r, y_r, z_r) be any point within rth subdivision δV_r. Consider the sum $\sum_{r=1}^{\infty} f(x_r, y_r, z_r)\delta V_r$. The limit of this sum, if it exists, as $n \to \infty$ and $\delta V_r \to 0$ is called triple integral of $f(x,y,z)$ over the region V and is denoted by $\iiint f(x,y,z)\,dV$. It can also be expressed as the repeated integral $\int_{x_1}^{x_2} \int_{y_1}^{y_2} \int_{z_1}^{z_2} f(x,y,z)\,dx\,dy\,dz$.

3.7.4 FOURIER SERIES

If numbers $a_0, a_1, \ldots a_n, b_0, b_1, \ldots b_n$ are derived from a function f by means of Euler–Fourier formulae,

$$a_n = \frac{1}{\Pi}\int_{-\Pi}^{\Pi} f(x)\cos nx\,dx \quad (n=0,1,2,\ldots) \quad (3.155a)$$

$$b_n = \frac{1}{\Pi}\int_{-\Pi}^{\Pi} f(x)\sin nx\,dx \quad (n=1,2,\ldots) \quad (3.155b)$$

Then, the series $a_0 + \sum_{n=1}^{\infty}(a_n \cos nx + b_n \sin nx)$ is called Fourier series.

3.7.4.1 Dirichlet's Condition

Assumptions for expansion in a Fourier's series can be listed as:
Given function $f(x)$ is assumed to be defined and single valued in the given range $(-l, l)$
$f(x)$ is periodic outside $(-l, l)$ with period $2l$

Series $a_0 + \sum_{n=1}^{\infty}(a_n \cos nx + b_n \sin nx)$ is uniformly convergent, so that term-by-term integration of the series is possible
Even and odd function of x (sine series and cosine series)
Case 1: when $f(x)$ is an even function.

$$f(x) = a_0 + \sum_{n=1}^{\infty} a_n \cos nx \qquad (3.156a)$$

where, $a_0 = \frac{1}{\Pi}\int_0^{\Pi} f(x)\,dx \qquad (3.156b)$

$$a_n = \frac{2}{\Pi}\int_0^{\Pi} f(x)\cos nx\,dx \qquad (3.156c)$$

$$b_n = 0 \qquad (3.156d)$$

Case 2: when $f(x)$ is an odd function.

$$f(x) = \sum_{n=1}^{\infty} b_n \sin nx \qquad (3.157a)$$

where $b_n = \frac{2}{\Pi}\int_0^{\Pi} f(x)\sin nx\,dx \qquad (3.157b)$

$$a_0 = 0 \qquad (3.157c)$$

$$a_n = 0 \qquad (3.157d)$$

3.7.4.2 Fourier Series in the Interval (a, b)

If function $f(x)$ satisfies Dirichlet's conditions in the interval (a, b), then

$$f(x) = a_0 + \sum_{n=1}^{\infty}\left(a_n \cos\frac{2n\Pi x}{b-a} + b_n \sin\frac{2\Pi nx}{b-a}\right) \qquad (3.158a)$$

where $a_0 = \frac{1}{b-a}\int_a^b f(x)\,dx \qquad (3.158b)$

$$a_n = \frac{2}{b-a}\int_a^b f(x)\cos\frac{2n\Pi x}{b-a}\,dx \qquad (3.158c)$$

$$b_n = \frac{2}{b-a}\int_a^b f(x)\sin\frac{2n\Pi x}{b-a}\,dx \qquad (3.158d)$$

3.7.5 SERIES

3.7.5.1 Sequences

A set of numbers $u_1, u_2, \ldots u_n$ formed according to some definite rule is called a sequence. A sequence is generally denoted by $\{u_n\}$. A sequence is called a finite sequence, if it has a finite number of terms and if it has infinite number of terms, then it is called infinite sequence.

3.7.5.2 Series

Sum of the corresponding terms of the sequence $u_1, u_2, u_3 \ldots$ i.e., $u_1 + u_2 + u_3 + \ldots$ is called series. A series is called a finite or infinite according as the corresponding sequence is finite or infinite. Infinite series $u_1 + u_2 + u_3 + \cdots + u_n + \cdots$ is denoted by $\sum u_n$ and sum of the first n terms series is denoted by S_n.

$$S_n = u_1 + u_2 + u_3 + \cdots + u_n \qquad (3.159)$$

A series $u_1 + u_2 + u_3 + \cdots$ is called positive term series, if $u_n > 0 \,\forall\, n$.

A series $u_1 - u_2 + u_3 \cdots$ is called alternative series, if $u_n > 0 \,\forall\, n$.

The infinite series $u_1 + u_2 + u_3 + \cdots$ is said to be convergent, if $\lim_{n\to\infty} S_n = \lim_{n\to\infty} (u_1 + u_2 + u_3 + \cdots + u_n)$ is finite. It is said to be divergent, if $\lim_{n\to\infty} S_n = \lim_{n\to\infty} (u_1 + u_2 + u_3 + \cdots + u_n) = +\infty$ or $-\infty$. It becomes an oscillatory series, if $\lim_{n\to\infty} S_n = \lim_{n\to\infty} (u_1 + u_2 + u_3 + \cdots + u_n)$ is not defined or oscillates between two limits.

3.7.5.3   Series Tests

3.7.5.3.1   Leibnitz's Test (Alternating Series Test)

Infinite series $u_1 - u_2 + u_3 - u_4 + \cdots$ is convergent, if

$$u_1 > u_2 > u_3 > u_4 \cdots$$

and

$$u_n \to 0 \text{ as } n \to \infty$$

3.7.5.3.2   Geometric Series Test

Geometric series $a + ar + ar^2 + \cdots$ is

　　Convergent, if $|r| < 1$

　　Divergent, if $r \geq 1$

　　Oscillatory, when $r \leq -1$

　　Other series tests are as follows: Hyperharmonic or p-series test, Comparison test, D'Alembert's ratio test, Raabe's test, Gauss's test, Logarithmic test, De Morgan and Bertrand's test, Higher Logarithmic test, Cauchy's Root test, Cauchy's condensation test, Cauchy's Integral test, Kummer's test, details of which can be had from the Bibliography provided.

3.7.6   Vector

Vector is completely specified by its magnitude and direction. Vector is represented by a directed line segment as shown in Figure 3.15. Thus $\overline{PQ}$ represents a vector, where magnitude is the length PQ and direction is from P (starting point) to Q (end point).

3.7.6.1   Unit Vector

A vector of unit magnitude is called unit vector. Unit vector corresponding to vector A is written as $\overline{A}$.

3.7.6.2   Null Vector

A vector of zero magnitude which can have no direction associated with it is called zero (or null) vector and is denoted by **0**—a thick zero.

3.7.6.3   Product of Two Vectors

3.7.6.3.1   Scalar or Dot Product

Scalar or dot product of two vectors A and B is defined as scalar $ab \cos\theta$, where θ is the angle between A and B.

$$\text{Mathematically,} A \cdot B = \cos\theta \tag{3.160}$$

3.7.6.3.2   Vector or Cross Product

Vector or cross product of two vectors A and B is defined as a vector $ab \sin\theta$, where θ is the angle between A and B as shown in Figure 3.16.

$$\text{Mathematically,} A \times B = ab \sin\theta\, N \tag{3.161}$$

N is a unit vector normal to the plane A and B.

3.7.6.4   Differentiation of Vector

If a vector R varies continuously as a scalar variable changes, then R is said to be function of t and is written as $R = f(t)$.

In scalar calculus, we define derivative of vector function $R = f(t)$ as

$$\frac{dR}{dt} \text{ or } \frac{df}{dt} \text{ or } F'(t) = \lim_{\delta t \to 0} \frac{F(t + dt) - F(t)}{dt}$$

If $F(t)$ has constant magnitude, then $F \cdot \dfrac{df}{dt} = 0$ (3.162)

for $F(t) \cdot F(t) = \left[F(t)\right]^2 = $ constant and $\dfrac{dF}{dt}$ is perpendicular to F

If $F(t)$ has constant or fixed direction, then $F \times \dfrac{dF}{dt} = 0$ (3.163)

3.7.6.5   Vector Calculus

Consider a vector operator ∇ (del) defined by

$$\nabla = i \frac{\partial}{\partial x} + j \frac{\partial}{\partial y} + k \frac{\partial}{\partial z} \tag{3.164}$$

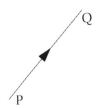

FIGURE 3.15   A vector represented by directed line segment.

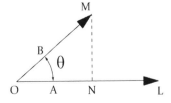

FIGURE 3.16   A vector or cross product of two vectors A and B, where θ is the angle between A and B.

3.7.6.5.1   Gradient

Del (∇) applied to vector function ∇f is defined as the gradient of the scalar point function f and written as grad f. Thus gradient f is defined as

$$\text{Grad } f = \nabla f = i \frac{\partial f}{\partial x} + j \frac{\partial f}{\partial y} + k \frac{\partial f}{\partial z} \qquad (3.165)$$

Grad f is a vector normal to surface f which is a constant and has a magnitude equal to the rate of change of f along this normal.

3.7.6.5.2   Directional Derivative

If δr denotes the length PP' and N' is a unit vector in the direction PP', then limiting value of $\dfrac{\delta f}{\delta r}$ as $\delta r \to 0$ is called directional derivative of f at P along the direction PP'.

$$\text{Since } \delta r = \frac{\delta N}{\cos \alpha} = \frac{\delta N}{N} \cdot N' \qquad (3.166)$$

$$\frac{\partial f}{\partial r} = \underset{\delta r \to 0}{\text{Lt}} \left[N \cdot N \frac{\delta f}{\delta n} \right] = \vec{N} \frac{\delta f}{\delta n} \vec{N} = N' \nabla f \qquad (3.167)$$

Thus directional derivative of f in the direction of N' is the resolved part of ∇f in the direction of N'.

$$\text{Since } \nabla f N' = |\nabla f| \cos \alpha \le |\nabla f| \qquad (3.168)$$

Therefore, ∇f gives the maximum rate of change of f.

3.7.6.5.3   Vector Identities

$$\text{grad}(U + V) = \text{grad}\, U + \text{grad}\, V,$$

$$\text{i.e., } \quad \nabla(U + V) = \nabla U + \nabla V \qquad (3.169)$$

$$\text{div}\left(\vec{A} + \vec{B} \right) = \text{div } A + \text{div } B$$

$$\text{i.e., } \quad \nabla \cdot A = \nabla \cdot B \qquad (3.170)$$

$$\text{Curl}\left(\vec{A} + \vec{B} \right) = \text{Curl } A + \text{Curl } B$$

$$\text{i.e., } \quad \nabla \times (A + B) = \nabla \times A + \nabla \times B \qquad (3.171)$$

$$\nabla \cdot (U\vec{A}) = (\nabla U) \cdot A + U(\nabla \cdot A) \qquad (3.172)$$

$$\nabla \times (U\vec{A}) = (\nabla U) \times A + U(\nabla \times A) \qquad (3.173)$$

$$\nabla \times (\vec{A} \times \vec{B}) = B \cdot (\nabla \times A) - A \cdot (\nabla \times B) \qquad (3.174)$$

$$\nabla \cdot (\vec{A} \times \vec{B}) = (B \cdot \nabla)A - B(\nabla \cdot A) - (A \cdot \nabla)B + A(\nabla \cdot B) \qquad (3.175)$$

3.7.6.6   Divergence

The divergence of a continuously differentiable vector point function F is denoted by Divergence F and is define by the equation

$$\text{div } F = \nabla F = i \cdot \frac{\partial F}{\partial x} + j \cdot \frac{\partial F}{\partial y} + k \cdot \frac{\partial F}{\partial z} \qquad (3.176)$$

if $F = fi + \phi j + \psi k$, then

$$\text{div } F = \nabla F = \left(i \cdot \frac{\partial}{\partial x} + j \cdot \frac{\partial}{\partial y} + k \cdot \frac{\partial}{\partial z} \right)(fi + \phi j + \psi k) \qquad (3.177)$$

$$\text{div } F = \left(i \cdot \frac{\partial f}{\partial x} + j \cdot \frac{\partial \varphi}{\partial y} + k \cdot \frac{\partial \psi}{\partial z} \right) \qquad (3.178)$$

3.7.6.7   Curl

The curl of a continuously differentiable vector point function F is defined by the equation

$$\text{Curl } F = \nabla \times F = i \times \frac{\partial F}{\partial x} + j \times \frac{\partial F}{\partial y} + k \times \frac{\partial F}{\partial z} \qquad (3.179)$$

if $F = fi + \phi j + \psi k$, then

$$\nabla \times F = \left(i \cdot \frac{\partial}{\partial x} + j \cdot \frac{\partial}{\partial y} + k \cdot \frac{\partial}{\partial z} \right) \times (fi + \phi j + \psi k)$$

$$= \begin{vmatrix} i & j & k \\ \dfrac{\partial}{\partial x} & \dfrac{\partial}{\partial y} & \dfrac{\partial}{\partial z} \\ f & \phi & \psi \end{vmatrix} \qquad (3.180)$$

$$= i\left(\frac{\partial \psi}{\partial y} - \frac{\partial \phi}{\partial z} \right) + j\left(\frac{\partial \psi}{\partial y} - \frac{\partial f}{\partial z} \right) + k\left(\frac{\partial \phi}{\partial x} - \frac{\partial f}{\partial y} \right)$$

3.7.7   Line Integral

Consider a continuous vector function $F(R)$ which is defined at each point of curve C in space. Divide C into n parts as shown in Figure 3.17 at the points $P_0, P_1, \ldots, -P_{i-1}, P_i, \ldots P_n$. Let their position vectors be $R_0, R_1, \ldots, R_{i-1}, R_i, \ldots, R_n$. Let U_i be the position vector of any point on the arc $-P_{i-1}P_i$. Now consider the sum as $S = \displaystyle\sum_{i=0}^{n} F(U_i) \cdot \delta R$ where $\delta R_i = R_i - R_{i-1}$.

Limit of this sum, as $n \to \infty$ in such a way that $\delta R_i \to 0$, provided it exists, is called tangential line integral of $F(R)$ along C and is symbolically written as

$$\int_C F(R) \cdot dR \text{ or } \int_C F \cdot \frac{dR}{dt} dt$$

When path of integration is a closed curve, then this is denoted by using $\oint$ in place of $\int$.

If $\quad F(R) = f(x, y, z)i + \phi(x, y, z)j + \psi(x, y, z)k \quad$ and
$dR = i\,dx + j\,dy + k\,dz$ then

$$\int_C F(R) \cdot dR = \int_C \left(f\,dx + \phi\,dy + \psi\,dz \right) \qquad (3.181)$$

Two other types of line integral are $\int_C F \times dR$ and $\int_C f\,dR$ which are both vectors.

3.7.8 Surfaces

A surface S may be represented by $F(x, y, z) = 0$. The parametric representation of S is of the form

$$R(u, v) = x(u, v)i + y(u, v)j + z(u, v)k \qquad (3.182)$$

and continuous function $u = \phi(t)\, v = \psi(t)$ of a real parameter represents a curve C on this surface S. Parametric representation of the circular cylinder $x^2 + y^2 = a^2, -1 \le z \le 1$ (radius a and height 2) is

$$R(u, v) = a\cos ui + a\sin uj + vk \qquad (3.183)$$

where parameters u and v vary in the rectangle $0 \le u \le 2\pi$ and $-1 \le v \le 1$. Also $u = t, v = bt$ represent a circular helix on this circular cylinder. The equation of the circular helix is

$$R = a\cos ti + a\sin tj + btk \qquad (3.184)$$

Differentiating $R = R(u, v)$, with respect to t, we get

$$\frac{dR}{dt} = \frac{\partial R}{\partial u} \cdot \frac{du}{dt} + \frac{\partial R}{\partial v} \cdot \frac{dv}{dt} \qquad (3.185)$$

The vectors $\left(\dfrac{\partial R}{\partial u} \right)$ and $\left(\dfrac{\partial R}{\partial v} \right)$ are tangential to S at P and determine tangent plane of S at $P \cdot N$

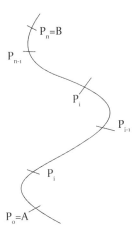

FIGURE 3.17 A continuous vector function $F(R)$ defined at each point of curve C in space which is divided into n parts at the points $P_0, P_1, \dots, -P_{i-1}, P_i, \dots P_n$.

$\left(\dfrac{\partial R}{\partial u} \right) \times \left(\dfrac{\partial R}{\partial v} \right) (\ne 0)$ gives a normal vector N of S at P.

3.7.8.1 Smooth Surfaces

If S has a unique normal at each of its points whose direction depends continuously on the points of S, then the surface S is called a smooth surface. If S is not smooth but can be divided into infinitely many smooth portions, then it is called piecewise smooth surface. e.g., surface of sphere is smooth while the surface of a cube is piecewise smooth.

3.7.8.2 Orientable and Nonorientable Surfaces

A surface S is said to be orientable or two sided if the positive normal direction at any direction p of S can be continued in a unique and continuous way to the entire surface. If positive direction of the normal is reversed as we move around a curve on S passing through P, then the surface generate is nonorientable.

3.7.8.3 Surface Integral

Consider a continuous function $F(R)$ and a surface S. Divide S into a finite number of subsurfaces. Let the surface element surrounding any point $P(R)$ be δS which can be regarded as a vector; its magnitude being the area and its direction that of the outward normal to the element.

Consider the sum $\sum F(R) \cdot \delta S$, where the summation extends all over the subsurfaces. The limit of this sum as the number of subsurfaces tends to infinity and the area of each subsurface is tend to zero, is called normal surface integral of $F(R)$ over S and is denoted by $\int_S F \cdot dS$ or $\int_S F \cdot N\,dS$, where N is unit outward normal at P to S.

Other types of surface integrals are $\int_S F \times dS$ and $\int_S f\,ds$ which are both vectors.

Notation: Only one integral sign when there is one differential (say dR or dS) and two (or three) signs when there are two (or three) differentials.

3.7.8.4 Flux across a Surface

If F represents velocity of a fluid particle, then total outward flux of F across a closed surface S is the surface integral $\int_S F \cdot dS$.

When flux of F across every surface S in a region E vanishes, F is said to be a solenoid vector point function in E.

F could equally well be taken as any other physical quantity, e.g., gravitational force, electric force, and magnetic force.

3.7.9 Volume Integral

Consider a continuous vector $F(R)$ and surface S enclosing the region E.

Divide E into finite number of subregions $E_1, E_2, \dots E_n$. Let δv_i be the volume of subregion E_i enclosing any point whose position vector is R_i.

Consider the sum $V = \sum_{i=1}^{n} F(R_i)\delta v_i$

The limit of this sum as $n \to \infty$ in such a way that $\delta v_i \to 0$ is called volume integral of $F(R)$ over E and is symbolically written as $\int_E F\,dv$.

If $F(R) = f(x,y,z)i + \phi(x,y,z)j + \psi(x,y,z)k$

$\Rightarrow \quad \delta v = \delta x\,\delta y\,\delta z$

then $\int_R F\,dv = i\iiint_E f\,dx\,dy\,dz + j\iiint_E \phi\,dx\,dy\,dz$

$$+ k\iiint_E \psi\,dx\,dy\,dz \qquad (3.186)$$

3.7.10 STROKE'S THEOREM

The close line integral of any vector $\bar{A}$ integral over any close curve C is always equal to the surface integral of the curl of vector A integrated over the surface area S which is enclosed by the closed curve C.

$$\oint_C (\bar{A}) \cdot (\overline{dl}) = \iint_S (\nabla \times \bar{A})\,ds \qquad (3.187)$$

3.7.11 GREEN THEOREM

Let $\phi(x,y), \psi(x,y), \phi y$ and ψx be continuous in a region E of the xy plane bound by a closed curve C, then

$$\int_C \phi\,dx + \psi\,dy = \iint_E \left(\frac{\partial \psi}{\partial x} - \frac{\partial \phi}{\partial y}\right)dx\,dy \qquad (3.188)$$

This equation can be extended to regions which may be divided into finite number of subregions such that to boundary of each is cut almost in two points by any line parallel to either axis apply in equation (i) to each of the subregion and adding result, the surface integral combines into an integral over the whole region, the line integral over the common boundary cancel, whereas remaining line integrals combine into the line integral over the external curve C. This theorem converts line integrals around a closed curve into double integrals and is a special case of Stroke's theorem.

3.7.12 GAUSS'S DIVERGENCE THEOREM (RELATION BETWEEN SURFACE AND VOLUME INTEGRALS)

"Volume integral of a vector field F taken over a any volume V is equal to the surface integral of F taken over the closer surface surrounding the volume V."

Mathematically, for any vector field $\bar{F}$,

$$\int_S F \cdot N\,d_s = \int_E \operatorname{div} F d_v \qquad (3.189)$$

where N is the unit external vector

If $F(R) = f(x,y,z)i + \phi(x,y,z)j + \psi(x,y,z)k \qquad (3.190)$

Then, $\iint (f\,dy\,dz + \phi\,dz\,dx + \psi\,dx\,dy)$

$$= \iiint_E \left(\frac{\partial f}{\partial x} + \frac{\partial \phi}{\partial y} + \frac{\partial \psi}{\partial z}\right)dx\,dy\,dz \qquad (3.191)$$

4 Engineering Mechanics

4.1 STATICS OF PARTICLE: FORCES ON A PLANE

A force can be depicted by its magnitude, direction, and point of application. A force system or a system of forces refers to several forces acting on a body.

4.1.1 Types of System of Forces

4.1.1.1 Coplanar Force System

Forces which lie on a single plane are called coplanar forces or coplanar force system which is shown in Figure 4.1.

4.1.1.2 Concurrent Force System

When all the forces pass through a common point, the force system is known as concurrent force system and forces are known as concurrent forces which are shown in Figure 4.2.

4.1.1.3 Collinear Force System

If the forces are acting in a single straight line as shown in Figure 4.3, this force system is known as collinear forces.

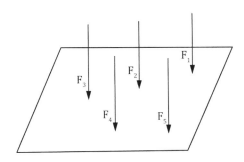

FIGURE 4.1 Coplanar force system.

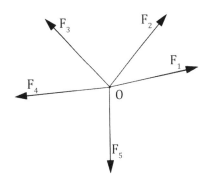

FIGURE 4.2 Concurrent force system.

4.1.1.4 Parallel Force System

Figure 4.4 shows the parallel forces which are forces whose lines of action are parallel to each other. Parallel forces are of two types:

  a. Like parallel forces
     These are forces which are parallel to each other with their lines of action in the same direction. In Figure 4.4, F_2, F_3, and F_4 are like parallel forces.
  b. Unlike parallel forces
     These are forces which are parallel to each other with their lines of action in different directions. In Figure 4.4, F_1 and F_2 are unlike parallel forces.

4.1.2 Resultant of Forces

A single force which can be substituted for a set of forces in a force system and cause the same external effect is known

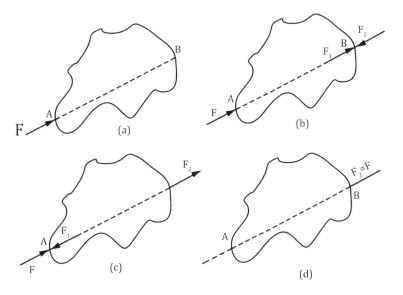

FIGURE 4.3 (a–d) Collinear force system.

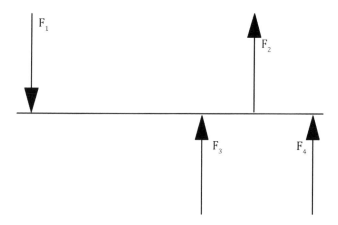

FIGURE 4.4   Parallel force system.

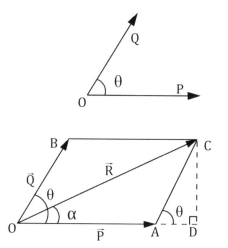

FIGURE 4.6   Parallelogram law of forces.

as the resultant of coplanar forces. Resultant of two forces is shown in Figure 4.5. The process of determining the resultant force is called the composition of forces. The various methods for finding out the resultant forces are parallelogram law, triangle law, and polygon law of forces.

4.1.2.1   Parallelogram Law of Forces

If two forces acting at a point are represented in magnitude and direction by the two adjacent sides of a parallelogram, then their resultant can be represented in magnitude and direction by the diagonal passing through the point. Consider two forces $\vec{P}$ and $\vec{Q}$ acting at a point O and inclined at an angle θ as shown in Figure 4.6. The forces $\vec{P}$ and $\vec{Q}$ can be represented in magnitude and direction by the sides OA and OB of a parallelogram OACB as shown in Figure 4.6. The resultant of the forces $\vec{P}$ and $\vec{Q}$ is the diagonal OC of the parallelogram represented as $\vec{R}$. The magnitude of the resultant is

$$R = \sqrt{\left(P^2 + Q^2 + 2PQ\cos\theta\right)} \tag{4.1a}$$

The direction of the resultant is

$$\alpha = \tan^{-1}\left(\frac{Q\sin\theta}{P + Q\cos\theta}\right) \tag{4.1b}$$

If P and Q are perpendicular, $\theta = 90°$. Then,

$$R = \sqrt{\left(P^2 + Q^2\right)} \tag{4.2a}$$

$$\alpha = \tan^{-1}\left(\frac{Q}{P}\right) \tag{4.2b}$$

4.1.2.2   Triangle Law of Forces

The law states that if two forces acting at a point are represented in magnitude and direction by the two adjacent sides of a triangle taken in order, then the resultant of the forces in magnitude and direction is given by the closing side of the triangle taken in the reverse order.

In Figure 4.7a, forces F_1 and F_2 act at an angle θ. In Figure 4.7b, $\overrightarrow{OA}$ and $\overrightarrow{AB}$ represent the forces F_1 and F_2 in magnitude and direction. The closing side $\overrightarrow{OB}$ of the triangle taken in the reverse order represents the resultant R of the forces P and Q. The magnitude and the direction of R can be found by using sine and cosine laws of triangles.

4.1.2.3   Polygon Law of Forces

Polygon law of forces is an extension of triangle law of forces and is used to determine the resultant of more than two concurrent forces. It states that if a number of forces acting simultaneously at a point in a plane are represented in magnitude and direction by the sides of a polygon taken in order, then the resultant of all these forces may be represented in magnitude and direction by the closing side of the polygon taken in opposite order.

Consider the forces F_1, F_2, F_3, and F_4 are acting at a point as shown in Figure 4.8a. Starting from the point O, $\overrightarrow{OA}$ represents the force F_1 in magnitude (using suitable scales) and direction. From the tip A, $\overrightarrow{AB}$ is drawn representing the force F_2. Similarly, $\overrightarrow{BC}$ represents the force F_3, and $\overrightarrow{CD}$ represents force F_4. The starting point O is joined to the end point D giving $\overrightarrow{OD}$ in opposite order. $\overrightarrow{OD}$ represents the resultant force R in magnitude and direction as shown in Figure 4.8b.

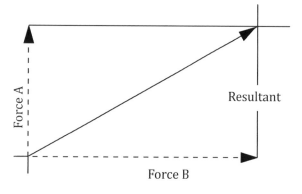

FIGURE 4.5   Resultant of two forces.

$$R = F_1 + F_2 + F_3 + F_4 \tag{4.3}$$

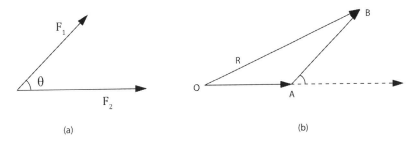

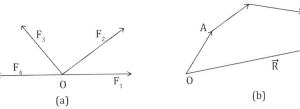

(a)

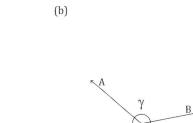

FIGURE 4.7 (a and b) Triangle law of forces.

FIGURE 4.8 (a and b) Polygon law of forces.

4.1.2.4 Resultant of Coplanar Concurrent Force Systems

Resultant of a number of coplanar concurrent forces is given by

$$R = \sqrt{\left(\sum F_x^2 + \sum F_y^2\right)} \qquad (4.4)$$

where $\sum F_x$ and $\sum F_y$ are the sum of components of all the forces acting along two mutually perpendicular X and Y directions.

4.1.3 Equilibrium of Coplanar Forces

A number of forces acting on a particle are said to be in equilibrium when their resultant force is zero. If the resultant force is not equal to zero, then the particle can be brought to rest by applying a force equal and opposite to the resultant force. Such a force is called equilibrant.

4.1.3.1 Lami's Theorem

Lami's theorem is an equation which relates the magnitudes of three coplanar, concurrent, and noncollinear forces, which keeps an object in static equilibrium, with the angles directly opposite to the corresponding forces. According to the theorem,

$$\frac{A}{\sin \alpha} = \frac{B}{\sin \beta} = \frac{C}{\sin \gamma} \qquad (4.5)$$

where A, B, and C are the magnitudes of three coplanar, concurrent, and noncollinear forces, which keep the object in static equilibrium. α, β, and γ are the angles directly opposite to the forces A, B, and C, respectively, as shown in Figure 4.9.

FIGURE 4.9 Lami's theorem.

4.1.3.2 Conditions for the Equilibrium of Coplanar Concurrent Force Systems

Resultant of a number of coplanar concurrent forces is given by Equation 4.4. For the resultant force to be zero, both $\sum F_x$ and $\sum F_y$ must be zero. Therefore, the equations of equilibrium are

$$\sum F_x = 0 \qquad (4.6a)$$

$$\sum F_y = 0 \qquad (4.6b)$$

4.2 MOMENT OF A FORCE

A force produces a twisting effect about an axis perpendicular to the plane which contains the line of action of the force. The measure of this capacity of the force is called as moment. The moment is calculated as the product of the force and the perpendicular distance from the axis to the line of action of the force. The moment center is defined as the intersection of the plane and the axis. The moment arm is the perpendicular distance from the moment center to the line of action of the force. Conventionally, clockwise moments are assumed to be positive and vice versa.

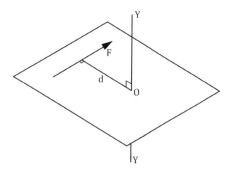

FIGURE 4.10   Moment of a force.

Figure 4.10 shows the details of the moment of a force. If O is the moment center and d is the moment arm, the moment M of force F about point O will be equal to the product of F and d. The moment acts clockwise and hence the figure shows a positive moment.

$$M = Fd \qquad (4.7)$$

4.2.1   VARIGNON'S THEOREM

It was the French mathematician Pierre Varignon (1654–1722) who formulated the Varignon's Theorem. The theorem was put forth in 1687 in his book *Projet d' unè nouvelle mècha-nique*. Varignon's theorem states that the moment of a resultant of two concurrent forces about any point is equal to the algebraic sum of the moments of its components about the same point. The principle of moments states that the moment of the resultant of a number of forces about any point is equal to the algebraic sum of the moments of all the forces of the system about the same point.

From Figure 4.11,

$$R \times r = F_1 \times r_1 + F_2 \times r_2 \qquad (4.8)$$

where F_1, F_2 are the component forces, R is the resultant force, and r, r_1, r_2 are perpendicular distances to the line of action of forces R, F_1, and F_2, respectively.

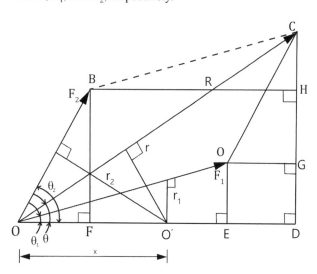

FIGURE 4.11   Varignon's theorem.

4.3   COUPLE

A couple (also called as force couple or pure moment) is a system of forces that has no resultant force but definitely has a resultant moment. The resultant moment of a couple is called a torque.

A couple consists of two equal forces that are parallel to each other and acting in opposite direction. Referring Figure 4.12, the magnitude of the couple can be computed using Equation 4.9. If F is the magnitude of two forces and d is the moment arm or the perpendicular distance between the forces,

$$\text{Magnitude of couple} = Fd \qquad (4.9)$$

4.4   STATICS OF PARTICLE: NONCOPLANAR CONCURRENT FORCES IN SPACE

Some forces do not lie in the same plane; however, their line of action may sometimes pass-through a single point. Such forces are called noncoplanar concurrent forces.

The magnitude of a force F in space as shown in Figure 4.13 is

$$F = \sqrt{\left(F_x^2 + F_y^2 + F_z^2\right)} \qquad (4.10)$$

Components of a force in space as shown in Figure 4.14 are

$$F_x = F \cos\theta_x \qquad (4.11a)$$

$$F_y = F \cos\theta_y \qquad (4.11b)$$

$$F_z = F \cos\theta_z \qquad (4.11c)$$

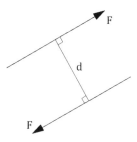

FIGURE 4.12   Couple.

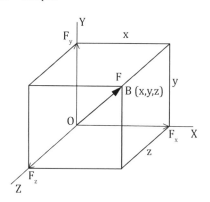

FIGURE 4.13   Noncoplanar concurrent forces system.

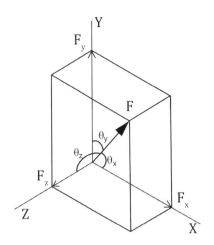

FIGURE 4.14 Components of a force in space.

Their direction cosines are

$$\cos\theta_x = \frac{F_x}{F} \qquad (4.12a)$$

$$\cos\theta_y = \frac{F_y}{F} \qquad (4.12b)$$

$$\cos\theta_z = \frac{F_z}{F} \qquad (4.12c)$$

4.4.1 RESULTANT OF CONCURRENT FORCE SYSTEMS IN SPACE

Components of the resultant are

$$R_x = \sum F_x \qquad (4.13a)$$

$$R_y = \sum F_y \qquad (4.13b)$$

$$R_z = \sum F_z \qquad (4.13c)$$

Magnitude of the resultant is

$$R = \sqrt{\left(R_x^2 + R_y^2 + R_z^2\right)} \qquad (4.14)$$

4.4.2 EQUILIBRIUM OF CONCURRENT SPACE FORCES

For equilibrium of concurrent space forces, the resultant of all the forces must be equal to zero.

$$\sum F_x = 0 \qquad (4.15a)$$

$$\sum F_y = 0 \qquad (4.15b)$$

$$\sum F_z = 0 \qquad (4.15c)$$

Also, the sum of moments produced due to the forces should be equal to zero.

$$\sum M_x = 0 \qquad (4.16a)$$

$$\sum M_y = 0 \qquad (4.16b)$$

$$\sum M_z = 0 \qquad (4.16c)$$

4.5 STATICS OF RIGID BODY: NONCONCURRENT, COPLANAR FORCES ON A PLANE RIGID BODY

The force system in which lines of action of individual forces lie in the same plane but act at different points of applications as shown in Figure 4.15 are called as coplanar nonconcurrent force system.

4.5.1 CONDITIONS OF EQUILIBRIUM OF COPLANAR NONCONCURRENT FORCE SYSTEMS

A rigid body is said to be in equilibrium when the net effect of external forces acting on it is zero. The necessary and sufficient conditions for the equilibrium of a rigid body can be expressed analytically as

$$\sum F_x = 0 \qquad (4.17a)$$

$$\sum F_y = 0 \qquad (4.17b)$$

$$\sum M = 0 \qquad (4.17c)$$

4.5.2 FREE BODY DIAGRAM

The free body diagram is a graphical illustration of an object or a body along with all the external forces acting on it. A sketch showing the physical situation of a system consisting of several bodies is called a space diagram. For a system consisting of several bodies to be in equilibrium, each body of the system considered separately must be in equilibrium. Figure 4.16a shows a case of mowing a lawn using a lawnmower. Figure 4.16b shows its free body diagram representing the forces acting on the lawnmower.

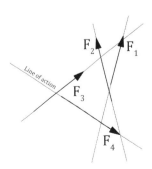

FIGURE 4.15 Coplanar nonconcurrent forces system.

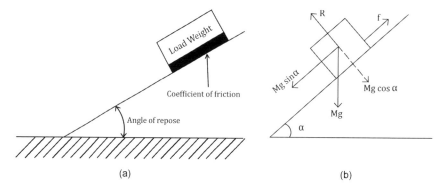

FIGURE 4.16   Free body diagram.

4.5.3 PRINCIPLE OF TRANSMISSIBILITY

It states that the conditions of equilibrium or conditions of motion of a rigid body will remain unchanged if force acting at a given point of the rigid body is replaced by a force of the same magnitude and same direction, but acting at a different point, provided that the two forces have the same line of action.

Consider a rigid body under the action of a force F applied at A and acting along AB as shown in Figure 4.17. Two equal and opposite forces applied at B will not change the condition of the rigid body. Now the removal of force at A and the force at B which is opposite to the force at A will not change the condition of the rigid body.

4.5.4 PRINCIPLE OF SUPERPOSITION OF FORCES

This principle states that the combined effect of force system acting on a particle or a rigid body is the sum of the effects of individual forces.

4.6 TYPES OF SUPPORTS

4.6.1 ROLLER SUPPORTS

Some supports are provided with rollers on the surface where they rest. They are free to rotate as well as displace along the surface. These supports are called as roller supports and are

shown in Figure 4.18. They are in the form of bearing pads commonly placed at one end of long bridge slabs.

4.6.2 HINGED SUPPORTS

Some supports are provided with hinges at the surface of their rest. They are capable of resisting forces acting in any direction of the plane. These supports are called as hinged support and such a support is shown in Figure 4.19. However, they do not provide resistance to rotation. They are the supports used to provide rotation to the doors.

4.6.3 FIXED SUPPORTS

Fixed supports are also known as rigid supports as they can restrain both rotation and translation. A fixed support as shown in Figure 4.20 can resist vertical and horizontal forces as well as

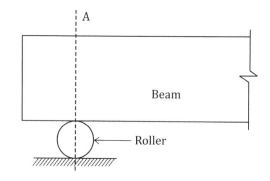

FIGURE 4.18   Roller support.

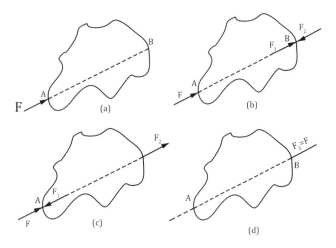

FIGURE 4.17   (a–d) Principle of transmissibility.

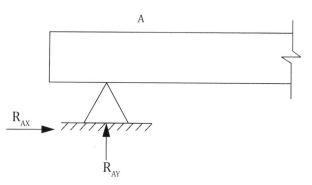

FIGURE 4.19   Hinged support.

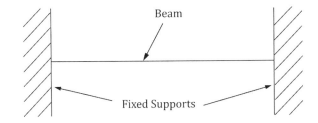

FIGURE 4.20 Fixed support.

moment. For a structure to be stable at least one fixed support is necessary. A common example of a fixed support is the high mast lamp set into a concrete base. The steel reinforcement of a Reinforced Concrete (RC) beam is embedded in an RC column to produce a fixed support as shown in Figure 4.20. Similarly, all the welded joints in a steel structure are examples of fixed supports.

4.6.4 Pinned Supports

Pinned supports can resist both vertical and horizontal forces as shown in Figure 4.21. They are also called as hinged supports. They are not capable of resisting moments. It allows rotation but restricts translation in all directions. Even then, rotation is allowed in one direction only and is restricted in all the other directions. The best example of a pinned support is the knee in our body. It allows rotation in only one direction and resists lateral movements. In practical cases, ideal cases of pinned supports are rarely found. Beams supported on walls or connected to other steel beams can be considered as pinned.

4.7 TYPES OF BEAMS

Horizontal slender structural members are called beams. The loads are transferred horizontally along their length to the supports. At the supports, the loads are then resolved into vertical forces. Beams can resist transverse vertical loads, shear forces, and bending moments. Beams can be classified into different types based on the various types of support given to the beams.

4.7.1 Simply Supported Beam

In some beams, the ends of a beam rest freely on the supports. Such beams are called as simply (freely) supported beams shown in Figure 4.22.

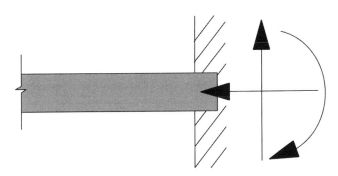

FIGURE 4.21 Pinned support.

FIGURE 4.22 Simply supported beam.

4.7.2 Fixed Beam

A beam which is fixed at both ends is called as a fixed beam or "built-in beam" or "encastre beam" and is shown in Figure 4.23.

4.7.3 Cantilever Beam

A beam fixed at one end and free at the other is called a cantilever beam which is shown in Figure 4.24.

4.7.4 Continuously Supported Beam

A beam with more than two supports is called a continuously supported beam which is shown in Figure 4.25.

4.7.5 Overhanging Beams

Some beams are freely supported at two points and the beam may be extending beyond the supports at either one end or both the ends. Such beams are called overhanging beams shown in Figure 4.26.

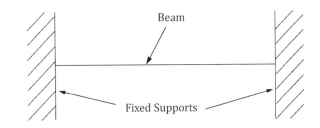

FIGURE 4.23 Fixed beam.

FIGURE 4.24 Cantilever beam.

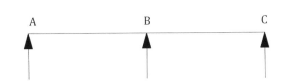

FIGURE 4.25 Continuously supported beam.

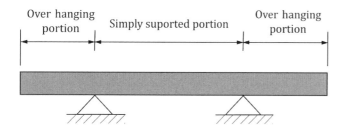

FIGURE 4.26   Overhanging beam.

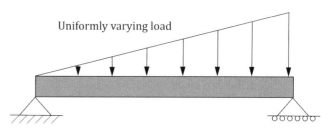

FIGURE 4.29   Uniformly varying load.

4.8   TYPES OF LOADS

4.8.1   Point Loads

Loads applied at certain points in a beam are said to be point loads or concentrated loads shown in Figure 4.27.

4.8.2   Uniformly Distributed Load

It is a type of load which has the same intensity of load over a certain length of the beam shown in Figure 4.28. The total load will be $w \times x$, where w is the intensity of load and x is the loaded length.

4.8.3   Uniformly Varying Load

When the load varies uniformly from one point to another point on the beam, it is called uniformly varying load shown in Figure 4.29. When the intensity of load varies from zero at one end to w/unit length at the other end, over a length x, then the total load will be $\frac{1}{2} \times x \times w$ = the area of the triangle.

4.8.4   Support Reactions

When loads are applied, reactions develop at the supports. A beam is kept in equilibrium by the system of forces consisting of applied loads and reactions. The nature of reactions depends upon the type of support and types of loads. The equations for equilibrium shown in Eqn. 4.17 can be applied to the beams and frames for calculating the reactions at the supports.

4.9   FRICTION

Friction is defined as the resistive force acting in opposite direction at the surface of a body which tends to move or it moves. Frictional forces always act tangentially at the point of contact. Friction may be categorized into two groups—(a) static friction; (b) kinetic friction/dynamic friction.

4.9.1   Static Friction

Even when there is no motion between a body and the surface on which it moves, the body experiences a frictional force under the action of the external force acting on the body. This is called as static friction.

4.9.2   Kinetic Friction/Dynamic Friction

The frictional force experienced by a body when it is in motion is called kinetic friction. There are two types of kinetic friction which are sliding friction and rolling friction.

4.9.2.1   Sliding Friction

When a body slides over another body or on a surface, sliding friction occurs. This is shown in Figure 4.30. The resistive force to sliding is known as sliding frictional force.

4.9.2.2   Rolling friction

When a body rolls over another body or surface, the resistive force is known as rolling friction force shown in Figure 4.31.

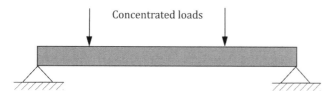

FIGURE 4.27   Point load.

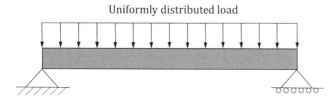

FIGURE 4.28   Uniformly distributed load.

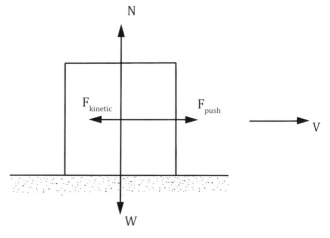

FIGURE 4.30   Sliding friction.

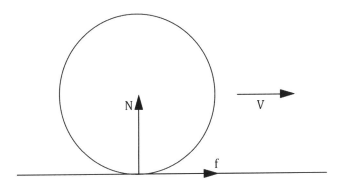

FIGURE 4.31   Rolling friction.

4.9.3   Laws of Friction

The force of friction depends on certain assumptions specified in the laws of friction which are in close agreement with experiments.

4.9.3.1   Law 1

When two bodies are in contact, the direction of forces of friction on one of them at its point of contact is opposite to the direction in which the point of contact tends to move relative to the other.

4.9.3.2   Law 2

The magnitude of frictional force that can be exerted between two surfaces is limited. If the magnitude of forces acting on the body is large enough to exceed this, motion will occur. Thus, limiting friction is defined as the friction exerted when equilibrium breaks and one body slides on another body or on a surface.

4.9.3.3   Law 3

The ratio of the limiting friction to the normal reaction between two surfaces depends on the material property of the surfaces. It does not depend on the magnitude of the normal reaction. This ratio is denoted by μ and is called as coefficient of friction. Thus, if the normal reaction is R, the limiting friction is μR.

4.9.3.4   Law 4

The amount of limiting friction does not depend on the contact area between two surfaces and their shape if the normal reaction is unaltered.

4.9.3.5   Law 5

When a body moves, the direction of friction is opposite to the direction of relative motion. It is independent of velocity. There exists a constant ratio between the magnitude of the frictional force and the normal reaction.

4.9.4   Coefficient of Friction

The coefficient of friction is defined as the ratio of the frictional force to the normal force. This coefficient depends on the material properties of the contact surfaces. A rougher surface will have a higher coefficient.

4.9.5   Calculation of Frictional Force on a Body

Consider Figure 4.32,

N = total reaction perpendicular to the contact surface
f = frictional force
W = weight of the body = mg
m = mass of the body
g = acceleration due to gravity
μ = coefficient of friction
R = resultant of f and N
ϕ = angle of friction

By definition,

$$f = \mu N \qquad (4.18a)$$

From Figure 4.32,

$$\tan\phi = \frac{f}{N} \qquad (4.18b)$$

$$\tan\phi = \mu \qquad (4.18c)$$

Considering the equilibrium of the body, from Equation 4.15b,

$$N = mg \qquad (4.18d)$$

Then, frictional force,

$$f = \mu mg \text{ (from Equation 4.18a)} \qquad (4.18e)$$

If external force, $F < \mu mg$ then, there is no motion in the body and the body will remain so until the frictional force of the body is equal to external force. When $\mu mg > F$ then, body slides over the surface.

4.9.6   Angle of Friction

Just at the time of sliding, the angle between the normal reaction and the resultant reaction is called the angle of static friction as shown in Figure 4.32.

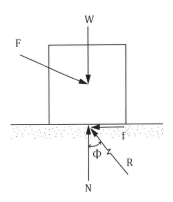

FIGURE 4.32   Forces on a body.

From Equation 4.18c,

$$\phi = \tan^{-1} \mu \qquad (4.19)$$

4.9.7 ANGLE OF REPOSE

Angle of repose is defined as the angle of the inclined plane with horizontal at which the body is in the condition of just sliding as shown in Figure 4.33. In bodies resting on an inclined plane, the angle of friction is always equal to the angle of repose.

4.9.8 CONE OF FRICTION

Figure 4.34 shows an inverted cone with a semicentral angle, α, which is equal to limiting frictional angle, ϕ. This cone is called the cone of friction.

4.9.9 WEDGE FRICTION

Wedges are small pieces of materials with a triangular or trapezoidal cross section as shown in Figure 4.35. Wedges are advantageous in heavy weightlifting, for slightly adjusting the position of a body, etc. Compared to the weight to be lifted, the weight of the wedge is very small. Hence, it is usually neglected. The problems on wedges are usually the problems

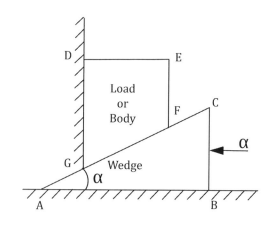

FIGURE 4.35   Wedge friction.

related to the equilibrium of bodies resting on inclined planes. These problems are therefore solved by applying Lami's theorem or by equilibrium method.

4.9.10 LADDER FRICTION

A ladder is used for climbing on the walls. It basically consists of two long uprights which are connected by a number of crossbars called as rungs. They act as the steps while climbing.

A ladder AB is shown in Figure 4.36. Its end A rests on the ground and end B leans against a wall. The following forces act on the ladder.

- Self-weight of the ladder, W, acts downward at its midpoint
- Normal reaction, R_b, and frictional force, $F_b = \mu R_b$, are the two forces at the end B. The ladder may slip downward. Therefore, the frictional force will act upward. In the case of a smooth wall, the frictional force will be equal to zero as $\mu = 0$.
- Normal reaction, R_a, and frictional force, $F_a = \mu R_a$, are the forces at the end A. While slipping, since the ladder has a tendency to move away from the wall, the frictional force will act toward the wall.

Applying equilibrium conditions, given in Equation 4.15, the unknown forces can be determined.

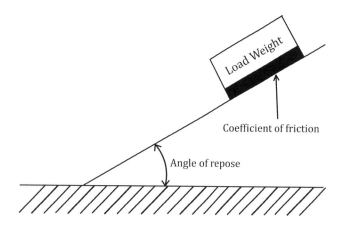

FIGURE 4.33   Angle of repose.

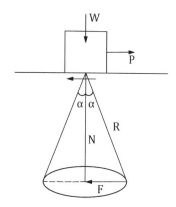

FIGURE 4.34   Cone of friction.

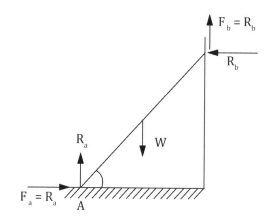

FIGURE 4.36   Ladder friction.

4.10 PROPERTIES OF SURFACES AND SOLIDS

4.10.1 CENTER OF GRAVITY

The center of gravity (CG) is a point where the whole weight of a body acts. The earth attracts all the particles in a body toward its center and this force of attraction is called as the weight of the body. These forces exerted by the particles may be considered to act along parallel lines. There exists a point in the body, through which the resultant of all such parallel forces acts. This point, through which the whole resultant or the weight of the body acts is known as the CG and is shown in Figure 4.37. It may be noted that each body has only one CG.

4.10.2 CENTROID

As CG is for 3D bodies, centroid is for 2D plane figures (like circle, triangle, quadrilateral, etc.). They have only areas, but no mass. The center of area of such figures is known as centroid shown in Figure 4.38.

4.10.3 AXIS OF REFERENCE

For calculating the location of the CG of a body, an assumed axis known as the axis of reference is used with respect to which computations are made. For plane figures, this axis of reference is usually taken as the bottommost line in the figure for calculating Y (distance measured vertically) and the leftmost line of the figure for calculating X (distance measured horizontally) as shown in Figure 4.39.

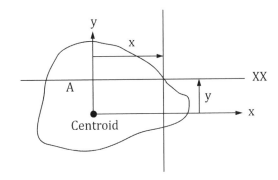

FIGURE 4.38 Centroid.

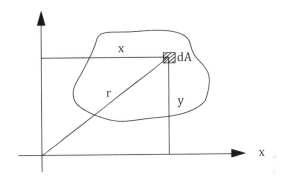

FIGURE 4.39 Axis of reference.

4.10.4 AXIS OF SYMMETRY

Axis of symmetry is a line or axis, which divides the given line, area, or volume into two equal and identical parts.

4.10.5 CG OF SOLID BODIES

Consider a lamina with a definite area. Its plane consists of a number of particles with masses m_1, m_2, m_3, etc., and hence weights m_1g, m_2g, m_3g, etc., with coordinates (x_1, y_1), (x_2, y_2), (x_3, y_3), etc. Taking moments about O, with the reference axis OX in Figure 4.37, the following equations are obtained.

$$\bar{x} = \frac{\sum_{i=1}^{n} m_i x_i}{\sum_{i=1}^{n} m_i} \qquad (4.20a)$$

$$\bar{y} = \frac{\sum_{i=1}^{n} m_i y_i}{\sum_{i=1}^{n} m_i} \qquad (4.20b)$$

4.10.6 CENTROID OF AREA

The method used for finding out the centroid of a plane figure and the CG of a body is the same. The distance of centroid of an area from Y-axis and X-axis is given by Equations 4.21a and b, respectively, in which dA is an elemental area, x is the distance

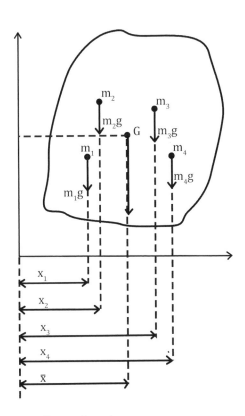

FIGURE 4.37 Center of gravity.

of this elemental area from Y-axis, and y is the distance of elemental area dA from X-axis as shown in Figure 4.40.

$$\bar{x} = \frac{\int x\,dA}{\int dA} \qquad (4.21a)$$

$$\bar{y} = \frac{\int y\,dA}{\int dA} \qquad (4.21b)$$

A composite area can be divided into a number of plane figures like triangle, square, rectangle, circle, semicircle, etc., for which areas and locations of centroids of each area are known. The distances of centroid of the composite area from the Y axis and X axis are given by Equations 4.22a and b, where $a_1, a_2, \ldots\ldots$ are areas of each section, $x_1, x_2, \ldots\ldots\ldots$ are the distances of centroid of $a_1, a_2, \ldots\ldots..$ from the Y-axis, and $y_1, y_2, \ldots\ldots\ldots$ are the distances of centroid of $a_1, a_2, \ldots\ldots..$ from the X-axis as shown in Figure 4.41.

$$\bar{x} = \frac{\int xa}{\int a} = \frac{\sum(xa)}{\sum a} = \frac{a_1 x_1 + a_2 x_2 + \cdots}{a_1 + a_2 + \cdots} \qquad (4.22a)$$

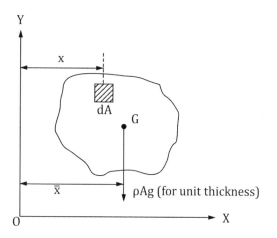

FIGURE 4.40   Centroid of a plane figure.

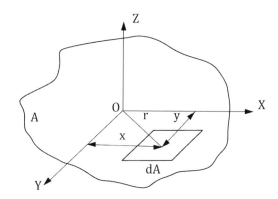

FIGURE 4.41   Centroid of composite area.

$$\bar{y} = \frac{\int ya}{\int a} = \frac{\sum(ya)}{\sum a} = \frac{a_1 y_1 + a_2 y_2 + \cdots}{a_1 + a_2 + \cdots} \qquad (4.22b)$$

4.10.7   THEOREMS OF PAPPUS–GULDINUS

These theorems relate a surface of revolution to its generating curve and a volume of revolution to its generating area.

4.10.7.1   Theorem 1

The area of the surface generated by revolving a plane curve about a nonintersecting axis in the plane of the curve is equal to the product of the length of the curve and the distance travelled by the centroid of the curve while the surface is being generated. The entire area generated by the curve

$$A = \int 2\pi y\,dL = 2\pi \int (y\,dL) = 2\pi \bar{y} L \qquad (4.23)$$

where $2\pi \bar{y}$ is the distance traveled by the centroid of the curve of length L as shown in Figure 4.42.

4.10.7.2   Theorem 2

The volume of a body generated by revolving a plane area about a nonintersecting axis in the plane of the area is equal to the product of area and the distance traveled by the centroid of the plane area while the body is being generated. The entire volume generated by A

$$V = \int 2\pi y\,dA = 2\pi \int y\,dA = 2\pi \bar{y} A \qquad (4.24)$$

where $2\pi \bar{y}$ is the distance traveled by the centroid of area A (see Figure 4.43).

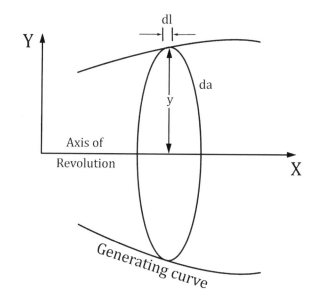

FIGURE 4.42   Pappus–Guldinus Theorem 1.

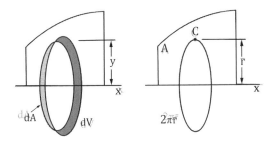

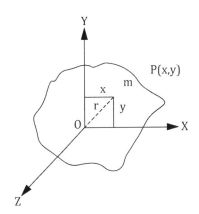

FIGURE 4.43 Pappus–Guldinus Theorem 2.

4.10.8 MOMENT OF INERTIA

The moment of inertia is a measure of the distribution of the mass or area of an object relative to a given axis. For solid bodies, it is called as mass moment of inertia, and for plane figures, it is called as area moment of inertia or second moment of area. For computing the mass moment of inertia, the body is broken down into very small bits of mass dM. The masses of these bits are multiplied by the square of the distance to the x (and y) axis and then summed up for the entire object. The area moment of inertia is the product of area and the square of its moment arm about a reference axis. The area moment of inertia is obtained by adopting the same procedure after breaking down the object into very small bits of area dA.

4.10.8.1 Radius of Gyration

The radius of gyration indicates the distribution of the components of an object around an axis. In terms of the mass moment of inertia, it is measured as the perpendicular distance from the axis of rotation to a point mass (of mass, m) which gives an equivalent inertia to the original object (of mass, m). Mathematically, the radius of gyration is the root mean square distance of the parts of the object from either its center of mass or a given axis. The radius of gyration is given by the following formula:

$$k = \sqrt{\frac{I}{A}} \qquad (4.25)$$

where I is the second moment of area and A is the total cross-sectional area.

4.10.8.2 Perpendicular Axis Theorem

The theorem of perpendicular axis for a plane laminar body states that the moment of inertia of a plane lamina about an axis perpendicular to the plane of the lamina is equal to the sum of the moments of inertia of the lamina about any two mutually perpendicular axes in its own plane and intersecting each other at the point where the perpendicular axis passes through it. If I_{XX} and I_{YY} are the moments of inertia of a plane lamina about the perpendicular axis XX and YY, respectively, which lie in the plane of the lamina and intersect each other at O, then the moment of inertia I_{ZZ} of the lamina about an axis passing through O and perpendicular to its plane is given by Equation 4.26 (see Figure 4.44).

$$I_{ZZ} = I_{XX} + I_{YY} \qquad (4.26)$$

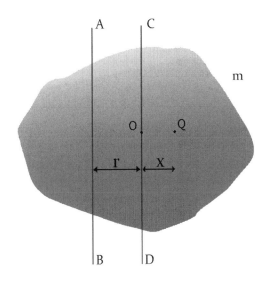

FIGURE 4.44 Perpendicular axis theorem.

4.10.8.3 Parallel Axis Theorem

The theorem of parallel axis states that the moment of inertia of a rigid body about any axis is equal to its moment of inertia about a parallel axis through its center of mass plus the product of the mass of the body and the square of the perpendicular distance between the two parallel axes.

If I_G is the moment of inertia of a plane lamina of area A, about its centroidal axis in the plane of the lamina, then the moment of inertia about any axis AB which is parallel to the centroidal axis and at a distance r from the centroidal axis is given by (see Figure 4.45),

$$I_{AB} = I_G + Ar^2 \qquad (4.27)$$

The parallel axis theorem is used for finding the rectangular moment of inertia about an axis different from the centroidal axis of the section. The centroidal rectangular moments of inertia, $\overline{I_x}$ and $\overline{I_y}$, are determined using Equations 4.28a and b where y and x are the distances of an elemental area dA from XX and YY axes, respectively.

$$\overline{I_x} = \int_A y^2 \, dA \qquad (4.28a)$$

FIGURE 4.45 Parallel axis theorem.

$$\overline{I_y} = \int_A x^2 \, dA \qquad (4.28b)$$

Using the parallel axis theorem, the moments of inertia about the axes XX and YY parallel to the centroidal axes x and y, I_x and I_y can be determined using Equations 4.29a and b:

$$I_x = \overline{I_x} + Ay^2 \qquad (4.29a)$$

$$I_y = \overline{I_y} + Ax^2 \qquad (4.29b)$$

The product of inertia is given using Equation 4.29c.

$$I_{xy} = Axy \qquad (4.29c)$$

4.10.8.4 Moment of Inertia of Composite Areas

The composite area can be divided into a set of simple areas. Moment of inertia of each area about its centroidal axis can be calculated. The centroidal moments of inertia of a cross section composed of several simpler geometric parts are computed using parallel axis theorem. For this, first, break the cross section into several simple shapes. Then, calculate the individual moments of inertia. Next, compute the moments of inertia about the centroidal axes of the entire section using parallel axis theorem. Finally, sum up the moment of inertia of each simple area to get the moment of inertia of the composite area.

4.10.9 INSTANTANEOUS CENTER

The instant center of rotation is also called the instantaneous center or the instantaneous velocity center, or the instant center. It is the point fixed to a body which undergoes planar movement that has zero velocity at a particular instant of time. At this instant, the velocity vectors of the trajectories of other points in the body generate a circular field around this point which is identical to what is generated by a pure rotation.

The properties of the instantaneous center are as follows:

  i. The magnitude of the velocity of any point on a body is proportional to its distance from the instantaneous center and is equal to the angular velocity times the distance.
  ii. The direction of the velocity of any point on a body is perpendicular to the line joining that point and the instantaneous center.

The above properties are used to locate the instantaneous center of a body.

Case 1: When the direction of velocities of any two points on the body is known.

In this case, the point of intersection of lines drawn perpendicular to the direction of velocities will be the instantaneous center.

Case 2: When the direction of velocities is parallel and magnitudes are unequal.

In this case, the point of intersection of the line joining the tip of velocity vectors and either the line joining the points or extension of the line joining the points will be the instantaneous center.

4.11 KINETICS OF PARTICLE

Particle kinetics deals with the study of the movement of the particles and forces which cause this movement. In kinetics of particle, only the translational motion of the particle is considered.

4.11.1 WORK

In mechanics, work is said to be done whenever the point of application of a force is moved along the line of action of that force or along the line of action of the component of that force. The work done by a force F during a differential displacement ds of its point of application is $F\cos\theta\,ds$, where θ is the angle between F and ds as shown in Figure 4.46.

$F\cos\theta$ is the force in the direction of displacement. Alternatively, the work done may be interpreted as the force multiplied by the displacement component $ds\cos\theta$ in the direction of the force.

$$\text{Work done, } dW = F \cdot ds \cos\theta \qquad (4.30)$$

4.11.1.1 Work Done by Friction Force

Work is done in displacing a body with constant velocity. When a body is pushed through a horizontal floor, work is done against friction shown in Figure 4.47.

$$\text{Work } W = \mu_k mg \; s \qquad (4.31)$$

4.11.1.2 Work Done by a Spring

$$W = \frac{1}{2}k\left(x_i^2 - x_f^2\right) \qquad (4.32)$$

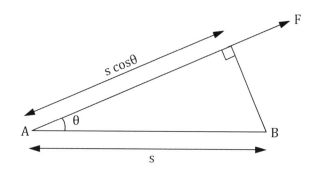

FIGURE 4.46 Work done by a force.

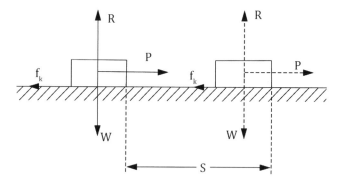

FIGURE 4.47 Work done by friction force.

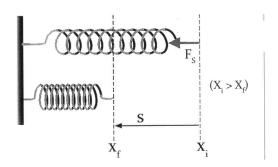

FIGURE 4.48 Work done by a spring.

where

    k = spring constant

    x_i = initial length of spring (see Figure 4.48)

    x_f = final length of spring (both lengths are in stressed condition)

4.11.1.3 Work Done by Torque

$$W = T \times \theta \qquad (4.33)$$

where

    T = torque (in N-m)

4.11.1.4 Work Done by a Force on a Moving Body

It is defined as the product of the force and the distance moved in the direction of the force.

$$W = F \cdot s \qquad (4.34)$$

For moving objects, the quantity of work divided by time (power) has to be calculated. Thus, at any instant, the rate of the work done by a force (measured in Joules/second, or Watts) can be determined. This is computed as the scalar product of the force (a vector) and the velocity vector of the point of application which is termed as instantaneous power. The small amount of work δW that occurs over an instant of time dt is calculated as

$$\delta W = F \cdot ds = F \cdot V dt \qquad (4.35)$$

where "$F \cdot V$" is the Power over the instant dt. Work is computed as the sum of such small amounts of work integrated over the trajectory of the point.

$$W = \int_{t_1}^{t_2} (F \cdot V dt) = \int_c F \cdot ds \qquad (4.36)$$

where c is the trajectory from $\mathbf{x}\,(t_1)$ to $\mathbf{x}\,(t_2)$. This calculation can be generalized for a constant force that is not directed along the line. Then, the dot product $F \cdot ds = F \cos \theta\, ds$, where θ is the angle between the force vector and the direction of movement, i.e.,

$$W = \int_c F \cdot ds = F \cdot s \cos\theta \qquad (4.37)$$

4.11.1.5 Work Done by a Variable Force

Work of a force is the line integral of its scalar tangential component along the path of its application point.

$$W = \int_a^b F(s)\,ds \qquad (4.38)$$

4.11.2 POWER

The rate at which work is done or the rate at which energy is used or transferred is termed as power.

$$\text{Power} = \frac{\text{work done}}{\text{time}} \qquad (4.39)$$

The SI unit for power is Watt denoted as W. A power of 1W means that work is done at the rate of 1 J/s. If work is done by a machine moving at a speed v against a constant force of resistance, F, work done per second equals FV, the same as power.

$$\text{power} = \text{force} \times \text{velocity}$$

$$P = F \times V \qquad (4.40)$$

4.11.3 ENERGY

Energy is the capacity of a body to do a work. The SI unit is Joules denoted as J. Depending on the status of the body, that is, whether it is at rest or in motion, the energy possessed by a body can be classified as potential energy (PE) and kinetic energy (KE).

4.11.3.1 Potential Energy

The PE of a body is the energy possessed by a body at rest. It is defined as the amount of work it would do if it were to move from its current position to the standard position.

$$\text{Potential Energy} = m \times g \times h \qquad (4.41)$$

where

    m is the mass of the body, g is the acceleration due to gravity, and h is the height.

4.11.3.2 Kinetic Energy

KE is described as the energy due to motion. The KE of a body may be defined as the amount of work it can do before being brought to rest.

$$\text{Kinetic Energy} = \frac{1}{2}mv^2 \qquad (4.42)$$

where m is the mass of the body and v is the velocity with which the body moves.

4.11.4 WORK-ENERGY PRINCIPLE

The principle of work and KE, also known as the work–energy principle, states that the work done by all the forces acting on a particle equals the change in the KE of the particle. That is, the work done, W, by the resultant force on a particle equals the change in the particle's KE,

$$W = \Delta KE = \frac{1}{2}mv^2 - \frac{1}{2}m\,u^2 \qquad (4.43)$$

where u and v are the initial and final velocities, respectively, and m is the mass of the body.

4.11.5 IMPULSE AND MOMENTUM

The principle of impulse and momentum is derived from Newton's second law, $F = ma$. This principle relates force, mass, velocity, and time, and is suitably used for solving problems where large forces act for a very small time.

When a large force acts over a short period of time, that force is called an impulsive force. The impulse of a force F acting over a time interval t_1 to t_2 is defined by $\int_{t_1}^{t_2} F \cdot dt$.

The product of mass and velocity is called momentum. i.e., the resultant force acting on a body is equal to the rate of change of momentum of the body.

$$\text{Impulse} = \text{final momentum} - \text{initial momentum} \qquad (4.44a)$$

$$Ft = m(V_2 - V_1) \qquad (4.44b)$$

4.11.5.1 Law of Conservation of Momentum

When the sum of the impulses due to external forces is zero, the momentum of the system remains constant or is conserved.

$$\text{Final Momentum} = \text{Initial Momentum} \qquad (4.45a)$$

It states, *The total momentum of two bodies remains constant after their collision or any other mutual action.*
Mathematically,

$$m_1u_1 + m_2u_2 = m_1v_1 + m_2v_2 \qquad (4.45b)$$

m_1 and m_2 are the masses of the first and second body, respectively; u_1 and u_2 are their initial velocities; v_1 and v_2 are their final velocities.

4.11.6 COLLISION OF ELASTIC BODIES

Collision of two bodies occurs in a very small interval of time. At this time, the two bodies exert a very large force on each other which is called as impact. The common normal to the surfaces of two bodies in contact when they collide with each other is called the line of impact. If the line of impact passes through the mass centers of the colliding bodies, then the impact is called central impact. If the velocities of the two bodies before collision are along the line of impact, the impact is called direct impact; otherwise, it is called indirect or oblique impact.

4.11.6.1 Newton's Law of Collision of Elastic Bodies

Consider two bodies A and B having a direct impact as shown in Figure 4.49a. Let u_1 and u_2 be the initial velocity of the first and second body, respectively, and v_1 and v_2 be the final velocity of the first and second body, respectively. The impact will take place only if u_1 is greater than u_2. Therefore, the velocity of approach will be equal to $(u_1 - u_2)$. After impact, the separation of the two bodies will take place, only if v_2 is greater than v_1. Therefore, the velocity of separation will be equal to $(v_2 - v_1)$.

It states, *When two moving bodies collide with each other, their velocity of separation bears a constant ratio to their velocity of approach.*

$$\text{Mathematically, } v_2 - v_1 = e \times (u_1 - u_2) \qquad (4.46)$$

where e = constant of proportionality called the coefficient of restitution and its value lies between 0 and 1. If $e = 0$, the two bodies are inelastic and if $e = 1$, the two bodies are perfectly elastic.
Notes:

1. Before or after impact, if the two bodies are moving in the same direction, then the velocity of approach or separation is the difference of their velocities.
2. But if the two bodies are moving in the opposite directions, then the velocity of approach or separation is the algebraic sum of their velocities.
3. The above formula holds well under the assumed conditions (i.e., $u_1 > u_2$ and $v_2 > v_1$).

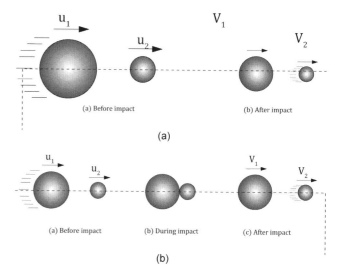

FIGURE 4.49   (a and b) Newton's law of collision of elastic bodies.

4.11.7 TYPES OF COLLISION

When two bodies collide with one another, they are said to have an impact. The two types of impacts are direct impact and indirect (or oblique) impact.

4.11.7.1 Direct Collision of Two Bodies

The line of impact of the two colliding bodies is the line joining the centers of these bodies and passes through the point of contact or point of collision. Before the impact, if the two bodies are moving along the line of impact, the collision is called as direct impact. Now consider the two bodies A and B having a direct impact as shown in Figure 4.49b. Let m_1 = mass of the first body, u_1 = initial velocity of the first body, v_1 = final velocity of the first body, and m_2, u_2, v_2 = corresponding values for the second body. According to the law of conservation of momentum,

$$m_1u_1 + m_2u_2 = m_1v_1 + m_2v_2 \qquad (4.47a)$$

The loss of KE, during impact, is determined as the difference between the KEs of the system.

$$\text{Loss of kinetic energy during impact} = \left[\frac{1}{2}m_1u_1^2 + \frac{1}{2}m_2u_2^2\right]$$

$$-\left[\frac{1}{2}m_1v_1^2 + \frac{1}{2}m_2v_2^2\right]$$

$$(4.47b)$$

4.11.7.2 Direct Impact of a Body with a Fixed Plane

Let u be the initial velocity of the body, v is the final velocity of the body, and e is the coefficient of restitution. Even after impact, the fixed plane will not move. Thus the velocity of approach is equal to (u) and velocity of separation is equal to (v). The Newton's Law of collision of Elastic Bodies also holds good for this type of impact. i.e., $v = eu$. If a body falls from a height, h, then the velocity with which the body impinges on the floor,

$$u = \sqrt{2gh} \qquad (4.48)$$

If a body is first projected upwards from the ground with some initial velocity, it will reach the greatest height and will then return to the ground with the same velocity, with which it was projected upward. From Figure 4.50, let h_0 = height from which the ball is dropped, h_1 = height after the first rebound, and h_2 = height after the second rebound. The velocity with which the ball impinges on the floor,

$$u = \sqrt{2gh_0} \qquad (4.49a)$$

and the velocity with which the ball rebounds first time

$$v = \sqrt{2gh_1} \qquad (4.49b)$$

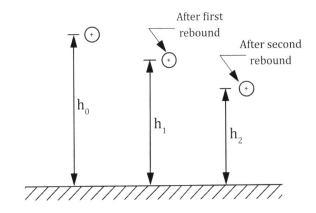

FIGURE 4.50 Direct impact of a body with a fixed plane.

Similarly, the velocity with which the ball impinges after the first rebound

$$u_1 = v = \sqrt{2gh_1} \qquad (4.49c)$$

and velocity with which the ball rebound the second time

$$v_1 = \sqrt{2gh_2} \qquad (4.49d)$$

Similarly, the velocity with which the ball impinges after the second rebound,

$$u_2 = v_1 = \sqrt{2gh_2} \qquad (4.49e)$$

During the first impact

$$v = eu \qquad (4.50a)$$

$$\sqrt{2gh_1} = e \times \sqrt{2gh_0} \qquad (4.50b)$$

During the second impact

$$v_1 = eu_1 \qquad (4.50c)$$

$$\sqrt{2gh_1} = e \times \sqrt{2gh_2} \qquad (4.50d)$$

4.11.7.3 Indirect Impact of Two Bodies

Before the impact, if the two bodies are not moving along the line of impact, the collision is called an indirect (or oblique) impact as shown in Figure 4.51. Now the law of conservation of momentum may be applied in the amended form in this case also. i.e., total initial momentum along the line of impact = total final momentum along the line of impact

$$m_1u_1\cos\alpha_1 + m_2u_2\cos\alpha_2 = m_1v_1\cos\theta_1 + m_2v_2\cos\theta \quad 4.51$$

where m_1 and m_2 are the masses of the first and second body, respectively; u_1 and u_2 are their initial velocities; v_1 and v_2 are the final velocities of the first and second body, respectively.

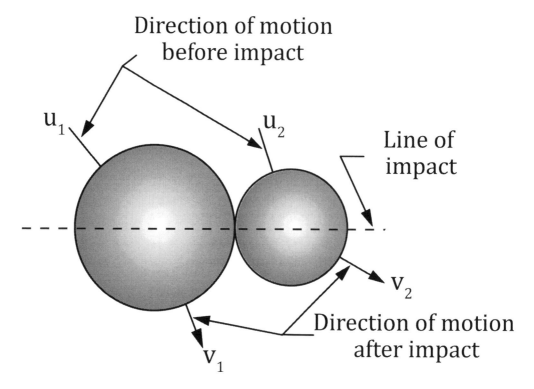

FIGURE 4.51   Indirect impact of two bodies.

4.11.7.4 Indirect Impact of a Body with a Fixed Plane

The Newton's Law of Collision of Elastic Bodies also holds good for this impact which is shown in Figure 4.52. i.e.,

$$v\cos\theta = eu\cos\alpha \qquad 4.52a$$

In this impact also, the principle of momentum (i.e., equating the initial momentum and the final momentum) is not applied, since the fixed plane has infinite mass. The components of initial and final velocities at right angles to the line of impact are same, i.e.,

$$u\sin\alpha = v\sin\theta \qquad 4.52b$$

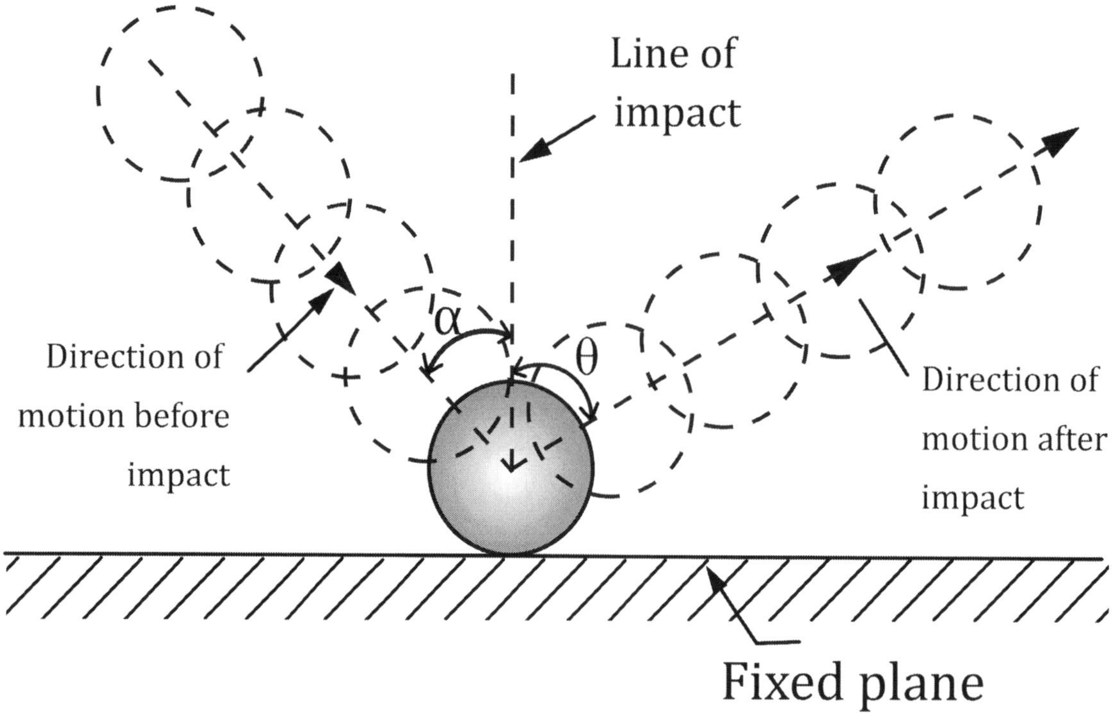

FIGURE 4.52   Indirect impact of a body with a fixed plane.

4.12 KINEMATICS OF A PARTICLE

Kinematics is the study of the geometry of motion of particles, rigid bodies, etc. In kinematics, the forces associated with these motions are not considered.

4.12.1 PLANE MOTION

When all the parts of the body move in a parallel plane, then the rigid body is said to perform plane motion. The plane motion may be classified into different types of motions.

4.12.1.1 Translation

Translational motion is the motion by which a body shifts from one point in space to another as shown in Figure 4.53. There are two types of translation: (1) rectilinear motion and (2) curvilinear motion. Rectilinear motion is that type of motion where the body moves in a straight line while curvilinear motion defines the motion of an object moving in a curved path.

4.12.1.2 Rotation

A rotation is a circular movement of an object around a center point as shown in Figure 4.54. There are two types of rotational motion: (1) fixed axis rotation and (2) rotation about a fixed point. Fixed axis rotation is a type of rotation where the

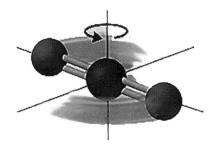

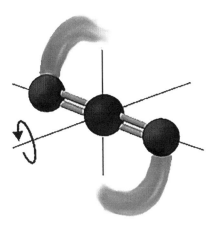

FIGURE 4.54 Rotational motion.

body rotates about a fixed axis which may or not be its centroidal axis. Here the angular velocity vector does not change direction. In the rotation about a fixed point, the body rotates about a fixed point and there is a change in the direction of the angular velocity vector.

4.12.1.3 General Plane Motion

General planar motion is a type of motion which includes for simultaneous rotational and translational motion in a 2-D plane as shown in Figure 4.55.

4.12.2 KINETICS OF RIGID BODIES UNDER COMBINED TRANSLATIONAL AND ROTATIONAL MOTION

The Kinetic Energy (KE) of a rigid body is the sum of the KE in translation and the KE in rotation.

The total KE of the body = KE in translation + KE in rotation

$$KE = \frac{1}{2}mv^2 + \frac{1}{2}I\omega^2 \qquad (4.53)$$

4.12.3 PRINCIPLE OF CONSERVATION OF ENERGY

The sum of KE and PE of a rigid body under the action of conservative forces, such as gravitational force, spring force, elastic force, remains constant.

$$(KE + PE)_{initial} = (KE + PE)_{final} \qquad (4.54)$$

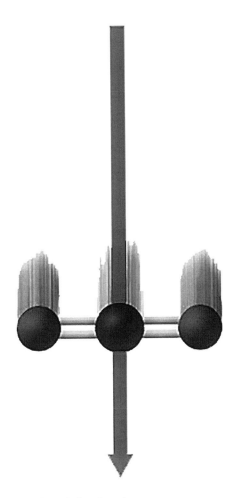

FIGURE 4.53 Translational motion.

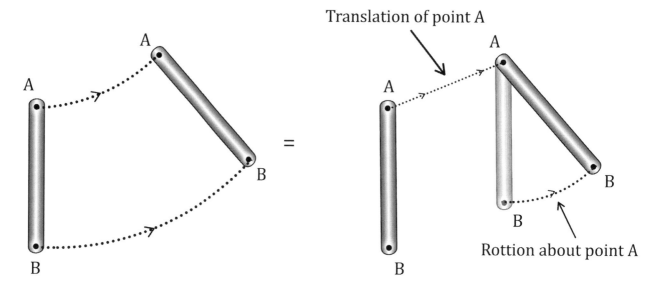

FIGURE 4.55 General plane motion.

5 Mechanics of Structures and Their Analysis

5.1 SIMPLE STRESS

5.1.1 INTRODUCTION

Any type of structural material can be classified as elastic, plastic, and rigid materials. When subjected to external loading, elastic materials undergo deformation. However, the elastic deformation vanishes when the loads are removed. Materials which undergo continuous deformation when loaded are called plastic materials. The plastic deformation can be considered as permanent deformation. The plastic materials do not regain their original dimensions when the loading is removed. Materials which do not undergo any deformation when loaded are called rigid materials. Structural members are always designed such that they do not exceed the elastic condition under working loads.

5.1.2 STRESS

Stress in a material is defined as the force of resistance offered against deformation. The load is termed as the external force acting on the body. The distinction between stress and load is that the load is often applied to the body, while stress is developed within the material of the body. Stress depends on the material properties of the body on which load is applied. Intensity of stress over a given section is the unit force of that portion.

$$\sigma = \frac{P}{A} \tag{5.1}$$

where p is the normal load extended and A is the area where the load is applied.

A material is capable of offering the following types of stresses.

i. **Normal stress**: If the force applied in the section plane is normal, the correlating stress is termed normal stress. Normal stress can be compressive or tensile. Members subjected to compressive force are under compressive stress and are called as compression members whereas members subjected to pure tensile force is under tensile stress and are called as tension members. The nature of compressive and tensile forces is shown in Figure 5.1.

ii. **Shear stress**: Shear stress or shearing stress also known as tangential stress is caused by forces, which are parallel to force-resistant area. It differs from the normal stress mainly in the way in which it acts

on a body. If F is the subsequent shearing force that passes through the center of area A, it is sheared as is shown in Figure 5.2,

$$\tau = \frac{F}{A} \tag{5.2}$$

iii. **Bearing stress**: It is the pressure of contact between different bodies which is shown in Figure 5.3. The difference between bearing pressure and compressive stress is that while the former one is external pressure acting on the body, the latter is an internal stress caused by compressive forces.

5.1.3 STRESS ON AN OBLIQUE PLANE DUE TO AXIAL LOADING

As shown in Figure 5.4, a section with normal plane at an angle of θ is drawn. The shearing stress and normal stress on this section can be calculated as:

$$\text{Normal component, } F = P\cos\theta \tag{5.3}$$

$$\text{Tangential component, } V = P\sin\theta \tag{5.4}$$

$$\text{Normal stress, } \sigma = \frac{P}{A}\cos^2\theta \tag{5.5}$$

$$\text{Shear stress, } \tau = \frac{P}{A}\sin\theta\cos\theta \tag{5.6}$$

Highest shear stress or maximum shear strength occurs on planes, which makes an inclination of 45° or 175° with the normal cross section.

5.2 SIMPLE STRAIN

The strain is characterized as the change in length to the original length ratio. The transition in length is caused by the force applied.

$$\varepsilon = \frac{\delta}{L} \tag{5.7}$$

where δ is the deformation in length and L is the initial length, thus ε is the strain which is dimensionless. From Figure 5.5,

$$\text{Tensile strain} = \frac{\text{Increase in length}}{\text{Initial length}} \tag{5.8}$$

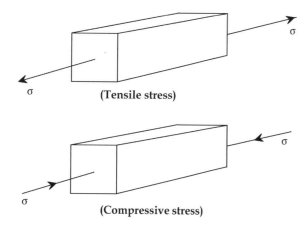

FIGURE 5.1 Normal stress.

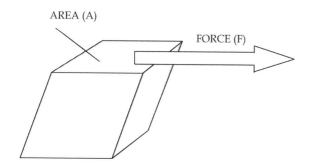

FIGURE 5.2 Shear stress.

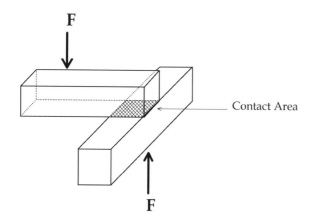

FIGURE 5.3 Bearing stress.

$$\text{Compressive strain} = \frac{\text{Decrease in length}}{\text{Initial length}} \qquad (5.9)$$

5.2.1 Stress–Strain Diagram

The stress–strain curve of a particular material represents the distinction of the stress developed in the material with respect to the changes in the strain occurring in it. An example is shown in Figure 5.6. It can be considered as unique because it is different for different materials. It is found out by noting the amount of deformation (strain) experienced by the material at distinct intervals of normal loading (stress).

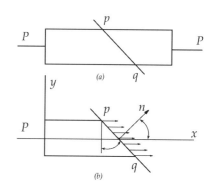

FIGURE 5.4 Axial loading stress on an oblique plane.

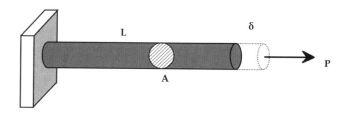

FIGURE 5.5 Simple strain.

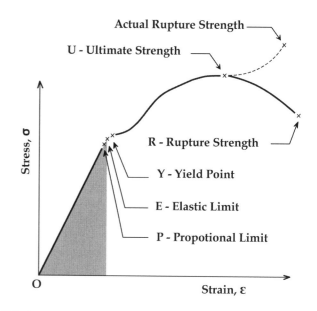

FIGURE 5.6 Stress–strain diagram.

5.2.1.1 Proportional Limit

The proportional limit is the highlight of the stress–strain curve. The stress values remain proportional to the strain to this point. In Figure 5.6, point *P* is referred to as the proportional limit or the proportionality limit. Hooke's law is justifiable only up to this point. Hooke's law states that stress is directly proportional to strain within the proportional limit or

$$\sigma = E\varepsilon \qquad (5.10)$$

The proportionality constant is termed as the Young modulus or the Elasticity modulus and is indicated by letter E. Its value is equal to the slope of the stress–strain curve within the proportional limit.

5.2.1.2 Elastic Limit

There is a restriction above which the material does not return to its present form if the load is eliminated. This limit is called as the elastic limit. It implies the maximum possible stress at a stage where no permanent deformation arises after the load is fully removed.

5.2.1.3 Elastic and Plastic Ranges

Figure 5.6 shows the elastic and plastic ranges in a material. The region between O and P is the elastic range and plastic range is the region between P and R.

5.2.1.4 Yield Point

After the elastic range, there exists a point at which the material will yield or elongate appreciably without any increase in load. This point is termed as yield point.

5.2.1.5 Ultimate Strength

After the yield point, the stress–strain diagram rises up to a maximum ordinate. This is called as the ultimate strength which indicates the maximum stress that the material can withstand before failure.

5.2.1.6 Rupture Strength

If a material is stressed beyond its ultimate strength, the stress–strain diagram will move downward and finally stop at a point, where the material will break. The strength of the material at failure is termed as the rupture strength or the break strength.

5.2.1.7 Modulus of Resilience

Resilience is the capability of a material to absorb energy without permanent deformation. The work on a unit amount of material, as the force tends to increase steadily from O to P, is called the resilience modulus. Its unit is Nm/m³. The resilience modulus is calculated as the area below the stress curve from the origin O to the elastic limit E (shown by the shaded region in Figure 5.6).

5.2.1.8 Modulus of Toughness

The capacity of any material to absorb energy without failure is called toughness. Work on a unit volume of material as the force tends to increase progressively from O to R is called a toughness modulus. Its unit is Nm/m³. Modulus of rupture is computed as the area of the whole stress–strain curve from O to R as shown in Figure 5.6.

5.2.2 Axial Deformation

If a steel rod has uniform cross-sectional area, the loading is axial, and the stress is within the proportional limit, the axial deformation may be computed using Equation 5.11.

$$\delta = \frac{PL}{AE} = \frac{\sigma L}{E} \qquad (5.11)$$

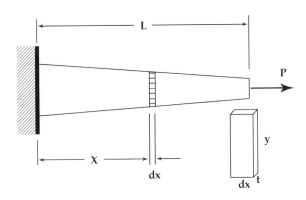

FIGURE 5.7 Axial deformation.

If the cross-sectional area of the bar is not uniform throughout the entire length, as shown in Figure 5.7, the axial deformation is calculated by such a differential length and integration.

$$\delta = \frac{P}{E} \int_0^L \frac{dx}{L} \qquad (5.12)$$

If a long rod L is suspended vertically from one end, the total elongation is attained due to its self-weight:

$$\delta = \frac{\rho g L^2}{2E} = \frac{MgL}{2AE} \qquad (5.13)$$

where ρ=density of the rod in kg/m³, M=total mass of the rod in kg, A=cross-sectional area of the rod in m², L=length of the rod in m, and g=9.81 m/s².

The ratio of the steady force (P) acting on an elastic body to the resulting displacement (δ) is called stiffness of the material.

$$k = \frac{P}{\delta} \qquad (5.14)$$

5.2.3 Shearing Deformation

Shearing deformation is caused by shearing forces, which is shown in Figure 5.8. An element subject to shear will not alter its length but will alter its shape. Shear strain is the change in angle at the corner of the original rectangular element:

$$\gamma = \frac{\delta_s}{L} \qquad (5.15)$$

The ratio of the shear stress τ to the shear strain γ is termed as the modulus of elasticity in shear or modulus of rigidity and is represented by G.

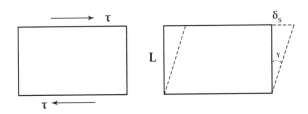

FIGURE 5.8 Shearing deformation.

$$G = \frac{\tau}{\gamma} \tag{5.16}$$

The shear deformation and the shear force applied are related as:

$$\delta_s = \frac{VL}{A_s G} = \frac{\tau L}{G} \tag{5.17}$$

where V=shearing force acting on an area A_s.

5.2.4 POISSON'S RATIO

Consider a bar which is loaded with a tensile force, as seen in Figure 5.9. The ratio of the lateral strain to the longitudinal strain is termed as the Poisson's ratio.

$$\upsilon = -\frac{\varepsilon_y}{\varepsilon_x} = -\frac{\varepsilon_z}{\varepsilon_x} \tag{5.18}$$

where ε_x = strain in the x-direction; ε_y and ε_z=the strains in the perpendicular directions y and z, respectively. The negative sign in the equation shows that the transverse dimension reduces with an increase in the linear dimension.

5.2.5 BIAXIAL DEFORMATION

Consider an element that is simultaneously subjected to tensile stresses, σ_x and σ_y, in the x and y directions, respectively. The strain in the x and y directions will be $\frac{\sigma_x}{E}$ and $\frac{\sigma_y}{E}$ accordingly. As the tensile stresses are acting at the same time, the stress in the y direction results in lateral contraction on the x direction having magnitude $-\upsilon\varepsilon_y$ or $-\frac{\upsilon\sigma_y}{E}$. The final strain in the x and y directions is given in Equations 5.19 and 5.20, respectively.

$$\varepsilon_x = \frac{\sigma_x}{E} - \upsilon\frac{\sigma_y}{E} \tag{5.19a}$$

$$\text{or } \sigma_x = \frac{(\varepsilon_x + \upsilon\varepsilon_y)E}{1 - \upsilon^2} \tag{5.19b}$$

$$\text{and } \varepsilon_y = \frac{\sigma_y}{E} - \upsilon\frac{\sigma_x}{E} \tag{5.20a}$$

$$\text{or } \sigma_y = \frac{(\varepsilon_y + \upsilon\varepsilon_x)E}{1 - \upsilon^2} \tag{5.20b}$$

5.2.6 TRIAXIAL DEFORMATION

When three mutually perpendicular normal stresses σ_x, σ_y, and σ_z act on an element at the same time and are accompanied by strains ε_x, ε_y, and ε_z accordingly,

$$\varepsilon_x = \frac{1}{E}\left[\sigma_x - \upsilon(\sigma_y + \sigma_z)\right]$$

$$\varepsilon_y = \frac{1}{E}\left[\sigma_y - \upsilon(\sigma_x + \sigma_z)\right]$$

$$\varepsilon_z = \frac{1}{E}\left[\sigma_z - \upsilon(\sigma_x + \sigma_y)\right] \tag{5.21}$$

In the above equations, tensile stresses and elongation are considered as positive. Negative sign is considered for compressive stresses and contraction.

5.2.7 THERMAL STRESS

Changes in temperature cause a body to change its dimensions. It may either contract or expand. If α is the thermal expansion coefficient in m/m°C, L is the length of the body (m), T_f and T_i are the final and initial temperatures, respectively, in °C, the amount of resulting change, δ_T, is given by

$$\delta_T = \alpha L(T_f - T_i) = \alpha L \Delta T \tag{5.22}$$

If temperature deformation occurs freely, no load or stress is induced in the structure. Otherwise, an internal stress, known as thermal stress, will be developed. As given in Figure 5.10, the thermal stress can be attained for a homogeneous rod positioned between unyielding supports as:

$$\sigma = E\alpha\Delta T \tag{5.23}$$

where σ=thermal stress in MPa and E=elasticity modulus of the rod in MPa.

When the wall yields for x as demonstrated in Figure 5.11, the following equations will be used:

$$\delta_T = x + \delta_p \tag{5.24}$$

$$\alpha L \Delta T = x + \frac{\sigma L}{E} \tag{5.25}$$

where σ is the thermal stress. It is to be noted that the rod will be in tension if the temperature reduces to the normal, and in compression, if the temperature rises above normal.

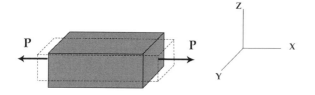

FIGURE 5.9 Concept of Poisson's ratio.

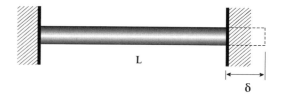

FIGURE 5.10 Rod subjected to thermal stress.

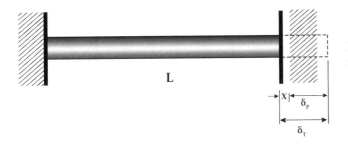

FIGURE 5.11 Wall yielding due to thermal stress.

5.3 STATICALLY DETERMINATE MEMBERS

A structure is regarded as the assembly of a number of components, such as walls, beams, slabs, columns, foundations, etc., that stay in balance. It is usually categorized in two categories as determinate and indeterminate or redundant structures. Determinate structures are assessed using fundamental equilibrium equations. Examples of determinate structures include cantilever beams, simply supported beams, overhanging beams, three hinged arches, etc.

5.3.1 DETERMINACY AND STABILITY

A specified structure is regarded to be externally determinate when the total number of reaction components is equal to the available number of equilibrium equations. In other words, determinacy is ascertained as:

$$r = 3n, \text{ statically determinate} \quad (5.26a)$$

$$r > 3n, \text{ statically determinate} \quad (5.26b)$$

where n is the total parts of structural members and r is the total unknown number of reactive force and moment components.

5.3.2 STATICALLY INDETERMINATE MEMBERS

A structure is regarded to be statically indeterminate when the number of independent equations of equilibrium is less than the internal resisting forces or the reactive forces over a cross section.

Examples of indeterminate structures include fixed beams, continuous beams, fixed arches, two hinged arches, portals, multistorey frames, etc.

5.3.3 TRUSS STRUCTURES

A truss is a structure consisting of bars or links, presumed to be linked to the joints by frictionless pins and organized in such a way that the area only within the limits of the structure is segmented by the bars into geometric figures that are normally triangles. Internal determinacy is a kind of indeterminacy associated with trusses. The overall shape of a truss is a triangle. If j is the estimated number of joints, i is the estimated number of links, and r is the minimal number of external determinacy/stability reactions required,

$$i + r = 2j \quad (5.27)$$

In this case, the total number of unknowns will be equal to the total number of equations available (at joints).

The conditions for determinacy are as follows:

   i. If $i + r = 2j$, the truss is stable and internally determinate.
   ii. If $i + r > 2j$, the truss is stable but internally indeterminate.
   iii. If $i + r < 2j$, the truss is unstable.

A structure will have determinacy and indeterminacy only if the structure is stable.

5.4 RELATIONSHIP BETWEEN ELASTIC CONSTANTS

Elastic constants represent the stress–strain relationship within the elastic limit. The modulus of elasticity or Young modulus (E), Bulk modulus (K), and Modulus of rigidity or shear modulus (N, C, or G) are the various kinds of elastic constants. Young's modulus E is the slope of the stress–strain curve in uniaxial tension. Lateral to longitudinal strain in uniaxial tensile stress is regarded as Poisson's ratio (v). It is therefore dimensionless and typically ranges from 0.2 to 0.49 and for most metals is around 0.3. The bulk modulus assesses the solid resistance to changes in volume even without any massive variation in shape or form. It is usually large (bigger than E). The shear modulus assesses its resistance to shear deformations preserving volume and is characterized as the relationship between shear stress and shear strain. Generally, its value is rather smaller than E. The relationship between the elasticity modulus (E), the shear modulus (G), and the Poisson ratio is as follows:

$$G = \frac{E}{2(1+v)} \quad (5.28)$$

The bulk modulus of elasticity K

$$K = \frac{E}{3(1-2v)} = \frac{\sigma}{\Delta V/V} \quad (5.29)$$

where V = the volume and ΔV = the volume change. The ratio $\Delta V/V$ = volumetric strain and is expressed as

$$\frac{\Delta V}{V} = \frac{\sigma}{K} = \frac{3(1-2v)}{E} \quad (5.30)$$

5.5 PRINCIPAL PLANES AND PRINCIPAL STRESSES

The maximum and minimum normal stresses which can occur in a material are called the principal stresses. The major principal stress is the maximum normal stress, whereas the minimum normal stress is referred to as minor principal stress. The planes upon which normal stress reaches its maximum and minimum value are, respectively, known as major and minor principal planes. Shear stress on principal planes is zero.

5.5.1 PRINCIPAL AXIS

The principal axis is the axis in which the inertial moment is maximum or minimum and the inertial product is zero. The principal plane is a stress property, whereas the principal axis is a moment of inertia property. There is no relationship between the principal axis and the principal plane.

5.5.2 MAXIMUM AND MINIMUM NORMAL STRESS

The rotation of the stress condition of a stress element can give stresses to any angle shown in Figure 5.12. The angle at which the maximum (or minimum) normal stress is developed can be obtained as follows:

Equation for the stress transformation in x or y direction is

$$\sigma_n = \frac{\sigma_x + \sigma_y}{2} + \frac{\sigma_x - \sigma_y}{2} \cos 2\theta + \tau_{xy} \sin 2\theta \quad (5.31a)$$

where σ_n is the normal stress acting on the inclined plane. To maximize (or minimize) the stress, the derivative of σ_n with respect to the rotation angle θ is equated to zero and the inclination angle is computed from Equation 5.31b.

$$\tan 2\theta_p = \frac{2\tau_{xy}}{\sigma_x - \sigma_y} \quad (5.31b)$$

The angle θ_p as shown in Figure 5.13 can be replaced back into the rotation stress equation to get the original maximum and minimum stress values. These stresses are regarded as the principal stresses, σ_1 and σ_2, which are generally expressed as,

$$\sigma_{1,2} = \frac{\sigma_x + \sigma_y}{2} \pm \sqrt{\left(\frac{\sigma_x - \sigma_y}{2}\right)^2 + \tau_{xy}^2} \quad (5.32)$$

Shear stress, τ_n, will become zero if the stress element is rotated by θ_p.

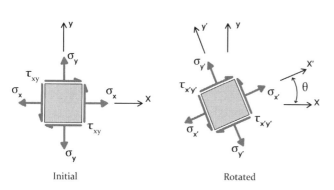

FIGURE 5.12 Rotating stresses from x–y coordinate system to new x'–y' coordinate system.

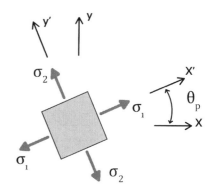

FIGURE 5.13 Principal stresses, σ_1 and σ_2, at principal angle, θ_p.

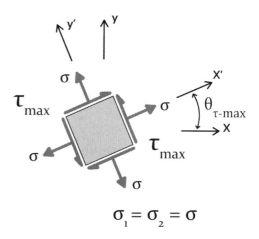

$$\sigma_1 = \sigma_2 = \sigma$$

FIGURE 5.14 Maximum shear stresses, τ_{max}, at angle, $\theta_{\tau-max}$.

5.5.3 MAXIMUM SHEAR STRESS

Shear stress will have a maximum value at a given angle $\theta_{\tau-max}$ as shown in Figure 5.14. If τ_n is the shear stress acting on the inclined plane,

$$\tau_n = -\frac{\sigma_x - \sigma_y}{2} \sin 2\theta + \tau_{xy} \cos 2\theta \quad (5.33)$$

Maximum shear stress is computed as:

$$\tau_{max} = \sqrt{\left(\frac{\sigma_x - \sigma_y}{2}\right)^2 + \tau_{xy}^2} \quad (5.34)$$

5.6 MOHR'S CIRCLE METHOD

The Mohr circle is a two-dimensional pictorial portrayal of the Cauchy stress tensor transformation law. The stress elements acting on a rotated coordinate system that acts on a different plane passing through a point can be visually exhibited and can be determined using the Mohr circle. The magnitudes of normal stress (σ_n) as abscissa and that of shear stress as ordinate, (τ_n) of each point on the circle, acting on the rotated coordinate system are shown in Figure 5.15. This can also be demonstrated as, when the axes represent the

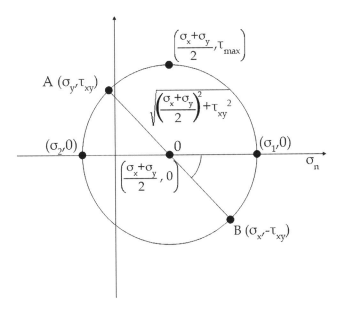

FIGURE 5.15 Mohr's circles for a three-dimensional state of stress.

principal axes of stresses, the Mohr's circle is the locus of points that at all inclinations truly represent the stress state on individual planes.

5.6.1 Equation of the Mohr Circle

A two-dimensional infinitesimal material element around a material point P, as shown in Figure 5.16, with a unit area in the direction parallel to the y-z plane, can be regarded to generate the Mohr circle equations for the two-dimensional cases of plane stress and plane strain shown in Equations 5.35–5.37.

$$\left[\sigma_n - \frac{1}{2}(\sigma_x + \sigma_y)\right]^2 + \tau_n^2 = \left[\frac{1}{2}(\sigma_x - \sigma_y)\right]^2 + \tau_{xy}^2 \quad (5.35)$$

$$\left(\sigma_n - \sigma_{avg}\right)^2 + \tau_n^2 = R^2 \quad (5.36)$$

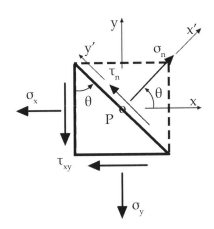

FIGURE 5.16 Stress components in a plane passing through a continuum point under plane stress conditions.

where

$$R = \sqrt{\left[\frac{1}{2}(\sigma_x - \sigma_y)\right]^2 + \tau_{xy}^2} \quad (5.36(i))$$

and

$$\sigma_{avg} = \frac{1}{2}(\sigma_x + \sigma_y) \quad (5.36(ii))$$

This represents the equation of a circle (the Mohr circle) in the form

$$(x - a)^2 + (y - b)^2 = r^2 \quad (5.37)$$

with radius $r = R$ centered at a point with coordinates $(a, b) = (\sigma_{avg}, 0)$ in the (σ_n, τ_n) system.

In the convention on physical-space signs, normal stresses acting outside the action plane are regarded positive (tension) and normal stresses acting inside the action plane are deemed negative (compression). Shear stresses in the positive direction of an axis acting on the positive faces of the material element are perceived positive. Shear stresses on the negative faces of the material element are also termed positive in the negative direction of the axis

5.7 COMBINED STRESSES

5.7.1 Stresses Developed through Axial Load Combinations and Bending Moments

The simple stress developed when a member is axially loaded

$$\sigma = \frac{P}{A} \quad (5.38)$$

The stress for a lateral load (bending) member is given by

$$\sigma = \frac{My}{I} \quad (5.39)$$

The stresses in quoted equations are normal for the cross section of the member. The resulting stress behaving on the section due to the combined action of the axial load P and the bending moment M is determined by the arithmetical sum of the direct stress and the flexural stress. The stress equation is usually written in the form:

$$\sigma = \pm\frac{P}{A} \pm \frac{My}{I} \quad (5.40)$$

5.7.2 Direct Shear in Association with Torsion

If a shaft with a circular cross section is exposed to a torque at any point, the shearing stress due to the applied torque is administered by

$$\tau = \frac{Tr}{J} \quad (5.41)$$

When a direct shear force acts on a beam of any section at any point, the shear stress is exerted by

$$\tau = \frac{V}{Ib}A'y' \tag{5.42}$$

The consequent shearing stress due to direct shear and torsion is given as follows by the algebraic sum:

$$\tau = \frac{Tr}{J} \pm \frac{V}{Ib}A'y' \tag{5.43}$$

The positive or negative signs in Equation 5.43 depends on the shearing stresses that act along the same line of action or in opposite directions.

5.7.3 STRESSES PRODUCED DUE TO COMBINED ACTION OF BENDING AND TORSION

When a shaft is bent and twisted, it is optimal to convey the subsequent direct and shear stresses as the moment and the torque applied. If a bending moment M and torque T act on a rod together, the stresses on a component at the top of the rod are as shown in Figure 5.17. Those on the lower surface, except compressive, are the same.

$$\text{The normal stress, } \sigma = \frac{My}{I} = \frac{M}{\frac{\pi}{32}d^3} \tag{5.44a}$$

$$\text{The shear stress, } \tau = \frac{Tr}{J} = \frac{T}{\frac{\pi}{16}d^3} \tag{5.44b}$$

For solid shaft,

$$\text{Maximum principal stress, } \sigma = \frac{1}{2}\left[\sigma_x + \sqrt{\sigma_x^2 + 4\tau_{max}^2}\right] \tag{5.45a}$$

$$\text{Maximum shear stress, } \tau = \frac{1}{2}\sqrt{\sigma_x^2 + 4\tau_{max}^2} \tag{5.45b}$$

5.8 PHYSICAL PROPERTIES OF MATERIALS AND THEIR MEASURING PARAMETERS

5.8.1 BRITTLENESS

It is a material property which indicates instant failure without significant deformations. A broken material does not yield and fails without the actual structure being notified.

5.8.2 DUCTILITY

Ductility of the material can be described as its ability to undergo large deformations before failure. Ductile materials exhibit the phenomenon of yielding. Ductility is measured by the length, elongation percentage, and the tension test area reduction percentage.

5.8.3 ELASTICITY

It is a property of the material, in which the stresses developed due to the applied load will disappear when the load is removed.

5.8.4 HARDNESS

Hardness of a material refers to its resistance to scratching, abrasion, or indentation.

Initial tangent modulus is the tangent line slope at the start of the stress–strain chart as can be seen in Figure 5.18.

5.8.5 MALLEABILITY

It is a property of the material by which it can be rolled or beaten up into thin sheets.

5.8.6 MODULUS OF RESILIENCE

The resilience module is characterized as the total energy that can be absorbed per unit volume without permanently distorting the system. It is measured by the integration of the stress–strain curve within the elastic limit. In uniaxial tension,

$$U_r = \frac{\sigma_y^2}{2E} \tag{5.46}$$

where U_r=resilience modulus, σ_y=yield strength, and E=the Young's modulus of elasticity.

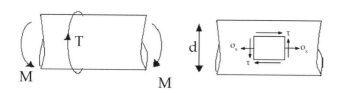

FIGURE 5.17 Combined stresses due to bending and torsion.

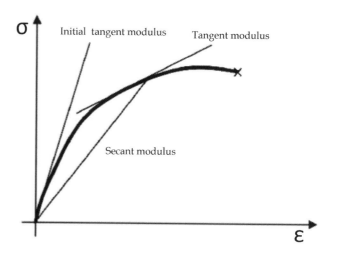

FIGURE 5.18 Slope of stress–strain curve.

5.8.7 PLASTICITY

It is a property of the material, in which the material undergoes permanent change in shape without failure under the applied load.

5.8.8 PROOF RESILIENCE

Proof resistance is characterized as the total energy absorbable within in the elastic limit without permanent distortion.

5.8.9 RELATIVE DENSITY

Relative density is the ratio of material density to water density.

5.8.10 RESILIENCE

Resilience implies the capacity of a material to absorb energy when it is elastically deformed and release energy when it is unloaded.

5.8.11 SECANT MODULUS

Secant module is the stress-to-strain ratio in a stress-strain diagram at any point in the curve. It is the line slope from its origin to any point on a curve of stress–strain as shown in Figure 5.18.

5.8.12 SPECIFIC MODULUS OF ELASTICITY

Specific modulus of elasticity is the ratio of modulus of elasticity to relative density of a material.

5.8.13 STIFFNESS

The capacity of a material to prevent the deformations is called stiffness. Modulus of elasticity, secant modulus, etc., are the measures of stiffness.

5.8.14 TENACITY

It refers to the resistance of rupture of a material due to tearing.

5.8.15 TOUGHNESS

Toughness is characterized as the capacity of a material to absorb energy and deform without fracturing plastically.

5.9 SHEAR FORCES AND BENDING MOMENTS

Beams are usually long cuboid members. The applied loads are generally perpendicular to the axis of beam. Bending and shearing are produced due to the transverse loading in the beam. Axial forces are produced in the beam only when the loads are not perpendicular to the beam.

5.9.1 TRANSVERSE LOADING

Forces extended perpendicular to the longitudinal axis of a member are named as cross loads. It bends and deflects the member from its previous position. It in turn will lead to the development of internal tensile and compressive strains along with the change in the curvature of the member.

5.9.2 SHEAR FORCE DIAGRAM (SFD) AND BENDING MOMENT DIAGRAM (BMD)

SFD and BMD are two analytical tools that are used in structural analysis. They are used as an aid in structural design. At a certain point in the structural element, they give the real value of shear force and bending moment. In order to design a structural member (its type, size, and material) for a specified set of loads by preventing any structural failure, these diagrams can be conveniently used. The deflection of a beam either using moment area method or the conjugate beam method can be calculated by diagrams of shear force and bending moment.

5.9.3 SIGN CONVENTION

When the external forces acting on the beam usually tend to shear off the beam at that point, as shown in Figure 5.19, it is said that the shear is positive at any given point. At any point in a beam, the bending moment is also positive when the external forces acting on the beam generally tend to bend the beam at that point as shown in Figure 5.20. If the tendency of the bending moment is to bend so that it generates convexity over the center line, this bending moment is termed as hogging bending moment, as seen in Figure 5.21. It is quite possible that the bending moment at a section may bend the member at the section so as to produce concavity above the center line. Such a bending

FIGURE 5.19 Normal positive shear force convention.

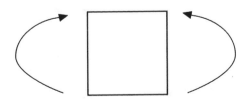

FIGURE 5.20 Normal bending moment convention.

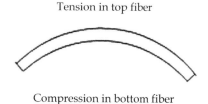

Tension in top fiber

Compression in bottom fiber

FIGURE 5.21 Hogging bending.

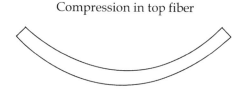

Compression in top fiber

Tension in bottom fiber

FIGURE 5.22 Sagging bending.

moment is called a sagging bending moment as shown in Figure 5.22.

5.9.4 SFD AND BMD FOR DIFFERENT CANTILEVER AND SIMPLY SUPPORTED BEAMS

Table 5.1 shows the loading diagram, shear force, and BMDs for the cantilever, simply supported, and overhanging types of beams for various loading patterns at different loading positions.

5.9.5 RELATION BETWEEN LOAD, SHEAR, AND BENDING MOMENT

The rate of change in the bending moment, M, relative to the distance, x, is equal to the shear force, or the slope of the moment diagram at that point is the shear shown in Figure 5.36. The rate of change of the shear force, V, is equal to the shear force with respect to the distance, or the slope of the moment diagram at that point is the shear at that point as shown in Figure 5.36.

$$\frac{dV}{dx} = \text{Load} \qquad (5.47)$$

5.10 BENDING STRESSES IN BEAMS

At a segment of a beam, the bending moment appears to bend or deflect the beam, and the internal stress resists bending. The bending process deters when each cross section establishes full resistance to the bending moment. This resistance is called bending stress by the internal stresses, and the relevant theory is called the theory of bending.

5.10.1 ASSUMPTIONS IN THE THEORY OF SIMPLE BENDING

The assumptions made in the simple bending hypothesis are as follows:

*   The beam material is absolutely perfectly homogeneous (of the same type throughout) and isotropic (in all directions of the same elastic properties).

TABLE 5.1
SFD and BMD for Various Cases of Loading on Beams

Type of Beam	Type of Load	Position of Load	Loading Diagram SFD BMD
Cantilever beam having length 'l'	A concentrated load W	At the free end	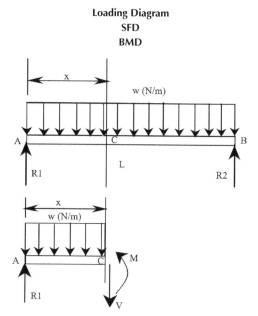

(Continued)

TABLE 5.1 (*Continued*)

SFD and BMD for Various Cases of Loading on Beams

Type of Beam	Type of Load	Position of Load	Loading Diagram SFD BMD
	An uniformly distributed load of *w* per unit run	Over the whole length	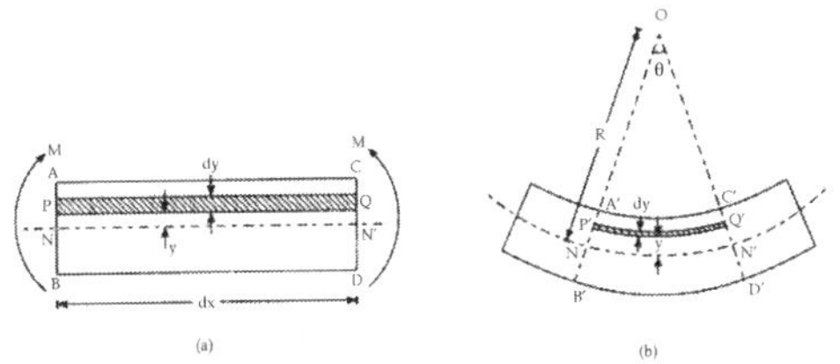
	A uniformly varying load	Zero at the free end to *w* per unit run at the fixed end	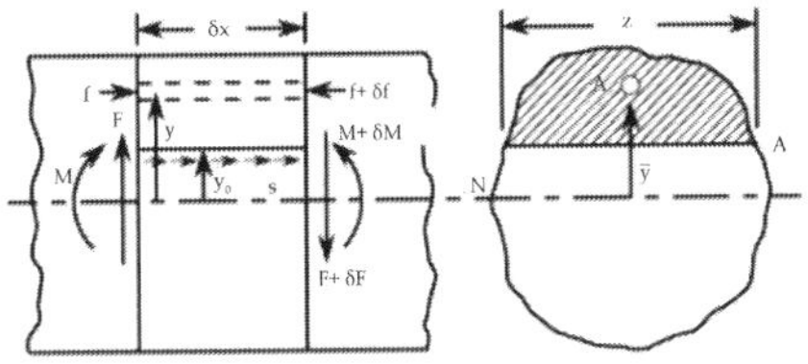
Simply supported beam of span *l*	A concentrated load *W*	At mid-span	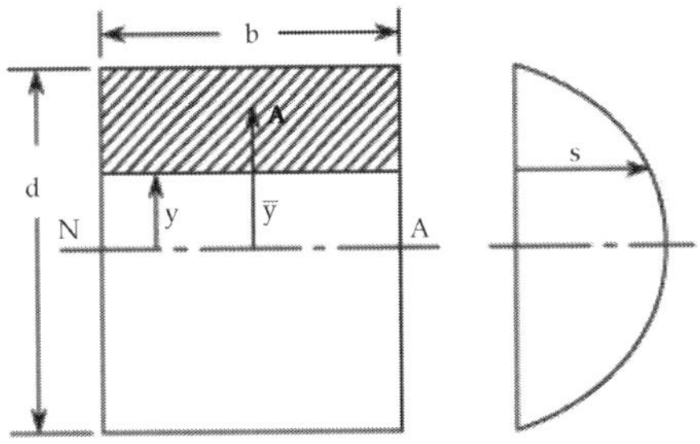
	A concentrated load *W*	Placed eccentrically on the span	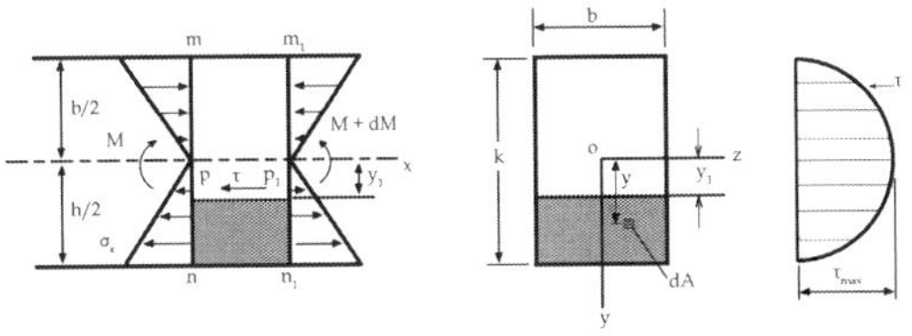
	An uniformly distributed load *w* per unit run	Over the whole length	
		Over part of its span For a certain distance from one end	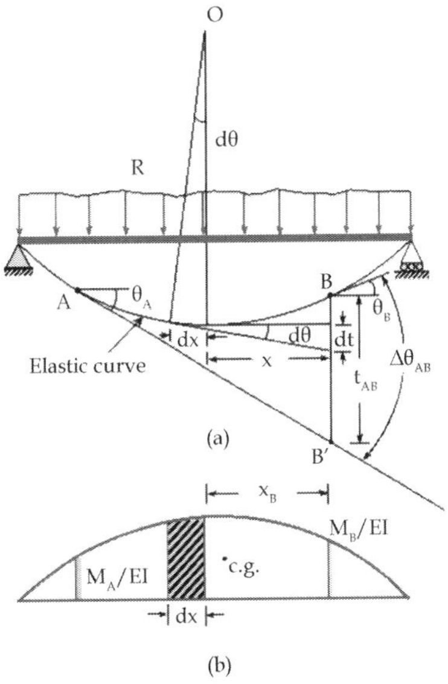

(*Continued*)

TABLE 5.1 (*Continued*)
SFD and BMD for Various Cases of Loading on Beams

Type of Beam	Type of Load	Position of Load	Loading Diagram SFD BMD
		On an intermediate part of the span	
	A uniformly varying load	Zero at each end to *w* per unit run at the mid span	
		Zero at one end to *w* per unit run at the other end	
	A couple	At the mid-span	
Overhanging beam with overhang at one end	An uniformly distributed load of w per unit run	Over the whole length	
Overhanging beam with equal overhangs at either end	An uniformly distributed load of *w* per unit run	Over the whole length	

- The beam material is stressed within its elastic limits and therefore complies with the Hooke's law.
- The transverse sections that remain plane before bending remain plane after bending.
- Each layer of the beam is able to expand or contract individually, above or below the layer.
- The value of E in tension and compression is the analogous.

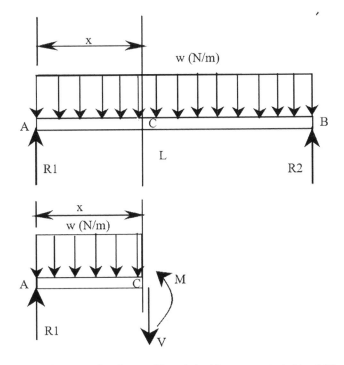

FIGURE 5.23 Cantilever of length *l* with a concentrated load *W* at the free end.

5.10.2 Theory of Simple Bending

Figure 5.37a shows the cross section of a beam. On bending, the layers above NN' have been compressed and those below have been stretched as shown in Figure 5.37b. The magnitude of the compression or elongation depends on the position of NN'. This layer NN', which is neither compressed nor stretched, is known as neutral plane or neutral layer. This bending theory is termed as the theory of simple bending.

Now consider a layer PQ at a distance y from the neutral axis of the beam NN'. Let this layer be compressed to $P'Q'$ after bending as shown in Figure 5.37b. Let M be the moment acting on the beam, θ be the angle subtended at the center by the arc, and R be the radius of curvature of the beam. The strain developed in the layer is proportional to its distance from the neutral axis.

$$\text{Stress, } f = \text{strain} \times \text{modulus of elasticity}$$

$$= e \times E = \frac{y}{R} \times E = y \times \frac{E}{R} \qquad (5.48a)$$

Since E and R are constants in this expression, stress is directly proportional to y in any section, i.e., distance of the point from neutral axis.

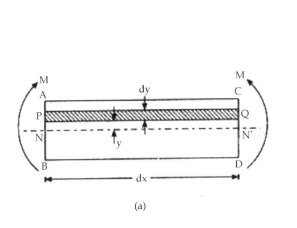

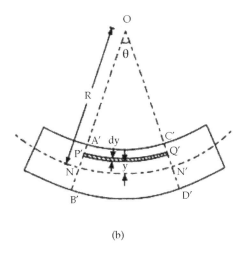

(a) (b)

FIGURE 5.24 Cantilever of length *l* with a uniformly distributed load of *w* per unit run over the whole length (a and b).

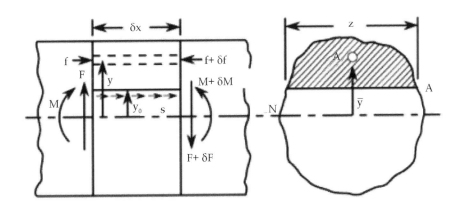

FIGURE 5.25 Cantilever of length *l* carrying a load of uniform intensity from zero at the free end to w per unit at the fixed end.

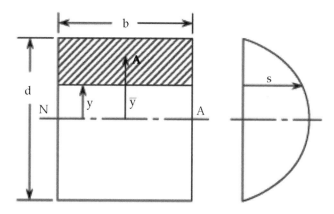

FIGURE 5.26 Simply supported beam of span *l* with a concentrated load in the middle span.

The above expression can also be written as:

$$\frac{f}{y} = \frac{E}{R} \tag{5.48b}$$

5.10.3 Position of Neutral Axis

The neutral axis of a section is the neutral layer intersection line with any normal beam cross section. The stresses are compressive on one side of the neutral axis, while they are tensile on the other. There will be no stress of any kind on the neutral axis.

5.10.4 Moment of Resistance

The compressive and tensile stresses on either side of the neutral axis form a pair, regarded as the resistance moment, which must be equal in value to the external moment *M*, so that it can resist the external bending moment.

So the equation for bending becomes

$$\frac{M}{I} = \frac{f}{y} = \frac{E}{R} \tag{5.48c}$$

where *I* is the moment of inertia of the cross section of the beam.

5.10.5 Section Modulus

The relation for finding out the stress on the extreme fibers of the section is

$$\frac{M}{I} = \frac{f}{y} \tag{5.49a}$$

$$M = \frac{f}{y} \times I = f \times \frac{I}{y} \tag{5.49b}$$

The stress in a fiber is proportional to its distance from the center of gravity. If y_{max} is the distance between the center of gravity of the section and the extreme fiber of the stress, then

$$M = f_{max} \times \frac{I}{y_{max}} = f_{max} \times z \tag{5.49c}$$

$$\text{where } z = \frac{I}{y_{max}} \tag{5.49d}$$

z is termed as modulus of section or section modulus.

Therefore, the above equation becomes

$$M = f \times z \tag{5.49e}$$

where *f* denotes the maximum stress, tensile, or compressive in nature.

5.11 SHEARING STRESSES IN BEAMS

When bending moment, *M*, and shear force, *V*, together act on a cross section of a beam, then it will be subjected to nonuniform bending. Figure 5.38 shows two transverse sections at a distance δx apart. The shearing forces at the sections are *F* and $F+\delta F$ while the bending moments are *M* and $M+\delta M$.

Let *s* be the complementary shear stress and therefore the transverse shear stress at a distance y_0 from the neutral axis is as shown in Figure 5.39.

$$s = F\frac{A\bar{y}}{zI} \tag{5.50}$$

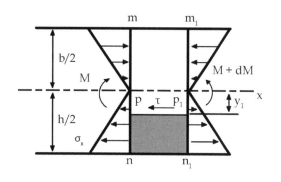

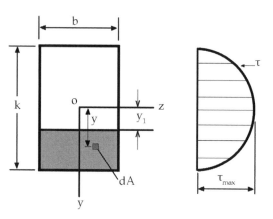

FIGURE 5.27 Simply supported beam with a concentrated load placed eccentrically on the span

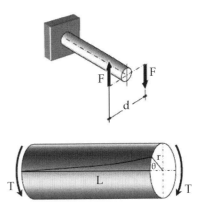

FIGURE 5.28 Simply supported beam subject to uniform load [UDL] throughout the entire length.

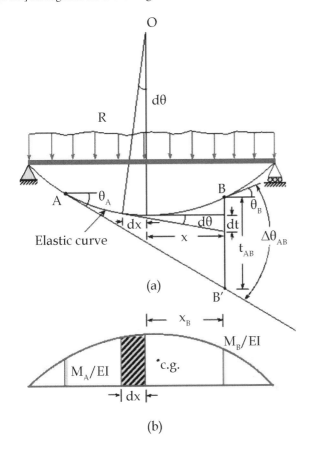

(a)

(b)

FIGURE 5.29 Simply supported beam with an UDL for a certain distance from one end. (a) Beam. (b) M/EI diagram.

FIGURE 5.31 Simply supported beam carrying a load, the intensity of which varies from zero at each end to w per unit runs in the middle of the span.

FIGURE 5.32 Simply supported beam with a load of uniform intensity from zero at one end to w per unit run at the other end.

FIGURE 5.33 Simply supported beam with a couple at midspan.

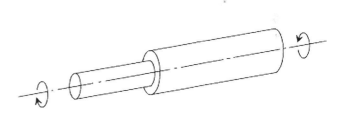

FIGURE 5.34 Beam with overhang at one end and carrying an UDL over the whole length.

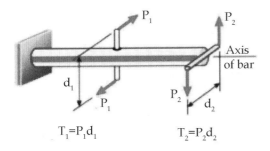

FIGURE 5.30 Simply supported beam carrying an UDL on an intermediate part of the span.

It must be noted that Z is the actual section width at the position where s is computed and I is the total moment of inertia at the neutral axis. As shown in Figure 5.39, there will be a parabolic variation of shear stress with y. The maximum shear force occurs at the neutral axis and is given by:

$$s = \frac{3F}{2bd} \qquad (5.51)$$

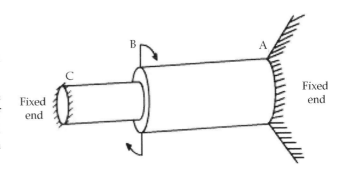

FIGURE 5.35 Simply supported beam with equivalent overhangs and UDL of w per unit run over the whole length.

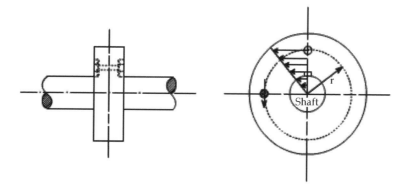

FIGURE 5.36   Relation between load, shear, and bending moment.

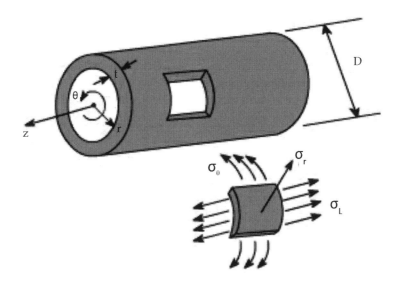

FIGURE 5.37   Theory of simple bending.

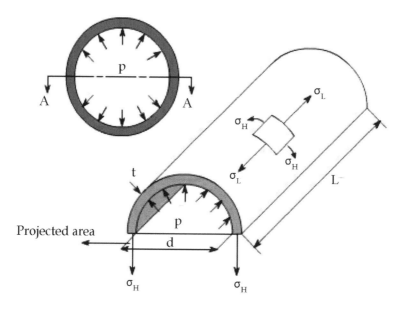

FIGURE 5.38   Transverse section subjected to shear stress.

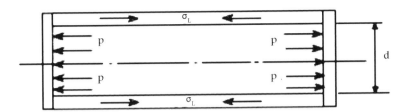

FIGURE 5.39 Parabolic variation of shear stress.

If $\dfrac{F}{bd}$ is represented as the mean stress, then

$$s = 1.5 \times s_{\text{mean}} \tag{5.52}$$

5.11.1 Shear Flow

The shear flow in each section is the product of the shear stress and the correlating width. Shear flow, q, refers to the longitudinal force per unit length transmitted throughout the section.

$$q = \tau \times z = \frac{V}{I} \times Q \tag{5.53}$$

where q is the shear flow, V is the shear force at a given section, I is the moment of inertia at the neutral axis, b is the width of section, and Q be the first moment of area about the neutral axis $= \int y\, dA$. Figure 5.40 shows a rectangular beam in pure bending.

$$\text{Shear stress, } \tau = \frac{VQ}{Ib} \tag{5.54}$$

For the rectangular section displayed in Figure 5.40:

$$\tau = \frac{V}{2I}\left(\frac{h^2}{4} - y_1^2\right) \tag{5.55}$$

Shear stresses vary quadratically from the neutral axis to the distance y_1. The maximum shear stress on the neutral axis is nil on the top and underside surfaces of the beam. The max shear stress for a rectangular cross section is as follows:

$$\tau = \frac{3V}{2A_{\text{max}}} \tag{5.56}$$

For a circular cross section:

$$\tau = \frac{4V}{3A_{\text{max}}} \tag{5.57}$$

5.11.2 Built-Up Beams

The built-up beams are made of two or more pieces of materials combined to form a single solid beam. Shear flow, f, in such beams at critical locations is offered by

$$f = \frac{VQ}{I} \tag{5.58}$$

5.12 DEFLECTIONS

A beam or a frame deflects when it is loaded. If the deflection exceeds the permissible value, the structure will not look aesthetic and may result in psychological upsetting of the occupants. This also may cause cracking in the materials of the structure.

The extent of the deflection of a member under a specified load is related directly to the slope of the deflected member's shape under that load. Deflection is measured by integrating the function that characterizes the member's slope under this load arithmetically.

5.12.1 Correlation between Slope, Deflection, and Radius of Curvature

As shown in Figure 5.37b, a small portion PQ of the beam that is bent into an arc is considered. In Cartesian coordinates, the radius of curvature is given by the relation.

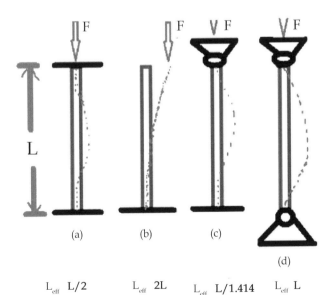

FIGURE 5.40 Rectangular beam in pure bending. (a) Both end fixed. (b) One end fixed other end free. (c) One end fixed other pinned. (d) Both ends pinned.

$$R = \frac{\left[1+\left(\dfrac{dy}{dx}\right)^2\right]^{\frac{3}{2}}}{\dfrac{d^2y}{dx^2}} \qquad (5.59)$$

Substituting for R in Equation 5.48c, we get

$$M = EI\frac{d^2y}{dx^2} \qquad (5.60)$$

Since $\dfrac{dy}{dx}$ (represents the slope at any point in the center line of the bent beam) is an extremely small quantity, $\left(\dfrac{dy}{dx}\right)^2$ is much smaller still and hence can be neglected. The expression $\left\{1+\left(\dfrac{dy}{dx}\right)^2\right\}^{\frac{3}{2}}$ is, therefore, equal to unity. To be consistent with the sign conventions adopted, a negative sign is introduced and the equation of the bending of the beam can be represented as:

$$EI\frac{d^2y}{dx^2} = -M \qquad (5.61)$$

This equation is regarded as the elastic curve differential equation of the beam. The product EI is termed the beam's flexural rigidity. Table 5.2 shows the beam deflection formula for different types of beam with different loading conditions.

5.12.2 MOMENT AREA METHOD

The rotation, inclination, and deflection of beams and frames are derived from the moment area theorem. The theorem was formulated by Otto Mohr and disclosed later in 1873 by Charles Greene. In this technique, area of the BMD is considered for the measurement of the slope and/or the deflections along the axis of the beam or frame at any point. For the computation of the deflection, two theorems known as the moment area theorems are used. One theorem is used to compute the change of slope in the elastic curve between two points. While the other theorem is used to compute the vertical distance between a point on the elastic curve and a line tangent to the elastic curve at a second point (called tangential deviation). For this purpose, the bending moment chart for the beam is first drawn and then divided by the flexural rigidity (EI) to obtain the "M/EI" chart.

Theorem 1:

The change of slope between two points on the elastic curve corresponds to the area of the M/EI diagram between these two points as shown in Figure 5.42.

$$\theta_{AB} = \int_A^B \frac{M}{EI}dx \qquad (5.62)$$

where M is the moment, EI is the flexural rigidity, and θ_{AB} is the change of slope between points A and B of the elastic curve.

Theorem 2:

The vertical divergence of a point A on an elastic tangent curve extending from another point B equates to the moment of the area under the M/EI diagram between these two points (A and B). This moment is calculated on point A to evaluate the deviation from B to A.

$$t_{A/B} = \int_A^B \frac{M}{EI}xdx \qquad (5.63)$$

where $t_{A/B}$ is the deviation of tangent at point B as regards tangent at point A.

5.12.2.1 Sign Rules

1. The deviation is said to be positive at any point if the point is above the tangent.
2. When computed from the left tangent if θ is in the counterclockwise direction, the slope change is positive.

TABLE 5.2
Beam Deflection Formulae

Type of Beam	Slope at Free End	Deflection at Any Section in Terms of x	Maximum Deflection
1. Cantilever beam—concentrated load P at the free end			

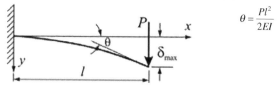

| | $\theta = \dfrac{Pl^2}{2EI}$ | $y = \dfrac{Px^2}{6EI}(3l-x)$ | $\delta = \dfrac{Pl^3}{3EI_{max}}$ |

(Continued)

TABLE 5.2 (*Continued*)
Beam Deflection Formulae

Type of Beam	Slope at Free End	Deflection at Any Section in Terms of x	Maximum Deflection
2. Cantilever beam—concentrated load P at any point 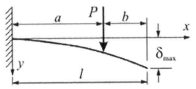	$\theta = \dfrac{Pa^2}{2EI}$	$y = \dfrac{Px^2}{6EI}(3a - x)$ for $0 < x < a$ $y = \dfrac{Px^2}{6EI}(3x - a)$ for $a < x < 1$	$\delta = \dfrac{Pa^2}{6EI}(3l - a_{max})$
3. Cantilever beam—uniformly distributed load ω (N/m) 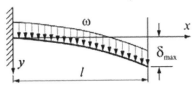	$\theta = \dfrac{wl^3}{6EI}$	$y = \dfrac{wx^2}{24EI}\left(x^2 + 6l^2 - 4lx\right)$	$\delta = \dfrac{wl^4}{8EI}_{max}$
4. Cantilever beam—uniformly varying load: maximum intensity ω_0 (N/m) 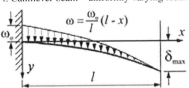	$\theta = \dfrac{w_0 l^3}{24EI}$	$y = \dfrac{w_0 x^2}{120lEI}\left(10l^3 - 10l^2 x + 5lx - x^3\right)$	$\delta = \dfrac{w_0 l^4}{30EI}_{max}$
5. Cantilever beam—couple moment M at the free end 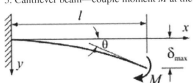	$\theta = \dfrac{ml}{EI}$	$y = \dfrac{Mx^2}{2EI}$	$\delta = \dfrac{Ml^2}{2EI}_{max}$
6. Beam simply supported at ends—concentrated load P at the center 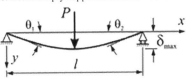	$\theta_1 = \theta_2 = \dfrac{Pl^2}{16EI}$	$y = \dfrac{Px}{12EI}\left(\dfrac{3l^2}{4} - x^2\right)$ for $0 < x < \dfrac{l}{2}$	$\delta = \dfrac{Pl^3}{48EI}_{max}$
7. Beam simply supported at ends—uniformly distributed load ω (N/m) 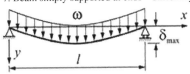	$\theta_1 = \theta_2 = \dfrac{wl^3}{24EI}$	$y = \dfrac{wx}{24EI}\left(l^3 - 2lx^2 + x^3\right)$	$\delta = \dfrac{5wl^4}{384EI}_{max}$
8. Beam simply supported at ends—couple moment M at the right end 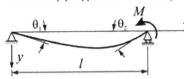	$\theta_1 = \dfrac{Ml}{6EI}$ $\theta_2 = \dfrac{Ml}{3EI}$	$y = \dfrac{Mlx}{6EI}\left(1 - \dfrac{x^2}{l^2}\right)$	$\delta = \dfrac{Ml^2}{9\sqrt{3}EI}_{max}$ at $x = \dfrac{l}{\sqrt{3}}$
9. Beam simply supported at ends—uniformly varying load: maximum intensity ω_0 (N/m) 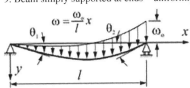	$\theta_1 = \dfrac{7w_0 l^3}{360EI}$ $\theta_2 = \dfrac{w_0 l^3}{45EI}$	$y = \dfrac{w_0 x}{360lEI}\left(7l^4 - 10l^2 x^2 + 3x^4\right)$	$\delta_{max} = 0.00652\dfrac{w_0 l^4}{EI}$ at $x = 0.519l$

5.12.3 Conjugate Beam Method

The conjugate beam method is an exceptionally versatile beam deflection calculation method. The correlation between the loading, shear, and bending moments is obtained by

$$\frac{d^2M}{dx^2} = \frac{dV}{dx} = -w(x) \tag{5.64}$$

where M is the bending moment, V is the shear force, and w (x) is the intensity of distributed load.

Similarly,

$$\frac{d^2v}{dx^2} = \frac{d\theta}{dx} = \frac{M}{EI} \tag{5.65}$$

It can be ascertained from the above two equations that if M/EI is the load on an imaginary beam, the consequent shear and moment in the beam are the slope and the displacement of the real beam. The imaginary beam is called the "conjugate beam" and its length is almost the same as original beam.

The conjugate beam method involves two major steps.

1. Set up an additional beam, called "conjugate beam."
2. Determine the "shearing forces" and "bending moments" in the conjugate beam.

Two concepts related to the conjugate beam are as follows:

Theorem 1:

The slope of the real beam at one point is arithmetically equal to the shear at the corresponding point of the conjugate beam.

Theorem 2:

The displacement of a point in the real beam is mathematically equal in value to the moment in the conjugate beam at the corresponding point. The slope of (the center line of) the actual real beam is exactly equal to the "shearing force" at the correlating cross section of the conjugate beam at any cross section. This slope is positive or anticlockwise when the "shearing force" is positive in the beam convention to rotate the beam element anticlockwise. The "bending moment" of the conjugate beam at any point provides the deflection of the real beam at that point (the center line of). This deflection is down if the "bending moment" is positive in the beam convention, probably causing top fibers in compression. The positive shear force and moment of bending are shown in Figure 5.41. The differences between real beams and conjugate beams are shown in Table 5.3.

5.13 INDETERMINATE STRUCTURES

A structure is statically indeterminate if the static equilibrium equations are inadequate to evaluate the internal forces and reactions on that structure. The fixed and continuous beams are called indeterminate beams.

The degree of indeterminacy of a structure can be characterized as the number of unknown forces exceeding the static equation. These unidentified extra forces are termed redundant forces. Support reactions or forces in internal members might be redundant. When redundancies are taken away from the structure, they become determinate.

For the analysis of indeterminate structures, the following two methods are available.

1. Force or Flexibility or Compatibility method
2. Displacement or Stiffness method

5.13.1 Force Method

In this method, the selected redundant reactions are removed from the structure so that the resulting structure is determinate. The deflections of the points, where the redundant are acting, are found out in the direction of redundant removed.

Now the external loads are eliminated and the redundant reactions are assumed to act as loads and the deflections for the same points are found out in terms of redundant. The resultant equations are equated to find out the values of the redundant. The method of consistent distortions (deformations), Castigliano's second theorem, and the three-moment theorem are force methods.

5.13.2 Displacement Method

In this method of analysis, the displacements of joints (may be rotation or translation) are selected as redundant and similar equations as in force method are written. Solution of these equations gives the displacement assumed. By substituting these displacements in original equations, the reaction components can be found out. Slope deflection method is a displacement method, whereas moment distribution method is a successive approximation method based on the same general theory as the displacement method. This method is of the greatest importance because it is the matrix analysis method which can be computerized for general usage.

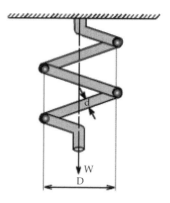

FIGURE 5.41 Positive shearing force and bending moment.

TABLE 5.3
Differences between Real Beam and Conjugate Beam

Real Beam	Conjugate Beam
Fixed support $v=0$ $\theta=0$	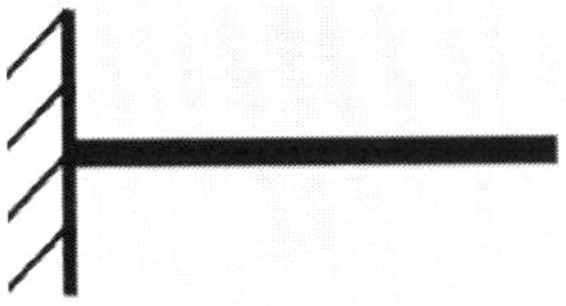 Free end $\bar{M}=0$ $\bar{Q}=0$ 
Free end $v \neq 0$ $\theta \neq 0$	 Fixed support $\bar{M} \neq 0$ $\bar{Q} \neq 0$ 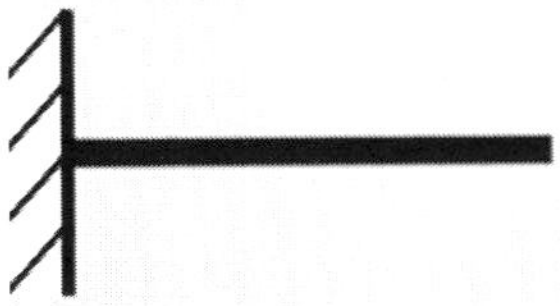
Hinged support $v=0$ $\theta \neq 0$	 Hinged support $\bar{M}=0$ $\bar{Q} \neq 0$ 
Middle support $v=0$ θ continue	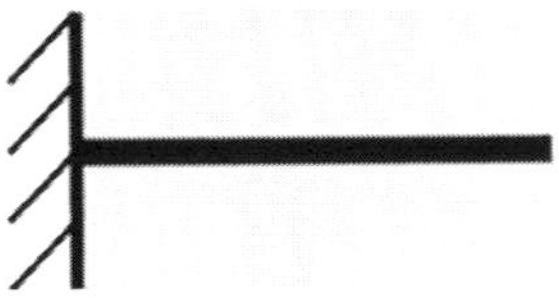 Middle hinge $\bar{M}=0$ $\bar{Q}$ continue 
Middle hinge v continue θ discontinue	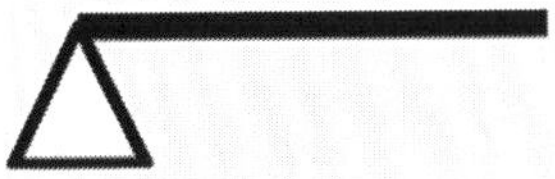 Middle support $\bar{M}$ continue $\bar{Q}$ discontinue 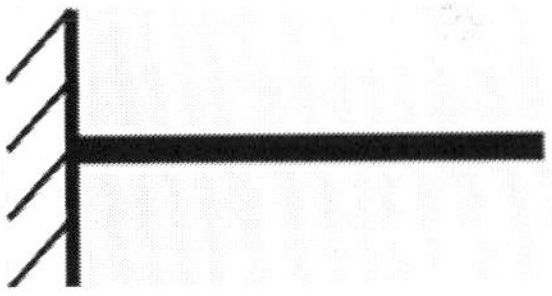

5.14 FIXED BEAMS

These are constrained at the supports, so that the slopes at the supports remain zero when the beam is loaded.

Consider a fixed beam, fixed at both the supports loaded with an arbitrary load. There are four unknowns V_A, M_A at support A and V_B, M_B at support B. Two equations of statics $\sum V = 0$ and $\sum M = 0$ are available. Since two more equations are required to solve this beam, the degree of redundancy is two. The boundary conditions for this are $\theta_A = 0$, $\theta_B = 0$, $\delta_A = 0$, and $\delta_B = 0$. Any two conditions will give us two equations. Three techniques are there for the solution of such beams. These are:

1. Method of superposition
2. Double integration method
3. Moment area method

5.14.1 METHOD OF SUPERPOSITION

This method consists of finding slopes and deflections separately for each load and reaction. The redundant reactions are selected first. Force–displacement relationships are developed. Deformation consistency is checked using compatibility equation.

5.14.2 DOUBLE INTEGRATION METHOD

This method is used to find out the deflections and rotations at various points. This method can be extended to solve the propped and fixed beams having one and two redundants, respectively. The origin is selected at the fixed end so as to obtain the constants of integration to be zero as both slope and deflection are zero at the fixed end. After writing the formulae of moments and integrating them, the boundary conditions are applied at the other end. For a fixed support, both rotation and deflection should be zero at the other end also.

5.14.3 MOMENT AREA METHOD

For structures primarily subjected to bending, the deflections can be found out by moment-area method, developed by Mohr. Using compatibility of displacement, the deflections of indeterminate structures can be easily found out.

In this technique, the area of the bending moment graphs is used to calculate the slope and the deflections along the beam axis or frame at certain points. For the measurement of deflection, two algebraic concepts known as the moment zone equations are used. One hypothesis is used to calculate the change in the slope of the elastic curve between two

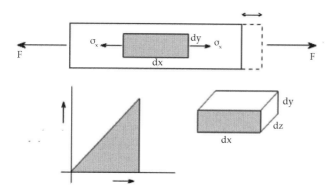

FIGURE 5.42 Mohr's first theorem.

points. The other theorem is just used to ascertain the vertical distance (called tangential deviation) between an elastic curve point and an elastic curve line tangent at a second point.

5.14.3.1 Mohr's First Theorem (Mohr I)

As seen in Figure 5.42, recognize the elastic curve of a loaded simple beam. Tangents are drawn on points A and B on the elastic curve. Total angle of the two tangents is indicated as $\Delta\theta_{AB}$. As to evaluate the value of $\Delta\theta_{AB}$, consider the augmental distinction in angle $d\theta$ over an infinitesimal segment dx found at a distance of x from point B. The bending equation gives the radius of the curvature and bending moment for each part of the beam.

$$\frac{M}{I} = \frac{E}{R} \tag{5.66}$$

where M denotes the bending moment, I is the moment of inertia, E is the modulus of elasticity, and R is the radius of curvature.

The elementary length dx and the variation in angle $d\theta$ are related as,

$$dx = d\theta \times R \tag{5.67}$$

The total change in rotation between A and B is thus

$$\int_A^B d\theta = \int_A^B \frac{M}{EI} dx \tag{5.68}$$

where M/EI is the curvature. Thus we have

$$d\theta_{BA} = \theta_B - \theta_A = \int_A^B \frac{M}{EI} dx \tag{5.69}$$

This is supposed to be

$$\left[\text{Change in slope} \right]_{AB} = \left[\text{Area of } \frac{M}{EI} \text{ diagram} \right]_{AB} \tag{5.70}$$

This is Mohr's First Theorem (Mohr I): It specifies that the slope change over any length of a bending member is equal to the curvature diagram area over that length.

5.14.3.2 Mohr's Second Theorem (Mohr II)

$$d\Delta = x \times d\theta \tag{5.71}$$

We have

$$\int_A^B d\Delta = \int_A^B \frac{M}{EI} \times x \times dx \tag{5.72}$$

$$\Delta_{BA} = \left[\int_A^B \frac{M}{EI} \times dx \right] \bar{x} = \text{first moment of } \frac{M}{EI} \text{ diagram about } B \tag{5.73}$$

This is easily interpreted as

$$\left[\text{Vertical Intercept} \right]_{BA} = \left[\text{Area of } \frac{M}{EI} \text{ diagram} \right]_{BA} \times$$

$$\left[\text{Distance from } B \text{ to centroid of } \left(\frac{M}{EI} \right)_{BA} \text{ diagram} \right] \tag{5.74}$$

This is Mohr's Second Theorem (Mohr II): The vertical intercept between one terminal and the tangent to the curve of another terminal is the first moment of the curvature diagram around the terminal where the intercept is quantified for an originally straight beam subject to bending moment.

From the above definition, two important things are to be noted:

- From the fundamental diagram, the vertical intercept will not give the deflection. It is the distance from the beam's deformed position to the tangent of the deformed beam shape at a different location. That is:

$$\Delta \neq \delta \tag{5.75}$$

- The moment of the curvature diagram should be taken at the point at which the vertical intercept is necessary. That is:

$$\Delta_{BA} \neq \Delta_{AB} \tag{5.76}$$

5.15 CONTINUOUS BEAMS

A continuous beam is a multispan beam with hinged support that is statically indeterminate. The end spans can be cantilever, freely supported or fixed. A reaction along the beam axis must be able to establish at least one of the supports of a continuous beam. These beam types are used in structures such as bridges and buildings, where the supports are strong enough for heavy loads.

5.15.1 THREE-MOMENT THEOREM

The theorem of three moments presented by Clapeyron in 1857 gives the relation between the support moments in a continuous beam. This theorem is usually developed for the beams of constant cross section between each pair of supports.

Let A, B, C be the three successive support points, the length of AB is indicated by l and BC by l', and the weight per unit of length in these segments by w and w'. The bending moments M_A, M_B, M_C can then be linked to the following three points:

$$M_A l + 2M_B(l + l') + M_C l' = \frac{1}{4}wl^3 + \frac{1}{4}w'(l')^3 \quad (5.77)$$

This equation can be written as

$$M_A l + 2M_B(l + l') + M_C l' = \frac{6a_1x_1}{l} + \frac{6a_2x_2}{l'} \quad (5.78)$$

where a_1 and a_2 are the bending moment area diagram due to vertical loads on AB and BC, respectively, x_1 is the distance from A to the center of the beam moment diagram AB, x_2 is the distance from C to the center of the beam moment area BC.

The second equation is more specific since it does not necessitate a uniform allocation of the weight of each segment.

The three-moment equation is given by

$$\frac{M_1L_1}{E_1I_1} + 2M_2\left(\frac{L_1}{E_1I_1} + \frac{L_2}{E_2I_2}\right) + \frac{M_3L_2}{E_2I_2}$$

$$= 6\left[\frac{\Delta A - \Delta B}{L_1} + \frac{\Delta C - \Delta B}{L_2}\right] - 6\left[\frac{A_1X_1}{E_1I_1L_1} + \frac{A_2X_2}{E_2I_2L_2}\right] \quad (5.79)$$

5.16 TORSION

Torsion is the twisting of a structural member when filled by pairs that rotate the longitudinal axis around it. The torque unit, T is N-m.

As shown in Figure 5.43,

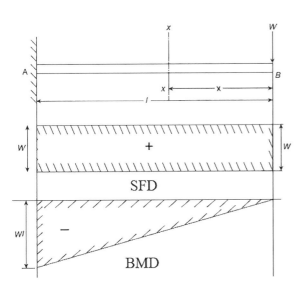

FIGURE 5.43 Torsion.

$$T_1 = P_1d_1 \quad (5.80)$$

$$T_2 = P_2d_2 \quad (5.81)$$

The couples T_1, T_2 are known as torques or twisting couples or twisting moments.

Consider a bar that is rigidly affixed to one end and twisted to the other end by a torque or twisting moment T equivalent to $F \times d$, which is introduced perpendicular to the bar axis, as seen in Figure 5.44. Such a bar is supposed to be in torsion.

5.16.1 TERMS RELATED TO TORSION

5.16.1.1 Torsional Shearing Stress (τ)

For a hollow or solid circular shaft subject to a twisting moment T, the torsional shear stress p from the center of the shaft is at a distance p

$$\tau = \frac{Tp}{J} \quad (5.82)$$

$$\tau = \frac{Tr}{J_{max}} \quad (5.83)$$

where r is the outer radius and J is the polar moment of inertia of the section.

For solid cylindrical shaft as shown in Figure 5.45:

$$J = \frac{\pi}{32}D^4 \quad (5.84)$$

$$\tau = \frac{16T}{\pi D_{max}^3} \quad (5.85)$$

For hollow cylindrical shaft as shown in Figure 5.46:

$$J = \frac{\pi}{32}\left(D^4 - d^4\right) \quad (5.86)$$

$$\tau = \frac{16TD}{\pi\left(D^4 - d^4\right)_{max}} \quad (5.87)$$

5.16.1.2 Angle of Twist

The angle θ through which the bar length L will twist is

$$\theta = \frac{TL}{JG} \text{ in radians} \quad (5.88)$$

where T=the torque in Nmm; L=the shaft length in mm; J=the polar moment of inertia in mm⁴; G=shear modulus in Mpa; D and d=diameter in mm; and r=the radius in mm.

5.16.1.3 Power Transmitted by the Shaft

If a twisting moment T is acted on a shaft that rotates at a constant angular velocity (in radians per second); the transmitted power of the shaft is offered by

$$P = T\omega = 2\pi Tf \quad (5.89)$$

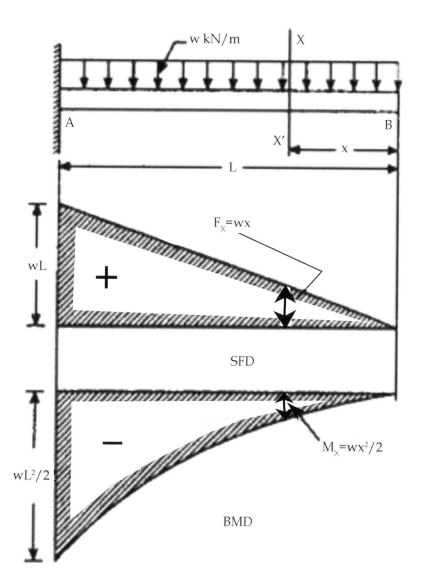

FIGURE 5.44 Bar in torsion.

where T=the torque in Nm; f=the number of revolutions per second; and P=the power in watts.

5.16.1.4 Equivalent Torque

σ_1 and σ_2 for the combined bending and twisting case are expressed by the relations:

$$\sigma_1, \sigma_2 = \frac{16}{\pi d^3} \left[M \pm \sqrt{M^2 + T^2} \right] \tag{5.90}$$

$$\tau = \frac{16}{\pi d^3} \sqrt{M^2 + T^2} \frac{16}{\pi d^3 e_{max}} \tag{5.91}$$

where $\sqrt{M^2 + T^2}$ is regarded as the equivalent torque, which would create the same maximum shear stress as pure torsion by acting alone.

$$\text{Hence, } T_e = \sqrt{M^2 + T^2} \tag{5.92}$$

5.16.2 Composite Shafts

A composite shaft is made up of shafts of two different diameters. The composite shaft may experience twist when it is subjected to the torque. This twist can be calculated based on the position of the torque applied, i.e., when the shafts are in series and parallel.

5.16.2.1 Composite Shafts in Series

If two or more shafts of various materials, diameters, or basic shapes are linked together, as seen in Figure 5.47, so that each shaft has the same torque, the shafts are said to be linked in series, and the composite shaft thus generated is called a series–connected shaft.

The equilibrium of the shaft, in this case, requires a torque "T," which is to be same throughout both the parts.

In these instances, the composite shaft strength is achieved by individually taking into consideration each shaft element by applying the theory of torsion to each. The composite shaft

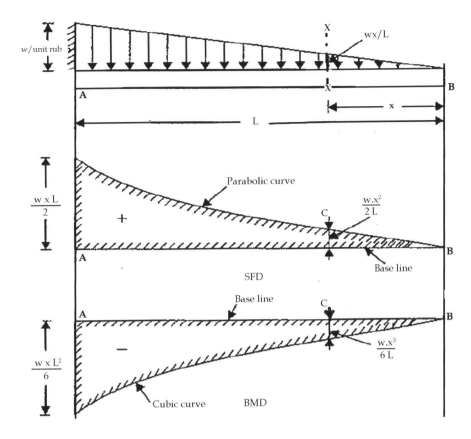

FIGURE 5.45 Solid cylindrical shaft.

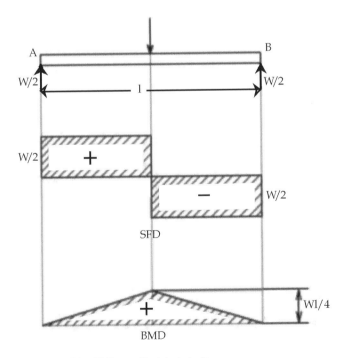

FIGURE 5.46 Hollow cylindrical shaft.

is therefore weaker than its weakest component. If the relative dimensions of different parts are needed, a solution is usually achieved by matching the torque in each shaft, for example, for two series shafts.

$$T = \frac{G_1 J_1 \theta_1}{L_1} = \frac{G_2 J_2 \theta_2}{L_2} \qquad (5.93)$$

It is appropriate in some applications to ensure that the twist angle in each shaft is equal, i.e., $\theta_1 = \theta_2$, so that for similar materials in each shaft.

$$\frac{J_1}{L_1} = \frac{J_2}{L_2} \qquad (5.94)$$

Or

$$\frac{L_1}{L_2} = \frac{J_1}{J_2} \qquad (5.95)$$

At the free end the total angle of twist must be the sum of angles $\theta_1 = \theta_2$ over each cross section.

5.16.2.2 Composite Shafts in Parallel

If two or more shafts are rigidly fixed together, as shown in Figure 5.48, to share the applied torque, the composite shaft so developed is said to have been connected in parallel.

For parallel connection,

$$\text{Total Torque, } T = T_1 + T_2 \qquad (5.96)$$

In this case, the twist angle of each portion is equal and

$$\frac{T_1 L_1}{G_1 J_1} = \frac{T_2 L_2}{G_2 J_2} \qquad (5.97)$$

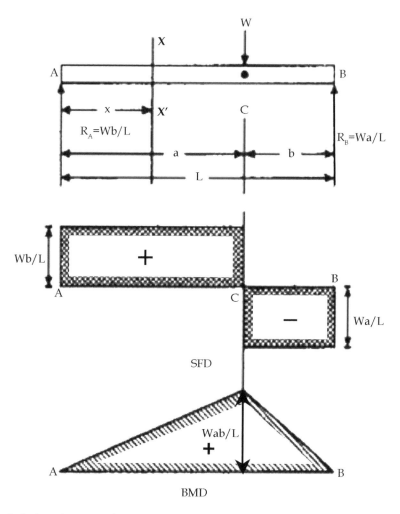

FIGURE 5.47   Composite shafts in series connection.

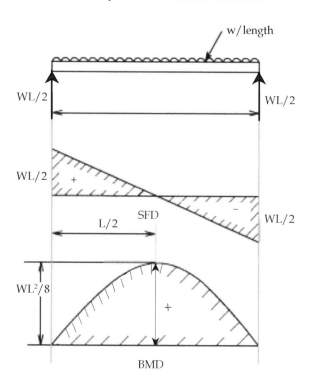

FIGURE 5.48   Composite shafts in parallel connection.

for equal lengths (as is normally the case for parallel shafts)

$$\frac{T_1}{T_2} = \frac{G_1 J_1}{G_2 J_2} \quad (5.98)$$

This type of configuration is statically indeterminate, since the applied torque apportioned to each segment is unknown. In each part of the composite shaft, two formulae are used in terms of torques, and the maximum shear stress in each part can then be found from the relationships shown in Equations 5.99 and 5.100.

$$\tau_1 = \frac{T_1 R_1}{J_1} \quad (5.99)$$

$$\tau_2 = \frac{T_2 R_2}{J_2} \quad (5.100)$$

5.16.3   Shaft Couplings

The bolts fail in shear, in the case, especially of shaft couplings. In this case, the coupling's torque capacity is evaluated as follows. The shear stress in any bolt is presumed to be uniform and the distance from its center to the coupling center as shown in Figure 5.49 is governed. The torque capacity of the coupling may therefore be determined by

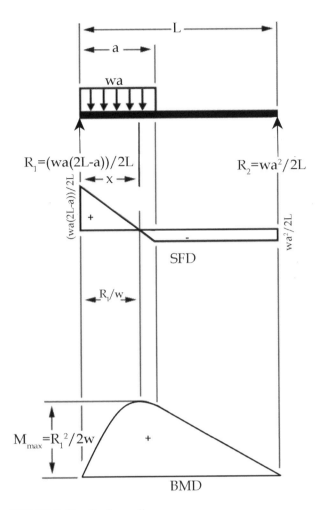

FIGURE 5.49   Shaft coupling.

$$T = \left(\frac{\pi}{4} d_b^2 \right) \tau_b' rn \qquad (5.101)$$

where

   d_b is the diameter of bolt; τ_b' is the maximum shear stress in bolt; n is the no. of bolts; r is the distance from center of bolt to that of coupling.

5.17   THIN CYLINDERS

A pressure vessel is being used for pressurized fluid storage. It is usually spherical or cylindrical in shape. There are various types of pressure vessels used for technical purposes. Boilers, gas tanks, metal tires, and pipelines are included. A thin pressure vessel therefore has a thickness to the internal radius ratio of not more than 1/10. Simply consider a long circular cross-sectional cylinder, as shown in Figure 5.50, with internal radius "r" and a constant wall thickness "t."

5.17.1   FAILURE OF THIN CYLINDERS

A thin-walled cylinder may completely fail if its internal pressure is unusually high. It could completely fail to burst along

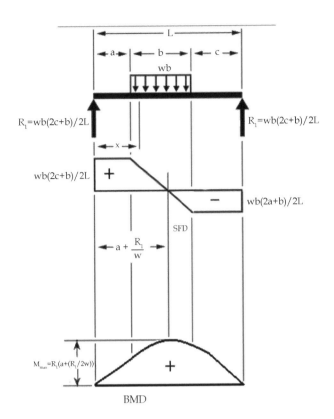

FIGURE 5.50   Thin cylinder with cylindrical polar system.

a path along the circumference of the cylinder. It will simply fail under ordinary circumstances by bursting down a path parallel to the axis. This means that the axial stress is considerably below the hoop stress.

5.17.2   APPLICATIONS OF THIN CYLINDERS

Liquid storage tanks or containers, boilers, water pipes, submarine hulls, and certain aircraft components are prominent examples for cylinders and spheres with thin walls.

   The assumptions made for the evaluation of the thin-walled cylinders are as follows:

   • There is no shear stress in the wall.
   • The longitudinal and hoop stresses do not vary across the wall.
   • Radial stresses that act normally on the curved plane of the isolated element are trivially small compared to two other stresses, particularly if t/Ri is less than 1/10.

5.17.3   THIN CYLINDERS SUBJECTED TO INTERNAL PRESSURE

When an internal pressure is applied to a thin-walled cylinder, three main mutually perpendicular stresses are formed in the cylindrical materials.

   • Circumferential or hoop stress
   • Radial stress
   • Longitudinal stress

5.17.3.1 Hoop Stress on Thin Cylinders

The stress which is developed to resist the bursting effect of the applied pressure is known as Hoop stress. It is calculated by taking equilibrium of the cylinder. In Figure 5.51, one half of the cylinder is shown. This cylinder is pressurized internally p.

Combined total force on one half of the cylinder due to the "p" internal pressure

$$= p \times \text{projected area} = p \times d \times L \quad (5.102)$$

where d is the inner diameter and L is the cylinder length.

The total resistant force due to hoop stress σ_H developed in the cylinder walls

$$= 2 \times \sigma_H \times L \times t \quad (5.103)$$

where t is the wall thickness.

Since $\sigma_H \times L \times t$ is the force in one wall of the half cylinder, equating both the two equations

$$\text{Circumferential or hoop Stress}, \sigma_H = \frac{pd}{2t} \quad (5.104)$$

5.17.3.2 Radial Stress on Thin Cylinders

The circumferential and longitudinal stresses are usually much higher for pressure vessels. Since the magnitude of radial stresses is very small in thin cylinders, they are usually neglected.

5.17.3.3 Longitudinal Stress on Thin cylinders

Now consider the same figure again but with the vessel having closed ends and containing a fluid under a gauge pressure p as shown in Figure 5.52. The cylindrical walls will then have both longitudinal and circumferential stress.

$$\text{Total cylinder end force due to internal pressure} = p \times \frac{\pi d^2}{4} \quad (5.105)$$

$$\text{Area of material surface resisting this force} \approx \pi \times d \times t \quad (5.106)$$

Hence the longitudinal stress,

$$\sigma_L = \frac{pd}{4t} \quad (5.107)$$

5.18 THICK CYLINDERS

If the thickness to the internal radius ratio of a cylinder exceeds 1/10, it is considered as a thick pressure vessel. For thick cylinders such as guns, high-pressure hydraulic pipes etc., the stress variation across the thickness is very significant, because of the wall thickness being relatively high.

Stress in thick shell is calculated by Lame's theorem. The assumptions of the theorem are based solely on the following:

- The cylinder material is homogeneous and isotropic.
- The cylindrical plane sections perpendicular to the longitudinal axis stay plane under the pressure.

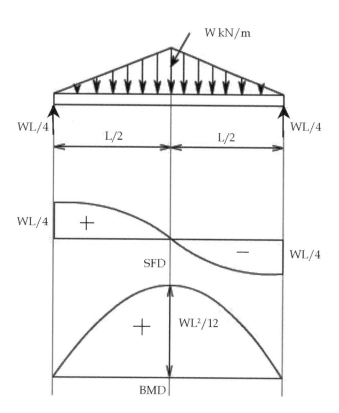

FIGURE 5.51 Hoop stress on thin cylinders.

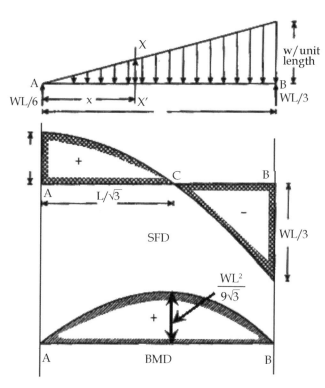

FIGURE 5.52 Longitudinal stress on thin cylinders.

Hoop stress is calculated as:

$$\sigma_h = \frac{b}{r^2} + a \qquad (5.108)$$

Radial stress is calculated as:

$$\sigma_r = \frac{b}{r^2} - a \qquad (5.109)$$

Here, a and b are constant and r is the distance measured from center of shell.

If cylinder has closed ends, longitudinal stress over the cross section is given as

$$\sigma_l = \frac{pr_1^2}{r_2^2 - r_1^2} \qquad (5.110)$$

5.19 COLUMNS AND STRUTS

The compressive load which a structural member can hold can be split into two distinct categories; obviously it depends on the relative length and cross-sectional dimension. Short, thick members are usually referred to as columns and usually fail to crush when the yield stress of the compression material exceeds its yield. Long, slender columns are usually referred to as struts and can fail to buckle before the compression yield stress is achieved. Because of the following fundamental reasons, the buckling occurs.

a. The strut might not initially be completely straight.
b. The load cannot be applied precisely along the strut axis.
c. One part of the material can result more easily in compression than others due to a lack of uniformity in the properties of the material all throughout the strut.

Any member's resistance to bending is calculated by its flexural rigidity EI and the quantity I that can be written as

$$I = Ak^2 \qquad (5.111)$$

where I is the inertial moment of area; A is the cross-sectional area; k is the radius of gyration.

Therefore, the radius of gyration is offered by the load per unit area which the member can hold. There will be two principal moments of inertia, if the least of these is taken then the ratio, called slenderness ratio, is given by

$$\frac{l}{k} \text{ i.e. } \frac{\text{length of member}}{\text{least radius of gyration}} \qquad (5.112)$$

Its number value implies if the member falls into the column or strut class.

5.19.1 EULER'S THEOREM OF COLUMNS

In this hypothesis, the following assumptions are made:

- Initially, the column is straight, and the load applied is axial.
- The column material is homogeneous, linear, and isotropic.
- The column length is very large in comparison with its cross-sectional dimensions.
- The column will have a uniform cross section.
- The axial compression shortening of the column is not regarded.
- The self-weight of the column is ignored.
- The ends of the column are free of friction.

There are four types of columns, depending on the end conditions as shown in Figure 5.53. They are as follows:

i. Both ends hinged
ii. Both ends fixed
iii. One end fixed and the other end hinged
iv. One end fixed and the other end free.

Euler's formula:

$$F = \frac{\pi^2 EI}{(KL)^2} \qquad (5.113)$$

Here, F is the load under which a column just buckles, E is the elasticity modulus of column material, I is the least moment of inertia, K is the effective factor of length, and L is the total column length.

Effective column length depends on the end-support conditions of the columns:

a. For both ends, pinned effective length is L. So, $K = 1$
b. For both ends fixed, this is $L/2$, so $K = \frac{1}{2}$

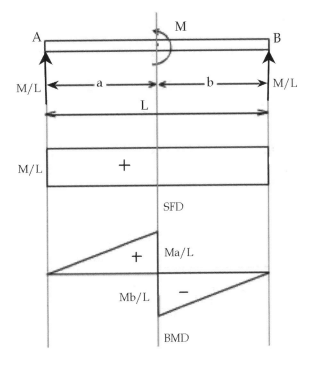

FIGURE 5.53 Types of columns.

c. For one end fixed and the other free, this is $2L$ so, $K=2$

d. For one end fixed and the other hinged, this is $L/1.414$, so $K=1/1.414$

Rankine Gordon-Formula:

$$\frac{1}{\text{critical load}} = \frac{1}{\text{crushing load}} + \frac{1}{\text{buckling load}} = \frac{1}{P_c} + \frac{1}{P_e}$$

(5.114)

where P_e=Euler's Load=F.

P_c=Crushing strength of the column material

Or,

$$P_{cr} = \frac{P_c}{1+a\left(\frac{L_e}{k}\right)^2}$$

(5.115)

can determine the crushing load by multiplying the crushing strength with the cross-sectional area of the column. Euler

formula can be usually used to determine buckling load. Rankine's formula applies to all column types, both short and long columns.

5.20 SPRINGS

A spring can be characterized as an elastic member, whose primary role is to distort or deflect the load applied; it regains its natural form when the load is set to release. Springs are also categorized as units with energy absorption. Springs store energy and slowly or quickly reinstate it; obviously it depends on the application of the load.

5.20.1 Closed Coiled Helical Springs

A helical spring is formulated as a circular cylindrical helix by winding a wire. These can be axial loads or torsional loads. The angle of the horizontal coil is called the helix angle (α). If α is smaller than 18, closely coiled helical springs are recognized. Simply consider a closed coiled spring with an axial load W, as illustrated in Figure 5.54. Every coil can be considered to lie in a plane at the right angle to the helix axis.

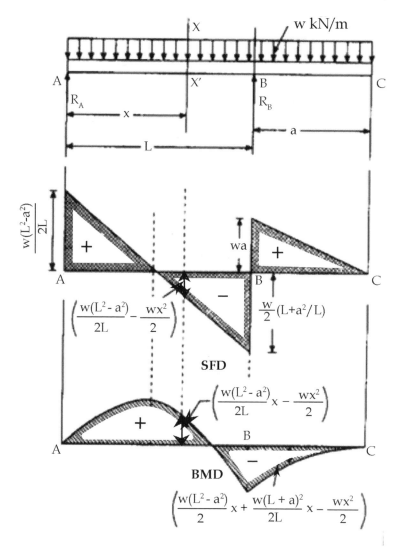

FIGURE 5.54 Closed coiled spring.

The wire that actually renders the helical spring close is under torsion.

5.20.1.1 Under Axial Load, W

Since the angle of the helix is fairly small, the action on any cross section is a pure torque and the bending and shear effects can be ignored. The torque value is given by:

$$\text{Torque} = W \times \frac{D}{2} \qquad (5.116)$$

where W is the axial load and D is the mean coil diameter.

The spring stiffness

$$\mu = \frac{W}{x} = \frac{Gd^4}{8D^3n} \qquad (5.117)$$

The strain energy

$$U = \frac{1}{2}Wx \qquad (5.118)$$

where x is the deflection of the spring.

Which by replacing in terms of τ from Equation 5.117 can be reduced to

$$U = \left(\frac{\tau^2}{4G}\right) \times \text{volume} \qquad (5.119)$$

5.20.1.2 Under Axial Torque Load, T

This results in a pure bending moment of magnitude T in all cross sections. The total strain energy is given by

$$U = \frac{T^2l}{2EI} = \frac{T^2\pi Dn}{2E \times \frac{\pi d^4}{64}} = \frac{32T^2Dn}{Ed^4} \qquad (5.120)$$

However, if T causes one end of the spring to rotate out through an angle φ around the axis relative to the other end, then

$$U = \frac{1}{2}T\varphi \qquad (5.121)$$

Substituting this in Equation 5.120

$$\varphi = \frac{Tl}{EI} = \frac{64TDn}{Ed^4} \qquad (5.122)$$

$$\text{Maximum bending stress} = \frac{T \times \frac{d}{2}}{\frac{\pi d^4}{64}} = \frac{32T}{\pi d^3} \qquad (5.123)$$

Strain energy: The energy of the strain is characterized as the energy stored in a material when the work on the material is carried out. In the particular instance of a spring, the energy of the strain is caused by bending and apparently given by the expression.

$$U = \frac{32T^2Dn}{Ed^4} \qquad (5.124)$$

5.20.2 OPEN COILED HELICAL SPRING

The spring's coils are not close to each other. The helix angle is larger than 18. Consider an open helical spring under axial load.

Let α be the angle of the helix, then the length of the wire

$$l = \frac{\pi Dn}{\cos\alpha} \qquad (5.125)$$

The total strain energy due to bending and twisting is given by:

$$U = \frac{\left[\left(\frac{WD}{2}\right)\cos\alpha + T\sin\alpha\right]^2 l}{2GJ} + \frac{\left[T\cos\alpha - \left(\frac{WD}{2}\right)\sin\alpha\right]^2 l}{2EI} \qquad (5.126)$$

5.21 STRAIN ENERGY

Sometimes strain energy is called internal work, to differentiate it from external work "W." Consider a simple bar confined to a tensile force F with a small element of dx, dy, and dz dimensions as seen in Figure 5.55. The energy of strain U is the area under the triangle.

$$\frac{U}{\text{volume}} = \frac{1}{2}\frac{\sigma_x^2}{E} \qquad (5.127)$$

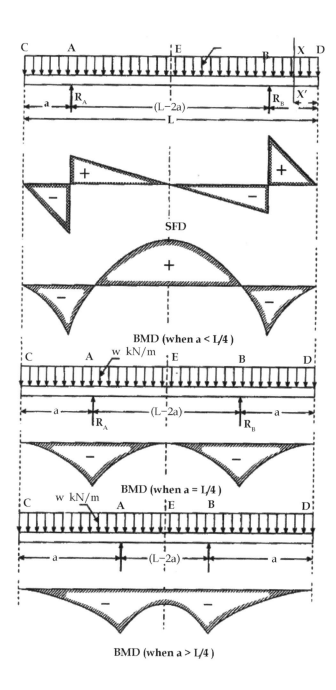

FIGURE 5.55 Strain energy.

6 Principles of Surveying

6.1 BASIC PRINCIPLES OF SURVEYING

Surveying is the methodology used to determine exactly the relative position of points above or below the surface of the earth using direct or indirect distance, direction, and elevation measurements. It also includes points determination with predetermined angular and linear measurements.

Leveling is a branch of the survey whose object is

- to actually find the elevations of points for a specified or presumed data, and
- to set points at a previously given elevation or totally different elevations for a specified or presumed date.

The first operation is vital for the design of the works, whereas the second operation is crucial for all types of engineering work. Leveling deals with vertical plane measurements.

6.1.1 OBJECTIVES OF SURVEYING

The main purpose of the survey is to prepare a map showing the relative positions of the objects on the earth's surface. The map is drawn to some appropriate scale. It shows a country's natural characteristics such as villages, towns, rivers, roads, railways, etc. Maps might also include particulars of various engineering works, such as roads, railways, canals of irrigation, etc.

6.1.2 NEED OF SURVEYING

- To make a topographical map, it shows the country's hills, valleys, rivers, villages, cities, etc.
- To aid in preparation of a cadastral map showing the boundaries and other properties of the field houses.
- To plan and prepare an engineering map for details such as roads, railways, canals, and so on.
- To start preparing a military map displaying roads and railways, communications with various parts of the country
- To help in making the contour map and help to decide the capacity of the reservoirs and tunnels, find the perfect possible route for roads, railroads, etc.
- To aid in preparation of archeological maps as well as for locations where there may be ancient relics.
- To help in sketching out of a geological map showing regions with underground resources.

6.1.3 GENERAL PRINCIPLE OF SURVEYING

- Working from whole to part.
- To fix the position of new stations with at least two readings (linear or angular) from the fixed point of reference.

The first principle is that the entire area is first enclosed by major stations and major survey lines. The area is then fragmented into several parts by establishing well-conditioned triangles; the most well-conditioned triangle is an almost equilateral triangle. The main lines of the survey are measured with a standard chain very precisely. Then the triangle sides are calculated.

According to the second principle, the new stations should always be fixed by at least two measurements from fixed reference points. Linear measurements refer to horizontal distance which is measured by chain or tape. Angular measurements refer to the magnetic bearing or horizontal angle taken by a prismatic compass or theodolite.

6.1.4 SCALES

The area that is surveyed is vast, and therefore, plans are made to some scale. Scale is the fixed ratio that each distance on the plane has a corresponding ground distance. The following methods can represent scale:

- One centimeter on the plan represents a number of meters on the ground, e.g., 1 cm = 10 m. This scale type is called the engineering scale.
- 1 unit of length on the plan symbolizes a number of units of the same length on the ground, e.g., 1/1000. This map distance ratio to the subsequent ground distance unbiased of the measuring units is referred to as the representative factor (RF). For instance, 1 cm = 50 m, RF = $1/(50 \times 100) = 1/5000$.

The two above-mentioned scales are also called numerical scales. The four major types of scales that are used for measurements for different tasks are plain scale, diagonal scale, Vernier scale, chord scale.

6.1.4.1 Plain Scale

Plain scale is the one on which it is possible to measure only two dimensions. For example, measurements such as units and lengths, meters, and decimeters, etc. Six different plain scales in metric are used by engineers, architects, and surveyors.

6.1.4.2 Diagonal Scale

The diagonal scale is the one on which three dimensions like meters, decimeter, and centimeter can be measured; units, tens and hundreds; yards, feet and inches, etc. A short length is partitioned into several parts using the similar triangle principle in which the sides are proportional, as shown in Figure 6.1. Figure 6.1 shows 1-1 for 1/10 PQ, 2-2 for 2/10 PQ,... 9-9 for 9/10 PQ.

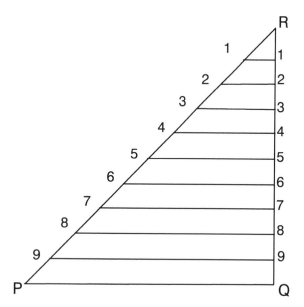

FIGURE 6.1   Diagonal scale.

6.1.4.3   Vernier Scale

The Vernier scale is a device used to quantify the marginal portion of one of the smallest graded divisions. It is normally a small auxiliary scale that slides along the primary scale. Lowest Vernier count is the difference between the smallest division in the main division and the smallest division in the Vernier.

6.1.4.4   Scale of Chords

Chord scale is used to calculate an angle and is marked on a rectangular protractor or a standard wooden box scale.

6.2   CLASSIFICATION OF SURVEYING

6.2.1   Primary Classification

Surveys are classified primarily as plane surveys and geodesic surveys.

6.2.1.1   Plane Surveying

The earth's shape is spheroidal, so the surface of earth is obviously curved. However, the curvature of the earth is not taken into account in the plane survey. This is because plane surveys are performed only in a small area. The surface of the earth is therefore regarded as a plane. In this survey, a line that joins two points is considered straight. The three-point triangle is considered to be a plane triangle and the angles of the triangle are presumed to be plane angles. Plane surveys are carried out by state agencies such as the Department of Irrigation, the Department of Railways, etc. Plane survey is carried out on an area of less than 250 km².

6.2.1.2   Geodetic Surveying

The curvature of the earth is considered in geodetic surveys and extends over a wide area. The line connecting two points is regarded a curved line. The three-point triangle is considered spherical and the angles of the triangle are therefore presumed to be spherical angles. Geodetic surveys are carried out by the Survey Department of India and cover an area exceeding 250 km².

6.2.2   Secondary Classification

Instrument-based:

- Chain surveying
- Compass surveying
- Plane table surveying
- Theodolite surveying
- Tacheometric surveying
- Photographic surveying
- Modern surveying

Methods-based:

- Triangulation surveying
- Traverse surveying

Object based:

- Geological surveying
- Mine surveying
- Archaeological surveying
- Military surveying

Based on field nature:

- Land surveying
- Marine surveying
- Astronomical surveying

6.3   VALUES OF A QUANTITY

6.3.1   True Value of a Quantity

The real value is the value of a quantity that is completely free of errors. It is never discoverable and indeterminate.

6.3.2   Observed Value of a Quantity

The value observed is the value of a quantity obtained from field measurement after correction for all observation-related errors.

6.3.3   Most Probable Value of a Quantity

The value that is most obviously true than any other value is called a quantity's most probable value. This is most likely free, but not entirely free, from mistakes. The most probable value for direct observations with the same weight is the arithmetic mean. When observing unequal weights directly, the most probable value is the weights; the most probable value is the weighted arithmetical mean.

6.3.4 Principle of Least Square

According to the principle of least square, the most probable value of the quantity observed from a given set of observations is that for which the sum of error squares is a minimum.

6.4 ERRORS

6.4.1 Sources of Errors

Survey errors can occur from three main sources:

6.4.1.1 Instrumental Errors

Survey errors may be caused by imperfection or incorrect adjustment of the measuring instrument. An angle measuring instrument, for example, may be out of control or a tape may be too long. Such mistakes are called instrumental errors.

6.4.1.2 Personal Errors

Error can also occur because of a lack of perfection of human sight in observation and of touch in instruments manipulation. For example, there may be an error in reading a level or reading an angle in a theodolite's circle. Such errors are referred to as personal errors.

6.4.1.3 Natural Errors

Surveying errors can also happen due to differences in phenomena such as temperature, humidity, gravity, wind, refraction, and magnetic decay. If they are not properly monitored during measurement, the findings seem to be incorrect. A tape can be 20 m at 20°C, for example, but its length will alter if the field temperature is drastically different.

6.4.2 Types of Surveying Errors

Ordinary survey errors in all the survey work classes can be categorized as follows:

6.4.2.1 Mistakes

These happen only during manual recording. Errors result from inexperience, inattention, carelessness, poor judgment, or confusion in the observer's mind. They do not follow any math (probability law) rules and can be small or large, positive or negative. They cannot be quantified. Examples of mistakes are as follows:

1. False recording, e.g., Write 69 instead of 96
2. Counting eight for three

6.4.2.2 Accidental Errors

Errors in the survey may occur due to unavoidable circumstances such as possible variations in atmospheric conditions that are completely beyond the control of the observer. Errors in surveys due to imperfection in instruments of measurement and even imperfection of eyesight fall into this class. They can be positive and signs can change. They are not accountable.

6.4.2.3 Systematic or Cumulative Errors

A cumulative or systematic error is an error that always has the same size and signs under the same conditions. A systematic error often obeys a certain mathematical or physical law and it can be corrected and applied. These mistakes are of a constant nature and are regarded positive or negative, because the result is large or small. Their effect is therefore cumulative. These are the examples:

- Faulty line alignment
- An instrument is not properly leveled
- An instrument is not correctly adjusted

6.4.2.4 Compensating Errors

- Error in compensation typically occurs in either direction, i.e., sometimes the error tends to be positive and sometimes negative and thus offset each other. They tend sometimes in one direction and sometimes in the other, i.e., the obvious result is statistically likely to be large or small. They should be managed in accordance with mathematical laws of probability because they obey the laws of chance. Some instances are as follows:
  - Inaccurate centering
  - Inaccurate object bisection

6.4.3 Most Probable Error

The most probable error is defined as that quantity which when added to and subtracted from, the most probable value fixes the limits within which it is an even chance the true value of the measured quantity must lie.

The probable error of a single observation is calculated using the equation,

$$E_s = \pm 0.6745 \sqrt{\frac{\sum v^2}{n-1}} \tag{6.1}$$

The probable error of the mean of a number of observations of the same quantity is calculated from the equation:

$$E_s = 0.6745 \sqrt{\frac{\sum v^2}{n(n-1)}} = \frac{E_m}{\sqrt{n}} \tag{6.2}$$

where E_s is the probable error of single observation, v is the difference between the mean of the series and the single observation, E_m is the probable error of the mean, and n is the number of observations in the series.

6.5 VERTICAL CONTROL

Vertical control networks are a number of points on which precise heights or elevations have been determined. Vertical checkpoints are usually called bench marks.

6.6 MEASUREMENT OF DISTANCE

6.6.1 Methods of Linear Surveying

Linear survey methods can be subdivided into three heads as direct measurement, optical measurement, and electronic measurement.

6.6.1.1 Direct Measurement

In this method of surveying, distances on the surface of the earth are actually measured using chains, tapes, etc.

The following methods are generally employed for linear measurements:

- **Pacing**: The distance can be determined by pacing if approximate results are required. It consists of walking upon a line and counting the number of paces (800 mm) the needed distance can be obtained by multiplying the number of paces by the average pace length. The pace length varies with the following:
  - individual, age, physical condition, height
  - land nature (uphill and downhill)
  - the country's slope, and
  - the pacing speed
- **Passometer**: It is a pocket tool used to record the number of paces automatically. It should be carried vertically or hung from a button in a waistcoat pocket. The mechanism is operated by the body's movement and strain.
- **Pedometer**: It is like a passometer. The key difference is that it records the distance traveled by those who carry it. The distance is measured by an indicator. It is equipped with a knob or stud, which can be carried in the same way as the passometer when the release indicator is pressed to zero.
- **Odometer**: It calculates the approximate distances and can be connected to any vehicle's wheel. It records the number of wheel revolutions. When the wheel circumference is known, the distance traversed can be achieved by multiplying the number of revolutions by the wheel circumference.
- **Speed meter**: The locomotive speed meter is used for approximate distance measurement. It results better than pacing, provided the route is smooth.
- **Perambulator**: It is used for rapid distance measurement. It is made up of a single wheel with a fork and a handle. It is wheeled along the line that is desirable in length.
- **Time measurement**: The distance is determined approximately by the travel time. If you know the average time per km for a walker or a horse or a light ray as in modern instruments, you can easily calculate the distance traversed.
- **Chaining**: A more accurate method called chaining is used to calculate the distance using a chain, rope, or a tape. A chain is used for ordinary precision work. Where high precision is needed, a steel tape is used.

6.6.1.2 Measurement by Optical Means

In this technique, observations are made via a telescope and distances are ascertained by analysis such as in tacheometer or triangulation.

6.6.1.3 Electronic Method

Distances are measured using instruments that rely on the propagation, reflection, and subsequent reception of radio or light waves in these linear surveying methods. The various tools used with electronic methods are as follows:

- Geodimeter
- Tellurometer
- Decca navigator
- Lambda position fixing system

The distance calculated in the case of the geodimeter is based on the light wave propagation. The other three instruments use radio waves to measure distance.

6.7 CONTOURING

A contour is an invisible line of nearly constant elevation on the surface of the ground. This can be regarded as the intersection line between the level surface and the soil surface. The intersection line of the water surface of a lake or lake with the encircling ground, for example, represents a contour line.

6.7.1 Terms related to Contouring

- **Contour line**: A line with equal elevation points is termed a contour line. It enables depiction of terrain relief in a two-dimensional plan or map.
- **Contour gradient**: An imaginary line on the surface of the earth with a constant horizontal inclination (slope) is termed the contour gradient. The inclination of a contour gradient is usually given as an upward gradient or a downward gradient. It is found as the relation between the vertical height and the horizontal distance specified.
- **Contour interval**: The lateral distance between two successive contours is an interval of contours. For instance, if the different consecutive contours are 100, 98, 96 m, etc. Contour intervals for flat country are usually tiny, e.g., 0.25, 0.5, 0.75 m, and so on. There is a greater steep slope in the hilly region, e.g., 5, 10, 15 m, etc.

6.7.2 Characteristics of Contours

The main features of the contour lines that assist to draw or read a contour map are as follows:

- The fluctuation of the vertical distance between each of the two contour lines is presumed to be uniform.
- The horizontal distance between each of the two contour lines implies the slope amount and differs

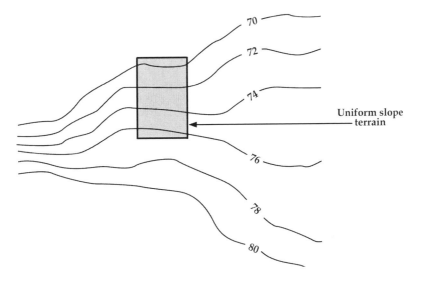

FIGURE 6.2 Contours of terrain having different types of slope.

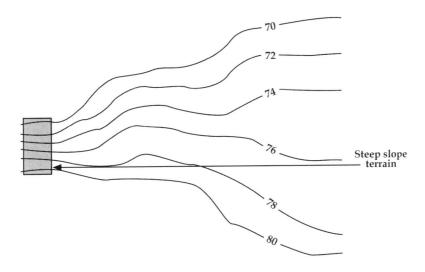

FIGURE 6.3 Contours of terrain having different types of slope.

inversely with the slope amount. Therefore, the contours are equally spaced for uniform slopes as in Figure 6.2, for steep slope contours as in Figure 6.3.

- At any point in the contour, the steepest slope of the terrain is depicted along the normal contour at that point as in Figure 6.4. They are perpendicular to the mountain ridges and valley lines traversing these lines.
- Contours do not actually pass through buildings like permanent structures as in Figure 6.5.
- Contours of various elevations cannot be crossed (exceptions are caves and overhanging cliffs).
- Contours of various heights cannot mobilize to form a single contour (an exception is the vertical cliff).
- Contour lines cannot start or end on the plan.
- A contour line should be completely closed but not inherently within the map limits.
- A map's closed contour line symbolizes depressions or hills as in Figure 6.6. A full set of ring contours

with higher values shows a hill, while the lower value shows a depression (without an outlet).

- Contours deflect uphill on lines of the valley and downhill on lines of ridge. U-shaped contour lines

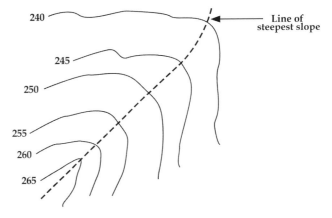

FIGURE 6.4 Steepest slope of a terrain indicated at contour lines.

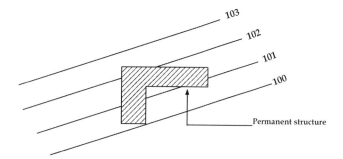

FIGURE 6.5   Contours across permanent structure.

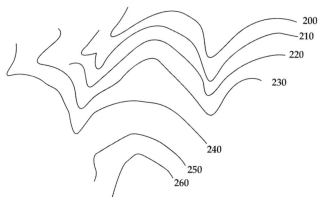

FIGURE 6.7   Valley and ridge lines.

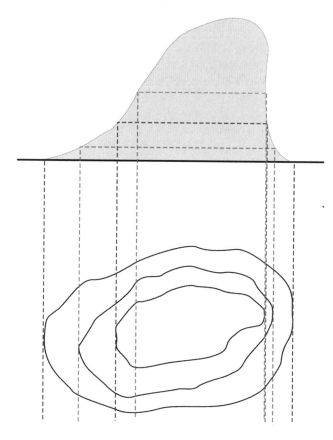

FIGURE 6.6   Hill and its contour.

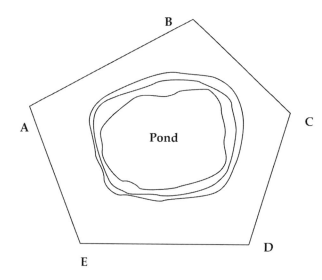

FIGURE 6.8   Closed traverse.

cross a ridge and cross a valley at the right angles in V-shaped form. The concavity in the contour lines is in the case of a ridge toward the higher ground and in the case of a valley toward the lower ground as in Figure 6.7.

• Contours have no sharp turns.

6.8   TRAVERSE SURVEYING

Traversing is the type of survey in which the framework for a number of connected survey lines and the directions and lengths of the survey lines are calculated by using an angle calculating instrument and a tape. Whenever the lines form a circuit that finishes at the base point, that they really are simply regarded to as a closed crossing. If the circuit finishes

anywhere else, it is said to be an open crossing. A completely closed crossing is appropriate for the location of lakes, forests, etc., and for huge areas. Figure 6.8 shows a closed traverse. The open crossing is appropriate for the survey of a long narrow strip of land for a road, canal, or coastline. Figure 6.9 shows an open traverse.

6.8.1   PROCEDURE FOR TRAVERSE CALCULATIONS

6.8.1.1   Balancing Angles of Closed Traverses

$$\sum \text{interior angles} = (n-2) \times 180 \qquad (6.3)$$

where n is the number of interior angles. If a is the tiny division visible on the instrument,

$$\text{Maximum error} = \sqrt{n} \times a \qquad (6.4)$$

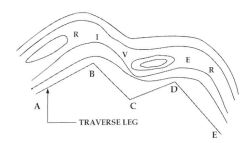

FIGURE 6.9   Open traverse.

6.8.1.2 Closure of Latitudes and Departures

- The mathematical total sum of all latitudes must be nil or latitude distinction between the initial and final points of control.
- The algebraic sum of all departures shall be equal to zero percent or the departure distinction between the initial and final control points.

6.9 HYDROGRAPHIC SURVEYING

Hydrographic survey is the survey branch that deals specifically with the quantification of water bodies. It is the art of describing the underwater levels, contours, and characteristics of seas, gulfs, rivers, and lakes.

In one or more of the following conditions, hydrographic surveys are actually carried out.

1. Measurement of sea coast tides
2. Setting the depth of the bed by sounding
3. Finding of current direction

6.9.1 Horizontal Controls

The primary horizontal control is mainly developed in an extensive survey by running theodolite and tape traverses in front of the triangulation station. The traverse lines run approximately to the shore lines.

6.9.2 Vertical Controls

Vertical controls depends on a series of bench marks developed by spirit leveling near the shoreline, which are used to set and check tide gages, etc.

6.9.3 Sounding and the Methods Employed in Sounding

The depth measurement below the water surface is known as sounding. The datum or the horizontal line is the surface of water, the level of which continuously goes on changing with time. The aim of producing soundings is thus to identify the structure of the subaqueous source. Soundings are required for the following purposes:

i. Making nautical charts for navigation;
ii. Calculation of silting or scouring areas and determination of volume of dredged material;
iii. Conducting subaqueous inquiries to gather data necessary to build, develop, and improve port facilities.

6.9.3.1 Sounding Boat

A row-boat for sounding should be stable and sufficiently roomy. For quiet water, a flat bottom boat is more suitable, but for rough water, round-bottomed boat is more suitable.

6.9.3.2 Fathometer

An instrument regarded as a fathometer is used for sounding the ocean. It is an electronic device used to quantify the time expected to travel to the bottom end of the water and back by the sound (impulses). The time of travel is transformed to depth in either digital or graphical fathometer. It is termed echo sounder too.

6.10 CURVES

Curves are typically used on highways and railways when the direction of motion needs to be changed. A curve can be circular, parabolic, or spiral and always be tangential in both directions. Furthermore, circular curves are subdivided into three classes: (1) simple, (2) compound, and (3) reverse.

6.10.1 Simple Curve

A simple curve is that consisting of a single circle arc. Both straight lines are tangential. A compound curve comprises two or more simple arcs which turn in the same direction and at common points of tangent. A reverse curve is the one consisting of two circular arches in the same or distinct radii with their centers on the various sides of the fairly common tangent. Both arches therefore bend in various directions with a normal tangent at junction end. Figure 6.10 shows the fairly typical layout of the single curve.

6.10.1.1 Elements of a Simple Curve

- **Point of intersection (PI)**: The intersection point indicates the intersection of the back and forward tangents. The surveyor shows that it is one of the preliminary crossing stations.
- **Intersecting angle (I)**: The intersection angle at the intersection point is the deflection angle. The surveyor calculates its real value from the preliminary angles of the crossing station or determines it on the field.
- **Radius (R)**: It is the circle radius of which the curve is an arc.
- **Point of curvature (PC)**: Curvature point is the point at which the circular curve begins. At this point, the back tangent is tangent to the curve.

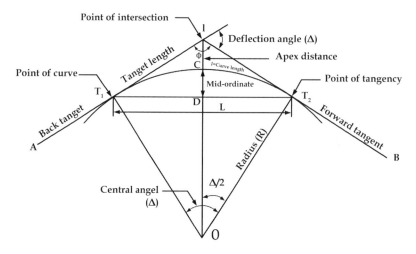

FIGURE 6.10 Simple curve.

- **Point of tangency (*PT*):** Tangency point is the point at which the curve finishes. At this point, the forward tangent is tangent to the curve.
- **Degree of curve (*D*):** The degree of curve defines the "strength" or "flatness" of the curve. There are two widely used definitions for the curve degree, the arc, and the chord definition. The relationship between radius (*R*) and curve degree (*D*) may be specified as

$$R = 5730/D \qquad (6.5)$$

- **Length of curve (*l*):**

$$l = R\Delta \qquad (6.6)$$

where Δ is in radians.

$$l = \frac{\pi R}{180°}\Delta \qquad (6.7)$$

where Δ is in degrees.

$$\text{Tangent length: } T = R\tan\frac{\Delta}{2} \qquad (6.8)$$

$$\text{Length of the long chord}(L): L = 2R\sin\frac{1}{2}\Delta \qquad (6.9)$$

$$\text{Apex distance}(E): E = R\left(\sec\frac{\Delta}{2} - 1\right) \qquad (6.10)$$

$$\text{Mid-ordinate}(M): M = R\left(1 - \cos\frac{\Delta}{2}\right) \qquad (6.11)$$

6.10.1.2 Methods of Setting Out of Single Circular Curve

- By offsets or long chord ordinate.
- Gradually bisecting arcs or chords.
- By tangent offsetting.
- Offsets produced from the chord.

6.10.2 Compound Curve

The typical compound curve layout is shown in Figure 6.11.

The elements of a compound curve are point of curvature (*PC*), point of tangency (*PT*), point of intersection (*PI*), point of compound curves (*PCC*), tangent length of the first curve (*T₁*), tangent length of the second curve (*T₂*), vertex of first

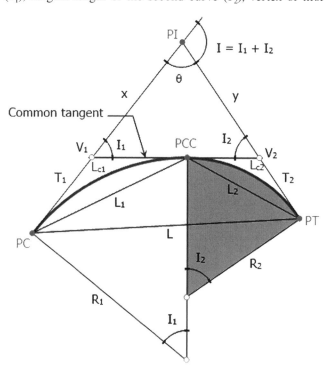

FIGURE 6.11 Compound curve.

curve (V_1), vertex of second curve (V_2), central angle of first curve (I_1), central angle of second curve (I_2), angle of intersection (I) = $I_1 + I_2$, curve length of first curve (Lc_1), curve length of second curve (Lc_2), chord length of first curve (L_1), and chord length of second curve (L_2).

- The common tangent measurement length from V_1 to $V_2 = T_1 + T_2$
- $\Theta = 180° - I$
- X and Y can be found from the triangle $V_1 - V_2 - PI$
- L can be found from the triangle $PC - PCC - PT$

Tangent lengths can be computed as

$$t_s = R_s \tan\frac{\Delta_s}{2} \qquad (6.12)$$

$$t_L = R_L \tan\frac{\Delta_L}{2} \qquad (6.13)$$

$$T_s = (t_s + t_L)\frac{\sin\Delta_l}{\sin\Delta} + t_S \qquad (6.14)$$

$$T_L = (t_s + t_L)\frac{\sin\Delta_s}{\sin\Delta} + t_L \qquad (6.15)$$

$$\text{where } \Delta = \Delta_s + \Delta_L \qquad (6.16)$$

Length of curves can be computed as

$$l_s = \frac{\pi R_s \Delta_s}{180} \qquad (6.17)$$

$$l_L = \frac{\pi R_L \Delta_L}{180} \qquad (6.18)$$

$$l = l_s + l_L \qquad (6.19)$$

6.10.3 Transition Curves

The centrifugal forces start developing during the turn, so that the motor vehicle and its entire contents are instantly subject to centrifugal forces. The velocity of the vehicle is greater at the curvature, and the effort on vehicles and drivers to change from tangent to curve is greater. If transition curves are not provided, drivers tend to form their own transition curves by relocating laterally in their travel lane and often in the adjacent lane, that is, not only risky for them but also for other road users. Figure 6.12 shows a typical transition curve.

6.10.3.1 Requirement of Transition Curve

The major requirements of a transition curve are as follows:

- It should be straight to tangential
- It should meet the circular curve tangentially
- Curvature at the origin should be zero.

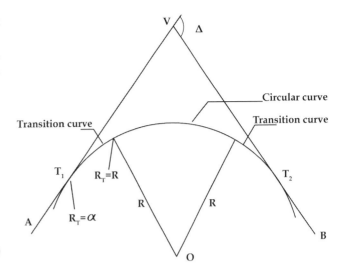

FIGURE 6.12 Transition curves.

- Curvature on the junction of the circular curve should be the same.
- Rate of increase of curvature = rate of increase of super elevation.

6.11 EARTHWORKS

One of the main aims of land survey is the quantifying of the area of the tract surveyed and the volume of earthwork. The land area in the aircraft survey is the area estimated on a horizontal plane. The main goal of the survey is to calculate the areas and the volumes. The lands are usually of irregular shaped polygons.

6.11.1 Computation of Areas

Graphical methods are those used to acquire the necessary data from the plan calculations. In this particular case, the actual figure area is found as a whole or the skeleton areas and the irregular strips are individually found. The skeleton areas are first subdivided into several regular polygons by means of survey lines. Then the area can be calculated as: (1) measurement of the survey skeleton area plus; (2) assessment of the area between skeleton survey lines and boundaries. The instrumental method is used to ascertain the area of a map by means of a planimeter. It is the quickest method and generates precise results compared to other methods. There are two kinds of planimeter—polar planimeter Amsler and planimeter rolling.

6.11.1.1 Areas of Skeleton

Areas of skeleton are determined from geometry as follows:

1. **By dividing into triangles**: Splitting the figure into a fraction of triangles was the most reliable method. The sides of each triangle are calculated or the base and elevation are scaled and its area is located. Area of a triangle forming a skeleton is computed as

$$\text{Area} = \sqrt{s(s-a)(s-b)(s-c)} \qquad (6.20)$$

where a, b, c are the side of triangle and semiperimeter, $s = (a + b + c)/2$

$$\text{Area} = \frac{1}{2} \times \text{base} \times \text{height} \quad (6.21)$$

2. **By division into square**: A piece of tracing paper ruled out into squares in this technique is positioned over the drawing, each assigning it to a certain number of square meters or square centimeters. The number of the entire square is calculated and the area is ascertained. The sections of the fractured square are calculated for the entire square and the broken square.

3. **By splitting into trapezoids or by sketching parallel lines and transferring them to rectangles**: The length of the rectangles is attained by placing the tracing paper over the plane and area is computed according to:

$$\text{Required area} = \{\text{length of rectangles}\}$$

$$\times \{\text{constant distance common breadth}\} \quad (6.22)$$

$$\text{Area of Trapezium} = \frac{1}{2}(\text{sum of parallel sides})$$

$$\times \text{distance between them} \quad (6.23)$$

$$= \frac{1}{2}\,(a + b) \times h$$

6.11.1.2 Considering the Area along Boundaries

The area is computed by any of the techniques, e.g., the mid-ordinate method, the average ordinate method, the trapezoidal rule, or the Simpson's rule.

6.11.1.2.1 Midpoint-Ordinate Rule

The rules clearly state that if the sum of all ordinates in the middle of each division is multiplied by the length of the base line with the ordinates.

$$\text{Area} = \Delta = \text{Average ordinate} \times \text{Length of base}$$

$$\text{Area} = \frac{O_1 + O_2 + \ldots + O_n}{n} L = (O_1 + O_2 + \ldots + O_n)d = d\sum O \quad (6.24)$$

where O_1, O_2,$\ldots$ = the ordinates at the mid-points of each division; $\sum O$ = sum of the mid-ordinates; n = number of divisions; L = base line length = nd; d = each divisions distance.

6.11.1.2.2 Average Ordinate Rule

The rule says that the average number of ordinates taken in each of the equal length divisions multiplies by baseline length divided by number of ordinates.

$$\text{Area} = \Delta = \text{Average ordinate} \times \text{Length of base}$$

$$\text{Area} = \frac{O_0 + O_1 + \ldots + O_n}{n + 1} L = \frac{L}{(n + 1)}\sum O \quad (6.25)$$

where O_0 = ordinate at one base end; O_n = ordinate at the other end of the base fragmented into n equal divisions; O_1, O_2,$\ldots$ = ordinates at each divisions end.

6.11.1.2.3 The Trapezoidal Rule

The boundaries between the ends of the ordinates are presumed to be straight in trapezoidal rule. The areas between the base and the irregular border lines are therefore deemed to be trapezoids

$$\Delta = \left(\frac{O_0 + O_n}{2} + O_1 + O_2 + \ldots + O_{n-1}\right)d \quad (6.26)$$

Add the end offset average to the middle offset sum. Multiply the total amount thus achieved by the common distance between the ordinates in order to attain the needed area.

6.11.1.2.4 Simpson's Rule

The limits between the ends of the ordinates are presumed to form an arc of parabola in Simpson's rule. Therefore, the rule of Simpson is almost always referred to as a parabolic rule. It states clearly that there must be a sum of first and last ordinates. Add the remaining unusual ordinates twice and the sum of the remaining ordinates four times. Multiply the common distance between the ordinates that gives the necessary area by 1/3 of the total amount.

$$\Delta = \frac{d}{3}\Big[(O_0 + O_n) + 4(O_1 + O_3 + \ldots + O_{n-1})$$

$$+ 2(O_2 + O_4 + \ldots + O_{n-2})\Big] \quad (6.27)$$

6.11.2 Computation of Volume

Volumes are computed from cross sections or contours. The first two methods are commonly used for the calculation of earth work while the third method is generally adopted for the calculation of reservoir capacities.

6.11.2.1 Method of Cross Sections

Cross sections are well ideally suited to the calculation of pipeline quantities, roads, canals, dams, etc. For most prevalent cross-sectional cases, typically used equations are given below.

6.11.2.1.1 Horizontal Ground (Figure 6.13)

Structures made by man usually have constant slopes

$$w = \frac{b}{2} + m \cdot h \quad (6.28)$$

$$\text{Area} = h \cdot \frac{[2w + b]}{2} = h(b + m \cdot h) \quad (6.29)$$

6.11.2.1.2 Sloping Ground with Constant Cross-Fall (Figure 6.14)

The ground is normally not flat. If the cross-fall is constant,

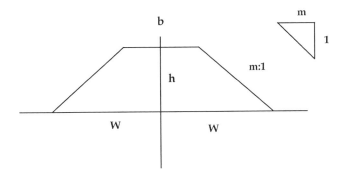

FIGURE 6.13 Horizontal ground.

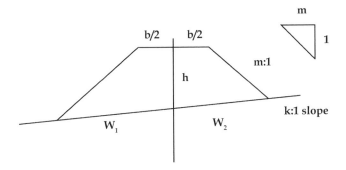

FIGURE 6.14 Two-level section.

$$w_1 = \left(\frac{b}{2} + m \cdot h\right)\left[\frac{k}{k-m}\right] \qquad (6.30)$$

$$w_2 = \left(\frac{b}{2} + m \cdot h\right)\left[\frac{k}{k+m}\right] \qquad (6.31)$$

$$\text{Area} = \frac{\left[\left(\frac{b}{2} + m \cdot h\right)(w_1 + w_2) - \frac{b^2}{2}\right]}{2m} \qquad (6.32)$$

This is a two-level section, as two points define the cross-fall.

6.11.2.1.3 Side Hill Two-Level Section (Figure 6.15)

In this specific case, the ground slopes transversely and the slope of the ground reduces the level of formation so that one part of the area is in the cutting and the other part is in the embankment.

$$w_1 = \frac{b}{2} + \frac{rs}{r-s}\left(h + \frac{b}{2r}\right) \qquad (6.33)$$

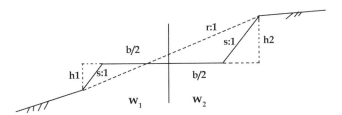

FIGURE 6.15 Side hill two-level section.

$$w_2 = \frac{b}{2} + \frac{rs}{r-s}\left(\frac{b}{2r} - h\right) \qquad (6.34)$$

$$h_1 = \left(h + \frac{w_1}{r}\right) \qquad (6.35)$$

$$h_2 = \left(h - \frac{w_2}{r}\right) \qquad (6.36)$$

$$\text{Area in cutting, } A_1 = \frac{1}{2}\left[\frac{\left(\frac{b}{2} + rh\right)^2}{r-s}\right] \qquad (6.37)$$

$$\text{Area in filling, } A_2 = \frac{1}{2}\left[\frac{\left(\frac{b}{2} - rh\right)^2}{r-s}\right] \qquad (6.38)$$

6.11.2.1.4 Three-Level Section (Figure 6.16)

In this particular case, the cross-slope of the ground is not formal and r_1:1 on one side and r_2:1 on the other side of the cross-sectional center line.

$$w_1 = \frac{r_1 s}{r_1 - s}\left(h + \frac{b}{2s}\right) \qquad (6.39)$$

$$w_2 = \frac{r_2 s}{r_2 + s}\left(h + \frac{b}{2s}\right) \qquad (6.40)$$

$$\text{Area, } A = \frac{b}{4}(h_1 + h_2) + \frac{h}{2}(w_1 + w_2) \qquad (6.41)$$

6.11.2.1.5 Multilevel Section (Figure 6.17)

In this case, the coordinate system is adopted to work out the cross-sectional area and the points of change of level are given coordinates.

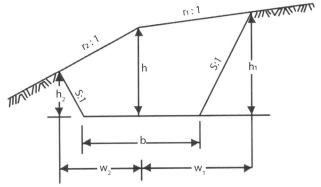

FIGURE 6.16 Three-level section.

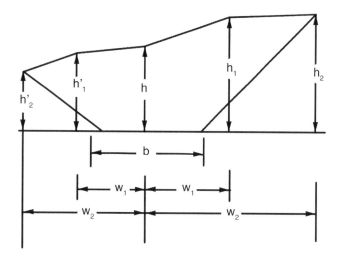

FIGURE 6.17 Multi-level section.

$$A = \frac{1}{2}\left[\sum F - \sum D\right] \quad \text{if } SF > SD \qquad (6.42)$$

$$A = \frac{1}{2}\left[\sum D - \sum F\right] \quad \text{if } SD > SF \qquad (6.43)$$

where SF = sum of the products of the coordinates connected by full lines

SD = sum of the product s of the coordinates joined by dotted lines

$$\sum F = w_1 h_2 + w_1^1 h_2^1 \qquad (6.44)$$

$$\sum D = h(w_1 + w_2) + h_1 w_2 + h_1^1 w_2^1 + \frac{b}{2}\left(h_2 + h_2^1\right) \quad (6.45)$$

After obtaining the cross-sectional areas as above, the volume of earthwork is worked out by applying any one of the following two formulae:

- Trapezoidal formula
- Prismoidal formula

6.11.2.2 Trapezoidal Formula

This is known as average and area rule, and it is presumed that a solid is a trapezoid between two consecutive cross sections.

$$\text{Volume } V = d\left(\frac{A_1 + A_n}{2} + A_2 + A_3 + \ldots + A_{n-1}\right) \quad (6.46)$$

where A_1, A_2, A_3,…, A_n are the equispaced cross-sectional areas, d is the distance between two consecutive areas, and V is the volume of earth work.

6.11.2.3 Prismoidal Formula

A prismoid is regarded as a solid whose end faces are located in parallel planes and comprise two polygons, not inherently of the same kind of number of sides; the longitudinal faces are plane surfaces between the end planes. The longitudinal faces can be triangles, trapezia, and parallelograms.

If d is the prismoid length quantified perpendicular to the two parallel planes at the end, A_1 is the cross-sectional area of one end plane, A_2 is the cross-sectional area of the other end plane, A_m is the cross-sectional area of the middle part of the plane between and parallel to the end planes, V is the prismoid volume,

$$V = \frac{d}{6}\left(A_1 + A_2 + 4A_m\right) \qquad (6.47)$$

6.11.3 Methods of Contours for Volume Computation

In this method, the contour plan is studied in detail and profiles of cross sections are drawn with respect to the contour gradient, and the earthwork volumes between adjacent cross sections are calculated using a trapezoidal rule.

6.11.4 Prismoidal Correction for Volume

The term prismoidal correction is used to indicate the difference in volume obtained by the application of trapezoidal formula and prismoidal for the same volume. For a level section, if A_1 and A_2 are the cross-sectional areas of the two sections at distances d apart with formation width b, side slope s:1 and with h and H heights at the center,

$$\text{Prismoidal correction, } C_p = \frac{ds}{6}\left(h - H\right)^2 \quad (6.48)$$

6.11.5 Curvature Correction for Volumes

The correction for curvature can be applied in the following two ways.

6.11.5.1 Equivalent Areas

The corrected area of each section is obtained by the following expression:

$$\text{Equivalent area of section} = A\left(1 \pm \frac{e}{R}\right) \quad (6.49)$$

where A is the area of section, e is the centroid of area, and R is the radius of curvature. Use + sign if the centroid is the center of curvature on the opposite side of the center line and − sign vice versa.

6.11.5.2 Pappus Theorem

$$\text{Curved volume } V_c = \frac{d}{4}\left(A_1 + A_2\right)\left(1 \pm \frac{e}{R}\right) \quad (6.50)$$

where A_1 and A_2 are the cross-sectional areas of end sections, d is the distance between the sections, R is the radius of curvature, and e is the mean distance of the centroid.

6.12 GEODETIC SURVEYING

The curvature of the earth is kept in mind in geodetic surveys and a greater degree of accuracy is attained in linear and angular observations. Geodetic surveys can be carried out over large areas and lines linking two points on the earth's surface are regarded as arches. Triangulation is employed in geodetic surveys.

6.12.1 TRIANGULATION

The system consists of a number of interlinked triangles in triangulation, whereby the length of only one line titled as the base line and the angles of the triangles are computed quite exactly. Knowing the length of one side and the three angles, it is possible to calculate the length of the other two sides of each triangle. The apexes of the triangles are called the stations of triangulation and the entire figure is called the system of triangulation or the figure of triangulation.

6.12.1.1 Objectives of Triangulation

The main goal of triangulation is to offer a number of stations whose relative and absolute positions are exactly established, both horizontally and vertically. From these stations, a much more comprehensive survey is carried out. Further goals are provided below:

1. To assess exact points locations in engineering works.
2. To maintain precise photogrammetric survey control.
3. Maintaining accurate plane and geodetic survey of vast areas.

6.12.1.2 Classification of Triangulation System

Triangulation surveys are classified as first order (primary), second order (secondary), and third order (tertiary) triangulation based entirely on the scope and intent of the survey and the degree of accuracy required.

The triangulation of first order is of the greatest order and is used either to know the figure of earth or to provide the accurate control points interlinked to the secondary triangulation. The primary system of triangulation covers a large area (normally the entire country). The general requirements for primary triangulation are set out in Table 6.1.

The triangulation of the second order comprises a number of points in the primary triangulation. At close intervals, the stations are fixed so that the sizes of the triangles developed are tinier than the primary triangulation. The tools and techniques used are not of the same greatest refinement. The particular regulations for secondary triangulation are set out in Table 6.2.

The triangulation of the third order comprises a number of points fixed in the secondary triangulation and forms the instant control for detailed engineering and other inquiries. The triangle sizes are significantly smaller than the secondary triangulation, and the instrument can be used with moderate accuracy. The third-order triangulation parameters are given in Table 6.3.

6.12.2 CURVATURE AND REFRACTION

The error due to curvature comes into play in leveling, mostly because the horizontal line and level line do not coincide in the case of long distances. The level is a curved line parallel to the surface level, but the horizontal line goes straight. This implies that the vertical distance of the target from the level line will be greater than the distance from the horizontal line that we compute. This is shown in Figure 6.18. Refraction is the phenomenon by which when light travels from a denser

TABLE 6.1
General Specifications of Primary Triangulation

1	Average triangle closure	Less than 1 second
2	Maximum triangle closure	Not more than 3 seconds
3	Length of base line	5–15 km
4	Length of the sides of triangles	30–150 km
5	Actual error of base	1 in 300,000
6	Probable error of base	1 in 1,000,000
7	Discrepancy between two measures of a section	10 mm km
8	Probable error or computed distance	1 in 60,000 to 1 in 250,000
9	Probable error in astronomic azimuth	0.5 seconds

TABLE 6.2
General Specifications of Secondary Triangulation

1	Average triangle closure	3 seconds
2	Maximum triangle closure	8 seconds
3	Length of base line	1.5–5 km
4	Length of the sides of triangles	8–65 km
5	Actual error of base	1 in 150,000
6	Probable error of base	1 in 500,000
7	Discrepancy between two measures of a section	20 mm km
8	Probable error or computed distance	1 in 20,000 to 1 in 50,000
9	Probable error in astronomic azimuth	2.0 seconds

TABLE 6.3

Specifications for Third-Order Triangulation

1	Average triangle closure	6 seconds
2	Maximum triangle closure	12 seconds
3	Length of base line	0.5–3 km
4	Length of the sides of triangles	1.5–10 km
5	Actual error of base	1 in 75,0000
6	Probable error of base	1 in 250,000
7	Discrepancy between two measures of a section	25 mm km
8	Probable error or computed distance	1 in 5,000 to 1 in 20,000
9	Probable error in astronomic azimuth	5 seconds

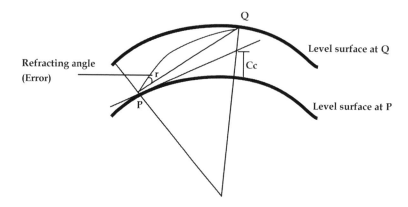

FIGURE 6.18 Curvature and refraction.

medium to a lighter medium, it distracts itself from the normal media to the plane of the media. This hypothesis must be taken into account when measuring distances in the case of geodesic surveys. The corrections for refraction and curvature are determined using the following formulae:

$$\text{Curvature correction, } C_c = -0.07849D^2 \text{ m} \quad (6.51)$$

$$\text{Refraction correction, } C_r = 0.01121D^2 \text{ m} \quad (6.52)$$

$$\text{Combined correction } C = C_c + C_r = -0.06728D^2 \text{ m} \quad (6.53)$$

where D is in km.

6.12.3 Intervisibility and Height of Stations

The stations should be fixed on the highest possible ground for the indivisibility of two stations.

6.12.3.1 Distance between Stations

If the intervening ground is completely free of any obstacles, the distance of an identifiable horizon can be calculated by the formula by a station of documented altitude just above the datum and as well as the altitude of the signal that can only be seen at a fixed distance.

$$h = \frac{D^2}{2R}(1 - 2m) \quad (6.54)$$

where h is the station altitude above a datum, D is the distance to the visible horizon, R is the mean radius of the earth, and M is the co-efficient of refraction (0.07 for sight over land and 0.08 for sights over water). If the D and R values are replaced in the correct units, the h value of $m = 0.07$ is provided by

$$h = 0.574D^2 \quad (6.55)$$

where h is in feet and D is in miles and

$$h = 0.06728D^2 \quad (6.56)$$

where h in m and D in km.

6.12.3.2 Relative Elevations of Stations

If D is the known distance in kilometers between two stations A and B, h_1 is the known station elevation "A" above the datum, h_2 is the required elevation "B" over the datum, D_1 is the distance in km from "A" to the tangency point, and D_2 is the distance in km from "B" to the tangency point, then,

$$h_1 = 0.06728D_1^2 \quad (6.57)$$

$$D_1 = \sqrt{\frac{h_1}{0.06728}} = 3.8553\sqrt{h_1} \quad (6.58)$$

where D_1 is in km and h_1 is in m.

Knowing D_1, D_2 is given by

$$D_2 = D - D_1 \qquad (6.59)$$

Knowing D_2, h_2 is calculated from the relation

$$h_2 = 0.06728 D_2^2 \qquad (6.60)$$

6.13 PHOTOGRAMMETRY

Photogrammetry is a method used for calculating the three-dimensional photographic coordinates. It could be used for calculation purposes. Triangulation is the basic principle that is used by photogrammetry. Pictures from at least two various locations are taken in this method. The aim of taking photos from more than two points is to develop line of sight.

6.13.1 TERRESTRIAL PHOTOGRAMMETRY

In terrestrial photogrammetry maps, photographs or terrestrial photogrammetry are taken from terrestrial (or ground) images for calculation purposes. The principle of terrestrial photogrammetry was improved upon and perfected by Capt. Deville, then Surveyor General of Canada in 1888. In terrestrial photogrammetry, photographs are taken and maps are compiled from the photographs.

6.13.2 AERIAL PHOTOGRAMMETRY

Aerial photogrammetry maps are made from air images (photos taken from air). Aerial photography platforms include fixed-wing aircraft, helicopters, rockets, parachutes, surveillance drones, pigeons, balloons, kites, stand-alone telescopes, and poles. Mounted cameras can be sparked automatically or remotely; pictures taken by a photographer by hand held.

6.13.3 ORTHOPHOTOS

Vertical pictures are sometimes used to generate orthophotos. It is regarded as orthophoto maps, that are geometrically "rectified" images so that they can be used as a map. An orthophoto is a simulation of a photo taken from an infinite distance, looking straight to nadir. The perspective must obviously be deleted and the differences in the terrain rectified. Orthophotos are frequently used in geographical information systems (GISs), such as mapping agencies to develop maps. Then when the pictures are aligned or "registered" that they just can be widely adopted with documented real world coordinates. Huge sets of orthophotos, generally extracted from various sources and fragmented into "tiles," are prevalently used in online map systems such as Google Maps. OpenStreetMap allows to derive new map data using analogous orthophotos.

6.14 MODERN SURVEYING EQUIPMENT

With current modern survey tools, survey work is accurate, faster, and less stressful. Electronic Distance Measurements (EDMs), digital theodolites, digital levels, Global Positioning System (GPS), data collectors, and total station equipment are included in modern equipment. The length and angle of the total stations are gauged between two points, which is an apparent vital survey function.

6.14.1 EDM INSTRUMENTS

Direct quantification of distances and directions can be attained by using electronic instruments that depend on the spread, reflection, and reception of light waves or radio waves. They can be categorized as infrared wave instruments, light wave instruments, and microwave instruments in three types.

6.14.1.1 Infrared Wave Instruments

These instruments gauge distances aiding modulated infrared waves with amplitude. Prisms attached to the target are used to reflect the waves at the end of the line. The range of such a tool shall be 3 km and the precision shall be ±10 mm. Distomat is a very tiny, compact EDM, especially used for construction and other civil engineering works with a distance of less than 500 m. To calculate the distance, the instrument should be pointed to the reflector, touch a key, and note the readings.

6.14.1.2 Light Wave Instruments

These tools calculate distances based on the moduled propagation of light waves. The accuracy about such a tool varies from 0.5 to 5 mm/km and ranges from almost 3 km. The light wave instrument, for instance, is a geodimeter based solely on the spread of modulated light waves. It is preferable for observation at night and necessitates a prism system at the end of the line to reflect the waves. It was programmed by E. Bergestand from the Swedish Geographical Survey, in collaboration with the Swedish M/s AGA, the major manufacturer.

6.14.1.3 Microwave Instruments

These instruments just use higher frequency radio waves, which Dr. T.L. Wadley devised in South Africa as early as 1950. Tellurometer was the first successful microwave electronic distance measurement equipment. The name derives from the Latin word "tellus," meaning Earth. The range of these tools is up to 100 km which can be used throughout the day and during night. For a microwave instrument, an EDM requires high-frequency radio waves (microwaves) to calculate the distance is a tellurometer. It is a very portable instrument and can be used with a battery of 12–24 V. Two tellurometers are needed to calculate the distance, one stationed at each end of the line with two highly qualified persons to make observations. One device is used as a master device and another as a remote device. Each technician is offered with a voice facility (interaction facility) to communicate during the calculation.

6.14.2 Total Station

It is a lightweight, compact, and completely integrated electronic device with the combined ability of an EDM and an angular calculating instrument like wild theodolite. It is attached to a tripod and sends an infrared signal to a prism. The signal is returned to the total station. The time needed to reach the prism and back of the signal gets to decide the distance from one point to the other. This offers all the relevant information to measure this distance.

The following processes can be conducted using a total station.

- Measuring distance
- Angular measurement
- Processing of data
- Digital point details display
- Data storage is an electronic field book

The key characteristics of the station are as follows:

1. All tasks of the keyboard control are governed by the operating keyboard.
2. The digital panel showcases the distance, angle, height, and coordinates values of the observed point in which the reflector (target) is established.
3. Remote height objects can be obviously read directly by the heights of certain unreachable objects such as towers. The microprocessor supplied in the instrument randomly assigns the adjustment of the curvature of the earth and the mean refraction.
4. The traversing program, the reflector coordinates, and the angle or bearing of the reflector can be safely stored and recalled for the next instrument setting.
5. Setting for distance, direction, and height, when a general direction and horizontal distance are actually entered in place in order to retrieve the point on the ground by using a target, the instrument exhibits the angle through which the theodolite is to be turned and the distance through which the reflector is to relocate.

6.14.3 Automatic Level

An automatic level is a unique leveling device containing an optical compensator that establishes a line of sight or collimation even though the instrument is marginally tilted, e.g., NAK2 Wild Automatic Level.

6.15 MODERN SURVEYING METHODS

6.15.1 Remote Sensing

Remote sensing is the science of gathering information about an object or a phenomenon without physical contact with the object and therefore in contrast to observation on site. In modern use, the word usually relates to the use of aerial sensor systems for the tracking and categorization of objects on Earth (both on the surface and in the atmosphere and oceans) using propagated signals (e.g., radiation of electromagnetic). It can be divided into active remote sensing (when an aircraft or satellite first emits a signal) and passive remote sensing (e.g., Sunlight) when data are simply reported. Remote sensors gather data by sensing the earth's energy. Such sensors can be installed on aircraft or satellites. Remote sensors can be active or passive. Active sensors just use internal stimuli to gather earth data. A laser-beam remote sensing system, for instance, projects a laser on the earth's surface and quantifies the time it takes for the laser to reflect back to its sensor. Passive sensors react to external stimuli, in contrast to active sensors. Thereafter record natural energy reflected or radiated from the surface of the earth. By far the most prevalent radiation source observed by passive sensors is the sunlight reflected.

6.15.2 Geographical Information System

A GIS is a scheme capable of capturing, store, manipulate, evaluate, handle, and exhibit all types of spatial or geographical information. The word implies any data system that combines geographical information edits, analyzes, stores, shares, and displays. GIS applications are tools that allow users to create customizable queries (user-created searches), formatting map information, examine spatial information, and report the results of all these activities. The science of geographical information is the science behind geographical concepts, applications, and systems. GIS has several applications referring to planning, engineering, transport/logistics, telecommunications, insurance, business, and management. GIS and location intelligence programs can thus serve as the basis for many location-enabled services based on analysis and visualization. By using location as a key index variable, GIS can relate any unrelated data. Locations or extents in the space-time of the Earth can be reported as occurrence dates and x, y, and z coordinates, portraying longitude, latitude, and elevation. All spatial-temporal locations and references to extent based solely on Earth must ideally be linked to each other and ultimately to a "real" physical location or extent. This major feature of the GIS has begun to open up new possibilities for scientific research.

6.15.3 Global Positioning System

This program was created by the United States Defense Department (DoD) and is termed the Time and Global Positioning System (NAVSTAR GPS) navigation system or obviously the GPS system. The United States Air Force has stationed 24 satellites at an elevation of 20,200 km above the surface of the earth. The satellites have all been placed so that at least four satellites can be seen from any point on Earth. To retrieve the position of any point on the ground, the user needs a GPS receiver. The receiver functions the earned signals from the satellite and calculates the position (latitude and longitude) and point elevation in relation to the data. GPS has been quickly adapted for surveying, as it can explicitly give the position (Latitude, Longitude, and Height) all without

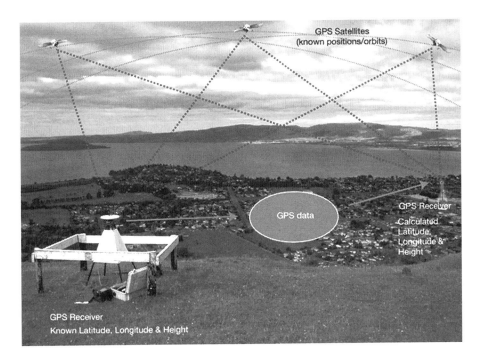

FIGURE 6.19 Positioning of GPS point.

calculating angles and distances between intermediate points. In the readily available receivers, the GPS receiver defines its position (Latitude, Longitude, and Height) almost instantly with an unease of a few meters from the satellite data. This data may include a summary of the shifting positions of the satellites (its orbit) and the transmission time of the data. Figure 6.19 shows a typical GPS point positioning. Figure 6.20 shows a typical geodetic GPS receiver.

6.15.3.1 GPS Baseline

Frequently, the GPS receivers used in surveys are much more complicated and expensive than that used in daily life. The two frequencies transmitted by GPS satellites are used. These receivers have a separate, high-quality antenna. A GPS baseline uses two GPS receivers of survey quality, one at each end of the line to be quantified. At the same time, they gather data from the same GPS satellites. The length of these concurrent observations differs with the length of the line and the precision required, but usually is 1 hour or more. When data from both points are coupled later, a custom software computes the distinction in position (latitude, length, and height) between the two points. Figure 6.21 shows GPS Baseline Measurement.

6.15.3.2 Kinematic GPS

There are several differences in this type of GPS survey, but it is essentially analogous to the basic GPS method, except that while one GPS receiver continues to remain in a specific position (base station), the other moves between points and only one has to be at each point for several seconds. Adjustments to the GPS data (solely based on the given position of the base station and its position calculated from the GPS) can be transmitted instantly from the receiver at the base station to the receiver at the other end of the line (the remote). The remote

station's position can be calculated and saved in a few seconds. For transmission of corrections, radios or mobile phones can be used. However, this method can give the basic method a comparable accuracy, and this method is usually limited to a distance of about 20 km. Figure 6.22 shows GPS Real-Time Baseline Measurements.

FIGURE 6.20 Geodetic GPS receivers.

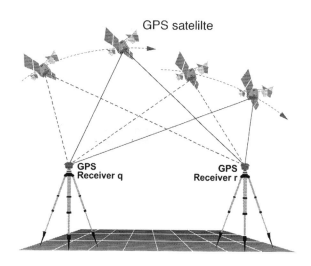

FIGURE 6.21   GPS baseline measurement.

6.15.3.3   Continuously Operating Reference Stations

A GPS receiver of survey quality could be permanently installed in a desirable location with a known location, which can be used as the starting point for any GPS calculation in the area. One or more GPS survey quality receivers can be used to concurrently retrieve GPS data at any necessary point. These data can be used to compute the positions in unique combination with the GPS data collected from the Continuously Operating Reference Station (CORS). If more than one CORS is easily available, the unknown position with respect to these multiple known positions can be computed, offering more faith in the outcomes. The duration of the inferences totally depends on the distance from the CORS, but it is usually 1 or 2 hours.

6.15.3.4   Heights from GPS

GPS automatically gives height, latitude, and longitude as it is a three-dimensional system. However, the height will be above the conceptual surface of the earth used for measurements, regarded as the ellipsoid (so the height is known as the ellipsoidal height). It will not be above the mean sea level. Due to the differing density of the earth, the significant difference between ellipsoidal height and MSL height can be large (up to 100 m) and uneven.

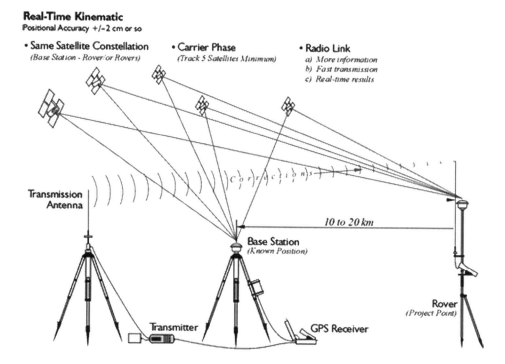

FIGURE 6.22   GPS real-time baseline measurements.

7 Building Materials

7.1 PRINCIPAL PROPERTIES OF BUILDING MATERIALS

Building materials have an important role to play in this modern age of technology. The rapid advancement of constructional methods and changes in the organization of the building industry may influence the choice of materials.

7.1.1 PHYSICAL CHARACTERS

- Density is the mass of a unit volume of homogeneous material.
- Specific weight also known as the unit weight is the weight per unit volume of material.
- Porosity is indicative of other major properties of material, such as bulk density, heat conductivity, durability, etc.
- Void ratio is defined as the ratio of volume of voids to the volume of solids.
- Hygroscopicity is the property of a material to absorb water vapor from air.
- Water absorption denotes the ability of the material to absorb and retain water.
- Weathering resistance is the ability of a material to endure alternate wet and dry conditions for a long period without considerable deformation and loss of mechanical strength.
- Water permeability is the capacity of a material to allow water to penetrate under pressure. Materials like glass, steel, and bitumen are impervious.
- Frost resistance denotes the ability of a water-saturated material to endure repeated freezing and thawing with considerable decrease of mechanical strength.
- Heat conductivity is the ability of a material to conduct heat.
- Thermal capacity is the property of a material to absorb heat described by its specific heat.
- Chemical resistance is the ability of a material to withstand the action of acids, alkalis, sea water, and gases.
- Durability is the ability of a material to resist the combined effects of atmospheric and other factors.

7.1.2 MECHANICAL PROPERTIES

- Strength is found from tests on standard cylinders, prisms, and cubes—smaller for homogeneous materials and larger for less homogeneous ones.
- Hardness is the ability of a material to resist penetration by a harder body. Mohs scale is used to find the hardness of materials.
- Elasticity is the ability of a material to restore its initial form and dimensions after the load is removed.
- Plasticity is the ability of a material to change its shape under load without cracking and to retain this shape after the load is removed.

7.2 STRUCTURAL CLAY PRODUCTS

Typical structural clay products are building brick, paving brick, terra-cotta facing tile, roofing tile, and drainage pipe. Structural clay products are made from 35% to 55% clays or argillaceous (clayey) shale, 25% to 45% quartz, and 25% to 55% feldspar. Colors can range from buff and other light shades of brown through red to black.

7.2.1 PROPERTIES

The properties exhibited by structural clay products are determined by particle size, firing temperature, and ultimate microstructure. Lower firing temperatures are employed— typically in the range of 1050°C–1100°C (approximately 1925°F–2000°F). Because of the presence of large and small particles in their microstructures, fired clay products have relatively high compressive strengths.

7.3 ROCKS AND STONES

Stone is a "naturally available building material" which has been used from the early age of civilization. Stones used for civil engineering works can be classified geologically, physically, and chemically.

Geological classification: Igneous, sedimentary, and metamorphic rocks.

i. **Igneous rocks**: These rocks are formed by cooling and solidifying of the rock masses from their molten magmatic condition of the material of the earth. Trap and basalt belong to this category.

ii. **Sedimentary rocks**: Due to weathering action of water, wind and frost, existing rocks disintegrates. The disintegrated material is carried by wind and water; the water being the most powerful medium. Sand stones, lime stones, mud stones, etc., belong to this class of rock.

iii. **Metamorphic rocks**: Previously formed igneous and sedimentary rocks undergo changes due to metamorphic action of pressure and internal heat. For example, due to metamorphic action, granite becomes gneiss, trap and basalt change to schist and laterite, lime stone changes to marble, sand stone becomes quartzite, and mud stone becomes slate.

Physical classification: Based on the structure, rocks are classified as Stratified rocks, Unstratified rocks, and Foliated Rocks

i. **Stratified rocks**: These rocks have a layered structure. Sand stones, lime stones, slate, etc., are examples.
ii. **Unstratified rocks**: These rocks are not layered. They possess crystalline and compact grains. Granite, trap, marble, etc., are examples.
iii. **Foliated rock**: These rocks have a tendency to split along a definite direction.

Chemical classification: Chemically, the rocks may be classified into Siliceous rocks, Argillaceous rocks, and Calcareous rocks.

i. **Siliceous rocks**: The main content of these rocks is silica.
ii. **Argillaceous rocks**: The main constituent of these rocks is argil, i.e., clay. These stones are hard and durable but they are brittle.
iii. **Calcareous rocks**: The main constituent of these rocks is calcium carbonate. Limestone is a calcareous rock.

7.3.1 Requirements of Good Building Stones

i. **Strength**: The stone should be able to resist the load coming on it.
ii. **Durability**: Stones selected should be capable of resisting adverse effects of natural forces.
iii. **Hardness**: The stone used in floors and pavements should be able to resist abrasive forces.
iv. **Toughness**: Building stones should be tough enough to sustain stresses developed due to vibrations.
v. **Specific gravity**: The specific gravity of good building stone is between 2.4 and 2.8.
vi. **Porosity and absorption**: Building stone should not be porous.
vii. **Dressing**: Giving required shape to the stone is called dressing.
viii. **Appearance**: In case of the stones to be used for face works, where appearance is a primary requirement, its color and ability to receive polish are major factors.
ix. **Seasoning**: Good stones should be free from the quarry sap. Laterite stones should not be used for 6–12 months after quarrying.

7.3.2 Tests on Stones

The common tests conducted on stones are Crushing strength test, Water absorption test, Abrasion test, Impact test, and Acid test. The details of these tests are available in standard codes.

7.3.3 Common Building Stones

i. **Basalt and trap**: The structure is medium to fine grained and compact. Their weight varies from 18 to 29 kN/m^3. The compressive strength varies from 200 to 350 N/mm^2. They are used as pavement.
ii. **Granite**: Granites are also igneous rocks. The color varies from light gray to pink. The structure is crystalline, fine to coarse-grained. Specific gravity is from 2.6 to 2.7 and compressive strength is 100–250 N/mm^2. They are used as coarse aggregates in concrete.
iii. **Sand stone**: These are sedimentary rocks, and hence stratified. They consist of quartz and feldspar. They are found in various colors like white, grey, red, buff, brown, yellow, and even dark gray. The specific gravity varies from 1.85 to 2.7 and compressive strength varies from 20 to 170 N/mm^2. Its porosity varies from 5% to 25%.
iv. **Slate**: These are metamorphic rocks. They are composed of quartz, mica, and clay minerals. The color varies from dark gray, greenish gray, purple gray to black. The specific gravity is 2.6–2.7. Compressive strength varies from 100 to 200 N/mm^2.
v. **Laterite**: It is a metamorphic rock. It is having porous and spongy structure. It contains high percentage of iron oxide. Its specific gravity is 1.85 and compressive strength varies from 1.9 to 2.3 N/mm^2.
vi. **Marble**: This is a metamorphic rock. Its specific gravity is 2.65 and compressive strength is 70–75 N/mm^2. It is used for facing and ornamental works.
vii. **Gneiss**: It is a metamorphic rock. It is having fine to coarse grains. Alternative dark and white bands are common. The specific gravity varies from 2.5 to 3.0 and crushing strength varies from 50 to 200 N/mm^2.
viii. **Quartzite**: Quartzites are metamorphic rocks. The structure is fine to coarse grained and often granular and branded. The specific gravity varies from 2.55 to 2.65. Crushing strength varies from 50 to 300 N/mm^2.

7.4 WOOD AND WOOD PRODUCTS

Wood has been used as a building material for thousands of years. Pieces of wood that is smaller than 127 mm (5″) wide by 127 mm thick (regardless of length) are generally referred to as lumber. These pieces are machine-planed and sawn to fit certain dimensional specifications (e.g., 50.8 mm × 101.6 mm (2″ × 4″), 50.8 mm × 203.2 mm (2″ × 8″, etc.)) and are primarily used in residential construction. Pieces of wood over 127 mm (5″) wide by 127 mm thick (regardless of length) are referred to as timber, and any timber pieces that exceed 203.2 mm (8″) wide by 203.2 mm (8″) thick are referred to as beams. Another type of wood commonly used in construction is known as engineered wood. As its name implies, engineered wood is the product of a more intricate fabrication process in which various wood strands, fibers, veneers, or other forms of wood are glued.

7.4.1 Problems of Using Wood as a Building Material

i. **Shrinkage and swelling of wood**: Wood is a hygroscopic material. This means that it will adsorb moisture.

ii. **Deterioration of wood**: Biotic agents include decay and mold fungi, bacteria, and insects. Abiotic agents include sun, wind, water, certain chemicals, and fire.

iii. **Biotic deterioration of wood**: Biological deterioration of wood due to attack by decay fungi, wood boring insects, and marine borers during its processing and in service has technical and economical importance.

iv. **Fungi**: It is necessary to give some short information about fungi agents to take measures against the wood deterioration.

Physiological requirements of wood destroying and wood inhabiting fungi:

a. A favorable temperature: The temperature must be 25°C–30°C for optimum growth.

b. An adequate supply of oxygen: Oxygen is essential for the growth of fungi.

c. Moisture: Generally wood will not be attacked by the common fungi at moisture contents below the fiber saturation point (FSP).

d. Nutrients: Wood has carbon content. That means, wood is a very suitable nutrient for fungi. Because fungi derive their energy from oxidation of organic compounds. Decay fungi wood rotters can use polysaccharides while stain fungi evidently require simple forms such as soluble carbohydrates, proteins, and other substances present in the parenchyma cell of sapwood. Additionally, the presence of nitrogen in wood is necessary for the growth of fungi.

e. Insects: Termites, carpenter ants, and marine borers.

   i. Termites: Subterranean termites damage wood that is untreated, moist, in direct contact with standing water, soil, and other sources of moisture.

   Dry wood termites attack and inhabit wood that has been dried to moisture contents as low as 5%–10%.

   Carpenter ants: Carpenter ants do not feed on wood.

   They attack most often wood in ground contact or wood that is intermittently wetted.

   ii. Carpenter bees: They cause damage to unpainted wood.

   iii. Marine borers: They rapidly destroy wood in salt water and brackish water.

7.4.2 Minimizing the Problems of Wood

1. **Coating**: Coating provides protection to wood used both indoors and outdoors. But coating does not totally prevent changes in moisture content. Coating with solid color or pigmented stains protects wood against ultraviolet rays. Deteriorating paint film actually increases the decay hazard.

2. **Drying**: Generally wood will not be attacked by the common fungi at moisture content below the FSP. FSP for different wood lies between 20% and 35%, and 30% is accepted generally: Fungi cannot attack wood used indoor and in heated rooms, since the equilibrium moisture content is much more below than FSP. One of the most effective ways to prevent degradation of wood is to thoroughly dry it and keep it dry. The last case is very important since even wood that has been kiln dried will readily regain moisture if placed in a humid environment.

   Wood can be dried in air or in some type of dry kiln. Air drying alone is not sufficient for wood items which are used in heated rooms. Therefore kiln drying is necessary. Kiln drying has many advantages: One of them is the killing of staining or wood destroying fungi or insects that may be attack the wood and lower its grade. Wood that will be used indoor needs only to be dried to provide for long-term protection against rot.

3. **Treating with wood preservatives**: Some of the wood preservatives may harm humans and other creatures. For this reason, if wood is used outdoor in situations where it is often wet or in close proximity to liquid water, then wood must be treated with wood-preserving chemicals to achieve long-term durability.

   Wood preservatives are divided into two groups: Water-borne and oil-borne chemicals. About 75% of wood that is commercially treated today is treated with water-borne salts, and chromated copper arsenate is the compound used in treating for the greatest volume of wood.

   Oil-based or oil-borne preservatives are generally used for treating of wood used outdoors in industrial applications, such as ties, piling, and poles. In a serious situation, wood is treated with water-borne preservatives.

4. **Remedial treatment**: Retreatment of wood window frames, door frames, and wood timber and beams is sometimes carried out by drilling holes in areas where decay has begun and filling these holes with a suitable treating compound. Treating compound in the form of solid rods is mostly preferred since it provides a slow release of active ingredients. Retreatment of wood used in ground contact must be realized by application of pastes and wrapping with preservative-impregnate bandages.

7.4.3 Seasoning of Timber

This is a process by which moisture content in a freshly cut tree is reduced to a suitable level.

   i. **Natural seasoning**: It may be air seasoning or water seasoning. On about 300 mm high-platform timber balks are stacked as shown in Figure 7.1. A well-seasoned timber contains only 15% moisture. This is a slow but a good process of seasoning. Water seasoning is carried out on the banks of rivers. The thicker end of the timber is kept pointing upstream side. After a period of 2–4 weeks the timber is taken out. Then timber is stalked in a shed with free air circulation.

   a. Artificial seasoning: In this method timber is seasoned in a chamber with regulated heat, controlled humidity, and proper air circulation. Seasoning can be completed in 4–5 days only. Different ways of artificial seasoning are given here.

    i. Boiling: In this method timber is immersed in water and then water is boiled for 3–4 hours. Then it is dried slowly.

    ii. Kiln seasoning: Kiln is an airtight chamber. Timber to be seasoned is placed inside it. Then fully saturated air with a temperature 35°C–38°C is forced in the kiln. The heat gradually reaches inside timber. Then relative humidity is gradually reduced and temperature is increased, and maintained till desired degree of moisture content is achieved.

      The kiln used may be stationary or progressive. In progressive kiln the carriages carrying timber travel from one end of kiln to other end gradually. The hot air is supplied from the discharging end so that temperature increase is gradual from charging end to discharging end. This method is used for seasoning on a larger scale.

   iii. Chemical seasoning: In this method, the timber is immersed in a solution of suitable salt. Then the timber is dried in a kiln.

   iv. Electrical seasoning: In this method high-frequency alternate electric current is passed through timber. However, it is a costly process.

7.5 MATERIALS FOR MAKING CONCRETE

Concrete is a composite material composed of coarse aggregate bonded together with fluid cement which hardens over time. In Portland cement concrete when the aggregate is mixed together with the dry cement and water, they form a fluid mass that is easily molded into shape. Major characters:

1. Has high compressive strength.
2. It is a corrosion-resistance material
3. It hardens with age.
4. It is more economical
5. It binds rapidly with steel, and as it is weak in tension, the steel reinforcement is placed in cement concrete at suitable places to take up the tensile stresses.
6. It has a tendency to shrink when:
   a. There is initial shrinkage of cement concrete which is mainly due to the loss of water through forms, absorption by surfaces of forms, etc.
   b. The shrinkage of cement concrete occurs as it hardens. This tendency of cement concrete can be minimized by proper curing of concrete.
7. It has a tendency to be porous
8. It forms a hard surface, capable of resisting abrasion.

7.5.1 Materials Used in RCC Work

1. **Cement**: Most of the cement concrete work in building construction is done with ordinary Portland cement at present. But other special varieties of cement such as rapid hardening cement and high alumina cement are used under certain circumstance.

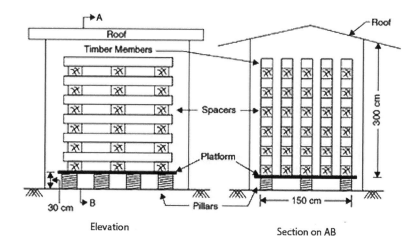

Elevation   Section on AB

FIGURE 7.1 Natural seasoning.

2. **Aggregates**: Aggregates are bound together by means of cement. The aggregates are classified into two categories: Fine and coarse. The material which passes through BIS test sieve no. 480 is termed as a fine aggregate. Usually, the natural river sand is used as a fine aggregate. But at places, where natural sand is not available economically, the finely crushed stone may be used as a fine aggregate. The material which is retained on BIS test sieve no. 480 is termed as a coarse aggregate. The broken stone is generally used as a coarse aggregate.

3. **Steel**: The diameters of bars vary from 5 to 40 mm. Sometimes square bars are used as steel reinforcement. For road slabs and such other constructions, the reinforcement may also consist of sheets of rolled steel of suitable thickness.

4. **Water**: The water should be clean and free from harmful impurities such as oils, alkali, acid, etc. In general, the water which is fit for drinking should be used for making concrete.

7.5.2 Types of Concrete

1. **High-strength concrete**: Concrete with a compressive strength of 40 MPa.

2. **High-performance concrete**: Describes concretes which may have better lifespan, lifespan in corrosive environments, or many other parameters.

3. **Lightweight concrete**: It is made by using small, lightweight aggregates, such as by adding foaming agents to the mix of concrete.

4. **Self-consolidating concrete**: It is also called self-compacting concrete.

5. **Sprayed concrete**: It is sprayed directly onto a surface, and is used in infrastructure projects and to repair old, cracked concrete surfaces. It is technically called as shotcreting or *guniting*.

6. **Water-resistant concrete**: Water-resistant concretes are engineered to have fine particle cement replacements that do not allow water to pass through.

7. **Micro reinforced concretes**: They contain small steel, fiberglass, or plastic fibers that change the characters.

7.6 MORTARS

Mortar is a workable paste used to bind building blocks such as stones, bricks, and concrete masonry units together, fill and seal the irregular gaps between them, and sometimes add decorative colors or patterns in masonry walls. The mortars are classified into the following five categories:

1. **Lime mortar**: The fat lime shrinks to a great extent and hence it requires about 2–3 times its volume of sand. The lime should be slaked before use. This mortar is unsuitable for water-logged areas or in damp situations. It possesses good cohesiveness with other surfaces and shrinks very little. It is sufficiently durable, but it hardens slowly. It is generally used for lightly loaded above-ground parts of buildings.

2. **Surkhi mortar**: The powder of surkhi should be fine enough to pass BIS No. 9 sieve and the residue should not be more than 10% by weight. The surkhi mortar is used for ordinary masonry work of all kinds in foundation and superstructure. But it cannot be used for plastering or pointing since surkhi is likely to disintegrate after some time.

3. **Cement mortar**: Depending upon the strength required and importance of work, the proportion of cement to sand by volume varies from 1:2 to 1:6 or more. The proportion of cement with respect to sand should be determined with due regard to the specified durability and working conditions. The cement mortar is used where a mortar of high strength and water-resisting properties is required such as underground constructions, water-saturated soils, etc.

4. **Gauged mortar**: This process is known as the gauging. It makes lime mortar economical, strong, and dense. The usual proportion of cement to lime by volume is about 1:6–1:8. It is also known as the composite mortar or lime-cement mortar, and it can also be formed by the combination of cement and clay. This mortar may be used for bedding and for thick brick walls.

5. **Gypsum mortar**: Prepared from gypsum-binding materials and anhydrite-binding materials.

7.6.1 Properties of a Good Mortar

i. It should be capable of developing good adhesion with the building units such as bricks, stones, etc.

ii. It should be capable of developing the designed stresses.

iii. The joints formed by mortar should not develop cracks and they should be able to maintain their appearance for a sufficiently long period.

7.6.2 Preparation of Cement Mortar

Hand mixing: This method is generally used when a small quantity of mortar is required. First a nonporous platform is prepared in the site. The sand is measured in the required proportion and laid on the platform. The required quantity of cement is then evenly spread over the sand. After this, both ingredients are mixed in dry state by overturning with shovels 2–3 times until its color becomes uniform. A depression is formed at the center of the mix. Water is added, and the whole mass is again turned 2–3 times to get a uniform mix of required consistency.

Machine mixing: This method is used when large quantity of mortar is required continuously. In this method mixing of ingredients is done in a machine which is known as mixer machine.

7.6.3 PRECAUTIONS IN USING MORTAR

1. **Consumption of mortar**: After preparation, the mortar should be consumed as early as possible. The lime mortar should be consumed within 36 hours after its preparation and it should be kept wet or damp. The cement mortar should be consumed within 30 minutes after adding water.
2. **Frost action**: The setting action of mortar is affected by the presence of frost.
3. **Sea water**: It helps in preventing too quick drying of the mortar. However, it is not advisable to use sea water in making pure lime mortar or surkhi mortar because it will lead to efflorescence.
4. **Soaking of building units**: The building units should be soaked in water before mortar is applied. If this precaution is not taken, the water of mortar will be absorbed by the building units and the mortar will become weak.
5. **Sprinkling of water**: The water may be sprinkled for about 7–10 days. The exposed surfaces are sometimes covered to give protection against sun and wind.
6. **Workability**: The joints should be well formed and the excess mortar from the joints should be neatly taken off by a trowel. The surfaces formed by mortar for building units to rest should be even.

7.6.4 TESTS FOR MORTAR

1. **Crushing test**: The brick work is crushed in a compression testing machine and the load is noted down. Then the crushing strength is obtained as load divided by cross-sectional area.
2. **Tensile strength test**: The mortar prepared is placed in a mould of briquette which has central cross-sectional area as 38 mm×38 mm. After curing the briquette (Figure 7.2) is pulled under the grips of tensile-testing machine. The ultimate load noted. Then the tensile strength of mortar is load divided by the central cross-sectional area.

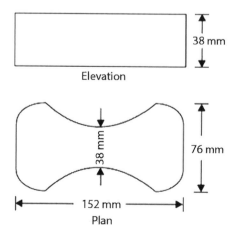

FIGURE 7.2   Briquette mould.

3. **Adhesive test**: Two bricks are joined together with mortar to be tested as shown in Figure 7.3. The upper brick is suspended from an overhead support. A board is hung from the lower brick. Then weights are added to the board till the bricks separate. The adhesive strength is the load divided by area of contact.

7.7 PAINTS, ENAMELS, VARNISHES

The paints are coatings of fluid materials, and they are applied over the surfaces of timber and metals. The varnishes are transparent or nearly transparent solutions of resinous materials and they are applied over the painted surfaces. The distempers are applied over the plastered surfaces.

7.7.1 PAINTING

Following are the objectives of painting a surface:

1. It protects the surface from weathering effects of the atmosphere and actions by other liquids, fumes, and gases.
2. It prevents decay of wood and corrosion in metal.
3. It gives even surface for easy cleaning.

7.7.2 CHARACTERISTICS OF AN IDEAL PAINT

1. It should possess a good spreading power.
2. The paint should be such that it can be easily and freely applied on the surface.
3. The paint should be such that it dries in reasonable time and not too rapidly.
4. The paint should be such that its color is maintained for a long time.
5. The paint should not affect the health of workers during its application.
6. The surface coated with paint should not show cracks when the paint dries.

7.7.3 PIGMENT VOLUME CONCENTRATION NUMBER (PVCN)

$$PVCN = \frac{V_1}{V_1 + V_2} \qquad (7.1)$$

where V_1 = volume of pigment in the paint
V_2 = volume of nonvolatile vehicle or carrier in the paint.

The higher the value of PVCN the lower will be the durability and gloss of the paints.

7.7.4 COMPONENTS OF PAINT

1. **Bases**: A base is a solid substance in a fine state of a division and it forms a bulk of paint. Bases for paints:

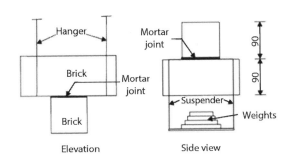 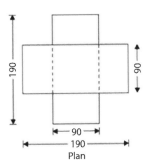

FIGURE 7.3   Adhesive test.

a.   White lead
b.   Red lead
c.   Oxide of zinc or zinc white
d.   Lithophone

2. **Vehicles**: The vehicles are the liquid substances which hold the ingredients of paint in liquid suspension. They are required mainly for two reasons.
   i.   To make it possible to spread the paint evenly and uniformly
   ii. To provide a binder for the ingredient of a paint
   Vehicles for Paint:
   a.   Linseed oil
        i.   Raw linseed oil
        ii.   Boiled linseed oil
        iii. Pale-boiled linseed oil
        iv.   Double-boiled linseed oil
        v.   Stand oil
   b.   Tung oil
   c.   Poppy oil
   d.   Nut oil

3. **Driers**: These substances accelerate the process of drying. Driers are cobalt, lead, manganese, etc. They are used for the following purposes:
   i.   To bring down the cost of paint.
   ii.   To improve the durability of paint.
   iii. To modify the weight of paint.
   iv.   To prevent shrinkage and cracking
        The litharge, red lead, and sulfate of manganese can also be used as driers. The litharge is the most commonly used drier, the proportion being 1/8 kg to 5.0 L of oil.

4. **Coloring pigment**: (Table 7.1) Following are the divisions of the coloring pigment:

i.   Natural earth colors such as ochres, umbers, iron oxides, etc.
ii. Precipitates such as Prussian blue, chrome green, chrome yellow, etc.
iii. Metal powders such as aluminum powder, bronze powder, copper powder, zinc powder, etc.

5. **Solvents**: The function of a solvent is to make the paint thin. The most commonly used solvent is the spirit of the turpentine.

7.7.5 Types of Paint

1. **Aluminum paint**: The spirit or oil evaporates and a thin metallic film of aluminum is formed on the surface. The advantages of an aluminum paint are as follows:
   a.   It is visible in darkness.
   b.   It resists heat to a certain degree.
   c.   It possesses a high covering capacity. A liter of paint can cover an area of about 200 m².
   d.   It gives good appearance to the surface.
   e.   It is impervious to the moisture.
   f.   It possesses high electrical resistance.
   g.   The aluminum paint is widely used for painting gas tanks, marine piers, etc.

2. **Anticorrosive paint**: This paint essentially consists of oil and a strong drier. A pigment such as chromium oxide or lead or red lead or zinc chrome is taken and after mixing it with some quantity of very fine sand, it is added to the paint. The advantages of anticorrosive paint are as follows:
   a.   It is low cost.
   b.   It lasts for a long duration.
   c.   The appearance of the paint is black.

3. **Asbestos paint**: This is a peculiar type of paint.

4. **Bituminous paint**: This paint is prepared by dissolving asphalt or mineral pitches or vegetable bitumen in any type of oil or petroleum. A variety of bituminous paints is available. The paint presents a black appearance and it is used for painting iron.

5. **Cellulose paint**: It is useful where a shorter drying time is required due to it being an "air drying" paint.

TABLE 7.1
Coloring Pigments for Paint

Black	Graphite, lamp black, ivory black
Green	vegetable black
Red	Chrome green, copper sulfate
Yellow	Carmine, red lead, vermilion red
	Chrome yellow, raw sienna, yellow ochre, zinc chrome

Cellulose paint will require approximately 4–5 coats as opposed to a 2k paint requiring 2–3 coats to achieve a full and proper finish.

6. **Cement paint**: Following are the advantages of cement paints:
   a. It requires less skill and time for applying cement water paints and the applying implements can be cleaned with water only.
   b. The preparation of surfaces is easier in a cement paint system as it is not necessary to remove the previous coats of cement paints.
   c. They become an integral part of the substrata and add to its strength.
   d. They can be applied over new and damp walls which cannot be painted over with oil paints until they are sufficiently dried.

7.7.6 Defects in Painting

Following are the usual defects which are found in the painting work:

1. **Blistering**: This defect is caused by the water vapor which is trapped behind the painted surface. The formation of bubbles under the film of paint occurs in this defect. It may occur from various causes such as imperfect seasoning of timber, excess oil used in final coat, etc.
2. **Bloom**: In this defect, the formation of dull patches occurs on the finished polished surface. It is due to the defect of paint or bad ventilation.
3. **Fading**: The gradual loss of color is known as the fading and it is mainly due to the effect of sunlight.
4. **Flaking**: A small portion of the painted surface is sometimes seen loose. It is known as the taking and is due to poor adhesion.
5. **Flashing**: Sometimes the glossy patches are seen on the painted surface. This is known as the flashing and it is mainly due to poor workmanship, cheap paint, or weather actions.
6. **Grinning**: When the final coat of paint has not sufficient opacity, the background is clearly seen. This is known as the grinning.
7. **Running**: This defect occurs when the surface to be painted is too smooth.
8. **Sagging**: When a vertical or inclined surface is too thickly painted, the defect of sagging occurs.

7.7.7 Varnish

Varnish is a solution of resin in oil, turpentine or alcohol. Varnish is applied:

i. To the painted surface to increase its brilliance and to protect it from the atmospheric action
ii. To the unpainted wooden surface to brighten appearance of the grains of wood.

Composition of Varnishes:
1. **Resins**: Commonly used resins are copal mastic, amber gum, and lac.
2. **Solvents**: Boiled linseed oil is used to dissolve copal or amber, turpentine oil for common resin or mastic, methylated spirit for lac.
3. **Driers**: Litharge or lead acetate is the commonly used drier.

7.7.8 The Qualities of a Good Varnish

1. It should be dry quickly.
2. It should be uniform and pleasant looking on drying.
   Different kinds of Varnishes:
   1. Oil varnish: These are made by dissolving hard resins like amber or copal in oil.
   2. Turpentine varnish: These are made from soft resins like mastic.
   3. Spirit varnish: Varnishes in which spirit is used as a solvent as known as spirited varnish or French polish.
   4. Water varnish: They consist of lac dissolved in hot water with borax, ammonia, potash, or soda just enough to dissolve the lac.

7.7.9 Distempering

The main object of applying distemper to the plastered surfaces is to create a smooth surface. The distempers are available in the market under different trade names.

7.7.9.1 Properties
1. On drying, the film of distemper shrinks. Hence it leads to cracking and flaking, if the surface to receive distemper is weak.
2. The coatings of distemper are usually thick and they are more brittle than other types of water paints.
3. They exhibit poor workability.
4. They prove to be unsatisfactory in damp locations such as kitchen, bathroom, etc.

7.7.10 Ingredients of a Distemper

Distemper is composed of base, carrier, coloring pigments, and size. For the base, whiting or chalk is used and for the carrier, water is used. The distempers are available in powder form or paste form. As the water dries, the oil makes a hard surface which is washable. It should be remembered that most of the manufacturers of readymade distempers supply the directions for use of their products. These directions are to be strictly followed to achieve good results.

7.8 TAR, BITUMEN, ASPHALT

7.8.1 Tar

The tar is a dark black liquid with high viscosity.

1. **Coal tar**: Coal tar is a brown or black liquid of extremely high viscosity. Coal tars are complex and variable mixtures of phenols, polycyclic aromatic hydrocarbons, and heterocyclic compounds.
2. **Mineral tar**: This variety of tar is obtained by distilling the bituminous shale.
3. **Wood tar**: There are two types: hardwood tars, derived from such woods as oak and beech; and resinous tars, derived from pine wood, particularly from resinous stumps and roots.

7.8.2 Bitumen

The bitumen is the binding material which is present in asphalt. It is obtained by partial distillation of crude petroleum. It is chemically a hydrocarbon. It is insoluble in water, but it completely dissolves in carbon bisulfide, chloroform, benzol, coal tar, and oil of turpentine.

7.8.3 Forms of Bitumen

Depending upon the temperature and other factors, various types of bitumen are found and used throughout the world.

1. **Cutback bitumen**: Cut-back bitumen is the one which is prepared with the addition of a volatile oil to reduce the thickness of the binder.
2. **Fluxed bitumen**: Fluxed bitumen is the bitumen which is prepared by the addition of relatively non-volatile oils to reduce the viscosity of the binder.
3. **Modified bitumen**: Modified bituminous binders are those whose properties such as cohesive strength, adhesive property, etc., are improved by adding suitable oils.

7.8.4 Properties of Bituminous Materials

1 **Adhesive**: Binds together all the components
2 **Water-proof**: Bitumen is insoluble in water and can serve as an effective sealant
3 **Strong**: Has good cohesive strength

7.8.5 Asphalt

It is black or brownish black in color. It remains in solid state at low temperature and becomes liquid at a temperature of about 50°C–100°C.

7.8.6 Forms of Asphalt

1. **Mastic asphalt**: Mastic asphalt is a type of asphalt which differs from dense graded asphalt (asphalt concrete) in that it has a higher asphalt/bitumen (binder) content, usually around 7%–10% of the whole aggregate mix, as opposed to rolled asphalt concrete, which has only around 5% added asphalt/bitumen.
2. **Asphalt emulsion**: A number of technologies allow asphalt/bitumen to be mixed at much lower temperatures. These involve mixing with petroleum solvents to form "cutbacks" with reduced melting point, or mixtures with water to turn the asphalt/bitumen into an emulsion. Asphalt emulsions contain up to 70% asphalt/bitumen and typically less than 1.5% chemical additives. There are two main types of emulsions with different affinity for aggregates, cationic, and anionic.
3. **Cutback asphalt**: "Cutback asphalt" means asphalt that has been liquefied by blending with a diluent of petroleum solvents or any other diluent that contains Volatile Organic Compounds (VOC).

7.9 MISCELLANEOUS MATERIALS

7.9.1 Abrasives

Abrasives are those materials used in operations such as grinding, polishing, lapping, honing, pressure blasting, or other similar process. Abrasives come in different particle or grit sizes depending on how much material needs to be removed. Abrasives can either be bonded or coated. The abrasives on sandpaper are coated on paper that can be held by the hand. The person can control the smoothing strength of the sandpaper by adjusting the pressure and speed of the hand movement.

7.9.2 Adhesives

An adhesive is a material used for holding two surfaces together. An adhesive must wet the surfaces, adhere to the surfaces, develop strength after it has been applied, and remain stable. The adhesives materials allow joint substrates with different geometries, sizes, and composition. With the adhesive we can joint glass, plastics, metals, and ceramics.

7.9.3 Asbestos

The asbestos is a naturally occurring fibrous mineral substance. It is composed of hydrous silicates of calcium and magnesium. Two types of asbestos are crocidolite and amosite.

Crocidolite: Crocidolite is known colloquially as blue asbestos and is a member of the Amphibole group. The needle-like fibers are the strongest of all asbestos fibers and have a high resistance to acids.

Amosite: Amosite is also known as brown asbestos and is, like crocidolite, a member of the Amphibole group. Its harsh, spiky fibers have good tensile strength and resistance to heat.

7.9.4 Cork

Cork is an impermeable, buoyant material, a prime-subset of bark tissue that is harvested for commercial use primarily from the Cork Oak.

Properties: The properties of cork derive naturally from the structure and chemical composition of its extremely strong, flexible cell membranes, which are water-proof and airtight.

    i. **Lightness**: Cork is light and floats on water.
    ii. **Elasticity and resiliency**: The cell membranes of cork are highly flexible, making it both compressible and elastic.
    iii. **Impermeability**: The presence of suberin renders cork impermeable to both liquids and gases.
    iv. **Insulation and fire-retardant qualities**: The value of cork is further enhanced by its low conductivity of heat, sound, and vibration.
    v. **Resistance to wear**: Remarkably resistant to wear and has a high friction coefficient.
    vi. **Hypoallergenic properties**: It helps protect against allergies and does not pose a risk to asthma sufferers.

7.9.5 Fly Ash

Fly ash is a byproduct from burning pulverized coal in electric power generating plants. During combustion, mineral impurities in the coal (clay, feldspar, quartz, and shale) fuse in suspension and float out of the combustion chamber with the exhaust gases. As the fused material rises, it cools and solidifies into spherical glassy particles called fly ash. Fly ash is collected from the exhaust gases by electrostatic precipitators or bag filters.

7.9.6 Gypsum

Calcium sulfate, commonly known as natural gypsum, is found in nature in different forms, mainly as the dihydrate ($CaSO_4 \cdot 2H_2O$) and anhydrite ($CaSO_4$).

The manufacturer receives quarried gypsum and crushes the large pieces before any further processing takes place. Synthetic gypsum, commonly known as the FGD (Flue Gas Desulfurization) gypsum or DSG (desulfurized) gypsum, may also be used in the production of gypsum board. This product is primarily derived from coal-fired electrical utilities which have systems in place to remove sulfur dioxide from flue gasses. These systems capture the sulfur dioxide by passing the gasses through scrubbers that contain limestone (calcium carbonate) which absorbs and chemically combines with the sulfur dioxide to form pure calcium sulfate, or gypsum.

7.10 METALS AND ALLOYS

All metals may be classified as ferrous or nonferrous. A ferrous metal has iron as its main element. A metal is still considered ferrous even if it contains less than 50% iron, as long as it contains more iron than any other one metal. A metal is nonferrous if it contains less iron than any other metal.

An alloy is a substance composed of two or more elements. Therefore, all steels are an alloy of iron and carbon, but the term "alloy steel" normally refers to steel that also contains one or more other elements. For example, if the main alloying element is tungsten, the steel is a "tungsten steel" or "tungsten alloy." If there is no alloying material, it is a "carbon steel."

7.10.1 Ferrous Metals

Steel: Plain carbon steel is a ferrous metal that consists of iron and carbon. Carbon is the hardening element. Steel is produced in a variety of melting furnaces, such as open-hearth, Bessemer converter, crucible, electric-arc, and induction.

Cast iron: Cast iron is produced by melting a charge of pig iron, limestone, and coke in a cupola furnace. It is a brittle and hard metal with above average levels of wear resistance. The desirable properties of cast iron are less than those of carbon steel because of the difference in chemical makeup and structure. Cast iron differs from steel mainly because its excess of carbon (more than 1.7%) is distributed throughout as flakes of graphite, causing most of the remaining carbon to separate.

Iron ore: Iron ore is smelted with coke and limestone in a blast furnace to remove the oxygen and earth foreign matter from it. Limestone is used to combine with the earth matter to form a liquid slag.

Wrought iron: Wrought iron is one of the ferrous metals which is an alloy that is almost pure iron. It is made from pig iron in a puddling furnace and has a carbon content of less than 0.08%.

Gray cast iron: If the molten pig iron is permitted to cool slowly, the chemical compound of iron and carbon breaks up to a certain extent. Much of the carbon separates as tiny flakes of graphite scattered throughout the metal.

White cast iron: White cast iron is very hard and brittle, often impossible to machine, and has a silvery white fracture.

Malleable cast iron: Malleable cast iron is made by heating white cast iron from 1400 to 1700°F (760 and 927°C) for about 150 hours in boxes containing hematite ore or iron scale. This heating causes a part of the combined carbon to change into the free or uncombined state. This can be welded and brazed. Any welded part should be annealed after welding.

Steel: A form of iron, steel is one of the ferrous metals that contain less carbon than cast iron, but considerably more than wrought iron. The carbon content is from 0.03% to 1.7%.

Low-carbon steel (carbon content up to 0.30%): This steel is soft and ductile, and can be rolled, punched, sheared, and worked when either hot or cold.

Medium-carbon steel (carbon content ranging from 0.30% to 0.50%): This steel may be heat-treated after fabrication.

High-carbon steel (carbon content ranging from 0.50% to 0.90%): This steel is used for the manufacture of drills, taps, dies, springs, and other machine tools and hand tools that are heat treated after fabrication to develop the hard structure necessary to withstand high shear stress and wear.

7.10.2 Nonferrous Metals

Aluminum: Aluminum is a lightweight, soft, low strength metal which can easily be cast, forged, machined, formed, and welded. It is suitable only in low-temperature applications, except when alloyed with specific elements.

Cobalt: Cobalt is a hard, white metal similar to nickel in appearance, but has a slightly bluish cast.

Copper: Copper is a reddish metal, is very ductile and malleable, and has high electrical and heat conductivity. It is used as a major element in hundreds of alloys.

- Beryllium copper contains 1.50%–2.75% beryllium. It is ductile when soft, but loses ductility and gains tensile strength when hardened.
- Nickel copper contains 10%, 20%, or 30% nickel. Nickel alloys have moderately high to high tensile strength, which increases with the nickel content. They are moderately hard, quite tough, and ductile. They are very resistant to the erosive and corrosive effects of high velocity sea water, stress corrosion, and corrosion fatigue.

Lead: Lead is a heavy, soft, malleable metal with low melting point, low tensile strength, and low creep strength. It is resistant to corrosion from ordering atmosphere, moisture, and water, and is effective against many acids.

Magnesium: Magnesium is an extremely light metal, is white in color, has a low melting point, excellent machinability, and is weldable. Welding by either the arc or gas process requires the use of a gaseous shield.

Nickel: Nickel is a hard, malleable, ductile metal. As an alloy, it will increase ductility and has no effect on grain size.

Monel metal: Monel metal is a nickel alloy of silver-white color containing about 67.00% nickel, 29.00%–80.00% copper, 1.40% iron, 1.00% manganese, 0.10% silicon, and 0.15% carbon.

Tin: Tin is a very soft, malleable, somewhat ductile, corrosion-resistant metal having low tensile strength and high crystalline structure. It is used in coating metals to prevent corrosion.

Zinc: Zinc is a medium low strength metal having a very low melting point.

Alloys: An alloy is a metal (parent metal) combined with other substances (alloying agents), resulting in superior properties such as strength, hardness, durability, ductility, tensile strength, and toughness.

Aluminum alloy: An aluminum alloy is a chemical composition where other elements are added to pure aluminum in order to enhance its properties, primarily to increase its strength. These other elements include iron, silicon, copper, magnesium, manganese, and zinc at levels that combined may make up as much as 15% of the alloy by weight.

Copper alloys: Copper is used in a wide range of products due to its excellent electrical and thermal conductivity, good strength, good formability, and resistance to corrosion. Pipe and pipe fittings are commonly manufactured from these metals due to their corrosion resistance. They can be readily soldered and brazed, and many can be welded by various gas,

arc, and resistance methods. They can be polished and buffed to almost any desired texture and luster.

One of the most important properties of copper is its ability to fight bacteria.

Brass: Brass is mainly an alloy that consists of copper with zinc added. Brasses can have varying amounts of zinc or other elements added. If the zinc content of the brass ranges from 32% to 39%, it will have increased hot-working abilities but the cold-working will be limited.

7.10.3 OTHER BRASS ALLOYS

Tin brass: This is an alloy that contains copper, zinc, and tin. This alloy group would include admiralty brass, naval brass, and free-machining brass.

Bronze: Bronze is an alloy that consists primarily of copper with the addition of other ingredients. Bronze is characterized by its dull-gold color.

7.10.4 OTHER BRONZE ALLOYS

Phosphor bronze (or tin bronze): This alloy typically has a tin content ranging from 0.5% to 1.0%, and a phosphorous range of 0.01% to 0.35%.

Aluminum bronze: This has an aluminum content range of 6%–12%, an iron content of 6% (max), and a nickel content of 6% (max). These combined additives provide increased strength, combined with excellent resistance to corrosion and wear.

Silicon bronze: This is an alloy that can cover both brass and bronze (red silicon brasses and red silicon bronzes). They typically contain 20% zinc and 6% silicon.

Nickel brass (or nickel silver): This is an alloy that contains copper, nickel, and zinc. The nickel gives the material an almost silver appearance.

Copper nickel (or cupronickel): This is an alloy that can contain anywhere from 2% to 30% nickel. This material has a very high corrosion-resistance and has thermal stability.

Nickel alloys: Nickel has been used in alloys that date back to the dawn of civilization. Chemical analysis of artifacts has shown that weapons, tools, and coins contain nickel in varying amounts. Nickel and nickel alloys are used for a wide variety of applications, the majority of which involve corrosion resistance and/or heat resistance.

Low-expansion alloys: Nickel was found to have a profound effect on the thermal expansion of iron.

Electrical-resistance alloys: Several alloy systems based on nickel or containing high nickel contents are used in instruments and control equipment to measure and regulate electrical characteristics or are used in furnaces and appliances to generate heat.

Types of resistance alloys containing nickel include:

- Cu-Ni alloys containing 2%–45% Ni
- Ni-Cr-Al alloys containing 35%–95% Ni

Types of resistance heating alloys containing nickel include:

- Ni-Cr alloys containing 65%–80% Ni with 1.5% Si
- Ni-Cr-Fe alloys containing 35%–70% Ni with 1.5% Si+1% Nb

Soft magnetic alloys: Two broad classes of magnetically soft materials have been developed in the Fe-Ni system. The high-nickel alloys (about 79% Ni with 4%–5% Mo; bal Fe) have high initial permeability and low saturation induction.

7.11 CERAMIC MATERIALS

Ceramics are classified as inorganic and nonmetallic materials that are essential to our daily lifestyle. This category of materials includes things like tile, bricks, plates, glass, and toilets.

Clay products: Structural clay products, ceramic products intended for use in building construction. Typical structural clay products are building brick, paving brick, terra-cotta facing tile, roofing tile, and drainage pipe. These objects are made from commonly occurring natural materials, which are mixed with water, formed into the desired shape, and fired in a kiln in order to give the clay mixture a permanent bond.

Tiles: A tile is a manufactured piece of hard-wearing material such as ceramic, stone, metal, or even glass, generally used for covering roofs, floors, walls, showers, or other objects such as tabletops. Thinner tiles can be used on walls than on floors.

Characteristics of a good tile are as follows:

- i. It should be regular in shape and size.
- ii. It should be sound, hard, and durable.
- iii. It should be well burnt.
- iv. It should fit in properly, when placed in position.
- v. It should give an even and compact structure when seen on its broken surface.
- vi. It should possess uniform color.

Types of common tiles:

Drain tiles: The purpose of the drain tile is to collect water around your basement foundation and deliver it to your sump crock so it can be pumped up and away from your home. It is basically a pipe with holes or perforations and stone surrounding it.

Flooring tiles: Flooring Tiles in India have over the years gained popularity over the cemented and concrete flooring. Tiles are available in different patterns, designs, and utility options. Various flooring tiles are:

Terrazzo tiles: Terrazzo tiles are available in the market with sizes varying from 20 cm×20 cm to 30 cm×30 cm and thickness varying between 2 and 3 cm.

Chequered tiles: The chequered tiles are similar to the terrazzo tiles except that they have groves of about 2.5 cm in both directions. The method of laying such tiles is similar to that of laying other tiles.

Glazed tiles: Glazed tiles are available in the market having either a glossy finish or a mat finish.

They are usually available in sizes of 10 cm×10 cm, 15 cm×15 cm, or 10 cm×20 cm in white or different colors.

Vitrified tiles: Vitrified tiles are increasingly being adopted for construction of floors in residential buildings. Vitrified tiles are available in glossy or mat finish having different design and patterns. Sizes up to 60 cm×60 cm are also available and accordingly each tile covers a large area.

PVC tiles: PVC tiles are being used for flooring in residential as well as nonresidential buildings. PVC tiles are usually square in shape having sides of 30, 60, or 90 cm.

Roof tiles: Roof tiles are designed mainly to keep out rain and are traditionally made from locally available materials such as terracotta or slate. A large number of shapes (or "profiles") of roof tiles have evolved.

Encaustic tiles: These are ceramic tiles in which the pattern or figure on the surface is not a product of the glaze but of different colors of clay. They are usually of two colors but a tile may be composed of as many as six.

Terracotta: Terracotta is a ceramic made out of fire-baked clay and can be used as a building material.

7.12 POLYMERIC MATERIALS

7.12.1 CHARACTERISTICS OF POLYMERS

The majority of manufactured polymers are thermoplastic, meaning that once the polymer is formed it can be heated and reformed over and over again. This property allows for easy processing and facilitates recycling.

7.13 POLYMER FIBER COMPOSITES

Fiber-reinforced polymer (FRP), also Fiber-reinforced plastic, is a composite material made of a polymer matrix reinforced with fibers. The fibers are usually glass, carbon, or aramid, although other fibers such as paper or wood or asbestos have been sometimes used. Properties of the fiber composites depend on:

- i. The properties of the fiber
- ii. The geometry and orientation of the fibers in the composite

Manufacturing FRP composites: FRP composite manufacturing can be an energy-intensive process with high heat and pressure needed to bond the composite material together.

Fiber fabrication: High temperatures are required in the manufacturing of both carbon and glass fibers. The fibers can be woven into a fabric or formed into a tape, depending upon specifications for the component to be manufactured. In some cases, the long fibers are arranged in one direction or the fibers are chopped short and set in multiple directions.

Component production: There are many different ways to make composite parts. Typically, the strong, stiff, reinforcing fibers are combined with the polymer either before or during part fabrication. These parts are made by layering the composite material over a mold in the final shape of a part (much like the layering of plywood) and then heated under pressure.

7.14 GEOSYNTHETICS

Geosynthetics are man-made polymer-based materials or natural fiber-based, which facilitate cost-effective building, environmental, transportation, and other construction projects. Most geosynthetics are made from synthetic polymers of polypropylene (PP), polyester, or polyethylene or of natural fibers like coir, jute, hemp, sisal, etc. Geosynthetic products available today include but are not limited to geowebs, geogrids, geonets, geomeshes, geocomposites, and geotextiles.

Geotextile is a permeable geosynthetic made of textile materials. They are used to provide one or more of the following functions: separation, reinforcement, filtration, drainage, or liquid barrier.

i. **Filtration**: Geosynthetics can be used as filters to prevent soils from migrating into the adjacent material.

ii. **Drainage**: Geotextiles or geocomposites can be used as drainage, or conduit, by allowing water to drain from or through low-permeability soils.

iii. **Separation**: Geosynthetics can be used as a separator to separate the two dissimilar materials and prevent them from mixing.

iv. **Reinforcement**: Geogrids or geotextiles can be used as reinforcement to increase shear strength of soils, thereby providing a more competent structural material.

v. **Erosion control**: Geosynthetics can be used to minimize the movement of soil particles due to flow of water.

Different types of geosynthetics are given below:

i. **Geotextiles**: Geotextiles (Figure 7.4) are indeed textiles in a traditional sense but consist of synthetic fibers or natural ones like coir, jute, cotton, wool,

and silk. There are two types of geotextiles. They are woven and nonwoven geotextiles.

ii. **Geogrid**: Geogrids (Figure 7.5) are plastics formed into a very open net-like configuration. Single or multilayer materials are usually made from extruding and stretching high-density polyethylene or by weaving or knitting the PP.

iii. **Geonets**: Geonets (Figure 7.6) are stacked crisscrossing polymer strands that provide in-plane drainage. The geonets are all made of polyethylene.

iv. **Geocomposites**: Geocomposites (Figure 7.7) are geotextile filters surrounding a geonet. Some of the functions of the geocomposites are as blanket drains, panel drains, edge drains, and wick drains.

v. **Geomembranes**: Geomembranes (Figure 7.8) are impervious thin sheets of rubber or plastic material primarily used for linings and covers of liquid- or solid-storage impoundments. Thus the primary function is always as a liquid or vapor barrier.

FIGURE 7.5 Geogrid.

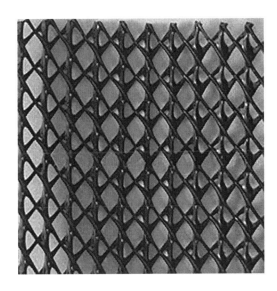

FIGURE 7.6 Geonets.

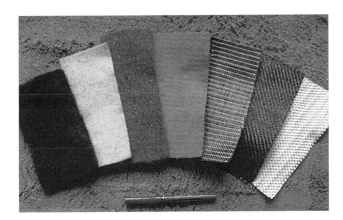

FIGURE 7.4 Geotextiles.

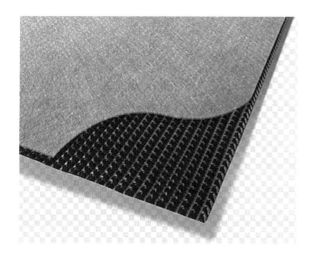

FIGURE 7.7 Geocomposites.

FIGURE 7.8 Geomembranes.

They are relatively impermeable when compared to soils or geotextiles.

vi. **Geosynthetic clay liners**: Geosynthetic clay liners (Figure 7.9) include a thin layer of finely ground bentonite clay. The clay swells and becomes a very effective hydraulic barrier when wetted

FIGURE 7.9 Geosynthetic clay liners.

FIGURE 7.10 Geofoam.

- Geofoam: (Figure 7.10) Geofoam is a newer category of the geosynthetic product. It is a generic name for any foam material utilized for geotechnical applications.
- It is used within soil embankments built over soft, weak soils.
- Used under roads, airfield pavements, and railway track systems which are subjected to excessive freeze–thaw conditions.

vii. **Geopipe**: Another significant product which has been adopted as a geosynthetic is the plastic pipe (Figure 7.11). The specific polymer resins that are

FIGURE 7.11 Geopipe.

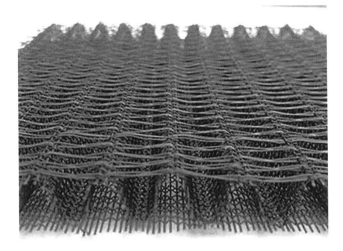

FIGURE 7.12   Turf reinforcement mats.

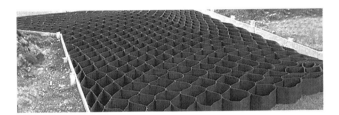

FIGURE 7.13   Geocell.

used in the manufacturing of plastic pipes are high-density polyethylene (HDPE), polyvinyl chloride (PVC), PP, polybutylene, acrylonitrile butadiene styrene, and cellulose acetate butyrate.

viii. **Turf reinforcement mats**: Turf reinforcement mats (Figure 7.12) are three-dimensional structures composed of fused polymer nettings, randomly laid monofilaments, or yarns woven or tufted into an open and dimensionally stable mat. Erosion protection can be increased by applying these Mats, which can provide more protection compared to that of plants grown normally.

ix. **Geocell**: Geocells (Figure 7.13) were manufactured from a novel polymeric alloy called Neoloy. The geocell with a higher elastic modulus has stiffness of the reinforced base and a higher bearing capacity. Geocells made from HDPE are found to be significantly better in stiffness, ultimate bearing capacity, and reinforcement geocells.

7.15   SOIL STABILIZERS

Soil stabilization is a method of improving soil properties by blending and mixing other materials. The components of stabilization technology include soils and or soil minerals and stabilizing agent or binders (cementitious materials).

7.15.1   STABILIZING AGENTS

These are hydraulic (primary binders) or nonhydraulic (secondary binders) materials that when in contact with water or in the presence of pozzolanic minerals react with water to form cementitious composite materials.

1. **Soil stabilization with cement**: The soil stabilized with cement is known as soil cement. The cementing action is believed to be the result of chemical reactions of cement with siliceous soil during hydration reaction. The important factors affecting the soil–cement are nature of soil content, conditions of mixing, compaction, curing, and admixtures used. The quantity of cement for a compressive strength of 25–30 kg/cm^2 should normally be sufficient for tropical climate for soil stabilization. Lime, calcium chloride, sodium carbonate, sodium sulfate, and fly ash are some of the additives commonly used with cement for cement stabilization of soil.

2. **Soil stabilization using lime**: Lime changes the nature of the adsorbed layer and provides pozzolanic action. Plasticity index of highly plastic soils is reduced by the addition of lime with soil. Normally 2%–8% of lime may be required for coarse grained soils and 5%–8% of lime may be required for plastic soils.

3. **Stabilization with bitumen**: Bituminous materials when added to a soil, it imparts both cohesion and reduced water absorption. Depending upon the above actions and the nature of soils, bitumen stabilization is classified in following four types:
   * Sand bitumen stabilization
   * Soil bitumen stabilization
   * Water-proofed mechanical stabilization, and
   * Oiled earth

4. **Chemical stabilization of soil**: The vapor pressure gets lowered, surface tension increases, and rate of evaporation decreases. The freezing point of pure water gets lowered and it results in prevention or reduction of frost heave. Depressing the electric double layer, the salt reduces the water pickup and thus the loss of strength of fine grained soils. Calcium chloride acts as a soil flocculent and facilitates compaction. Frequent application of calcium chloride may be necessary to make up for the loss of chemical by leaching action. For the salt to be effective, the relative humidity of the atmosphere should be above 30%. Sodium chloride is the other chemical that can be used for this purpose with a stabilizing action similar to that of calcium chloride.

5. **Electrical stabilization of clayey soils**: This is done by electro-osmosis. This is an expensive method of soil stabilization and is mainly used for drainage of cohesive soils.

6. **Soil stabilization by grouting**: This method is not useful for clayey soils because of their low permeability. This is a costly method for soil stabilization.

This method is suitable for stabilizing buried zones of relatively limited extent. The grouting techniques can be classified as following:
   - Clay grouting
   - Chemical grouting
   - Chrome lignin grouting
   - Polymer grouting, and
   - Bituminous grouting

7. **Soil stabilization by geotextiles and fabrics**: Geotextiles are porous fabrics made of synthetic materials such as polyethylene, polyester, nylons, and PVC. Woven, nonwoven and grid form varieties of geotextiles are available.

7.16 SUSTAINABLE CONSTRUCTION MATERIALS

A sustainable material is one that

- does not deplete nonrenewable (natural) resources,
- whose use has no adverse impact on the environment.

We can preserve natural resources in many ways:

- avoiding using scarce materials, such as peat and weathered limestone,
- creating less waste,
- using less; using reclaimed, rather than new materials,
- using renewable materials (crops),

We can reduce the impact by

- using materials with low embodied.
- reducing transport of materials and associated fuel, emissions, and road congestion.

Five sustainable building materials that could transform construction:

1. **Wool bricks**: Developed by Spanish and Scottish (Figure 7.14) researchers with an aim to "obtain a composite that was more sustainable, non-toxic, using abundant local materials that would mechanically improve the bricks' strength," these wool bricks are exactly what the name suggests. Simply by adding wool and a natural polymer found in seaweed to the clay of the brick, the brick is 37% stronger than other bricks, and more resistant to the cold wet climate often found in Britain.

   Solar tiles: (Figure 7.15) Traditional roof tiles are either mined from the ground or set from concrete or clay—all energy-intensive methods. Once installed, they exist to simply protect a building from the

FIGURE 7.14 Wool bricks.

FIGURE 7.15 Solar tiles.

elements despite the fact that they spend a large portion of the day absorbing energy from the sun.

Sustainable concrete: While 95% of a building's CO_2 emissions are a result of the energy consumed during its life, there is much that can be done to reduce that 5% associated with construction. Crushed glass can be added, as can wood chips or slag—a byproduct of steel manufacturing. While these changes are not radically transforming concrete, by simply using a material that would have otherwise gone to waste, the CO_2 emissions associated with concrete are reduced (Figure 7.16).

2. **Paper insulation**: (Figure 7.17) It is made from recycled newspapers and cardboard, paper-based insulation is a superior alternative to chemical foams. Paper insulation can be blown into cavity walls, filling every crack and creating an almost draft-free space.

3. **Triple-glazed windows**: In fact, super-efficient windows (Figure 7.18) would better describe this particular building material. The three layers of glass do a better job of stopping heat from leaving the building,

FIGURE 7.16 Sustainable concrete.

FIGURE 7.18 Triple-glazed windows.

with fully insulated window frames further contributing. In most double-glazed windows, the gas argon is injected between each layer of glass to aid insulation, but in these super-efficient windows, krypton—a better, but more expensive insulator—is used. In addition to this, low-emissivity coatings are applied to the glass, further.

FIGURE 7.17 Paper insulation.

8 Building Construction Technology and Management

8.1 BASICS OF CONSTRUCTION TECHNOLOGY

The construction sector is currently undergoing a change process to adapt to the new market situation. In this changed scenario, management processes need to be improved in order to increase competitiveness. This chapter looks at studies from a variety of perspectives on these processes. The chapter identifies success factors for architectural practices and differences between project management systems from a strategic approach. A process-oriented approach examines management models that can improve quality, reduce production times, and minimize costs by providing examples of their application to building processes, ceramic coatings, and Construction and Demolition (C&D) waste. Finally, the chapter presents a person-centered approach, with examples of worker-focused studies, job promotion systems, architectural career experiences, and the qualities that a project leader should have. Users are also examined to integrate their requirements into the design process.

8.1.1 What is Construction Technology?

Construction technology is used for the creation, design, and installation of structures and their different components. The art of building houses, skyscrapers, hospitals, and bridges is included. In addition, the management of construction technology refers to the design, coordination, and successful implementation of such structures. Construction Technology and Management, a branch of civil engineering, gives students an insight into the scientific principles of construction, understanding of the behavior of materials, and the basics of structural mechanism. Construction technologists should have good knowledge of various types of construction materials and test procedures to ensure quality control.

8.1.2 Scope of Construction Technology and Management

- Graduates/post-graduates can work as a construction planning engineer and a site engineer in any construction projects in arena of Residential buildings, Cooperatives, huge housing projects, commercial buildings, and hospitals.
- They also estimate the overall project costs and conduct negotiations regarding the elimination of unnecessary scope and cost.
- As managers of quality control, they enforce quality control to reduce potential defects and poor manufacturing.

- Construction managers also optimize and speed up design and planning of works.

8.1.3 Impact of Construction Technology

Construction essentially means building new buildings, residential areas, offices, shopping malls, restaurants, and more. It has a positive and a negative impact on the nation's growth.

8.1.3.1 Positive Construction Impact for a Worker

1. **High pay scale**: Constructors are mostly paid high compared to other workers. In some ways, you are the highest paid worker if you can become a union leader.
2. **Less working hours**: Most IT companies are committed to working for at least 8 hours a day. In the construction sector, however, you are likely to leave earlier than other office workers.
3. **No need to go formal**: In construction, you do not have to dress up like a coat and a suit as a professional. You can wear and do your job comfortably.

8.1.3.2 Positive Impact on Surroundings

1. Construction gives new and attractive lives to those people who still dream of living in their own home.
2. Shopping malls collect a number of items, such as food, clothes, and many other items under a single roof every day. This saves time for people who do not have sufficient time to shop for their everyday items.
3. New office space renovates the office area and transforms your old office into a new look. New building sites open more scope for business and also contribute to business growth.

8.1.3.3 Positive Impact on the Country

1. **Brings tourism**: New buildings, wonderful offices, our national heritage are all part of the building. The construction of new buildings attracts tourism and contributes to the country's economic growth.
2. **Improvement in social conditions**: The building of new social communities gives the social gatherings isolated space. This space allows people to connect, share their views, and sometimes also enjoy events.
3. **Regional benefits**: Building temples, gurudwaras, mosques, and churches help people to learn more about the country's various religions. This helps parents to provide their children with religious knowledge from the beginning of their lives.

8.1.3.4 Negative Impact of Construction for a Worker

1. **Wastes time**: Without a doubt, you go home early and spend time with your family and friends, but this wastes almost half of your day, which could have been used to do more work and earn more for your family.
2. **Dangerous work**: Construction is a dangerous job and the lives of workers are always at stake. Other jobs such as IT, electronics, electrical, etc., are not.

8.1.3.5 Impact on Environment: Cutting of Trees

Construction sites need clearance for trees, forests, and other occupied areas to meet the needs and requirements of customers. This is destroying many lives and also affecting the environment.

8.1.3.6 Wrapping Up

The construction of new buildings, companies, offices, etc., is always intended to improve the company. In view of its positive effects, negatives can be ignored to help a nation advance effectively.

8.2 PLANNING FOR AND CONTROLLING CONSTRUCTION

The building process includes materials, equipment, subcontractors, project owners, and inspectors who must work together to finish the job. The project manager is the person responsible for supervising this process and for facilitating coordination and communication. The project manager not only ensures that the building is completed on schedule and within a specified budget but also that it is delivered on schedule. In order to successfully manage these tasks, he or she must rely on project management techniques and systems to help plan and monitor projects.

8.2.1 Community Development

The development of the Community is a process in which the members of the Community take collective action and generate solutions to common problems. The well-being of the Community (economic, social, environmental, and cultural) often develops from this type of collective action at grassroots level. The term Community development covers a variety of communities, projects, groups, initiatives, and activities. This process aims to empower and support people in a community to work together to identify the problems affecting their community, identify the causes, plan, and implement the plan. Empowerment and participation are the key elements in this approach to community development. Participation is the engagement of people in their planning, decision-making, and problem-solving.

Effective development of the community should be

- a long-term effort
- well-planned

- equitable and inclusive
- integrated and holistic into the larger picture
- initiated and supported by members of the community
- advantageous to the community
- based on experience leading to best practices

8.2.2 Managing Community Development

Management is an organization of people and resources to achieve the goals of a group. People in community development groups need to work effectively together to achieve what they are planning to do. It must permit and facilitate project work to be carried out in a way that is participatory and reliable, coordinated and flexible, indicative and in contact with members.

8.2.3 The Role of Management

The ultimate responsibility of any management committee is to ensure that the objectives of an organization are met. The Management Committee assumes a number of roles. It is up to each group to determine the roles that its management committee wants to play. They often include the following:

- developing objectives and an activity program to represent the general membership.
- implementing policies agreed to at general meetings.
- policy decisions or changes to be made.
- to ensure that the organization meets its legal responsibilities.
- to ensure that the organization has appropriate legal structures.
- to ensure that the organization carries out its employer responsibilities.
- guarantee accountability for all financial resources.
- to ensure regular ongoing planning and evaluation of the activities of the organizations.
- ensuring that all organization members receive the training they need.
- answering problems.
- represent external groups of the organization, e.g., the media.

8.2.4 Teamwork

"Working together" is important for the development of the Community. Teamwork is therefore essential in community development organizations. Each part of the organization, management committee, staff, and participants form a separate team and part of the larger group. Every group has a role, e.g., the role of the management committee is to manage, the task of the staff is to carry out, and the participants have to participate in certain activities.

8.2.5 Accountability

The management committee of a community group must be accountable to its employees, volunteers, participants, and especially the community it is intended to represent. This means that:

a. the use of money and resources is open.
b. be open to critique and questioning.
c. taking the group's responsibility.
d. to build trust in the groups to which the Management Committee is responsible.
e. take the time to look at the entire management committee as a group and question how it implements its own goals and objectives for community development.
f. ensuring the capacity, ability, and representation of the committee to do the job.

8.3 CONSTRUCTION SAFETY

Measures are to be taken to translate this social concern into concrete programs of action—legislative, administrative, or educational. Construction is a relatively hazardous undertaking.

8.3.1 SAFETY IMPORTANCE

The cost of accidents is costly. However, economic costs are not the only reasons why a contractor should be aware of the safety of construction. The reasons why safety is considered include are as follows:

1. **Humanitarian concern**: When an accident occurs, it is difficult to quantify economically the resulting suffering of injured workers and their families. The contractor must never overlook this even if he has accident insurance.
2. **Economic reasons**: Even if a contractor has insurance, the cost of accidents comes out of his own pocket by increasing insurance premiums. Moreover, there are other indirect costs related to accidents. The direct and indirect accident costs may be:
   Direct Cost:
   a. Costs of medical care for injured
   b. Compensation costs for workers
   c. Premium insurance increases
   d. Replacement of damaged equipment and material in accidents
   e. Reparation and clean-up facility
   f. Legal counsel fees
   Cost indirect:
   a. Operational slowdown
   b. Moral decrease affecting productivity
   c. The time lost by injured workers and fellow workers was productive
   d. Accident-related administrative work
   e. Loss of customer faith
   f. Work slowdown requires overtime
3. **Laws and regulations**: Under various acts and laws, the employer should take care of the safety of the employee. Infringement of these laws will be punishable.
4. **Organizational image**: A good safety record can lead to higher morality and productivity and greater loyalty for employees. It will also improve the public image of the company and thus facilitate the acquisition of negotiated jobs.

8.3.2 DESIGNING FOR SAFETY

Safety in the construction of a project is also largely influenced by decisions taken during the design and planning process. Some designs or construction plans are inherently difficult and dangerous to implement if other comparable plans reduce the likelihood of accidents considerably. Technology choice can also be crucial in determining the safety of a workplace. Safeguards built into machines can notify or prevent injuries to operators. Proper material choices also influence the safety of building. The education of workers and managers in proper procedures and hazards can directly affect the safety of the workplace. The realization of the high costs associated with construction injuries and illnesses is an important motivation for awareness and training. Prequalification of security contractors and subcontractors is another important way of improving safety.

8.3.3 ROLE OF VARIOUS PARTIES IN DESIGNING FOR SAFETY

1. **Designer**: During the planning phase, architects, engineers, and designers should take due account of the safety of the workers who are subsequently employed in the construction of such structures. They should exercise adequate care and should not include anything in the design that requires the use of harmful structural procedures and undue dangers.
2. **Employer**: The employers should provide and maintain buildings, planning, equipment, and work passages and organize the work in a way that protects workers from accident risks. While buying machinery, he should ensure the equipment's specifications comply with the different safety regulations. Proper instructions on safety requirements should be given to all workers. The employer should establish a proper signaling system for any hazard.
3. **Workers**: At the start of the job, the workers should inspect their workplaces and the equipment to be used by them and report any dangerous defect to their supervisor or any competent person. They should use all safeguards and safety devices made available for their safety and not interfere with equipment for which they have not been duly authorized to operate or use. Workers should not be allowed to sleep or rest in dangerous places such as scaffolding, running machines or vehicles, heavy equipment or near fires, and dangerous substances. Workers should use protective equipment for their safety and make themselves conversant with safety instructions.

4. **Manufacturer or Dealer**: The manufacturer and dealers of equipment should ensure that the machines, vehicles, and other equipment used in construction industry comply with the safety laws and regulations and that these machines are as safe as possible. In case of toxic, corrosive, explosive flammable liquids, and other dangerous items, the manufacturers should ensure that these are not sent to the user without adequate instructions.

8.4 THE CONSTRUCTION PROCESS

In general, the construction industry focuses on producing a single and unique end product. Building projects developed clearly in a sequential or linear manner. The general steps involved are as follows:

- The identification of the need for a facility by the owner.
- Development of initial feasibility and cost projections.
- Finalization of conceptual design by a professional designer.
- Preparation of an approximate cost estimate.
- Preparation of the final design documents that define the project fully for construction purposes.
- Advertisement of the project on the basis of the final design documents.
- Invitation of construction quotations.
- Selection of a builder based on proposals and quotations received.
- Establishment of a work contract.
- The building process of the facility is initiated.
- Completion of the work.
- Allotment of a test period to determine whether the facility operates as planned in complex projects. This period is typical for industrial projects and is known as the start of a project.

8.4.1 INITIATING THE PROJECT

The aim of the execution of a project process is to recognize the work to be done to supply the products actually required. Before deciding to continue the project, this understanding is necessary. A process is defined as an operating system for designing, developing, and producing something like a project. The start of a project process checks whether the project is viable and the start of a project aims to build the right foundation for the project so that all stakeholders know what the project will achieve. The alternative would be to allow projects to start without knowing anything of the following: planning, milestones, cost, and quality. It is like building a house on a foundation. Starting a project can be a big investment for a company, but planning and running the rest of the project is a must. During the initiation of a project, the project manager will create a collection of management products to show how to manage the project, the costs, how to check quality, how to plan, how to communicate, etc. A project

process is a series of actions, modifications, or operations that achieve the end result cost, schedule, and technical performance goals. It is also a course or time span in which something is created—a continuous movement or progression. The purchase process begins with a customer who needs a facility. When looking to buy construction items, the nature of the risk is influenced by this process. Since the purchased item is to be produced in construction, there are many complex issues that may lead to a failure to complete the project in a workable and professional manner. Figure 8.1 shows how a project has been initiated by the authority to organize a project.

8.4.2 DESIGNING THE PROJECT

During this phase, detailed design work is developed for the foundation, floor plans, structure, mechanical systems, electrical systems, interior and exterior finish, and landscaping.

Review requirements: The first step is to review the requirements for the project.

Establish a budget: The cost estimates need to be converted into a detailed budget that can be used to guide the design process.

Reconsider site characteristics: Additional soil samples may be taken to determine the type of foundation that is needed.

Review preliminary plans: If modifications need to be made, the owner is consulted before work proceeds.

Prepare detailed drawings and specifications: Drawings describe the size and shape. Specifications are written descriptions of the type and quality of materials and workmanship needed for the project.

The construction process can be divided into the following stages.

1. **Conceptual stage**: The client identifies the need for the construction item and appoints and briefs consultants to study the client's requirement, and proposes a design outline to evaluate the efficacy of the project.
2. **Design stage**: The project concept is further developed and documentation on production and contracts is prepared and bids are called.
3. **Construction stage**: Production programs and construction on site are prepared.
4. **Operation and maintenance stage**: The completed building or works are maintained, prepared, or altered during its lifetime.

8.4.3 DESIGN METHODOLOGY

Although the conceptual design process can be formal or informal, it can be characterized by a number of measures: formulation, analysis, search, decision, specification, and modification. However, these actions are highly interactive at the beginning of the development of a new project as shown in Figure 8.2. Many redesign iterations are expected to refine

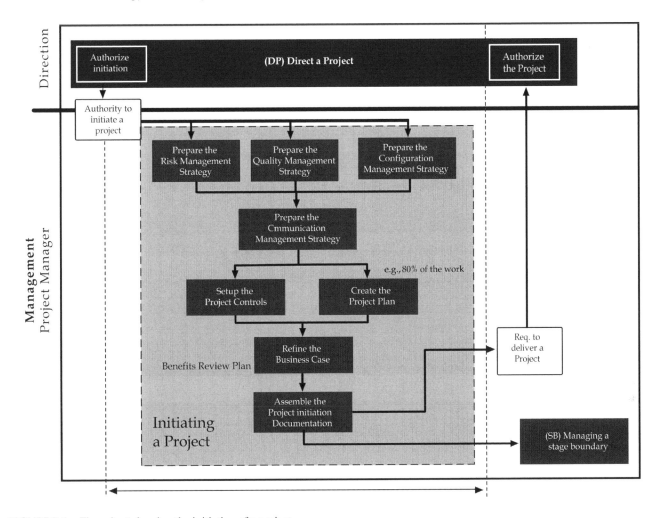

FIGURE 8.1 Flow chart showing the initiation of a project.

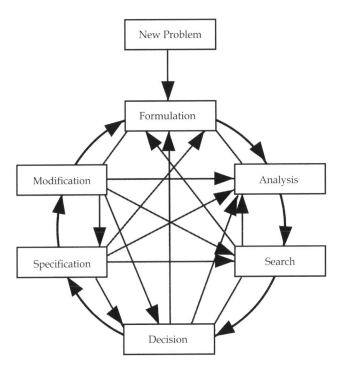

FIGURE 8.2 Conceptual design process.

the functional requirements, design concepts, and financial constraints; although at this stage the analytical tools used to solve the problem can be very crude.

The following can be described as the series of actions taken during the conceptual design process:

- Formulation refers, in general terms, to the definition or description of a design problem by synthesizing ideas describing alternative facilities.
- Analysis refines the definition or description of the problem by separating important information from peripheral information and combining the essential details. As part of the analysis, interpretation and prediction are normally required.
- Search involves the collection of a set of potential solutions to perform the functions specified and meet user requirements.
- Decision means that each of the possible solutions is assessed and compared with the alternatives until the best solution is reached.
- The specification is to describe the solution chosen in a form that contains sufficient details for its implementation.

- Modification refers to a solution change or re-design if a solution is found to be desirable or a new information is discovered during the design process.

8.4.4 Functional Design

The purpose of the functional design of a proposed facility is to treat the facility as a complex system of interrelated spaces. The spaces are systematically organized according to the functions to be performed in these spaces to serve a collection of needs. The arrangement of physical spaces can be regarded as an iterative design process to find an appropriate floor plan to facilitate the movement of persons and goods related to the intended operations. A designer often relies a lot on a heuristic approach, i.e., the implementation of selected rules or strategies to encourage research to find a solution. The heuristic approach used for the arrangement of spatial arrangements for installations is generally based on the following considerations:

- identification of goals and task constraints,
- the current status of each task during the iterative design process,
- assessment of differences between current and target status, and
- means of targeting the search efforts on the basis of past experiences.

The procedure for searching for the objectives can therefore be recycled iteratively to make trade-offs and thereby improve the solution of spatial layouts.

8.4.5 Preliminary Design

The first phase of the design process is preliminary design. The project planner will coordinate a number of meetings with the users and the design team for the collection of information. What can be expected is given in the following:

- The user group shall appoint a departmental administrator to serve as a department-planner link.
- Users communicate effectively the specific needs/requirements and the design team will conduct field research on the layout and impact of the existing areas in question, including the building systems.
- The design team creates schemes based solely on information and field investigation collected from users.
- Schemes will be reviewed and refined by all stakeholders accordingly. This project phase defines design and overall layout parameters.
- Any changes to the agreed and authorized scope, work schedule, or budget of the project shall be endorsed by the planner and the relevant administrative bodies with all the consequences understood and accepted by the source of the project funding.

8.4.6 Design Development

Design development phase includes the intense development of all the major project design components. Quite often it is the user who ensures that all the user-specific requirements are met by the design.

- During the design development phase, the planner will work with the design team to keep the scope in line with the initial approval plus any changes incorporated during the design development phase.
- The design team must receive input from a number of different groups and coordinate with them.
- At the end of the design development phase, an estimate will be prepared to verify that the project remains within the budget.
- **Swing space**: Before starting construction, if the project area is occupied, the occupants must be relocated to a third location. The temporary location or "swing space" must be clearly identified pretty early in the design process, so that movements can be made in a timely manner in order to start the construction as planned.
- **Equipment**: By the beginning of design development, all equipment requirements from the preliminary design phase documentation on new and existing equipment will have been clearly established. The equipment schedule, which documents the key requirements of all equipment, will be updated when necessary during the design development. Any changes must be timely in order to avoid procurement delays.

8.4.7 Construction Documents and Bidding

The main objective of this work phase is to verify that all the information is on the drawings and specifications so that the project can be tendered.

The planner will begin the transition of project management to a Project Manager within the Facilities Office. The project manager will supervise the implementation of the project from building to occupancy. The project will be tendered to a Constructor who will then provide a Guaranteed Maximum Price to be matched with the construction budget of the project before construction begins.

Early demolition: Demolition is often carried out as a prerelease of construction as part of this phase of the work. This helps the design team to identify field conditions more clearly. Early demolition identifies potential problems early. This ensures that the problems are dealt with during the project design phase itself and not during the construction phase when changes are quite expensive and affects the schedule. Early demolition thus streamlines the construction process and makes the construction offers more competitive, as the offers are more accurate and generally lower with less unknown contingency.

8.4.8 Types of Contractors

The construction phase of the project involves the project implementation, as described in the contract documents (building drawings and specifications). The project manager enters into a contract with the contractor on the basis of the agreed work price. It is important to choose the contractor(s). There are various types of contractors that all have their own expertise and are suitable for various types of projects. The contractor, on request, provides a portfolio of previous jobs, giving an idea of his level of expertise. A modern building requires the know-how of dozens of contractors to complete it. Although many people have skills in various fields, union regulations often prevent them from working in more than one area. The general contractor, who acts as a liaison between the customer and the building activity, is responsible for effective coordination of the activities of the various trades.

a. **General contractor**: A general contractor supervises the construction project from start to finish. This can involve the creation of a building from the ground up or a renovation project involving any trade combination. General contractors are often carpenters, as carpentry tends to involve many other businesses.

b. **Electrician**: Electricians install breaker boxes, cables, and fixtures and deal with power companies to turn off grid power so new systems can be securely hooked up. Electricians are also involved in the installation of alternative and decentralized solar, wind, and geothermal energy systems.

c. **Plumber**: Plumbers are responsible for building water systems, including the connection of a house to a well or municipal water mains, the installation of water heating equipment, and the construction and maintenance of septic and drainage systems.

d. **Heating and ducting**: Heating and ducting contractors install and operate furnaces, boilers, and extensive piping and ducting to heat a building. They can work with electricians and plumbers because their systems are interlocked.

e. **Drywaller**: Drywallers install drywall on framing and smooth wall tapes, spacers, and sand. On a well-run work site, drywallers are called in as soon as the plumbers and electricians have finished their work, so that the work can continue quickly without the different trades stumbling over each other as they try to do their jobs.

f. **Painter**: Painters start their work as soon as the drywaller has finished the drywall sanding. They apply finishing coats to the tastes of the customer.

g. **Finish carpenter**: The final subcontractor working on the interior of a building is the finish carpenter. Finishing carpenters install door and window fittings, baseboards, and other interior fittings, which can include crown molding and wainscoting.

h. **Other contractors**: Other contractors may include pool builders, masons, roofers, excavators, land-scrapers, cabinet makers, and interior design consultants, depending on the nature of the work being carried out.

8.4.9 Types of Contracts

A summary of different methods for implementing construction would include the following basic approaches and contracting procedures.

a. **Lump-sum contract**: This is a traditional method in which the owner, having selected an Architect/Engineer (A/E), has a set of definitive documents defining the scope of work required. The A/E will define such requirements in the documents prepared by him. These documents must be complete in all details and project definition, to make it possible for the owner to seek lump-sum bids. This procedure has several advantages. They are as follows:

   i. Before the construction begins, the total construction cost is defined and a definite contractual commitment is established between the owner and the contractor.
   ii. Contractual relations are clearly defined between all parties, the owner and A/E, the owner and the contractor.
   iii. The scope and limits of the project are defined before construction begins.
   iv. The best price for work obtained through a competitive process with agreed work costs.

   The definitiveness of lump-sum contract makes it the established method of contracting and in many instances the only method utilized, because of the statutory requirements. There are, however, distinct disadvantages which cannot be ignored. They are as follows:

   i. The lack of time flexibility in this method could lead to a delay in the implementation of the project, which could result in an increase in prices due to an escalation.
   ii. This method requires the competence of the appointed A/E rather than other methods, since all decisions concerning materials and systems are taken and fixed during the design phase. With other methods allowing more flexibility, it is easier to implement changes or revisions to effect cost reduction or necessary modifications to the initial design concept during later stages of work, whether design or construction.

b. **Cost plus fixed fee contract**: All costs in accordance with predetermined standards or in accordance with specific regulations are reimbursed to the contractor by the owner, except that the contractor is paid a fixed amount representing a profit. This amount is fixed at the time of awarding the contract. The variation in the lump sum fee is the percentage fee in which the fee is calculated as a percentage of the total costs incurred by the contractor in the installation. Although the initial scheme may be arrived at by using the percentage calculation, it is normally set as a fixed sum.

c. **Cost plus bid fee contract**: In this method, an owner will issue a request for proposals to selected number of contractors. The bid fee is established prior to contractor selection and is incorporated in the contract between the parties.

d. **Guaranteed maximum contract**: This method is often used as either a direct substitute for a lump-sum or cost-plus or modification of either contracting system. No owner wishes or is in a position to issue a blank cheque for building a new facility. If the budget is exceeded, new appropriations must be sought. The contractor agrees to carry out the work within a certain price ceiling before the start of the construction. This is the agreed maximum or guaranteed price. If the ceiling is exceeded, the excess costs shall be borne by the contractor. In contrast, if the work is done below the ceiling, the savings are shared by the owner and contractor on a predetermined ratio.

e. **Negotiated contract (competitive and noncompetitive)**: The end result of the negotiated contract can be any form of contracting previously described, lump-sum, or cost-plus. A typical situation might encompass a preselected list of contractors being invited to submit proposals describing:
   • Past experience ability in specific fields
   • Availability of personnel to carry out work within the time frame required
   • Estimate price (or quote)
   • Compensation method proposed, including the amount of fee

   The above factors would form the basis for the selection of a contractor, leading to the contract negotiation, as well as the funding source for the work. Again, depending on the context, there is an option for competitive selection from a limited and preselected group of contractors or noncompetitive negotiations.

f. **Unit-price contract**: This type of contract is based on estimates of the quantity of the work items. The costs per unit of each item are bid by the contractor and the owner is responsible for the estimated quantity of these items. The works are performed in line with the contract documents. The total construction cost can only be determined by the owner after the project has been completed. The contractor is obliged to perform at his quoted unit prices the quantities of the work actually required in the field, whether they are greater or lesser than the owners' estimate. This method is also called K-2 form of contract.

g. **Design build**: As the term implies, this approach establishes a single administrative, management, and professional responsibility for the two separate functions of design and construction. The owner enters into one agreement for both. The method of contracting can be any of the traditional methods or modifications previously described.

h. **Turn-key contracts**: As with the design build method, turn-key construction utilizes a single contract for all functions. There is one administrative, management, and professional responsibility for design and construction. There is a single party under contract to an owner to fulfill these functions in addition to other function that may be necessary to implement a project. These may include site selection, land acquisition, and financing all tasks that may be necessary to make the turn-key complete. As the term implies, it is an abbreviation for "turn the key" comprising all the functions required to enable an owner to turn the key (open the door) and start operating his newly developed and acquired facility.

8.4.10 Managing Construction Projects

Managing Construction Projects proposes new ways of approaching about project management in construction, to analyze the skills required to manage uncertainty, and to offer techniques to think about the challenges involved. Project management is essentially an organizational innovation that identifies a person or a small team responsible for ensuring that the project mission is effectively delivered to the clients. The activities involved are as follows:

   i. investigation,
   ii. planning,
   iii. execution/construction, and
   iv. operation of the complex system.

During the course of all these processes, several complicated and difficult issues may arise. The engineer has to take important decisions on such issues on intuitions, insight, perceptions, and experience. The goal can be achieved with the help of different groups of people working together in an organized manner resorting to easy management practices.

8.4.11 Preparing the Site

In preparing the site for construction works, geotechnical reporting, site clearing, excavation, grading, and compaction are included. In order to build an excellent project, the condition of the project site including the surface and surface condition must be carefully examined and evaluated. The overall assessment of the site may involve evaluating the present and installation of underground services, stipulating the appropriate basis depending on the recommendation of the geotechnical report, anticipating the groundwater level, the grading quantity required for proper drainage to remove water from the structure, whether the site is difficult to excavate or not, the depth of freezing. In order to build the structure in accordance with the design, the excavation volume must be estimated accurately and appropriate drainage, structural elevations, and layout must be carried out with immense accuracy. Stages or steps to prepare the construction project site are as follows:

- geotechnical report on the soil properties of the site
- building site clearing and excavation
- project site grading
- compaction of the project site

8.4.12 Geotechnical Report Related to Site Soil Properties

The geotechnical report communicates the condition of the project site and the design and construction recommendations. Therefore, a geotechnical report on the soil of the site is a must to understand the properties and condition of the soil of the project site. This report usually describes the soil property and recommends it. It is produced on the basis of a series of soil tests. The structure type determines the sampling method, the test type, and the number of tests required. The geotechnical reports provide information on the type of foundation suitable for the site, settlements and related recommendations, liquefaction possibilities, slope stability, groundwater level, soil bearing capacity, hazards associated with excavation, soil strength, soil classification, and much more. These valuable data properly define the properties and behavior of the soil in the future. If the project site is prone to earthquakes, the geotechnical report should include the necessary tests and recommendations.

8.4.13 Construction Site Clearing and Excavation

Clearing and excavation is part of the larger work that is done in the preparation of building projects. After the structure has been precisely laid, the excavation work begins and the soil is removed to the required depth in which the structure is built. At the project site, different types of machineries are used for excavating and transporting soil. The selection of the type of machinery used for excavation is based on the type of soil, the length of the distance to be transported, the load capacity of the soil, and the accessibility of the site. For example, blasting, drilling, and machineries such as boulders, backhoe, shovels, and scooper are used to dig and transport blasted and drilled materials when rocks are present.

8.4.14 Grading of Construction Project Site

Grading on a designed and built site is very important in order to force water away from the structure. The International Building Code (IBC 2009) provides the recommendations necessary to create the correct classification. The Code states that the grade slope should be at least one vertical unit to 20 horizontal units, that is, to say 5% at a distance of 3 m measured perpendicular to the wall. If a horizontal distance of 3 m due to physical hindrances is not available, other options must be used, such as swales and impervious surfaces, for which the lowest slope should be 2% if it is within the 3 m limit. Under certain conditions, the IBC code allows for a minimum slope of one unit horizontally to 48 units vertically. If the construction site is not flat, the appropriate cutting and filling must be carried out and the cutting and filling volume must be determined by the lowest level of the structure. During grading, final soil settlement should be considered. The ground under the foundation in the construction of the site must be compacted to the required degree, which is 90% of the maximum dry density according to the IBC code. Compaction of soil layers that support loads is essential because it reduces settlement and thus prevents unwanted incidents. Tamping, vibration, and rolling are types of loads used for compact soil layers. Several compaction machines are used at the construction site, such as a smooth wheel roller, sheep foot roller, rubber tire, screwdriver, and compactor plate. The compaction of the soil not only improves the sharpness but also decreases the soil permeability and compression. To achieve a target degree of compaction, optimum moisture content should be provided. Silt and clay are considerably sensitive in this respect, as incorrect moisture content leads to a failure of the process and poor compaction is achieved, which is usually unwanted. The optimum moisture content and compaction percentage are determined in the laboratory by either the standard modified compaction test or the Proctor modified test.

8.4.15 Building the Project

Building the project begins with the foundation and ends when the owner approves final payment to the contractor.

Laying the foundation: The type and size of the foundation are based on the size, weight, contents of the building, and the characteristics of the soil. Slab foundations are made with steel-reinforced concrete and are used most often on single-storey buildings.

Framing and enclosing: Once the foundation of the building is completed, the framework of the building is completed. Framing supports the entire structure. Houses often have wood frames. Many buildings have steel frames. On high-rise buildings, lower floors are enclosed before upper floors, allowing interior work to begin.

Installing building mechanical systems: Once a structure is enclosed, work begins on the building systems: plumbing, heating, ventilating, air-conditioning, electrical, communication, and safety and security systems. Plumbing begins first. Tubs and shower bases are also installed during this phase because walls and floor material must be fitted around these fixtures. The building sewer and water supply piping are also connected to the water and sewer mains at this time. The installation of heating, ventilation, and air-conditioning

(HVAC) systems can also begin after the building is enclosed. The electrical power system is usually installed after the plumbing pipes and HVAC ducts. In this way the electrical wiring will not be run in locations that interfere with the plumbing pipes and HVAC ducts.

Finishing the building interior exterior work: Installing trim around roof edges, windows, and exterior doors. Covering exterior walls with siding, brick, or stone. Building porches, patios, and decks. Installing driveways and walks.

Installing landscaping: Landscaping adds value to new structures. Following a landscape plan, workers will place plants, build retaining walls or raised beds, and create lawn areas. Irrigation system pipes and sprinkler heads are installed before planting. Grass areas are planted with sod or seed.

8.4.16 COMPLETING THE FINAL INSPECTION

Numerous inspections take place during the construction process. The plumbing and electrical systems must be inspected and approved by the appropriate inspectors before walls and ceilings are enclosed. The appropriate inspectors complete final inspections of the plumbing, HVAC, and electrical systems plus interior and exterior finish work. These inspections ensure that the work has been done according to the code. Any items that need to be changed, repaired, or replaced are added to a punch list and given to the contractor responsible for correction.

After the completion of the project, the superintendent notifies the architect or engineer that all is ready for a final inspection. The owner's representative, selected personnel from the architect's or engineer's office including the clerk of the works, and selected personnel from the general contractor's office, including the superintendent, make a complete inspection of the facility. Any correction is listed. This list is called a "punch list."

Final inspection involves five steps:

1. Reviewing standards in the contract, drawings, and specifications.
2. Making an inspection based on these standards.
3. Creating a punch list for defects that must be corrected by responsible party.
4. Making corrections—inspectors approve corrected items.
5. Signing approval forms after final inspection.

8.4.17 CLOSING THE CONTRACT

Closing the contract: Once the project is completed, and the requirements of the contract have been met, the owner makes final payment and the contract is closed.

Approvals: Building officials issue a completion certificate once all code-related items have been corrected in the punch list. The date on the certificate of completion is the date from which warranties become effective.

Final payment: The last step in the construction process is when the owner makes the final payment to the contractor.

The contract is then closed. Closing the contract also releases the contractor from any bonds required by the contract. The contractor's responsibility for the project is now limited to the responsibilities stated in the contractor's warranty.

The contractor must request the release of the final payment after all the deficiencies noted in the punch list have been dealt with. There is also a request for the release of retention funds. Final payment disputes can make or break a project. In order to make sure that the project tends to lean toward the former, it is really important to stick to the financial side—well before the work starts.

Building projects payment basically consists of two main areas, progress and final. The first payment, payment for progress, is made during construction and in view of further work. The second category, "final payment," includes the reimbursement of the remaining amounts due by the end of the work of a contractor or a subcontractor, as the name suggests.

8.5 CONSTRUCTION TOOLS AND EQUIPMENT

8.5.1 TOOLS

There are a number of construction tools for construction works such as concrete, brick masonry, leveling, wood works, floor works, plate works, laying of bricks, plastering, etc. Each building tool is necessary to achieve good results throughout the entire project. In addition to these tools, employees should also use certain safety tools to prevent unforeseen accidents.

Some important tools and their applications are listed below:

Bolster: Bolster is like chisel, which is used for cutting bricks. Its cutting edge is wider than the brick width. It is useful for cutting bricks accurately.

Boning stalks: Boning stalks are made of wood and T-shaped. It is used to level the excavated trench over its entire length. To level the trench surface, a minimum of three boning rods are used.

Brick hammer: Brick hammer is used for cutting the bricks and for pushing the bricks out of the line.

Bump cutter: Bump cutter is used to level concrete floors, foundations, and so on. It is called screed, too.

Chain lewis and Pin lewis: Chain lewis and pin lewis are two different tools used to lift heavy stones, particularly in stone masonry construction.

Chisel: Chisel is generally used in woodworking and must be useful for removing bumps of concrete or excess concrete from the hardened surface.

Circular screw: Circular screw for cutting wooden boards, frames, etc. It is used in less time when accurate cutting is necessary. It is safer than what you have seen.

Crowbar: Crowbar is used to dig the ground and to remove the ground roots, nails, and so on.

Digging bar: Digging bar is pin-shaped solid metal rod at the bottom. It is also used to dig hard ground surfaces.

End frames: Their use of the line and pins is similar. However, instead of pins, L-shaped frames are used at the end of the thread, which keep the bricks working efficiently and correctly aligning.

Float: Float is made of wood that smoothes the plastered surface of the concrete. It has a grip on the top and a smooth wooden surface on the bottom.

Hand screw: Hand screw is used for cutting wood materials such as doors, windows, panels, etc.

Head Pan: Head Pan is made of iron to lift the excavated soil or cement or concrete into the workplace, etc. More commonly used in building sites.

Hoe: Hoe is also used for excavating the soil, but the metal plate has an acute angle to the wooden handle in this case.

Line and pins: Line and pins are a thread with two solid metal rods connected to pin points. It is used to align the course of bricks while laying bricks.

Mason's square: The square of Mason is used to obtain a perfect right angle at the corner of the wall of masonry. It is the shape of "L." The first course is properly arranged using the Mason square and the first remaining layers of bricks are laid.

Measuring box: Measuring box is used to measure the amount of sand and aggregate used in concrete production. It has fixed dimensions, so that the aggregate does not have to be weighed at all times. The measuring box's general size is 300 mm × 300 mm × 400 mm (length × width × depth). The volume of the measuring box is usually 1 ft^3, making it easy to measure the concrete or mortar ratio.

Plumb Bob: Plumb Bob is used to verify structure verticality. It contains a solid metal bob that is attached to the thread end. Inn surveys are also used to level the instrument position.

Plumb rule: The plumb rule is used to verify the vertical wall line whether or not it is perfectly vertical. It contains a wood panel with uniform edges. In its center is provided a groove in which a plumb bob is located. When the rule is placed vertically with the wall, the plumb bob should be in the groove line.

Putty knife: Putty knife is used to decrease the thickness of the finish when the finish is thicker.

Spade: Spade is used to dig the ground for base trenches and so on. At the end of the long wooden handle, it contains metal plates.

Spirit level: Spirit level is made from wood or hard plastic in the middle with a bubble tube. The bubble tube is partially filled with alcohol. The air bubble is therefore formed in it. In brick masonry, the spirit level is used to check the surface level. The level of the spirit is placed on the surface and the bubble is controlled. The surface is lifted when the bubble settles in the tube in the center of the tube.

Trowel: Trowel is used in small quantities for lifting and applying cement mortar. It is made from steel and wooden handle for holding. The ends of the trowel can be pointed or wearing bulls.

Wheel barrow: Wheel barrow is used to carry bulk material weights such as cement, sand, concrete mix, etc. It contains one or two wheels at the front and two handles at the back of the wheel barrow.

8.5.2 Building Construction Equipment

Mechanization of building operations increases building productivity and results in saving of time and better quality of work. Some important building equipment are as follows:

Batching plants for concrete: In order to ensure uniformity of mix proportions in a concrete mix, batching plants are used. The cement is ordinarily measured by the number of bags (50 kg) and aggregates are batched by volume.

Concrete mixers: The essential feature of the ordinary batch mixer is a revolving drum fitted with blades to stir the materials. The mixers may be tilting and nontilting types. Tilting mixer is more efficient. The output per hour is affected by length of the mixing period and the time taken to discharge and charge the drum.

Wheel barrows: It is the most basic and most commonly used equipment for transporting materials on any construction work. Usually, wheel barrows will be of two nominal capacities, 60 L and 80 L.

Hoist: Chain hoist, platform hoist, and skip hoist are the three hoists mostly used for building operations.

Concrete pumps: Pumping concrete through steel pipe lines is one of the most satisfactory methods of transporting concrete. It completely eliminates the use of the barrows, skips, hoists, towers, buckets, etc. A considerable saving in time and labor is affected. The equipment consists of a heavy duty single-acting horizontal piston pump.

Concrete vibrators: Most of the high-frequency vibrators now available give at least 3600 impulses/minute and some of them produce frequencies of two or more times this number. They may be electrically driven or may be operated from a petrol engine or air compressor. Concreting vibrators are of four types: internal or spud or needle vibrator, surface vibrators, external or form vibrator, and vibrating table.

8.6 CONSTRUCTION SCHEDULING

Construction planning is a graphical representation showing the phasing rate of construction tasks with the start and finish dates and the sequential relationship between the various activities or operations in a project so that the work can be carried out in an orderly and efficient manner.

8.6.1 Purpose of Scheduling

Many decisions taken during the estimation process relate to the project completion schedule. In fact, a preliminary schedule is sometimes prepared during the estimate process. Once the contract has been awarded, the project will be guided by a much more comprehensive schedule. A schedule lists the sequence for the full implementation of tasks. Managers use the schedule to decide when employees, equipment, and materials are required. Adding daily progress information to the planned schedule warns managers of tasks that do not proceed as planned. Corrective steps should be taken to restore the project to schedule. The schedule can also be used to track money expenditure. Summary charts show how much money was spent in comparison with what was planned. As with everyday progress, the early identification of financial problems enhances the problem-solving skills.

8.6.2 METHODS OF SCHEDULING

Scheduling can be carried out using various methods depending on the project size. The methods used are as follows: (1) Bar charts or Gantt charts, (2) Milestone charts, (3) Network Analysis.

Bar charts: These types of charts were introduced by Henry Gantt around 1900 A.D. The bar chart comprises two coordinate axes, one of which symbolizes the work or tasks to be held out and the other of which actually represents the time, where each bar reflects one project particular task or occurrence. The start and end of each bar represents the start and finishing time of the activity, and the bar symbolizes the time required to achieve the activity. The given steps are involved in drawing up the bar chart:

- Split the project into a number of activities
- List the activity
- Find the connection between these activities. Systematically organize the activities
- Evaluate working quantity and the time needed
- Draw on the scale shown in Figure 8.3. Figure 8.3 indicates a bar chart for a project with seven different activities, namely, A, B, C, D, E, F, and G, to be finished in a timely and successful manner. The time required to complete these activities is 11, 6, 11, 8, 6, 9, and 16 time units.

Milestone charts: A change over the original bar chart is the milestone chart. In each activity, some major events are to be carried out in order to finish the activity. Such important events are referred to as milestones and represented by a square or circle. These events are instantly recognizable across the main bar that truly represents the activity. It was

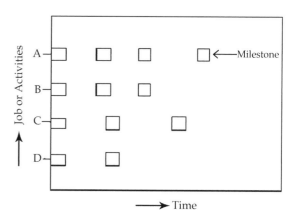

FIGURE 8.4   Milestone chart.

noticed that the details will be lacking if a particular activity represented by a bar is very long. However, if the activity is divided into a number of subactivities or major events, each of which can be acknowledged during the project's progress, it can be easily controlled and there is also some interrelation between the established activities. Let us consider the milestone chart shown in Figure 8.4. The activity or job A is divided into 4 key events or milestones, Job B into 3 milestones, and so on. Each milestone can be considered to be a specific event along the job and is represented by a square.

Network analysis: There are two methods of network analysis—Critical Path Method (CPM) and Project Evaluation and Review Technique (PERT).

- CPM: In this method, a project's activities and events are shown as a flow chart or network. The analysis of the network consists of paths, each of which shows a number of activities in order. The longest path to complete is the critical path (CP). This path requires the minimum time required to complete the project. A second path can be critical if the CP can be shortened. You can create more than one network for a project. Figure 8.5 shows a CPM network for a basketball hoop project. Every arrow is a task. The circles with their numbers are events. At the beginning and at the end of the event circle, the number identifies a task. If two or more arrows point to the same event, the next task cannot start until all the previous tasks are finished. Dashed lines show links but do not take longer. Many people can work simultaneously in major projects without interfering with one another. Working on several tasks at the same time reduces the total project duration. If only one task was carried out at a time, the project's duration would increase. Use the sequence below to build a CPM network. List each activity.
- Determine the sequence in which the work can be carried out
- Draw a diagram of the network
- Estimate how long each activity needs to be completed
- Determine the CP

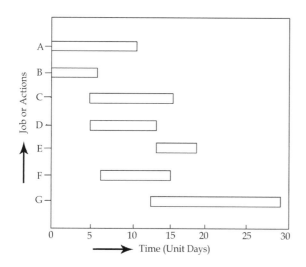

FIGURE 8.3   Bar chart.

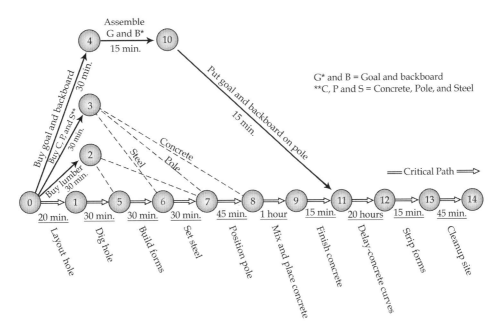

FIGURE 8.5 CPM network.

CPM software for helping planners build networks is available. The software totally reminds the planner of possible missed steps and calculates the critical route. Printers and plotters generate network printouts, tables, and graphs which elucidate CPM data. So once the job has been started, progress report data are entered into the computer and the software recalculates the CP. Managers are instantly notified to schedule problems. CPM network shows that different staff performs different tasks. Simultaneously note the special symbol used to show the CP.

PERT: To date, the analysis focused on determining the CP, slack events, and floats of activity. For this intent, that we really used single time estimates of the duration of the activity although previously three times for each activity were established. Now we consider the project duration variability. The discrepancy of the PERT analysis is measured by the standard variance or its square root. The variance of a number set is the average square difference between the numbers in the set and their average arithmetic.

8.6.3 Scheduling Workers

A schedule allows the contractor to identify the number and type of employees required for a project. The general contractor either directly hires workers or hires a subcontractor who hires the employees. If the project necessitates employees to be hired, the contractor contracts the suitable union and enquiries a certain number of employees. Construction companies normally recruit a number of employees, who are called regular employees; in practice, construction companies commonly have many projects in progress at any time. Contractors prefer to keep their regular employees consistent, as regular employees tend to be trustworthy employees.

8.6.4 Obtaining Materials

During the estimation process, the list of materials needed for the project is created. The contractor refers to the schedule to determine if different materials are required at the workplace. The contractor places orders for the materials based on this information. The goal is to obtain the materials when necessary. They should not arrive too early and should certainly not arrive. Small project materials can normally be purchased immediately, as they are normally available locally. Materials for major projects however must be ordered well in advance. For example, structural steel framing for a building must be manufactured specifically for the project and then shipped to the work site. It can take months for this process. Materials are obtained from stock or purchased as needed. Stock material is the building company's own and stored materials. Building companies often have two types of materials in stock: frequently used materials and materials from previous projects. Receipts or invoices supplied with the materials are submitted to the project manager to keep accurate records of the materials used at work.

8.6.5 Obtaining Equipment

The contractor utilizes the timeframe to ascertain if different types of equipment are required at the site. Contractors get equipment by buying, renting, leasing, or contracting a subcontractor's services.

8.6.6 Obtaining Permits

Licenses are written documents that give a company the permission to build, remodel, or repair. It is the general contractor's responsibility to apply for permits. In general, subcontractors are required to obtain the special permits they need to complete their work. Licenses are granted by local administration. Several types of permits are required at a building site. This permit enables construction to begin and maintain the inspection schedule for the foundation, structure, and finishing work. Specific licenses for plumbing, electrical, and HVAC systems are required. Permits may also be required for road entrances and for water, sewer, and natural gas connections.

8.7 BEGINNING CONSTRUCTION

8.7.1 SITE PREPARATION

The preparation of the site involves the demolition or wreckage of buildings and other structures, the clearing of construction sites, and the sale of materials from demolished buildings. The preparation of the site also involves blasting, testing drilling, landfilling, leveling, moving earth, excavation, drainage, and other land preparation.

8.7.1.1 Establishing Site Preparation

Constructing boundary fencing is temporarily used at building sites to mark clearing and grading limits and to describe areas to be safeguarded. Building borders contain and reduce disturbed areas, protect trees and vegetation, and prevent invasion of other sensitive areas. Construction border fencing is not a border fencing property but marks the perimeter of the construction disturbance area.

The process of marking the position of building on the selected site is called setting out of the building. For setting out the foundation lines, the engineer has to prepare a foundation plan to a convenient scale with all the measurements and then transfer these lines to ground. The setting of building on the ground may be done by center line method.

8.7.1.2 Providing Access to Site

Construction sites can be dangerous areas and access should be granted to authorized personnel only. Risks to unauthorized staff include:

- Falling materials or tools
- Falling into trenches
- Falling from height
- Being struck by moving vehicles and plants
- Standing on sharp objects
- Electricity contacts or hazardous materials
- Staining, noise, and vibration

Building sites may also be vulnerable to vandalism, robbery, arson, protests, suicides, etc. Construction sites, however, pose a challenge in securing access as

- their nature and layout are subject to frequent modifications;
- a large number of contractors, suppliers, consultants, and so on require access;

- they often find themselves in highly populated areas;
- user access to neighboring sites or parts of the site itself may need to be maintained; and
- there can be time pressures to quickly complete the work.

8.7.1.3 Clearing the Site

The clearing of the site is the process to clear the vegetation and surface soil of the building site. A successful site clearing process involves several steps. When the area to be cleared is designated, the first step is to remove the vegetation. This starts with undergrowth. Large vegetation, such as trees and shrubs, must be cleaned after the undergrowth has been cleared. The trees are cut to leave tall stumps that can be removed more easily. Stumps are removed by machinery. Any trees and shrubs within 30 ft of the building site must also be cleaned.

8.7.1.4 Locating a Structure

A detailed investigation and observation of ground conditions and stability is vital to evaluate whether a site is ideal for a building. Inadequate bearing capacity of the soil can lead to minor or major building failures.

Preliminary survey: This conducts a preliminary site survey including general landforms, flooding, evidence of risk of landslide or subsidence, types of soil for load-bearing capacity, drainage and runoff, water table and presence of natural springs or waterlogged soils, proximity of the site or proposed construction of excavations or exposed banks, expansive clay presence, prior use of the site, such as buried structures, contamination, earthworks, and uncompacted filling. Proof of good soil actually includes where foundations of adjacent buildings show any signs of settlement or inadequate bearing, no proof of landslides in the area, no direct evidence of buried services, and no organic soil, turf, or soft clay.

8.7.1.5 Locating Temporary Buildings

A "temporary building" is a building or structure that can be used up to 12 months. Boat shelters, storage containers, sales offices, greenhouses, or any other similar type of portable building or structure may be included. Three sets of drawings shall be submitted by individuals who wish to obtain a permit to construct or locate a "temporary building":

- Site plan showing the location of the "temporary building" including all buildings or structures on the site, access to firefighting, moisturizer, parking, services.
- Building floor plan showing the size and use of all areas including, where applicable, ramps and washrooms for disabled persons.
- Elevations showing the size and location of all external openings—concerning spatial separation.
- Details of the wall/floor construction to determine the fire-resistance requirements, wherever applicable.

- Structural details signed by a professional anchoring engineer and any additional structural work, including appropriate cover letters. Further information may be required for addressing cooking equipment, propane cylinders location, etc.
- If the building contains plumbing devices, drainage, vents, and water piping plans drawn and sized in accordance with plumbing codes are required.

8.7.1.6 Securing the Site

An unsecured construction site is a playground for vandals and thieves. Copper pipes, air-conditioning units, water heaters, shingles, and even trailers left on construction sites (among other supplies, tools, and assets) may be a temptation for them. The safety of the building site should be a priority between the construction equipment on site and the investment in the construction project.

Tips to ensure a secure building site:

Store equipment properly: Keep all unused equipment and tools secure. Building vehicles and keys should be locked when not in use. Locks should be mounted on fences and sheds.

Authorized personnel only: Limit access to the site and only provide the necessary personnel with site keys. Provide only one access point to the job site if possible. This allows you to monitor the traffic from and to the site closely. All workers on the site should be able to access the site with identification and credentials.

Use proper lighting: Maintain well-illuminated construction site and areas around all trailers, access points, and storage areas. Motion-sensitive lighting is ideal, particularly in areas outside the public eye. Lights are often all it takes to dissuade criminals.

Install adequate fencing: Close the entire building site properly and show signs "No Trespassing" around the site. A fence must not only deter a criminal from accessing the building site but should also prevent anyone who has unauthorized access to the site from removing anything. Barbed or electric wire and an alarm also help to dissuade theft. If you cannot close the whole site at least, install proper fencing around construction trailers and storage areas.

Manage site and equipment effectively: Complete your inventory and enter or mark your company identification numbers on each property. Allocate accountability to a particular person or team for all the property assets on the project site.

8.7.2 Earthwork and Foundation

Earthwork is the modifying of earth in order to make it suitable for construction. The different forms of earthworks include as follows:

a. **Stabilization**: The soil stabilization process improves the engineering properties of soil and therefore makes it more stable. It is required when the soil is not suitable for the intended purpose. Stabilization in its true sense includes compaction, preconsolidation, drainage, and many other processes of this nature.

b. **Loosening**: For earth works if the soil is hard, it should be loosened in order to make the work easier and better. Hence, the hard earth must be loosened either by blasting or air hammering.

c. **Excavation**: For the construction work of a foundation, earthwork in excavation to the needed depth is mandated. Different soil layers may be found during excavation; dewatering may sometimes be necessary. The sides of the trenches are kept vertical to a depth of 2 m from the bottom in soft or firm soils. The excavation profiles must be enlarged for more depths by allowing steps of 50 cm on either side after every 2 m from the bottom. Excavation can also be carried out to provide a 1:4 slope (1 horizontal: 4 vertical). The natural drainage of the area shall be maintained during the excavation. Excavation is conducted from top to bottom. Undermining or cutting is not permitted.

In water-logged areas, excavation is made depending on the depth of the excavation, depth of the water table, and other factors. Different methods like building sand drains, building deep wells, chemical stabilization, electro-osmosis, etc., are used to dig foundations in water-logged areas.

8.7.3 Damp Proofing of Foundation Walls

A cover on the outside of the foundation wall prevents the movement of water and water vapor through the wall. To keep cracks tight, the walls should be strengthened and concreting should be done. A low water-cement ratio mix should be used in order to reduce the shrinkage. Damp proofing is done as an external coating on the exterior side of the wall. New concrete should be cured and dried for 7–14 days, and some admixtures can be applied to green concrete.

Another simple treatment is spraying of an asphaltic material with a roller or a brush. Some other spray-on treatments require a priming coat and one or more final layers of a liquid-rubber, elastomeric coating. Bentonite waterproofing panels are also adopted for very wet foundations. These panels are made of sandwiched bentonite clay in carton. This clay is very expansive and forms an impenetrable barrier against infiltration when it comes into contact with water.

8.7.4 Building the Super Structure

Super structure is the part of the structure above the ground level. Columns and walls are generally constructed in super structure. The important parts of super structure are as follows.

8.7.4.1 Floor

A floor is usually the lower horizontal surface of a room or the underlying structure.

Functional floor requirements are as follows:

1. The main function is to support the occupants, the furnishings, and equipment inside the building.
2. A good floor should have adequate fire resistance, sound insulation, thermal insulation, and humidity insulation.

The various types of floor structures are as follows:

a. **Solid floor**: It is made of reinforced or plain concrete. The thickness of the slab varies depending on the load to be carried by the floor and the bearing capacity of the ground.
b. **Suspended floor**: It may be constructed of wood, reinforced concrete, or steel. It is actually a ground floor with an empty space (void) beneath the structure. It can be made of precast concrete panels, cast in-situ with reinforced concrete or timber joists.
c. **Timber floor**: It is a light self-weight and a dry construction form. Due to its light weight, it is economical, especially when the loads imposed are small.
d. **Concrete floor**: It has high strength and good fire resistance. Different types are in-situ solid concrete floors, in-situ hollow block floors, and precast floors.

8.7.4.2 Walls

The vertical members of a building that enclose, separate, and protect the interior spaces are called walls. There are two types of walls: (1) load bearing wall and (2) nonload bearing wall. A load-carrying wall is part of the building's structure which carries loads and transfers it downward. A nonload bearing wall is only a partition which divides a building's different spaces.

8.7.4.3 Roof and Ceiling Framing

The roof is the top part of a building and is built to protect the building from rain, snow, hail, direct sunlight from the wind, etc. Depending on the form and the material used, there are different types of roof. A ceiling is an overhead inner surface covering a room's upper limits. It is not usually regarded as a structural element but as a finished surface that hides the underside of the roof structure or the floor above. Ceilings can be tastefully decorated.

Basic roof design types are as follows:

1. **Flat**: must have a slight drainage slope
2. **Shed**: meeting with two slopes on a ridge; two walls stretch to the ridge
3. **Hip**: two gables, a pyramid may be regarded as a hip roof
4. **Gambrel**: the typical barn roof has four slopes in one direction
5. **Mansard**: two gambrels, basically what the hip is to gable is to the gambrel.

Pitched or sloping roof design: A roof is referred to as a pitch, if its slope exceeds 10°. In areas with heavy rainfall, a steep roof quickly throws off the rain, while a less steep roof,

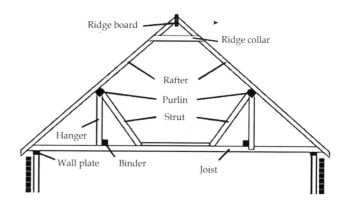

FIGURE 8.6 Typical section of a pitched roof.

not more than 35°–40°, preserves a useful insulating blanket of snow during the cold season, but allows the water to run freely. The different types of pitched roof are: (1) lean-to roof, (2) couple roof, (3) couple close roof, (4) collar beam roof, (5) collar and tie roof-double or purlin roof, (6) king post truss, (7) queen post truss, (8) mansard truss, (9) bel-fast truss, (10) steel truss, and (11) composite truss. The parts of a roof truss are shown in Figure 8.6.

Flat or terraced roof design: A roof is flat when its slope is less than 10°. Flat roofs are traditionally used in hot climates where the accumulation of water is not an issue. In the northern climate, they were generally unknown before the end of the last century. Commonly used flat roofs are: (1) Madras terrace roof, (2) Bengal terrace roof, (3) reinforced brick cement roof, (4) reinforced cement concrete roof, and (5) filler slab.

Building gable roofs with trusses: The gable roof, also known as the saddle roof, is easy to design and cheap to construct. Its symmetrical shape is attractive, efficiently sheds rainwater and snow and gives maximum space in the attic. It is a good choice in climates where snowfall occurs, but it may not work in areas with high winds

8.7.4.4 Installing Fascia and Sheathing

Fascia is an architectural term for a vertical frieze or band that forms the outer surface of a cornice that is visible to an observer. Sheathing is the board or panel material used in residential and commercial building floor, wall, and roof assemblies. The primary purpose of sheathing is to form a surface on which other materials can be introduced in any application.

8.8 INSTALLATION OF PLUMBING AND SANITARY FITTINGS

Plumbing is the science of creating and maintaining sanitary conditions in man-made buildings. It includes installing, repairing and servicing the pipes, fittings and equipment needed to supply water and to remove liquid waste from water.

8.8.1 Plumbing Systems

Plumbing system is used in buildings for water supply. It supplies water to the kitchen and toilet through the piping system.

The drainage system is used to eliminate human waste through a well-organized drainage pipeline network. For distribution system pipes, galvanized iron (GI), copper, high-density polyethylene, and chlorinated polyinyl chloride (CPVC) are generally used. Nowadays, mainly CPVC plastic pipes are used because they are not rusted, light weight, easy to install and maintain, and are economical. It includes fixtures and fixture traps; soil and waste pipes; ventilation pipes; building drain and construction sewer; and storm drainage pipes.

8.8.1.1 Piping System

A piping system is a configuration of various components with a solid joint technique to haul fluid from real source to final destination. The various components combined are defined as pipe components. They are pipe, pipe fittings, flanges, gaskets, bolts, and valves.

Two-pipe system: This method provides an ideal solution where the fixture cannot be closely fixed. Figure 8.7 shows the line diagram of a two-pipe system. Two pipes are provided in this system. One pipe collects foul soil from washrooms, while the second pipe collects foul water from the kitchen, bathrooms, washrooms, rain water, etc. The soil pipes (pipes carrying the soil waste) are connected directly to the drain, while the waste pipes (pipes carrying foul water) are connected via the trapped gully. All traps seen in this system are ventilated completely. Two-pipe system has separate stacks for carrying sewage and waste water.

a. Soil stack: All fixtures which carry human excreta, urine, or obnoxious wastes are connected to a separate stack known as soil stack, which is connected directly to a sewer line through a manhole. Each connected to the soil pipe must be provided with a trap.

b. Antisiphonic pipe: When fixtures are connected one on top of other as in tall buildings, high rate of flow from WCs (Water Closets) from upper floors may induce negative pressure and induce siphonage when the flow passes the fixtures in lower floors resulting in loss of trap seal.

This can be prevented by connecting the crown of the trap to a separate stack which terminates at roof level. Traps on all floors are connected to this pipe. This stack called as the antisiphonage pipe provides access to atmospheric air eliminating the possibility of generating negative pressure thereby maintaining hydraulic equilibrium in the system.

c. Waste stack: Fittings used for washing and cleanup, e.g., wash basins, bath tubs, showers, drinking water foundations, kitchen sinks and water from air conditioning machines, etc., are connected to a separate stack called waste stack. Many of these fixtures do not have their own traps but may discharge over a floor trap within the bathroom or kitchen.

One pipe system: Only one main pipe is provided in this system, which collects both foul soil waste and unfoul waste from the buildings. The main pipe is connected directly to the drainage system. If this system is provided in multistorey buildings, the washing blocks of different floors are placed over each other in such a way that the waste water discharged from the various units can be transported through short drains. This system is shown by line diagram in Figure.8.8. All traps of WCs, sinks of basins, etc., are fully ventilated

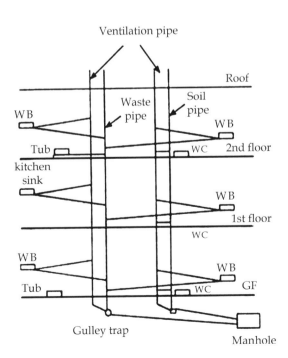

FIGURE 8.7 Two-pipe system.

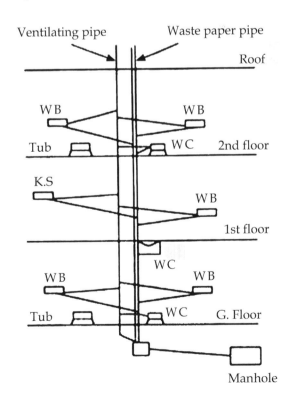

FIGURE 8.8 One-pipe system.

with the ventilation pipe connected. But all traps and waste pipes are totally exempt.

Single-stack system: This is similar to a single-pipe system; the only difference is that there is no ventilation in the traps. Figure 8.9 illustrates this system.

Single-stack partially ventilated system: This system is located between a single pipe and a single-stack system. Only one pipe is provided in this system to collect all kinds of foul and foul waste water. Only the water closet traps are ventilated by a relief vent pipe. This system is shown in Figure 8.10.

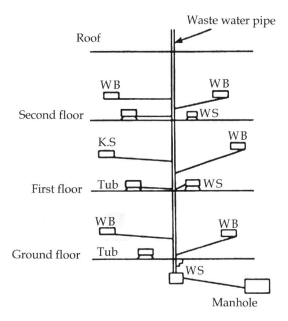

FIGURE 8.9 Single-stack system.

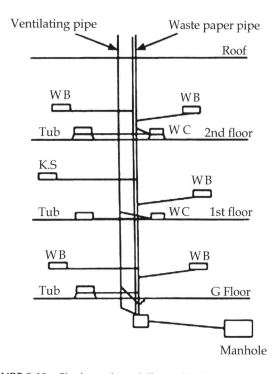

FIGURE 8.10 Single-stack partially ventilated system.

8.8.1.2 Materials for Residential and Light Commercial Piping System

PVC pipe: Polyvinyl chloride (PVC) pipe, by far the most commonly used pipe in residential homes, is the white pipe commonly used in plumbing applications. Easy and versatile with a variety of fittings and sizes, PVC is ideal for most applications in hot and cold water. PVC functions well for home plumbing piping because over time it does not rust or corrode. It is also easy to work with, because it does not require welding or metal-working and is a cheap option for your home. The main disadvantage of the PVC pipe is that it cannot be used for hot water. PVC will warp when exposed to hot water.

CPVC pipe: CPVC pipes are made of PVC supplemented with additional chlorine. It brings with it all the benefits of PVC additional durability. CPVC does not degrade with exposure to hot water and is safe to drink water. However, it is marginally more expensive than PVC and splits if it can freeze. CPVC could be utilized in many other implementations in which copper piping is used, and for a number of reasons, it is rated as higher option. CPVC is generally being used in applications where PVC properties are desired, but PVC cannot be used, including hot water supply, water supply, hot water drainage, waste and water pipes.

Copper pipe: Copper piping is the benchmark for most domestic plumbing applications since 1960s. It has long life and durability. It tolerates heat well and is exceptionally corrosion-resistant. Copper is not degraded with water and is therefore safe for drinking water use. However, this is one of the most expensive materials for piping and is at high risk of robbery on jobs or in vacant houses.

PEX pipe: PEX is an extremely flexible piping option that can literally be snaked through a home and bent around corners when needed. It does not likely require glue and in freezing conditions holds better than CPVC just because the materials can expand and contract. PEX is usually connected by stab-in or compression fittings necessitating a special tool. PEX can be split into existing pipes, including copper pipes, making it an excellent choice for adding and refurbishing. PEX is sufficiently durable for hot water applications but cannot be directly connected to the hot water heater. It must be connected to an 18-inch section of copper or other hot water-safe piping for hot water supply lines. Due to its flexibility and durability, the PEX pipe is excellent for retrofitting an older home, snaking through walls in a remodel, working in low-ventilation areas where glue is dangerous

Galvanized pipe: Galvanized piping is a zinc-coated steel or iron piping to help prevent rust and corrosion. Galvanized metal is frequently used in construction, but galvanized steel pipes can be used for plumbing. This specific type of pipe is absolutely best utilized for water lines because gas lines can corrode the zinc and harm the pipe or block the whole system. Galvanized pipes can last up to 100 years even in highly corrosive conditions. For the following applications, galvanized pipes are typically used: water supply lines, outdoor applications, underground applications, brass pipes. Brass piping is, after all, an option for certain applications.

8.8.2 Sanitary Fittings

Sanitary fittings are frequently being used in buildings to collect and eliminate the house drain. Sanitary fittings that are used in buildings are as follows:

Wash basins: Pottery or white glazed earthware or enameled iron is typically made of wash basins, etc. They are also made from pressed steel or plastic, sometimes. There are two types of wash basins—the back flat and the back angle. Normal wash basin is attached to the wall on brackets, while a pedestal is mounted on a wall-mounted pedestal. They are offered in various shapes and sizes. Standard sizes are 630 mm×450 mm, 550 mm×400 mm, and 450 mm×300 mm. Standard wash basin angle sizes are 600 mm×480 mm, 400 mm×400 mm. It has an oval bowl at the top with overflowing slot. The waste pipe is supplied at the bottom of the bowl with a metallic strainer.

Sinks: A sink is a rectangular tank used for the washing of utensils and glass goods in the kitchen or laboratory. These can be made from glazed earthenware, inoxidable steel, or pressed steel. The sink usually has an outlet of about 40 mm diameter. The pipe discharges water over a trap on the floor. The mouth of the outlet pipe is given with brass or nickel grating to prevent the entry of solid substances. Common cooking sinks size is 600 mm×400 mm×150 mm, 600 mm×450 mm×250 mm, 750 mm×450 mm×250 mm. Common sizes are 400 mm×250 mm×150 mm, 450 mm×300 mm×150 mm, and 600 mm×400 mm×200 mm.

Bath tubs: Bath tubs are usually made of enamel, porcelain, or plastic coated iron or steel. They may be parallel or confined sides. The usual size of the tub is 1.70–1.85 m; width: 0.7–0.75 m; depth: 0.6 m.

Water closets (W/C): A water closet is a sanitary installation engineered to directly obtain human feces and try to pass through a trap to the septic tank or underground sewer. It is probably connected to a cupboard flush and discharges the human excretion into the ground pipe. There are three types of water closets: (1) Indian type, (2) European type, (3) Anglo-Indian type.

Flushing cisterns: Flushing is achieved by storing the water required for each flush in a cistern placed above the WC. Cisterns placed about 1800 mm above the floor level provide enough pressure for efficient flushing of wash-down squatting WCs through a 32 mm diameter flush pipe. Some designs have low-level cisterns. An inverted bell with a mechanism to lift it momentarily by a lever or chain is installed over the flush pipe terminating at the high water level in a cistern. Operation of the lever creates a partial vacuum inside the bell thus enabling the atmospheric pressure to push the water in the cistern into the flush pipe by siphonic action. Siphonic cisterns operate only when the cistern is full of water and hence do not waste water by overflow.

8.9 INSTALLATION OF HVAC AND COMMUNICATION SYSTEMS

Equipment for HVAC (Heating, ventilation, and air conditioning) is used to heat and/or cool residential, commercial, or industrial buildings. The HVAC system may also provide fresh outdoor air to dilute indoor air contaminants such as inhabitant odors, volatile organic compounds emitted from indoor furnishings, chemicals used for cleaning, etc. A carefully designed system provides a reasonably comfortable indoor environment throughout the year. An HVAC system must be properly designed, sized, and installed for proper operation. A proper HVAC system offers a better indoor environment and minimizes operating costs. In the energy-efficient home planning process, everything must be done to control the heating and cooling load on the home before the HVAC system is designed.

8.9.1 Temperature Control

The most basic control system is a thermostat for heating and cooling. Programmable thermostats, also known as setback thermostats, can save households great energy. These programmable thermostats adjust the temperature setting automatically if people sleep or are not at home.

8.9.2 Humidity Control

Humidity is becoming more and more an issue for builders and owners. High indoor moisture leads to the growth of mold and mildew in the building. There are several methods to control indoor moisture. The easiest (and most costly) method is to connect a moisture system to an electric heater. When the humidity in the building exceeds the humidity level, the heater is switched on. The extra heat causes the air-conditioning system to run longer and eliminate more humidity.

8.9.3 Cleaning Air

Control of air purity is one of the functions of air-conditioning. It consists of removing unwanted particulate or gaseous matter from the air supplied to the conditioned space. The types of air-cleaning problems encountered in air conditioning can usually be classified with respect to the kind of dirt to be removed. Lint, course dirt, fine dirt, smoke, fumes, smog, smudging dirt, bacteria, etc., are the usual contaminants. Air cleaning is done by air cleaners which may be of following types.

a. **Impingement filters**: In impingement filters, impurities are trapped and retained by passing air through surfaces coated with viscous fluid. Expanded metal, glass fiber, steel wool, bronze or copper wool, hemp fibers are some of the materials used for the surfaces. The density of packing determines the air resistance through the filter. Density of packing goes on increasing toward the air-leaving side. They are made in units about 50 cm square.

b. **Dry strainer filter**: They have collecting surfaces made of cellulose, cloth, felt, glass fiber, or similar material. Their effectiveness in removing impurities depends upon the physical properties of the medium used. The air velocity through the filter varies from 3 to 15 m/mt.

c. **Electrostatic precipitation**: They are of recent origin and have proved very efficient. They consist of two parts—a charging chamber and a precipitating chamber. In first chamber, the dust particles are highly charged, and in second chamber, the charged particles are repelled by electric current and dust particles get clung to the plates.

d. **Air washers**: They consist of a GI steel casing with a door. Sprays are fitted at the intake to the casing. The air after passing through the casing becomes saturated. The water drops containing dust particles adhere to the scrubber plates and are washed down to the bottom of the washer. Air washers serve two purposes. They clean the air and also adjust the humidity. The air speed through washers usually range between 150 and 180 m/mt. The resistance of the air washer varies from 5 to 12.5 mm water gauge. The quantity of water required for spray nozzles varies from 480 to 800 L/1000 m³ of air.

8.10 ELECTRICAL POWER SYSTEM

An electrical power system is a network of electrical components used for power transmission, storage, and use. The grid that supplies power to an expanded area is an example of an electric power system. An electrical grid power system can be widely divided into the generators supplying power, the transmission system supplying power from the generating centers to the load centers, and the distribution system supplying power to nearby homes and industries. Most of these systems rely on three-phase AC power—the standard for large-scale transmission and distribution of power throughout modern times. In aircraft, electrical rail systems, ocean liners and automobiles, specialized power systems which do not always rely on three-phase AC power are found. An electrical supply system consists of three main components, i.e., the power plant, the transmission lines, and the system. Electricity is produced in power plants located in favorable locations, normally very away from consumers. It is then transmitted to load centers over long distances using conductors known as transmission lines. It is finally distributed through a distribution network to a large number of small and large consumers. The electrical supply system can be classified widely in (1) DC or just AC Overhead system (2) or underground system. Three-phase, three-wire AC. The system is universally adopted as an economic proposal for the generation and transmission of electricity. However, electrical power is distributed in 3 phases, 4-wire AC system. The underground system costs more than the overhead system.

8.11 LANDSCAPING

Landscaping is a process which changes a land area in one or all three categories: Plants—the addition of decorative, eatable, indigenous, or other plant types. Terrain—the shape of the land is altered by grading, backfilling, mounting, terracing, etc. Structures—building of fences, patio coverings, walls, decks, planters raised, or other building features. For aesthetic and economic reasons, humans manipulated the land. Adding plants, modifying the existing terrain and building structures are all part of landscaping. Landscaping today refers to the design, layout, and construction of gardens that improve the physical appearance and generate helpful space for outdoor activities around a house or building. Landscape is a task that brings science and art together. With horticultural knowledge and familiarity with landscape design elements and principles, a landscape expert can help you transform your property. The landscape architect, designer, or contractor with whom you choose to work guides you through the landscape design and construction phase. Landscape plans transform ideas into visuals and ensure that scale and layout are considered before construction. Your landscape designer may develop a plan for your yard that fits your needs and gives remedies to specific problems of landscaping, such as slopes, wind, sun, or space.

9 Concrete Technology

9.1 FRESH CONCRETE

Concrete remains in its fresh state from the time it is mixed until it sets. It is during this time that the concrete is handled, transported, placed, and compacted. Properties of concrete in its fresh state are very important because it influences the quality of the hardened concrete. Fresh concrete is that stage of concrete in which concrete can be moulded and it is in plastic state. This is also called "Green Concrete."

9.1.1 PROPERTIES OF FRESH CONCRETE

9.1.1.1 Consistency

Consistency of a concrete mix is a measure of the stiffness or fluidity of the mix. It indicates the easiness of flow of concrete. Slump test is commonly used to measure consistency of concrete.

9.1.1.2 Setting of Concrete

The hardening of concrete before its hydration is known as setting of concrete. It is the transition process of changing of concrete from plastic state to hardened state.

Following are the factors that affect the setting of concrete:

1. Water cement (w/c) ratio
2. Suitable temperature
3. Cement content
4. Type of cement
5. Fineness of cement
6. Relative humidity
7. Admixtures
8. Type and amount of aggregate

9.1.1.3 Workability

The workability of a concrete mix is the relative ease with which concrete can be placed, compacted, and finished without separation or segregation of the individual materials. As the strength of concrete is adversely and significantly affected by the presence of voids in the compacted mass, it is vital to achieve a maximum possible density. Presence of voids in concrete reduces the density and greatly reduces the strength, 5% of voids can lower the strength by as much as 30%. Slump test can be used to find out the workability of concrete.

Factors affecting workability of concrete are as follows:

i. **Water content or w/c ratio**: More the w/c ratio more will be workability of concrete. Addition of water increases the interparticle lubrication. High water content results in a higher fluidity and greater workability but reduces the strength of concrete. Increased water content also results in bleeding.

ii. **Amount and type of aggregate**: More the amount of aggregate less will be workability.
- Using smooth and round aggregate increases the workability. Workability reduces if angular and rough aggregate is used.
- In case of greater size of aggregate, less water is required to lubricate it. So extra water is available for workability.
- Angular aggregates increase flakiness or elongation and thus reduces workability. Round smooth aggregates require less water and less lubrication and greater workability in a given w/c ratio.
- Porous aggregates require more water compared to nonabsorbant aggregates for achieving same degree of workability.

iii. **Aggregate cement ratio**: More the aggregate cement ratio, lesser the workability and the paste will be stiff.

iv. **Weather conditions**
1. Temperature: If temperature is high, evaporation increases, thus workability decreases.
2. Wind: If wind is moving with greater velocity, the rate of evaporation increases and this reduces the amount of water and ultimately reduces workability.

v. **Admixtures**: Chemical admixtures can be used to increase workability. Use of air entraining agent produces air bubbles which acts as a sort of ball bearing between particles and increases mobility, workability, and decreases bleeding and segregation. The use of fine pozzolanic materials also has better lubricating effect and more workability.

vi. **Sand to aggregate ratio**: If the amount of sand is more, the workability will reduce because sand has more surface area. More contact area causes more resistance.

9.1.1.4 Bleeding and Segregation in Concrete

Bleeding in concrete is sometimes referred as water gain. It is a particular form of segregation, in which some of the water from the concrete comes out to the surface of the concrete, being of the lowest specific gravity among all the ingredients of concrete.

When the surface is worked up with the trowel, the aggregate goes down and the cement and water come up to the top surface. This formation of cement paste at the surface is known as "*Laitance*." In such a case, the top surface of slabs and pavements will not have good wearing quality. This laitance formed on roads produces dust in summer and mud in rainy season.

Segregation can be defined as the separation of the constituent materials of concrete. A good concrete is one in which all the ingredients are properly distributed to make a homogeneous mixture. There are considerable differences in the sizes and specific gravities of the constituent ingredients of concrete. Therefore, it is natural that the materials show a tendency to fall apart.

Segregation may be of three types:

a. Coarse aggregate separating out or settling down from the rest of the matrix.
b. Cement paste separating away from coarse aggregate.
c. Water separating out from the rest of the material being a material of lowest specific gravity.

9.1.1.5 Hydration in Concrete

Concrete derives its strength by the hydration of cement particles. The hydration of cement is a process which continues for a long time. Initially, the rate of hydration is fast. It continues over a very long period at a decreasing rate day by day. For the hydration to happen continuously, extra water must be added to recoup the loss of water on account of absorption and evaporation. Therefore, the curing can be considered as creation of a favorable environment during the early period for uninterrupted hydration. The desirable conditions are suitable temperature and ample moisture.

9.1.1.6 Air Entrainment

Air entrainment reduces the density of concrete and consequently reduces the strength. Air entrainment is used to produce a number of effects in both the plastic and the hardened concrete. These include:

a. Resistance to freeze–thaw action in the hardened concrete.
b. Increased cohesion, reducing the tendency to bleed and segregation in the plastic concrete.
c. Compaction of low workability mixes including semidry concrete.
d. Stability of extruded concrete.
e. Cohesion and handling properties in bedding mortars.

9.2 RHEOLOGY OF CONCRETE

Rheology measurements on concrete indicate that it is reasonable to approximate the concrete flow behavior using a Bingham model. The Bingham model, along with other common rheological models, is shown in Figure 9.1. Shear yield stress (y-axis intercept), τ_o, indirectly measures interparticle friction, while the viscosity μ (slope of the line) depends on the rheology of the paste and the volume fraction of aggregates. Other types of flow behavior are also possible, as indicated in Figure 9.1. "Shear thinning" means that the rate of increase of shear stress slows down with increasing shear strain rate; the opposite is true for "shear thickening" behavior. Thixotropy defined as the property exhibited by certain gels of becoming fluid when stirred or shaken and returning to the semisolid state upon standing. This is considered equivalent to the pseudoplastic behavior.

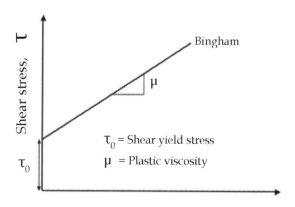

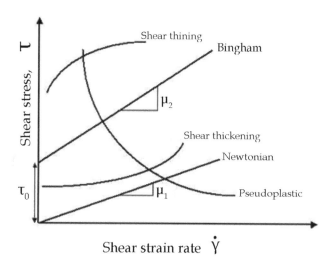

FIGURE 9.1 Common rheological model for fluids.

9.3 HARDENED CONCRETE

Fully cured, hardened concrete must be strong enough to withstand the structural and service loads which will be applied to it and must be durable enough to withstand the environmental exposure for which it is designed. If concrete is made with high-quality materials and is properly proportioned, mixed, handled, placed, and finished, it will be the strongest and durable building material.

9.3.1 PROPERTIES OF HARDENED CONCRETE

9.3.1.1 Strength

Compressive strength is the characteristic strength of concrete. Because, concrete is strong in compression but relatively weak in tension and bending. Compressive strength of concrete is measured in N/m^2. Compressive strength mostly depends upon amount and type of cement used in concrete mix, the water-cement ratio as well as the mixing, placing, and curing methods. Concrete tensile strength ranges from 7% to 12% of compressive strength. Both tensile strength and bending strength can be increased by adding reinforcement.

9.3.1.2 Creep

Deformation of concrete structure under sustained load is defined as concrete creep. Long-term pressure or stress on concrete can make it change its shape. This deformation usually occurs in the direction in which the force is applied.

9.3.1.3 Durability

Durability might be defined as the ability to maintain satisfactory performance over an extended service life. The design service life of most buildings is often 30 years, although buildings often last 50–100 years. Most concrete buildings are demolished due to obsolescence rather than deterioration. Appropriate ingredients, mix proportions, finishes, and curing practices can be adopted to increase the durability of concrete.

9.3.1.4 Shrinkage

Shrinkage is the decrease in volume of concrete caused by either drying or chemical changes.

9.3.1.5 Modulus of Elasticity

The modulus of elasticity of concrete depends on the modulus of elasticity of the concrete ingredients and their mix proportions.

9.3.1.6 Water Tightness

Water tightness is also called as the impermeability of concrete. It is directly related to the durability of concrete. The lesser the permeability, the more the durability of concrete.

9.3.2 Factors Affecting Properties of Hardened Concrete

9.3.2.1 W/C Ratio

W/c ratio is defined as the weight of water to the weight of cement in concrete mix. For normal concrete construction, a usual value of 0.45 is used. If the w/c is greater than 0.45, the excess water will remain as pore water in concrete. When concrete dries out it is filled with air and reduce the concrete density. As a result, it reduces the strength and durability of concrete.

9.3.2.2 Type and Amount of Cement

The primary factor affecting concrete compressive strength is the amount of cement. The higher the cement content, the greater the tendency for shrinkage cracks to form while the concrete is curing and hardening. Types of cement also have a great impact on the properties of hardened concrete.

9.3.2.3 Type and Amount of Aggregate

The strength of concrete mostly depends upon the strength of aggregate. The coarse aggregates act like bone in concrete. Increasing the amount of aggregate at a constant cement content reduces the concrete strength. Strength of concrete can also be affected by the type of aggregate. Angular and rough surface of concrete increases the concrete strength. Rough surface of aggregates makes strong bond between cement paste and aggregates. Round and smooth surface of aggregates decrease the strength.

9.3.2.4 Weather Condition

Increase of water temperature, either at mixing stage or during curing, affects the rate of strength gain of concrete. In cold climates, exterior concrete is exposed to repeated freeze–thaw cycles which can affect the strength of concrete adversely. Concrete expands and contracts with changes in moisture content. Excessive shrinkage can cause concrete to crack.

9.4 PRESTRESSED CONCRETE

In prestressed concrete, predetermined engineering stresses are given to members to counteract stresses that occur when the prestressed concrete unit is subjected to service loads. High tensile strands are stretched between abutments at each end of long casting beds. Concrete is then poured into the forms encasing the strands. As the concrete sets, it bonds to the tensioned steel. When the concrete reaches a specific strength, the strands are released from the abutments. This compresses the concrete, arches the member, and creates a built-in resistance to service loads.

9.5 PROPORTIONING OF CONCRETE MIXES

The process of adding relative proportions of cement, sand, coarse aggregate, and water, so as to obtain a concrete of desired quality, is known as the proportioning of concrete. The proportions of coarse aggregate, cement, and water should be such that the resulting concrete has the following properties:

1. When concrete is fresh, it should have enough workability so that it can be placed in the formwork economically.
2. The concrete must possess maximum density, or in the other words, it should be strongest and most watertight.
3. The cost of materials and labor required to form concrete should be minimum.

The determination of the proportions of cement, aggregates, and water to obtain the required strengths shall be made as follows:

a. By designing the concrete mix (design mix concrete), or
b. By adopting nominal mix (nominal mix concrete).

9.5.1 Types of Mixes

1. **Nominal mixes**: These mixes of fixed cement-aggregate ratio which ensures adequate strength are termed nominal mixes.

2. **Standard mixes**: The nominal mixes of fixed cement-aggregate ratio (by volume) vary widely in strength and may result in under- or over-rich mixes. For this reason, the minimum compressive strength has been included in many specifications. These mixes are termed standard mixes.

IS 456-2000 has designated the concrete mixes into a number of grades as M10, M15, M20, M25, M30, M35, and M40. In this designation, the letter M refers to the mix and the number to the specified 28-day cube strength of mix in N/mm². The mixes of grades M10, M15, M20, and M25 correspond approximately to the mix proportions (1:3:6), (1:2:4), (1:1.5:3), and (1:1:2), respectively.

3. **Designed mixes**: In these mixes, the performance of the concrete is specified by the designer but the mix proportions are determined by the producer of concrete, except that the minimum cement content can be laid down.

9.5.2 MIX PROPORTION DESIGNATIONS

The common method of expressing the proportions of ingredients of a concrete mix is in the terms of parts or ratios of cement, fine, and coarse aggregates. For example, a concrete mix of proportions 1:2:4 means that cement, fine, and coarse aggregate are in the ratio 1:2:4 or the mix contains one part of cement, two parts of fine aggregate, and four parts of coarse aggregate. The proportions are either by volume or by mass. The w/c ratio is usually expressed in mass.

9.5.3 METHODS OF PROPORTIONING CONCRETE

9.5.3.1 Arbitrary Method

The general expression for the proportions of cement, sand, and coarse aggregate is 1:n:2n by volume.

- 1:1:2 and 1:1.2:2.4 for very high strength
- 1:1.5:3 and 1:2:4 for normal works
- 1:3:6 and 1:4:8 for foundations and mass concrete works

Recommended Mixes of Concrete:

- M10—1:3:6
- M15—1:2:4
- M20—1:1.5:3
- M25—1:1:2

9.5.3.2 Fineness Modulus Method

The term fineness modulus is used to indicate an index number which is roughly proportional to the average size of the particle in the entire quantity of aggregates. The fineness modulus is obtained by adding the percentage of weight of the material retained on the following sieve and divided by 100. The coarser the aggregates, the higher the fineness modulus.

9.5.3.3 Minimum Void Method

The quantity of sand used should be such that it completely fills the voids of coarse aggregate. Similarly, the quantity of cement used should be such that it fills the voids of sand, so that for a dense mix, the minimum void is obtained. In actual practice, the quantity of fine aggregate used in the mix is about 10% more than the voids in the coarse aggregate and the quantity of cement is kept as about 15% more than the voids in the fine aggregate.

9.5.3.4 Maximum Density Method

Percentage of material finer than diameter d (by weight) is given by Equation 9.1

$$P = 100\sqrt{\left(\frac{d}{D}\right)} \qquad (9.1)$$

where D = maximum size of aggregate (i.e., coarse aggregate).
d = maximum size of fine aggregate

A box is filled with varying proportions of fine and coarse aggregates. The proportion which gives heaviest weight is then adopted.

9.5.3.5 W/C Ratio Method

According to Abram's w/c ratio law, the strength of well-compacted concrete with good workability is dependent only on w/c ratio. The optimum w/c ratio for the concrete of required compressive strength is decided from graphs and expressions developed from various experiments. If w/c ratio is less than 0.4–0.5, complete hydration will not occur.

- Some practical values of w/c ratio for structure reinforced concrete
  0.45 for 1:1:2 concrete
  0.5 for 1:1.5:3 concrete
  0.5–0.6 for 1:2:4 concrete.

Thumb rules for deciding the quantity of water in concrete:

i. Weight of water = 28% of the weight of cement + 4% of the weight of total aggregate
ii. Weight of water = 30% of the weight of cement + 5% of the weight of total aggregate

9.6 PRODUCTION OF CONCRETE

Concrete is a mixture of two components: aggregates and paste. The paste, comprising cement and water, binds the aggregates (usually sand and gravel or crushed stone) into a rocklike mass as the paste hardens.

Portland cement is made from a combination of a calcareous material (usually limestone) and of silica and alumina found as clay or shale. In lesser amounts, it can also contain iron oxide and magnesia. Aggregates, which comprise 75% of concrete by volume, improve the formation and flow of cement paste and enhance the structural performance of concrete. Aggregates can also be classified according to the type of rock they consist of: basalt, flint, and granite, among others. A wide range of chemicals called admixtures are added to cement to act as plasticizers, superplasticizers, accelerators, set retarders, dispersants, and water-reducing agents. Accelerators which are used to reduce the setting time, include calcium chloride or aluminum sulfate and other acidic materials. Plasticizing or superplasticizing agents increase the fluidity of the fresh cement mix with the same w/c ratio, thereby improving the workability of the mix as well as its ease of placement. Examples of plasticizers are polycarboxylic acid materials while those of superplasticizers are sulfanated melamine formaldehyde or sulfanated naphthalene formaldehyde condensates. Set retarders are used to delay the setting of concrete. These include soluble zinc salts, soluble borates, and carbohydrate-based materials. Gas forming admixtures, powdered zinc, or aluminum in combination with calcium hydroxide or hydrogen peroxide, are used to form aerated concrete by generating hydrogen or oxygen bubbles that become entrapped in the cement mix.

9.6.1 Manufacturing Process

First, the cement (usually Portland cement) is measured. Next, the other ingredients—aggregates (such as sand or gravel), admixtures (chemical additives), any necessary fibers—are measured and dry-mixed together. Then the required proportion of water is added to the dry mix and mixed thoroughly to form the concrete. The concrete is then shipped to the work site and placed, compacted, and cured.

9.6.2 Transport to Work Site

- Once the concrete mixture is ready, it is transported to the work site. There are many methods of transporting concrete, including wheelbarrows, buckets, belt conveyors, special trucks, and pumping. Pumping transports large quantities of concrete over large distances through pipelines using a system consisting of a hopper, a pump, and the pipes. Pumps come in several types—the horizontal piston pump with semi-rotary valves and small portable pumps called squeeze pumps. A vacuum provides a continuous flow of concrete, with two rotating rollers squeezing a flexible pipe to move the concrete into the delivery pipe.

9.6.3 Placing and Compacting

- Once at the site, the concrete must be placed and compacted. These two operations are performed almost simultaneously. Placing must be done so that segregation of the various ingredients is avoided and full compaction is achieved eliminating the air bubbles. Vibrators are used to densify the concrete.

9.6.4 Curing

- Once it is placed and compacted, the concrete must be cured before it is finished to make sure that it does not dry too quickly. As the cement solidifies, the concrete shrinks. To minimize this problem, concrete must be kept damp during the initial days it requires to set and harden.

9.6.5 Quality Control

Concrete manufacturers expect their raw material suppliers to supply a consistent, uniform product. At the cement production factory, the proportions of the various raw materials that go into cement must be checked to achieve a consistent kiln feed, and samples of the mix are frequently examined using X-ray fluorescence analysis. Quality control charts are widely used by the suppliers of ready-mixed concrete and by the engineer on site to continually assess the strength of concrete.

9.7 UNDERWATER CONCRETING

The most common method of handling concrete under water is by tremie. A tremie consists essentially of a vertical steel pipeline which is topped by a hopper. It is so long as to reach from a working platform above water to the lowest point of the underwater formwork.

During the underwater placing, it must be ensured that concrete should not be washed out by the water. The slump value lies within the range of 150–250 mm. Antiwashout admixtures, which enable the concrete to flow during pumping and make the concrete highly viscous when it is at rest, are also added. Concrete mix should contain high amount of cementitious materials that lie within the range about 360–400 kg/m^3. Underwater concrete should also contain 15%–20% pozzolans as they improve the fluidity and in turn workability of concrete.

9.8 CONCRETING UNDER EXTREME CLIMATIC CONDITIONS

Below freezing temperatures, freshly placed concrete may be damaged by the formation of ice within its pore structure. In very hot weather the concrete may stiffen prematurely, preventing it from being compacted and finished properly. Sometimes even thermal cracking can occur.

9.8.1 Hot Weather Concreting

High temperatures result in rapid hydration of cement, increased evaporation of mixing water, greater mixing water demand, and large volume changes resulting in cracks. Any operation of concreting done at atmospheric temperature above 400°C may be put under hot weather concreting. The effect of hot weather may be as follows:

a. **Accelerated setting**: A higher temperature of fresh concrete results in a more rapid hydration of cement and leads to reduced workability/accelerated setting. This reduces the handling time of concrete.

b. **Reduction in strength**: Concrete mixed, placed, and cured at higher temperature normally develops higher early strength than concrete produced and cured at normal temperature, but at 28 days or later, the strength is generally lower.

c. **Increased tendency to crack**: Rapid evaporation may cause plastic shrinkage and cracking and subsequent cooling of hardened concrete would introduce tensile stresses.

d. **Rapid evaporation of water during curing period**: It is difficult to retain moisture for hydration and maintain reasonably uniform temperature conditions during the curing period.

e. **Difficulty in control of air content in air-entrained concrete**: It is more difficult to control air content in air-entrained concrete. This adds to the difficulty of controlling workability. For a given amount of air-entraining agent, hot concrete will entrain less air than concrete at normal temperatures. It is desirable to limit the maximum temperature of concrete as 35°C to keep margin for increase in temperature during transit.

9.8.2 COLD WEATHER CONCRETING

Any concreting operation done at a temperature below 5°C is termed as cold weather concreting. The effect of cold weather concreting may be as follows:

a. **Delayed setting**: When the temperature is falling to about 5°C or below, the development of strength of concrete is retarded compared with that at normal temperature. Thus, the time period for removal of form work has to be increased.

b. **Freezing of concrete at early stage**: The permanent damage may occur when the concrete in fresh stage is exposed to freeze before certain prehardening period. Concrete may suffer irreparable loss in its properties to an extent that compressive strength may get reduced to 50% of what could be expected for normal temperature concrete.

c. **Stresses due to temperature differentials**: Large temperature differentials within the concrete member may promote cracking and affect its durability adversely. It is a general experience that large temperature differentials within the concrete member may promote cracking and have a harmful effect on the durability. Such differentials are likely to occur in cold weather at the time of removal of form insulation.

d. **Repeated freezing and thawing of concrete**: If concrete is exposed to repeated freezing and thawing after final set and during the hardening period, the final qualities of the concrete may also be impaired. In view of above, it is desirable to limit the lowest temperature of concrete as 5°C.

9.9 SPECIAL CONCRETES AND HIGH PERFORMANCE CONCRETES

9.9.1 LIGHT WEIGHT CONCRETE

One of the main advantages of conventional concrete is the self-weight of concrete. Density of normal concrete is of the order of 2200–2600 kg/m³. This self-weight will make it to some extent an uneconomical structural material. Only method for making concrete light is by inclusion of air. This is achieved by (1) replacing original mineral aggregate by light weight aggregate, (2) by introducing gas or air bubble in mortar, and (3) by omitting sand fraction from concrete. This is called no-fine concrete. Light weight aggregate includes pumice, saw dust rice husk, thermocole beads, formed slag, etc. Self-weight of light weight concrete varies from 300 to 1850 kg/m³. Light weight concrete has low thermal conductivity and high fire resistance.

9.9.2 AERATED CONCRETE

It is made by introducing air or gas into a slurry composed of Portland cement. No fine concrete is made up of only coarse aggregate, cement, and water. This type of concrete is used for load-bearing cast in situ external walls for building. They are also used for temporary structures because of low initial cost and can be reused as aggregate.

9.9.3 HIGH-DENSITY CONCRETE

Heavy-density concrete uses heavy natural aggregates such as barites or magnetite or manufactured aggregates such as iron or lead shot. The density achieved will depend on the type of aggregate used. Typically using barites, the density will be in the region of 3500 kg/m³, which is 45% greater than that of normal concrete, while with magnetite the density will be 3900 kg/m³, or 60% greater than normal concrete. Very heavy concretes can be achieved with iron or lead shot as aggregate is 5900 and 8900 kg/m³, respectively.

9.9.4 MASS CONCRETE

Mass concrete is defined in ACI as "any volume of concrete with dimensions large enough to require that measures be taken to cope with generation of heat from hydration of the cement and attendant volume change to minimize cracking." The one characteristic that distinguishes mass concrete from other concrete work is thermal behavior. Because the cement–water reaction is exothermic by nature, the temperature rise within a large concrete mass, where the heat is not quickly dissipated, can be quite high.

9.9.5 READY-MIX CONCRETE

Ready-mix concrete has cement, aggregates, water, and other ingredients, which are weigh-batched at a centrally located plant. This is then delivered to the construction site in truck-mounted transit mixers and can be used straight away without any further treatment. This results in a precise mixture, allowing specialty concrete mixtures to be developed and implemented on construction sites.

9.9.6 POLYMER CONCRETE

The impregnation of monomer and subsequent polymerization is the latest technique adopted to reduce inherent porosity of concrete and increase strength and other properties of concrete. There are mainly three types of polymer concrete.

9.9.6.1 Polymer-Impregnated Concrete

It is a precast conventional concrete cured and dried in oven or by dielectric heating from which the air in the open cell is removed by vacuum. Then a low-viscosity monomer is diffused through the open cell and polymerized by using radiation, application of heat, or by chemical initiation.

Mainly the following types of monomers are used:

a. Methyl methacrlylate
b. Acrylonitrile
c. t-butyl styrene
d. Other thermoplastic monomer

9.9.6.2 Polymer Cement Concrete

Polymer cement concrete is made by mixing cement, aggregate, water, and monomer. Such plastic mixture is cast in moulds, cured dried, and polymerized. The monomers that are used in polymer cement concrete are

1. Polyster-styrene
2. Epoxy-styrene
3. Furans
4. Vinyldene chloride

9.9.6.3 Polymer Concrete

Polymer concrete is an aggregate bound with a polymer binder instead of Portland cement as in conventional concrete. The main technique in producing PC is to minimize void volume in the aggregate mass so as to reduce the quantity of polymer needed for binding the aggregate. This is achieved by properly grading and mixing the aggregate to attain maximum density and minimum voids.

9.9.7 SHOTCRETE

Shotcrete is a process where concrete is projected or "shot" under pressure using a feeder or "gun" onto a surface to form structural shapes including walls, floors, and roofs. The surface can be wood, steel, polystyrene, or any other surface that concrete can be projected onto. The surface can be troweled smooth while the concrete is still wet.

It is defined as a mortar conveyed through a hose and pneumatically projected at high velocity on to a surface. There are mainly two different methods, namely, wet mix and dry mix process. In wet mix process, the material is conveyed after mixing with water.

9.9.8 PREPACKED CONCRETE

In constructions where the reinforcement is very complicated or where certain arrangements like pipe, opening, or other arrangements are incorporated, this type of concreting is adopted. One of the methods of concrete process in which mortar is made in a high-speed double drum and grouting is done by pouring on prepacked aggregate. This is mainly adopted for pavement slabs.

9.9.9 VACUUM CONCRETE

Concrete poured into a framework that is fitted with a vacuum mat to remove water not required for setting of the cement; in this framework, concrete attains its 28-day strength in 10 days and has a 25% higher crushing strength. The elastic and shrinkage deformations are considerably greater than for normal-weight concrete.

9.9.10 PUMPED CONCRETE

Pumped concrete must be designed to that it can be easily conveyed by pressure through a rigid pipe of flexible hose for discharge directly into the desired area. Pozzocrete use can greatly improve concrete flow characteristics making it much easier to pump, while enhancing the quality of the concrete and controlling costs.

9.9.11 HIGH-PERFORMANCE CONCRETE

High-performance concrete is a concrete mixture, which possess high durability and high strength when compared to conventional concrete. This concrete contains one or more of cementitious materials such as fly ash, silica fume or ground granulated blast furnace slag, and usually a super plasticizer. This enhances the strength, durability, and workability qualities to a very high extent.

American Concrete Institute defines High Performance Concrete as "A concrete which meets special performance and uniformity requirements that cannot always be achieved routinely by using only conventional materials and normal mixing, placing and curing practices." The main ingredients of high-performance concrete are cement (with little C_3A as possible because the lower amount of C_3A, the easier to control the rheology, and lesser the problems of cement-super plasticizer compatibility), fine aggregate (both river sand and crushed stones), coarse aggregate, water, chemical admixtures like plasticizers, super plasticizers, retarders, or air-entraining agents and mineral admixtures like fly ash, silica fumes, carbon black powder, or anhydrous gypsum-based mineral additives.

High-performance concrete can be used in severe exposure conditions where there is a danger to concrete by chlorides or sulfates or other aggressive agents as they ensure very low permeability. High-performance concrete is mainly used to increase the durability of concrete.

10 Reinforced Concrete Structures

10.1 FUNDAMENTALS OF REINFORCED CONCRETE

Concrete has gained so much importance and popularity because of the development of reinforced concrete. Introducing the reinforcing bars in concrete makes the concrete an excellent composite building material wh can resist significant amount of tensile stresses/strains also. Steel bars embedded in the tension zone of concrete make it able to resist tension.

10.1.1 DESIGN PHILOSOPHIES FOR DESIGN OF REINFORCED CONCRETE STRUCTURES

There are three design philosophies for the design of reinforced concrete and prestressed concrete.

1. **Working stress method (WSM):**
   It is the most traditional method for the design of reinforced concrete, structural steel, and timber. Basic assumptions in this method are:
   a. The material behaves in a linear elastic manner.
   b. Adequate safety can be ensured by restricting the stresses in the materials that are induced due to the application of working load/service loads. This assumption also introduces the concept of factor of safety which is expressed as:

   $$\text{Factor of safety} = \frac{\text{strength of material}}{\text{allowable stress in the material}}$$

2. **Ultimate load method (ULM):**
   In this method, the nonlinear stress–strain behavior of steel and concrete is accounted for and stresses induced in the structure at the verge of failure at ultimate loads are considered.
   The safety in the design of structure is taken care by the concept of load factor which is expressed as:

   $$\text{Load factor} = \frac{\text{ultimate or design load}}{\text{working or service load}}$$

   This concept makes it possible to assign different factor of safety to different types of loads and can be suitably combined.

3. **Limit state method (LSM):**
   This method takes into account the safety at ultimate load and serviceability at service loads. LSM employs different safety factors at ultimate and service loads. Limit state is the state of "about to collapse" or "impending failure" beyond which the structure is not of any practical use, i.e., either the structure collapses or becomes unserviceable. In LSM, two types of limit states are defined which are:

   a. Limit state of collapse:
      This limit state deals with the strength of the structure in terms of collapse, overturning, sliding, buckling, etc. Various limit states of collapse are Flexure, Compression, Shear, and Torsion.
   b. Limit state of serviceability:
      This limit state deals with the deformation of the structure to such an extent that the structure becomes unserviceable due to excessive deflection, cracks, vibration, leakage, etc. Various limit states of serviceability are deflection, excessive vibrations, corrosion, and cracking.

10.1.2 BASIC DEFINITIONS

i. Characteristic strength and characteristics load:
   Characteristic strength:
   Characteristic strength of a material is that strength below which not more than 5% of the test results are expected to fail. For concrete, this value is known as characteristic strength of concrete, and for steel, this value is known as yield strength of steel.
   Characteristic strength of concrete:
   Characteristic strength of concrete is denoted by f_{ck} (N/mm^2) and its value is different for different grades of concrete, e.g., M15, M25, etc. In the symbol M used for designation of concrete mix refers to the mix and the number refers to the specified characteristic compressive strength of 150 mm size cube at 28 days expressed in N/mm^2.
   Characteristic strength of steel:
   The characteristic value shall be assumed as the minimum yield stress or 0.2% proof stress. The characteristic strength of steel designated by symbol f_y (N/mm^2).
   Characteristic load:
   The characteristic load is that load which has 95% probability of not being exceeded during the life time of the structure.
ii. Partial safety factor for materials:
   The design strength of concrete and steel is obtained by dividing the respective characteristics strength by appropriate partial safety factors for concrete and steel.
iii. Partial safety factor for loads:
   Various load combinations for dead loads (DL), live load (LL), and earthquake load are computed and multiplied by a factor called partial safety factor for assessing the ultimate load.

10.2 DESIGN OF SINGLY REINFORCED SECTIONS

The LSM is the modification of the ULM. This takes into account the drawbacks of ultimate load theory (ULT). In LSM, the attention is given to the "acceptable limit" before failure as against the ULT which gives the load factor on failure load of the structure. In the method of design based on LSM, the structure shall be designed to withstand safely all loads liable to act on it throughout its life and also satisfy the serviceability requirements. The acceptable limit for the safety and serviceability requirements before failure occurs is called limit state.

The design values are derived from the characteristic values through the use of partial safety factor, one for material strength and the other for loads are used.

10.2.1 LIMITING DEPTH OF NEUTRAL AXIS

An expression for the depth of the neutral axis at the ultimate limit state can be easily obtained from the strain diagram in Figure 10.1.

Considering similar triangles,

$$X_u/d = 0.0035/\left(0.0035 + \left(0.87 f_y/E_s\right) + 0.002\right) \quad (10.1)$$

When the maximum strain in tension reinforcement is equal to $\left(0.87 f_y/E_s\right) + 0.002$, then the value of neutral axis will be X_{umax}.

$$X_{umax}/d = 0.0035/\left(0.0035 + \left(0.87 f_y/E_s\right) + 0.002\right) \quad (10.2)$$

The values of X_{umax} for different grades of steel, obtained by applying Equation 10.2, are listed in Table 10.1.

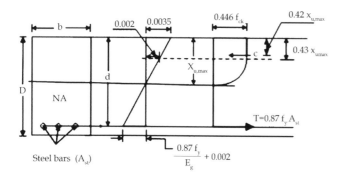

FIGURE 10.1 Strain diagram and neutral axis.

TABLE 10.1
Limiting Depth of Neutral Axis for Different Grades of Steel

Grades of Steel	Fe 250	Fe 415	Fe 500
X_{umax}/d	0.53	0.48	0.46

The limiting depth of neutral axis X_{umax} corresponds to the so-called balanced section, i.e., a section that is expected to result in a "balanced" failure at the ultimate limit state in flexure. If the neutral axis depth is less than X_{umax}, then the section is under-reinforced (resulting in a "tension" failure), whereas if exceeds X_{umax}, it is over-reinforced (resulting in a "compression" failure).

10.2.2 ANALYSIS OF SINGLY REINFORCED RECTANGULAR SECTIONS

Analysis of a given reinforced concrete section at the ultimate limit state of flexure implies the determination of the ultimate moment M_u of resistance of the section. This is easily obtained from the couple resulting from the flexural stresses shown in Figure 10.2.

$$M_u = C_u \times z = T_u \times z \quad (10.3)$$

where C_u and T_u are the resultant (ultimate) forces in compression and tension, respectively, and z is the lever arm.

$$T_u = f_{st} \times A_{st} \quad (10.4)$$

where $f_{st} = 0.87 f_y$ for $X_u \leq X_{umax}$ and the line of action of T_u corresponds to the level of the centroid of the tension steel.

10.2.2.1 Concrete Stress Block in Compression

In order to determine the magnitude of C_u and its line of action, it is necessary to analyze the concrete stress block in compression. As ultimate failure of a reinforced concrete beam in flexure occurs by the crushing of concrete, for both under- and over-reinforced beams, the shape of the compressive stress distribution ("stress block") at failure will be, in both cases, as shown in Figure 10.3. The value of C_u can be computed, once the compressive stress in concrete is uniform at $0.477 f_{ck}$ for a depth of $3X_u/7$, and below this it varies parabolically over a depth of $4X_u/7$ to zero at the neutral axis (Figure 10.3).

For a rectangular section of width b,

$$C_u = 0.477 f_{ck}\, b\left[\left(3X_u/7\right) + \left(2/3 \times \left(4X_u/7\right)\right)\right]$$

Therefore $C_u = 0.36\, f_{ck}\, b\, X_u \quad (10.5)$

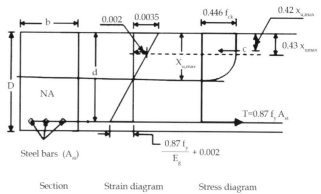

FIGURE 10.2 Couple resulting from the flexural stresses.

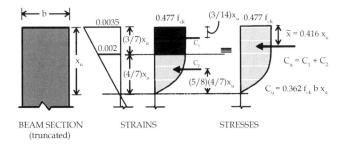

FIGURE 10.3 Compressive stress distribution at failure.

Also, the line of action of C_u is determined by the centroid of the stress block, located at a distance x' from the concrete fibers subjected to the maximum compressive strain. Accordingly, considering moments of compressive forces C_u, C_1, and C_2 (Figure 10.3) about the maximum compressive strain location,

$$0.36 \cdot f_{ck} \cdot b \cdot X_u \cdot x' = 0.477 \cdot f_{ck} \cdot b \cdot X_u \Big[(3/7)(1.5X_u/7)$$

$$+ (2/3)(4/7)\Big(X_u - \big((5/8)(4X_u/7)\big)\Big)\Big]$$

(10.6)

Solving $x' = 0.416 X_u$.

10.2.2.2 Depth of Neutral Axis

For any given section, the depth of the neutral axis should be such that $C_u = T_u$, satisfying equilibrium of forces. Equating $C_u = T_u$, with expressions for C_u and T_u given by Equations 10.5 and 10.4, respectively.

$$X_u = 0.087 \cdot f_y \cdot A_{st}/0.36 \cdot f_{ck} \cdot b$$

valid only if resulting $X_u \leq X_{umax}$ (10.7)

10.2.2.3 Ultimate Moment of Resistance

The ultimate moment of resistance M_u of a given beam section is obtainable from Equation 10.3. The lever arm z for the case of the singly reinforced rectangular section (Figures 10.2 and 10.3) is given by

$$Z = d - 0.416X_u$$ (10.8)

Accordingly, in terms of the concrete compressive strength,

$$M_u = 0.36 \cdot f_{ck} \cdot b \cdot X_u (d - 0.416X_u) \quad \text{for all } X_u \quad (10.9)$$

Alternatively, in terms of the steel tensile stress,

$$M_u = f_{st} \times A_{st} (d - 0.416X_u) \quad \text{for all } X_u \quad (10.10)$$

with $f_{st} = 0.87 f_y$ for $X_u \leq X_{umax}$

10.2.2.4 Limiting Moment of Resistance

The limiting moment of resistance of a given (singly reinforced, rectangular) section corresponds to the condition, defined by Equations 10.2. From Equation 10.9, it follows that

$$M_{u,\lim} = 0.36 \cdot f_{ck} \cdot b \cdot X_{umax} (d - 0.416X_{umax}) \quad (10.11)$$

$$M_{u,\lim} = 0.36 \cdot f_{ck} \cdot \big(X_{umax}/d\big)\big(1 - \big(0.416 X_{umax}/d\big)\big) b \, d^2$$

(10.11a)

10.2.2.5 Safety at Ultimate Limit State in Flexure

The bending moment expected at a beam section at the ultimate limit state due to the factored loads is called the factored moment M_u. For the consideration of various combinations of loads (DLs, LLs, wind loads, etc.), appropriate load factors should be applied to the specified "characteristic" loads, and the factored moment M_u is determined by structural analysis.

The beam section will be considered to be "safe," if its ultimate moment of resistance M_u is greater than or equal to the factored moment M_u. In other words, for such a design, the probability of failure is acceptably low. At ultimate failure in flexure, the type of failure should be a tension (ductile) failure. For this reason, the designer has to ensure that $X_u \leq X_{umax}$ whereby it follows that, for a singly reinforced rectangular section, the tensile reinforcement percentage p_t should not exceed $p_{t\lim}$, and the ultimate moment of resistance M_u should not exceed $M_{u,\lim}$.

10.2.3 MODES OF FAILURE: TYPES OF SECTION

A reinforced concrete member is considered to have failed when the strain of concrete in extreme compression fiber reaches its ultimate value of 0.0035. At this stage, the actual strain in steel can have the following values:

- Equal to failure strain of steel $\big(f_y/1.5E_s + 0.002\big)$ corresponding to balanced section.
- More than failure strain corresponding to under reinforced section.
- Less than failure strain corresponding to over reinforced section.

  Thus for a given section, the actual value of X_u/d can be determined from Equation 10.7. Three cases may arise.
  Case 1: X_u/d equal to the limiting value X_{umax}/d: Balanced section
  Case 2: X_u/d less than the limiting value: under-reinforced section
  Case 3: X_u/d more than the limiting value: over-reinforced section

- **Balanced section**: Section in which tension steel also reaches yield strain simultaneously as the concrete reaches the failure strain in bending is called "Balanced Section."
- **Under-reinforced section**: Section in which tension steel also reaches yield strain at loads lower than the load at which concrete reaches the failure strain in bending is called "Under-Reinforced Section."
- **Over-Reinforced section**: Section in which tension steel also reaches yield strain at loads higher than the load at which concrete reaches the failure strain in bending is called "Over-Reinforced Section."

10.2.4 COMPUTATION OF MOMENT OF RESISTANCE

S. No.	Types of Problems	Data Given	Data Determine	
1	Identify the type of section as balance, under-reinforced, or over-reinforced	Grade of concrete and steel, size of beam and reinforcement provided	If $X_u/d=X_{umax}/d$ implies balanced If $X_u/d<X_{umax}/d$ implies under-reinforced If $X_u/d>X_{umax}/d$ implies over reinforced $X_u/d=0.087 \cdot f_y \cdot A_{st}/0.36 \cdot f_{ck} \cdot b \cdot d$	
			Fe	X_u/d
			250	0.53
			415	0.48
			500	0.46
2	Calculate moment of resistance (MR)	Grade of concrete and steel, size of beam and reinforcement provided	If $X_u/d=X_{umax}/d$ implies balanced $MR=M_{u,lim}=0.36 \cdot X_u/d(1-0.42 \cdot X_u/d) \cdot b \cdot d^2 \cdot f_{ck}$ If $X_u/d<X_{umax}/d$ implies under reinforced $MR=M_u=0.87 \cdot f_y \cdot A_{st} \cdot d(1-(A_{st} \cdot f_y/b \cdot d \cdot f_{ck}))$ Or $MR=M_u=0.87 \cdot f_y \cdot A_{st} \cdot d(1-0.42 \cdot X_u/d)$ If $X_u/d>X_{umax}/d$ implies over-reinforced revise the depth	
3	Design the beam. Find out the depth of beam "D" and reinforcement required "A_{st}"	Grade of concrete and steel, width of beam, BM or loading on the beam with the span of the beam	Design as a "Balanced Design" For finding "d," use the equation, $MR=M_{u,lim}=0.36 \cdot X_u/d(1-0.42 \cdot X_u/d) \cdot b \cdot d^2 \cdot f_{ck}$; for finding "$A_{st}$," use the equation, $MR=M_u=0.87 \cdot f_y \cdot A_{st} \cdot d(1-(A_{st} \cdot f_y/b \cdot d \cdot f_{ck}))$ OR $M.R=M_u=0.87 \cdot f_y \cdot A_{st} \cdot d(1-0.42 \cdot X_u/d)$	

where

d = effective depth of beam in mm

b = width of beam in mm

X_u = depth of actual neutral axis in mm from extreme compression fiber

$X_{u,max}$ = depth of critical neutral axis in mm from extreme compression fiber

A_{st} = area of tensile reinforcement

f_{ck} = characteristic compressive strength of concrete in MPa

f_y = characteristic strength of steel in MPa

$M_{u,lim}$ = limiting MR of a section without compression reinforcement

10.2.5 DESIGN TYPE OF PROBLEMS

The designer has to make preliminary plan layout including location of the beam, its span and spacing, estimate the imposed and other loads from the given functional requirement of the structure. The DLs of the beam are estimated assuming the dimensions b and d initially. The bending moment, shear force, and axial thrust are determined after estimating the different loads. Let us assume that the imposed and other loads are given. Therefore, the problem is such that the designer has to start with some initial dimensions and subsequently revise them, if needed. The following guidelines are helpful to assume the design parameters initially.

- **Selection of breadth of the beam b:**

    Normally, the breadth of the beam b is governed by: (1) proper housing of reinforcing bars and (2) architectural considerations. It is desirable that the width of the beam should be less than or equal to the width of its supporting structure like column or the wall, etc. Practical aspects should also be kept in mind. It has been found that most of the requirements are satisfied with b as 150, 200, 230, 250, and 300 mm. Again, width-to-overall depth ratio is normally kept between 0.5 and 0.67.

- **Selection of depths of the beam d and D:**

    The effective depth has a major role to play in satisfying (1) the strength requirements of bending moment and shear force and (2) deflection of the beam. The initial effective depth of the beam, however, is assumed to satisfy the deflection requirement depending on the span and type of the reinforcement.

    The basic values of span to effective depth ratios for spans up to 10 m are as follows:

    Cantilever 7

    Simply supported 20

    Continuous 26

    For spans above 10 m, the above values may be multiplied with 10/span in meters, except for cantilevers where the deflection calculations should be made. Further, these ratios are to be multiplied with the modification factor depending on reinforcement percentage and type. The total depth D can be determined by adding 40–80 mm to the effective depth.

- **Selection of the amount of steel reinforcement A_{st}:**

    The amount of steel reinforcement should provide the required tensile force T to resist the factored moment M_u of the beam. Further, it should also satisfy the minimum and maximum percentages of reinforcement requirements. The minimum reinforcement "A_s" is provided for creep, shrinkage, thermal, and other environmental requirements irrespective of the strength requirement. The minimum reinforcement A_s to be provided in a beam depends on the f_y of steel and it follows the relation:

$$A_s/b \cdot d = 0.85/f_y$$

The maximum tension reinforcement should not exceed $0.04 \, b \, D$, where D is the total depth.

Besides satisfying the minimum and maximum reinforcement, the amount of reinforcement of the singly reinforced beam should normally be 75%–80% $p_{t,\lim}$. This will ensure that strain in steel will be more than $(0.87f_y/E_s + 0.002)$ as the design stress in steel will be $0.87f_y$.

Moreover, in many cases, the depth required for deflection becomes more than the limiting depth required to resist $M_{u,\lim}$. Thus, it is almost obligatory to provide more depth. Providing more depth also helps in the amount of steel which is less than that required for $M_{u,\lim}$. This helps to ensure ductile failure. Such beams are designated as under-reinforced beams.

- **Selection of diameters of bar of tension reinforcement**:

    Reinforcement bars are available in different diameters such as 6, 8, 10, 12, 14, 16, 18, 20, 22, 25, 28, 30, 32, 36, and 40 mm. Some of these bars are less available. The selection of the diameter of bars depends on its availability, minimum stiffness to resist while persons walk over them during construction, bond requirement, etc. Normally, the diameters of main tensile bars are chosen from 12, 16, 20, 22, 25, and 32 mm.

- **Selection of grade of concrete**:

    Besides strength and deflection, durability is a major factor to decide on the grade of concrete. M20 is the minimum grade under mild environmental exposure and also other grades of concrete under different environmental exposures.

- **Selection of grade of steel**:

    Normally, Fe 250, 415, and 500 are used in reinforced concrete work. Mild steel (Fe 250) is more ductile and is preferred for structures in earthquake zones or where there are possibilities of vibration, impact, blast, etc.

10.3 DESIGN OF DOUBLY REINFORCED SECTIONS

Concrete has a very good compressive strength and almost negligible tensile strength. Hence, steel reinforcement is used on the tensile side of concrete. Thus, singly reinforced beams reinforced on the tensile face are good both in compression and tension. However, these beams have their respective limiting moments of resistance with specified width, depth, and grades of concrete and steel. The amount of steel reinforcement needed is known as $A_{st,\lim}$. Problem will arise, therefore,

if such a section is subjected to bending moment greater than its limiting moment of resistance as a singly reinforced section.

There are two ways to solve the problem. First, we may increase the depth of the beam, which may not be feasible in many situations. In those cases, it is possible to increase both the compressive and tensile forces of the beam by providing steel reinforcement in compression face and additional reinforcement in tension face of the beam without increasing the depth (Figure 10.4). The total compressive force of such beams comprises (1) force due to concrete in compression and (2) force due to steel in compression. The tensile force also has two components: (1) the first provided by $A_{st,\lim}$ which is equal to the compressive force of concrete in compression. The second part is due to the additional steel in tension—its force will be equal to the compressive force of steel in compression. Such reinforced concrete beams having steel reinforcement both on tensile and compressive faces are known as doubly reinforced beams.

10.3.1 Basic Principle

The moment of resistance Mu of the doubly reinforced beam consists of (1) $M_{u,\lim}$ of singly reinforced beam and (2) M_{u2} because of equal and opposite compression and tension forces (C_2 and T_2) due to additional steel reinforcement on compression and tension faces of the beam (Figure 10.5). Thus, the moment of resistance M_u of a doubly reinforced beam is

$$M_u = M_{u,\lim} + M_{u2} \tag{10.12}$$

$$M_{u,\lim} = 0.36 \cdot f_{ck} \cdot \left(X_{umax}/d\right)\left(1 - \left(0.416 X_{umax}/d\right)\right)b \, d^2 \tag{10.13}$$

Also, $M_{u,\lim}$ can be written as

$$M_{u,\lim} = 0.87 A_{st,\lim} \, f_y \left(d - 0.42 x_{u,\max}\right)$$

$$= 0.87 p_{t,\lim}\left(1 - \left(0.42 X_{u\max}/d\right)\right)b \cdot d \cdot f_y \tag{10.14}$$

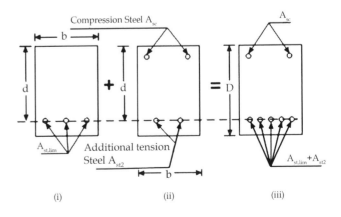

For (i) >> $M_{u1} = M_{u,\lim}$
For (ii) >> M_{u2} = Due to A_{sc} and A_{st2}
For (iii) >> $Mu = M_{u,\lim} + M_{u2}$

FIGURE 10.4 Doubly reinforced beam.

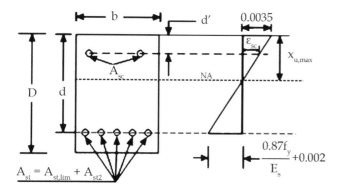

$A_{st} = A_{st,lim} + A_{st2}$

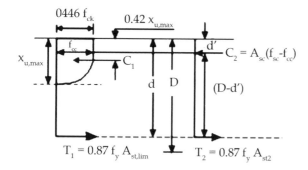

$T_1 = 0.87\, f_y\, A_{st,lim}$ $T_2 = 0.87\, f_y\, A_{st2}$

(i)   Beam Cross Section
(ii) Strain Diagram
(iii) Force Diagram of Beam $M_{u,lim}$
(iv) Force Diagram of Beam M_{u2}

FIGURE 10.5   Stress, strain, and force diagram of doubly reinforced beam.

The additional moment M_{u2} can be expressed in two ways (Figure 10.5): considering (1) the compressive force C_2 due to compression steel and (2) the tensile force T_2 due to additional steel on tension face. In both the equations, the lever arm is $(d-d')$. Thus, we have

$$M_{u2} = A_{sc}\left(f_{sc} - f_{cc}\right)\left(d - d'\right) \tag{10.15}$$

$$M_{u2} = A_{st2}\left(0.87 f_y\right)\left(d - d'\right) \tag{10.16}$$

where

A_{sc} = area of compression steel reinforcement
f_{sc} = stress in compression steel reinforcement
f_{cc} = compressive stress in concrete at the level of centroid of compression steel reinforcement
A_{st2} = area of additional steel reinforcement

Since the additional compressive force C_2 is equal to the additional tensile force T_2, we have

$$A_{sc}\left(f_{sc} - f_{cc}\right) = A_{st2}\left(0.87 f_y\right) \tag{10.17}$$

Any two of the three equations (Equations 10.15–10.17) can be employed to determine A_{sc} and A_{st2}.

The total tensile reinforcement A_{st} is then obtained from:

$$A_{st} = A_{st1} + A_{st2} \tag{10.18}$$

where $A_{st2} = p_{t,lim}\left(b \cdot d/100\right) = M_{u,lim}/\left(0.87 \cdot f_y\left(d - 0.42\right)\right)$

$$\tag{10.19}$$

10.3.2   Determination of F_{SC} and F_{CC}

It is seen that the values of f_{sc} and f_{cc} should be known before calculating A_{sc}. The following procedure may be followed to determine the value of f_{sc} and f_{cc} for the design type of problems (and not for analyzing a given section). For the design problem the depth of the neutral axis may be taken as $x_{u,max}$ as shown in Figure 10.5. From Figure 10.5, the strain at the level of compression steel reinforcement ε_{sc} may be written as:

$$\varepsilon_{sc} = 0.0035\left(1 - \left(d'/X_{umax}\right)\right) \tag{10.20}$$

The stress in compression steel f_{sc} is corresponding to the strain ε_{sc} of Equation 10.20 and is determined for (a) mild steel and (b) cold worked bars Fe 415 and 500 as given below:

a. **Mild steel Fe 250**
   The strain at the design yield stress of 217.39 N/mm^2 (f_d=0.87f_y) is 0.0010869 (= 217.39/E_s). The f_{sc} is determined from the idealized stress–strain diagram of mild steel after computing the value of ε_{sc} from Equation ix as follows:
   i. If the computed value of $\varepsilon_{sc} \leq 0.0010869$, $f_{sc} = \varepsilon_{sc} E_s = 2 \,(10^5)\, \varepsilon_{sc}$ N/mm^2
   ii. If the computed value of $\varepsilon_{sc} > 0.0010869$, $f_{sc} = 217.39$ N/mm^2

b. **Cold worked bars Fe 415 and Fe 500**
   From the stress-strain diagram of these bars, it shows that the stress is proportional to strain up to a stress of $0.8f_y$. The stress–strain curve for the design purpose is obtained by substituting f_{yd} for f_y up to $0.8f_{yd}$. Thereafter, from $0.8f_{yd}$ to f_{yd}, Table A of SP-16 gives the values of total strains and design stresses for Fe 415 and Fe 500. Table 10.2 below presents these values as a ready reference here.

TABLE 10.2

Values of ε_{sc} and f_{sc}

Stress Level	Fe 415		Fe 500	
	Strain ε_{sc}	Stress f_{sc} (N/mm^2)	Strain ε_{sc}	Stress f_{sc} (N/mm^2)
$0.80f_{yd}$	0.00144	288.7	0.00174	347.8
$0.85f_{yd}$	0.00163	306.7	0.00195	369.6
$0.90f_{yd}$	0.00192	324.8	0.00226	391.3
$0.95f_{yd}$	0.00241	342.8	0.00277	413
$0.975f_{yd}$	0.00276	351.8	0.00312	423.9
$1.0f_{yd}$	0.0038	360.9	0.00417	434.8

The above procedure has been much simplified for the cold worked bars by presenting the values of f_{sc} of compression steel in doubly reinforced beams for different values of d'/d only taking the practical aspects into consideration. In most of the doubly reinforced beams, d'/d has been found to be between 0.05 and 0.2. Accordingly, values of f_{sc} can be computed from Table 10.2 after determining the value of ε_{sc} from Equation 10.20 for known values of d'/d as 0.05, 0.10, 0.15, and 0.2. Table F of SP-16 presents these values of f_{sc} for four values of d'/d (0.05, 0.10, 0.15 and 0.2) of Fe 415 and Fe 500. Table 10.3 below, however, includes Fe 250 also whose f_{sc} values are computed down in along with those of Fe 415 and Fe 500. This table is very useful and easy to determine the f_{sc} from the given value of d'/d. The table also includes strain values at yield which are explained below:

i. The strain at yield of Fe 250 = Design yield stress/E_s

$$= 250/1.15 \times 200000$$
$$= 0.0010869$$

Here, there is only elastic component of the strain without any inelastic strain.

ii. The strain at yield of Fe 415 = Inelastic strain + Design yield stress/E_s
$$= 0.002 + 415/(1.15 \times 20,000)$$
$$= 0.0038043$$

iii. The strain at yield of Fe 500 = 0.002 + 415/ (1.15 × 20,000)

$$= 0.0041739$$

10.3.3 Minimum and Maximum Steel

In compression:

There is no stipulation in standard codes regarding the minimum compression steel in doubly reinforced beams. However, hangers and other bars provided up to 0.2% of the whole area of cross section may be necessary for creep and shrinkage of concrete. Accordingly, these bars are not considered as compression reinforcement. From the practical aspects of consideration, therefore, the minimum steel as compression reinforcement should be at least 0.4% of the area of concrete in compression or 0.2% of the whole cross-sectional area of the beam so that

the doubly reinforced beam can take care of the extra loads in addition to resisting the effects of creep and shrinkage of concrete.

The maximum compression steel shall not exceed 4% of the whole area of cross section of the beam.

- **In tension**:

The minimum amount of tensile reinforcement shall be at least (0.85 bd/f_y) and the maximum area of tension reinforcement shall not exceed (0.04 bD).

It has been stipulated that the singly reinforced beams shall have A_{st} normally not exceeding 75%–80% of $A_{st,lim}$ so that X_u remains less than $X_{u,max}$ with a view to ensuring ductile failure. However, in the case of doubly reinforced beams, the ductile failure is ensured with the presence of compression steel. Thus, the depth of the neutral axis may be taken as $X_{u,max}$ if the beam is over-reinforced. Accordingly, the A_{st1} part of tension steel can go up to $A_{st,lim}$ and the additional tension steel A_{st2} is provided for the additional moment $M_u - M_{u,lim}$. The quantities of A_{st1} and A_{st2} together form the total A_{st}, which shall not exceed 0.04 bD.

10.3.4 Types of Problems and Steps of Solution

The doubly reinforced beams have two types of problems: (1) design type and (2) analysis type.

- **Design type of problem**

In the design type of problems, the given data are b, d, D grades of concrete and steel. The designer has to determine A_{sc} and A_{st} of the beam from the given factored moment. These problems can be solved by two ways: (1) use of the equations developed for the doubly reinforced beams, named here as direct computation method; (2) use of charts and tables of SP-16.
  a. Direct computation method
    1. To determine $M_{u,lim}$ and $A_{st,lim}$ from Equations 10.13 and 10.18, respectively.
    2. To determine M_{u2}, A_{sc}, A_{st2}, and A_{st} from Equations 10.12, 10.15, 10.17, and 10.18, respectively.
    3. To check for minimum and maximum reinforcement in compression and tension as explained earlier.

- **Analysis type of problems**

In the analysis type of problems, the data given are $b, d, d', D, f_{ck}, f_y, A_{sc}$, and A_{st}. It is required to determine the moment of resistance M_u of such beams. These problems can be solved (1) by direct computation method and (2) by using tables of SP-16.
  a. Direct computation method
    1. To check if the beam is under-reinforced or over-reinforced.

First, X_{umax} is determined assuming it has reached limiting stage using X_{umax}/d coefficients given in standard codes. The

TABLE 10.3

Values of f_{sc} for Different Values of d'/d

f_y (N/mm²)	0.05	0.10	0.15	0.20	Strain at Yield
			d'/d		
250	217.4	217.4	217.4	217.4	0.0010869
415	355	353	342	329	0.0038043
500	412	412	395	370	0.0041739

strain of tensile steel ε_{st} is computed from $\varepsilon_{st} = \varepsilon_c \, (d - X_{umax})/X_{umax}$ and is checked if ε_{st} has reached the yield strain of steel:

$$\varepsilon_{st} \text{ at yield} = f_y/(1.15E) + 0.002 \quad (10.20a)$$

The beam is under-reinforced or over-reinforced if ε_{st} is less than or more than the yield strain.

2. To determine $M_{u,\lim}$ from Equation 10.14 and $A_{st,\lim}$ from the $p_{t\lim}$.
3. To determine A_{st2} and A_{sc} from Equations 10.17 and 10.18, respectively.
4. To determine M_{u2} and M_u from Equations 10.15 and 10.12, respectively.

10.4 SHEAR IN REINFORCED CONCRETE

Shear is one of the major considerations in the design of reinforced structures. Any reinforced concrete member which is subjected to bending moments is usually accompanied by transverse shear, axial forces, and torsion. The shear associated with varying bending moments is called flexural shear.

The simplified method/conventional method of shear design of reinforced concrete members assumes that plane sections remain plane before and after the bending. In this method, the shear design is achieved by transverse shear reinforcement, and on the other hand, the longitudinal reinforcement is designed for axial forces and flexure moments.

10.4.1 MODES OF FAILURE

For beams with low span-to-depth ratio or inadequate shear reinforcement, the failure can be due to shear. A failure due to shear is sudden as compared to a failure due to flexure. The following five modes of failure due to shear are identified.

1. Diagonal tension failure
2. Shear compression failure
3. Shear tension failure
4. Web crushing failure
5. Arch rib failure

The occurrence of a mode of failure depends on the span-to-depth ratio, loading, cross section of the beam, amount, and anchorage of reinforcement.

1. Diagonal tension failure:
    In this mode, an inclined crack propagates rapidly due to inadequate shear reinforcement (Figure 10.6).
2. Shear compression failure:
    There is crushing of the concrete near the compression flange above the tip of the inclined crack (Figure 10.7).
3. Shear tension failure:
    Due to inadequate anchorage of the longitudinal bars, the diagonal cracks propagate horizontally along the bars (Figure 10.8).

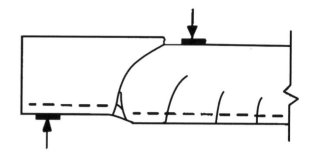

FIGURE 10.6   Diagonal tension failure.

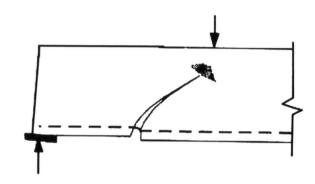

FIGURE 10.7   Shear compression failure.

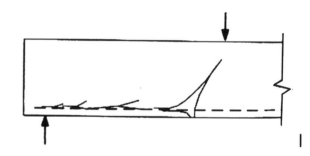

FIGURE 10.8   Shear tension failure.

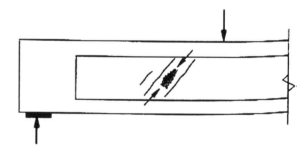

FIGURE 10.9   Web crushing failure.

4. Web crushing failure:
    The concrete in the web crushes due to inadequate web thickness (Figure 10.9).
5. Arch rib failure:
    For deep beams, the web may buckle and subsequently crush (Figure 10.10). There can be anchorage failure or failure of the bearing.

The objective of design for shear is to avoid shear failure. The beam should fail in flexure at its ultimate flexural strength.

FIGURE 10.10 Arch rib failure.

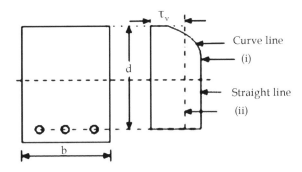

Note:
(i)   Actual distribution
(ii)   Average distribution

(a)

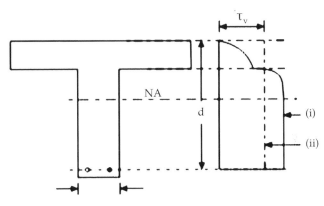

Note:
(i)   Actual distribution
(ii)   Average distribution

(b)

FIGURE 10.11 Distribution of shear stress and average shear stress. (a) Rectangular beam. (b) T beam.

Hence, each mode of failure is addressed in the design for shear. The design involves not only the design of the stirrups but also limiting the average shear stress in concrete, providing adequate thickness of the web and adequate development length of the longitudinal bars.

10.4.2 SHEAR STRESS

The distribution of shear stress in reinforced concrete rectangular, T and L-beams of uniform and varying depths depends on the distribution of the normal stress. However, for the sake of simplicity the nominal shear stress τ_v is considered, which is calculated as follows:

i. In beams of uniform depth (Figure 10.11)

$$\tau_v = V_u / b*d \qquad (10.21)$$

where
V_u=shear force due to design load
b=breadth of rectangular beams and breadth of the web bw for flanged beams
d=effective depth
ii. In beams of varying depth

$$\tau_v = V_u \pm \left(M_u * \tan \beta / d\right) / b*d \qquad (10.22)$$

where
M_u=bending moment at the section
β=angle between the top and the bottom edges

The positive sign is applicable when the bending moment M_u decreases numerically in the same direction as the effective depth increases, and the negative sign is applicable when the bending moment M_u increases numerically in the same direction as the effective depth increases.

10.4.3 DESIGN SHEAR STRENGTH OF REINFORCED CONCRETE

Recent laboratory experiments confirmed that reinforced concrete in beams has shear strength even without any shear reinforcement. This shear strength (τ_c) depends on the grade of concrete and the percentage of tension steel in beams. On the other hand, the shear strength of reinforced concrete with the reinforcement is restricted to some maximum value τ_{cmax} depending on the grade of concrete. These minimum and maximum shear strengths of reinforced concrete are given below:

- **Design shear strength without shear reinforcement**
- The design shear strength of concrete τ_c for different grades of concrete with a wide range of percentages of positive tensile steel reinforcement can be got from standard codes. It is worth mentioning that the reinforced concrete beams must be provided with the minimum shear reinforcement even when τ_v is less than τ_c given in Table 10.4.

    In Table 10.4, As is the area of longitudinal tension reinforcement which continues at least one effective depth beyond the section considered except at support where the full area of tension reinforcement may be used.
- **Maximum shear stress τ_{cmax} with shear reinforcement**
- The maximum shear stress of reinforced concrete in beams τ_{cmax} is given in Table 10.5. Under no circumstances, the nominal shear stress in beams τ_v shall exceed τ_{cmax} given in Table 10.5 for different grades of concrete.

TABLE 10.4
Design Shear Strength of Concrete, τ_c in N/mm²

	Grades of Concrete				
100 A_s/bd	M20	M25	M30	M35	M40 and Above
≤0.15	0.28	0.29	0.29	0.29	0.30
0.25	0.36	0.36	0.37	0.37	0.38
0.50	0.48	0.49	0.50	0.50	0.51
0.75	0.56	0.57	0.59	0.59	0.60
1.00	0.62	0.64	0.66	0.67	0.68
1.25	0.67	0.70	0.71	0.73	0.74
1.50	0.72	0.74	0.76	0.78	0.79
1.75	0.75	0.78	0.80	0.82	0.84
2.00	0.79	0.82	0.84	0.86	0.88
2.25	0.81	0.85	0.88	0.90	0.92
2.50	0.82	0.88	0.91	0.93	0.95
2.75	0.82	0.90	0.94	0.96	0.98
≥3.00	0.82	0.92	0.96	0.99	1.01

TABLE 10.5
Maximum Shear Stress, τ_{cmax}, in N/mm²

Grades of Concrete	M20	M25	M30	M35	M40 and Above
τ_{cmax} (N/mm²)	2.8	3.1	3.5	3.7	4.0

10.4.4 Critical Section for Shear

For beams generally subjected to uniformly distributed loads or where the principal load is located further than $2d$ from the face of the support, where d is the effective depth of the beam, the critical sections depend on the conditions of supports as shown in Figures 10.12a–c and are mentioned below.

i. When the reaction in the direction of the applied shear introduces tension (Figure 10.12a) into the end region of the member, the shear force is to be computed at the face of the support of the member at that section.

ii. When the reaction in the direction of the applied shear introduces compression into the end region of the member (Figures 10.12b and c), the shear force computed at a distance d from the face of the support is to be used for the design of sections located at a distance less than d from the face of the support.

10.4.5 Enhanced Shear Strength of Sections Close to Supports

Figure 10.13 shows the shear failure of simply supported and cantilever beams without shear reinforcement. The failure plane is normally inclined at an angle of 30° to the horizontal. However, in some situations, the angle of failure is more steeper either due to the location of the failure section closed to a support or for some other reasons. Under these situations, the shear force required to produce failure is increased.

Such enhancement of shear strength near a support is taken into account by increasing the design shear strength of concrete to ($2d\tau_c/a_v$) provided that the design shear stress at the face of the support remains less than the value of τ_{cmax} given in Table 10.5. In the above expression of the enhanced shear strength,

d = effective depth
τ_c = design shear strength of concrete before the enhancement as given in Table 10.4
a_v = horizontal distance of the section from the face of the support

Similar enhancement of shear strength is also to be considered for sections close to point loads. It is evident from the

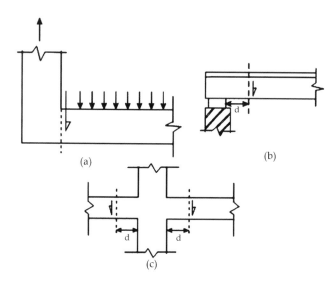

FIGURE 10.12 Support conditions for loacting factored shear force (a–c).

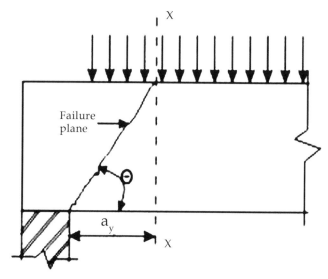

FIGURE 10.13 Shear failure near supports.

expression ($2d\tau_c/a_v$) that when a_v is equal to $2d$, the enhanced shear strength does not come into picture. Further, to increase the effectivity, the tension reinforcement is recommended to be extended on each side of the point where it is intersected by a possible failure plane for a distance at least equal to the effective depth, or to be provided with an equivalent anchorage.

10.4.6 MINIMUM SHEAR REINFORCEMENT

Minimum shear reinforcement has to be provided even when τ_v is lesser than τ_c given in Table 10.4. The amount of minimum shear reinforcement, is given below.

The minimum shear reinforcement in the form of stirrups shall be provided such that:

$$A_{sv}/b\,S_v \geq 0.4/0.87 f_y \qquad (10.23)$$

where

A_{sv}=total cross-sectional area of stirrups effective in shear

S_v=stirrup spacing along the length of the member

b=breadth of rectangular beams and breadth of the web b_w for flanged beams

f_y=characteristic strength of stirrup reinforcement in N/mm^2 which shall not be taken greater than 415 N/mm^2

The above provision is not applicable for members of minor structural importance such as lintels where the maximum shear stress calculated is less than half the permissible value.

The minimum shear reinforcement is provided for the following:

   i. Any sudden failure of beams is prevented if concrete cover bursts and the bond to the tension steel is lost.

   ii. Brittle shear failure is arrested which would have occurred without shear reinforcement.

   iii. Tension failure is prevented which would have occurred due to shrinkage, thermal stresses, and internal cracking in beams.

   iv. To hold the reinforcement in place when concrete is poured.

   v. Section becomes effective with the tie effect of the compression steel.

Further, the maximum spacing of shear reinforcement measured along the axis of the member shall be not more than $0.75\,d$ for vertical stirrups and d for inclined stirrups at $45°$, where d is the effective depth of the section. However, the spacing shall not exceed 300 mm in any case.

10.4.7 DESIGN OF SHEAR REINFORCEMENT

When τ_v is more than τ_c given in Table 10.4, shear reinforcement shall be provided in any of the three following forms:

   a. vertical stirrups,

   b. bent-up bars along with stirrups, and

   c. inclined stirrups.

In the case of bent-up bars, it is to be seen that the contribution toward shear resistance of bent-up bars should not be more than 50% of that of the total shear reinforcement.

The amount of shear reinforcement to be provided is determined to carry a shear force V_{us} equal to

$$V_{us} = V_u - \tau_c bd \qquad (10.24)$$

where b is the breadth of rectangular beams or b_w in the case of flanged beams.

The strengths of shear reinforcement V_{us} for the three types of shear reinforcement are as follows:

   a. Vertical stirrups:

$$V_{us} = 0.87 f_y A_{sv} d / S_v \qquad (10.25)$$

   b. For inclined stirrups or a series of bent-ups at different cross sections:

$$V_{us} = \left(0.87\, f_y A_{sv} d / S_v\right) \cdot (\sin\alpha + \cos\alpha) \qquad (10.26)$$

   c. For single bar or single group of parallel bars, all bent-ups at the same sections.

$$V_{us} = 0.87 f_y\, A_{sv} S_v \sin\alpha \qquad (10.27)$$

where

A_{sv}=total cross-sectional area of stirrup legs or bent-up bars within a distance S_v,

S_v=spacing of stirrups or bent-up bars along the length of the member,

τ_v=nominal shear stress,

τ_c=design shear strength of concrete,

b=breadth of rectangular beams and breadth of the web b_w for flanged beams,

f_y=characteristic strength of stirrup or bent-up reinforcement in N/mm^2 which shall not be taken greater than 415 N/mm^2,

d=effective depth, and

α=angle between the inclined stirrup or bent-up bars and the axis of the member, not less than 450.

The following two points are to be noted:

• The total shear resistance shall be computed as the sum of the resistance for the various types separately, where more than one type of shear reinforcement is used.

• The area of stirrups shall not be less than the minimum specified in standard codes.

10.4.8 Shear Reinforcement for Sections Close to Supports

The total area of the required shear reinforcement A_s is obtained from:

$$A_s = a_v b \left(\tau_v - 2 d \tau_c / a_v \right) / 0.87 f_y \text{ and } \geq 0.4 \, a_v b / 0.87 f_y \quad (10.28)$$

For flanged beams, b will be replaced by b_w, the breadth of the web of flanged beams.

This reinforcement should be provided within the middle three quarters of a_v, where a_v is less than d and horizontal shear reinforcement will be effective than vertical.

The following method is for beams carrying generally uniform load or where the principal load is located further than $2d$ from the face of support. The shear stress is calculated at a section, a distance d from the face of support. The value of τ_c is calculated in accordance with Table 10.4 and appropriate shear reinforcement is provided at sections closer to the support. No further check for shear at such sections is required.

10.5 BOND, DEVELOPMENT LENGTH, AND SPLICING OF REINFORCEMENT

The bond between steel and concrete is very important and essential so that they can act together without any slip in a loaded structure. With the perfect bond between them, the plane section of a beam remains plane even after bending. The length of a member required to develop the full bond is called the anchorage length. The bond is measured by bond stress. The local bond stress varies along a member with the variation of bending moment. The average value throughout its anchorage length is designated as the average bond stress. In our discussion, the average bond stress will be used.

Thus, a tensile member has to be anchored properly by providing additional length on either side of the point of maximum tension, which is known as "Development length in tension." Similarly, for compression members also, we have "Development length L_d in compression."

It is worth mentioning that the deformed bars are known to be superior to the smooth mild steel bars due to the presence of ribs. In such a case, it is needed to check for the sufficient development length L_d only, rather than checking both for the local bond stress and development length as required for the smooth mild steel bars.

10.5.1 Design Bond Stress τ_{BD}

a. Definition

The design bond stress τ_{bd} is defined as the shear force per unit nominal surface area of reinforcing bar. The stress is acting on the interface between bars and surrounding concrete and along the direction parallel to the bars.

This concept of design bond stress finally results in additional length of a bar of specified diameter to be provided beyond a given critical section. Though,

TABLE 10.6

Design Bond Stress for Plain Bars in Tension

Grades of Concrete	M20	M25	M30	M35	M40 and Above
Design bond stress τ_{bd} (N/mm^2)	1.2	1.4	1.5	1.7	1.9

the overall bond failure may be avoided by this provision of additional development length L_d, slippage of a bar may not always result in overall failure of a beam. It is, thus, desirable to provide end anchorages also to maintain the integrity of the structure and thereby, to enable it carrying the loads.

b. Design bond stress—values

The local bond stress varies along the length of the reinforcement while the average bond stress gives the average value throughout its development length. This average bond stress is still used in the WSM. However, in the LSM of design, the average bond stress has been designated as design bond stress τ_{bd}. (Table 10.6).

For deformed bars, these values shall be increased by 60%. For bars in compression, the values of bond stress in tension shall be increased by 25%.

10.5.2 Development Length

Development length prevents the bar from pulling out in tension or pushing it in compression. When the required development length cannot be provided due to certain restrictions or other considerations then in that case bends, hooks or mechanical anchorages are provided to supplement with an equivalent embedment length.

a. A single bar

Figure 10.14a shows a simply supported beam subjected to uniformly distributed load. Because of the maximum moment, the A_{st} required is the maximum at $x = L/2$. For any section 1-1 at a distance $x < L/2$, some of the tensile bars can be curtailed. Let us then assume that section 1-1 is the theoretical cut-off point of one bar. However, it is necessary to extend the bar for a length L_d.

Figure 10.14b shows the free body diagram of the segment AB of the bar. At B, the tensile force T trying to pull out the bar is of the value $T = (\pi \varphi^2 \sigma_s / 4)$, where φ is the nominal diameter of the bar and σ_s is the tensile stress in bar at the section considered at design loads. It is necessary to have the resistance force to be developed by τ_{bd} for the length L_d to overcome the tensile force. The resistance force $= \pi \varphi (L_d) (\tau_{bd})$.

Equating the two, we get:

$$\pi \varphi (L_d)(\tau_{bd}) = \left(\pi \varphi^2 \sigma_s / 4 \right) \quad (10.29)$$

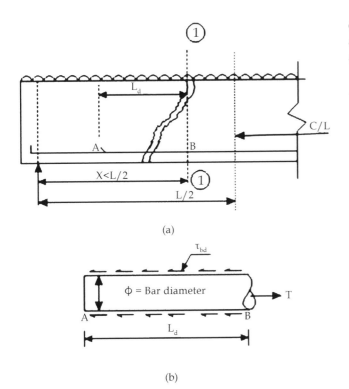

FIGURE 10.14   Development length of bars. (a) Beam showing L_d of a bar. (b) Free body diagram of segment AB.

thus gives

$$L_d = \varphi \sigma_s / 4 \tau_{bd} \qquad (10.30)$$

The above equation is used to determine the development length of bars.

Similarly, other sections of the bar should have the required L_d as determined for such sections. For bars in compression, the development length is reduced by 25% as the design bond stress in compression τ_{bd} is 25% more than that in tension. Following the same logic, the development length of deformed bars is reduced by 60% of that needed for the plain round bars. It is to be noted that the consequence of stress concentration at the lugs of deformed bars has not been taken into consideration.

b. Bars bundled in contact

The respective development lengths of each of the bars for two, three, or four bars in contact are determined following the same principle. The development length of each bar of bundled bars shall be that for the individual bar, increased by 10 per cent for two bars in contact, 20 per cent for three bars in contact, and 33 per cent for four bars in contact.

- In addition to single bar, bars may be arranged in pairs in contact or in groups of three or four bars bundled in contact
- Bundled bars shall be enclosed within stirrups or ties to ensure the bars remaining together.
- Bars larger than 32 mm diameter shall not be bundled, except in columns.

Curtailment of bundled bars should be done by terminating at different points spaced apart by not less than 40 times the bar diameter except for bundles stopping at support.

10.5.3   CHECKING OF DEVELOPMENT LENGTHS OF BARS IN TENSION

1. At least one-third of the positive moment reinforcement in simple members and one-fourth of the positive moment reinforcement in continuous members shall be extended along the same face of the member into the support, to a length equal to $L_d/3$.
2. Such reinforcements of 1 above shall also be anchored to develop its design stress in tension at the face of the support, when such member is part of the primary lateral load resisting system.
3. The diameter of the positive moment reinforcement shall be limited to a diameter such that the L_d computed for $\sigma_s = f_d$ does not exceed the following:

$$L_d \le M_1/V + L_0 \qquad (10.31)$$

where
M_1 = moment of resistance of the section assuming all reinforcement at the section to be stressed to f_d.
$f_d = 0.87 f_y$
V = shear force at the section due to design loads.
L_0 = sum of anchorage beyond the center of support and the equivalent anchorage value of any hook or mechanical anchorage at simple support. At a point of inflexion, L_0 is limited to the effective depth of the member or 12φ whichever is greater.
φ = diameter of bar

It has been further stipulated that M_1/V in the above expression may be increased by 30% when the ends of the reinforcement are confined by a compressive reaction.

10.5.4   REINFORCEMENT SPLICING

Splicing is required when the bars available are shorter than that required (due to nonavailability of longer bars owing to manufacturing or transportation problem), when there is a change in the bar diameter, etc. Splicing is done to transfer the axial load from one bar to another with the same line of action. Splicing introduces stress concentrations in the surrounding concrete. To avoid such stress concentrations, splices in flexural members should not be at sections where the bending moment is more than 50% of the moment of resistance and not more than half the bars shall be spliced at a section. Splicing is done by lapping of bars welding of bars or with mechanical connections.

1. **Lap splices**:
   The following are the salient points:
   - They should be used for bar diameters up to 36 mm.

- They should be considered as staggered if the center-to-center distance of the splices is at least 1.3 times the lap length.
- The lap length including anchorage value of hooks for bars in flexural tension shall be L_d or 30φ, whichever is greater. The same for direct tension shall be $2L_d$ or 30φ, whichever is greater.
- The lap length in compression shall be equal to L_d in compression but not less than 24φ.
- The lap length shall be calculated on the basis of diameter of the smaller bar when bars of two different diameters are to be spliced.
- Lap splices of bundled bars shall be made by splicing one bar at a time and all such individual splices within a bundle shall be staggered.

2. **Strength of welds**

The strength of welded splices and mechanical connections shall be taken as 100% of the design strength of joined bars for compression splices.

For tension splices, such strength of welded bars shall be taken as 80% of the design strength of welded bars. However, it can go even up to 100% if welding is strictly supervised and if at any cross section of the member not more than 20% of the tensile reinforcement is welded. For mechanical connection of tension splice, 100% of design strength of mechanical connection shall be taken.

10.6 CONTINUOUS BEAMS

Beams are made continuous over the supports to increase structural integrity. A continuous beam provides an alternate load path in the case of failure at a section. In regions with high seismic risk, continuous beams and frames are preferred in buildings and bridges. A continuous beam is a statically indeterminate structure.

The advantages of a continuous beam as compared to a simply supported beam are as follows.

- For the same span and section, vertical load capacity is more.
- Mid-span deflection is less.
- The depth at a section can be less than a simply supported beam for the same span. Else, for the same depth the span can be more than a simply supported beam.
  - The continuous beam is economical in material.
- There is redundancy in load path.
  - Possibility of formation of hinges in case of an extreme event.
- Requires less number of anchorages of tendons.
- For bridges, the number of deck joints and bearings is reduced.
  - Reduced maintenance.

There are several disadvantages for a continuous beam as compared to a simply supported beam.

- Difficult analysis and design procedures.
- Difficulties in construction, especially for precast members.
- Increased frictional loss due to changes of curvature in the tendon profile.
- Increased shortening of beam, leading to lateral force on the supporting columns.
- Secondary stresses develop due to time dependent effects like creep and shrinkage, settlement of support, and variation of temperature.
- The concurrence of maximum moment and shear near the supports needs proper detailing of reinforcement.
- Reversal of moments due to seismic force requires proper analysis and design.

10.6.1 ANALYSIS OF CONTINUOUS BEAM

The analysis of continuous beams is based on elastic theory. For prestressed beams, the following aspects are important.

- Certain portions of a span are subjected to both positive and negative moments. These moments are obtained from the corresponding moment diagram.
- The beam may be subjected to partial loading and point loading. The envelop moment diagrams are developed from "pattern loading." The pattern loading refers to the placement of LLs in patches only at the locations with positive or negative values of the influence line diagram for a moment at a particular location.
- For continuous beams, prestressing generates reactions at the supports. These reactions cause additional moments along the length of a beam.

The analysis of a continuous beam is illustrated to highlight the aspects stated earlier. The bending moment diagrams for the following load cases are shown schematically in the following figures:

1. DL
2. LL on every span
3. LL on a single span.

For moving point loads as in bridges, first the influence line diagram is drawn. The influence line diagram shows the variation of the moment or shear for a particular location in the girder, due to the variation of the position of a unit point load. The vehicle load is placed based on the influence line diagram to get the worst effect. An influence line diagram is obtained by the Müller-Breslau Principle.

The placement of LL is as follows:

1. LL in all the spans.
2. LL in adjacent spans of a support for the support moment. The effect of LL in the alternate spans beyond is neglected.
3. LL in a span and in the alternate spans for the span moment.

The envelop moment diagrams are calculated from the analysis of each load case and their combinations. The analysis can be done by moment distribution method or by computer analysis.

The envelop moment diagrams provide the value of a moment due to the external loads. It is to be noted that the effect of prestressing force is not included in the envelop moment diagrams. Figure 10.15 shows typical envelop moment diagrams for a continuous beam.

In the above diagrams, M_{max} and M_{min} represent the highest and lowest values (algebraic values with sign) of the moments at a section, respectively. Note that certain portions of the beam are subjected to both positive and negative moments. The moment from the envelop moment diagrams will be represented as the M_0 diagram. This diagram does not depend on whether the beam is prestressed or not.

10.7 TORSION IN REINFORCED CEMENT CONCRETE (RCC) ELEMENTS

Loads acting normal to the plane of bending in case of beams and slabs gives rise to bending moments and shear force. However, when the loads act away from the plane of bending will induce torsional moment along with bending moment and shear. Space frames (Figure 10.16a), inverted L-beams as in supporting sunshades and canopies (Figure 10.16b), beams curved in plan (Figure 10.16c), edge beams of slabs (Figure 10.16d) are some of the examples where torsional moments are also present.

Skew bending theory and space-truss analogy are some of the theories developed to understand the behavior of reinforced concrete under torsion combined with bending moment and shear. These torsional moments are of two types:

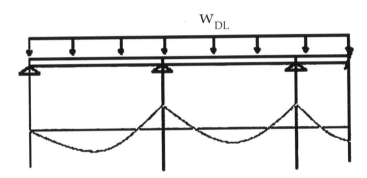

Moment diagram for DL

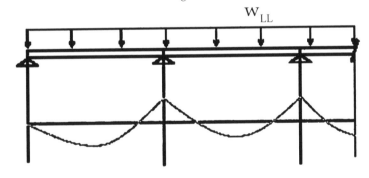

Moment diagram of LL on every span

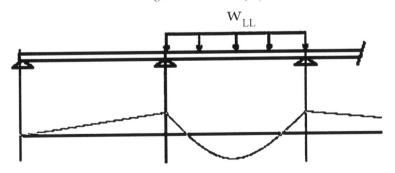

Moment diagram of LL on one span

FIGURE 10.15 Bending moment diagrams.

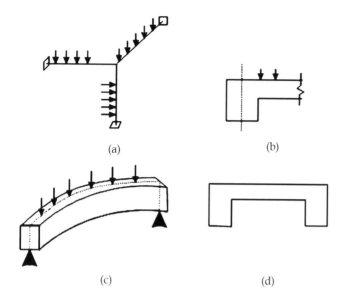

(a)　　　　　　　　　　　　(b)

(c)　　　　　　　　　　　　(d)

FIGURE 10.16 Beams under combined bending, shear and torsion. (a) Space frame. (b) Inverted L-beam. (c) Beam curved in plan. (d) Edge beam in slab.

　　i. primary or equilibrium torsion, and
　　ii. secondary or compatibility torsion.

The primary torsion is required for the basic static equilibrium of most of the statically determinate structures. Accordingly, this torsional moment must be considered in the design as it is a major component.

The secondary torsion is required to satisfy the compatibility condition between members. However, statically indeterminate structures may have any of the two types of torsions. Minor torsional effects may be ignored in statically indeterminate structures due to the advantage of having more than one load path for the distribution of loads to maintain the equilibrium. This may produce minor cracks without causing failure. However, torsional moments should be taken into account in the statically indeterminate structures if they are of equilibrium type and where the torsional stiffness of the members has been considered in the structural analysis. It is worth mentioning that torsion must be considered in structures subjected to unsymmetrical loadings about axes.

10.7.1 Analysis for Torsional Moment in a Member

We know that the bending moments are distributed among the sharing members with the corresponding distribution factors proportional to their bending stiffness EI/L where E is the elastic constant, I is the moment of inertia, and L is the effective span of the respective members. In a similar manner, the torsional moments are also distributed among the sharing members with the corresponding distribution factors proportional to their torsional stiffness GJ/L, where G is the elastic shear modulus, J is the polar moment of inertia, and L is the effective span (or length) of the respective members.

10.7.2 Approach of Design for Combined Bending, Shear and Torsion

The longitudinal and transverse reinforcements are determined taking into account the combined effects of bending moment, shear force, and torsional moment. Two empirical relations of equivalent shear and equivalent bending moment are given. These fictitious shear forces and bending moment, designated as equivalent shear and equivalent bending moment, are separate functions of actual shear and torsion, and actual bending moment and torsion, respectively. The total vertical reinforcement is designed to resist the equivalent shear V_e and the longitudinal reinforcement is designed to resist the equivalent bending moment M_{e1} and M_{e2}, respectively. These design rules are applicable to beams of solid rectangular cross section. However, they may be applied to flanged beams by substituting b_w for b.

10.7.3 Critical Section

Sections located less than a distance d from the face of the support are to be designed for the same torsion as computed at a distance d, where d is the effective depth of the beam.

10.7.4 Shear and Torsion

a. The equivalent shear, a function of the actual shear and torsional moment, is determined from the following empirical relation:

$$V_e = V_u + 1.6\left(T_u/b\right) \tag{10.32}$$

where
V_e = equivalent shear
V_u = actual shear
T_u = actual torsional moment
b = breadth of beam

b. The equivalent nominal shear stress τ_{ve} is determined from

$$\tau_{ve} = V_e/bd \tag{10.33}$$

However, τ_{ve} shall not exceed τ_{cmax} given in standard codes.

c. Minimum shear reinforcement is to be provided, if the equivalent nominal shear stress τ_{ve} obtained from Equation 10.33 does not exceed τ_c given in standard codes.

d. Both longitudinal and transverse reinforcement shall be provided, if τ_{ve} exceeds τ_c and is less than, τ_{cmax}.

10.7.5 Reinforcement in Members Subjected to Torsion

1. Reinforcement for torsion shall consist of longitudinal and transverse reinforcement.

2. The longitudinal flexural tension reinforcement shall be determined to resist an equivalent bending moment M_{e1} as given below.

$$M_{e1} = M_u + M_t \tag{10.34}$$

where M_u=bending moment at the cross section

$$M_u = (T_u/1.7)(1 + D/b) \tag{10.35}$$

T_u=torsional moment
D=overall depth of the beam
b=breadth of the beam

3. The longitudinal flexural compression reinforcement shall be provided if the numerical value of M_t as defined above in Equation 10.34 exceeds the numerical value of M_u. Such compression reinforcement should be able to resist an equivalent bending moment M_{e2} as given below.

$$M_{e2} = M_t - M_u \tag{10.36}$$

The M_{e2} will be considered as acting in the opposite sense to the moment M_u.

4. The transverse reinforcement consisting of two legged closed loops (Figure 10.17) enclosing the corner longitudinal bars shall be provided having an area of cross section A_{sv} given below:

$$A_{sv} = T_u S_v / b_1 d_1 (0.87 f_y) + V_u S_v / 2.5 \, d_1 (0.87 f_y) \tag{10.37}$$

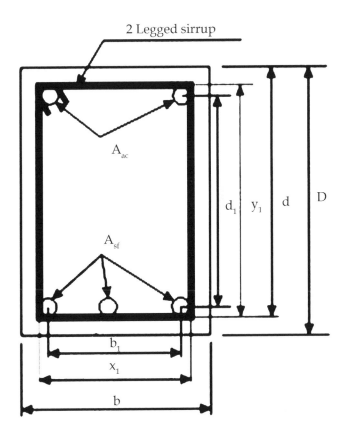

FIGURE 10.17 Stirrups in beams.

However, the total transverse reinforcement shall not be less than the following:

$$A_{sv} = (\tau_{ve} - \tau_c) b S_v / (0.87 f_y) \tag{10.38}$$

where
T_u=torsional moment
V_u=shear force
S_v=spacing of stirrup reinforcement
b_1=center-to-center distance between corner bars in the direction of the width
d_1=center-to-center distance between corner bars
b=width of the member
f_y=characteristic strength of stirrup reinforcement
τ_{ve}=equivalent shear
τ_c = shear strength of concrete as per standard codes

10.7.6 REQUIREMENT OF REINFORCEMENT

Beams subjected to bending moment, shear, and torsional moment should satisfy the following requirements:

a. **Tension reinforcement**
    The minimum area of tension reinforcement should be governed by

$$A_s / b \cdot d = 0.85 / f_y$$

where
A_s=minimum area of tension reinforcement
b=breadth of rectangular beam or breadth of the web of T-beam
d=effective depth of the beam
f_y = characteristic strength of reinforcement in N/mm^2
    The maximum area of tension reinforcement shall not exceed 0.04 bD, where D is the overall depth of the beam.

b. **Compression reinforcement**
    The maximum area of compression reinforcement shall not exceed 0.04 bD. They shall be enclosed by stirrups for effective lateral restraint.

c. **Side face reinforcement**
    Beams exceeding the depth of 750 mm and subjected to bending moment and shear shall have side face reinforcement. However, if the beams are having torsional moment also, the side face reinforcement shall be provided for the overall depth exceeding 450 mm. The total area of side face reinforcement shall be at least 0.1% of the web area and shall be distributed equally on two faces at a spacing not exceeding 300 mm or web thickness, whichever is less.

d. **Transverse reinforcement**
    The transverse reinforcement shall be placed around the outermost tension and compression bars.

They should pass around longitudinal bars located close to the outer face of the flange in T- and I-beams.

e. **Maximum spacing of shear reinforcement**

The center-to-center spacing of shear reinforcement shall not be more than $0.75d$ for vertical stirrups and d for inclined stirrups at $45°$, but not exceeding $300\,mm$, where d is the effective depth of the section.

f. **Minimum shear reinforcement**

This has been discussed in Section 10.4.6.

g. **Distribution of torsion reinforcement**

The transverse reinforcement shall consist of rectangular close stirrups placed perpendicular to the axis of the member. The spacing of stirrups shall not be more than the least of x_1, $(x_1+y_1)/4$ and $300\,mm$, where x_1 and y_1 are the short and long dimensions of the stirrups (Figure 10.17).

Longitudinal reinforcements should be placed as close as possible to the corners of the cross section.

h. **Reinforcement in flanges of T- and L-beams**

For flanges in tension, a part of the main tensile reinforcement shall be distributed over the effective flange width or a width equal to one-tenth of the span, whichever is smaller. For effective flange width greater than one-tenth of the span, nominal longitudinal reinforcement shall be provided to the outer portion of the flange.

Linear interpolation may be done for intermediate values.

11 Steel Structures

11.1 STEEL AS A STRUCTURAL MATERIAL

Steel has been used as a structural material for the construction of buildings, bridges, towers, and other structures since the late 19th century. It exhibits desirable physical properties that make it one of the most versatile structural materials in use. For example, the Forth Bridge of UK, which was constructed between 1882 and 1890, uses about 55,000 tons of steel. However, in order to get the best out of steel, structures should be properly designed and protected to resist fire and corrosion. Steel is essentially a noncombustible material; however when it is heated to a high temperature of about 500°C, its strength and stiffness are tremendously reduced. As per the International Building Code, steel should be enveloped in sufficient fire-resistant materials, thereby increasing the overall cost of steel structures. When steel comes in contact with water it can corrode, creating potentially dangerous structures. Therefore adequate measures should be taken to prevent lifetime corrosion of steel by providing paint coatings.

11.2 PLASTIC ANALYSIS AND DESIGN

Plastic analysis is defined as the analysis based on the ultimate load the structure will have to carry unlike the conventional elastic analysis in which the maximum strength is assumed to be the load at which the structure first attains its yield point stress. It is the analysis of the inelastic material beyond the elastic limit from the stress–strain diagram. Actually, the ultimate load is found from the strength of steel in the plastic range, which is how the term "plastic" has occurred. The concept of factor of safety is however retained. In the case of elastic design, the allowable stress is equal to the yield stress divided by the factor of safety, whereas in plastic design, the working load is equal to the ultimate load divided by the factor of safety. A steel rigid frame attains its ultimate strength through the formation of "plastic hinges." These hinges in turn form under overload because of the ability of structural steel to deform plastically after the yield point is reached.

11.2.1 BASICS OF PLASTIC ANALYSIS

Plastic analysis is based on the idealization of stress–strain curve as perfectly plastic. In this analysis, it is assumed that the width-thickness ratio of plate elements is small so that local buckling does not occur; in other words, the section can be classified as perfectly plastic. With these assumptions, it can be said that the section reaches its plastic moment capacity and then it will be subjected to considerable moment at applied moments.

Consider a simply supported beam subjected to a point load W at its mid-span as shown in Figure 11.1.

The stress distribution across any cross section is linear as shown in Figure 11.2a. As the load W gradually increases, the bending moment at every section increases and the stresses also increase. At a section close to the support, where the bending moment is maximum, the stresses in the extreme fibers reach the yield stress. The corresponding moment at this state is called the first **yield moment** M_y of the cross section. But this does not imply failure of the beam as it continues to take additional load. As the load continues to increase, more and more fibers reach the yield stress and the stress distribution is as shown in Figure 11.2b. Eventually, the whole of the cross section reaches the yield stress and the corresponding stress distribution is as shown in Figure 11.2c. The moment corresponding to this state is known as **plastic moment** of the cross section which is denoted by M_p.

The magnitude of the moment M_p may be computed directly from the stress distribution diagram shown in Figure 11.2c.

If f_y is the yield stress, A is the area above the neutral axis, and $\bar{y}$ is the distance measured from the neutral axis up to the center of gravity of that area,

$$M_p = f_y \times A \times \bar{y} \qquad (11.1)$$

The ratio of the plastic moment M_p to the yield moment M_y is known as the **shape factor**, F, as it depends on the shape of the cross section.

$$\text{Shape factor, } F = \frac{M_p}{M_y} \qquad (11.2)$$

Thus, for a rectangular cross section shape factor, F will be 1.5. For an I-section, the value of the shape factor is about 1.12.

11.2.2 PRINCIPLES OF PLASTIC ANALYSIS

Plastic analysis must always satisfy the following three conditions:

i. **Mechanism condition**: The ultimate or collapse load is reached when a mechanism is formed. The number of plastic hinges formed should be just sufficient to form a mechanism. This is also called as continuity condition.

ii. **Equilibrium condition**: All the equilibrium conditions, i.e., the summation of all forces and moments should be equal to zero $\left[\sum F_x = 0, \sum F_y = 0, \sum M_{xy} = 0 \right]$.

iii. **Yield condition**: This condition is also known as plastic moment condition. The bending moment at any section of the structure should not be more than the fully plastic moment of the section ($M \leq M_p$).

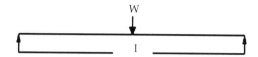

FIGURE 11.1   Upheld bar exposed to a point load W at its mid-range.

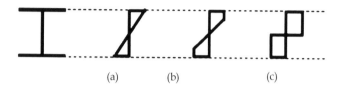

FIGURE 11.2   Plastic analysis. (a) at M_y. (b) $M_y < M < M_p$. (c) M_p.

11.2.2.1   Collapse Mechanisms

When a system of loads is applied to an elastic body, it will deform and will show a resistance against deformation. Such a body is known as a **structure**. On the other hand, if no resistance is set up against deformation in the body, then it is known as a **mechanism**. Various types of independent mechanisms are identified to enable prediction of possible failure modes of a structure.

    i. Beam mechanism

        Figure 11.3 shows a simply supported beam and a fixed beam with corresponding mechanisms.

   ii. Panel or sway mechanism

Figure 11.4 shows a panel or sway mechanism for a portal frame fixed at both ends.

  iii. Gable mechanism

        Figure 11.5 shows the gable mechanism for a gable structure fixed at both the supports.

  iv. Joint mechanism

        Figure 11.6 shows a joint mechanism. It occurs at a joint where more than two structural members meet.

11.2.2.2   Combined Mechanism

Various combinations of independent mechanisms can be made depending upon whether the frame is made of strong beam and weak column combination or strong column and weak beam combination. The one shown in Figure 11.7 is a combination of a beam and sway mechanism. Failure is triggered by formation of hinges at the bases of the columns and the weak beam developing two hinges. This is illustrated by the right hinge being shown on the beam, in a position slightly away from the joint.

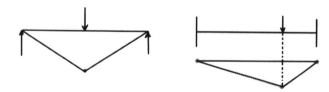

FIGURE 11.3   Beam mechanism.

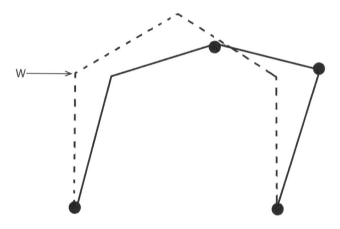

FIGURE 11.5   Gable mechanism.

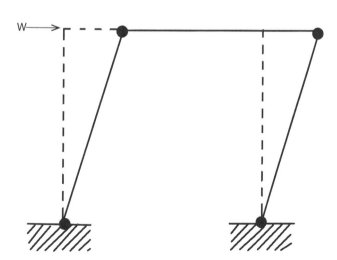

FIGURE 11.4   Panel or sway mechanism.

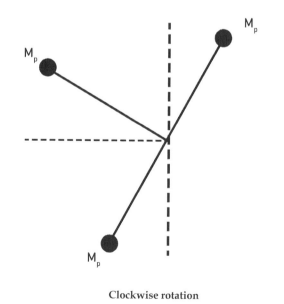

Clockwise rotation

FIGURE 11.6   Joint component.

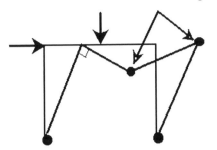

Two hinges developed on the beam

FIGURE 11.7 Consolidated system.

From the above examples, it is seen that the number of hinges needed to form a mechanism equals the statistical redundancy of the structure plus one.

11.2.2.3 Number of Independent Mechanisms

If N is the number of possible plastic hinges, r is the number of redundancy and n is the number of possible independent mechanisms, then

$$n = N - r \qquad (11.3)$$

This is found out to avoid any possible omission of a combined mechanism. After the number of independent mechanisms is found out, all possible combinations are made in such a way that the external work is maximum or the internal work is minimum, so as to obtain the lowest possible load.

11.2.2.4 Theorems of Plastic Analysis

i. Static or lower bound theorem

For a given frame and loading if there exists any distribution of bending moment throughout the frame, which is both safe and statically admissible with set of loads P, the value of the load P must be less than or equal to the collapse load P_u, $(P \le P_u)$. A load computed on the basis of an assumed equilibrium moment diagram in which the moments are not greater than M_p is less than or at best equal to the true ultimate load. Hence the static method represents the lower limit to the true ultimate load and has a maximum factor of safety. The equilibrium and yield conditions are satisfied in the static theorem.

ii. Kinematic or upper bound theorem

For a given frame subjected to a set of loads P, the value of P which is found to correspond to any assumed mechanism must be greater than or equal to the collapse load P_u $(P \ge P_u)$. A load computed on the basis of an assumed mechanism will always be greater than or equal to the true ultimate load. Hence the kinematic theorem represents an upper limit to the true ultimate load and has a smaller factor of safety. Equilibrium and continuity conditions are satisfied in the kinematic theorem.

iii. Uniqueness theorem

For a given frame and loading at least one safe and statically admissible bending moment distribution can be found, and if in this distribution, the bending moment is equal to the fully plastic moment at sufficient cross section to cause the failure of the frame as a mechanism, due to rotation of plastic hinges, then the corresponding load will be equal to the collapse load $(P = P_u)$. All the three conditions are satisfied; equilibrium, continuity, and yield conditions are satisfied in this theorem.

11.2.2.5 Methods of Plastic Analysis

1. Static method of plastic analysis

In this method, the free and bending moment diagrams are drawn for the structure first and then a combined bending moment diagram is drawn in such a way that a mechanism is formed. After that from the equilibrium equation, the collapse load is found out. It is also checked that the bending moment at any section is not more than the fully plastic moment at any section.

2. Kinematic method of plastic analysis

In this method, first of all the possible places of plastic hinges are located. Then the possible independent and combined mechanisms are determined and the collapse load is found out using the principle of virtual work. A bending moment diagram corresponding to the collapse load is also drawn. Here also it is checked that the bending moment at any section is not more than the fully plastic moment at any section.

The static method of analysis is usually more difficult for complicated frames and finding the correct equilibrium equation is illusive. In that case, the kinematic method is more practical.

11.3 INTRODUCTION TO LIMIT STATE DESIGN

Limit states are the acceptable limits for the safety and serviceability requirements of the structure before failure occurs.

Limit states are mainly of two types:

i. Limit state of strength
ii. Limit state of serviceability

11.3.1 LIMIT STATE OF STRENGTH

Limit state of strength also known as limit sate of collapse or ultimate limit state deals with the strength and stability of structures subjected to the maximum design loads out of the possible combinations of several types of loads. Therefore, this limit state ensures that neither any part nor the whole structure should collapse or become unstable under any combination of expected overloads. Limit state of strength ensures the safety of the structure against the following:

i. Stability against overturning:

The structure as a whole or part of it should not overturn, slide, or lift off its seat. It should be proportioned or designed under factored load to ensure this.

ii. Stability against sway:

The whole structure including the portions between expansion joints should be adequately stiff against sway. Sway stability can be provided by braced frames, joint rigidity or by utilizing shear walls, staircases, and lift cores.

iii. Fatigue:

It is the weakening of the structure caused by repeatedly applied loads or when it is subjected to cyclic loading. Welded connections significantly influence the fatigue strength of steel structures than bolted connections.

iv. Brittle fracture:

It is the sudden failure of the structure under service conditions without any warning. The thickness of the material should be limited and the abrupt changes of section and stress concentrations should be avoided.

11.3.2 Limit State of Serviceability

Limit state of serviceability deals with deflection and cracking of structures under service loads, durability under working environment during their anticipated exposure conditions during service, stability of structures as a whole, fire resistance, etc.

i. Deflection

It is the main serviceability limit state that must be considered in the design. It may adversely affect the appearance or effective use of the structure. It may even lead to the improper functioning of services or equipment. When the deflection is excessive, trusses, beams, and girders should be precambered.

ii. Vibrations

It can cause discomfort to people, damages to the structure, and can limit the functional effectiveness of the structure. Floor systems susceptible to vibrations such as open floor areas free of partitions should be given special considerations, so as ensure that such vibrations are acceptable for the intended use and occupancy.

iii. Dynamic effects

Dynamic effects of live loads, impact loads, a sudden gust of wind, vibrations caused by machines, and oscillation caused by harmonic resonance should all be given suitable provisions in the design. They may sometimes be serious and require a proper evaluation and due consideration.

iv. Corrosion and durability

Steel work is critically affected by corrosion in exposed situations, accelerated by sea water, smoke, soot, acid or alkaline vapors, etc., and as a result, there will be loss of surface and durability. Durability of the steel structures should be ensured by following the recommendations of the code.

v. Ponding

Ponding is the unwanted pooling of water on flat roofs and can occur even in the presence of roof drains. It can accelerate the deterioration of many materials including seam adhesives in single ply roof systems, steel equipment supports, and particularly roofing asphalts. Adequate roof slope, precambering of beams, etc., can be adopted to prevent the effects of ponding.

vi. Fire resistance

All hot-rolled steel structural sections have some inherent fire resistance, and it is a function of the size of the section, its degree of exposure to the fire, and the load it carries. The strength of the structural steel decreases substantially at high temperatures. Therefore in steel buildings, steel members should be provided with some form of fire protection.

11.3.3 Partial Safety Factors

Structures should be designed with loads obtained by multiplying the characteristic loads with suitable factors of safety depending on the nature of loads or their combinations, and the limit state being considered. These factors of safety for loads are termed as partial safety factors (γ_f) for loads. Thus, the design loads are calculated as

Design load, (F_d) = Characteristic load (F)

$$\times \text{Partial safety factor for load } (\gamma_f) \quad (11.4)$$

The values of γ_f for loads in two limit states are given in Table 11.1 for different combinations of loads.

γ_f considers the variability of loading and allows for the following:

i. Possibility of unfavorable load deviation from the characteristic value.
ii. Possibility of inaccurate assessment of load.
iii. Uncertainty in the assessment of effects of the load.
iv. Uncertainty in the assessment of the limit states being considered.

Design strength is calculated by dividing the characteristic strength further by the partial safety factor for the material (γ_m), where γ_m depends on the material and the limit state being considered. Thus,

Design strength of the material

$$= \frac{\text{Characterisitic strength of the material}, f}{\text{Partial safety factor of the material}, \gamma_m} \quad (11.5)$$

γ_m considers the variability of material strength and allows for the following:

i. Possibility of unfavorable deviation of material strength from characteristic value as they may

TABLE 11.1

Partial Safety Factors for Loads, γ_f

Load Combinations	Limit State of Collapse			Limit State of Serviceability (For Short-Term Effects Only)		
	DL	IL	WL	DL	IL	WL
DL+IL	1.5	1.5	1.0	1.0	1.0	—
DL+WL	1.5/0.9[a]	—	1.5	1.0	—	1.0
DL+ IL+ WL	1.2	1.2	1.2	1.0	0.8	0.8

Notes

[a] This value is to be considered when stability against overturning or stress reversal is considered.

When considering earthquake effects, substitute EL for WL.

For the limit states of serviceability, the values of γ_f given in this table are applicable for short term effects only. While assessing the long-term effects due to creep, the dead load and that part of the live load likely to be permanent may be considered. Abbreviations

DL = Dead Load, IL = Imposed Load (Live Load), WL = Wind Load, EL = Earthquake Load.

initially vary appreciably and further vary with time due to creep, corrosion, and fatigue.

ii. The method of analysis is often subjected to appreciable errors.

iii. Possibility of unfavorable variation of member sizes.

iv. Possibility of unfavorable reduction in member strength due to fabrication and tolerances.

v. Uncertainty in the calculation of strength of the material.

vi. Mode of failure (ductile/brittle).

11.3.4   Design Criteria

The reliability of the design can be expressed as:

$$\text{Design action } (Q_d) \leq \text{Design strength } (S_d) \quad (11.6)$$

The left-hand side of the equation refers to the load effects on the structure and the right-hand side refers to the resistance or capacity of the structure.

The design action Q_d can be expressed as:

$$Q_d = \sum_k \gamma_f Q_{ck} \quad (11.7)$$

where

Q_{ck} = characteristic load (action)

γ_f = partial safety factor for loads

The design strength S_d can be expressed as

$$S_d = S_u / \gamma_m \quad (11.8)$$

where

S_u = ultimate strength

γ_m = partial safety factor for materials

11.4   SIMPLE CONNECTIONS—RIVETED, BOLTED, AND PINNED CONNECTIONS

Steel sections are usually manufactured to certain standard lengths, as they are governed by rolling, transportation, and handling restrictions. However when used in structures, most of the steel structural members will have to span great lengths and enclose large three-dimensional spaces. Hence connections are structural members used for joining different members of a structural steel framework. Connections can depend on the type of loading, strength and stiffness of the members, economy, and difficult or ease of erection.

11.4.1   Riveted Connections

Rivet is a round rod which holds two metal pieces together permanently. They are made from mild steel bars with yield strength ranging from 220 to 250 N/mm². A rivet consists of a head and a body. The body of rivet is termed as shank. The head of rivet is formed by heating the rivet rod and upsetting one end of the rod by running it into the rivet machine. The rivets are manufactured in different lengths to suit different purposes. The size of rivet is expressed by the diameter of the shank. The length of the rivet should be sufficient enough to form the second head.

The process by which two plates are joined together by the use of rivets is known as riveting. Holes are drilled in the plates to be connected at the appropriate places. For driving the rivets, they are heated till they become red hot and are then placed in the hole. Keeping the rivets pressed from one side, a number of blows are applied and a head at the other end is formed. When the hot rivet so fitted cools it shrinks and presses the plates together. These rivets are known as hot driven rivets. The hot driven rivets of 16, 18, 20 and 22 mm diameter are used for the structural steel works.

Some rivets are driven at atmospheric temperature. These rivets are known as cold-driven rivets. The cold-driven rivets need larger pressure to form the head and complete the driving. The small size rivets ranging from 12 to 22 mm in diameter may be cold-driven rivets. The strength of rivet increases in the cold driving. The use of cold-driven rivets is limited because of equipment necessary and inconvenience caused in the field.

The head of the rivets can be of different shapes as shown in Figure 11.8a–f. Snap head is usually used in structural steel construction. Sometimes it is necessary to flatten the rivet heads so as to provide sufficient clearance. A rivet head which has the form of a truncated cone is called a countersunk head. The snap and pan heads form a projection beyond the plate face, and where this is an objection as in bearings. A countersunk head is employed when continuity between plates or plate and masonry is necessary.

11.4.1.1 Types of Rivet Joints

Riveted joints are mainly of two types, namely, lap joints and butt joints. Two plates are said be connected by a lap joint when the connected ends of the plates lie in parallel planes. Lap joints may be further classified according to the number of rivets used and the arrangement of rivets adopted. The different types of lap joints are shown in Figure 11.9, and they are as follows:

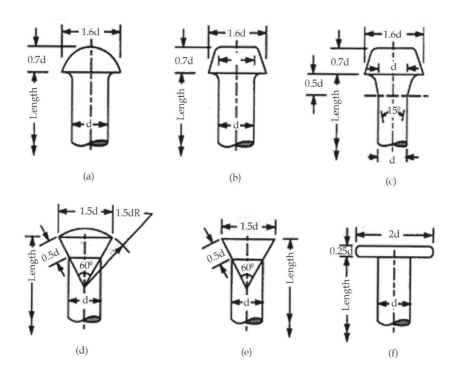

FIGURE 11.8 Various shapes of rivet head. (a) Snap head. (b) Pan head. (c) Pan head with tapered neck. (d) Round counter sunk head 60°. (e) Flat counter sunk head 60°. (f) Flat head.

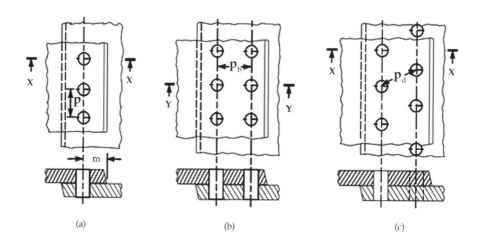

FIGURE 11.9 Types of lap joints. (a) Single riveted lap joint. (b) Double riveted lap joint (chain riveting). (c) Double riveted lap joint (zig zag riveting).

i. single-riveted lap joint
ii. double-riveted lap joint

Double-riveted lap joints can be either chain-riveted lap joint or zigzag-riveted lap joint.

On the other hand, in a butt joint, the connected ends of the plates lie in the same plane. The abutting ends of the plates are covered by one or two cover plates or strap plates. Butt joints may also be classified into single-cover butt joint and double-cover butt joint. In single-cover butt joint, the cover plate is provided on one side of the main plate. In case of double-cover butt joint, cover plates are provided on either side of the main plate. Butt joints are also further classified according to the number of rivets used and the arrangement of rivets adopted. Figure 11.10 shows the different types of butt joints.

11.4.2 BOLTED CONNECTIONS

A bolt may be defined as a metal pin with a head at one end and a threaded shank at the other end to receive a nut. Steel washers are usually provided under the bolt so as to distribute the clamping pressure on the bolted member and also to prevent the threaded portion of the bolt from bearing on the connecting pieces. The holes made for placing the bolts in the joints may be either drilled or punched.

Bolted connections can be classified in the following ways:

i. Based on the resultant force transferred, they are referred to as
   a. Concentric connections: when the load passes through the center of gravity (CG) of the section, e.g., axially loaded tension and compression members.
   b. Eccentric connections: when the load is away from the CG, e.g., bracket connections. Seat connections.
   c. Moment-resisting connections: when joints are subjected to moments, e.g., beam to column connections in framed structures.
ii. Based on the type of force, they are classified as
   a. Shear connections: when the load is through shear, e.g., lap joint and butt joint.

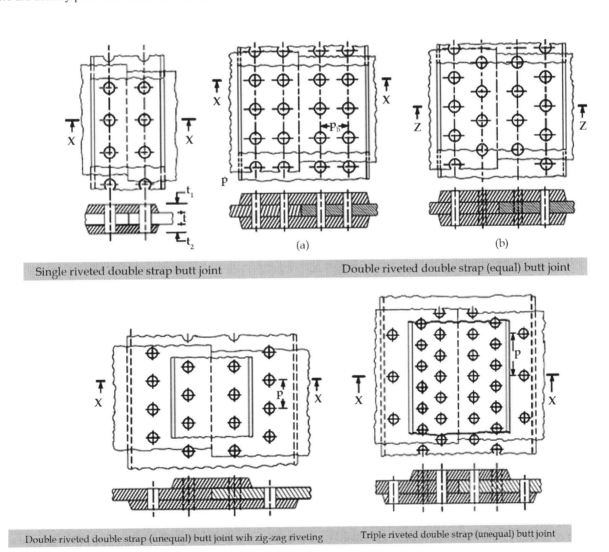

Single riveted double strap butt joint

Double riveted double strap (equal) butt joint

Double riveted double strap (unequal) butt joint wih zig-zag riveting

Triple riveted double strap (unequal) butt joint

FIGURE 11.10 Types of butt joints. (a) Chain riveting. (b) zig-zag riveting.

b. Tension connections: when the load is transferred by tension on bolts, e.g., hanger connection.

c. Combined shear and tension connections: when an inclined member is to be connected to a column through bracket, e.g., connections of bracings.

iii. Based on the force mechanism

a. Bearing type: when the bolts bear against the holes to transfer the load, e.g., slip-type connections.

b. Friction type: when the force is transferred by friction between the plates due to tensioning of bolts, e.g., slip-critical connection.

11.4.2.1 Types of Bolts

The most common type of bolts used to connect structural elements are as follows:

1. **Unfinished bolts**

They are also called ordinary, black, rough, or common bolts. They are least expensive. They are primarily used for light structures under static load and secondary members such as small trusses, purlins, etc. They are recommended for connections subjected to impact load, vibration, and fatigue. They are forged from low carbon rolled steel circular rods, permitting large tolerances.

2. **Turned bolts**

These are similar to unfinished bolts, with the differences that the shank of these bolts is formed from a hexagonal rod. The surfaces of the bolts are prepared carefully and are machined to fit in the hole. Tolerances allowed are very small. These bolts have high shear and bearing resistance as compared to unfinished bolts. However, these bolts are obsolete nowadays.

3. **Ribbed bolts**

These are also called fluted bolts. The head of the bolt is like a rivet head. The thread and nut are provided on the other end of the shank. From the shank core, longitudinal ribs project out making the diameter of the shank more than the diameter of the hole. These ribs cut grooves into the connected members while tightening and ensure a tight fit. These bolts have more resistance to vibrations as compared to ordinary bolts. The permissible stresses for ribbed are same as that for rivets.

4. **High-strength bolts**

These bolts are called friction grip bolts. These are made from bars of medium carbon heat treated steel. Their high strength is achieved through quenching and tempering processes or by alloying steel. Steel washers of hard steel or carburized steel are provided to evenly distribute the clamping pressure on the bolted member and to prevent the threaded portion of the bolt from bearing on the connecting pieces.

11.4.2.2 Types of Bolted Joints

There are two types of bolt joints subjected to axial forces, lap joints and butt joints.

In lap joints, the two members are overlapped and connected together using bolts. It can be either single-bolted lap joint or double-bolted lap joint. The joints are provided in the same manner as in riveted connections (refer Figure 11.9). Since the center of gravity of load in each of the members does not lie in the same line, the load in the lap joint will have eccentricity. This might cause undesirable bending and the bolts may fail in tension. To minimize this effect, at least two bolts should be provided in a line.

In butt joints, the two members to be connected are placed end to end with additional plate/plates placed on either one or both sides, called cover plates, and connected to the main plates using bolts. If the cover plate is provided on one side of the main plates, it is called single-cover butt joint and if it is provided on both the sides it is called double-cover butt joint. The joints are provided in the same manner as in riveted connections (refer Figure 11.10).

11.4.2.3 Bearing-Type Connections

In bearing-type connections, it is assumed that the load to be transferred is larger than the frictional resistance caused by tightening the bolts. As a result, the members slip a little over each other, placing the bolts in shear and bearing. The number of bolts required for making the connection is given by load divided by strength of the bolt. The strength of the bolt is the minimum of strength of bolts in shear, bearing, and tension.

i. **Shearing strength of bolts**

The resistance of a bolt to shear is called the nominal capacity of a bolt in shear and is denoted by V_{nsb}. It is given by

$$V_{nsb} = \frac{f_{ub}}{\sqrt{3}} \left(n_n A_{nb} + n_s A_{sb} \right) \quad (11.9)$$

The nominal shear capacity of a bolt for long joints will be lesser and is modified and expressed as

$$V_{nsb} = \frac{f_{ub}}{\sqrt{3}} \left(n_n A_{nb} + n_s A_{sb} \right) \beta_{lj} \beta_{lg} \beta_{pkg} \quad (11.10)$$

where
f_{ub} = ultimate tensile stress of the bolt
n_n = number of shear planes with threads intercepting the shear plane
n_s = number of shear planes without threads intercepting the shear plane
β_{lj} = a reduction factor to allow for the overloading of the end bolts that occur in long connections
β_{lg} = a reduction factor to allow for the effect of large grip length
β_{pkg} = a reduction factor to account for packing plates in excess of 6 mm

For the safety of joint in shear, the strength of bolt V_{sb}, i.e., the maximum factored shear force that the bolt can carry,

$$V_{sb} \leq \frac{V_{nsb}}{\gamma_{mb}} \qquad (11.11)$$

where γ_{mb} = partial safety factor for the material of bolt = 1.25.

Therefore, the strength of the bolt in shear can be written as

$$V_{sb} = \frac{V_{nsb}}{\gamma_{mb}} = \frac{f_{ub}}{\sqrt{3}\gamma_{mb}}(n_n A_{nb} + n_s A_{sb})\beta_{lj}\beta_{lg}\beta_{pkg} \qquad (11.12)$$

ii. Bearing strength of bolt

The nominal bearing strength of the bolt is given by

$$V_{npb} = 2.5 k_b d t f_u \qquad (11.13)$$

where

k_b = smaller of $\dfrac{e}{3d_0}$, $\dfrac{p}{3d_0} - 0.25$, $\dfrac{f_{ub}}{f_u}$, and 1.0

d_0 = diameter of the hole

e, p = end and pitch distances of the fasteners along bearing direction

f_{ub} = ultimate tensile stress of the bolt

f_u = ultimate tensile stress of the plate in MPa

d = nominal diameter of the bolt in mm, and,

t = aggregate thickness of the connected plates experiencing bearing strength in the same direction

For the safety if the joint in bearing, the bearing strength of the bolt

$$V_{pb} \leq \frac{V_{npb}}{\gamma_{mb}} \qquad (11.14)$$

where γ_{mb} = the partial safety factor for the material of bolt = 1.25.

Therefore, the bearing strength of the bolt can be written as

$$V_{pb} = 2.5 k_b d t \frac{f_u}{\gamma_{mb}} \qquad (11.15)$$

iii. Tensile strength of the bolt

The nominal tensile capacity of the bolt is given by

$$T_{nb} = 0.9 f_{ub} A_{nb}$$
$$\prec f_{yb} A_{sb} \frac{\gamma_{mb}}{\gamma_{m0}} \qquad (11.16)$$

where

f_{ub} = ultimate tensile stress of bolt

f_{yb} = yield stress of bolt

A_{nb} = net tensile stress area of bolt

A_{sb} = shank area of bolt

The bolt is safe in tension if the factored tension force

$$T_{db} \leq \frac{T_{nb}}{\gamma_{mb}} \qquad (11.17)$$

where

γ_{mb} = the partial safety factor for the material of bolt = 1.25

γ_{m0} = partial safety factor for the material resistance governed by yielding = 1.0

Therefore, the tensile strength of the bolt can be written as

$$T_{db} = 0.9 \frac{f_{ub}}{\gamma_{mb}} A_{nb} \qquad (11.18)$$

iv. Tensile strength of plate

If the tensile strength of the plate is less than the tensile load on the plate, the plate fails in tensile through rupture. The pattern of bolts can be chain or staggered (zig-zag). The net area will be more in the case of staggered pattern.

$$\text{Net area, } A_n = (B - nd_h t) \text{ for chain bolting} \qquad (11.19)$$

$$A_n = \left[B - nd_h + \sum_{i=1}^{m} \frac{p_{si}^2}{4g_i} \right] t \text{ for staggered bolting} \qquad (11.20)$$

The tensile strength of the plate is given by

$$T_n = 0.9 A_n \frac{f_u}{\gamma_{m1}} \qquad (11.21)$$

where

f_u is the ultimate tensile stress of material in MPa.

A_n = net effective area in mm²

γ_{m1} = partial safety factor = 1.25

v. Strength and efficiency of the joint

Efficiency of a bolted joint (η) also known as the percentage strength of the joint is the ratio of the strength of the joint to the strength of the main member expressed as a percentage.

$$\eta = \frac{\text{strength of bolted joint/pitch length}}{\text{strength of solid plate/pitch length}} \times 100 \qquad (11.22)$$

vi. Combined shear and tension for bearing-type connections

The following interaction equation can be used to check the safety of the connection,

$$\left(\frac{V_b}{V_{db}}\right)^2 + \left(\frac{T_b}{T_{db}}\right)^2 \leq 1.0 \qquad (11.23)$$

where

V_b = factored shear force on bolt
V_{db} = design shear capacity
T_b = factored tensile force on bolt
T_{db} = design tension capacity

11.4.2.4 Slip-Critical Connection

When the connection length is limited or when the member forces are large, it is most suitable to use High-Strength Friction Grip (HSFG) bolts. At serviceability load these do not slip and the joints are called slip-resistant connections, although they may slip at ultimate loads and behave like bearing-type connections.

  i. **Shear strength of HSFG bolts**
    The nominal shear capacity of a bolt is given by

$$V_{nsf} = \mu_f n_e K_h F_o \qquad (11.24)$$

where

μ_f = slip factor
n_e = number of interfaces offering frictional resistances to slip
K_h = 1.0 for fasteners in clearance holes
    = 0.85 for fasteners in oversized and short slotted holes and for fasteners in long slotted holes perpendicular to slot
    = 0.7 for fasteners in long slotted holes loaded parallel to slot
F_o = minimum bolt tension at installation = $A_{nb}f_o$
A_{nb} = net area of bolt at the threads
f_o = proof stress = $0.7f_{ub}$
f_{ub} = ultimate tensile stress of bolt

For the joint to be safe the factored design shear force,

$$V_f \leq \frac{V_{nsf}}{\gamma_{mf}} \qquad (11.25)$$

where

γ_{mf} = 1.1 for slip resistance designed at service load = 1.25 for slip resistance designed at ultimate load

Therefore, the shear strength of the bolt can be written as

$$V_{df} = \frac{V_{nsf}}{\gamma_{mf}} \qquad (11.26)$$

  ii. **Bearing strength of HSFG bolts**
    Bearing only occurs in HSFG bolts after slip takes place. If the slip is critical, the slip resistance

has to be checked. But if slip is not critical and limit state method is used at ultimate limit state, HSFG bolts will slip into bearing and need a check. Since at ultimate limit sate HSFG bolts slip into bearing, the bolt may deform due to high local bearing stresses between the bolt and the plate.

    The design bearing capacity of bolt V_{npb} can be determined using

$$V_{npb} = 2.5k_b dt f_u \qquad (11.27)$$

t = summation if thickness of all the connected plates experiencing bearing stress in the same direction

  iii. **Tensile strength of HSFG bolts**
    The nominal tensile strength of HSFG bolts subjected to factored tensile force is determined in a way similar to that of the black bolts.

$$T_{nf} = 0.9 f_{ub} A_{nb}$$

$$\prec f_{yb} A_{sb} \frac{\gamma_{m1}}{\gamma_{m0}} \qquad (11.28)$$

The factored design tensile force

$$T_f \leq \frac{T_{nf}}{\gamma_{mf}} \qquad (11.29)$$

Therefore, the tensile strength of bolt can be written as

$$T_{df} = \frac{T_{nf}}{\gamma_{mf}} \qquad (11.30)$$

where

γ_{m1} = partial safety factor for material resistance governed by ultimate stress = 1.15
γ_{m0} = partial safety factor for material resistance governed by yield = 1.10
γ_{mf} = partial safety factor for the material of bolts = 1.25
A_{nb} = net tensile stress area
A_{sb} = shank area of the bolt
f_{ub} = ultimate tensile stress of bolt

  iv. **Combined shear and tension for slip critical connections**

The interaction equation similar to that for black bolts is used in this case also,

$$\left(\frac{V_f}{V_{df}}\right)^2 + \left(\frac{T_f}{T_{df}}\right)^2 \leq 1.0 \qquad (11.31)$$

where

V_f = applied factored shear force at design load
V_{df} = design shear capacity
T_f = externally applied factored tension at design load
T_{df} = design tensile strength

11.4.3 PIN CONNECTIONS

When two structural members are connected by means of a cylindrical shaped pin, the connection is called a pin connection. Pins are manufactured from mild steel bars with diameters ranging from 9 mm to 330 mm. Pin connections are provided when hinged joints are required, i.e., for the connection where zero moment of free rotation is desired. Pins are provided in the following cases:

1. Tie rod connections, water tanks, and elevated bins
2. As diagonal bracing connections in beams and columns
3. Truss bridge girders
4. Hinged arches
5. Chain-link cables suspension bridges

The different types of pins used for making connections are dilled pin, undrilled pin, and forged steel pin. To make a pin connection, one end of the bar is forged like a fork and a hole is drilled in this portion. The other end of the bar to be connected is also forged and an eye is made. A hole is drilled into it in such a way that it matches with the hole on the fork end bar. The eye bar is inserted in the jaws of the fork end and a pin is placed. Both the forged ends are made octagonal for a good grip. The pin in the joint is secured by means of a cotter pin or screw, as shown in Figure 11.11.

11.4.3.1 Shear Capacity

The shear capacity of a pin is as follows:

i. If the rotation is not required and the pin is not intended to be removed,

$$\text{shear capacity} = 0.6 f_{yp} A \qquad (11.32)$$

ii. If the rotation is required or if the pin is intended to be removed,

$$\text{shear capacity} = 0.5 f_{yp} A \qquad (11.33)$$

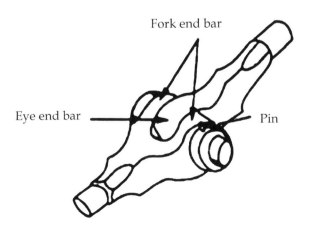

FIGURE 11.11 Knuckle joint.

where
f_{yp} = the design strength of the pin, and
A = the cross sectional area of the pin.

11.4.3.2 Bearing Capacity

The bearing capacity of a pin is as follows:

i. If the rotation is not required and the pin is not intended to be removed,

$$\text{bearing capacity} = 1.5 f_{yp} dt \qquad (11.34)$$

ii. If the rotation is required or if the pin is intended to be removed,

$$\text{bearing capacity} = 0.8 f_{yp} dt \qquad (11.35)$$

where f_{yp} = lower of the design strength of the pin and connected part.

11.4.3.3 Flexural Capacity

Flexure is most critical in case of pins. The members jointed by pins are separated by some distance because of the following reasons:

a. To prevent friction
b. To allow for bolt heads, if the member is built up
c. To facilitate painting

Due to these reasons, large bending moments are generated and therefore the pin diameters are governed by flexure.
The moment capacity of a pin is as follows:

i. If the rotation is not required and the pin is not intended to be removed,

$$\text{moment capacity} = 1.5 f_{yp} Z \qquad (11.36)$$

ii. If the rotation is required or the pin is intended to be removed,

$$\text{moment capacity} = 1.0 f_{yp} Z \qquad (11.37)$$

where f_{yp} = design strength of the pin and Z is section modulus of the pin.

A pin may be subjected to shear, bearing, and flexure, of which flexure is critical. Assuming that the pin is cylindrical shaped beam and is not intended to be removable, if M_u is the ultimate moment, diameter d is given by:

$$d = \left[21.33 \frac{M_u}{\pi f_{yp}} \right]^{\frac{1}{3}} \qquad (11.38)$$

11.4.4 Simple Welded Connections

Welding is the process of joining two structural members by creating a strong metallurgical bond between them by heating or pressure or both. It is distinguished from other forms of mechanical connections, such as riveting or bolting, which are formed by friction or mechanical interlocking. It is one of the oldest and reliable methods of joining.

11.4.4.1 Types of Welds

The two most common types of welds are fillet welds and groove welds (or butt welds). The other types of welds are plug weld, slot weld, spot weld, etc.

Groove welds are the most reliable form of welding. They are provided when the members to be jointed are lined up. They require edge preparations and are thus costly. Single V, U, J, etc., are cheaper to form, but require double the weld metal than double-grooved joints. The choice between single and double grooves depends on the cost of preparation offset by saving in weld metal.

Fillet welds are the most used type of weld. They are easy to make, require less material preparation, and easier to fit than groove welds. But for a given amount of weld material, they are not as strong and can cause greater concentration of stress. They are mostly provided when two members in different planes are to be jointed which are frequently met within structures. They are weaker than groove welds.

Slot and plug welds are mainly used to supplement fillet welds, when the required length of fillet weld cannot be provided. They stich the different parts of members together. They are expensive and also tensile forces are poorly transmitted. These welds are mostly avoided because the penetration of the welds cannot be ascertained and the welds are difficult to inspect.

11.4.4.2 Weld Symbols

Knowledge of welding symbols is essential for a site engineer to be able to read the drawings. They save a lot of space as the descriptive notes can be avoided. The type, size, positions, welding process, etc., of the welds are conveyed through the drawings. The symbolic representation includes elementary symbols along with supplementary symbols, a means of showing dimensions or some complementary indications. Table 11.2 depicts the elementary symbols for welds.

Supplementary symbols characterize the external surface of the weld and they complete the elementary symbols. Supplementary symbols are shown in Table 11.3.

Apart from the symbols, the methods of representation also include the following:

- An arrow line for each joint.
- A dual reference line, consisting of two parallel lines, one continuous, and the other dashed.
- A certain number of dimensions and conventional signs.

The location of welds is classified on the drawings by specifying the position of the arrow line, position of the reference line, and the position of the symbol. An example is shown in Figure 11.12.

The position of arrow line with respect to the weld has no special significance. The arrow line joins one end of the continuous reference line such that it forms an angle with it and shall be completed by an arrowhead or a dot. The reference line is a straight line drawn parallel to the bottom edge of the drawing. The symbol is placed either above or beneath the reference line. The symbol is placed on the continuous line side if the weld is on the arrow side of the joint, and it is placed on the dashed line side if the weld is on the other side of the joint.

11.4.4.3 Welding Process

Welding is the process by which the parts that are to be connected are heated and fused, with supplementary molten metal at the joint.

In the shielded metal arc welding process, *current* arcs across a gap between the electrode and the base metal are heating the connected parts and depositing part of the electrode into the molten base metal. A special coating on the electrode vaporizes and forms a protective gaseous shield, preventing the molten weld metal from oxidizing before it solidifies. The electrode is moved across the joint, and a weld bead is deposited, its size depending on the rate of travel of the electrode. As the weld cools, impurities rise to the surface, forming a coating called *slag* that must be removed before the member is painted or another pass is made with the electrode.

For shop welding, an automatic or semi-automatic process is usually used. Foremost among these is the submerged arc welding. In this process, the end of the electrode and the arc are submerged in a granular flux that melts and forms a gaseous shield. Other commonly used processes for shop welding are *gas shielded metal arc, flux cored arc,* and *electro-slag welding*.

11.4.4.4 Weld Defects

Good welding techniques, standard electrodes, and proper joint preparation are the basic tools to achieve a sound weld. However, defects are inevitable and knowledge of these is essential to minimize them. Some of the common defects found in the welds are as follows:

i. **Incomplete fusion**: It is the failure of the base metal to get completely fused with the weld metal which is caused by rapid welding and also because of the presence of foreign materials on the surface to be welding.

ii. **Incomplete penetration**: It is the failure of the weld metal to penetrate the complete depth of joint. It is normally found with single v_{ee} and bevel joints and also because of large size electrodes.

iii. **Porosity**: It occurs due to voids or gas pockets entrapped in the weld while cooling. It results in stress concentration and reduced ductility of the metal. Porosity is not a problem normally because each void is spherical and not a notch, and therefore, a smooth flow of stress around the void without any measurable loss in strength occurs.

iv. **Slag inclusions**: These are metallic oxides and other solid compounds which are sometimes found as elongated or globular inclusions. Since they are

TABLE 11.2

Elementary Weld Symbol

No	Designation	Illustration	Symbol
1	Butt weld between plates with raised edges (the raised edges being melted down completely)		
2	Square butt weld		
3	Single-V butt weld	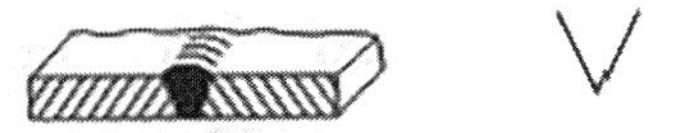	
4	Single-bevel butt weld		
5	Single-V butt weld with broad root face		
6	Single-bevel butt weld with broad root face		
7	Single-U butt weld (parallel or sloping sides		
8	Single-U butt		
9	Backing run; back or backing weld		
10	Fillet weld		
11	Plug weld; plug or slot weld		
12	Spot weld	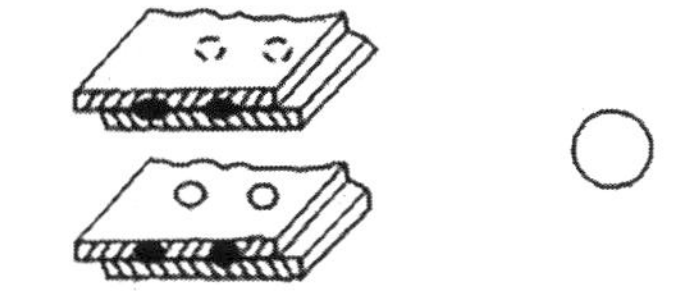	
13	Seam weld	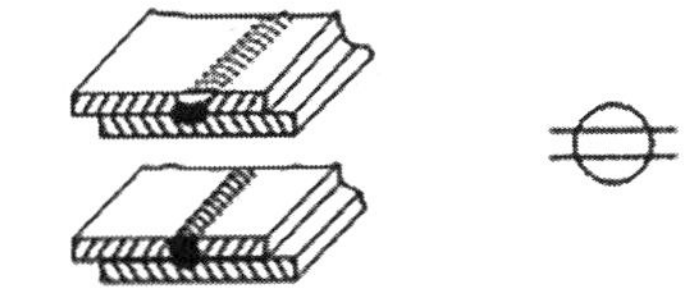	

lighter than the molten materials, they float and rise to the weld surface from where they are removed after cooling of the weld.

v. **Cracks**: Cracks can be divided into hot and cold. Hot cracks occur due to the presence of sulfur, carbon, silicon, and hydrogen in the weld metal. Whereas cold cracks are formed when phosphorus and hydrogen are trapped in the hollow spaces of the metal structure. Formation of cracks can be eliminated by preheating of the metal to be welded.

vi. **Undercutting**: It is the local decrease of the thickness of the parent metal at the weld toe. This is caused by the use of excessive current or a very long arc. An undercut may result in loss of gross cross section and acts like a notch.

11.4.4.5 Inspection of Welds

A poor weld leads to a collapse; therefore, proper inspection of the weld is necessary. Some of the materials for inspecting welds are as follows:

TABLE 11.3

Supplementary Weld Symbols

Form of weld	Illustration	BS symbol
Flat (flush) single-V butt weld		
Convex double-V butt weld		
Concave filled weld		
Flat (flush) single-V butt weld with flat (flush) backing run		

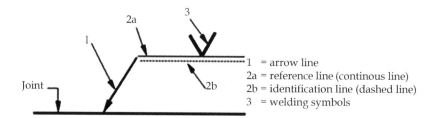

1 = arrow line
2a = reference line (continous line)
2b = identification line (dashed line)
3 = welding symbols

FIGURE 11.12 Representation of weld.

i. **Magnetic particle method**: Iron fillings are spread over the weld and it is then subjected to an electric current. The fillings form patterns which are interpreted to local surface cracks.

ii. **Dye penetration method**: This method is used to estimate the depth of a crack. A dye is applied over the weld surface and then the surplus is removed. A dye absorber is placed over the weld which oozes the dye giving an idea of the depth of the crack.

iii. **Ultrasonic method**: In this method, ultrasonic sound waves are sent through the weld. Defects like flaws, blow holes, etc., affect the time interval of sound transmission identifying the defect.

iv. **Radiography**: X-rays or gamma rays are used to locate defects. This method is used in groove welds only. It cannot be used in fillet welds because the parent material will also form part of the projected picture.

11.4.5 Design of Welds

11.4.5.1 Design of Butt Welds

For butt welds the most critical form of loading is tension applied in the transverse direction. The butt weld is normally designed for direct tension or compression. However, a provision is made to protect it from shear. Design strength value is often taken the same as the parent metal strength. For design purposes, the effective area of the butt-welded connection is taken as the effective length of the weld times the throat size. Effective length of the butt weld is taken as the length of the continuous full size weld. The throat size is specified by the effective throat thickness. For a full-penetration butt weld, the throat dimension is usually assumed as the thickness of the thinner part of the connection. Even though a butt weld may be reinforced on both sides to ensure full cross-sectional areas, its effect is neglected while estimating the throat dimensions. Such reinforcements often have a negative effect, producing stress concentration, especially under cyclic loads.

Intermittent butt welds are used to resist shear only and the effective length should not be less than four times the longitudinal space between the effective length of welds not more than 16 times the thinner part. They are not to be used in locations subjected to dynamic or alternating stresses. Some modern codes do not allow intermittent welds in bridge structures.

For butt welding parts with unequal cross sections, say unequal width, or thickness, the dimensions of the wider or thicker part should be reduced at the butt joint to those of the smaller part. This is applicable in cases where the difference in thickness exceeds 25% of the thickness of the thinner part or 3.0 mm, whichever is greater. The slope provided at the joint for the thicker part should not be steeper than one in five (Figure 11.13a and b). In instances, where this is not practicable, the weld metal is built up at the junction

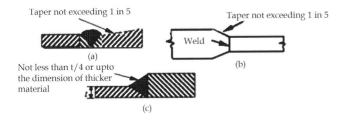

FIGURE 11.13 Tapering limits of weld (a–c).

equal to a thickness which is at least 25% greater than the thinner part or equal to the dimension of the thicker part (Figure 11.13c). Where reduction of the wider part is not possible, the ends of the weld shall be returned to ensure full throat thickness.

The design strength of the butt weld in tension or compression is governed by yield,

$$T_{dw} = \frac{f_y L_w t_e}{\gamma_{mv}} \quad (11.39)$$

where

f_y = smaller of yield stress of the weld (f_{yw}) and the parent metal (f_y) in MPa
L_w = effective length of the weld in mm
t_e = effective throat thickness of the weld in mm
γ_{mv} = partial safety factor
    = 1.25 for shop welding
    = 1.5 for site welding

The design strength of the butt weld in shear is also governed by yield,

$$V_{dw} = \frac{f_{yw1} L_w t_e}{\gamma_{mv}} \quad (11.40)$$

where

f_{yw1} = smaller of shear stress of weld $f_{yw1}/\sqrt{3}$ and the parent metal $f_y/\sqrt{3}$
γ_{mv} = partial safety factor as above
f_y = yield stress of the weld (MPa)

11.4.5.2 Design of Fillet Welds

Fillet welds are broadly classified into side fillets (Figure 11.14a) and end fillets (Figure 11.14b). When a connection with end fillet is loaded in tension, the weld develops high strength and the stress developed in the weld is equal to the value of the weld metal. But the ductility is minimal. On the other hand, when a specimen with side weld is loaded, the load axis is parallel to the weld axis. The weld is subjected to shear and the weld shear strength is limited to just about half the weld metal tensile strength. But ductility is considerably improved. For intermediate weld positions, the value of strength and ductility show intermediate values.

A simple approach to design is to assume uniform fillet weld strength in all directions and to specify a certain throat stress value. The average throat thickness is obtained by dividing the applied loads summed up in vector form per unit length by the throat size. This method is limited in usage to cases of pure shear, tension, or compression (Figure 11.15). For the simple method, the stress is taken as the vector sum of the force components acting in the weld divided by the throat area.

The design stress of a fillet weld

$$f_{wd} = f_{wn}/\gamma_{mw} \quad (11.41)$$

where

$$f_{wn} = \text{nominal strength of the fillet weld} = f_u/\sqrt{3} \quad (11.42)$$

The design strength of a fillet weld is based on its throat area and is given by

$$P_{dw} = L_w t_t \frac{f_u}{\sqrt{3}\gamma_{mw}} \quad (11.43)$$

$$P_{dw} = L_w KS \frac{f_u}{\sqrt{3}\gamma_{mw}} \quad (11.44)$$

where
L_w = effective length of the weld in mm
t_t = throat thickness in mm
S = size of the weld in mm
K = a constant whose value depends upon the angle between tension fusion faces

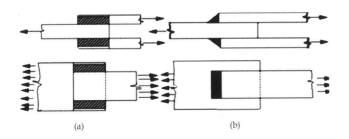

FIGURE 11.14 Types of fillet welds (a and b).

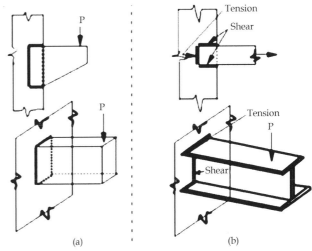

FIGURE 11.15 Shear and tension in fillet weld (a and b).

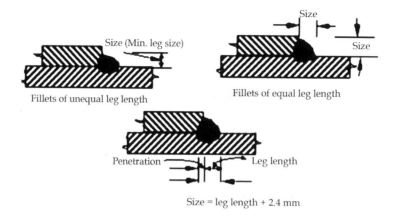

FIGURE 11.16   Size of fillet weld.

f_u = smaller of the ultimate strength of the weld and the parent material in N/mm²

P_{dw} = design strength of weld in N

γ_{mv} = partial safety factor
   = 1.25 for shop welding
   = 1.5 for site welding

The design strength shall be reduced appropriately for long joints as prescribed in the code.

The size of a normal fillet should be taken as the minimum leg size (Figure 11.16).

For a deep penetration weld, the depth of penetration should be a minimum of 2.4 mm. Then the size of the weld is minimum leg length plus 2.4 mm. The size of a fillet weld should not be less than 3 mm or more than the thickness of the thinner part joined. Minimum size requirement of fillet welds is given below in Table 11.4. Effective throat thickness should not be less than 3 mm and should not exceed $0.7t$ and $1.0t$ under special circumstances, where t is the thickness of thinner part.

Values of K for different angles between tension fusion faces are given in Table 11.5. Fillet welds are normally used for connecting parts whose fusion faces form angles between 60° and 120°.

The actual length of the weld is taken as the length having the effective length plus twice the weld size. Minimum effective length should not be less than four times the weld size. When a fillet weld is provided to square edge of a part, the weld size should be at least 1.5 mm less than the edge thickness (Figure 11.17a). For the rounded toe of a rolled section, the weld size should not exceed 3/4 thickness of the section at the toe (Figure 11.17b).

11.4.5.3   Design of Plug and Slot Welds

In certain instances, the lengths available for the normal longitudinal fillet welds may not be sufficient to resist the loads. In such a situation, the required strength may be built up by welding along the back of the channel at the edge of the plate if sufficient space is available. This is shown in Figure 11.18a. Another way of developing the required strength is by providing slot or plug welds. Slot and plug welds (Figure 11.18b)) are generally used along with fillet welds in lap joints.

The specifications to be adhered are as follows:

i. width or diameter should be $\geq 3t$ and also ≥ 25 mm;
ii. corner radius in slotted hole should be $\geq 1.5t$ and also ≥ 12 mm;
iii. clear distance between holes should be $\geq 2t$ and also ≥ 25 mm;

where "t" = thickness of plate having a hole or slot.

The strength of a plug or slot weld is calculated by considering the allowable stress and its nominal area in the shearing plane. This area is usually referred to as the faying surface and is equal to the area of contact at the base of the slot or plug. The length of the slot weld can be obtained from the following:

TABLE 11.4
Minimum Size of First Run or of a Single Run Fillet Weld

Thickness of Thicker Part (mm)	Minimum Size (mm)
$t \leq 10$	3
$10 < t \leq 20$	5
$20 < t \leq 32$	6
$32 < t \leq 50$	8 (First run) 10 (Minimum size of fillet)

TABLE 11.5
Value of K for Different Angles between Fusion Faces

Angle between Fusion Faces	60°–90°	91°–100°	101°–106°	107°–113°	114°–120°
Constant K	0.70	0.65	0.60	0.55	0.50

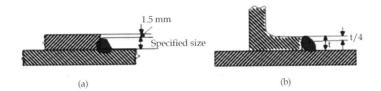

FIGURE 11.17   Weld at square edge and rounded toe (a and b).

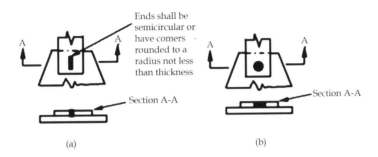

FIGURE 11.18   Slot and plug welds (a and b).

$$L = \frac{\text{Load}}{(\text{Width})\,\text{Allowable stress}} \qquad (11.45)$$

11.5   TENSION MEMBERS

A tension member is designed as a structural member subjected to tensile force in a direction parallel to its longitudinal axis. A tension member is also called a **tie member** or simply a tie. The stress in an axially loaded tension member is given by

$$f = \frac{P}{A} \qquad (11.46)$$

where

  P is the magnitude of load, and
  A is the cross-sectional area normal to the load.

The stress in a tension member is uniform throughout the cross section except near the point of application of load, and at the cross section with holes for bolts or other discontinuities, etc.

11.5.1   Types of Tension Members

The types of structure and method of end connections determine the type of a tension member in structural steel construction. Tension members used may be broadly grouped into four groups.

  1. Wires and cables
  2. Rods and bars
  3. Single structural shapes and plates
  4. Built-up members
    i. Wires and cables
        Cables are flexible tension members with one or more groups of wires or strands or rope. A strand consists of wires placed helically about a center wire. A rope consists of strands laid helically about

a center strand. Wires and cables are mainly used for derricks, hoists, guy wires, rigging slings, and hangers for suspension bridges. The advantages of wires and cables are their flexibility and strength. Figure 11.19 shows the composition of a rope.

  ii. Rods and bar
        They are hot rolled small tension members. They are used in bracing systems or in very light structures such as towers. The rods and bars have the disadvantage of inadequate stiffness resulting in noticeable sag under the self-weight. The round bars with threaded ends are used with pin-connections at the ends instead of threads. The ends of rectangular bars or plates are enlarged by forging and bored to form eye bars. The eye bars are used with pin connections.

  iii. Structural shapes and plates
        If some amount of rigidity is required or reversal of load occurs, then structural shapes are used. The different kinds of single structural shapes used as tension members are shown in Figures 11.20. The angle sections are considerably more rigid than the wire ropes, rods, and bars. Occasionally, I-sections are used as tension members. I-sections have more rigidity, and single I-sections are more economical than built-up sections.

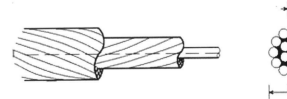

FIGURE 11.19   Composition of rope.

FIGURE 11.20   Structural shapes (a–g).

Double structural shapes are also used as tensions members such as double-angle section and double-channel section.

iv. Built-up sections

Two or more than two members are used to form built-up members. When the single-rolled steel section cannot furnish the required area, then built-up sections are used. They are mainly used in truss girder bridges. In built-up sections, the dashed line indicates a lateral system such as a lacing or battening, which makes the different components acts as one section.

The angle sections are placed back-to-back on two sides of a gusset plate (Figure 11.21a). When both the angle sections are attached on the same side of the gusset, as seen in Figure 11.21b, then built-up section has eccentricity in one plane and is subjected to tension and bending simultaneously. Four angle sections as shown in Figure 11.21c are also used in the two-plane trusses. For angle sections connected by plates as shown in Figure 11.21d are used as tension members in bridge girders.

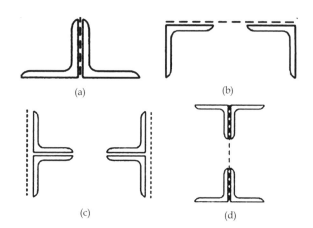

FIGURE 11.21   Angle sections (a–d).

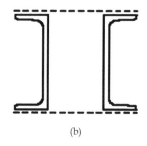

FIGURE 11.22   Channel sections (a and b).

A built-up section may be made of two channels placed back-to-back with a gusset in between them. Such sections are used for medium loads in a single plane-truss. In two-plane trusses, two channels are arranged at a distance with their flange turned inward (Figure 11.22a). It simplifies the transverse connections and also minimizes lacing. The flanges of two channels are kept outward, as in the case of chord members or long span girders, in order to have greater lateral rigidity (Figure 11.22b).

11.5.2 NET CROSS SECTIONAL AREA

A tension member is designed for its net sectional area at the joint. When a tension member is spliced or joined to a gusset plate by rivets and bolts, the gross cross-sectional area is reduced by rivet holes. If no holes are made in a tension member for its end connection, the gross cross-sectional area is effective in resisting the tension.

i. **For plates**

The presence of a hole in a plate subjected to tension produces a stress concentration at the hole edge in the elastic range but the stress distribution becomes uniform at yielding or at ultimate stress as shown in Figure 11.23. The effective area of the cross section of a plate at the hole is the net area.

In a plate, if the holes are in a single line (Figure 11.24a), the net sectional area is obtained by multiplying the net effective width b_e with the thickness t.

$$A_n = b_e t \qquad (11.47)$$

If there are "n" number of holes in a line, then

$$b_e = b - n d_0 \qquad (11.48)$$

where d_0 = diameter of the hole.

If the holes are staggered at close pitch (Figure 11.24b), the net effective area is obtained by multiplying the net effective width b_e with the thickness t of the plate where the net effective width b_e is obtained using the following expression:

$$b_e = b - n d_0 + n_1 \frac{p^2}{4g} \qquad (11.49)$$

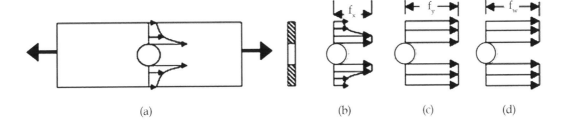

FIGURE 11.23   Stress distribution in plate with hole (a–d).

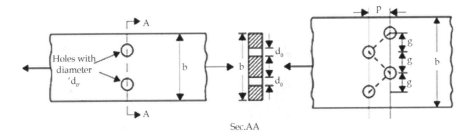

Sec.AA

FIGURE 11.24   Orientation of holes in a plate.

where
p = staggered pitch
g = gauge
n = number of holes
n_1 = number of inclined lines between the holes
d_0 = diameter of the holes

Again, the net section area of the plate is given by

$$A_n = b_e t \qquad (11.50)$$

ii. **For single angle connected by one leg only as shown in Figure 11.25**
   The net effective area is given by,

$$A_{net} = A_1 + A_2 k_1 \qquad (11.51)$$

where
A_1 = net cross-sectional area of the connected leg
A_2 = gross cross-sectional area of the unconnected leg

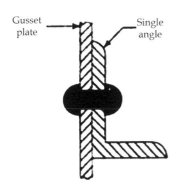

Gusset plate   Single angle

FIGURE 11.25   Single angle connected by one leg.

and,

$$k_1 = \frac{3 \times A_1}{3 \times A_1 + A_2} \qquad (11.52)$$

The area of a leg of an angle

= thickness of angle X (length of leg − ½ X thickness of leg)
$$(11.53)$$

iii. **For two angles placed back-to-back (or a single Tee) connected by only one leg of each angle (or by the flange of a Tee) to the same side of gusset plate as shown in Figure 11.26.**
   The net effective area is given by

$$A_{net} = A_1 + A_2 k_1 \qquad (11.54)$$

where
A_1 = net sectional area of the connected legs (or flange of the Tee)
A_2 = area of the outstanding legs (or web of the Tee)

and

$$k_1 = \frac{5 A_1}{5 A_1 + A_2} \qquad (11.55)$$

The area of a web of a Tee

= thickness of web (depth − ½ thickness of flange)
$$(11.56)$$

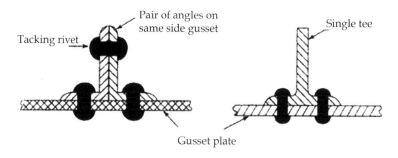

FIGURE 11.26 Two angles or tee placed back-to-back and one leg each connected to the gusset plate.

iv. **For double angles or Tees carrying direct tension placed back-to-back and connected to each side of gusset plate or to each side of a rolled section as shown in Figure 11.27.**

The net effective area is given by

$$A_{\text{net}} = \text{Gross area} - \text{Deduction for holes} \quad (11.57)$$

provided that angles or tees are tack riveted along their length at a pitch not exceeding 1 m.

11.5.3 Design of Tension Members

The tension members can sustain loads up to the ultimate load where they fail to rupture at a weakest section. However, the yielding of the gross cross section of the member may take place before the rupture at the weakest (net) section. Plates and other rolled structure may also fail by block shear, especially at bolted end connections. The factored design tension in the member (T) should be less than or equal to the design strength of the member (T_d) which is the least of the design strength due to the yielding of the gross cross section T_{dg}, the design strength due to the rupture at the net section T_{dn}, and the design strength due to the block shear T_{db}.

11.5.3.1 Design Strength Due to Yielding

Although steel tension members can sustain loads up to the ultimate load without failure, the elongation of the members at this load would be nearly 10%–15% of the original length and the structure supported by the member would become unserviceable. Hence, in the design of tension members, the yield load is usually taken as the limiting load. The corresponding design strength in member under axial tension is given by

$$T_{dg} = \frac{A_g f_y}{\gamma_{m0}} \quad (11.58)$$

where

f_y = yield strength of the material in MPa
A_g = gross area of the cross section
γ_{m0} = partial safety factor for the yielding = 1.1

11.5.3.2 Design Strength Due to Rupture

11.5.3.2.1 Flats/Plates

Since only a small length of the member adjacent to the smallest cross section at the holes would stretch a lot at the ultimate stress, and the overall member elongation need not be large, as long as the stresses in the gross cross section are below the yield stress. Hence, the design strength as governed by net cross section at the hole, T_{dn}, is given by

$$T_{dn} = 0.9 A_n f_u / \gamma_{m1} \quad (11.59)$$

where

f_u = ultimate strength of the material
A_n = net area of the cross section
γ_{m1} = partial safety factor for the rupture = 1.25

11.5.3.2.2 Single Angle (Figure 11.28)

The rupture of a single angle connected through one leg and affected by shear lag is given by

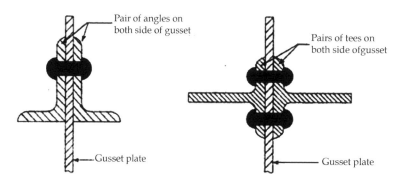

FIGURE 11.27 Double angles or tees placed back-to-back and connected to each side of gusset plate.

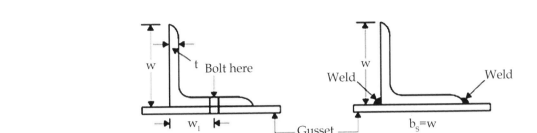

FIGURE 11.28 Single angle.

$$T_{dn} = \frac{0.9 A_{nc} f_u}{\gamma_{m1}} + \frac{\beta A_{go} f_y}{\gamma_{m0}} \qquad (11.60)$$

where

$$\beta = 1.4 - 0.076 \left(\frac{w}{t}\right)\left(\frac{f_y}{f_u}\right)\left(\frac{b_s}{L_c}\right) \qquad (11.61)$$

$$\text{but } 0.7 \le \beta \le \frac{f_u \gamma_{m0}}{f_y \gamma_{m1}} \qquad (11.62)$$

in which

w = width of the outstanding leg
b_s = shear lag width as shown in Figure 11.28
L_c = length of the end connection, i.e., distance between the outmost bolts or length of the weld along the direction of load
A_{nc} = net sectional area of the connected leg
A_{go} = gross area of the outstanding leg
t = thickness of the angle

11.5.3.2.3 Other Sections

The rupture strength of double-angles, tee, channel, I-section, etc., is also affected by shear-lag and may be calculated using the above equation. For the calculation of β, b_s should be taken as the distance from the farthest edge of the outstanding leg to the nearest bolt or weld line in the connecting leg of a section. Double angles should be tack welded or bolted along their length at a spacing not exceeding 1000 mm so that they act as one integral section.

11.5.3.3 Design Strength Due to Block Shear

A tension member may fail along end connection due to block shear as shown in Figure 11.29.

The corresponding design strength can be evaluated using the following equations. The block shear strength, T_{db}, at an end connection is taken as the smaller of

$$T_{db} = V_{bdg} + T_{bdn} = \frac{A_{vg} f_y}{\gamma_{m0}\sqrt{3}} + \frac{0.9 A_{tn} f_u}{\gamma_{m1}} \qquad (11.63)$$

or

$$T_{db} = V_{bdn} + T_{bdg} = \frac{0.9 A_{vn} f_u}{\gamma_{m1}\sqrt{3}} + \frac{A_{tg} f_y}{\gamma_{m0}} \qquad (11.64)$$

where A_{vg}, A_{vn} are the minimum gross and net area in shear along the bolt line parallel to the line of transmitted force, respectively (along 1-2-3 as shown in Figure 11.29); A_{tg}, A_{tn} = minimum gross and net area in tension from the bolt hole to the edge of a plate or between the bolt holes, perpendicular to the line of force, respectively (along 3-4-5 as shown in Figure 11.29); f_u, f_y are the ultimate and yield stress of the material of the plate, respectively.

Equation 11.63 accounts for the shearing strength (V_{bdg}) on the gross area A_{vg} at yielding and the tensile strength due to rupture (T_{bdn}) on the net area A_{tn}. Equation 11.64 accounts for the shearing strength on the net area A_{vn} and the tensile strength due to yielding (T_{bdg}) on the gross area A_{tg}.

11.5.4 Lug Angles

The lug angle is a short length of an angle section, as shown in Figure 11.30, used at a joint to connect the outstanding leg

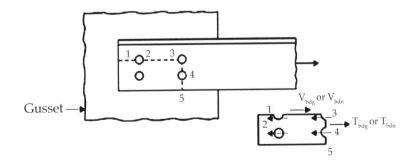

FIGURE 11.29 Block shear.

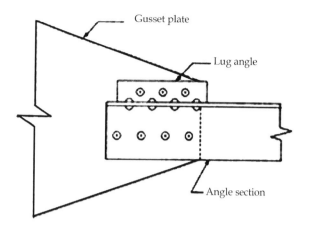

FIGURE 11.30 Lug angle.

of a member, thereby reducing the length of the joint. A lug angle is provided at the beginning of a joint so that it can be effective in sharing load.

11.6 COMPRESSION MEMBERS

A structural member loaded axially in compression is generally called a compression member. That is, the loads are applied on the longitudinal axis through the centroid of the member cross section, and the load over the cross-sectional area gives the stress on the compressed member. Vertical compression members in buildings are called columns, posts or stanchions. A compression member in roof trusses is called struts and in a crane is called a boom. A compression member may be called short or long (Figure 11.31), depending on the slenderness ratio ($\lambda = KL/r$), where KL is the effective length and r is the least radius of gyration of the cross section. For small values of KL/r, a compression member undergoes only simple compression and is called a short column. For larger values of KL/r, a compression member undergoes buckling, i.e., side-way deflection and is called a long

column. The loading capacity of a short column is determined by strength limit of the material, and therefore, they fail by crushing or yielding. The strength of a column of intermediate size is limited by its degree of inelasticity. A long column is constrained by the elastic limit and they fail by buckling. The knowledge of the buckling or the elastic stability of the compression members is extremely important for the design of compression members. The design of long columns is based on Euler's buckling theory.

11.6.1 EULER'S BUCKLING THEORY

Consider an initially straight prismatic long column with both the ends pinned subjected to an axial compression P as shown in Figure 11.32a.

The critical load of the column is given by

$$P_{cr} = \frac{\pi^2 EI}{L^2} \quad (11.65)$$

where
E = modulus of elasticity of the material of the column,
I = moment of inertia of the cross section of the column, and
L = length of column between the hinged ends.

If σ_{cr} is the crushing stress,

$$\sigma_{cr} = \frac{\pi^2 E}{\left(\dfrac{KL}{r}\right)^2} \quad (11.66)$$

$$\sigma_{cr} = \frac{\pi^2 E}{\lambda^2} \quad (11.67)$$

where λ is the slenderness ratio and KL is the effective length of the column in which K is a factor known as the effective

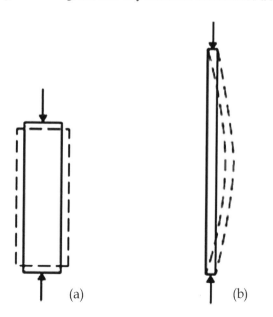

FIGURE 11.31 Compression members (a and b).

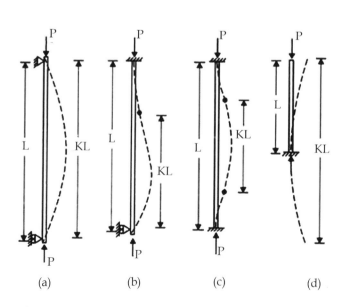

FIGURE 11.32 Buckling of long column (a–d).

TABLE 11.6
Effective Length Factor (K) for Different End Conditions

End Conditions of Column	K
Both ends pinned (Figure 11.32a)	1.0
One end fixed, other end pinned (Figure 11.32b)	0.707
Both ends fixed (Figure 11.32c)	0.5
One end fixed, other end free (Figure 11.32d)	2

length factor. The values of K for different end conditions of column are given in Table 11.6.

11.6.2 Types of Sections

Various types of sections may be used as single sections or in combination with either the same or different cross sections. Rods (Figure 11.33a) and bars (Figure 11.33b) withsatnd very little compression when the length is more and hence they are used for lengths less than 3 m only. But solid sections have much smaller radius of gyration and hence it is not efficient. Hence tubular sections, as shown in Figure 11.33c and d, which are more efficient are used in roof trusses and towers.

Single angle sections (Figure 11.33e) are also used as compression members nowadays, mostly in towers and also as a bracing member in plate girder bridges and in large built-up columns. Double angles placed back-to-back, as shown in Figure 11.33f, or with legs spread, as shown in Figure 11.33g

are most suited for trusses. When the legs are spread, the least radii of gyration are same as that when placed back-to-back. The stiffness of double-angle sections with legs spread is greater than when placed back-to-back but it is uneconomical because of the additional cost of lacing and riveting. Two angle sections can also be used in the form of a star (Figure 11.33h) and is the most effective because of its approximately equal radii of gyration in two directions. The box section, as shown in Figure 11.33i, is formed with welded connections. Four angle sections can be used as shown in Figure 11.33e–g). When the load is small, the built-up section shown in Figure 11.33j is used and for higher loads, the section shown Figures 11.33k is preferred. The section shown in Figures11.33l is used for moderate loads and large length, i.e., when the least radius of gyration required is large but it is uneconomical due to excessive lacing. The connecting systems can be replaced by plates on one side or on more than one side as shown in Figure 11.33m–o. Such sections are normailly provided in bridge trusses.

Two channels can be placed to form sections as shown in Figure 11.33(p–r). When the channels are placed back-to-back, they result in a small value of radius of gyration about y-y axis and are more often recommended. When two channels are placed face-to-face, it provides a larger value of radius of gyration as compared to the other two and therefore forms an ideal section for compression members.

Generally rolled section with additional intermediate support in the weak direction is used as column sections.

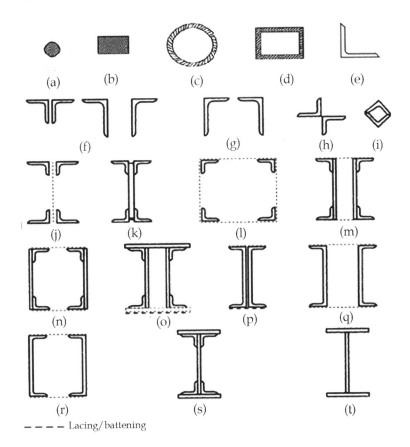

FIGURE 11.33 Typical compression members (a–t).

For larger loads, when rolled I-section does not suffice, an I-section is built up with plates and angles in case-riveted connections are done (Figure 11.33k and s). On the other hand, such a section is built up with plates only if welded connections are done (Figure 11.33t). Two I-sections laced together can also be used for higher loads.

11.6.3 Strength of Axially Loaded Compression Members

Maximum axial compression load permitted on a compression member,

$$P = \sigma_{ac} A \qquad (11.68)$$

where
P = axial compressive load (N)
σ_{ac} = allowable compressive stress (MPa)
A = effective cross-sectional area of the member mm^2

The average allowable compressive stress in the section is assumed. It should not be more than upper limit of the column formula specified by the code.
Permissible stress in axial compression (MPa):

$$\sigma_{ac} = 0.6 \frac{f_{cc} f_y}{\left[f_{cc}^n f_y^n \right]^{\frac{1}{n}}} \qquad (11.69)$$

where

$$f_{cc} = \text{elastic critical stress in compression} = \frac{\pi^2 E}{\lambda^2} \qquad (11.70)$$

f_y = yield stress of steel in MPa,
$\lambda = \dfrac{l_e}{r}$ = slenderness ratio of the member,

l_e = effective length of the member,
r = appropriate radius of gyration of the member,
E = modulus of elasticity = 200,000 MPa, and
n = a factor assumed as 1.4.

11.6.4 Effective Length of Compression Member (Table 11.7)

The actual length L of the compression member should be taken as the length from center-to-center of intersection of supporting members or the cantilevered length in the case of free-standing struts.

11.6.5 Maximum Slenderness Ratio

The slenderness ratio should not exceed the values given below in Table 11.8.

11.6.6 Angle Struts

Single-angle discontinuous struts connected by a single rivet or bolt may be designed for axial load only provided the compressive stress does not exceed $0.8\sigma_{ac}$. The value of σ_{ac} can be determined on the basis that the effective length l_e of the strut is from center-to-center of intersection at each end and r is the minimum radius of gyration. In no case, the $\dfrac{l_e}{r}$ ratio for single angle struts should exceed 180°. If a single discontinuous strut is connected by a weld or by two or more rivets or bolts in line along the angle at each end, it may be designed for axial load only provided the compression stress does not exceed σ_{ac} arrived at on the basis that l_e is taken as 0.85 times the length of the strut, center-to-center at each end, and r is the minimum radius of gyration.

TABLE 11.7
Effective Length of Members for Various End Conditions

Sl. No	Type	Effective Length of Member, l_e
1	Effectively held in position and restrained against rotation at both ends	$0.65L$
2	Effectively held in position at both ends restrained against rotation at one end	$0.80L$
3	Effectively held in position at both ends but not restrained against rotation	$1.0L$
4	Effectively held in position and restrained against rotation at one end and at the other end effectively restrained against rotation but not held in position	$1.2L$
5	Effectively held in position and restrained against rotation at one end and the other end partially restrained against rotation but not held in position	$1.5L$
6	Effectively held in position at one end but not restrained against rotation and at the other end restrained against rotation but not held in position	$2.0L$
7	Effectively held in position and restrained against rotation at one end but not held in position or restrained against rotation at the other end	$2.0L$

Note
L is the unsupported length of compression member.
For battened struts, the effective length should be increased by 10%.

TABLE 11.8
Equivalent Length for Various End Conditions

S. No	Type of Member	Slenderness Ratio
1	A member carrying compressive loads resulting from dead and superimposed loads	180
2	A member subjected to compressive loads resulting from wind/earthquake forces provided the deformation of such members does not adversely affect the stress in any part of the structure	250
3	A member normally carrying tension but subjected to reversal of stress due to wind or earthquake forces	350
4	Tension member (other than pretensioned member)	400

For double-angle struts which are discontinuous, back-to-back connected to both sides of the gusset or section by not less than two bolts or rivets in line along the angles at each end or by equivalent welding, the load may be regarded as applied axially. The effective length l_e in the plane of end gusset could be taken between 0.7 and 0.85 times the distance between the intersections depending on the restraint provided, the plane perpendicular to that of the end gusset, the effective length should be taken as equal to the distance between centers of intersections.

11.6.7 COMPRESSION MEMBERS COMPOSED OF BACK-TO-BACK COMPONENTS

A compression member composed of two angles, channels or tees, back to back, in contact or separated by a small distance should be connected together by riveting, bolting, or welding so that the slenderness ratio of each member between the connections is not greater than 40 nor greater than 0.60 times the most unfavorable slenderness ratio of the strut as a whole. In no case, the spacing of tacking rivets in a line exceeds 600 mm for such members.

For other types of built-up compression members, where cover plates are used, the pitch of tacking rivets should not exceed $32t$ or 300 mm, whichever is less, where t is the thickness of the thinner outside plate. Where plates are exposed to bad weather conditions, the pitch should not exceed $16t$ or 200 mm whichever is less.

The rivets, welds, and bolts in these connections should be sufficient to carry the shear force and bending moments, if any, specified for battened struts. The diameter of the connecting rivets should not be less than the minimum diameter given below in Table 11.9.

A minimum of two rivet or bolts should be used in each connection, one on line of each gauge mark, where the legs of the connected angles or tables of the connected tees are 125 mm wide or over, or where the webs of channel are 150 mm wide or over.

11.6.8 LACINGS AND BATTENS FOR BUILT-UP COMPRESSION MEMBERS

In a built-up section, the different components are connected together so that they act as a single column. Lacing is generally preferred in case of eccentric loads. Battening is normally used for axially loaded columns and in sections where the components are not far apart. Flat bars are generally used for lacing. Angles, channels, and tubular sections are also used for lacing of very heavily columns. Plates are used for battens.

11.6.8.1 Lacings

Lacing is the most common type of lateral system that is used in built-up columns. Different types of lacing systems used as compression members are shown in Figure 11.34. The simplest form of lacing is known as single lacing, which is shown in Figure 11.34a. Sometimes double lacing is preferred. Lacing bars are connected to the components of the column using bolts or welding. The connections are designed to transfer only the axial forces in the lacing bars to the main components, and hence, it is not moment resistant. At the end of the laced columns, tie plates (battens) are provided.

A lacing system should generally conform to the following requirements:

i. The compression member comprising two main components laced and tied should, where practicable, have a radius of gyration about the axis perpendicular to the plane of lacing not less than the radius of gyration at right angles to that axis.
ii. The lacing system should not be varied throughout the length of the strut as far as practicable.
iii. Cross (except tie plates) should not be provided along the length of the column with lacing system, unless all forces resulting from deformation of column members are calculated and provided for in the lacing and its fastening.
iv. The single-laced systems on opposite sides of the main components should preferably be in the same direction so that one system is the shadow of the other.
v. Laced compression members should be provided with tie plates at the ends of the lacing system and

TABLE 11.9
Minimum Diameter of Rivets

Thickness of Member	Minimum Diameter of Rivets
Up to 10 mm	16 mm
Over 10 mm up to 16 mm	20 mm
Over 16 mm	22 mm

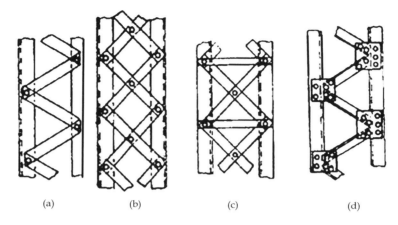

FIGURE 11.34 Lacing. (a) Single lacing. (b) Double lacing. (c) Double lacing with cross member (not recommended). (d) Single angle lacing with gussets.

at points where the lacing system are interrupted. The tie plates should be designed by the same method as followed for battens.

11.6.8.1.1   Guidelines for the Design of Lacing System

i. The angle of inclination of the lacing with the longitudinal axis of the column should be between 40° and 70°.

ii. The slenderness ratio $\lambda = \dfrac{l_e}{r}$ of the lacing bars should not exceed 145.

iii. The effective length l_e of the lacing bar should be according to Table 11.10 given below.

iv. For riveted or welded lacing system, $\left(\dfrac{L}{r_{e,\min}}\right)$ should not be greater than 50 or 0.7 times maximum slenderness ratio of the compression member as a whole, whichever is less. Here, $L =$ distance between the centers of connections of the lattice bars as shown in Figure 11.35, and $r_{e,\min} =$ the minimum radius of gyration of the components of the compression member.

v. Minimum width of lacing bars in riveted connection should be according to Table 11.11 given below.

vi. Minimum thickness of lacing bars:

t should not be less than $\dfrac{l}{40}$, for single lacing;

t should not be less than $\dfrac{l}{60}$, for double lacing

where $l =$ length between inner end rivets.

vii. The lacing of compression members should be designed to resist a transverse shear, $V = 2.5\%$ of the axial force in the member. The shear is divided equally among all transverse lacing systems in parallel planes. The lacing system should also be designed to resist additional shear due to bending if the compression member carries bending due to eccentric load, applied end moments, and/or lateral loading.

viii. The riveted connections may be made in two ways, as shown in Figure 11.36a and b.

ix. Welded connection

Lap joint: Overlap should not be less than ¼ times thickness of bar or member, whichever is less.

Butt joint: Full penetration butt weld or fillet weld on each side, lacing bar should be placed opposite to flange or stiffening component of main member.

11.6.8.2   Battens

Battening consists of provisions of battens along the length of a column as shown in Figure 11.37. Compression members composed of two main components battened should preferably have the components of the same cross section and symmetrically disposed about their x–x axis. The battens should be placed opposite to each-other at each end of the member and at points where the member is stayed in its length, and should as far as practicable, be spaced, and proportioned uniformly throughout. The effective length of columns should be increased by 10%.

TABLE 11.10
Effective Length of Lacing Bar

No	Type of Lacing	Effective Length, l_e
1	Single lacing, riveted at ends	Length between inner end rivets on lacing bar
2	Double lacing, riveted at ends	0.7 times the length between end rivets on lacing bars
3	Welded lacing	0.7 times the distance between inner ends or effective lengths of welds at ends

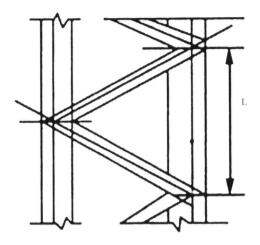

FIGURE 11.35   Length of lattice bar.

TABLE 11.11

Minimum Width of Lacing Bars

Nominal rivet diameter (mm)	22	20	18	16
Width of lacing bars (mm)	65	60	55	50

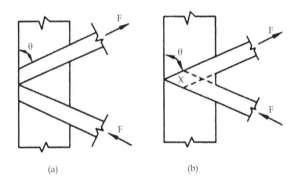

(a)   (b)

FIGURE 11.36   Riveted connection in lacing system (a and b).

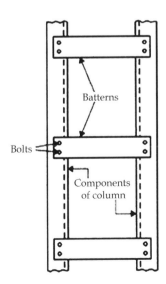

FIGURE 11.37   Battens.

11.6.8.2.1   Design Details of Battens

i. Spacing of batten C, from center-to-center of end fastening should be such that the slenderness ratio of the lesser main component, $\left(\dfrac{C}{r_{e,\min}}\right)$ should not be greater than 50 or 0.7 times the slenderness ratio of the compression member as a whole about x–x axis (parallel to battens) whichever is less.

ii. Effective depth, d, of the battens shall be taken as distance between end rivets or end welds.

$$d > \frac{3}{4}a, \text{ for intermediate batten}$$

$d > a$, for end batten.
$d > 2b$, for any batten
where
d = effective depth of batten
a = centroidal distance of members
b = width of members in the plane of battens

iii. Thickness of battens,

$$t > \left(\frac{l_b}{50}\right) \tag{11.71}$$

where l_b = distance between the innermost connecting line of rivets or welds.

11.6.8.2.2   Design of Battens

Battens should be designed to carry bending moment and shear arising from a transverse shear,

$$V = \frac{2.5}{100}P \tag{11.72}$$

where P = total axial load in the compression member.

Transverse shear V is divided equally between the parallel planes of battens. Battens and their connections to main components resist simultaneously a longitudinal shear,

$$V_1 = \frac{V \times C}{N \times S} \tag{11.73}$$

and moment,

$$M = \frac{V \times C}{2N} \tag{11.74}$$

due to transverse shear V,
where

C = spacing of battens
N = number of parallel planes of battens
S = minimum transverse distance between centroid of rivet group or welding.

The end connections should also be designed to resist the longitudinal shear force V_1 and the moment M.

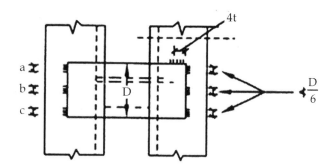

FIGURE 11.38   Welded connections in battens.

11.6.8.2.3   For Welded Connection (Figure 11.38)

1. Lap < $4t$
2. Total length of weld at edge of batten < $D/2$

   $a + b + c < D/2$

   let t = thickness of batten

   Length of continuous weld at each edge of batten < 1/3 of total length required.

   Return weld along transverse axis of the column < $4t$

   where t and D are the thickness and overall depth of battens, respectively.

11.7   BEAMS

A beam is a structural element that is capable of withstanding load primarily by resisting against bending. The bending force induced into the material of the beam as a result of the external loads, own weight, span, and external reactions to these loads is called a bending moment. Beams are characterized by their shape of cross section, their length, and their material.

In steel structures, the commonly used sections as beams are I, channel, tube, rectangular hallow sections. Built-up sections are also used when single sections are not able to resist the loads.

11.7.1   Behavior of Steel Beams

Laterally stable steel beams can fail only by (1) Flexure, (2) Shear, or (3) Bearing, assuming the local buckling of slender components does not occur. These three conditions are the criteria for limit state design of steel beams. Steel beams would also become unserviceable due to excessive deflection and this is classified as a limit state of serviceability.

The factored design moment, M, at any section, in a beam due to external actions shall satisfy

$$M \leq M_d \tag{11.75}$$

where M_d = design bending strength of the section.

11.7.1.1   Bending (Flexure)

The behavior of members subjected to bending is demonstrated in Figure 11.39.

This behavior can be classified under two parts:

- When the beam is adequately supported against lateral buckling, the beam failure occurs by yielding of the material at the point of maximum moment. The beam is thus capable of reaching its plastic moment capacity under the applied loads. Thus the design strength is governed by yield stress and the beam is classified as laterally supported beam.
- Beams have much greater strength and stiffness while bending about the major axis. Unless they are braced against lateral deflection and twisting, they are vulnerable to failure by lateral torsional buckling prior to the attainment of their full in-plane plastic moment capacity. Such beams are classified as laterally supported beam.

11.7.1.1.1   Beams Which Fail by Flexural Yielding

1. **Those which are laterally supported**

   The design bending strength of beams, adequately supported against buckling (laterally supported beams), is governed by yielding. The bending strength of a laterally braced compact section is the plastic moment M_p. If the section has a large shape factor, significant inelastic deformation may occur at service load, if the section is permitted to reach M_p at factored load. The limit of 1.5 M_y at factored load will control the amount of inelastic deformation for sections with shape factors greater than 1.5. This provision is not intended to limit the plastic moment of a hybrid section with a web yield stress lower than the flange yield stress. Yielding in the web does not result in significant inelastic deformations.

2. **Those which are laterally shifted**

   Lateral-torsional buckling cannot occur, if the moment of inertia about the bending axis is equal to or less than the moment of inertia out of plane. Thus, for shapes bent about the minor axis and shapes with $I_z = I_y$ such as square or circular shapes, the limit state of lateral-torsional buckling is not applicable and yielding is controlled provided the section is compact.

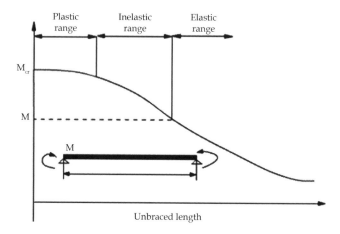

FIGURE 11.39   Members subjected to bending.

11.7.1.2 Shear

Let us take the case of an "I" beam subjected to the maximum shear force (at the support of a simply supported beam). The external shear "V" varies along the longitudinal axis "x" of the beam with bending moment as $V = d_m/d_x$. While the beam is in the elastic stage, the internal shear stresses τ, which resist the external shear, V, can be written as

$$\tau = \frac{VQ}{It} \tag{11.76}$$

where

V = shear force at the section,
I = moment of inertia of the entire cross section about the neutral axis,
Q = moment about neutral axis of the area that is beyond the fiber at which τ is calculated, and
t = thickness of the portion at which τ is calculated.

The above Equation 11.76 is plotted in Figure 11.40, which represents shear stresses in the elastic range. It is seen from the figure that the web carries a significant proportion of shear force and the shear stress distribution over the web area is nearly uniform.

Hence, for the purpose of design, we can assume without much error that the average shear stress as

$$\tau_{av} = \frac{V}{t_w d_w} \tag{11.77}$$

where

t_w = thickness of the web
d_w = depth of the web

The nominal shear yielding strength of webs is based on the Von Mises yield criterion, which states that for an unreinforced web of a beam, whose width-to-thickness ratio is comparatively small (so that web-buckling failure is avoided), the shear strength may be taken as

$$\tau_y = \frac{f_y}{\sqrt{3}} = 0.58 f_y \tag{11.78}$$

where f_y = yield stress.

The shear capacity of rolled beams V_c can be calculated as

$$V_c \approx 0.6 f_y t_w d_w \tag{11.79}$$

The girder webs will normally be subjected to some combination of shear and bending stresses. The most severe condition in terms of web buckling is normally the pure shear case. It follows that it is those regions adjacent to supports or the vicinity of point loads, which generally control the design. Shear buckling occurs largely as a result of the compressive stresses acting diagonally within the web, as shown in Figure 11.41 with the number of waves tending to increase with an increase in the panel aspect ratio c/d.

When $d/t_w \leq 67 \, \varepsilon$, where $\varepsilon = (250/f_y)^{0.5}$, the web plate will not buckle because the shear stress τ is less than critical buckling stress "τ_{cr}." The design in such cases is similar to the rolled beams here. Consider plate girders having thin webs with $d/t_w > 6 \, \varepsilon$. In the design of these webs, shear buckling should be considered. In a general way, we may have an unstiffened web, a web stiffened by transverse stiffeners and a web stiffened by both transverse and longitudinal stiffeners as shown in Figure 11.42.

11.7.1.3 Bearing

Beams can be either supported directly or other structural members such as steel stanchions, etc., or else they rest on concrete or masonry supports such as walls or pillars. In the latter case, the support is of a weaker material than steel and it becomes necessary to spread the load over a larger area, so that the bearing stress does not exceed a certain permissible value. This is achieved by the provision of a bearing plate.

The bearing stress in any part of a beam when calculated on the net area of contact shall not exceed the value of σ_p determined from the following formula:

$$\sigma_p = 0.75 f_y \tag{11.80}$$

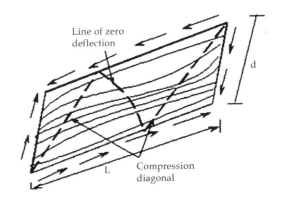

FIGURE 11.41   Shear buckling.

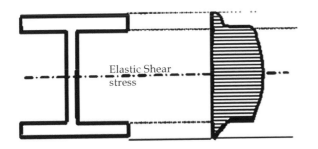

FIGURE 11.40   Shear stress in the elastic range.

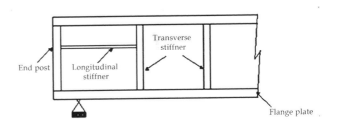

FIGURE 11.42   Stiffeners.

where
σ_p = maximum permissible bearing stress, and
f_y = yield stress of steel.

11.7.1.4 Deflection

When a beam is loaded, it deflects. The amount of maximum deflection depends upon the following:

i. span,
ii. moment of inertia of the section, I,
iii. distribution of the load,
iv. modulus of elasticity, E, and
v. support conditions.

A very common type of loading is uniformly distributed load (w) along a beam. The maximum deflection, caused at the mid-span of a simply supported beam, is given by:

$$\delta = \frac{5wL^4}{384EI} = \frac{5WL^3}{384EI} \tag{11.81}$$

where
W = total load = wL and
L = effective span of the beam.

In general, the maximum deflection of the beam is given by

$$\delta = K_l \frac{WL^3}{EI} \tag{11.82}$$

where co-efficient K_l depends upon the mode of distribution of load. In simply supported beam, for uniformly distributed load, $K_l = \dfrac{5}{384}$ while for concentrated load placed at mid-span, $K_l = \dfrac{1}{48}$.

Again, for maximum Bending Moment at the mid-span,

$$M = K_m WL \tag{11.83}$$

where K_m is a moment coefficient depending upon the distribution of load. For uniformly distributed load, $K_m = 1/8$ while for concentrated load at the mid-span, $K_m = 1/4$.

Again

$$\frac{M}{I} = \frac{f}{y} \tag{11.84}$$

or

$$M = \frac{f}{y} I \tag{11.85}$$

Taking $f = \sigma_{bc}$ (σ_{bc} is the bending stress at the extreme end fiber) and $y = d/2$ (where d = depth of the beam).

$$M = \frac{2\sigma_{bc}}{d} I \tag{11.86}$$

Therefore,

$$WL = \frac{2\sigma_{bc}I}{K_m d} \tag{11.87}$$

Substituting this value of WL in δ, we get,

$$\delta = \frac{K_l}{K_m} \frac{2L^2 \sigma_{bc}}{Ed} \tag{11.88}$$

$$\frac{\delta}{L} = \frac{2K_l}{K_m} \frac{L}{d} \frac{\sigma_{bc}}{E} \tag{11.89}$$

11.7.1.5 Other Beam Failure Criteria

1. **Web crushing**

This may occur under concentrated loads or at support point when deep slender webs are employed. A widely used method of overcoming web crushing problems is to use web cleats at support points, as shown in Figure 11.43.

2. **Shear buckling**

Thin webs subjected to predominant shear will buckle as shown in Figure 11.44. The maximum shear in a beam web is invariably limited to 0.7 times yield stress in shear. In addition, in deep webs, where shear buckling can occur, the average shear stress (p_v) must be less than the value calculated as follows:

$$p_v \leq \left(\frac{1000t}{D} \right)^2 \tag{11.90}$$

where
p = average shear stress in N/mm²
t and D are the web thickness and depth, respectively (in mm).

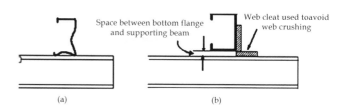

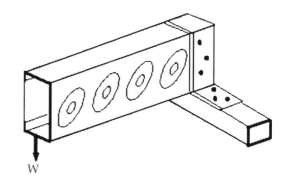

FIGURE 11.43 Web crushing. (a) Web crushing. (b) Cleats to avoid web crushing.

FIGURE 11.44 Web subjected to shear.

11.7.2 LATERALLY SUPPORTED BEAM

When the lateral support to the compression flange is adequate, the lateral buckling of the beam is prevented and the section flexural strength of the beam can be developed. The strength of I-sections depends upon the width-to-thickness ratio of the compression flange. When the width-to-thickness ratio is sufficiently small, the beam can be fully plastified and reach the plastic moment, such sections are classified as compact sections. However, provided section can also sustain the moment during the additional plastic hinge rotation till the failure mechanism is formed. Such sections are referred to as plastic sections. When the compression flange width-to-thickness ratio is larger, the compression flange may buckle locally before the complete plastification of the section occurs and the plastic moment is reached. Such sections are referred to as noncompact sections. When the width-to-thickness ratio of the compression flange is sufficiently large, local buckling of compression flange may occur even before extreme fiber yields. Such sections are referred to as slender sections.

The flexural behavior of such beams is presented in Figure 11.45. The section classified as slender cannot attain the first yield moment, because of a premature local buckling of the web or flange. The next curve represents the beam classified as "semi-compact" in which extreme fiber stress in the beam attains yield stress but the beam may fail by local buckling before further plastic redistribution of stress can take place toward the neutral axis of the beam. The curve shown as "compact beam" in which the entire section, both compression and tension portion of the beam, attains yield stress. Because of this plastic redistribution of stress, the member attains its plastic moment capacity (M_p) but fails by local buckling before developing plastic mechanism by sufficient plastic hinge rotation. The moment capacity of the section is calculated with low shear load. Low shear load is referred to the factored design shear force that does not exceed $0.6V_d$, where V_d is the design shear strength of cross section. The factored design moment is calculated as per:

$$M_d = \beta_b Z_p f_y / \gamma_{m0} \tag{11.91}$$

where

$\beta_b = 1.0$ for plastic and compact sections;

$\beta_b = Z_e/Z_p$ for semi-compact sections;

Z_e, Z_p = elastic and plastic section moduli of the cross section, respectively;

f_y = yield stress of the material, and

γ_{m0} = partial safety factor.

11.7.2.1 Holes in the Tension Zone

The fastener holes in the tension flange need not be considered provided that for the tension flange the following condition is satisfied. The presence of holes in the tension flange of a beam due to connections may lead to reduction in the bending capacity of the beam.

$$\left(\frac{A_{nf}}{A_{gf}}\right) = \left(\frac{f_y}{f_u}\right)\left(\frac{\gamma_{m1}}{\gamma_{m0}}\right)\Big/0.9 \tag{11.92}$$

where

A_{nf}/A_{gf} = ratio of net-to-gross area of the flange in tension,

f_y/f_u = ratio of yield and ultimate stress of the material, and

γ_{m1}/γ_{m0} = ratio of partial safety factors against ultimate to yield stress.

11.7.2.2 Shear Lag Effects

The simple theory of bending is based on the assumption that plane sections remain plane after bending. But, the presence of shear strains causes the section to warp. Its effect in the flanges is to modify the bending stresses obtained by the simple theory, producing higher stresses near the junction of a web and lower stresses at points away from it, as shown in Figure 11.46. This effect is called "shear lag." This effect is minimal in rolled sections, which have narrow and thick flanges and more pronounced in plate girders or box sections having wide thin flanges when they are subjected to high shear forces, especially in the vicinity of concentrated loads.

11.7.2.3 Biaxial Bending

When a plastic or compact section is subjected to biaxial bending, the section should be designed such that

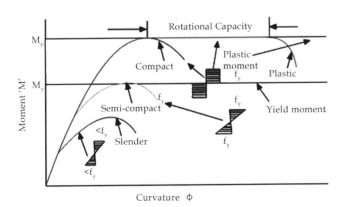

FIGURE 11.45 Flexural moment.

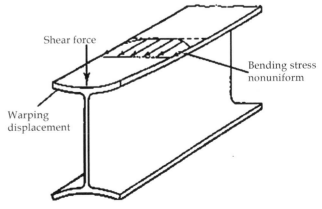

FIGURE 11.46 Shear lag.

$$\left(\frac{M_y}{M_{dy}}\right)^{\alpha_1} + \left(\frac{M_z}{M_{dz}}\right)^{\alpha_2} \leq 1.0 \qquad (11.93)$$

where M_y and M_z are the factored bending moments about the minor and major axes, respectively; M_{dy} and M_{dz} are the design bending strengths of the section about the minor and major axes, respectively. Also, α_1 and α_2 are constants. For I and channel sections, $\alpha_1 = 1.0$ and $\alpha_2 = 2.0$; for circular tubes, $\alpha_1 = 2.0$ and $\alpha_2 = 2.0$; for rectangular hallow sections, $\alpha_1 = \alpha_2 = 1.66$. For semi-compact section,

$$\frac{M_y}{M_{dy}} + \frac{M_z}{M_{dz}} \leq 1.0 \qquad (11.94)$$

11.7.3 Laterally Unsupported Beams

Under increasing transverse loads, a beam should attain its full plastic moment capacity. Two important assumptions made to achieve the ideal beam behavior are as follows:

- The compression flange of the beam is restrained from moving laterally.
- Any form of local buckling is prevented.

A beam experiencing bending about major axis and its compression flange not restrained against buckling may not attain its material capacity. If the laterally unrestrained length of the compression flange of the beam is relatively long then a phenomenon known as lateral buckling or lateral torsional buckling of the beam may take place and the beam would fail well before it can attain its full moment capacity. This phenomenon has close similarity with the Euler buckling of columns triggering collapse before attaining its squash load (full compressive yield load).

11.7.3.1 Lateral-Torsional Buckling of Beams

Lateral-torsional buckling is a limit-state of structural usefulness where the deformation of a beam changes from predominantly in-plane deflection to a combination of lateral deflection and twisting while the load capacity remains first constant, before dropping off due to large deflections. The various factors affecting the lateral-torsional buckling strength are as follows:

- Distance between lateral supports to the compression flange.
- Restraints at the ends and at intermediate support locations (boundary conditions).
- Type and position of the loads.
- Moment gradient along the length.
- Type of cross section.
- Nonprismatic nature of the member.
- Material properties.
- Magnitude and distribution of residual stresses.
- Initial imperfections of geometry and loading.

11.7.3.2 Design Bending Strength

The design bending strength of laterally unsupported beam as governed by torsional buckling is given by

$$M_d = \beta_b Z_p f_{bd} \qquad (11.95)$$

where f_{bd} = the design bending compressive stress.

The design bending compressive stress f_{bd} is expresses as

$$f_{bd} = \chi_{LT} f_y / \gamma_{m0} \qquad (11.96)$$

where χ_{LT} = bending stress reduction factor to account for lateral torsional buckling, given by

$$\chi_{LT} = \frac{1}{\left\{\phi_{LT} + \left[\phi_{LT}^2 - \lambda_{LT}^2\right]^{0.5}\right\}} \leq 1.0 \qquad (11.97)$$

$$\phi_{LT} = 0.5\left[1 + \alpha_{LT}\left(\lambda_{LT} - 0.2\right) + \lambda_{LT}^2\right] \qquad (11.98)$$

The imperfection parameter is given by

$$\alpha_{LT} = 0.21 \text{ for rolled steel section;}$$

$$\alpha_{LT} = 0.49 \text{ for welded steel section}$$

The nondimensional slenderness ratio, λ_{LT}, is given by

$$\lambda_{LT} = \sqrt{\beta_p Z_p f_y / M_{cr}} \leq \sqrt{1.2 Z_e f_y / M_{cr}}$$
$$= \sqrt{f_y / f_{cr,b}} \qquad (11.99)$$

where
M_{cr} = elastic critical moment,
$f_{cr,b}$ = extreme fiber compressive stress corresponding to elastic lateral buckling moment.

If the nondimensional slenderness ratio $\lambda_{LT} \leq 0.4$, then no allowance for lateral-torsional buckling is necessary. M_{cr}, the elastic lateral torsional buckling moment for simply supported, prismatic members with symmetric cross section, can be determined from

$$M_{cr} = \sqrt{\left\{\left(\frac{\pi^2 EI_y}{\left(L_{LT}\right)^2}\right)\left[GI_t + \frac{\pi^2 EI_w}{\left(L_{LT}\right)^2}\right]\right\}} = \beta_b Z_p f_{cr,b} \qquad (11.100)$$

$f_{cr,b}$ of nonslender-rolled steel sections can be calculated from

$$f_{cr,b} = \frac{1.1\pi^2 E}{\left(L_{LT}/r_y\right)^2}\left[1 + \frac{1}{20}\left(\frac{L_{LT}/r_y}{h_f/t_f}\right)^2\right]^{0.5} \qquad (11.101)$$

In the case of prismatic members made of standard rolled I-sections and welded doubly symmetric I-sections, M_{cr} is calculated by

$$M_{cr} = \frac{\pi^2 EI_y h_f}{2L_{LT}^2}\left[1 + \frac{1}{20}\left(\frac{L_{LT}/r_y}{h_f/t_f}\right)^2\right]^{0.5} \qquad (11.102)$$

where

I_t = torsional constant = $\Sigma b_i t_i^3/3$ for open section;

I_w = warping constant;

I_y, r_y = moment of inertia and radius of gyration, respectively, about the weaker axis;

L_{LT} = effective length for lateral torsional buckling;

h_f = center-to-center distance between flanges; and

t_f = thickness of flange.

11.7.3.3 Effective Length of Compression Flanges

The lateral restraints provided by the simply supported condition assumption in the basic case are the lowest and therefore the M_{cr} is also the lowest. It is possible, by other restraint conditions, to obtain higher values of M_{cr}, for the same structural section, which would result in better utilization of the section and thus, saving in weight of material. As lateral buckling involves three kinds of deformation, namely, lateral bending, twisting, and warping, it is feasible to think of various types of end conditions.

For the beam with simply supported end conditions and no intermediate lateral restraint, the effective length is equal to the actual length between the supports, when a greater amount of lateral and torsional restraints is provided at support. When the effective length is less than the actual length and alternatively the length becomes more when there is less restraint. The effective length factor would indirectly account for the increased lateral and torsional rigidities by the restraints.

11.8 MEMBERS UNDER COMBINED AXIAL LOAD AND MOMENT

Many members subjected to different types of stress resultants such as axial tensile force and axial compressive force are also subjected to bending moment along with transverse shear force.

11.8.1 GENERAL

As with combined bending moment and axial force in the elastic range, the plane of zero strain moves from the centroid so that the ultimate stress distribution appears as follows:

- To compute the modified plastic moment, M'_p, the stress blocks are divided into three components as shown in Figure 11.47.
- The outer pair is equal and opposite stress blocks which provide the moment.

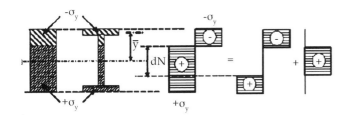

FIGURE 11.47 Modified plastic moment.

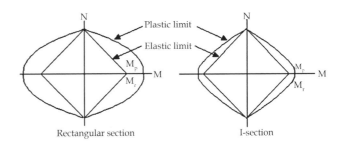

FIGURE 11.48 Interaction diagrams.

- The inner block acts in one direction, the area being balanced around the original neutral axis, defined by d_N. In the cases illustrated the inner block has a constant width, b or t_w, so that

$$M_p = 2A_p\bar{y}\sigma_y \tag{11.103}$$

$$N = bd_N\sigma_y \quad \text{(for rectangle)} \tag{11.104}$$

$$N = t_w d_N\sigma_y \text{ (for } I\text{-section)} \tag{11.105}$$

Interaction diagrams relating N and M can be drawn for initial yield and ultimate capacity as shown in Figure 11.48.

When the web and the flange of an I-beam have different yield strengths, it is possible to calculate the full plastic moment or the combined axial force and bending moment capacity taking into account the different yield strengths,

$$M_p = 2\sum_i A_i\bar{y}_i\sigma_{yi} \tag{11.106}$$

for all I segments with different yield stresses.

11.8.2 LOCAL CAPACITY CHECK

Any member subjected to bending moment and normal tension force should be checked for lateral torsional buckling and capacity to withstand the combined effects of axial load and moment at the points of greatest bending and axial loads. Figure 11.49 illustrates the type of three-dimensional

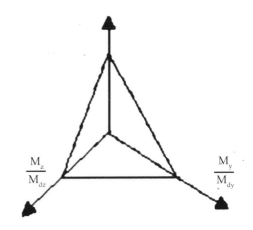

FIGURE 11.49 Three-dimensional interaction surface.

interaction surface that controls the ultimate strength of steel members under combined biaxial bending and axial force. Each axis represents a single load component of normal force N, bending about the y and z axes of the section (M_y or M_z) and each plane corresponds to the interaction of two components.

For plastic and compact sections, the design of members subjected to combined axial load and bending moment shall satisfy the following interaction relationship:

$$\left(\frac{M_y}{M_{ndy}}\right)^{u1}+\left(\frac{M_z}{M_{ndz}}\right)^{u2}\leq 1.0 \qquad (11.107)$$

where

M_y, M_z = design reduced flexural strength under combined axial force and the respective uniaxial moment acting alone,

N = factored applied axial force (Tension T, or Compression F),

N_d = design strength in tension (T_d) or in compression,

$$N_d = A_g f_y / \gamma_{m0} \qquad (11.108)$$

where

A_g = gross area of the cross section
$n = N/N_d$
α_1, α_2 = constants as given in Table in 11.12

The above interaction formulae $\left(\frac{M_y}{M_{ndy}}\right)^{u1}+\left(\frac{M_z}{M_{ndz}}\right)^{u2}\leq 1.0$ is a function of α_1 and α_2, and the values of α_1 and α_2 are in turn functions of the ratio $n = N/N_d$, where N and N_d are factored applied axial force and design strength in tension or compression.

It can be observed from Table 11.12 that for I-section or channel section, in case, the ratio $n = N/N_d$ is equal to 0.2, the value of α_1 becomes 1 and for $n = N/N_d > 0.2$, the value of α_1 is more than 1. As the value of α_1 increases above 1, the value of the component $\left(\frac{M_y}{M_{ndy}}\right)^{u1}$ gets reduced since the ratio $\left(\frac{M_y}{M_{ndy}}\right)^{u1}$ is always less than 1. The values of M_{ndy} and M_{ndz} are also proportionately reduced to accommodate the value of axial tension or compression depending upon type of

sections. The ratio $n = N/N_d$ is also directly related in reducing the bending strength, M_{ndy} and M_{ndz}.

11.8.3 MEMBERS SUBJECTED TO COMBINED BENDING AND AXIAL FORCES

For plastic and compact sections without bolts holes, the following approximations may be used while calculating/deriving the values of design reduced flexural strength M_{ndz} and M_{ndy} under combined axial force and the respective uniaxial moment acting alone, i.e., M_{ndz} and M_{ndy} acting alone. The value of reduced flexural strength of the section, either M_{ndz} or M_{ndy}, is directly proportional to the geometry of a particular section. We will now see how these values are changing depending upon geometry of a particular section:

i. **Plates**

For rolled steel plates irrespective of their thickness, the value of reduced flexural strength can be derived from the equation,

$$M_{nd} = M_d(1 - n_2) \qquad (11.109)$$

Here, the equation for reduced flexural strength is again a function of the ratio $n = N/N_d$, where N and N_d are factored applied axial force and design strength in tension or compression. For smaller values of the ratio n, the reduction in flexural strength is not significant since the reduction in flexural strength is directly proportional to square of the ratio n. It is obvious from the equation that as the ratio tends toward the value 1, the amount of reduction in flexural strength increases and for extreme case, when the ratio is equal to 1, the value of reduced flexural strength is zero, i.e., for this particular case no flexural or bending strength is available within the plate section. This situation also satisfies the condition,

$$\frac{N}{N_d}+\frac{M_y}{M_{dy}}+\frac{M_z}{M_{dz}}\leq 1.0 \qquad (11.110)$$

ii. **Welded I or H sections**

For welded I or H sections, the reduced flexural strength about the major axis can be derived from the equation,

$$M_{ndz} = M_{dz}(1 - n)/(1 - 0.5a) \leq M_{dz} \qquad (11.111)$$

and about the minor axis, from the equation

$$M_{ndy} = M_{dy}\left[1-\left(\frac{n-a}{1-a}\right)^2\right]\leq M_{dy} \qquad (11.112)$$

where

$$n = N/N_d \qquad (11.113)$$

TABLE 11.12

Constants α_1 and α_2

Section	α_1	α_2
I and Channel	$5n \geq 1$	2
Circular tubes	2	2
Rectangular tubes	$1.66/(1 - 1.13\,n^2) \leq 6$	$1.66/(1 - 1.13\,n^2) \leq 6$
Solid rectangle	$1.73 + 1.8\,n^2$	$1.73 + 1.8\,n^2$

and

$$a = (A - 2bt)/A \le 0.5 \qquad (11.114)$$

Here the reduction in flexural strength for major axis is linearly and directly proportional to the ratio n and inversely proportional to the factor a, which is a reduction factor for cross sectional area ratio. It should also be noted that for a particular sectional area A, as the width and/or thickness of the flange of I or H section increases, the factor "a" reduces which in turn increases the value of M_{ndz}.

For minor axis, the reduction in flexural strength is nonlinearly proportional to both the factors n and a, but as the value of the factor n increases considering other factor remaining unchanged, the value of M_{ndy} decreases, conversely as the value of the factor increases a, the value of M_{ndy} increases. For a particular case, when the numerical value of factor n is equal to 1 and the numerical value of the factor a is 0.5, the numerical value of M_{ndy} becomes zero. It can be observed that the factor a being the area ratio, it takes into account the effect of flange width and flange thickness. As the value of b or t_f increases, the value of the factor a reduces which in turn reduces further the value of design reduced flexural strength M_{ndy}.

iii. **Standard I or H sections**

For standard I or H sections, the reduced flexural strength about the major axis can be derived from the equation,

$$M_{ndz} = 1.11 M_{dz} (1 - n) \le M_{dz} \qquad (11.115)$$

and about the minor axis from the equation,

$$M_{ndz} = M_{ndy}, \quad \text{for } n \le 0.2 \qquad (11.116)$$

and

$$M_{ndy} = 1.56 M_{dy} (1 - n)(n + 0.6), \quad \text{for } n > 0.2 \quad (11.117)$$

Unlike welded I or H sections, here we do not find the factor a, but reduction in flexural strength for major axis is linearly and directly proportional to the ratio n. It is pertinent to note that for all cases, as the factor increases n, further reduction in reduced flexural strength of the member takes place. For a particular case, when the factor n becomes 1, the value of M_{ndz} reduces to zero.

For minor axis, no reduction in flexural strength takes place till the ratio $n = N/N_d$ is restricted to 0.2. When the value of n is more than 0.2, the reduction in flexural strength for minor axis is linearly and directly proportional to the ratio n. For a value of $n = 1$, the value of M_{ndy} reduces to zero.

iv. **Rectangular hollow sections and welded box sections**

When the section is symmetric about both axis and without bolt holes, the reduced flexural strength about the major axis can be derived from the equation,

$$M_{ndz} = M_{dz} (1 - n)/(1 - 0.5 a_w) \le M_{dz} \qquad (11.118)$$

and about the minor axis from the equation

$$M_{ndy} = M_{dy} (1 - n)/(1 - 0.5 a_f) \le M_{dy} \qquad (11.119)$$

As indicated in above equations, for rectangular hollow sections and welded box sections, the reduction in flexural strength for both the axes takes place in line with that of M_{ndz} for welded I or H sections as described earlier in (ii) above. The only variation is that the factor a is replaced either by a_w for M_{ndz} or by a_f for M_{ndy}.

v. **Circular hollow tubes without bolt holes**

The reduced flexural strength about both the axes can be derived from the equation,

$$M_{nd} = 1.04 M_d (1 - n^{1.7}) \le M_d \qquad (11.120)$$

For smaller values of the ratio n, the reduction in flexural strength is not significant since the reduction in flexural strength is directly proportional to the power of 1.7 for the ratio n. When the ratio n is equal to 1, the value of reduced flexural strength is zero, i.e., for this particular case no flexural or bending strength is available for the circular section. Usually, the points of greatest bending and axial loads are either at the middle or ends of members under consideration. Hence, the member can also be checked, conservatively, as follows:

$$\frac{N}{N_d} + \frac{M_y}{M_{dy}} + \frac{M_z}{M_{dz}} \le 1.0 \qquad (11.121)$$

where
N is the factored applied axial load in member under consideration,
$N_d = A_g f_y/\gamma_{m0}$, is the strength in tension,
M_z and M_y are the applied moment about the major and minor axes at critical region.

M_{dz} and M_{dy} are the moment capacity about the major and minor axes in the absence of axial load, i.e., when acting alone and A_g is the gross area of cross section. This shows that in point of time, the summation of ratios of various components of axial forces and bending moments (including bi-axial bending moments) will cross the limiting value of 1.

For semi-compact sections, when there is no high shear force semi-compact section design is satisfactory under combined axial force and bending, if the maximum longitudinal stress under combined axial force and bending f_x satisfies the following criteria:

$$f_x \leq f_y/\gamma_{m0} \qquad (11.122)$$

For cross section without holes, the above criteria reduces to

$$\frac{N}{N_d} + \frac{M_y}{M_{dy}} + \frac{M_z}{M_{dz}} \leq 1.0 \qquad (11.123)$$

where N_d, M_{dy}, and M_{dz} are as defined earlier.

11.8.4 OVERALL MEMBER STRENGTH CHECK

Members subjected to combined axial force and bending moment shall be checked for overall buckling failure considering the entire span of the member. This essentially takes care of lateral torsional buckling.

a. For bending moment and axial tension, the member should be checked for lateral torsional buckling to satisfy overall stability of the member under reduced effective moment M_{eff} due to tension and bending. The reduced effective moment, M_{eff}, can be calculated as per the equation

$$M_{eff} = \left[M - \psi T Z_{ec}/A \right] \leq M_d \qquad (11.124)$$

but in no case shall exceed the bending strength due to lateral torsional buckling M_d. Here M and T are factored applied moment and tension, respectively; A is the area of cross section; Z_{ec} is the elastic section modulus of the section with respect to extreme compression fiber and the factor is equal to 0.8 when tension and bending moments are varying independently or otherwise equal to 1. For extreme case, when the factor $\psi T Z_{ec}/A$ is equal to M, M_{eff} reduces to zero.

b. For bending moment and axial compression, when the member is subjected to combined axial compression and biaxial bending, the section should be checked to satisfy the generalized interaction relationship as per the equation

$$\frac{P}{P_d} + \frac{K_y M_y}{M_{dy}} + \frac{K_z M_z}{M_{dz}} \leq 1.0 \qquad (11.125)$$

Here K_y, K_z are the moment amplification factor about minor and major axis, respectively $\left(K_z = 1 - \dfrac{\mu_z P}{P_{dz}} \right)$ and $\left(K_y = 1 - \dfrac{\mu_y P}{P_{dy}} \right)$, where factors μ_z and μ_y are dependent on equivalent uniform moment factor, b, which in turn depends on the shape of the bending moment diagram between lateral bracing points in the appropriate plane of bending and nondimensional

slenderness ratio. P is the applied factored axial compression, M_y, M_z are the applied factor bending moments about minor and major axis of the member, respectively, and P_d, M_{dy}, M_{dz} are the design strength under axial compression, bending about minor and major axis, respectively, as governed by overall buckling criteria. The design compression strength, P_d, is the smallest of the minor axis (P_{dy}) and major axis (P_{dz}) buckling strength and the design bending strength (M_{dz}) about major axis is equal to (M_d), where M_d is the design flexural strength about minor axis, depending upon the lateral torsional buckling. For design bending strength about minor axis, M_{dy} is equal to M_d, where M_d is the design flexural strength about minor axis calculated using plastic section modulus for plastic and compact sections and elastic section modulus for semi-compact sections.

c. The factors are as designed below:
μ_z is the largest of μ_T and μ_{fz} as given below,

$$\mu_{LT} = 0.15\lambda_Y \beta_{MLT} - 0.15 \leq 0.90 \qquad (11.126)$$

$$\mu_{fz} = \lambda_z (2\beta_{Mz} - 4) + \left[\frac{Z_z - Z_{eZ}}{Z_{eZ}} \right] \leq 0.90 \qquad (11.127)$$

$$\mu_y = \lambda_y (2\beta_{My} - 4) + \left[\frac{Z_y - Z_{ey}}{Z_{ey}} \right] \leq 0.90 \qquad (11.128)$$

β_{My}, β_{Mz}, β_{MLT} = equivalent uniform moment factor, according to the shape of the bending moment diagram between lateral bracing points in the appropriate plane of bending.
λ_y, λ_z = nondimensional slenderness ratio about the respective axis.

11.9 COLUMN BASES AND CAPS

The columns are supported on the column bases. The column bases transmit the column load to the concrete or masonry foundation blocks. The column load is spread over large area on concrete or masonry blocks. The intensity of bearing pressure on concrete or masonry is kept within the maximum permissible bearing pressure. The column bases are of two types:

i. Slab bases
ii. Gusseted bases

The column load is first transmitted to the column footing through the column base. It is then spread over the soil through the column footing. The column footings are of two types: Independent footings and combined footings.

11.9.1 SLAB BASE

The slab base as shown in Figure 11.50 consists of cleat angles and base plate. The column end is faced for bearing over the whole area. The gussets (gusset plates and gusset angles) are

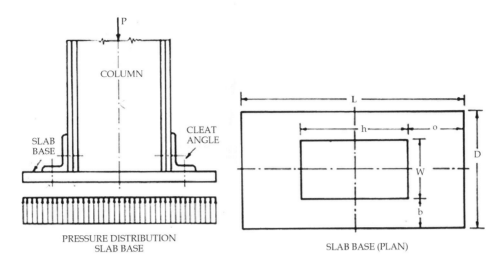

FIGURE 11.50 Slab base.

not provided with the column with slab bases. The sufficient fastenings are used to retain the parts securely in plate and to resist all moments and forces, other than the direct compression. The forces and moments arising during transit, unloading, and erection are also considered. When the slab alone distributes the load uniformly, the minimum thickness of a rectangular slab is derived as below.

The column is carrying an axial load P. Consider the load distributed over area $h \times w$ and under the slab over the area $L \times D$ as shown in Figure 11.50.

Let t = thickness of the slab

w = pressure or loading on the underside of the base

a = greater projection beyond column

b = lesser projection beyond column

σ_{bs} = allowable bending stress in the slab bases for all steels, it shall be assumed as 185 N/mm^2

Consider a strip of unit width.

$$\text{Along the } xx\text{-axis } M_{xx} = \left(\frac{wa^2}{2}\right) \qquad (11.129)$$

$$\text{Along the } yy\text{-axis } M_{yy} = \left(\frac{wb^2}{2}\right) \qquad (11.130)$$

If Poison ratio is adopted as ¼ the effective moment for width D

$$\text{Effective moment for width } L = \frac{w}{2}\left(a^2 - \frac{b^2}{4}\right) \qquad (11.131)$$

Since a is greater projection from the column, the effective moment for width D is more. Moment of resistance of the slab base of unit width

$$\text{M.R.} = \left(\frac{1}{6}1t^2\sigma_{bs}\right) \qquad (11.132)$$

$$\text{Therefore, } \left(\frac{1}{6}1t^2\sigma_{bs}\right) = \frac{w}{2}\left(a^2 - \frac{b^2}{4}\right) \qquad (11.133)$$

Hence the thickness of the slab base,

$$t = \left[\frac{a_w}{\sigma_{bs}}\left(a^2 - \frac{b^2}{4}\right)\right]^{\frac{1}{2}} \qquad (11.134)$$

For solid round steel column, where the load is distributed over the whole area, the minimum thickness of square base is given by

$$t = 10\left[\frac{90w}{16\sigma_{bs}}\left(\frac{B}{B-d_0}\right)\right]^{\frac{1}{2}} \qquad (11.135)$$

where

t = thickness of the plate in mm

w = total axial load in kN

B = length of the side of base of cap in mm

d_0 = diameter of the reduced end (if any) of the column in mm

σ_{bs} = allowable bending stress in steel (is adopted as 185 N/mm^2)

The allowable intensity of pressure on concrete may be assumed as 4 N/mm^2. When the slab does not distribute the load uniformly or where the slab is not rectangular, separate calculation shall be made to show that stresses are within the specified limits.

When the load on the cap or under the base is not uniformly distributed or where end of the column shaft is not machined with the cap or base, or where the cap or base is not square in plan, the calculations are made on the allowable stress of 185 N/mm^2 (MPa). The cap or base plate shall not be less than 1.50 $(d_0 + 75)$ mm in length or diameter.

The area of the shoulder (the annular bearing area) shall be sufficient to limit the stress in bearing, for the whole of the load communicated to the slab to the maximum value 0.75 f_y and resistance to any bending communicated to the shaft by the slab shall be taken as assisted by bearing pressures developed against the reduced and of the shaft in conjunction with the shoulder.

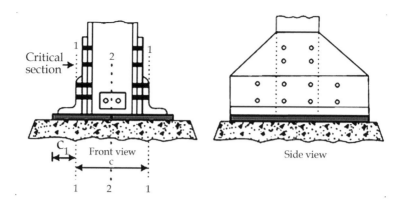

FIGURE 11.51   Gusset plates and angles placed on flanges.

11.9.2 GUSSET PLATE

When the load on the column section is too large or when the axial load is accompanied by bending moments, usually a gusset base is provided. It consists of a base plate, two gusset plates, and two gusset angles when bolted connections are made. The gusset plates and angles are placed on flanges as shown in Figure 11.51.

The upward pressure from below the base, as shown in Figure 11.52, causes bending of the gusset plates and puts their top edge in compression which may therefore buckle. This can be checked by limiting the width-to-thickness ratio for the gusset plate as below.

For the portion of gusset plate

welded to the column flange: $D \leq 29.3\varepsilon t_g$ (11.136)

For the outstand of the gusset plate from

the edge of the column flange: $S \leq 13.6\varepsilon t_g$ (11.137)

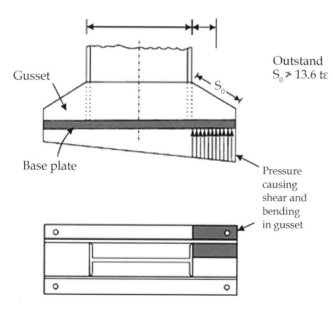

FIGURE 11.52   Upward pressure from below the base.

where $\varepsilon = \sqrt{\dfrac{250}{f_{yg}}}$ (11.138)

f_{yg} = design yield strength of gusset plate
t_g = thickness of the gusset plate

The gusset plates are designed to resist shear and bending. The moment in the gusset should not exceed the bending strength of the gusset plate.

Bending strength of gusset plate, $M_{dg} = \dfrac{f_{yg}}{\gamma_{m0}} Z_e$ (11.139)

where
  Z_e = elastic section modulus of the gusset plate
  γ_{m0} = partial safety factor = 1.0

11.10 PLATE GIRDER

A plate girder is a type of beam built from plates of steel that are either bolted or welded together. They are typically I-beams made from separate structural steel plates which are welded, bolted, or riveted together to form the deep vertical web with a pair of angles each edge to act as compression and tension flanges. Figure 11.53 shows the cross section of a plate girder.

Advantages of using plate girders in steel construction

There are number of advantages of using plate girders in bridge construction or other structures, some of which are as follows:

- Possessing the features of structure is simple.
- Provides convenient transport.
- Speedy erection.
- Heavy loading capacity.
- Great stability and long fatigue life.
- Being capable of an alternative span, loading capacity.

11.10.1 ELEMENTS OF PLATE GIRDER

The various elements of a plate girder are shown in Figure 11.54, and they are as follows:

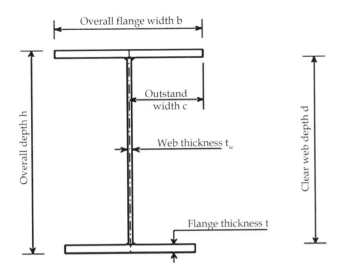

FIGURE 11.53 Cross section of a plate girder.

i. Web plate
ii. Flange plates and their curtailment
iii. Flange angles
iv. **Stiffeners**: bearing, transverse, and longitudinal
v. **Splices**: for web and flange
vi. End bearing

11.10.2   DESIGN COMPONENT OF PLATE GIRDER

The complete design of a plate girder consists of the following elements:

1. Calculation of external loads and estimation of self-weight
2. Calculations of shear force and bending moment
3. Deciding the depth of the plate girder
4. Design of web plate
5. Design of flange
6. Curtailment of flange plate
7. Design of connection between flange angle and web
8. Design of connection between the flange plates and angles
9. Design of stiffeners
10. Design of web splices
11. Design of flange splice
12. Design of end connection

11.10.3   SELF-WEIGHT AND ECONOMIC DEPTH

The depth which gives the minimum depth of the plate girder is called economic depth.

For practical and economic design, the depth-to-span ratio of girder is generally assumed between 1/8 and 1/12.

Overall depth, D: The vertical distance between outer faces of top and bottom flange is called overall depth.

Effective depth, d_e: The vertical distance between the center of gravity of the compression flange and tension flange is known as effective depth.

Clear depth, d_c: The vertical distance between the legs of angles of compression flange and tension flange is called clear depth.

11.10.3.1   Moment of Inertia of the Whole Section

Let M = maximum bending moment in the girder
   d = depth of web
   d_e = effective depth
   t_w = thickness of the web plate
   σ_b = allowable stress in bending
   A_f = area of flange plate
   l = length of plate girder
   γ_s = unit weight of steel

$$\text{If } A_f + \frac{1}{6} t_w d_e \rightarrow \text{Effective flange area} \qquad (11.140)$$

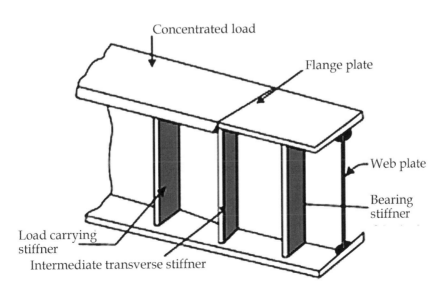

FIGURE 11.54   Elements of a pate girder.

$$\frac{1}{6}t_w \rightarrow \text{Web equivalent} \quad (11.141)$$

for compression flange,

$$M_{\text{compression}} = \left(A_f + \frac{1}{6}t_w d_e\right)d_e\sigma_b \quad (11.142)$$

For tension flange, the area for riveting has to be deducted. Assuming 75% of the web area acts in the bending, for tension flange

$$M_{\text{tension}} = \left(A_f + \frac{3}{4}\times\frac{1}{6}t_w d_e\right)d_e\sigma_b \quad (11.143)$$

Weight per meter run,

$$w = \left[1.76\left(\frac{M}{d_e\sigma_b} - \frac{1}{8}t_w d_e\right) + 1.6t_w d_e\right]\gamma_s \quad (11.144)$$

For minimum depth we have,

$$\frac{\partial w}{\partial d_e} = 0 \quad (11.145)$$

$$d_e \approx 1.1\sqrt{\frac{M}{t_w\sigma_b}} \quad (11.146)$$

For all practical purposes, $w = \dfrac{W}{300}$, where W is the total weight of the member.

11.10.4 Design of Web Plate

The web of a plate girder is designed for shear force. It is assumed that the web will resist the entire force.
Average shear force in the web,

$$\tau_{va,cal} = \frac{V}{d_w t_w} \prec \tau_{va} \quad (11.147)$$

where
V = maximum shear force on the girder
d_w = depth of the web plate
t_w = thickness of the web
τ_{va} = permissible average shear stress = 0.4 f_y (for unstiffened web)

11.10.5 Web Stiffeners

As per IS 800:1984, the provisions for web stiffeners are as follows:

1. $\dfrac{d_1}{t_w} \leq$ Lesser of, $\dfrac{816}{\sqrt{\tau_{va,cal}}}, \dfrac{1344}{\sqrt{f_y}}$ or 85 $\rightarrow$ no stiffener is required.

2. $\dfrac{d_2}{t_w} \leq$ Lesser of, $\dfrac{3200}{\sqrt{f_y}}$ and 200 $\rightarrow$ vertical stiffener is required.

3. $\dfrac{d_2}{t_w} \leq$ Lesser of, $\dfrac{4000}{\sqrt{f_y}}$ and 250 $\rightarrow$ vertical stiffener and one horizontal stiffener at a distance from the compression flange equal to two-fifth of the distance from compression flange to the neutral axis are provided.

4. $\dfrac{d_2}{t_w} \leq$ Lesser of, $\dfrac{6400}{\sqrt{f_y}}$ or 400 $\rightarrow$ same as condition (3) plus a horizontal stiffener at the neutral axis is provided.

where $d_2 = 2\times$clear distance from the compression flange angles or plates or tongue plates to the neutral axis.

In no case should the greater clear dimension of the web should exceed 270 t_w nor should the lesser clear dimension of the same panel exceed 180 t_w.

11.10.6 Design of Flange

Flanges are designed for resisting the maximum bending moment (M) over the beam.
If A_f = gross area of the flanges
d_w = depth of the web plate
D = total depth of the section
t_w = thickness of the web
σ_b = allowable stress in bending
Gross area of the web

$$A_w = t_w d_w \quad (11.148)$$

Gross area of the flange

$$A_f = \frac{M}{\sigma_b D} - \frac{A_w}{6} \quad (11.149)$$

11.10.7 Curtailment of Flange Plates

The girder section proportion is required only for a short distance near the point of maximum bending moment. This moment in a simple plate girder under uniformly distributed loads occurs near the center of the span and reduces toward the supports. So the bending stresses generated are resisted by the flange area. Therefore, it is possible to use a smaller flange area toward the supports. This reduction in the section may be achieved by cutting of the cover plates one by one until only the web plate and flange angles remain near the supports.

Generally the first flange cover plates are carried through the length of girder to serve the following purposes:

1. It ties together the outstanding legs of the flange angles.
2. A top flange with first cover plate curtailed may allow rain water to collect between the flange angles and the web, resulting in corrosion.

11.10.8 Web Splices

The plate girder web may be spliced for one or more of the following reasons:

i. The length of the plate girder may be large whereas the plates are available in limited lengths.
ii. Large-sized plates are difficult to handle, erect, and place as they may get twisted. It is also difficult to transport such plates.

Web splice plates are required only when plate girders are riveted. In the case of welded plate girders, the web plates may be butt-welded. There are mainly two types of splices—shear splice and moment splice, with the latter again divided into type I and type II. Moment splice-type I is much stronger than a shear splice.

i. **Shear splice**

Shear splice, which is shown in Figure 11.55, is the simplest type of splice which consists of a pair of splices plates, one on each side of the web.

Let σ_b = bending stress at the extreme fiber of the web plate

σ_{b1} = bending stress at the extreme fiber of the splice plate

A_w = area of the web plate

A_s = area of the splice plate

d = depth of the web plate to be spliced

d_s = depth of the splice plate

11.10.9   FLANGE SPLICES

i. **Flange angle splice**

Flange angles are spliced when angles of required length are not available. Flange angles are spliced in the following ways:

i. An angle splice on one side of the flange as shown in Figure 11.56a.
ii. One of the splice angles on the flange angle to be spliced and the other angle on the other flange angle as shown in Figure 11.56b.
iii. One angle splice on the flange angle to be spliced and one splice plate on the other flange angle as shown in Figure 11.56c.

Let P = force in splice plate or splice angle on uncut flange angle

V = shear force at the splice

R_v = strength of the rivet in single shear

n = number of rivets to resist shear force between web and flange

s = pitch of rivets

d_1 = effective depth of plate girder

$$n = \frac{P}{R_V - \dfrac{V}{2d_1}s} \qquad (11.150)$$

ii. **Flange plate splice**

Flange plates may have to be spliced if and when required. When the outermost cover plate is to be spliced, the splice plate area is kept equal to the area at the splice section the length of the splice plate is governed by the number of rivets to be provided which in turn is computed from the force at the splice section. If some inner flange plate is to be spliced, it should be at the theoretical cut-off point of the upper plate because the extended portion of the curtailed plate may be assumed to serve as the splice plate. It can also be spliced by providing a splice plate on the top most flange cover plate.

11.10.10   WELDED PLATE GIRDER

A welded plate girder is more efficient section than a riveted section as all the materials are effectively used to resist the load.

11.10.10.1   Web

The depth of the section is usually taken as 1/8 to 1/12 of the span. The web plate should be designed against local buckling. It should have adequate strength to carry shear force. The average shear force in the web is

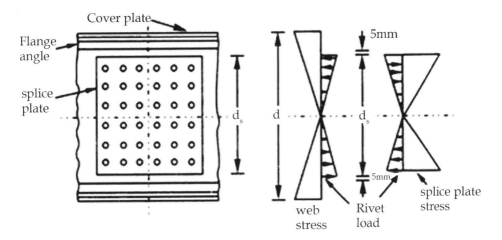

FIGURE 11.55   Shear splice.

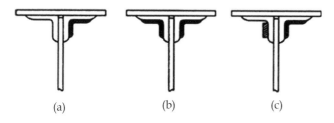

FIGURE 11.56 Flange angle splices (a–c).

$$\tau_{va,cal} = \frac{V}{d_w t_w} \prec \tau_{va} \qquad (11.151)$$

where

V = maximum shear force on the girder
d_w = depth of the web plate
t_w = thickness of the web
τ_{va} = permissible average shear stress = 0.4 f_y (for unstiffened web)

11.10.10.2 Flange

In the welded plate girder, since there is no hole in tension flange, the total effective flange area is given by

$$A_f = \frac{M}{\sigma_{bt} d_e} \qquad (11.152)$$

where

M = maximum bending moment in the girder
σ_{bt} = allowable tensile stress in bending
d_e = effective depth of web

11.10.10.3 Economic Depth

The average gross area of the section over the entire span can be expressed as

$$A_g = 2C_1 \left[\frac{M}{\sigma_{bc} d} - \frac{A_w}{6} \right] + C_2 d t_w \qquad (11.153)$$

where

C_1 = factor to account for reducing size of the flange at the region of lower than the maximum moment
C_2 = factor to account for the reducing web thickness at the region of reduce shear

For minimum area, we have

$$\frac{\partial A_g}{\partial d} = 0 \qquad (11.154)$$

If $C_1 = C_2 = 1$, then,

$$d = \sqrt{\frac{3M}{\sigma_b t_w}} \qquad (11.155)$$

If $C_1 = 0.7$ and $C_2 = 1$,

$$d = 1.35 \sqrt{\frac{M}{\sigma_b t_w}} \qquad (11.156)$$

11.10.10.4 Self-Weight of the Girder

Weight of the girder

$$W = A_g l \gamma_s \qquad (11.157)$$

where

A_g = gross area
l = length
γ_s = unit weight of steel

For $\dfrac{l}{d} = 15$, $\sigma_b = 165$ MPa, unit weight of steel, $\gamma_s = 77$ kN/m³, weight of the girder per meter run

$$w = \frac{W}{286} \text{ kN/m} \qquad (11.158)$$

11.10.10.5 Design of Flange

$$\text{The area of each flange} = A_f = \frac{M}{\sigma_s d} - \frac{A_w}{6} \qquad (11.159)$$

In the welded plate girder, the maximum thickness of the flange plate which is needed to resist the maximum bending moment is not required throughout the entire span, as the moment varies along the span.

11.10.10.6 Welds Connecting Flange with Web

The welds connecting the flange plates with web are designed to take the horizontal shear at that level.

If V is the shear force in N/mm, size of the weld,

$$S = \frac{V}{151} \text{ mm} \qquad (11.160)$$

The maximum effective size of the fillet weld,

$$S_{max} \approx \frac{2}{3} t_w \qquad (11.161)$$

11.10.10.7 Design of Intermediate Stiffeners

The size of the welds connecting web with stiffeners depend upon the following:

  i. Thickness of the web or stiffener
 ii. The requirements to withstand a minimum shear force between each component of the stiffeners and web are given below

$$V = \frac{125 t_w^2}{h} \qquad (11.162)$$

where h is the outstand of the plate in mm.

If the welding is done on both sides in pair, the effective length of each weld should not be less than four times the thickness of the stiffener. In case of one side, it should be greater than ten times the thickness of stiffener.

11.10.10.8 Design of Bearing Stiffener

Bearing stiffeners are provided at the location of the points of loads and at the ends. Their requirements are the same as in riveted girder.

11.11 ROOF TRUSSES

A roof truss is a framed structure formed by connecting various members at their ends to form a system of triangles. The members that carry compressive forces are called struts and those carrying tensile forces are called ties.

11.11.1 COMPONENTS OF A ROOF TRUSS

Figure 11.57 shows the different components of a roof truss. They are briefly described below:

i. **Principal rafter or top chord**

   The top inclined member of the truss extending from the eaves level to the ridge is known as principal rafter or top chord. They support the roof covering (sheeting) through purlins. They are mainly compression members. However, if the purlins are not supported at the panel points, they may also be subjected to bending moment and shear.

ii. **Principal (or main) tie or bottom chord**

   This is the bottom-most member of the truss, which is kept horizontal in most of the cases. It supports other members of the truss. It is usually in tension. However, if reversal of loads occurs due to wind, it may take compression also.

iii. **Ties**

   These are the tension members of the truss.

iv. **Struts**

   These are the compression members of the truss.

v. **Sag tie**

   A sag tie is the central vertical member of the truss used to reduce the moment due to self-weight in the long middle tie member and also to reduce its resulting deflection. It is provided only in long-span trusses.

vi. **Purlins**

   Purlins are structural members which are supported on the principal rafter, and which run transverse to the trusses. The span of the purlins is equal to the center-to-center spacing of the trusses. The purlins support the roof covering (sheeting) either directly or through common rafters. They usually made of either an angle section or a channel section and are therefore subjected to unsymmetrical bending.

vii. **Rafters (or common rafters)**

   If the spacing of the purlins is larger than the available length of sheeting, rafters may be provided to support the sheeting. Thus, the rafters are inclined beams, supported on the purlins. They are also known as common rafters to distinguish them from the principal rafter.

viii. **Ridge line**

   The ridge line or ridge is the line joining the vertices of the trusses. Special forms of ridges of roof covering material are usually available.

ix. **Eaves**

   The bottom edge of the inclined roof surface is called eaves.

x. **Panel points**

   These are the prominent points along the principal rafter, at which various members (i.e., ties and struts) meet. The distance of the principal rafter between any two panel points is termed as panel.

xi. **Roof covering**

   Roof coverings are mostly in the form of corrugated sheets of galvanized iron or asbestos cement. However, roof coverings of glass, fiber glass, or slates are also used. If the roof covering has smaller planer dimensions, subpurlins, supported on common rafters, are required.

xii. **Shoe angle**

   It is a supporting angle provided at the junction of the top and bottom chords of a truss. The reaction of the truss is transferred to the supports (common/wall) through the shoe angle. It is supported on the base plate.

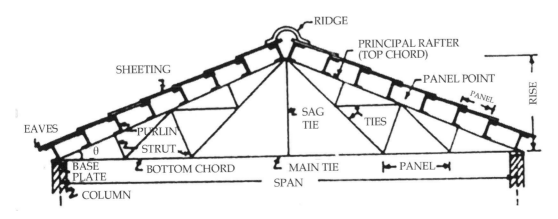

FIGURE 11.57 Components of a roof truss.

xiii. **Base plate, anchor plate, and anchor bolts**

Base plate supports the shoe angle of the truss through bolts. Base plate is anchored to the supporting column/wall through anchor bolts. It can take both downward as well as upward reactions from the truss.

Some of the prominent terms used in truss design are as follows:

Bay: Bay is defined as the distance between adjacent trusses.

Rise: It is defined as the distance from the highest point to the line-joining supports.

Span: It is the distance between centers of the end supports.

Pitch: For a symmetrical truss, it is defined as the ratio of the rise to spans. It is equal to half the slope of the truss and is expressed as $1/n$. For example, a pitch of ¼ corresponds to a slope of 1 in 2, or to a slope angle of 26½°.

Slope: Slope of a symmetrical truss is defined as the ratio of its rise to half the span. It is expressed as 1 in n or as degrees. It is thus numerically double the pitch. A pitch of ¼ corresponds to a slope of 1 in 2 or to a slope angle of 25½°.

11.11.2 Types of Roof Trusses

Some of the different types of steel roof trusses are shown in Figure 11.58. They are briefly described below:

i. **King Post Truss**

This type of truss is made out of wood most of the time, but it can also be built out of a combination of steel and wood. The King Post Truss spans up to 6 m.

ii. **Queen Post Truss**

The Queen Post Truss is designed to be a very reliable, simple, and versatile type of roof truss. It offers a good span, around 10 m, and has a simple design which makes it perfect for a wide range of establishments.

iii. **Howe Truss**

This type of truss is a combination of steel and wood. The diagonal web members are in compression and the vertical web members are in tension. The vertical members are manufactured out of steel which offers extra support and reliability. It has a very wide span, of about 6–30 m. This makes it versatile and very useful for a wide range of projects.

iv. **Fink Truss**

The Fink truss offers economy in terms of steel weight for short-span high-pitched roofs as the members are subdivided into shorter elements. There are many ways of arranging and subdividing the chords and internal members. This type of truss is commonly used to construct roofs in houses.

v. **Fan Truss**

Fan truss has a very simple design and is made out of steel. Fan trusses are a form of Fink roof truss. In Fan trusses, top chords are divided into small lengths in order to provide supports for purlins which would not come at joints in Fink trusses. It can be used for spans from 10 to 15m.

vi. **Pratt Truss**

A Pratt truss includes vertical members and diagonals that slope down toward the center, the opposite of the Howe truss. The interior diagonals are under tension under balanced loading and vertical elements under compression. If pure tension elements are used in the diagonals (such as eyebars) then crossing elements may be needed near the center to accept concentrated live loads as they traverse the span. It can be subdivided, creating Y- and K-shaped patterns. They are commonly used in long-span buildings ranging from 20 to 100 m in span.

vii. **Warren Truss**

In this type of truss, diagonal members are alternatively in tension and in compression. The Warren truss has equal length compression and tension web members, and fewer members than a Pratt truss. A modified Warren truss may be adopted where additional members are introduced to provide a node at purlin locations. Warren trusses are commonly used in long-span buildings ranging from 20 to 100 m in span. This type of truss is also used for the horizontal truss of gantry or crane girders

vii. **North Light Roof Truss**

The North Light Roof Truss is suitable for the larger spans that go over 20 m and get up to 30 m. It is generally more economical to change from a simple truss arrangement to one employing wide span lattice girders which support trusses at right angles. It consists of a series of trusses fixed to girders. The short vertical side of the truss is glazed so that when the roof is used in the Northern Hemisphere, the glazed portion faces north for the best light. It is also durable and used for industrial buildings, drawing rooms, etc.

viii. **Saw-Tooth Truss**

A variation of the North light truss is the saw-tooth truss which is used in multibay buildings. Similar to the North light truss, it is typical to include a truss of the vertical face running perpendicular to the plane of the saw-tooth truss. It is used for a span of 5–8 m.

ix. **Quadrangular Roof Trusses**

These trusses are used for large spans such as railway sheds and auditoriums. They may be considered as two trussed rafters held in place by the tie T at the center.

11.11.3 Geometry of the Roof Truss

In order to get a good structural performance, the ratio of span to truss depth should be chosen in the range of 10–15. To get an efficient layout of the truss members between the chords, the following is advisable:

- The inclination of the diagonal members in relation to the chords should be between 35° and 55°.
- Point loads should only be applied at nodes.

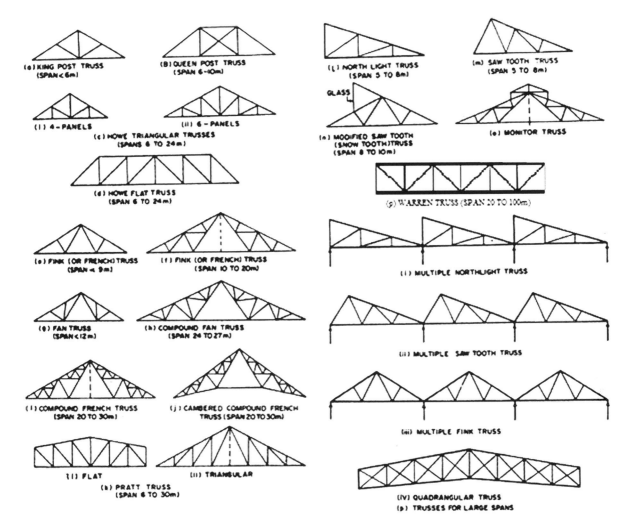

FIGURE 11.58 Types of steel roof truss.

- The orientation of the diagonal members should be such that the longest members are subject to tension (the shorter ones being subject to compression).

11.11.4 TRUSS MEMBER SECTIONS

The main criteria to be satisfied for selecting the sections are as follows:

- Sections should be symmetrical for bending out of the vertical plane of the truss.
- For members in compression, the buckling resistance in the vertical plane of the truss should be similar to that out of the plane.

For light building trusses, typical element cross sections are used. For large member forces, a good solution to use is given in the following:

- Chords having I sections, or a section made up of two channels.
- Diagonals formed from two battened angles.

Structural hollow sections, however, have higher fabrication costs and are only suited to welded construction.

11.11.5 TYPES OF CONNECTIONS

Members of trusses can be joined by riveting, bolting, or welding. Due to involved procedure and highly skilled labor requirement, riveting is not common these days. HSFG bolting and welding have become more common.

It may not always be possible to design connection in which the centroidal axes of the member sections are coincident as in apex connection shown in Figure 11.59a. Small eccentricities may be unavoidable and the gusset plates should be strong enough to resist or transmit forces arising in such cases without buckling as in support connection shown in Figure 11.59b.

11.11.6 LOADS ON ROOF TRUSSES

i. Dead load

Dead load on the roof trusses in single-storey industrial buildings consists of dead load of claddings and dead load of purlins, self-weight of the

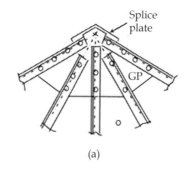
(a)

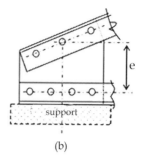
(b)

FIGURE 11.59 Types of connections. (a) Apex connection. (b) Support connection.

trusses in addition to the weight of bracings, etc. Further, additional special dead loads such as truss supported hoist dead loads, special ducting, and ventilator weight, etc., could contribute to roof truss dead loads. As the clear span length (column free span length) increases, the self-weight of the moment-resisting gable frames increases drastically. In such cases roof trusses are more economical. Dead loads of floor slabs can be considerably reduced by adopting composite slabs with profiled steel sheets.

ii. **Imposed load or live load**

The live load on roof trusses consists of the gravitational load due to erection and servicing as well as dust load, etc. The imposed load on sloping roofs with slopes up to and including 10° is to be taken as 750 N/m² of plan area. For roofs with slopes greater than 10°, an imposed load of 750 N/m² less 20 N/m² for every degree increase in slope over 10° subject to a minimum of 400 N/m² is to be taken for roof membrane. Additional special live loads such as crane live loads in trusses supporting monorails may have to be considered sometimes.

iii. **Snow load**

This load varies with the region in which the structure is built. Actual load due to snow will depend upon the shape of the roof and its capacity to retain the snow. In case of roofs with slope greater than 50°, snow load may be disregarded.

iv. **Wind load**

Wind load on the roof trusses, unless the roof slope is too high, would be usually uplift force perpendicular to the roof, due to suction effect of the wind blowing over the roof. Hence the wind load on roof truss usually acts opposite to the gravity load, and its magnitude can be larger than gravity loads, causing reversal of forces in truss members. It is the most critical load on an industrial building. The design wind pressure p_z is given by:

$$p_z = 0.6V_z^2 = 0.6(k_1k_2k_3V_b)^2 \qquad (11.163)$$

where
V_b = basic wind speed in m/s at 10 m height
k_1 = probability factor (or risk coefficient)

k_2 = terrain, height, and structure size factor
k_3 = topography factor

The wind force F acting in a direction normal to the individual structural element is

$$F = (C_{pe} - C_{pi})Ap_z \qquad (11.164)$$

where
C_{pe} = external pressure coefficient
C_{pi} = internal pressure coefficient

v. **Earthquake load**

Since earthquake load on a building depends on the mass of the building, earthquake loads usually do not govern the design of light industrial steel buildings. Wind loads usually govern. However, in the case of industrial buildings with a large mass located at the roof or upper floors, the earthquake load may govern the design.

11.11.7 Economical Spacing of Roof Trusses

When there is flexibility in the location of columns, the spacing of the columns and roof trusses should be economical. The economic spacing of trusses depends upon weight per square meter of floor area of trusses, purlins, column, roof coverings, etc., and their relative cost.

If C is the total cost, we have

$$C = T + P + R \qquad (11.165)$$

where
T = cost of truss, which is inversely proportional to the spacing of trusses
P = cost of purlins, the section of which depends upon the bending moment on it; since the bending moment is proportional to the square of the span of purlins
R = cost of roofing material, which is directly proportional to the spacing of trusses

$$T = 2P + R \qquad (11.166)$$

Hence the spacing should be such that the cost of truss is equal to twice the cost of purlins plus the cost of roof coverings.

The above expression is used to check the spacing of the roof trusses. In general, the economical spacing of trusses may range from 1/5 of span (for large spans) to 1/3 of span (for small spans).

11.11.8 Design of a Roof Truss

The design of a roof truss consists in selecting a suitable type of truss, estimation of loads and design of purlins, members of roof truss and their connections. Following are the steps followed in the design of roof trusses:

i. Depending upon the span, roofing material, lighting, etc., available, the type of truss is decided and a line diagram of the truss is prepared.

ii. Various loads acting over the roof truss are estimated.

iii. The purlins are designed and the loads acting at the panel points of the truss are computed.

iv. The roof truss is analyzed by any suitable method.

v. The compression members are designed. The principal rafter is designed as a continuous strut and the other compression members are designed as discontinuous struts.

vi. The tension members are designed. Single-angle tension members have twisting tendency and produce eccentric forces at the joints. Therefore, double-angle sections are preferred. A minimum of $50 \times 50 \times 6$ mm angle section is provided.

vii. The width of the truss members should be kept minimum as far as possible, because wide members have greater secondary stresses.

viii. If purlins are placed at intermediate points, i.e., between the joints of the top chord, the top chord will be subjected to moments. The member in such a case is designed for the direct stresses and moments.

ix. The members meeting at a joint are so proportioned that their centroidal axes intersect at one point (to avoid eccentric conditions).

x. The joints of the truss are designed.

12 Fluid Mechanics

12.1 PRESSURE AND ITS MEASUREMENT

A fluid is a substance which deforms continuously under the action of shearing forces. That is, if a fluid is at rest, there can be no shearing forces acting, and therefore, all forces in the fluid must be perpendicular to the planes upon which they act.

Pressure is the measure of a force on a specified area. However, depending on the application, there can be many different ways of interpreting pressure.

12.1.1 PRESSURE TERMINOLOGY

- Absolute pressure (Figure 12.1) is measured relative to absolute zero on the pressure scale, which is a perfect vacuum. (Absolute pressure can never be negative.) Absolute pressure is indicated by p and is identical to the familiar thermodynamic pressure.
- Gauge pressure (Figure 12.1) is measured relative to the local atmospheric pressure. Gauge pressure is thus zero when the pressure is the same as atmospheric pressure. (It is possible to have negative gauge pressure.) Gauge pressure is indicated by p_g and is related to absolute pressure as follows: $p_g = p - p_a$, where p_a is the local atmospheric pressure.
- Vacuum pressure (Figure 12.1) is also measured relative to the local atmospheric pressure, but is used when the gauge pressure is negative, i.e., when the absolute pressure falls below the local atmospheric pressure. (Positive vacuum pressure means that the gauge pressure is negative.) Vacuum pressure is indicated by p_{vacuum} and is related to absolute pressure as follows: $p_{vacuum} = p_a - p$, where p_a is the local atmospheric pressure.
- Differential pressure is the pressure difference between two media. Although most gauge pressures are technically a differential pressure sensor— measuring the difference between the medium and atmospheric pressure. A true differential pressure sensor is used to identify the difference between the two separate physical areas. For example, differential pressure is used to check the pressure drop—or loss—from one side of an object to the other.
- Sealed pressure uses a predetermined reference point, not necessarily vacuum. This allows for pressure measurement in locations that will vary with atmospheric changes.

12.1.2 UNITS OF PRESSURE

- **PSI (pounds per square inch)**: This is the unit of measure for one pound of force applied to one square inch of area. PSI is a typical unit of pressure in the United States.
- **BAR**: One bar equals the atmospheric pressure on the Earth at sea level. The BAR unit was created in Europe and is still commonly used there.
- **Pa (Pascal)**: One Pascal equals one Newton of pressure per square meter.
- **In Hg (inches of mercury)**: This is the pressure exerted by a one inch circular column of mercury, one inch tall, at gravity and 0°C (32°F). In Hg are typically used barometric pressure.
- **Torr**: This is the pressure exerted by a one millimeter tall circle column of mercury. It is also known as millimeter of mercury (mmHg). It is equal to 1/760 atmospheres.
- **atm (Standard atmosphere)**: This is a unit of pressure defined as 101,325 Pa (1.01325 bar). It is sometimes used as a reference or standard pressure. It is approximately equal to the atmospheric pressure at sea level.

12.1.3 PASCAL'S LAW

Pascal's law states that "the intensity of pressure at any point in a fluid at rest is the same in all direction."

If p_x = intensity of horizontal pressure on the element of the liquid, p_y = intensity of vertical pressure on the element of the liquid, p_z = intensity of pressure on the diagonal of the triangular element of the liquid, and θ = angle of the triangular element of the liquid (Figure 12.2), according to Pascal's law

$$p_x = p_y = p_z \qquad (12.1)$$

This applies to fluid at rest.

12.1.4 MEASUREMENT OF PRESSURE

In a stationary fluid the pressure is exerted equally in all directions and is referred to as the static pressure. In a moving fluid, the static pressure is exerted on any plane parallel to the direction of motion. If the fluid is flowing in a circular pipe, the measuring surface must be perpendicular to the radial direction at any point. The pressure connection, which is known as a piezometer tube (Figure 12.3), should be flush with the wall of the pipe so that the flow is not disturbed.

Manometer is a device for measuring fluid pressure consisting of a bent tube containing one or more liquid of different specific gravities. In using a manometer, generally a known pressure (which may be atmospheric) is applied to one end of the manometer tube and the unknown pressure to

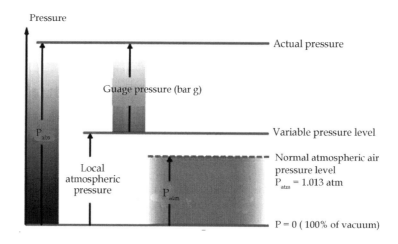

FIGURE 12.1   Pressure terminology.

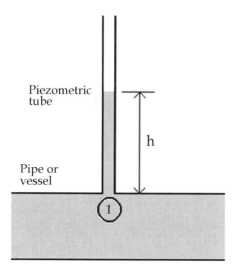

FIGURE 12.2   Forces on a fluid element.

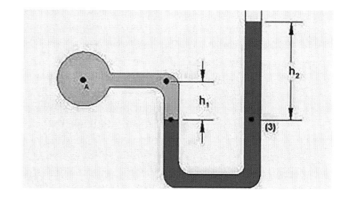

FIGURE 12.4   Simple U tube manometer.

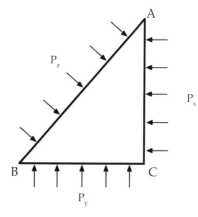

FIGURE 12.3   Piezometer.

be determined is applied to the other end. Various forms of manometers are as follows:

1. **Simple U-tube manometer**: It consists of glass tube bent in U-shape, one end of which is connected

to a point at which pressure is to be measured and the other end remains open to the atmosphere. The tube generally contains mercury or any other liquid whose specific gravity is greater (density=ρ_2) than the specific gravity of the liquid (density=ρ_1) whose pressure is to be measured. If g is the acceleration due to gravity and h_1 and h_2 are the heights of the liquids as shown in Figure 12.4.

For gauge pressure,

$$p = (\rho_2 g h_2 - \rho_1 g h_1) \tag{12.2}$$

For vacuum pressure,

$$p = -(\rho_2 g h_2 + \rho_1 g h_1) \tag{12.3}$$

2. **Inverted U-tube manometer**: Inverted U-tube manometer (Figure 12.5) is used for measuring pressure differences in liquids. The space above the liquid in the manometer is filled with air which can be admitted or expelled through the tap on the top, in order to adjust the level of the liquid in the manometer.

$$P_1 - P_2 = (\rho - \rho_m) g h \tag{12.4}$$

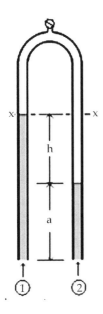

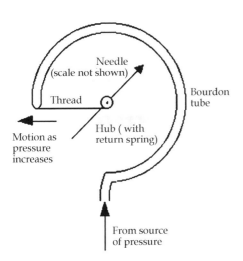

FIGURE 12.5   Inverted U-tube manometer.

If the manometric fluid is chosen in such a way that $\rho_m \ll \rho$ then,

$$P_1 - P_2 = \rho g h \qquad (12.5)$$

For inverted U-tube manometer the manometric fluid is usually air.

Pressure gauge is another device used for the measurement of pressure. The pressure to be measured is applied to a curved tube which is oval in cross section. Pressure applied to the tube tends to cause the tube to straighten out, and the deflection of the end of the tube is communicated through a system of levers to a recording needle. This gauge is widely used for steam and compressed gases. The pressure indicated is the difference between that communicated by the system to the external (ambient) pressure and is usually referred to as the gauge pressure. Bourdon gauge, a type of pressure gauge, is shown in Figure 12.6.

FIGURE 12.6   Bourdon gauge.

12.2   HYDROSTATIC FORCES ON SUBMERGED SURFACES

12.2.1   GENERAL SUBMERGED PLANE

Consider the plane surface shown in Figure 12.7.
Resultant force,

$$R = \sum p \delta A \qquad (12.6)$$

This resultant force will act through the center of pressure.

12.2.2   HORIZONTAL SUBMERGED PLANE

Resultant force = pressure × area of plane

$$R = pA \qquad (12.7)$$

12.2.3   INCLINED SUBMERGED SURFACE

This plane surface is totally submerged in a liquid of density ρ and inclined at an angle of θ to the horizontal as shown in Figure 12.8. Resultant force,

$$R = \rho g A \bar{x} \sin \theta \qquad (12.8)$$

12.2.4   VERTICAL SUBMERGED SURFACE

For a vertical wall, the pressure diagram can be drawn as shown in Figure 12.9. The area of this triangle represents the resultant force per unit width on the vertical wall.

$$\text{Resultant} = \text{Area of triangle} = \frac{1}{2}\rho g \qquad (12.9)$$

12.2.5   CURVED SUBMERGED SURFACE

In Figure 12.10, the liquid is resting on top of a curved base. The element of fluid ABC is equilibrium (as the fluid is at rest),

$$R = \sqrt{R_H^2 + R_Y^2} \qquad (12.10)$$

and acts through O at an angle of θ.
The angle the resultant force makes to the horizontal is

$$\theta = \tan^{-1}\left(\frac{R_Y}{R_H}\right) \qquad (12.11)$$

12.3   BUOYANCY AND FLOTATION

When a body is immersed in a fluid, an upward force is exerted by the fluid on the body. This upward force is equal to the weight of the fluid displaced by the body and is called the force of buoyancy or simply buoyancy.

Centre of Buoyancy:

It is defined as the point, through which the force of buoyancy is supposed to act. As the force of buoyancy is a vertical

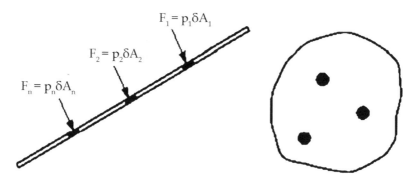

FIGURE 12.7   Forces on a general submerged plane.

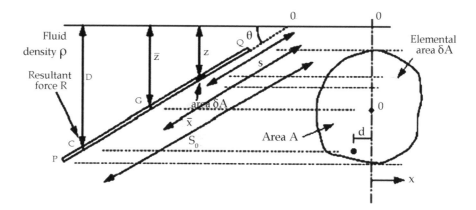

FIGURE 12.8   Resultant force and center of pressure on a submerged inclined surface in a liquid.

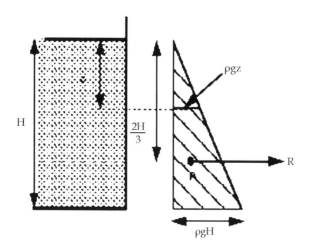

FIGURE 12.9   Pressure diagram for vertical wall.

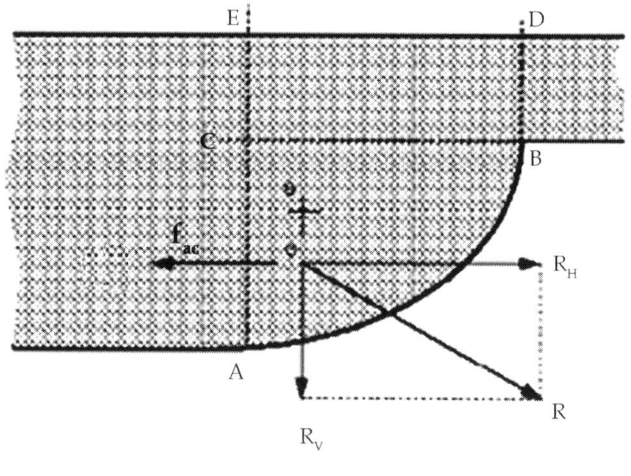

FIGURE 12.10   Resultant force on a curved submerged surface.

force and is equal to the weight of the fluid displaced by the body, the center of buoyancy will be the center of gravity of the fluid displaced.

Metacenter:

It is defined as the point about which a body starts oscillating when the body is tilted by a small angle. The metacenter may also be defined as the point at which the line of action of the force of buoyancy will meet the normal axis of the body when the body is given a small angular displacement. Consider a body floating in a liquid as shown in Figure 12.11a.

Let the body is in equilibrium and G is the center of gravity and B is the center of buoyancy. For equilibrium, both the points lie on the normal axis, which is vertical.

Metacentric Height:

The distance MG, i.e., the distance between the metacenter of a floating body and the center of gravity of the body, is called metacentric height.

Analytical Method for Metacenter Height:

Figure 12.11a shows the position of a floating body in equilibrium. The location of center of gravity and center of

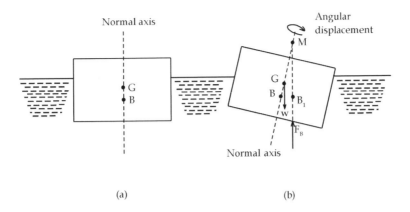

FIGURE 12.11 Buoyancy and metacentric height (a and b).

buoyancy in this position is at G and B. The floating body is given a small angular displacement in the clockwise direction. This is shown in Figure 12.11b, the vertical line through B_1 cuts the normal axis at M. Hence M is the metacenter and GM is metacentric height.

Flotation:

In flotation, the buoyant force equals the weight of the floating object and the volume of the object is always greater than the volume of water displaced. Flotation can be calculated using Archimedes' principle.

Archimedes' principle:

Archimedes' principle states that the buoyant force on a fluid is equal to the weight of the displaced fluid. To calculate the buoyant force, we use the equation buoyant force = density of fluid × volume of displaced fluid × acceleration due to gravity. In a completely submerged object, the volume of displaced fluid equals the volume of the object. If the object is floating, the volume of the displaced fluid is less than the volume of the object but the buoyant force = the weight of the object.

12.3.1 Conditions of Equilibrium of Floating and Submerged Bodies

12.3.1.1 Stability of a Submerged Body

a. **Stable equilibrium**: When $W = F_B$ and point B is above G, the body is said to be in stable equilibrium.

b. **Unstable equilibrium**: If $W = F_B$, but the center of buoyancy (B) is below center of gravity (G), the body is in unstable equilibrium as shown in Figure 12.12b. A slight displacement to the body, in the clockwise direction, gives the couple due to W and F_B also in the clockwise direction. Thus the body does not return to its original position and hence the body is in unstable equilibrium.

c. **Neutral equilibrium**: If $F_B = W$ and B and G are at the same point, as shown in Figure 12.12c, the body is said to be in Neutral Equilibrium.

12.3.1.2 Stability of Floating Body

a. **Stable equilibrium**: If the point M is above G (Figure 12.13).

b. **Unstable equilibrium**: If the point M is below G.

c. **Neutral equilibrium**: If the point M is at the center of gravity of the body, the floating body will be in neutral equilibrium.

12.4 FLUID KINEMATICS

Fluid kinematics deals with describing the motion of fluids without necessarily considering the forces and moments that cause the motion. In this chapter, several kinematic concepts related to flowing fluids are introduced. In addition, the material derivative and its role in transforming the conservation

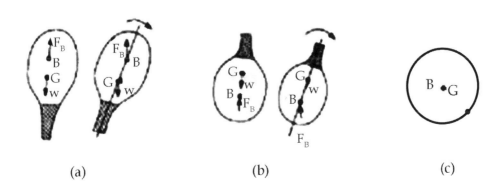

FIGURE 12.12 Stability of submerged body. (a) Stable equilibrium. (b) Unstable equilibrium. (c) Neutral equilibrium.

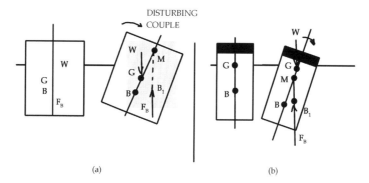

FIGURE 12.13 Stability of floating body (a and b).

equations from the Lagrangian description of fluid flow (following a fluid particle) to the Eulerian description of fluid flow (pertaining to a flow field). Fluid flow can be steady or unsteady, depending on the fluid's velocity:

1. Fluid flow can be *compressible or incompressible*, depending on whether you can easily compress the fluid. Liquids are usually nearly impossible to compress, whereas gases (also considered a fluid) are very compressible.
2. Liquid flow can be *viscous or nonviscous*. Viscosity is a measure of the thickness of a fluid, and very gloppy fluids such as motor oil or shampoo are called viscous fluids. Viscosity is actually a measure of friction in the fluid. When a fluid flows, the layers of fluid rub against one another, and in very viscous fluids, the friction is so great that the layers of flow pull against one another and hamper that flow.
3. Fluid flow can be *rotational or irrotational*. If, as you travel in a closed loop, you add up all the components of the fluid velocity vectors along your path and the end result is not zero, then the flow is rotational.

   Some flows that you may think rotational are actually irrotational. For example, away from the center, a vortex is actually an irrotational flow! You can see this if you look at the water draining from your bathtub. If you place a small floating object in the flow, it goes around the plug hole, but it does not spin about itself; therefore, the flow is irrotational.
4. **Laminar flow**: Occurs when the fluid flows in parallel layers, with no mixing between the layers, where the center part of the pipe flows the fastest and the cylinder touching the pipe is not moving at all. The flow is laminar when Reynolds number is less than 2300.
5. **Turbulent flow**: In turbulent flow occurs when the liquid is moving fast with mixing between layers. The speed of the fluid at a point is continuously undergoing changes in both magnitude and direction. The flow is turbulent when Reynolds number is greater than 4000.
6. **Transitional flow**: Transitional flow is a mixture of laminar and turbulent flow, with turbulence flow

in the center of the pipe and laminar flow near the edges of the pipe. Each of these flows behaves in different manners in terms of their frictional energy loss while flowing and have different equations that predict their behavior. The flow is transitional when Reynolds number is in between 2300 and 4000.

Rate of flow or discharge: The volume flow rate Q of a fluid is defined to be the volume of fluid that is passing through a given cross-sectional area per unit time.

Discharge,

$$Q = A \times V \tag{12.12}$$

where

$A =$ cross-sectional area of pipe
$V =$ average velocity of fluid across the section

12.4.1 CONTINUITY EQUATION

One of the fundamental principles used in the analysis of uniform flow is known as the Continuity of Flow. This principle is derived from the fact that mass is always conserved in fluid systems regardless of the pipeline complexity or direction of flow.

If steady flow exists in a channel and the principle of conservation of mass is applied to the system, there exists a continuity of flow, defined as "The mean velocities at all cross sections having equal areas are then equal, and if the areas are not equal, the velocities are inversely proportional to the areas of the respective cross sections." Thus if the flow is constant in a reach of channel, the product of the area and velocity will be the same for any two cross sections within that reach.

Consider two cross sections of a pipe as shown in Figure 12.14.

Let $V_1 =$ average velocity at cross section 1
$\rho_1 =$ density at section 1
$A_1 =$ area of pipe at section 1
And, V_2, ρ_2, A_2 are corresponding values at section 2.
According to law of conservation of mass,
Rate of flow at section 1 = rate of flow at section 2

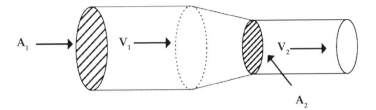

FIGURE 12.14   Flow of water through pipe of varying sections.

Or

$$\rho_1 A_1 V_1 = \rho_2 A_2 \qquad (12.13)$$

The above equation is applicable to the compressible as well as incompressible fluids and is called Continuity Equation. If the fluid is compressible, then

$$\rho_1 = \rho_2 \qquad (12.14)$$

And continuity equation reduces to

$$A_1 V_1 = A_2 \qquad (12.15)$$

12.4.1.1   Velocity and Acceleration

Velocity is an important basic parameter governing a flow field. Other field variables such as the pressure and temperature are all influenced by the velocity of the fluid flow. In general, velocity is a function of both the location and time. The velocity vector can be expressed in Cartesian coordinates as

$$V = V(x, y, z, t) \qquad (12.16)$$

$$V = u(x,\ y,\ z,\ t)i + v(x,\ y,\ z,\ t)j + w(x,\ y,\ z,\ t)k \qquad (12.17)$$

where the velocity components (u, v, and w) are functions of both position and time. That is, $u=u$ (x, y, z, t), $v=v$ (x, y, z, t), and $w=w$ (x, y, z, t). This velocity description is called the velocity field since it describes the velocity of all points, Eulerian viewpoint, in a given volume.

For a single particle, Lagrangian viewpoint, the velocity is derived from the changing position vector, or

$$V = \frac{d\mathbf{r}}{dt} = \frac{dx}{dt}i + \frac{dy}{dt}j + \frac{dz}{dt}k \qquad (12.18)$$

where $\mathbf{r}$ is the position vector ($\mathbf{r}=x\mathbf{i}+y\mathbf{j}+z\mathbf{k}$). Notice, the velocity is only a function of time since it is only tracking a single particle.

Acceleration is related to the velocity, and it can be determined once the velocity field is known. The acceleration is the change in velocity, δV, over the change in time, δt,

$$a = \frac{\delta V}{\delta t} = \frac{dV}{dt} \qquad (12.19)$$

The change in velocity must be track in both time and space. Using the chain rule of calculus, the change in velocity is,

$$dV = \frac{\partial V}{\partial x}\,dx + \frac{\partial V}{\partial y}\,dy + \frac{\partial V}{\partial z}\,dz + \frac{\partial V}{\partial t}\,dt \qquad (12.20)$$

Or

$$\frac{dV}{dt} = \frac{\partial V}{\partial x}\frac{dx}{dt} + \frac{\partial V}{\partial y}\frac{dy}{dt} + \frac{\partial V}{\partial z}\frac{dz}{dt} + \frac{\partial V}{\partial t} \qquad (12.21)$$

This can be simplified using u, v, and w, the velocity magnitudes in the three coordinate directions. In Cartesian coordinates, the acceleration field is

$$a = \frac{\partial V}{\partial t} + u\frac{\partial V}{\partial x} + v\frac{\partial V}{\partial y} + w \qquad (12.22)$$

This expression can be expanded and rearranged as

$$a = a_x i + a_y j + a_z k \qquad (12.23)$$

where

$$a_x = \frac{\partial u}{\partial t} + u\frac{\partial u}{\partial x} + v\frac{\partial u}{\partial y} + w\frac{\partial u}{\partial z} \qquad (12.24)$$

$$a_y = \frac{\partial v}{\partial t} + u\frac{\partial v}{\partial x} + v\frac{\partial v}{\partial y} + w\frac{\partial v}{\partial z} \qquad (12.25)$$

$$a_z = \frac{\partial w}{\partial t} + u\frac{\partial w}{\partial x} + v\frac{\partial w}{\partial y} + w\frac{\partial w}{\partial z} \qquad (12.26)$$

12.4.1.2   Velocity Potential Function

An irrotational flow is defined as the flow where the vorticity is zero at every point. It gives rise to a scalar function (ϕ) which is similar and complementary to the stream function (ψ). Let us consider the equations of irrotational flow and scalar function (ϕ). In an irrotational flow, there is no vorticity.

Here, ϕ is called as *velocity potential function* and its gradient gives rise to velocity vector. The knowledge ϕ immediately gives the velocity components. In Cartesian coordinates, the *velocity potential function* can be defined as,

$$\phi = \phi(x, y, z) \qquad (12.27)$$

It can be written as,

$$u\hat{i} + v\hat{j} + w\hat{k} = \frac{\partial \varphi}{\partial x}\hat{i} + \frac{\partial \varphi}{\partial y}\hat{j} + \frac{\partial \varphi}{\partial z}\hat{k} \qquad (12.28)$$

For a flow which is incompressible and irrotational,

$$\frac{\partial^2 \varphi}{\partial x^2} + \frac{\partial^2 \varphi}{\partial y^2} + \frac{\partial^2 \varphi}{\partial z^2} = 0 \qquad (12.29)$$

This is a Laplace equation.

Thus, any irrotational, incompressible flow has a velocity potential and stream function (for two-dimensional flow) that both satisfy *Laplace equation*. Conversely, any solution of *Laplace equation* represents both velocity potential and stream function (two-dimensional) for an irrotational, incompressible flow.

Properties of a velocity potential function:

1. If velocity potential exists, the flow should be irrotational.
2. If velocity potential satisfies the Laplace equation, it represents the possible steady, incompressible, irrotational flow.

12.4.1.3   Stream Function

The idea of introducing stream function works only if the continuity equation is reduced to two terms. For a steady, incompressible, plane, two-dimensional flow, this equation reduces to,

$$\frac{\partial u}{\partial x} + \frac{\partial v}{\partial y} = 0 \qquad (12.30)$$

Here, the striking idea of stream function works that will eliminate two velocity components *u* and *v* into a single variable. So, the *stream function* $\{\psi(x, y)\}$ relates to the velocity components in such a way that continuity equation is satisfied (Figure 12.15).

$$u = \frac{\partial \psi}{\partial y} \qquad (12.31)$$

$$v = -\frac{\partial \psi}{\partial x} \qquad (12.32)$$

In steady flow, a fluid particle will move along a streamline.

Equation of a stream line in a three-dimensional flow is given as

$$\frac{dx}{u} = \frac{dy}{v} = \frac{dz}{w} \qquad (12.33)$$

The following important points can be noted for stream functions;

1. The lines along which ψ is constant are called as streamlines. In a flow field, the tangent drawn at every point along a streamline shows the direction of velocity. So, the slope at any point along a streamline is given by,

$$\frac{dy}{dx} = \frac{v}{u} \qquad (12.34)$$

2. Infinite number streamlines can be drawn with constant ψ. This family of streamlines will be useful in visualizing the flow patterns. It may also be noted that streamlines are always parallel to each other.
3. The numerical constant associated to ψ represents the volume rate of flow.

12.4.2   Types of Flow

12.4.2.1   Eularian and Lagrangian Flow

Eulerian analysis uses the field concept, derived from the continuum assumption.

Lagrangian analysis involves following individual fluid particles as they move about and determining how fluid properties vary as a function of time.

12.4.2.2   Steady vs. Unsteady Flow

When the flowing particle velocity, temperature, and pressure do not change with respect to time is called steady flow. It should be noted that the conditions are different at different points, but at a particular point it is constant.

12.4.3   Streamlines, Streaklines, Pathlines

Streamline, pathline, streakline, and timeline form convenient tools to describe a flow.

12.4.3.1   Streamline

A streamline is one that drawn is tangential to the velocity vector at every point in the flow at a given instant and forms a powerful tool in understanding flows. This definition leads to the equation for streamlines.

$$\frac{du}{u} = \frac{dv}{v} = \frac{dw}{w} \qquad (12.35)$$

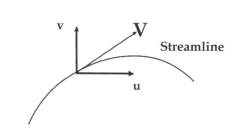

FIGURE 12.15   Velocity components along a streamline.

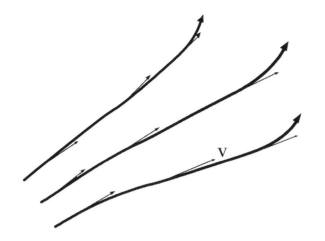

FIGURE 12.16   Streamlines.

where u, v, and w are the velocity components in x, y, and z directions, respectively, as sketched in Figure 12.16.

12.4.3.2   Stream Tube

Sometimes, as shown in Figure 12.17, we pull out a bundle of streamlines from inside of a general flow for analysis. Such a bundle is called stream tube and is very useful in analyzing flows. If one aligns a coordinate along the stream tube then the flow through it is one-dimensional.

12.4.3.3   Pathline

It is the line traced by a given particle. This is generated by injecting a dye into the fluid and following its path by photography or other means (Figure 12.18).

12.4.3.4   Streakline

It concentrates on fluid particles that have gone through a fixed station or point. At some instant of time the position of

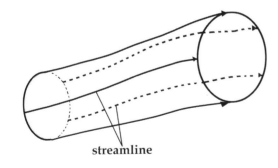

streamline

FIGURE 12.17   Stream tube.

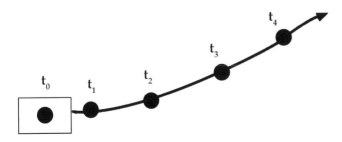

FIGURE 12.18   Pathlines.

all these particles is marked and a line is drawn through them. Such a line is called a streakline (Figure 12.19).

12.4.3.5   Timeline

It is generated by drawing a line through adjacent particles in flow at any instant of time. Figure 12.20 shows a typical timeline.

In a steady flow the streamline, pathline, and streakline all coincide. In an unsteady flow they can be different. Streamlines are easily generated mathematically while pathline and streaklines are obtained through experiments.

12.4.4   FREE AND FORCED VORTEX FLOW

Vortex flow is defined as the flow of a fluid along a curved path or the flow of a rotating mass of fluid is known as Vortex flow. The following are the two types of Vortex flow:

1. Free Vortex Flow
2. Forced Vortex Flow

12.4.4.1   Free Vortex Flow

When no external torque is required to rotate the fluid mass, that type of flow is called free vortex flow.

- There is no energy interaction between an external source and a flow or any dissipation of mechanical energy in the flow.
- Fluid mass rotates due to conservation of angular momentum.
- Velocity inversely proportional to the radius.

For a free vortex flow,

$$vr = \text{constant} \tag{12.36}$$

At the center ($r=0$) of rotation velocity approaches to infinite, that point is called singular point. The free vortex flow is irrotational and therefore also known as the irrotational vortex. In free vortex flow, Bernoulli's equation can be applied.

Examples:

- Flow of liquid through a hole provided at the bottom of the container
- Flow through kitchen sink
- Draining the bath tub
- Flow of liquid around a circular bend in pipe
- A whirlpool in a river
- The flow fields due to a tornado

12.4.4.2   Forced Vortex Flow

Forced vortex flow is defined as that type of vortex flow in which some external torque is required to rotate the fluid mass (Figure 12.21).

- To maintain a forced vortex flow, it required a continuous supply of energy or external torque.

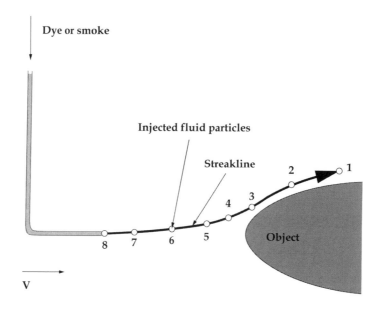

FIGURE 12.19   Streaklines.

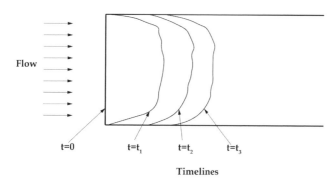

FIGURE 12.20   Timeline.

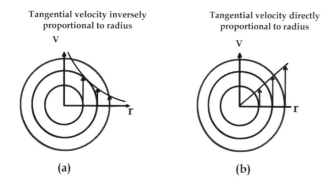

FIGURE 12.21   Free and forced vortex flow. (a) Free vortex. (b) Forced vortex.

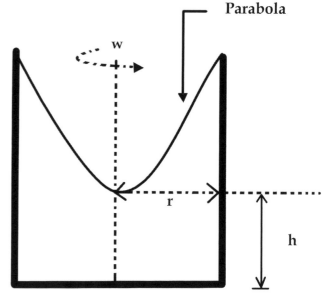

FIGURE 12.22   Vortex and tangential velocity.

$$v = r\omega \qquad (12.37)$$

where
ω = angular velocity
r = radius of fluid particle from the axis of rotation

- The surface profile of vortex flow is parabolic.

$$h = \frac{\omega^2 r^2}{2g} \qquad (12.38)$$

- In forced vortex total energy per unit weight increases with an increase in radius.
- Forced vortex is not irrotational; rather it is a rotational flow with constant vorticity 2ω.

- All fluid particles rotate at the constant angular velocity ω as a solid body. Therefore, a flow of forced vortex is called as a solid body rotation.
- Tangential velocity is directly proportional to the radius (Figure 12.22).

Examples:

- Flow of water through the runner of a turbine
- Flow of liquid through the passage of impeller of centrifugal pumps
- Rotation of water in a washing machine

12.5 DIMENSIONAL ANALYSIS

Dimensional analysis is a means of simplifying a physical problem by appealing to dimensional homogeneity to reduce the number of relevant variables. It is particularly useful for:

- presenting and interpreting experimental data;
- attacking problems not amenable to a direct theoretical solution;
- checking equations;
- establishing the relative importance of particular physical phenomena;
- physical modeling.

12.5.1 DIMENSIONS AND UNITS

Any physical situation can be described by certain familiar properties, e.g., length, velocity, area, volume, acceleration, etc. These are all known as dimensions. Dimensions are properties which can be measured. Units are the standard elements we use to quantify these dimensions. In dimensional analysis, we are only concerned with the nature of the dimension, i.e., its quality not its quantity. The following common abbreviations are used:
    Length=L; Mass=M; Time=T; Force=F; Temperature=Q.
    In this module, we are only concerned with L, M, T, and F (not Q). We can represent all the physical properties we are interested in with L, T and one of M or F (F can be represented by a combination of LTM).

Dimensional analysis is a method for reducing the number and complexity of experimental variables which affect a given physical phenomenon, by using a sort of compacting technique. If a phenomenon depends upon n dimensional variables, dimensional analysis will reduce the problem to only k dimensionless variables, where the reduction $n\text{-}k=1$, 2, 3, or 4, depending upon the problem complexity. Generally, $n\text{-}k$ equals the number of different dimensions which govern the problem. In fluid mechanics, the four basic dimensions are usually taken to be mass M, length L, time T, and temperature θ, or an $MLT\theta$ system for short. Sometimes one uses an $FLT\theta$ system, with force F replacing mass.

12.5.2 DIMENSIONAL HOMOGENEITY

Any equation describing a physical situation will only be true if both sides have the same dimensions. That is, it must be dimensionally homogenous.

For example, the equation which gives for flow over a rectangular weir is

$$Q = \frac{2}{3} B\sqrt{2g}H^{3/2} \qquad (12.39)$$

The SI units of the left hand side are m³/s. The units of the right-hand side must be the same. To be more precise, it is the dimension which must be consistent (any set of units can be used and simply converted using a constant). Writing the equation again in terms of dimensions,

$$L^3T^{-1} = L\left(LT^{-2}\right)^{1/2} L^{3/2} = L^3T^{-1} \qquad (12.40)$$

This property of dimensional homogeneity will be useful for

1. checking units of equations;

	Quantity	Common Symbol(s)	Dimensions
Geometry	Area	A	L^2
	Volume	V	L^3
	Second moment of area	I	L^4
Kinematics	Velocity	U	LT^{-1}
	Acceleration	a	LT^{-2}
	Angle	θ	1 (i.e., dimensionless)
	Angular velocity	ω	T^{-1}
	Quantity of flow	Q	L^3T^{-1}
	Mass flow rate	m	MT^{-1}
Dynamics	Force	F	MLT^{-2}
	Moment, torque	T	ML^2T^{-2}
	Energy, work, heat	E, W	ML^2T^{-2}
	Power	P	ML^2T^{-3}
	Pressure, stress	p, τ	$ML^{-1}T^{-2}$
Fluid properties	Density	ρ	ML^{-3}
	Viscosity	μ	$ML^{-1}T^{-1}$
	Kinematic viscosity	ν	L^2T^{-1}
	Surface tension	σ	MT^{-2}
	Thermal conductivity	k	$MLT^{-3}\theta^{-1}$
	Specific heat	c_p, c_v	$L^2T^{-2}\theta^{-1}$
	Bulk modulus	K	$ML^{-1}T^{-2}$

2. converting between two sets of units;
3. defining dimensionless relationships.

12.6 MODEL ANALYSIS

It is the study of models of actual machine or structure.

12.6.1 MODEL

It is the small-scale replica of the actual structure or machine. It is not necessary that models should be smaller than the prototypes (although in most of the cases it is), they may be larger than the prototypes.

Advantages:

- The performance of the machine can be easily predicted, in advance.
- With the help of dimensional analysis, a relationship between the variables influencing a flow problem in terms of dimensional parameters can be developed. This relationship helps in conducting tests on the model.
- The merits of alternative designs can be predicted with the help of model testing. The most economical and safe design may be, finally, adopted.

Three basic laws of similitude must be satisfied in order to achieve complete similarity between prototype and model flow fields.

1. **Geometric similarity**: Model and prototype must be the same in shape but can be different in size. All linear dimensions of the model are related to corresponding dimensions of the prototype by a constant length ratio, L_r. It is usually impossible to establish 100% geometric similarity due to very small details that cannot be put into the model. Modeling surface roughness exactly is also impossible.

$$L_r = \frac{L_p}{L_m} \qquad (12.41)$$

2. **Kinematic similarity**: Model and prototype flow fields are kinematically similar if the velocities at corresponding points are the same in direction and differ only by a constant factor of velocity ratio, V_r. This also means that the streamline patterns of two flow fields should differ by a constant scale factor.

$$V_r = \frac{V_p}{V_m} \qquad (12.42)$$

3. **Dynamic similarity**: Two flow fields should have force distributions such that identical types of forces are parallel and are related in magnitude by a constant factor of force ratio.

If a certain type of force, e.g., compressibility force, is highly dominant in the prototype flow, it should also be dominant in the model flow. If a certain type of force, e.g., surface tension force, is negligibly small in the prototype flow, it should also be small in the model flow. To establish dynamic similarity we need to determine the important forces of the prototype flow and make sure that the nondimensional numbers related to those forces are the same in prototype and model flows.

12.6.2 TYPE OF FORCES ACTING ON THE MOVING FLUID

1. **Inertial force**: it is equal to the mass and acceleration of the moving fluid.

$$F_i = \rho A \qquad (12.43)$$

2. **Viscous force**: it is equal to the shear stress due to viscosity and surface area of the flow. It presents in the flow problems where viscosity is having an important role to play.

$$F_v = \tau A \qquad (12.44)$$

3. **Gravity force**: product of mass and acceleration due to gravity.

$$F_g = \rho A L g \qquad (12.45)$$

4. **Pressure force**: product of pressure intensity and flow area.

$$F_p = p A \qquad (12.46)$$

5. **Surface tension force**: product of surface tension and the length of the surface of the flowing fluid.

$$F_s = \sigma d \qquad (12.47)$$

6. **Elastic force**: product of elastic stress and area of the flow.

$$F_e = K A \qquad (12.48)$$

12.6.3 DIMENSIONLESS NUMBERS

Reynolds number:

$$R_e = \frac{\text{Inertial force}}{\text{Viscous force}} = \frac{V \rho d}{\mu} \qquad (12.49)$$

Froude's number:

$$F_r = \sqrt{\frac{\text{Inertial force}}{\text{Gravity force}}} = \frac{V}{\sqrt{dg}} \qquad (12.50)$$

Euler's number:

$$E_u = \frac{\text{Pressure force}}{\text{Inertial force}} = \frac{p}{\rho V^2} \qquad (12.51)$$

Weber's number:

$$W_e = \frac{\text{Surface tension force}}{\text{Inertial force}} = \frac{\sigma}{\rho V^2 l} \qquad (12.52)$$

Mach's number:

$$M = \sqrt{\frac{\text{Inertial force}}{\text{Elastic force}}} = \frac{V}{\sqrt{K/\rho}} = \frac{V}{c} \qquad (12.53)$$

where c is the velocity of sound.

12.6.4 DYNAMIC SIMILARITY

For the dynamic similarity between the model and the prototype, the ratio of the corresponding forces acting at the corresponding points in the model and prototype should be equal. It means for dynamic similarity between the model and prototype, the dimensionless numbers should be the same for model and prototype. It is quite difficult to satisfy the condition that all the dimensionless numbers are the same for the model and prototype. Therefore, the model is designed on the basis of equating the dimensionless number which dominates the phenomenon.

Following are the dynamic similarity laws:

1. Reynolds model law
2. Froude's model law
3. Euler's model law
4. Weber model law
5. Mach model law

Reynolds model law: (pipe flow, submarines, aeroplane, etc.)

Froude's model law: (free-surface flow, jet from orifice or nozzle, etc.)

Euler's model law: (pressure force is a dominant force)

Weber model law: (surface tension is a dominant force)

Mach model law: (velocity of flow is comparable to the velocity of sound; compressible flow)

Classification of model:

- **Undistorted models**: are those models which are geometrically similar to their prototype. In other words, the scale ratio for the linear dimensions of the model and its prototype are the same.
- **Distorted models**: are those models which are geometrically not similar to its prototype. In other words, the scale ratio for the linear dimensions of the model and its prototype are not same.

For example river:

If the horizontal and vertical scale ratios for the model and the prototype are same then it is undistorted model. In this case the depth of the water in the model becomes very small which may not be measured accurately. Thus for such cases, distorted model is useful.

12.7 FLUID DYNAMICS

This section discusses the analysis of fluid in motion—fluid dynamics. The motion of fluids can be predicted in the same way as the motion of solids is predicted using the fundamental laws of physics together with the physical properties of the fluid.

It is not difficult to envisage a very complex fluid flow. Spray behind a car; waves on beaches; hurricanes and tornadoes; or any other atmospheric phenomenon are all examples of highly complex fluid flows which can be analyzed with varying degrees of success (in some cases hardly at all!). There are many common situations which are easily analyzed.

- **Uniform flow**: If the flow velocity is the same magnitude and direction at every point in the fluid, it is said to be uniform.
- **Nonuniform**: If at a given instant, the velocity is not the same at every point the flow is nonuniform. (In practice, by this definition, every fluid that flows near a solid boundary will be non-uniform—as the fluid at the boundary must take the speed of the boundary, usually zero. However, if the size and shape of the cross section of the stream of fluid are constant, the flow is considered uniform.)
- **Steady**: A steady flow is one in which the conditions (velocity, pressure, and cross section) may differ from point to point but DO NOT change with time.
- **Unsteady**: If at any point in the fluid, the conditions change with time, the flow is described as unsteady.

Combining the above we can classify any flow into one of four types:

1. **Steady uniform flow**: Conditions do not change with position in the stream or with time. An example is the flow of water in a pipe of constant diameter at constant velocity.
2. **Steady nonuniform flow**: Conditions change from point to point in the stream but do not change with time. An example is flow in a tapering pipe with constant velocity at the inlet—velocity will change as you move along the length of the pipe toward the exit.
3. **Unsteady uniform flow**: At a given instant in time the conditions at every point are the same, but will change with time. An example is a pipe of constant diameter connected to a pump pumping at a constant rate which is then switched off.
4. **Unsteady nonuniform flow**: Every condition of the flow may change from point to point and with time at every point; for example, waves in a channel.

In fluid dynamics, the Euler equations are a set of quasilinear hyperbolic equations governing adiabatic and inviscid flow. They are named after Leonhard Euler. The equations represent Cauchy equations of conservation of mass (continuity), and balance of momentum and energy, and can be seen as particular Navier–Stokes equations with zero viscosity and zero thermal conductivity. In fact, Euler equations can be obtained by linearization of some more precise continuity equations like Navier–Stokes equations in a local equilibrium state given by a Maxwellian. The Euler equations can be applied to incompressible and to compressible flow—assuming the flow velocity is a solenoidal field.

12.7.1 Euler's Equation

The Euler's equation for steady flow of an ideal fluid along a streamline is a relation between the velocity, pressure, and density of a moving fluid (Figure 12.23). It is based on the Newton's Second Law of Motion. The integration of the equation gives Bernoulli's equation in the form of energy per unit weight of the moving fluid.

It is based on the following assumptions:

- The fluid is nonviscous (i.e., the frictional losses are zero).
- The fluid is homogeneous and incompressible (i.e., mass density of the fluid is constant).
- The flow is continuous, steady, and along the streamline.
- The velocity of the flow is uniform over the section.
- No energy or force (except gravity and pressure forces) is involved in the flow.

Let,

- dA = cross-sectional area of the fluid element
- ds = length of the fluid element
- dW = weight of the fluid element
- P = pressure on the element at A
- $P+dP$ = pressure on the element at B
- v = velocity of the fluid element

$$\frac{dp}{\rho} + gdz + vdv = 0 \tag{12.54}$$

The above equation is known as Euler's equation of motion.

12.7.2 Bernoulli's Equation from Euler's Equation

It is given as follows:

If flow is incompressible, ρ is constant and

$$\frac{p}{\rho} + gz + \frac{v^2}{2} = \text{constant} \tag{12.55}$$

Or

$$\frac{p}{\rho g} + z + \frac{v^2}{2g} = \text{constant} \tag{12.56}$$

Or

$$\frac{p}{\rho g} + \frac{v^2}{2g} + z = \text{constant} \tag{12.57}$$

where

$\frac{p}{\rho g}$ = pressure energy per unit weight of fluid or pressure head.

$\frac{v^2}{2g}$ = kinetic energy per unit weight or kinetic head.

z = potential energy per unit weight or potential head

12.7.2.1 Bernoulli's Equation for Real Fluid

It is given as

$$\frac{p_1}{\rho g} + \frac{v_1^2}{2g} + z_1 = \frac{p_2}{\rho g} + \frac{v_2^2}{2g} + z_2 + h_L \tag{12.58}$$

where

h_L = loss of energy between points 1 and 2.

12.8 FLUID FLOW MEASUREMENTS

12.8.1 Practical Applications of Bernoulli's Equation

Bernoulli's equation finds wide applications in all types of problems of incompressible flow where there is involvement of energy considerations. The other equation, which is

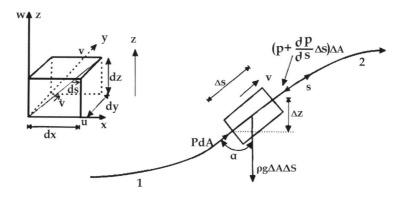

FIGURE 12.23 The Euler's equation for steady flow of an ideal fluid along a streamline.

commonly used in the solution of the problems of fluid flow, is the continuity equation. In this section, the applications of Bernoulli's equation and continuity equation will be discussed for the following measuring devices.

- Venturimeter
- Nozzle
- Orifice meter
- Pitot tube

12.8.1.1 Venturimeter (Figure 12.24)

- It is an instrument which is used to measure the rate of discharge in a pipeline and is often fixed permanently at different sections of the pipeline to measure the discharge.
- The principle of venturimeter was demonstrated by Italian physicist G.B. Venturi (1746–1822) in 1797, but it was first applied by C. Herschel (1842–1930) in 1887 to develop the device for measuring the discharge or rate of flow of fluid through pipes.
- The basic principle is that by reducing the cross-sectional area of the flow passage, a pressure difference is created and the measurement of pressure difference enables the determination of the discharge through pipes.

Construction:

It consists of three parts as shown in Figure 12.24.

i. An inlet section followed by a convergent cone
ii. A cylindrical throat
iii. A gradually divergent cone

Applying Bernoulli's equation at sections 1 and 2, we get theoretical discharge as follows:

$$Q_{th} = \frac{a_1 a_2 \sqrt{2gh}}{\sqrt{a_1^2 - a_2^2}} \qquad (12.59)$$

where

a_1 = area at section 1
a_2 = area at section 2

The above equation gives only the theoretical discharge because the loss of energy is not considered. But, in actual practice, there is always some loss of energy as the fluid flows and the actual discharge (Q) will always be less than the theoretical discharge. The actual discharge may be obtained by multiplying the theoretical discharge by a factor C_d, called coefficient of discharge, i.e.,

$$C_d = \frac{Q}{Q_{th}} \qquad (12.60)$$

For a given venturimeter, the cross-sectional areas of the inlet section and the throat, i.e., a_1 and a_2, are fixed.

$$Q_{act} = C_d \times \frac{a_1 a_2 \sqrt{2gh}}{\sqrt{a_1^2 - a_2^2}} \qquad (12.61)$$

- The coefficient of discharge of the venturimeter accounts for the effects of nonuniformity of the velocity distribution at sections 1 and 2.
- The coefficient of discharge of the venturimeter varies with the flow rate, viscosity of the fluid, and the surface roughness. But, in general, for the fluids of low viscosity, the value falls in the range of 0.95–0.98.
- The venture head (i.e., the pressure difference between sections 1 and 2) is usually measured by a manometer. If S_m and S are the specific gravities of the liquid in the manometer and liquid flowing in the venturimeter and x is the difference in the levels of two limbs of the manometer, then the expression for the venture head becomes

$$h = x \left[\frac{S_m}{S} - 1 \right] \qquad (12.62)$$

when $S_m > S$.

$$h = x \left[1 - \frac{S_m}{S} \right] \qquad (12.63)$$

when $S_m < S$.

- Venturimeter can also be used to measure the discharge through pipe, which is laid either in an inclined or in vertical position.

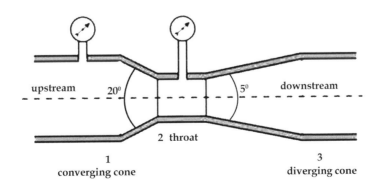

FIGURE 12.24 Venturimeter.

12.8.1.2 Nozzle

- A flow nozzle is also a device used for measuring the discharge through pipes. As shown in Figure 12.25, the flow nozzle consists of a streamlined convergent nozzle through which the fluid is gradually accelerated.
- It is essentially a venturimeter with divergent part omitted and hence the basic equations are the same as those for the venturimeter. Since there is no divergent cone on the downstream of throat of the nozzle, there is a greater dissipation of energy than the venturimeter.

The discharge coefficient for a nozzle can be given by the following empirical equation:

$$C_d = 0.99622 + 0.00059D - \frac{6.36 + 0.13D - 0.24\left(\dfrac{d}{D}\right)^2}{R_e}$$

(12.64)

where d and D are the diameter of the nozzle and the pipe, and R_e is the Reynolds number based on the diameter of the nozzle.

Cavitation:

When the pressure at any point in a liquid becomes equal to the vapor pressure of the liquid, the liquid vaporizes and forms bubbles. These bubbles have the tendency to break the continuity of the flow. Formation of vapor bubbles and their transport to regions of high pressure and subsequent collapse are known as "cavitation." It is quantified by a dimensionless number defined by:

$$\text{Cavitation number, } \sigma = \frac{p - p_v}{\left(\rho U_0^2 / 2\right)}$$

(12.65)

where p is the absolute pressure at the point under consideration, p_v is the vapor pressure of the liquid, U_0 is the reference velocity, and ρ is the density of the liquid.

12.8.1.3 Orifice Meter or Orifice Plate

An opening, in a vessel, through which the liquid flows out is known as orifice. This hole or opening is called an orifice, so long as the level of the liquid on the upstream side is above the top of the orifice. The typical purpose of an orifice is the measurement of discharge. An orifice may be provided in the vertical side of a vessel or in the base.

Types of Orifices:

Orifices can be of different types depending upon their size, shape, and nature of discharge. But the following are important from the subject point of view.

  A. According to size:
    i. Small orifice
    ii. Large orifice
  B. According to shape:
    i. Circular orifice
    ii. Rectangular orifice
    iii. Triangular orifice
  C. According to shape of edge:
    i. Sharp-edged
    ii. Bell-mouthed
  D. According to nature of discharge:
    i. Fully submerged orifice
    ii. Partially submerged orifice

Jet of water:

The continuous stream of a liquid, which comes out or flows out of an orifice, is known as the Jet of Water.

Vena-contract:

Consider an orifice is fitted with a tank. The liquid particles, in order to flow out through the orifice, move toward the orifice from all directions. A few of the particles first move downward, then take a turn to enter into the orifice and then finally flow through it.

The liquid particles loose some energy, while taking the turn to enter into the orifice. It can be observed that the jet, after leaving the orifice, gets contracted. The maximum contraction takes place at a section slightly on the downstream side of the orifice, where the jet is more or less horizontal. Such a section is known as *Vena-Contracta* as shown in Figure 12.26.

Hydraulic Coefficients:

The following four coefficients are known as hydraulic coefficients or orifice coefficients.

- Coefficient of contraction
- Coefficient of velocity

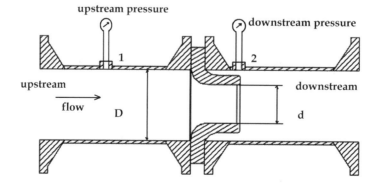

FIGURE 12.25 Nozzle.

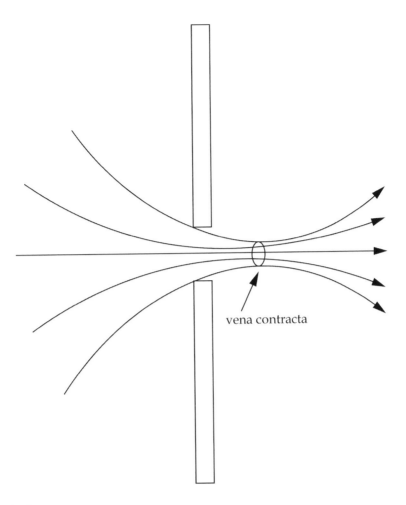

FIGURE 12.26   Vena-Contracta.

- Coefficient of discharge
- Coefficient of resistance

Coefficient of contraction:

The ratio of the area of the jet, at vena-contracta, to the area of the orifice is known as *coefficient of contraction*. Mathematically, coefficient of contraction,

$$C_c = \frac{\text{Area of jet at vena contracta}}{\text{Area of the orifice}} \quad (12.66)$$

The value of coefficient of contraction varies slightly with the available head of the liquid, size, and shape of the orifice. The average value of C_c is 0.64.

Coefficient of velocity:

The ratio of actual velocity of the jet, at vena-contracta, to the theoretical velocity is known as *coefficient of velocity*.

Mathematically, coefficient of velocity

$$C_v = \frac{\text{Actual velocity of the jet at vena contracta}}{\text{Theoretical velocity of the jet}} \quad (12.67)$$

The difference between the velocities is due to friction of the orifice. The value of coefficient of velocity varies slightly with the different shapes of the edges of the orifice. This value is

very small for sharp-edged orifices. For a sharp-edged orifice, the value of C_v increases with the head of water.

Coefficient of Discharge:

The ratio of actual discharge through an orifice to the theoretical discharge is known as *coefficient of discharge*. Mathematically, coefficient of discharge

$$C_d = \frac{\text{Actual discharge}}{\text{Theoretical discharge}} \quad (12.68)$$

$$C_d = C_v \times C_c \quad (12.69)$$

Thus the value of coefficient of discharge varies with the values of C_v and C_c. An average of coefficient of discharge varies from 0.60 to 0.64.

Coefficient of Resistance:

The ratio of loss of head in the orifice to the head of water available at the exit of the orifice is known as *coefficient of resistance*.

$$C_r = \frac{\text{Loss of head in the orifice}}{\text{Head of water}} \quad (12.70)$$

The loss of head in the orifice takes place, because the walls of the orifice offer some resistance to the liquid as it comes

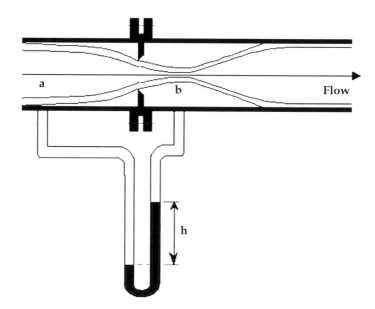

FIGURE 12.27 Orifice meter.

out. The coefficient of resistance is generally neglected, while solving numerical problems.

The orifice meter (Figure 12.27) consists of a flat orifice plate with a circular hole drilled in it. There is a pressure tap upstream from the orifice plate and another just downstream. There are three recognized methods of placing the taps. The coefficient of the meter will depend upon the position of taps. The principle of the orifice meter is identical with that of the venturimeter. The reduction of the cross section of the flowing stream in passing through the orifice increases the velocity head at the expense of the pressure head, and the reduction in pressure between the taps is measured by a manometer. Bernoulli's equation provides a basis for correlating the increase in velocity head with the decrease in pressure head.

$$Q = \frac{C_d a_0 a_1 \sqrt{2gh}}{\sqrt{a_1^2 - a_0^2}} \qquad (12.71)$$

where

C_d = coefficient of discharge for orifice meter
a_0 = area of orifice

The pressure recovery is limited for an orifice plate and the permanent pressure loss depends primarily on the area ratio.

For an area ratio of *0.5*, the head loss is about *70%–75%* of the orifice differential.

- The orifice meter is recommended for clean and dirty liquids and some slurry services.
- The rangeability is *4 to 1*
- The pressure loss is *medium*
- Typical accuracy is *2%–4%* of full scale
- The required upstream diameter is *10–30*
- The viscosity effect is *high*
- The relative cost is *low*

12.8.1.4 Pitot Tube

If a stream of uniform velocity flows into a blunt body, the stream lines take a pattern as shown in Figure 12.28.

Note how some move to the left and some to the right. But one, in the center, goes to the tip of the blunt body and stops. It stops because at this point the velocity is zero—the fluid does not move at this one point. This point is known as the *stagnation point*.

From the Bernoulli equation we can calculate the pressure at this point. Apply Bernoulli along the central streamline from a point upstream where the velocity is u_1 and the pressure p_1 to the stagnation point of the blunt body where velocity is zero, $u_2 = 0$. Also $z_1 = z_2$.

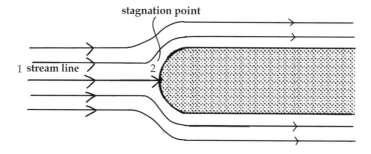

FIGURE 12.28 Streamlines around a blunt body.

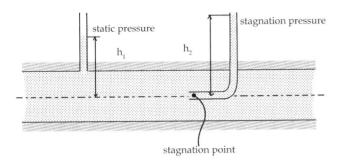

FIGURE 12.29 A Piezometer and a Pitot tube.

$$\frac{p_1}{\rho g} + \frac{u_1^2}{2g} + z_1 = \frac{p_2}{\rho g} + \frac{u_2^2}{2g} + z_2 \qquad (12.72)$$

$$p_2 = p_1 + \frac{1}{2}\rho \qquad (12.73)$$

This increase in pressure which brings the fluid to rest is called the *dynamic pressure.*

The blunt body stopping the fluid does not have to be a solid. It could be a static column of fluid. Two piezometers, one as normal and one as a Pitot tube within the pipe, can be used in an arrangement shown in Figure 12.29 to measure the velocity of flow.

$$u = \sqrt{2g(h_2 - h_1)} \qquad (12.74)$$

Pitot tubes (also called pitot-static tubes) are used, for example, to make airflow measurements in Heating, ventilation, and air conditioning (HVAC) applications and for aircraft airspeed measurements.

Static pressure is what is commonly called simply the pressure of the fluid. It is a measure of the amount that fluid pressure exceeds local atmospheric pressure. It is measured through a flat opening that is parallel with the fluid flow.

Stagnation pressure is also a measure of the quantity that fluid pressure exceeds local atmospheric pressure, but it includes the effect of the fluid velocity converted to pressure. It is measured through a flat opening that is perpendicular to the direction of fluid flow and facing into the fluid flow.

Dynamic pressure (also called velocity pressure) is a measure of the amount that the stagnation pressure exceeds static pressure at a point in a fluid. It can also be interpreted as the pressure created by reducing the kinetic energy to zero.

Velocity at any point,

$$v = C_v\sqrt{2gh} \qquad (12.75)$$

where
C_v = coefficient of pitot tube.

12.9 FLOW THROUGH PIPES

In general, the flow of liquid along a pipe can be determined by the use of Bernoulli Equation and Continuity Equation.

The former represents the conservation of energy, which in Newtonian fluids is either potential or kinetic energy, and the latter ensures that what goes into one end of a pipe must come out at the other end. However, as the flow moves down the pipe, losses due to friction between the moving liquid and the walls of pipe cause the pressure within the pipe to reduce with distance—this is known as head loss.

12.9.1 HEAD LOSS DUE TO FRICTION IN THE PIPE

Two equations can be used when the flow is either laminar or turbulent. These are given in the following:

12.9.1.1 Darcy's Equation for Round Pipes
For Round Pipes:

$$h_f = \frac{4flv^2}{2gd} \qquad (12.76)$$

where

h_f is the head loss due to friction (m);
l is the length of the pipe (m);
d is the hydraulic diameter of the pipe. For circular sections, this equals the internal diameter of the pipe (m);
v is the velocity within the pipe (m/s²);
g is the acceleration due to gravity;
f is the coefficient of friction.

12.9.1.2 Darcy's Equation for Noncircular Pipes

$$h_f = \frac{flv^2}{2g\mu} \qquad (12.77)$$

where μ = wetted area/wetted perimeter.

12.9.1.3 The Chezy Equation

$$v = C\sqrt{mi} \qquad (12.78)$$

where

$$i = \frac{h_f}{l} \qquad (12.79)$$

$$m = \frac{A}{P} \qquad (12.80)$$

or wetted area or wetted perimeter
and

$$C = \sqrt{\frac{2g}{f}} \qquad (12.81)$$

Reynolds number:

$$R_e = \frac{\text{Density} \times \text{Velocity} \times \text{Diameter of pipe}}{\text{Viscosity}} \qquad (12.82)$$

$$\text{Head loss} = 4f \frac{lV^2}{2dg} \qquad (12.83)$$

Note: f in the Darcy formula is not an empirical coefficient.
For Laminar flow,

$$f = \frac{16}{R} \qquad (12.84)$$

The changeover from laminar flow to turbulent flow occurs when $R = 2300$ and is independent of whether the pipe walls are smooth or rough.

12.9.1.4 Laminar Flow

When the flow is laminar it is possible to use the following equation to find the head loss.

The Poiseuille equation states that for a round pipe the head loss due to friction is given by

$$h_F = \frac{32 \mu l v}{\rho g d^2} \qquad (12.85)$$

12.9.1.5 Choice of Friction Factor f

The value of f must be chosen with care or else the head loss will not be correct. Assessment of the physics governing the value of friction in a fluid has led to the following relationships:

1. $h_f \propto L$
2. $h_f \propto v^2$
3. $h_f \propto 1/d$
4. h_f depends on surface roughness of pipes
5. h_f depends on fluid density and viscosity
6. h_f is independent of pressure

Consequently f cannot be a constant if it is to give correct head loss values from the Darcy equation. An expression that gives f based on fluid properties and the flow conditions is required.

12.9.1.6 Minor Energy (Head) Losses

1. Loss of head due to sudden enlargement
2. Loss of head due to sudden contraction
3. Loss of head at the entrance of a pipe
4. Loss of head at the exit of a pipe
5. Loss of head due to an obstruction in a pipe
6. Loss of head due to bend in the pipe
7. Loss of head in various pipe fittings
   1. Loss of head due to sudden enlargement (Figure 12.30):

   $$h_e = \frac{(V_1 - V_2)^2}{2g} \qquad (12.86)$$

   2. Loss of head due to sudden contraction (Figure 12.31):

   $$h_c = \frac{V_2^2}{2g} \left[\frac{1}{C_c} - 1 \right]^2 = \frac{kV_2^2}{2g} \qquad (12.87)$$

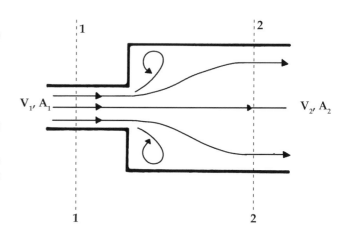

FIGURE 12.30 Loss of head due to sudden enlargement.

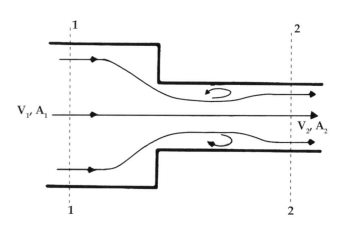

FIGURE 12.31 Loss of head due to sudden contraction.

where

$$k = \left[\frac{1}{C_c} - 1 \right]^2 \qquad (12.88)$$

If the value of C_c is assumed to be equal to 0.62, then

$$k = \left[\frac{1}{0.62} - 1 \right]^2 = 0.375 \qquad (12.89)$$

Then

$$h_c = \frac{kV_2^2}{2g} = 0.375 \frac{V_2^2}{2g} \qquad (12.90)$$

If the value of C_c is not given then the head loss due to contraction is taken as

$$h_c = 0.5 \frac{V_2^2}{2g} \qquad (12.91)$$

3. Loss of head at the entrance of a pipe:

$$h_i = 0.5 \frac{V^2}{2g} \qquad (12.92)$$

4. Loss of head at the exit of a pipe:

$$h_o = \frac{V^2}{2g} \qquad (12.93)$$

5. Loss of head due to an obstruction in a pipe:

$$= \frac{V^2}{2g} \left(\frac{A}{C_c (A-a)} - 1 \right)^2 \qquad (12.94)$$

6. Loss of head due to bend in the pipe:

$$h_b = \frac{kV^2}{2g} \qquad (12.95)$$

where k=coefficient of bend

7. Loss of head in various pipe fittings:

$$= \frac{kV^2}{2g} \qquad (12.96)$$

where k=coefficient of pipe fitting.

12.10 VISCOUS FLOW

12.10.1 REAL FLUIDS

The flow of real fluids exhibits viscous effect, that is, they tend to "stick" to solid surfaces and have stresses within their body.

According to Newton's law of viscosity, the shear stress, τ, in a fluid is proportional to the velocity gradient—the rate of change of velocity across the fluid path.

$$\tau \propto \frac{du}{dy} \qquad (12.97)$$

For a "Newtonian" fluid, we can write

$$\tau = \mu \frac{du}{dy} \qquad (12.98)$$

where the constant of proportionality, μ, is known as the coefficient of viscosity (or simply viscosity). We saw that for some fluids—sometimes known as exotic fluids—the value of μ changes with stress or velocity gradient.

12.10.2 LAMINAR AND TURBULENT FLOW

Inject a dye into the middle of the stream of a free-flowing water in a pipe, the type of flow can be classified as shown in Figures 12.32–12.34, as laminar flow, transitional flow, and turbulent flow, respectively.

The phenomenon was first investigated in the 1880s by Osbourne Reynolds in an experiment. He used a tank as shown in Figure 12.35 with a pipe taking water from the center into which injected a dye through a needle.

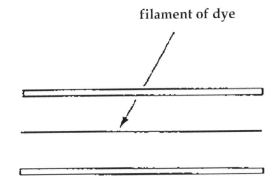

filament of dye

FIGURE 12.32 Laminar (viscous).

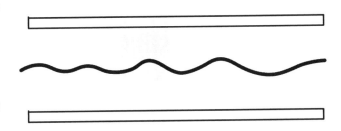

FIGURE 12.33 Transitional.

FIGURE 12.34 Turbulent.

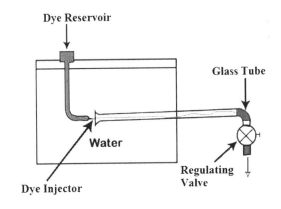

FIGURE 12.35 Reynolds number.

After many experiments, the following expression had been established

$$\frac{\rho u d}{\mu} \qquad (12.99)$$

where ρ=density, u=mean velocity, d=diameter, and μ=viscosity would help predict the change in flow type. If the value is less than about 2000 then the flow is laminar, if greater than 4000 then turbulent, and in between these then in the transition zone.

This value is known as the Reynolds number, R_e:

$$R_e = \frac{\rho u d}{\mu} \tag{12.100}$$

Laminar flow: $R_e < 2000$
Transitional flow: $2000 < R_e < 4000$
Turbulent flow: $R_e > 4000$

12.10.3 PRESSURE LOSS DUE TO FRICTION IN A PIPELINE

In a pipe with a real fluid flowing, at the wall there is a shearing stress retarding the flow, as shown in Figure 12.36.

Consider a cylindrical element of incompressible fluid flowing in the pipe, as shown in Figure 12.37.

The pressure at the upstream end is p, and at the downstream end, the pressure has fallen by Δp to $(p - \Delta p)$.

As the flow is in equilibrium, driving force = retarding force

$$\Delta p = \frac{\tau_w 4L}{d} \tag{12.101}$$

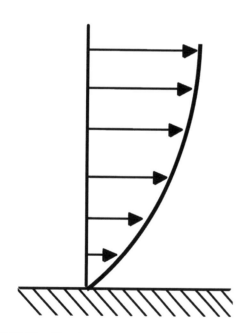

FIGURE 12.36 Shearing stress retarding the flow.

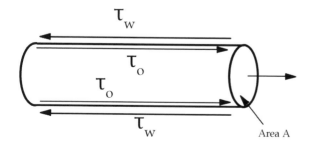

FIGURE 12.37 Cylindrical element of incompressible fluid.

Giving an expression for pressure loss in a pipe in terms of the pipe diameter and the shear stress at the wall on the pipe.

12.10.4 PRESSURE LOSS DURING LAMINAR FLOW IN A PIPE

In general the shear stress τ_w is almost impossible to measure. But for laminar flow it is possible to calculate a theoretical value for a given velocity, fluid, and pipe dimension.

In laminar flow the paths of individual particles of fluid do not cross, so the flow may be considered as a series of concentric cylinders sliding over each other—rather like the cylinders of a collapsible pocket telescope.

As before, consider a cylinder of fluid, length L, radius r flowing steadily at the center of a pipe as shown in Figure 12.38.

In equilibrium, the shearing forces on the cylinder equal the pressure forces.

$$\tau = \frac{\Delta p}{L} \frac{r}{2} \tag{12.102}$$

Discharge,

$$Q = \frac{\Delta p}{L} \frac{\pi d^2}{128 \mu} \tag{12.103}$$

This is the Hagen-Poiseuille equation for laminar flow in a pipe. It expresses the discharge Q in terms of the pressure gradient $\left(\frac{\partial P}{\partial x} = \frac{\Delta p}{L} \right)$, diameter of the pipe, and the viscosity of the fluid.

12.11 FLOW PAST IMMERSED BODIES

Whenever a body is placed in a stream, forces are exerted on the body. Similarly, if the body is moving in a stationary fluid, force is exerted on the body. Therefore, when there is a relative motion between the body and the fluid, force is exerted on the body.

Example: Wind forces on buildings, bridges etc., force experienced by automobiles, aircraft, propeller, etc.

12.11.1 FORCE EXERTED BY A FLOWING FLUID ON A BODY

Whenever there is relative motion between a real fluid and a body, the fluid exerts a force on a body and the body exerts equal and opposite force on the fluid. A body fully immersed in a real fluid may be subjected to two kinds of forces called drag force (F_D) and lift force (F_L).

12.11.1.1 Drag Force

The component of force in the direction of flow on a submerged body is called F_D.

12.11.1.2 Lift Force

The component of force in perpendicular to the flow is called the F_L. In a symmetrical body, moving through an ideal fluid

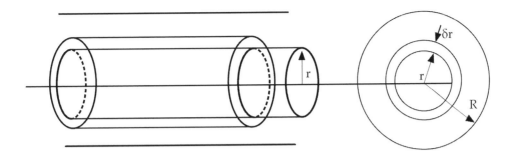

FIGURE 12.38 Laminar flow in a pipe.

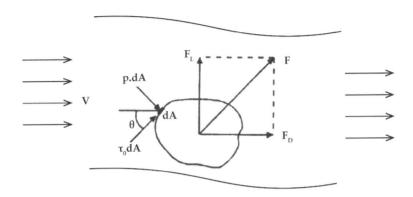

FIGURE 12.39 Force on immersed bodies.

(no viscosity) at a uniform velocity, the pressure distribution around the body is symmetrical, and hence the resultant force acting on the body is zero. However, real fluids such as air and water possess viscosity and if it is moved through these fluids at a uniform velocity, the body experiences a resistance to motion.

For the symmetrical body such as sphere and cylinder facing the flow is symmetrical, there is no F_L. For the production of F_L there must be symmetry of flow, but F_D exists always. It is possible to create drag without lift but impossible to create lift without drag.

12.11.1.3 Expressions for Drag and Lift

Consider a body held stationary in a stream of real fluid moving at a uniform velocity U. Let θ=inclination of the tangent to the small element dA with the direction of flow. Then the force acting on dA of the surface of the body can be considered to have two components τdA (called shear force) and pdA (called pressure force) acting along the directions tangential and normal to the surface, respectively. The tangential components are called shear force and the normal components are called pressure forces. The summation of component of the forces acting over the entire surface of the body in the direction of fluid flow is F_D and perpendicular to fluid flow is F_L (Figure 12.39).

For the body moving through a fluid density ρ at a uniform velocity U, the mathematical expression for the calculation of F_D and the F_L is given by

$$F_D = C_D A \qquad (12.104)$$

and

$$F_L = C_L A \qquad (12.105)$$

where

C_D=coefficient of drag
C_L=coefficient of lift
A=characteristics area
= area projected on a plane perpendicular to the relative motion of the fluid, in the case of calculating F_D
= area projected on a plane perpendicular to the direction of F_L, in the case of calculating F_L

Note:

- The F_L may exist even in ideal fluid by the presence of circulation.
- Real fluid also requires vortices or circulations around the body for producing lift.
- F_D is caused by inertia, viscosity, wave action of free surface, gravity.

12.11.1.4 Pressure Drag and Friction Drag

The contribution of pressure drag and friction drag to the total drag depends on the flowing parameters.

- Characteristics of fluid
- Shape of body
- Orientation of the body immersed in the fluid

12.11.1.5 Lifting Force

The lifting force acting on a body in a fluid flow can be expressed as

$$F_L = \frac{1}{2}C_L \rho V^2 A \qquad (12.106)$$

where

F_L=lifting force (N)
C_L=lifting coefficient
ρ=density of fluid (kg/m³)
v=flow velocity (m/s)
A=body area (m²)

12.11.1.6 Drag Force

The F_D acting on a body in fluid flow can be expressed as

$$F_D = \frac{1}{2}C_D \rho V^2 A \qquad (12.107)$$

where

F_D=drag force (N)
C_D=drag coefficient
ρ=density of fluid (kg/m³)
v=flow velocity (m/s)
A=body area (m²)

12.11.1.7 Thrust Power to Overcome Drag Force

Thrust power required to overcome the F_D can be calculated as

$$P = F_D V \qquad (12.108)$$

where

P=power (W)

12.11.2 Boundary Layer

Boundary layer is a very thin layer of fluid in the immediate vicinity of the wall (or boundary). When a real fluid flows past a solid boundary, there develops a thin layer very close to the boundary in which the velocity rapidly increases from zero at the boundary (due to no slip condition) to the nearly uniform velocity in the free stream. This region is called boundary layer. In this region, the effect of viscosity is predominant due to the high values of (du/dy) and most of the energy is lost in this zone due to viscous shear.

The layer of fluid which has its velocity affected by the boundary shear is called as boundary layer. A thin layer of fluid in the vicinity of the boundary, whose velocity is affected due to viscous shear, is called as the boundary layer.

12.11.2.1 Potential Flow or Irrotational Flow Region

The portion of the fluid outside the boundary layer where viscous effects are negligible is called potential flow or irrotational flow region. The flow in this region can be treated as Ideal Fluid Flow.

12.11.2.2 Factors Affecting the Growth of Boundary Layers

1. **Distance (x) from the leading edge**: Boundary layer thickness varies directly with the distance (x). More the distance (x), more is the thickness of the boundary layer.
2. **Free stream velocity**: Boundary layer thickness varies inversely as free stream velocity.
3. **Viscosity of the fluid**: Boundary layer thickness varies directly as viscosity.
4. **Density of the fluid**: Boundary layer thickness varies inversely as density.

12.11.2.3 Thicknesses of the Boundary Layer

1. **Boundary layer thickness**: It is the distance from the boundary in which the local velocity reaches 99% of the main stream velocity and is denoted by (δ). $y=(\delta)$ when $u=0.99\,U$.
2. **Displacement thickness ($\delta*$)**: It is defined as the distance perpendicular to the boundary by which the boundary will have to be displaced outward so that the actual discharge would be same as that of the ideal fluid past the displaced boundary. It is also defined as the distance measured perpendicular from the actual boundary such that the mass flux through this distance is equal to the deficit of mass flux due to boundary layer formation.
3. **Momentum thickness (θ)**: It is defined as the distance measured perpendicular from the actual boundary such that the momentum flux through this distance is equal to the deficit of the momentum flux due to the boundary layer formation.

$$\text{Momentum deficit} = \rho\big(b \cdot dy\big)\big(U - u\big)u \qquad (12.109)$$

$$\text{Total momentum deficit} = \text{Moment through thickness}(\theta) \qquad (12.110)$$

4. **Energy thickness ($\delta*$)**: It is the distance perpendicular to the boundary by which the boundary has to be displaced to compensate for the reduction in the kinetic energy of the fluid caused due to the formation of the boundary layer. Energy thickness is also defined as the distance measured perpendicular from the actual boundary such that the kinetic energy flux through this distance is equal to the deficit of kinetic energy due to the boundary layer formation.
5. **Shape factor (H)**: It is defined as the ratio of the displacement thickness to the momentum thickness.

$$H = \big(\delta^*/\theta\big) \qquad (12.111)$$

12.12 COMPRESSIBLE FLOW

Compressible flow (gas dynamics) is the branch of fluid mechanics that deals with flows having significant changes in fluid density. Gases, mostly, display such behavior. To distinguish between compressible and incompressible flow, the Mach number (the ratio of the speed of the flow to the speed of sound) must be greater than about 0.3 (since the density change is greater than 5% in that case) before significant compressibility occurs. The study of compressible flow is relevant to high-speed aircraft, jet engines, rocket motors, hyperloops, high-speed entry into a planetary atmosphere, gas pipelines, commercial applications such as abrasive blasting, and many other fields.

12.12.1 Thermodynamic Relations

1. **Equation of state**: It gives the relationship between the pressure, temperature, and specific volume of a gas.

$$p \forall = RT \qquad (12.112)$$

where
p = absolute pressure in kgf/m^2 or N/m^2
$\forall$ = specific volume
$R = G$As constant in J/kg K
T = absolute temperature

$$\frac{p}{\rho} = RT \qquad (12.113)$$

2. **Isothermal process**:

$$\frac{p}{\rho} = \text{constant} \qquad (12.114)$$

3. **Adiabatic process**:

$$\frac{p}{\rho^k} = \text{constant} \qquad (12.115)$$

where k = ratio of the specific heat at constant pressure to the specific heat at constant volume.

12.12.2 Mach Number

It may be seen that the speed of sound is the thermodynamic property that varies from point to point. When there is a large relative speed between a body and the compressible fluid surrounds it, then the compressibility of the fluid greatly influences the flow properties. Ratio of the local speed (V) of the gas to the speed of sound (a) is called as local Mach number (M).

$$M = \frac{V}{a} = \frac{V}{\sqrt{\gamma RT}} \qquad (12.116)$$

There are few physical meanings for Mach number:

a. It shows the compressibility effect for a fluid, i.e., $M < 0.3$ implies that fluid is incompressible.
b. It can be shown that Mach number is proportional to the ratio of kinetic to internal energy.
c. It is a measure of directed motion of a gas compared to the random thermal motion of the molecules.

As a rule of thumb, the compressible flow regimes are classified as below:

- $M < 0.3$ (incompressible flow)
- $M < 1$ (subsonic flow)
- $0.8 < M < 1.2$ (transonic flow)
- $M > 1$ (supersonic flow)
- $M > 5$ and above (hypersonic flow)

13 Engineering Hydrology

13.1 INTRODUCTION

The world's total water resources are estimated to be around 1.36×10^{14} ha m. 92.7% of this water is salty and is stored in oceans and seas. Only 2.8% of total available water is fresh water. Of this 2.8% fresh water, 2.2% is available as surface water and 0.6% as ground water. Of the 2.2% surface water, 2.15% is stored in glaciers and ice caps, 0.01% in lakes and streams, and the rest is in circulation among the different components of the Earth's atmosphere.

Of the 0.6% ground water only about 0.25% can be economically extracted. It can be summarized that less than 0.26% of fresh water is available for use by humans and hence water has become a very important resource.

Water is never stagnant (except in deep aquifers), it moves from one component to other component of the earth through various process of precipitation, runoff, infiltration, evaporation, etc. For a civil engineer, it is important to know the occurrence, flow, distribution, etc., as many structures have to be constructed in contact with water.

Hydrology:

Hydrology may be defined as applied science concerned with water of the Earth in all its states, their occurrences, distribution, and circulation through the unending hydrologic cycle of precipitation, consequent runoff, stream flow, infiltration and storage, eventual evaporation, and reprecipitation.

Hydrology concerns itself with three forms of water:

1. Above land as atmospheric water or precipitation
2. On land or surface as stored water or runoff
3. Below the land surface as ground water or percolation

13.1.1 IMPORTANCE OF HYDROLOGY

The importance of hydrology is seen in:

1. Design of hydraulic structures:

    Structures such as bridges, causeways, dams, spillways, etc., are in contact with water. Accurate hydrological predictions are necessary for their proper functioning. Due to a storm, the flow below a bridge has to be properly predicted. Improper prediction may cause failure of the structure. Similarly, the spillway in case of a dam which is meant for disposing excess water in a dam should also be designed properly; otherwise, flooding water may overtop the dam.

2. Municipal and industrial water supply:

    Growth of towns and cities and also industries around them is often dependent on fresh water availability in their vicinity. Water should be drawn from rivers, streams, and ground water. Proper estimation of water resources in a place will help planning and implementation of facilities for municipal (domestic) and industrial water supply.

3. Irrigation:

    Dams are constructed to store water for multiple uses. For estimating maximum storage capacity, seepage, evaporation, and other losses should be properly estimated. These can be done with proper understanding of hydrology of a given river basin and thus making the irrigation project a successful one. Artificial recharge will also increase the ground water storage. It has been estimated that ground water potential of Gangetic basin is 40 times more than its surface flow.

4. Hydroelectric power generation:

    A hydroelectric power plant needs continuous water supply without much variation in the stream flow. Variations will affect the functioning of turbines in the electric plant. Hence proper estimation of river flow and also flood occurrences will help to construct efficient balancing reservoirs and these will supply water to turbines at a constant rate.

5. Flood control in rivers:

    Controlling floods in a river is a complicated task. The flow occurring due to a storm can be predicted if the catchment characteristics are properly known. In many cases damages due to floods are high. Joint work of hydrologist and meteorologists in threatening areas may reduce damage due to floods. Flood plain zones maybe demarked to avoid losses.

6. Navigation:

    Big canals in an irrigation scheme can be used for inland navigation. The depth of water should be maintained at a constant level. This can be achieved by lock gates provided and proper draft to be maintained. If the river water contains sediments, they will settle in the channel and cause problems for navigation. Hence the catchment characteristics should be considered and sediment entry into the canals should be done.

7. Pollution control:

    It is an easy way to dispose sewage generated in a city or town into streams and rivers. If large stream flow is available compared to the sewage discharge, pollution problems do not arise as sewage gets diluted and flowing water also has self-purifying capacity. The problem arises when each of the flows is not properly estimated. In case sewage flow is high, it should be treated before disposal into a river or stream.

13.1.2 Hydrological Cycle

Water exists on the earth in gaseous form (water vapor), liquid, and solid (ice) forms, and is circulated among the different parts of the Earth mainly by solar energy and planetary forces. Sunlight evaporates sea water and this evaporated form is kept in circulation by gravitational forces of Earth and wind action. The different paths through which water in nature circulates and is transformed is called hydrological cycle.

13.1.2.1 Catchment or Descriptive Representation of Hydrological Cycle

Hydrological cycle is defined as the circulation of water from the sea to the land through the atmosphere back to the sea often with delays through process like precipitation, interception, runoff, infiltration, percolation, ground water storage, evaporation, and transpiration and also water that returns to the atmosphere without reaching the sea (Figure 13.1).

The hydrological cycle can also be represented in many different ways in diagrammatic forms as follows:

i. Horton's Qualitative representation (Figure 13.2)
ii. Horton's Engineering representation (Figure 13.3)

13.1.3 Some Important Definitions

1. **Precipitation**: It is the return of atmospheric moisture to the ground in solid or liquid form. Solid form—snow, sleet, snow pellets, hailstones. Liquid form—drizzle, rainfall.
2. **Infiltration**: Infiltration is the passage of water across the soil surface. The vertical downward movement of water within the soil is known as percolation. The infiltration capacity is the maximum rate of infiltration for the given condition of the soil. Obviously, the infiltration capacity decreases with time during/after a storm.
3. **Overland flow**: This is the part of precipitation which is flowing over the ground surface and is yet to reach a well-defined stream.
4. **Surface runoff**: When the overland flow enters a well-defined stream it is known as surface runoff.
5. **Interflow for subsurface flow**: A part of the precipitation which has in-filtered the ground surface may flow within the soil but close to the surface. This is known as interflow. When the interflow enters a well-defined stream, then only it is called runoff.
6. **Ground water flow**: This is the flow of water in the soil occurring below the ground water table. The ground water table is at the top level of the saturated zone within the soil and it is at atmospheric pressure. Hence it is also called phreatic surface. A portion of water may enter a well-defined stream. Only then it is known as runoff or base flow. Hence we say that runoff is the portion of precipitation which enters a well-defined stream and has three components, namely, surface runoff, interflow runoff, and ground water runoff or base flow.
7. **Evaporation**: This is the process by which state of substance (water) is changed from liquid state to vapor form. Evaporation occurs constantly from water bodies, soil surface, and even from vegetation. In short, evaporation occurs when water is exposed to atmosphere (during sunlight). The rate of evaporation depends on the temperature and humidity.
8. **Transpiration**: This is the process by which the water extracted by the roots of the plants is lost to the atmosphere through the surface of leaves and branches by evaporation. Hence it is also known as evapotranspiration.

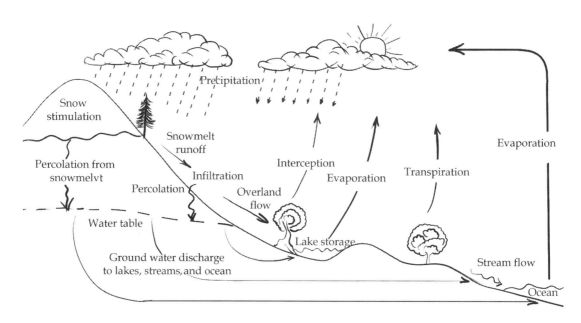

FIGURE 13.1 Hydrological cycle.

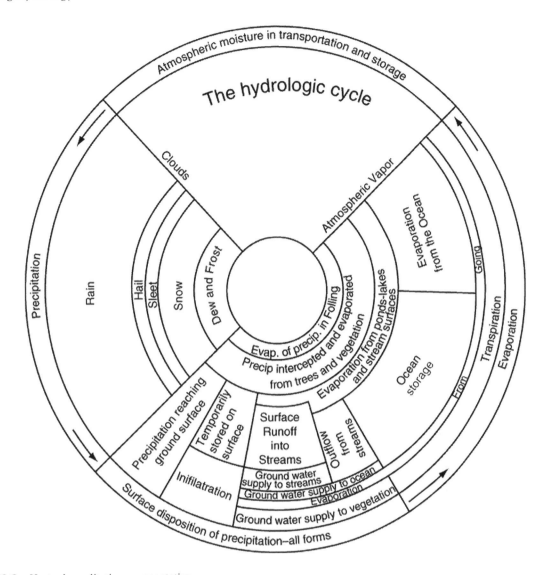

FIGURE 13.2   Horton's qualitative representation.

13.2   PRECIPITATION

It is defined as the return of atmospheric moisture to the ground in the form of solids or liquids. The following are the main characteristics of rainfall:

a. **Amount or quantity**: The amount of rainfall is usually given as a depth over a specified area, assuming that all the rainfall accumulates over the surface and the unit for measuring amount of rainfall is cm. The volume of rainfall = area×depth of rainfall (m³). The amount of rainfall occurring is measured with the help of rain gauges.

b. **Intensity**: This is usually the average of rainfall rate of rainfall during the special periods of a storm and is usually expressed as cm/hour.

c. **Duration of storm**: In the case of a complex storm, we can divide it into a series of storms of different durations, during which the intensity is more or less uniform.

d. **Aerial distribution**: During a storm, the rainfall intensity or depth, etc., will not be uniform over the entire area. Hence we must consider the variation over the area, i.e., the aerial distribution of rainfall over which rainfall is uniform.

13.2.1   Forms of Precipitation

Precipitation comes to earth's surface in different forms. Some of them are given below.

1. **Drizzle**: This is a form of precipitation consisting of water droplets of diameter less than 0.05 cm with intensity less than 0.01 cm/hour.

2. **Rainfall**: This is a form of precipitation of water drops larger than 0.05 cm diameter up to 0.5 cm diameter. Water drops of size greater than 0.5 cm diameter tend to break up as they fall through the atmosphere. Intensity varies from 0.25 to 0.75 cm/hour.

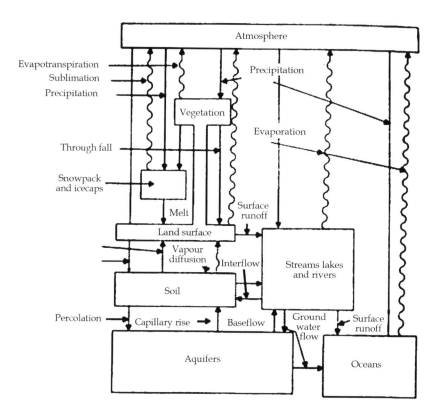

FIGURE 13.3 Horton's engineering representation.

3. **Glaze**: This is the ice coating formed when a drizzle or rainfall comes in contact with very old objects on the ground
4. **Sleet**: This occurs when rain drops fall through air which is below 0°C. The grains are transparent, round with diameter between 0.1 and 0.4 cm.
5. **Snow pellets**: These are white opaque round grains of ice. They are crystalline and rebound when falling onto the ground. The diameter varies from 0.05 to 0.5 cm.
6. **Snow**: This is precipitation in the form of ice crystals, usually a number of ice crystals combining to form snowflakes.
7. **Hails**: These are balls or irregular lumps of ice of over 0.5 cm diameter formed by repeated freezing and melting. These are formed by upward and downward movement of air masses in turbulent air currents.

13.2.2 NECESSARY CONDITIONS FOR OCCURRENCE OF PRECIPITATION

For precipitation to occur, moisture (water vapor) is always necessary to be present. Moisture is present due to the process of evaporation. There must also be some mechanism for large-scale lifting of moist, warm air so that there will be sufficient cooling. This will cause condensation (conversion of vapors) to liquid and growth of water drops.

Condensation nuclei, such as the oxides of nitrogen, salt crystals, carbon dioxide, silica, etc., must be present such that water vapor condenses around them. The conditions of electric charge in the cloud, size of water droplets or ice crystals, temperature and relative movement of clouds must be favorable so that the size of the condensed water drop increases and ultimately they begin to fall to the ground due to gravity. A drop of size 0.5 mm can fall through 2000 m in unsaturated air.

13.2.3 TYPES OF PRECIPITATION

One of the essential requirements for precipitation to occur is the cooling of large masses of moist air. Lifting of air masses to higher altitudes is the only large-scale process of cooling. Hence the types of precipitation based on the mechanism which causes lifting of air masses are as follows:

1. **Cyclonic precipitation**:
   This is the precipitation associated with cyclones or moving masses of air and involves the presence of low pressures. This is further subdivided into two categories:
   a. Nonfrontal cyclonic precipitation:
      In this, a low-pressure area develops. (Low-pressure area is a region where the atmospheric pressure is lower than that of surrounding locations.) The air from surroundings converges laterally toward the low-pressure area. This results in lifting of air and hence cooling. It may result in precipitation (Figure 13.4).
   b. Frontal cyclonic precipitation:
      FRONT is a barrier region between two air masses having different temperatures, densities,

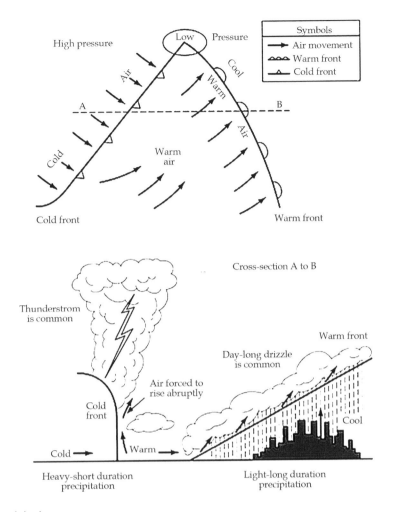

FIGURE 13.4 Cyclonic precipitation.

moisture, content, etc. If a warm and moist air mass moves upward over a mass of cold and heavier air mass, the warm air gets lifted, cooled, and may result in precipitation. Such a precipitation is known as warm front precipitation. The precipitation may extend for 500 km ahead of the front, i.e., the colder air region. If moving mass of cold air forces a warm air mass upward, we can expect a cold Front precipitation. The precipitation may extend up to 200 km ahead of the Front surface in the warm air.

2. **Convective precipitation**:

   This is due to the lifting of warm air which is lighter than the surroundings. Generally this type of precipitation occurs in the tropics where on a hot day, the ground surface gets heated unequally causing the warmer air to lift up and precipitation occurs in the form of high intensity and short duration.

3. **Orographic precipitation**:

   It is the most important precipitation and is responsible for most of heavy rains in India. Orographic precipitation is caused by air masses which strike some natural topographic barriers like mountains and cannot move forward and

hence the rising amount of precipitation. The greatest amount of precipitation falls on the windward side and leeward side has very little precipitation (Figure 13.5).

4. **Turbulent precipitation**:

   This precipitation is usually due to a combination of the several of the above cooling mechanisms. The change in frictional resistance as warm and moist air moves from the ocean onto the land surface may cause lifting of air masses and hence precipitation due to cooling. This precipitation results in heavy rainfall.

13.2.4 Water Budget Equation for a Catchment

The area of land draining into a stream at a given location is known as catchment area or drainage area or drainage basin or water shed. A catchment area is shown in Figure 13.6.

For a given catchment area in any interval of time, the continuity equation for water balance is given as:

$$(\text{Change in mass storage}) = (\text{mass in flow}) - (\text{mass outflow})$$

$$\Delta S = V_i - V_o \qquad (13.1)$$

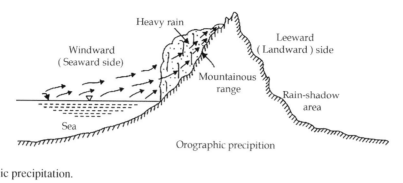

FIGURE 13.5 Orographic precipitation.

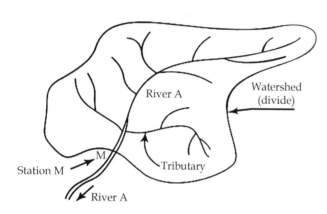

FIGURE 13.6 Catchment area.

The water budget equation for a catchment considering all process for a time interval Δ_t is written as:

$$\Delta S = P - R - G - E - T \qquad (13.2)$$

where
  Δ_s represents change in storage
  P = precipitation
  G = net ground water flowing outside the catchment
  R = Surface runoff
  E = evaporation
  T = transpiration

Storage of water in a catchment occurs in three different forms and it can be written as

$$S = S_S + S_m + S_g \qquad (13.3)$$

where
  S = storage,
  S_s = surface water storage,
  S_m = soil moisture storage,
  S_g = ground water storage.

Hence change in storage maybe expressed as

$$\Delta S = \Delta S_s + \Delta S_m + \Delta S_g \qquad (13.4)$$

The rainfall runoff relationship can be written as

$$R = P - L \qquad (13.5)$$

where
  R = surface runoff,
  P = precipitation,
  L = losses, i.e., water not available to runoff due to infiltration, evaporation, transpiration, and surface storage.

13.2.5 Rain Gauging (Measurement of Rainfall)

Rainfall is measured on the basis of the vertical depth of water accumulated on a level surface during an interval of time, if all the rainfall remained where it fell. It is measured in "mm." The instrument used for measurement of rainfall is called "Rain gauge."

These are classified as:

a. Nonrecording types
b. Recording types

a. **Nonrecording-type rain gauges**:
    These rain gauges are most widely adopted in India. They are known as nonrecording type because they do not record the rain but only collect the rain. Symon's rain gauge is the usual nonrecording type of rain gauge. It gives the total rainfall that has occurred at a particular period. It essentially consists of a circular collecting area 127 mm in diameter connected to a funnel. The funnel discharges the rainfall into a receiving vessel. The funnel and the receiving vessel are housed in a metallic container. The components of this rain gauge are shown in Figure 13.7. The water collected in the receiving bottle is measured by a graduated measuring jar with an accuracy of 0.1 ml. The rainfall is measured every day at 8:30 am IST, and hence, this rain gauge gives only depth of rainfall for previous 24 hours. During heavy rains, measurement is done 3–4 times a day. They can also be used to measure snowfall.
b. **Recording-type rain gauges**:
    These are rain gauges which can give a permanent, automatic rainfall record (without any bottle recording) in the form of a pen mounted on a clock driven chart. From the chart intensity or rate of rainfall in cm per hour or 6, 12 hours…, besides the total amount of rainfall can be obtained.

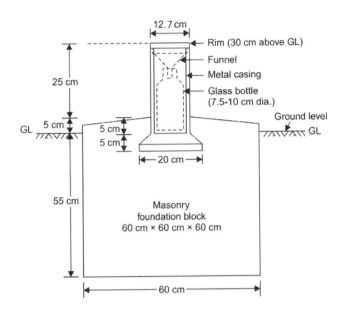

FIGURE 13.7 Symon's rain gauge.

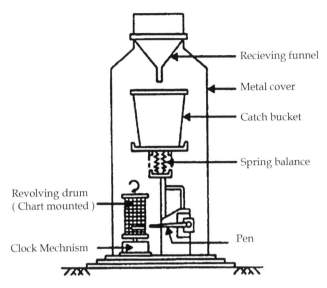

FIGURE 13.8 Weighing bucket rain gauge.

Advantages of recording rain gauges:
1. Necessity of an attendant does not arise.
2. Intensity of rainfall at any time as well as total rainfall is obtained, whereas nonrecording gauge gives only total rainfall.
3. Data from inaccessible places (hilly regions) can be continuously obtained once gauge is established.
4. Human errors are eliminated.
5. Capacity of gauges is large.
6. Time intervals are also recorded.

Disadvantages of recording rain gauges:
1. High initial investment cost.
2. Recording is not reliable when faults in gauge arise (mechanical or electrical) till faults are corrected.

13.2.5.1 Types of Recording or Automatic Rain Gauges

a. Weighing bucket rain gauge:
   This is the most common type of recording or automatic rain gauge adopted by Indian Meteorological Department. It consists of a receiving bucket supported by a spring or lever. The receiving bucket is pushed down due to the increase in weight (due to accumulating rain fall). The pen attached to the arm continuously records the weight on a clock-driven chart as shown in Figure 13.8. The chart obtained from this rain gauge is a mass curve of rainfall (Figure 13.9).

From the mass curve, the average intensity of rainfall (cm/hour) can be obtained by calculating the slope of the curve at any instant of time. The patterns as well as total depth of rainfall at different instants can also be obtained.

b. Tipping bucket rain gauge:
   This is the most common type of automatic rain gauge adopted by U.S. Meteorological Department. It consists of receiver draining into a funnel of 30 cm diameter as shown in Figure 13.10. The catch (rainfall) from funnel falls into one of the pair of small buckets (tipping buckets). These buckets are so balanced that when 0.25 mm of rainfall collects in one bucket, it tips and brings the other bucket into position. The tipping of the bucket is transmitted to an electricity-driven pen or to an electronic counter. This is useful in remote areas. The record from the tipping bucket type of rain gauge gives the data of the intensity of rainfall.

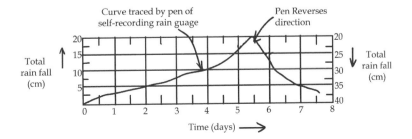

FIGURE 13.9 Mass curve of rainfall.

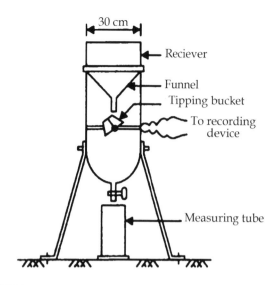

FIGURE 13.10 Tipping bucket rain gauge.

c. Siphon or float-type rain gauge:

This is also called integrating rain gauge as it depicts an integrated graph of rainfall with respect to time. The construction of this rain gauge is shown in Figure 13.11.

A receiver and funnel arrangement drains the rainfall into a container, in which a float mechanism

at the bottom is provided. As water accumulates, the float rises. A pen arm attached to the float mechanism continuously records the rainfall on a clock-driven chart and also produces a mass curve of rain fall. When the water level rises above the crest of the siphon, the accumulated water in the container will be drained off by siphonic action. The rain gauge is ready to receive the new rainfall.

13.2.5.2 Factors Governing Selection of Site for Rain Gauge Stations

1. The site for rain gauge station should be an open space without the presence of trees or any covering.
2. The rain gauge should be properly secured by fencing.
3. The site for rain gauge station should be a true representation of the area which is supposed to give rainfall data.
4. The distance of any object or fence from the rain gauge should not be less than twice the height of the object or fence and in no case less than 30 m.
5. The rain gauge should not be set upon the peak or sides of a hill, but on a nearby fairly level ground.
6. The rain gauge should be protected from high winds.
7. The rain gauge should be easily accessible to the observers at all times.

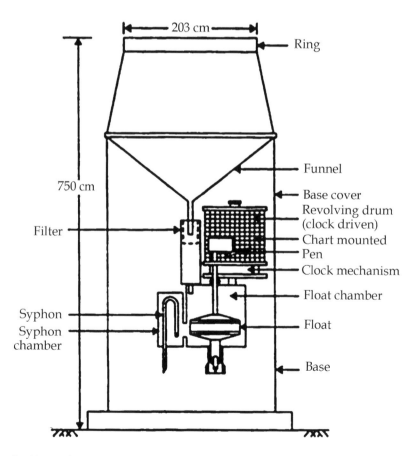

FIGURE 13.11 Siphon or float-type rain gauge.

Point rainfall: It is the total liquid form of precipitation or condensation from the atmosphere as received and measured in a rain gauge. It is expressed as so many "mm" of depth of water.

Rain gauge density or rain gauge network: Rainfall records are most important for hydrological investigations, so as well distributed network of rain gauge station within the catchment is essential. The rain gauge should be evenly and uniformly distributed within a given catchment. The total number of rain gauge in a given area should neither be too many, because it will increase the cost and nor be too low to give reliable result. A lesser density of rain gauge is used for studies of general storm and the high density of rain gauge is used for studies of rainfall pattern of a thunderstorm. Hence our aim is to provide an optimum density of gauges from practical consideration.

The Indian Standard (IS:4987-1968) recommends the following densities:

   i. In plain area = 1 stations/520 km^2
  ii. In region of average elevation of 1000 m = 1 station/ 260–390 km^2
 iii. In predominately hilly areas with heavy rainfall = 1 station/ 130 km^2

13.2.5.3   Optimum Number of Rain Gauges

If there are already some rain gauge stations in a catchment, the optimal number of stations that should exist to have an assigned percentage of error in the estimation of mean rainfall is obtained by statistical analysis as

$$N = \left(C_v/E\right)2 \qquad (13.6)$$

where N = optimal number of stations, E = allowable degree of error in the estimate of the mean rainfall, and C_v = coefficient of variation of the rainfall values at the existing n stations (in percent).

If there are m stations in the catchment, each recording rainfall values P_1, P_2, P_3, P_4, ... P_m in a known time, the coefficient of variation C_v is calculated as

$$C_v = \left(S_x/P_{avg}\right) \times 100 \qquad (13.7)$$

where
  S_x = standard deviation
  P_{avg} = mean of rainfall values of existing stations

13.2.6   Mean Precipitation over an Area

The rainfall measured by a rain gauge is called point precipitation because it represents the rainfall pattern over a small area surrounding the rain gauge station. However, in nature rainfall pattern varies widely. The average precipitation over an area can be obtained only if several rain gauges are evenly distributed over the area. But there is always limitation to establish several rain gauges. However, this drawback can be

overcome by adopting certain methods as mentioned below, which give fair results.

1. Arithmetic mean method:
    In this method to determine the average precipitation over an area, the rainfall data of all available stations are added and divided by the number of stations to give an arithmetic mean for the area. That is, if P_1, P_2, and P_3 are the precipitations recorded at three stations A, B, and C, respectively, then average precipitation over the area covered by the rain gauge is given by

$$P_{avg} = \left(P_1 + P_2 + P_3 + P_4 + ... P_n\right)/n \qquad (13.8)$$

This method can be used if the area is reasonably flat and individual gauge readings do not deviate from the mean (average). This method does not consider aerial variation of rainfall, noneven distribution of gauges, orographic influences (presence of hills), etc. This method can also be used to determine the missing rainfall reading from any station also in the given area.

2. Thiessen Polygon method:
    This is also known as weighted mean method. This method is very accurate for catchments having areas from 500 to 5000 km^2. In this method rainfall recorded at each station is given a weightage on the basis of the area enclosing the area. The procedure adopted is as follows:
    - The rain gauge station positions are marked on the catchment plan.
    - Each of these station positions is joined by straight lines.
    - Perpendicular bisectors to the previous lines are drawn and extended up to the boundary of the catchment to form a polygon around each station.
    - Using a planimeter, the area enclosed by each polygon is measured.
    - The average precipitation over an area is given as:

$$P_{avg} = \left[\left(P_1A_1 + P_2A_2 + P_3A_3 + P_4A_4 ... + P_nA_n\right)/\right.$$
$$\left.\left(A_1 + A_2 + A_3 + A_4 + ... A_n\right)\right] \qquad (13.9)$$

    where
      $P_1, P_2, P_3, P_4 ... P_n$ are rainfall amounts obtained from 1 to n rain gauge stations, respectively.
      $A_1, A_2, A_3, A_4 ... A_n$ are areas of polygons surrounding each station

3. Isohyetal method:
    Isohyets are imaginary line joining points of equal precipitation in a given area similar to contours in a given area, or isohyets are the contours of equal rainfall.
    In Isohyetal method for determining the average precipitation over an area, Isohyets of different

values are sketched in a manner similar to contours in surveying in a given area as shown in Figure 13.12. The mean (average) of two adjacent Isohyetal values is assumed to be the precipitation over the area lying between the two isohyets. To get the average precipitation over an area, the procedure to be followed is:

a. Each area between the isohyets is multiplied with the corresponding mean Isohyetal value (precipitation).
b. All such products are summed up.
c. The sum obtained from above is divided by the total area of the catchment (gauging area).
d. The quotient obtained from above represents average precipitation over gauging area.

13.3 ABSTRACTION FROM PRECIPITATIONS

When precipitation comes to the earth's surface, it produces runoff. The runoff is important for study to design the hydraulic structures, estimating flood, etc. All the precipitation that comes to the earth's surface does not contribute to runoff, some part of it disappear. Losses may occur due to the following reasons:

1. Evaporation
2. Transpiration
3. Infiltration
4. Interception
5. Depression storage
6. Watershed leakage

The first three contribute major amount of losses.

Precipitation − Surface run off = Total loss   (13.10)

where total loss = evaporation + transpiration + infiltration + interception + depression storage + watershed leakage.

13.3.1 INTERCEPTION

Interception may be defined as that amount of precipitation water which is intercepted by vegetative foliage, buildings, and other objects lying over the land surface. Interception does not reach the land surface but is returned back to the atmosphere by evaporation. This interception storage is the amount of precipitated water which wets the initially dried surface of the object lying above the ground surface. Various factors that affect interception are as follows:

i. Storm factor
ii. Plant factor
iii. Season of the year
iv. Prevailing wind

13.3.2 DEPRESSION STORAGE

A catchment area generally has many depressions of shallow depth and of varying size. When precipitation takes place, water runs toward these depressions and fills them before actual overland flow or run off toward a stream takes place. The depression storage is therefore defined as the water retained in these depressions/ditches. The size of depression may vary from micro to large capacity. These depressions form miniature reservoirs detaining water temporarily. The water stored in these depressions partly evaporates and partly

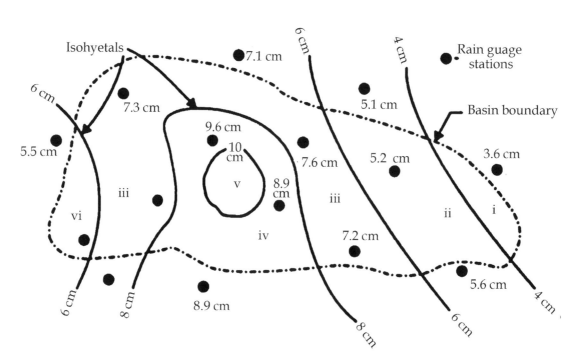

FIGURE 13.12   Isohyetal map.

infiltrates into the ground to meet the ground water reservoir, if any. The following relationship may be used for computing the depression storage.

$$V_{ds} = K\left[1 - e - P_e/K\right] \qquad (13.11)$$

where

V_{ds} = volume of water stored in surface depression
P_e = rainfall excess
K = depression storage capacity of the basin

The above relationship gives an exponentially decaying curve and attains value of K when the ratio P_e/K increases.

Following are the factors that affect depression storage:

   i. Land form
  ii. Soil characteristics
 iii. Topography
 iv. Antecedent precipitation index
  v. Man-made disturbance like terrace farming

13.3.3 WATERSHED LEAKAGE

Adjacent basins are separated by ridge lines, so that rainfall falling over a basin flows toward the drainage lines of the basin. Watershed leakage may be defined as flow of water from one basin to another basin or from one basin to the sea through major faults, fissures, or other geographical features. Due to these faults and fissures, underground hydraulic conduits so formed convey the discharge falling over a part of the catchment.

13.3.4 EVAPORATION

It is the process by which a liquid changes to gaseous state at the free surface through transfer of heat energy.

In an exposed water body like lakes or ponds, water molecules are in continuous motion with arrange of velocities (faster at the top and slower at the bottom). Additional heat on water body increases the velocities. When some water molecules posses' sufficient kinetic energy they may cross over the water surface. Simultaneously the water molecules in atmosphere surrounding the water body may penetrate the water body due to condensation. If the number of molecules leaving the water body is greater than the number of molecules arriving or returning, difference in vapor pressure occurs, leading to evaporation.

13.3.4.1 Dalton's Law of Evaporation

The rate of evaporation is a function of the difference in vapor pressure at the water surface and the atmosphere.

$$E \propto (e_w - e_a) \qquad (13.12)$$

$$E = k\,(e_w - e_a) \qquad (13.13)$$

E = daily evaporation

e_w = saturated vapor pressure of water at a given temperature
e_a = vapor pressure of air
k = proportionality constant

Considering the effect of wind, Dalton's law is expressed as

$$E = k'a(e_w - e_a)(a + b\,V) \qquad (13.14)$$

where

V = wind velocity in km/hour,
k', a and b are constants for a given area.

13.3.4.2 Measurement of Evaporation

In order to ensure proper planning and operation of reservoirs and irrigation systems, estimation of evaporation is necessary. However, exact measurement of evaporation is not possible. But the following methods are adopted as they give reliable results.

1. Pan measurement methods
2. Use of empirical formulae
3. Storage equation method
4. Energy budget method

1. **Pan measurement method**: Any galvanized iron cylindrical vessel of 1.2–1.8 m diameter and of 300 mm depth with opening at the top can be used as an evaporimeter or evaporation pan. During any interval of time, evaporation is measured as the drop in water level in the pan. Rainfall data, atmospheric pressure data, temperature, etc., should also be recorded.

It has been correlated that evaporation from a pan is not exactly the same as that taking place from a water body. Hence while using a pan measurement data for measuring evaporation from a lake or a water body, a correction factor has to be applied or multiplied by a pan coefficient.

$$\text{Pan coefficient } C_p = \frac{\text{Lake evaporation}}{\text{Pan evaporation}} \qquad (13.15)$$

The evaporation pans adopted in practice have a pan coefficient of 0.7–0.8.

The popularly used evaporation pans are
a. ISI standard pan or Class A pan
b. U..S Class A pan
c. Colorado sunken pan
d. U.S. Geological Survey floating pan

a. ISI standard pan or Class A pan: This evaporation pan should confirm to IS-5973:1976 and is also called Class A pan. It is also known as modified Class A pan. It consists of a circular copper vessel of 1220 mm effective diameter, 255 mm effective depth, and a wall thickness of 0.9 mm. A thermometer is assembled to record the variation in temperature. A wire mesh cover with hexagonal openings is provided at the top

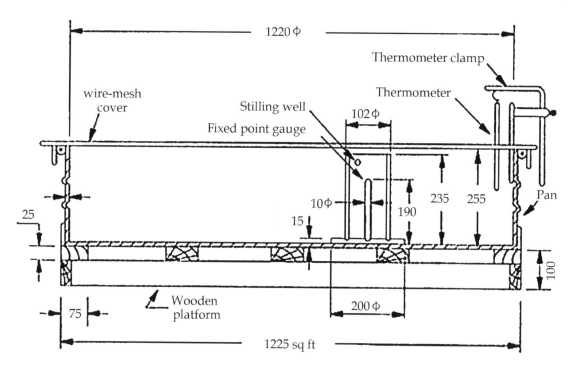

FIGURE 13.13   ISI standard pan or Class A pan.

to prevent entry of foreign matter. A fixed gauge housed in a stilling well as shown in Figure 13.13 is provided. During evaporation measurement a constant water level is maintained at the top level of fixed gauge. For this purpose water has to be added or removed periodically. The water-level measurements are done using micrometer hook gauge. The entire assembly is mounted on a level wooden platform. The evaporation for this pan is 14% less than U.S. Class A evaporation.

b. US Class A pan: The pan is made up of unpainted galvanized iron sheet and monel metal is used where corrosion poses a problem. It is mainly used by U.S. weather bureau.

c. Colorado sunken pan: This pan is made up of unpainted galvanized iron sheet and buried into the ground within 100 mm of top (Figure 13.14). The chief advantage of sunken pan is that radiation and aerodynamic characteristics are similar to those of lake. However, it has the following disadvantages:

   i. Difficult to detect leaks.

   ii. Extra care is needed to keep the surrounding area free from tall grass, dust, etc.

   iii. Expensive to install.

d. U.S. Geological Survey floating pan: This type of pan is kept free in water body, whose evaporation is found. With a view to simulate the characteristics by drum floats in the middle of a raft (4.25 m×4.87 m) is set afloat in a lake (Figure 13.15). The water level in the pan is kept at the same level as the lake leaving a rim of 75 mm. Diagonal baffles provided in the pan

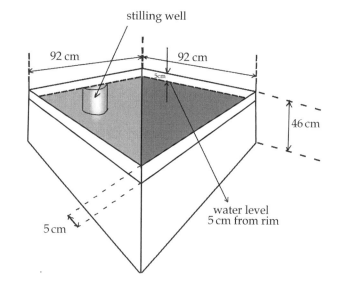

FIGURE 13.14   Colorado sunken pan.

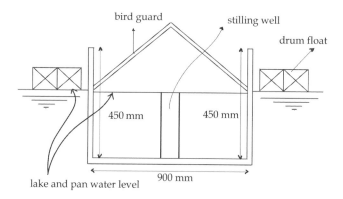

FIGURE 13.15   U.S. Geological Survey floating pan.

reduce the surging in the pan due to wave action. Its main disadvantages are as follows:

i. High cost of installation and maintenance
ii. Very difficult to take measurement

2. **Use of empirical formulae**: There are a large number of empirical equations for estimating the evaporation loss. Most of the equations are based on Dalton's law. However, we shall consider here the following two popular formulae:

a. Meyer's formula:

$$E = K_m(e_s - e_a)\left[1 + V_9/16\right] \qquad (13.16)$$

where

E = evaporation from water body in mm/day
e_s = saturation vapor pressure at the water surface temperature in mm of mercury
e_a = actual vapor pressure of overlying air at a specified height in mm of mercury
V_9 = monthly mean wind velocity in km/hour at a height of 9 m above the ground
K_m = coefficient accounting for various other factors
K_m = 0.36 for large deep waters
= 0.50 for small shallow waters

b. Rohwer's formula:

$$E = 0.771(1.465 - 0.000732\ P_a)(0.44 + 0.0733V_{0.6})(e_s - e_a) \qquad (13.17)$$

where

$V_{0.6}$ = mean wind velocity in km/hour at 0.6m above the ground level
P_a = mean barometric reading in mm of mercury

Thus Rohwer's formula takes into account a correction for the effect of pressure in addition to the wind speed effect. It is known that in the lower part of the atmosphere up to a height of about 500 m above the ground level, the wind velocity is assumed to follow the 1/7 power law as under

$$V_h = C\ h^{1/7} \qquad (13.18)$$

where V_h = wind velocity at height h above the ground.

3. **Water budget or water balance method**: This method balances all the incoming, outgoing, and stored water in a lake or reservoir over a period of time, using the following equation:

Inflow – Outflow = change in storage + evaporation loss

$$P + V_{iS} + V_{iG} = V_{oS} + V_{oG} + S + E \qquad (13.19)$$

where

P = precipitation
V_{iS} = surface water inflow

V_{iG} = ground water inflow
V_{oS} = surface water outflow
V_{oG} = ground water outflow
S = change in storage
E = evaporation loss

The above method does not give accurate results because it is very difficult to measure V_{iG} and V_{oG} for a lake or reservoir.

4. **Energy budget or energy balance equation**: This method uses the conservation of energy by incorporating all the incoming, outgoing, and stored energy of a lake or reservoir in the following form:

$$H_n = H_a + H_e + H_g + H_s + H_i \qquad (13.20)$$

where H_n = net heat energy received by water surface.

$$H_n = H_c(1 - r) - H_b \qquad (13.21)$$

$H_c(1 - r)$ = incoming solar radiation into a surface of reflection coefficient r
H_b = back radiation from the water body
H_a = sensible heat transfer from water surface to air
H_e = heat energy used up in evaporation

$$H_e = p_w L_a E \qquad (13.22)$$

p_w = density of water
L_a = latent heat of evaporation
H_g = heat flux into ground
H_s = heat stored in water body
H_i = net heat conducted out of the system by water flow

In the above equation, all the energy terms are in calories per square mm per day. If the time period is short, the terms H_s and H_i can be neglected as they are negligibly small. Then,

$$E = (H_n - H_g - H_s - H_i)/p_w\ L_a(1 + \beta) \qquad (13.23)$$

where β = Bowen's ratio.

13.3.5 TRANSPIRATION AND EVAPOTRANSPIRATION

Transpiration is the process in which water is lost through the living plants during the respiration process and back to the atmosphere. Plants also consume water for building tissue. It has been estimated that water lost by transpiration is about 800 times more than that required for tissue building. Hence transpiration losses must also be fairly accounted or measured.

The important factors which affect transpiration are as follows:

1. Atmospheric vapor pressure
2. Temperature
3. Wind

4. Light intensity
5. Characteristic of plant

One of the widely used laboratory methods is phytometer. It consists of a closed water tight tank with soil for plant growth. The soil is covered by a polythene sheet with the plant only exposed. The entire setup with the plant is weighed in the beginning (W_1) and at the end (W_2). The amount of water applied during plant growth (W) is noted. Water consumed by transpiration is given as

$$W_t = (W_1 + W) - W_2 \qquad (13.24)$$

The laboratory results are multiplied by a coefficient to obtain field results.

Transpiration ratio is defined as follows:

$$\text{Transpiration ratio} = \frac{\substack{\text{Total mass of water transpired by} \\ \text{the plant during its full growth}}}{\text{mass of dry matter produced}}$$
$$(13.25)$$

Transpiration ratio will increase if the water requirement of a crop increases. Transpiration ratio for wheat is 300–600 and for rice is 600–800.

13.3.5.1 Evapotranspiration or Consumptive Use of Water

Evapotranspiration or consumptive use of water by a crop is defined as the depth of water consumed by evaporation and transpiration during the crop growth, including water consumed by accompanying weed growth. Water deposited by dew or rainfall and subsequently evaporating without entering the plant system is part of consumptive use. When the consumptive use of the crop is known, the water use of large units can be calculated.

13.3.5.1.1 Factors Affecting Evapotranspiration
1. Evaporation which depends on humidity
2. Mean monthly temperature
3. Growing season of crop and cropping pattern
4. Monthly precipitation in the area
5. Irrigation depth or the depth of water applied for irrigation
6. Wind velocity in the locality
7. Soil and topography
8. Irrigation practices and method of irrigation

13.3.5.1.2 Basic Definitions
1. **Saturation capacity**: This can also be known as maximum moisture-holding capacity or total capacity is the amount of water required to fill all the pore spaces between soil particles by replacing all air held in pore spaces.
2. **Field capacity**: Field capacity is the moisture content of the soil after free drainage has removed most

of the gravity water. The concept of field capacity is extremely useful in arriving at the amount of water available in the soil for plant use. Most of the gravitational water drains through the soil before it can be used consumptively by plants.

3. **Permanent wilting point**: Permanent wilting point or wilting coefficient is that water content at which plants can no longer extract sufficient water from the soil for its growth. It is at the lower end of the available moisture range.
4. **Available moisture**: The difference in water content of the soil between field capacity and permanent wilting point is known as available moisture.

13.3.5.1.3 Potential Evapotranspiration and Actual Evapotranspiration

Evapotranspiration is the total loss of water from farm land as evaporation and from plants grown on it as transpiration. If sufficient moisture is always available to completely meet the needs of the plant, the resulting evapotranspiration is called potential evapotranspiration (PET). The real evapotranspiration occurring in a specific situation in the field is called actual evapotranspiration (AET).

At the moisture content in the soil corresponding to field capacity, the water supply to the plant is adequate and AET will be equal to PET, or in other words the ratio AET/PET will be equal to 1. With the reduction in available moisture in the soil, the ratio AET/PET decreases and finally AET will be zero at wilting point.

13.3.5.1.4 Determination of Evapotranspiration or Consumptive Use of Water

The time interval for supplying water to agricultural crops is a factor dependent on water requirement of crops, soil properties, as well as consumptive use. Hence accurate determination of consumptive use or evapotranspiration is very much essential. The methods of determining consumptive use are as follows:

i. Direct measurement method
ii. By use of empirical formulae

i. **Direct measurement method**:
   The consumptive use or evapotranspiration can be measured by five principal methods.
   1. Tank and Lysimeter methods
   2. Field experimental plots
   3. Soil moisture studies
   4. Integration method
   5. Inflow and outflow studies for large area
ii. **By use of empirical formulae**:
   The following are the some of the commonly used methods for computation of consumptive use:
   1. Blaney–Criddle method
   2. Penman method
   3. Hargreaves method
   4. Thornthwaite method

13.3.6 INFILTRATION

Infiltration may be defined as the downward movement of water from soil surface into the soil mass through the pores of the soil. When rain water falls on the ground, a small part of it is initially absorbed by the thin layer of soil so as to replenish the soil moisture deficiency, and therefore, excess water infiltrates downward to join ground water. Infiltration can also be defined as the entry or passage of water into the soil through soil surface. Once water enters into the soil, the passage of transmission of water in the soil, known as percolation, takes place, thus removing the water from near the surface to down below. Infiltration and percolation are directly interrelated. When percolation stops, infiltration also stops. The term infiltration was first used by Horton to describe the phenomenon of soaking water into the soil or absorbed by the soil. He defined it as the entry of water into the soil through the surface layer of the soil, vertically. Infiltration is a major process continuously affecting the magnitude, timing, and distribution of surface runoff at any measured cutlet of a basin.

13.3.6.1 Infiltration Capacity

The capacity of any soil to absorb water from rain falling continuously at an excessive rate goes on decreasing with time, until a minimum rate to infiltrate is reached. At any instant, infiltration capacity of a soil is the maximum rate at which water will enter the soil in a given condition.

13.3.6.2 Infiltration Rate

The infiltration rate at any instant is the rate at which water actually enters the soil during a storm and is equal to the infiltration capacity or the rainfall rate, whichever is less.

13.3.6.3 Infiltration Capacity Curve

Infiltration capacity curve for given soil formation graphically represents the variation of infiltration capacity with time during or a little after rainfall. Infiltration rate is maximum at

the beginning of the rainfall, and thereafter, infiltration rate decreases exponentially as shown in Figure 13.16.

13.3.6.4 Horton's Equation

Horton gave the following equation for finding infiltration rate (f_t) at any time period (t).

$$f_t = f_c + (f_0 - f_c) \qquad (13.26)$$

where

f_t = infiltration rate at any time t
f_c = constant infiltration rate at any time $t = T$
f_0 = infiltration rate in the beginning ($t = 0$)
k = a constant which depends on the soil and vegetation

13.3.6.5 Field Measurement of Infiltration Rate

Infiltration in the field can be measured with the help of two types of infiltrometers:

a. Single cylindrical or single-tube infiltrometers
b. Concentric double cylindrical or double-ring infiltrometers

a. Single cylindrical or single-tube infiltrometers:
   It consists of a hollow metal cylinder of 30 cm diameter and 60 cm length with both ends open. The cylinder is driven in the ground such that 10 cm of it projects above the ground (Figure 13.17). The cylinder is filled with water, such that a head of 7 cm within the infiltrometer is maintained above ground level. Due to infiltration of water, the water level in the cylinder will go on decreasing. Water is added to the cylinder, through graduated jar or burette, so as to maintain constant level. The volume of water added over a predetermined time interval gives the infiltration rate for that time interval. The observations are continued till almost uniform infiltration rate is obtained, which may take about 3–6 hours,

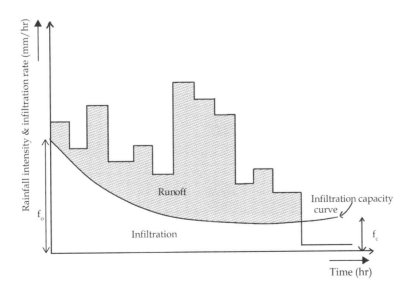

FIGURE 13.16   Infiltration capacity curve.

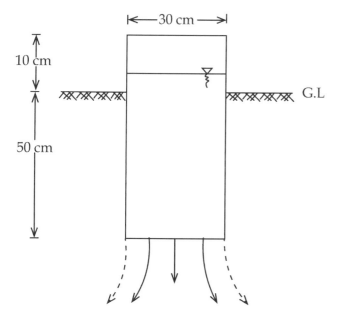

FIGURE 13.17 Single-cylindrical or single-tube infiltrometers.

the least depth of 15 cm. The diameter of the rings may vary from 25 to 60 cm (Figure 13.18). Water is applied in both the inner and outer rings to maintain a constant depth of about 5 cm. Water is replenished after the levels falls by about 1 cm. The water depth in the inner and outer rings should be kept the same during the observation period. However, the measurement includes the recording of volume of water added into the inner compartment, to maintain the constant water level and the corresponding elapsed time. As the purpose of the outer ring is to suppress the lateral percolation of water from the inner ring, the water added to it maintains the same depth as the inner ring. Observations are continued till constant infiltration rate is observed.

13.4 STREAM FLOW MEASUREMENT

A stream can be defined as a flow channel into which the surface runs off from a specified basin drains. Stream flow is the only part of the hydrological cycle that can be measured accurately. Hydrometer is the branch of science which deals with the measurement of water. Stream flow depends on rainfall characteristics, catchment characteristics, and climate. Stream flow is measured in units of discharge occurring at a specified time.

Continuous measurement of stream discharge is very difficult. Stream flow measurement can be done in two different ways.

1. Direct method
2. Indirect method

Direct method of measurement of discharge is very time consuming and a costly procedure. Hence, a two-step procedure

depending upon the type of soil. A plot of time in abscissa against rate of water added in mm/hour gives the infiltration capacity curve for the area. The major drawback is that infiltrated water percolates laterally at the bottom of the ring. Hence it does not truly represent the area through which infiltration takes place.

b. Concentric double cylindrical or double ring infiltrometers:

The above drawback of lateral percolation is rectified in the double-ring infiltrometer which consists of two concentric hollow rings driven into the soil uniformly without any tilt and disturbing the soil, to

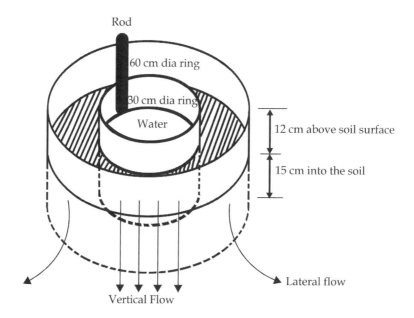

FIGURE 13.18 Concentric double-cylindrical or double-ring infiltrometers.

is followed. First, the discharge in a given stream is related to the elevation of the water surface (stage) through a series of careful measurements. In the next step, the stage of the stream is observed routinely in a relatively in expensive manner, and the discharge is estimated by previously determined stage discharge relationship. The observation stage is easy, inexpensive, and, if desired, continuous reading can also be determined.

Gauging site:

Measuring the area of cross section of river at a selected section. Gauging site must be selected with a case to ensure the stage–discharge curve is reasonably constant over a long time period. The ideal gauge site should follow following criteria:

1. The general cause of stream is straight for about 100 m downstream and 100 m upstream from the site.
2. The stream should not be braided at the gauge site and all the flow must be confined to single stream at all stages.
3. The stream bed should not be subjected to scour deposition of silt at the site.
4. The stream should be free from weed at the site.
5. Banks should be high enough to contain flood water.
6. The river should have a well-defined cross section which should not change in various seasons.
7. It should be easily accessible throughout the year.
8. The gauging site should be free from backwater effects in the channel.

13.4.1 STAGE

Stage of a river is defined as its water surface elevation measured above a particular datum. This datum can be mean sea level or any other arbitrary point. Stage measurement technique can be divided as:

a. manual gauges or nonrecording gauges
b. automatic gauges

a. **Manual gauges or nonrecording gauges**: They are of generally two types.
   1. Staff gauge:

      This is the simplest of stage measurement techniques. In this, elevation of water surface is measured with the help of a fixed graduated staff (Figure 13.19). It is fixed to a structure like abutment, pier, wall, etc. Sometimes it is not possible to read the entire range of water surface elevation of a stream by a single gauge (staff), and in such cases, staff is built in sections at different locations. Such gauges are called sectional gauges.

      When installing sectional gauges, care must be taken to provide an overlap between various gauges and refer all the gauges to some common datum. The staff is made by a durable material with a low coefficient of expansion with respect to temperature and moisture.
   2. Wire gauge:

      In this gauge, water surface elevation measurement is done above the surface such as from bridge. A weight is lowered by a reel to touch water surface and a mechanical counter measures the rotation of wheel, which is proportional to the length of wire paid out.

b. **Automatic gauges**:

   Automatic gauges take the reading itself and no man is required to take the reading. These are of two types.
   1. Float gauge:

      The float operated stage recorder which is in use. In this type, a float is balanced by means of a counterweight over the pulley of a recorder. A good instrument should have a large size float

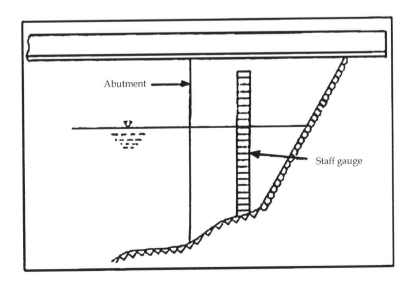

FIGURE 13.19 Staff gauge.

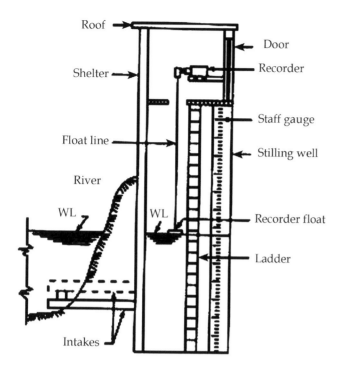

FIGURE 13.20   Float gauge.

and less friction over the pulley. To protect the float from debris and to reduce the effect of water wave on recording the data, stilling well is provided all—around the float type gauge (Figure 13.20). The instrument should be properly housed in a suitable enclosure to protect it from weather element. The installation of float-type gauge recorder is expensive.

2. Bubble gauge: In this type of gauge, compressed air or gas is made to bleed out at a very small rate through an outlet placed at the bottom of the river. In it a pressure gauge is provided, which measures the gas pressure, and by this pressure, we can calculate the water column above the instrument.

Stage hydrograph: The stage data are presented in the form of a plot of stage against chronological time, known as stage hydrograph. Stage data itself are of importance in the design of hydraulic structure and flood protection work.

13.4.2   MEASUREMENT OF VELOCITY

The measurement of velocity is an important aspect of many direct stream flow measurement techniques. It is done by two methods.

1. Current meter
2. Float method

Generally, a mechanical device called current meter is most commonly used for measuring the stream velocity.

The velocity measured by current meter is accurate. Approximate velocity is measured by floats.

1. **Current meter**: This instrument is used for measuring stream velocity and consists of a rotating element which rotates due to the reaction of stream current. The rotating element rotates with an angular velocity which is proportional to the stream velocity. These are of two types of current meters:
   i. Vertical axis meters
   ii. Horizontal axis meters

   i. Vertical axis meters: This instrument consists of a series of conical cups and those are mounted around a vertical axis. The cups are rotated in a horizontal plane. The normal range of velocity is from 0.15 to 4.0 m/s. The accuracy of this instrument is about 1.5% of the threshold value. This instrument cannot be used in situations where there are appreciable vertical components of velocity.

   ii. Horizontal axis meters: This type of meter consists of a propeller mounted at the end of horizontal shaft. They can measure the velocity in the range of 0.15–4.0 m/s according to the diameter of propeller. The accuracy of this instrument is about 1% of the threshold value.

   A current meter is designed in such a way that the rotation speed varies linearly with the stream velocity. A typical relationship is

   $$V = a\,N_s + b \qquad (13.27)$$

   where
   V = stream velocity at the instrument location (m/s)
   N_s = revolution per second of the meter
   a and b = constant of meter

A current meter measures the velocity at a point, but for the calculation of discharge, the mean velocity in each of the selected verticals is required. For mean velocity, we are required to find out the velocity at different depth. The commonly used methods for finding mean velocity are as follows:

a. One-point method: In this, the mean velocity is considered to be equal to the velocity measured at 0.6d below the free water surface. Here d is the depth of stream at that location. Used in shallow stream, depth up to 3 m.

b. Two-point method: In this method, mean velocity is taken as equal to the average of the velocity which is measured at 0.2d and 0.8d below the free water surface. Here d is the depth of stream at that location. Used in moderately deep stream.

c. Three-point method: In this, point velocity is observed at 0.2d, 0.6d, and 0.8d below the water surface. First the average velocity at 0.2d, 0.6d, and 0.8*d* is taken and this value is averaged with the velocity of 0.6*d*. Here *d* is the depth of stream at that location.

2. **Floats method**: A floating object on the surface of a stream floats with the surface velocity of stream and it can be measured by

$$V_s = S/t \qquad (13.28)$$

Here

V_s = surface velocity
S = distance travel
t = time taken by float to travel distance

This method is very useful in the following conditions:
  i.   small stream with rapidly changing water surface
  ii. for preliminary survey
  iii. small stream in flood

A simple float moving on stream surface is called surface float. Surface floats are affected by surface wind so they do not give the actual surface velocity.

For finding the average velocity over a depth, special type of float is used, in which part of float is under the water. Rod float gives average velocity.

In river, where the surface velocity at a depth of 0.5 m below the surface only can be measured, then the average velocity can be measured by

$$V_{\text{avg}} = k\, V_s \qquad (13.29)$$

where

V_{avg} = average velocity
k = reduction factor (value varies between 0.85 and 0.95)
V_s = surface velocity

13.4.3 Direct Method of Discharge Measurement

These are the following methods:

a. Area velocity method
b. Moving boat method

c. Dilution method
d. Electromagnetic method
e. Ultrasonic method

a. **Area velocity method**: This method of discharge measurement consists of measuring the cross-sectional area of the river at gauging site and measuring the velocity of flow throughout the cross-sectional area. For discharge estimation, the cross section is considered to be divided into a larger number of subsections. The average velocity in these subsections is measured. As the number of subsection increases, the accuracy of measuring discharge also increases.

The following guidelines should be followed for selecting the number of subsections.
  • The subsection width should not be more than 1/15 to 1/20 of the width of river.
  • The discharge in each subsection should be less than 10% of total discharge.
  • The difference of velocities in the adjacent subsection should not be more than 20%.

The method of midsections is applied for finding the total discharge. Depth of stream at various locations is measured by the sounding weights or sounding rod. An electroacoustic instrument called echo-depth recorder is used for measuring quick and accurate depth. In this, a high-frequency sound wave is send down by a source called transducer (which is kept immersed at water surface) and the echo reflected by the bed is also picked up by transducer. By comparing the time interval, the depth can be assessed. This is very useful in high-velocity streams and in deep streams for finding the depth (Figure 13.21).

b. **Moving boat method**: This method is similar to the conventional current meter method. In both the cases we apply the velocity area approach for determining the discharge. The basic difference in both the approaches is the procedure of data collection; in the conventional current meter method we use static approach and the data are collected at each observation point while the observer is stationary.

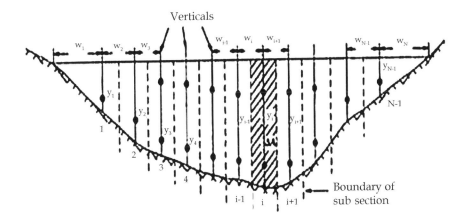

FIGURE 13.21   Section of stream of area velocity method.

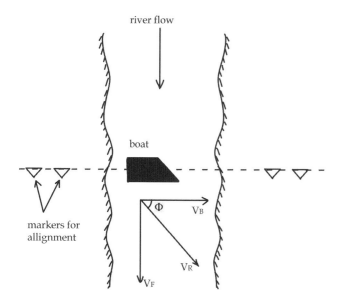

FIGURE 13.22   Moving boat method.

In the moving boat method, observer is on the boat in moving portion (Figure 13.22). In this method a special type of current meter is used which is free to move about vertical axis. This meter is towed in boat at a velocity V_b at right angle to the stream flow, and the stream velocity is V_f, if V_b and V_f are normal to each other then

$$V_b = V_R \cos \varnothing \qquad (13.30)$$

$$V_f = V_R \sin \varnothing \qquad (13.31)$$

V_R = resultant velocity of V_b and V_f
$\varnothing$ = angle between resultant velocity and boat velocity

This method generally has advantage over the conventional current meter method for the following conditions:

   i.   On very wide stream
   ii. When river is having straight stretch
   iii. In condition of unsteady flow

c. **Dilution method**: The dilution method of flow measurement is based on continuity principle and steady flow condition. The tracer with a concentration of C_t is injected into stream at a constant rate of Q_t m³/s. Let Q_S be the discharge in the stream which already contains the same tracer with a concentration of C_S. If the concentration of the tracer at the downstream section is C_m, then from the mass balance equation,

$$Q_S C_S + Q_t C_t = (Q_S + Q_t) \qquad (13.32)$$

Tracers generally used are of three types.
   1. Chemical
   2. Fluorescent dyes
   3. Radioactive material

The tracer used in dilution technique should have the following properties:
   • Tracer should not be absorbed by the channel boundary and vegetation and it should not chemically react with them.
   • Tracer should not be lost by evaporation.
   • Tracer should be nontoxic.
   • Tracer should not be very costly.
   • Tracer should be detected in a directive manner when it is in small concentrations.

The dilution method has the major advantage that the discharge is estimated directly in an absolute way. It is used for small turbulent streams, such as those in mountainous area.

d. **Electromagnetic method**: This method is based on the Faraday's principle; according to this an emf is induced in the conductor when it cuts normal magnetic field.

To produce the magnetic field, large coils are buried at the bottom of channel which carry current I to measure the produced voltage due to flow of water we provide the electrodes at the side of the channel; the discharge of system can be calculated by

$$Q = k_1 \left[\left(Ed/I \right) + k_2 \right] n \qquad (13.33)$$

where
d = depth of flow,
I = current in the coil,
E = system constant,
n, k_1, and k_2 = system constants.
This instrument can measure the discharge to an accuracy of 3%. The minimum detectable velocity is 0.0005 m/s. This method is especially suited for field situations where the cross-sectional properties can change with time due to weed growth, sedimentation.

e. **Ultrasonic method**: In this method, the average velocity is measured by ultrasonic signals. This method has the following advantages:
   • This is very rapid method with high accuracy.
   • This is suitable for automatic recording of data.
   • It can handle change in magnitude and direction of flow as in tidal river.
   • Its cost of installation does not depend on size of rivers.

13.4.4   INDIRECT METHOD OF DISCHARGE MEASUREMENT

This method includes such technique which finds the discharge with the help of relationship between discharge and depth. The broad classification of this method is as follows:

   a. Flow measuring structure
   b. Slope area method

a. **Flow measuring structure**: The conventional structures like weirs, flumes, etc., are used for measuring discharge. At these structures, discharge is a function of water surface elevation measured at a specified upstream location called control station.

$$Q = f(H) = kH^n \qquad (13.34)$$

where
Q = discharge
H = water surface elevation measured from specified datum
k = system constant
This equation can be used, when the downstream level is below a certain limiting water level known as modular limits and such flows that are independent of downstream water level are known as free flows.

If the tail water condition affects the flow, then the flow is known as submerged flow or drowned flow and the discharge under submerged flow is obtained by applying a reduction factor on the free flow discharge.

$$Q_S = K_1 Q \qquad (13.35)$$

where
Q_S = submerged flow discharge
K_1 = reduction factor
Q = free flow discharge
The flow measuring structure can be broadly considered under three categories:
  i.   Thin plate structure
  ii. Long bore weirs
  iii. Flumes
b. **Slope area method**: In this method, discharge is determined with the help of Manning's equation. This equation is used to relate the depths at either ends of river reach to the flowing discharge in that reach, and in this way by knowing the depth, discharge can be obtained. For this technique the following details are required.
  1. Cross-sectional properties at upstream and downstream including bed elevation.
  2. The value of Manning's coefficient n.
  3. The water surface elevation at both upstream and downstream.

$$V = \frac{1}{n} R^{2/3} S^{1/2} \qquad (13.36)$$

where
V = velocity of flow
n = Manning's coefficient
R = hydraulic mean depth
S = slope of bed of channel
The selection reach is the most important part of the slope area method. Following criteria should be followed.

  i.   The accuracy of high water marks must be good.
  ii. The reach should be straight and uniform so far as possible; if it is not possible then choose gradually contracting section.
  iii. The marked pole in water surface elevation must be large than the velocity head. It is good if fall is greater than 0.15 m.
  iv. As the length of reach increases the accuracy in the discharge measurement also increases. A reach length should be greater than 75 times the mean depth.

Stage–discharge relationship: A curve drawn between stream discharge and gauge height is called stage–discharge curve or rating curve, as shown in Figure 13.23. To determine discharge from rating curve only stage is required.

The discharge at a particular section depends upon the wide range of channel and flow parameter; the combined effect of these parameters is called control. If the relationship between stage and discharge does not change with time, then it is called permanent control. If the relationship changes with time, then it is called shifting control.

13.5 RUNOFF

The term runoff is used for water that is on the run or in a flowing state, in contrast to the water held in depression storage and water evaporated in the atmosphere. The runoff of a catchment area in any specified period is the total quantity of water draining into a stream or into a reservoir in that period. This can be expressed as (1) centimeters of water over a catchment or (2) the total water in cubic-meter or hectare-meters for a given catchment. Runoff is broadly divided into three types.

  1. Surface runoff
  2. Subsurface runoff
  3. Base flow

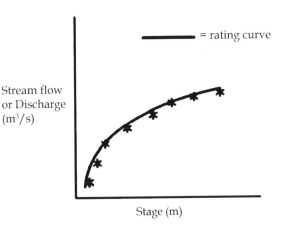

FIGURE 13.23 Stage–discharge curve or rating curve.

The rainfall is disposed off in the following manner:

1. Basin recharge
2. Direct runoff
3. Percolation down to ground water

1. **Basin recharge**: It consists of:
   a. Rain intercepted by leaves and stems of vegetation.
   b. Water held up in surface depression, commonly known as depression storage.
   c. Soil moisture held as capillary water in pore spaces of soil or as hygroscopic water absorbed on the surface of soil particles.
2. **Direct runoff**: Direct runoff is that water which reaches the stream shortly after it falls as rain. Direct runoff consists of:
   a. Overland flow or surface runoff
   b. Subsurface runoff or interflow
   Overland flow is that portion of water which travels across the ground surface to the nearest stream. However, if the soil is permeable, water percolates into it and when it becomes saturated flows laterally in the surface soil to a stream channel. The essential condition for interflow or influent stream is that the surface soil is permeable, but the subsoil is relatively impermeable so that water does not percolate deep to meet the ground water.
3. **Percolation down to ground water**: If the subsoil is also permeable, water percolates deep downward to meet the ground water. Much of the low water flow of rivers is derived from the ground water. Stream channels which are below the ground water table are called effluent streams.

13.5.1 COMPUTATION OF RUNOFF

The runoff from a catchment can be computed daily, monthly, or yearly. Following are some of the methods of computing the runoff.

1. By linear or exponential regression
2. By empirical equations and tables
3. By infiltration method
4. By unit hydrograph (see Section 13.6.4)
5. By rational method

It is difficult to obtain even a fairly approximate estimate of runoff because the various processes such as overland flow, base flow, infiltration, evaporation, etc., are highly irregular and complex. Thus none of the above methods can be considered as accurate. However, the Unit Hydrograph method is easier and is considered as the best among the methods mentioned.

1. **By linear or exponential regression**: Because of several factors affecting the runoff resulting from a given rainfall, the relationship between these two

is quite complex. However, we may use straight line regression for small- and medium-sized catchments and exponential form of regression for large catchments.

a. Straight line regression between P and R: The equation of linear regression line between observed values R and P is

$$R = aP + b \qquad (13.37)$$

where a and b are constants representing abstractions.

$$a = \frac{\left[N\left(\sum PR\right) - \left(\sum P\right)\left(\sum R\right) \right]}{N\left(\sum P^2\right) - \left(\sum P\right)^2} \qquad (13.38)$$

$$b = \frac{\sum R - a\sum P}{N} \qquad (13.39)$$

b. Exponential regression between P and R: For large catchments, it is advantageous to use the following exponential relationship

$$R = \beta \, m^P \qquad (13.40)$$

where β and m are constants. Taking logarithms on both sides, we get

$$\log_e R = P \log_e m + \log_e \beta \qquad (13.41)$$

which is a linear form, from which the values of coefficients β and m can be computed from observed sets of values of P and R.

2. **By empirical equations and tables**:
   a. Runoff coefficient
      The runoff and the rainfall can be interrelated by runoff coefficient, by the expression.

$$R = kP \qquad (13.42)$$

where
$R = $ runoff in cm
$P = $ rainfall in cm
$k = $ runoff coefficient
The runoff coefficient naturally depends upon all the factors which affect the runoff. This method is used only for small water control project and should be avoided for the analysis of major streams.
   b. Strange's tables and chart
      W.L. Strange gave tables and curves for runoff resulting from rainfall in the plains of South India. This tables and curves give runoff for daily rainfall and take into account three types

of catchments (good, average, and bad) and three surface conditions (dry, damp, and wet) prior to the rain.

c. Inglis's formula

C.C. Inglis gave the following formulae, derived from the data collected from 37 catchments in the Bombay Presidency.

For Ghat areas (Western Ghats)

$$R = 0.85\ P - 30.5 \tag{13.43}$$

where P and R in cm.

For non-Ghat areas (plain region)

$$R = 0.00394\ P^2 - 0.0701P \tag{13.44}$$

where P and R in cm.

d. Lacey's formula

$$R = \frac{P}{1 + \dfrac{304.8F}{P\ S}} \tag{13.45}$$

where
P and R in cm
S = catchment factor
F = monsoon duration factor

e. Khosla's formula

$$R_m = P_m - 0.48\ T_m \tag{13.46}$$

where
R_m = monthly runoff in cm
T_m = mean temperature in °C on the entire catchment
P_m = mean precipitation in cm

The temperature introduced in the formula takes into account various factors affecting losses by evaporation, transpiration, sunshine, and wind velocity.

3. **Runoff by infiltration method**:

Infiltration may be defined as the downward movement of water from soil surface, into the soil mass through the pores of the soil. When rain water falls on the ground, a small part of it is initially absorbed by the thin layer of soil so as to replenish the soil moisture deficiency, and therefore, excess water infiltrates downward to join ground water. The capacity of any soil to absorb water from rain falling continuously at an excessive rate goes on decreasing with time, until a minimum rate to infiltrate is reached. At any instant, infiltration capacity of a soil is the maximum rate at which water will enter the soil in a given condition. The infiltration rate at any instant is the rate at which water actually enters the soil during a storm, and is equal to the infiltration capacity or the rainfall rate, whichever is less.

The infiltration capacity of a soil can be determined experimentally by subjecting an experimental plot of rainfall rates in excess of infiltration capacity and by measuring the surface run off. For small area having uniform infiltration characteristics, the runoff volume can be estimated by subtracting infiltration from the design rainfall.

Infiltration index: Infiltration index is the average rate of loss such that the volume of rainfall in excess of that rate will be equal to the direct runoff. Estimating the runoff volume from large areas having heterogeneous infiltration loss and rainfall characteristics are made by use of infiltration indices. There are two types of infiltration indices.

1. W-index
2. Φ-index

The W-index is calculated from the expression:

$$W = (P - R - S)/t \tag{13.47}$$

where
P = precipitation
R = runoff
S = surface retention
t = duration of rainfall

The Φ-index is defined as the average rate of rainfall above which the rainfall volume equals the runoff volume (Figure 13.24). Alternatively, it is defined as the average rate of loss such that the volume of rainfall in excess of that rate will be equal to volume of direct runoff. The runoff volume above Φ-index is usually known as rainfall excess or effective rainfall.

$$\Phi\text{-index} = (P - R)/t \tag{13.48}$$

Thus W-index is essentially equal to Φ-index minus average rate of retention by interception and depression storage. It is the average infiltration rate during

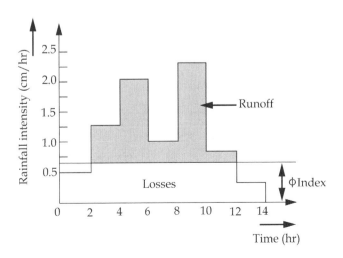

FIGURE 13.24 Φ-index and runoff.

the time rainfall intensity exceeds the capacity rate. The *W*-index is a refined version of *Φ*-index.

4. By unit hydrograph method
   See Section 13.6.4
5. By rational method
   The Rational method predicts the peak runoff according to the formula: Q = CiA, where C is a runoff coefficient, i is the rainfall intensity, and A is the subcatchment area. This formula is applicable to US or metric evaluation, as long as consistent units are employed.

13.6 HYDROGRAPHS

A hydrograph is the graphical representation of the discharge flowing in a river at the given location with the passage of time. A hydrograph is capable of representing discharge fluctuation in the river at a period. Hydrograph gives us the peak flow, which is important for the design of hydraulic structure at that site. The hydrograph result due to an isolated storm generally shows the singly peaked skew distribution of discharge and is known as storm hydrograph, flood hydrograph, or simply hydrograph. The hydrograph is the response of a given catchment to the rainfall input on the same catchment. Hydrograph consists of surface runoff, interflow, and base flow.

13.6.1 COMPONENTS OF HYDROGRAPH

A hydrograph consists of the following components as shown in Figure 13.25:

1. **Rising limb**: The rising limb is influenced by the storm characteristics. It is also known as concentration curve, represents increase in discharge due to gradual building up of storage in channels and over the catchment surface. As the rainfall continues, more and more flow from distant parts reaches the basin outlet, simultaneously the infiltration losses

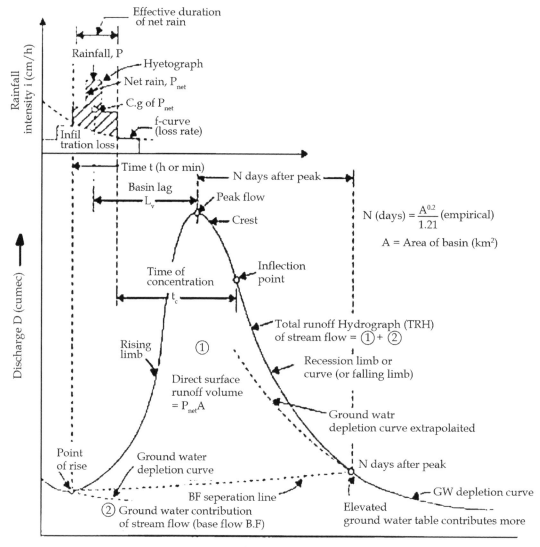

FIGURE 13.25 Storm hydrograph.

also decrease with time. Thus under a uniform storm over the catchment, the runoff increases rapidly.

2. **Crest segment**: The crest segment is one of the most important part of a hydrograph as it contains the peak flow. The point of inflection at the end of crest segment commonly assumed to mark the time at which surface in flow to the channel system ceases.

3. **Recession limb**: The recession limb extends from the point of inflection at the end of the crest segment to the commencement of the natural ground water flow. After the recession limb the withdrawal of water occurs from the storage built up in basin during the earlier phase of hydrograph. The starting point of recession limb represents the condition of maximum storage. Recession limb is independent of storm characteristics and depends entirely on the basin characteristics.

The recession of a storage can be expressed as

$$Q_t = Q_0 \; K_r^{(t-t_0)} \qquad (13.49)$$

where
Q_t = discharge at any time t
Q_0 = discharge at time $t = 0$
K_r = recession constant having value less than unity

13.6.1.1 Factors Affecting Flood or Storm Hydrograph

Various factors affecting the shape of the flood hydrograph are as follows:

1. Physiographic factors
2. Climate factors

Rising limb is affected by climatic factor and recession limb is affected by physiographic factors.

13.6.1.2 Time Parameters Used in Hydrograph Analysis

1. **Effective time duration** (t_{re}): It is the net duration of precipitation during which rainfall rates are more than infiltration rates.

2. **Lag time or basin lag** (t_l): It is the time interval between the center of mass of the net rainfall or rainfall excess and center of mass of runoff hydrograph. It is also taken as time lapse between the center of mass of effective rainfall and the peak of the hydrograph.

3. **Time to peak** (t_p): It is the time interval between the starting of the rising limb to the peak of the hydrograph.

4. **Time of concentration** (t_c): It is defined as the travel time of a water particle from the hydraulically most remote point in the basin to the outflow location. It is also taken as the time from the end of the net rainfall to the point of inflection of the falling limb of the hydrograph.

5. **Recession time** (T_r): It is the duration of direct runoff after the end of the effective rainfall duration. Thus, it is the difference in time between the end of effective rainfall and lowest point of the recession limb.

6. **Time base of hydrograph** (T_b): It is the time from which direct runoff begins to the time when it stops. It is the time from the beginning point to the end point of the direct runoff.

13.6.2 BASE FLOW SEPARATION

There are three methods of separation of base flow and direct runoff in a single peaked storm hydrograph as shown in Figure 13.26. The point A on the hydrograph which represents the beginning of the direct runoff can be easily identified in

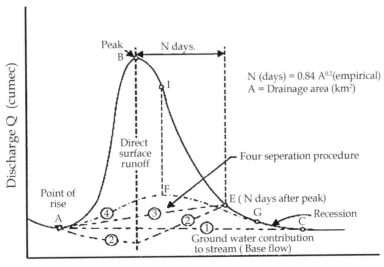

FIGURE 13.26   Base flow separation.

view of sharp change in the run off at that point. But the point D which marks the end of the direct runoff cannot be easily located. However, an empirical equation for the time interval N (days) from the peak (B) to the point D is:

$$N = 0.827 A^{0.2} \qquad (13.50)$$

where

A = area of drainage basin in km²

13.6.2.1 Method 1: Straight-Line Method

In this method, point A and D are joined by a straight line AD. This is the simplest method of separation.

13.6.2.2 Method 2: Two-Lines Method

This is most widely used method in which the base flow is seperated from direct runoff by two lines. The first line AE is obtained by extending the base flow curve, existing prior to the commencement of the surface runoff, till it intersects a vertical line, drawn from the peak in a point E. Point E (so obtained) and point D, located earlier, are joined by a straight line. The segments marked by two lines AE and ED demarcate the base flow and direct runoff.

13.6.2.3 Method 3: Curves Extension Method

This method consists of extending the base flow curve on recession side backward till it intersects a vertical line through the point of inflection in the point F. Thus, D′F is the curve extended backward. Points A and F are then joined by an arbitrary smooth curve. Thus the two segments marked by two curves AF and F′D demarcate the base flow and direct runoff. This method is preffered where the stream and ground water table are hydraulically connected and flow from ground water storage (the contribution of which may be significant) reaches the stream quickly. The surface runoff hydrograph obtained after the base flow separation is known as direct runoff hydrograph.

13.6.3 COMPUTATION OF DIRECT RUNOFF OR RAINFALL EXCESS FROM STORM HYDROGRAPH

13.6.3.1 Procedure

1. Find the ordinates of storm hydrograph representing total discharge Q at a given time interval, say t hours.
2. Separate the ground water flow. Find the ordinates of the base flow at the same time interval.
3. Find the ordinates of direct runoff by subtracting the ordinates of base flow from total discharge ordinates.
4. The direct runoff in depth of water (in cm) is found from the expression:

Direct runoff = Total volume of direct runoff/Area of basin

13.6.4 UNIT HYDROGRAPH

The concept of unit hydrograph initially called unit graph was proposed by L.K. Sherman.

A unit hydrograph is a hydrograph representing 1 cm of runoff from a rainfall of some unit duration and specific areal distribution. Unit duration refers to the duration of a runoff producing rainfall excess that results in a unit hydrograph.

13.6.4.1 Assumptions of Unit Hydrograph Theory

1. **Time invariance**: The runoff produced from a given drainage basin due to given effective rainfall shall always be the same irrespective of the time of its occurrence. The runoff response of a basin to a given effective rainfall is assumed to be time invariant.
2. **Linear response**: The direct runoff response to the rainfall is assumed to be linear.
3. Effective rainfall is uniformly disturbed over the catchment.
4. The rainfall intensity is constant during the storm period
5. The ordinates of direct runoff of common base time are directly proportional to the total amount of direct runoff represented by each hydrograph.
6. For a given drainage basin, the hydrograph of runoff due to a given period of rainfall reflects all the combined physical characteristics of the basin.

13.6.4.2 Uses of Unit Hydrograph

The unit hydrograph establishes a relationship between effective rainfall and direct runoff for a catchment. This relationship is very useful in study of the hydrology of a catchment, as

1. in the development of flood hydrograph for extreme rainfall magnitude
2. in extension of flood-flow records based on rainfall records.
3. in development of flood forecasting and warning system based on rainfall.

13.6.4.3 Limitations of Unit Hydrograph

1. Unit hydrograph assumes uniform distribution of rainfall over the catchment and the intensity of rainfall is assumed constant for the duration of rainfall excess. In practice these two conditions are never strictly satisfied.
2. Unit hydrograph method cannot be used for a catchment area greater than 5000 km² and less than 2 km².
3. Precipitation must be from rainfall only. Snowmelt runoff cannot be satisfactorily represented by unit hydrograph.
4. The catchment should not have large storage which affects the linear relationship between storage and discharge.

In the use of unit hydrograph 20% variation in time base and 10% variation in peak are expected in reproduction of result.

13.7 FLOODS

When the rain falls in heavier quantity, the discharge flowing in the river increases and causes the water to spill over and across the bank of river. Such spills may flood the adjoining areas. A flood is an unusually high stage in a river, normally the level at which the river overflows its banks and inundates the adjoining area.

Flood peak is a valuable data for the purpose of hydrologic design. At a given location, flood peaks vary from year to year in magnitude. The design of bridges, culvert, and spillways for dams and estimation of scour at a hydraulic structure, etc., are examples, where the flood peak values are required.

13.7.1 Types of Flood

Floods are mainly classified into two categories. They are:

1. Large area flood
2. Small area flood

1. **Large area flood**: This type of flood occurs due to storm of low intensity having a duration of a few days to several weeks. Sometimes, snow melt may also be reasons to cause large area floods. The time period of maximum total precipitation does not necessarily coincide with the time of occurrence of large area floods.
2. **Small area flood**: Storms of high intensity and duration of one day or less cause small area floods. Such flows cause good damage to agricultural land in the form of excessive soil erosion. This is the major cause of sedimentation in reservoir and river. In this type, the total flood volume is not much but such flood causes serious local area damage.

The importance of structure and economic development of surrounding area dictates the design criteria for choosing the flood magnitude.

13.7.2 Estimation of Flood Peak

To estimate the magnitude of a flood peak, the following alternative methods are available:

1. Rational method
2. Empirical method
3. Unit-hydrograph technique
4. Flood-frequency studies

The use of a particular method depends upon (1) the desired objective, (2) available data, and (3) the importance of the project.

1. Rational method

If a rainfall is applied to an impervious surface on the catchment surface, it will finally reach at a rate equal to the rate of rainfall. Initially, only a certain amount of water will reach the outlet, but after sometime, the water will reach the outlet from the entire area and in that case, the runoff rate would be equal to the rainfall rate. The time required to reach this condition is the time of concentration (T_C) (Figure 13.27). The peak value of the runoff is given by

$$Q_P = K \ P \ A \quad \text{for } t > T_C \quad (13.51)$$

where
Q_P = peak flood
K = coefficient of runoff
P = rainfall intensity
A = catchment area
In other words,

$$Q_P = \left(1/36\right) K \ P_C A \quad (13.52)$$

where
Q_P = peak rate of runoff in m³/s
K = coefficient of runoff
A = catchment area
P_c = mean rainfall intensity in cm/hour for a duration equal to time of concentration and a given frequency of occurrence.
If P cm of rainfall has fallen in T hours during an individual storm, then mean rainfall intensity is given by

$$P_C = P/T \quad (13.53)$$

2. Empirical method

The empirical formulae used for the estimation of the flood peak are essential regional formulae

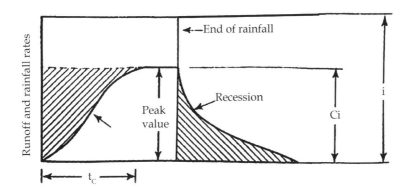

FIGURE 13.27 Estimation of flood peak by rational method.

based on statistical correlation of the observed peak and important catchment properties. Almost all the empirical formulae use the catchment affecting the flood peak and most of them neglect the flood frequency as a parameter. The empirical formulae are applicable only in the region from which they were developed. The maximum flood discharge Q_P from a catchment area A is given by

$$Q_P = f(A) \qquad (13.54)$$

3. Unit-hydrograph technique

The unit hydrograph technique can be used to predict the peak flood hydrograph if the rainfall producing the flood, infiltration characteristics of the catchment, and the appropriate unit hydrograph are available.

4. Flood frequency analysis

In this analysis, the predictions for the future floods are made on the basis of the available records of the past flood. These methods can be safely used to determine the maximum flood that is expected to a river with a given frequency, if sufficient past records are available. Prediction will be precise only if there is no appreciable change in the regime of the river during or after the period of records. The values of the annual maximum flood from a given catchment area for large number of successive years constitute a hydrologic data series called annual series. The data are arranged in decreasing order of magnitude and probability P of each event equal or exceed is calculated by the plotting position formula (Hazen's formula).

$$P = m/(N+1) \qquad (13.55)$$

where
m = order number of the event
N = total number of events in the data

13.8 FLOOD ROUTING

Flood routing is a process of estimation and determining the stage height, volume in storage, and rate of outflow from a reservoir or a stream reach for a given inflow hydrograph. Flood routing is the technique of determining the flood hydrograph at a section of a river by utilizing the data of flood flow at one or more upstream sections. Flood routing is a process of determining peak discharge or stage of outflow hydrograph, corresponding to a known inflow hydrograph.

13.8.1 Uses of Flood Routing

1. In estimation of design flood
2. In designing of reservoir
3. In designs of flood control structure
4. In determining adequacy of spillway

5. In predicting the behavior of river after a change has been done in channel section
6. In study of flood wave

13.8.2 Types of Flood Routing

On the basis of application, flood routing can be classified as

1. Reservoir routing
2. Channel routing

1. Reservoir routing:

In Reservoir routing, the effect of a flood wave entering a reservoir is studied to predict the variations of reservoir elevation and outflow discharge with time. This form of reservoir routing is essential (1) in the design of the capacity of spillways and other reservoir outlet structures, and (2) in the location and sizing of the capacity of reservoirs to meet specific requirements.

2. Channel routing:

In Channel routing, the change in the shape of a hydrograph as it travels down a channel is studied to predict the flood hydrograph at various sections of the reach. Information on the flood-peak attenuation and the duration of high-water levels obtained by channel routing is of utmost importance in flood-forecasting operations and flood-protection works.

13.8.3 Factors Used in Flood Routing

In all flood routing method, the following factors must be considered:

1. **Inflow hydrograph**: These hydrographs are simply runoff hydrograph.
2. **Outflow or spillway graph**: This is the depth-discharge relation of the reservoir spillway structure or of the downstream end of a reach. The rate of discharge is influenced by the hydraulic head and also by the type and size of spillway. In channel routing case the outflow for the reach above is also the inflow to the next reach below.
3. **Outflow hydrograph**: This hydrograph shows the spillway discharge as function of time. Determination of this hydrograph is very necessary.

13.8.4 Basic Equation Used in Flood Routing

A variety of routing methods are available and they can be broadly classified into two categories as: (1) hydrologic routing, and (2) hydraulic routing. Hydrologic-routing methods employ essentially the equation of continuity. Hydraulic methods, on the other hand, employ the continuity equation together with the equation of motion of unsteady flow. The basic differential equations used in the hydraulic routing, known as St. Venant equations, afford a better description of unsteady flow than hydrologic methods.

The equation of continuity used in all hydrologic routing as the primary equation is

$$I - Q = \frac{dS}{dt} \tag{13.56}$$

where

I = inflow rate
Q = outflow rate
S = storage

In small interval

$$\left[\frac{I_1 + I_2}{2}\right]\Delta t - \left[\frac{Q_1 + Q_2}{2}\right]\Delta t = S_1 - S_2 \tag{13.57}$$

Suffix 1 and 2 denote the beginning and end of the time interval Δt.

The time interval Δt should be sufficiently short so that the inflow and outflow hydrograph can be assumed to be straight line in that time interval. Time interval Δt must be shorter than the time of transit of the flood wave through the reach.

14 Water Resources Engineering

14.1 INTRODUCTION

Water resources engineering is the quantitative study of the hydrologic cycle—the distribution and circulation of water linking the earth's atmosphere, land, and oceans. The main sources of water supply are surface and ground water which have been used for a variety of purposes such as drinking, irrigation, hydroelectric energy, transport, recreation, etc. Often, human activities are based on the usual or normal range of river flow conditions. However, flows and storage vary spatially and temporally, and also they are finite (limited) in nature, i.e., there is a limit to the services that can be expected from these resources. Rare or extreme flows or water quality conditions outside the normal ranges will result in losses to river-dependent, human activities. Therefore, planning is needed to increase the benefits from the available water sources.

The purpose of water resources planning and management activities is to determine the following:

i. How can the renewable yet finite resources best be managed and used?
ii. How can this be accomplished in an environment of uncertain supplies, uncertain and increasing demands, and consequently of increasing conflicts among individuals having different interests in the management of a river and its basin?

14.1.1 NEED FOR PLANNING AND MANAGEMENT

Planning and management of water resources systems are essential due to following factors:

1. Severity of the adverse consequences of droughts, floods, and excessive pollution. These can lead to
   a. Too little water due to growing urbanization, additional water requirements, in stream flow requirements, etc. Measures should be taken to reduce the demand during scarcity times.
   b. Too much water due to increased flood frequencies and also increase in water requirements due to increased economic development on river floodplains.
   c. Polluted water due to both industrial and household discharges.
2. Degradation of aquatic and riparian systems due to river training and reclamation of flood plains for urban and industrial development, poor water quality due to discharges of pesticides, fertilizers, and wastewater effluents, etc.
3. While port development requires deeper rivers, narrowing the river for shipping purposes will increase the flood level.

4. River bank erosion and degradation of river bed upstream of the reservoirs may increase the flooding risks.
5. Sediment accumulation in the reservoir due to poor water quality.

Considering all these factors, the identification and evaluation of alternative measures that may increase the quantitative and qualitative system performance is the primary goal of planning and management policies.

14.2 WATER RESOURCES SUSTAINABILITY

14.2.1 DRIVING FORCES AND PRESSURES

Climate change and the hydrological variability of water's distribution and occurrence are natural driving forces that, when combined with the pressures from economic growth and major population change, make the sustainable development of water resources a challenge. The combination of these factors commonly results in increased water use, competition, and pollution in addition to highly inefficient water supply practices. Water management plans should consider the best existing practices and the most advanced scientific breakthroughs.

The scientific community has to convey more effectively its recommendations to decision-makers to enable the latter to develop and maintain multidisciplinary integrated approaches and solutions. Increased funding and resources need to be provided for the collection of detailed water data and information.

The roles and interdependencies of the different hydrological cycle components are often not fully appreciated. As a result, it is difficult to set up adequate protection and prevention strategies.

14.2.2 STATE OF OUR NATURAL WATER RESOURCES

All components of the hydrological cycle should be taken into account when developing water management plans. Each component has a specific role that must be better understood. For example, rain and snow directly supply terrestrial ecosystems and soil moisture is a unique water source for both agricultural development and terrestrial ecosystems. Furthermore, glacial melting has a strong influence on water availability in many nations and as a result more comprehensive global assessments are needed.

Methods are available to substantively predict annual variability in surface runoff and have created solutions to deal with it. However, overcoming the less predictable 5–10 year global cycles of distinctly lower and higher runoff remains a challenge. Groundwater resources could provide a valuable contribution to overcoming climate variability and meeting demands during extended dry periods. A surplus of surface water runoff during wet periods can be used to replenish aquifer systems.

However, we do not have enough data on groundwater and aquifer systems, especially in developing countries where the lack of adequate surface water resources is most extreme. This is particularly true in both Asia and Africa where there has been a dramatic reduction in water monitoring programs.

Water quality monitoring programs are inadequate or lacking in most developing nations; thus safeguarding human health is difficult. Despite two decades of increased international scientific attention and concern, attempts to collect, compile, and gain knowledge from consumption, pollution, and abstraction data and information at a global scale are still piecemeal and in relatively early stages of applicability.

14.2.3 IMPACTS

Poor quality water and unsustainable supplies limit national economic development and can lead to adverse health and livelihood conditions.

Landscape modifications further complicate our understanding of and ability to predict the impacts on water resources since these changes disrupt natural hydrological and ecosystem functioning.

A reasonable level of knowledge has been reached toward recognizing impacts on water quality and quantity from pollution and excessive groundwater and surface water withdrawals. The focus must now be on reducing these impacts. In most developing countries, specific and well-targeted programs are being funded to reduce impacts on water quality and quantity.

14.2.4 RESPONSES

Prevention strategies and new technologies that augment existing natural water resources, reduce demand, and achieve higher efficiency are part of the response to meet today's increasing demands on our available water resources.

To meet current and future water demands, increased attention should be given to precautionary approaches such as innovative uses of natural supplies and new technologies. In the past, methods like storing runoff in reservoirs, diverting flows from water-abundant to water-scarce regions, and extracting aquifer resources were resorted to provide ample water where and when it was needed. These methods are likely to remain part of most water resources development strategies. Nonconventional water resources, such as water reuse and desalination, are being increasingly used and new technologies such as artificial recharge are also becoming more and more common. Capturing rain at the source through rainwater harvesting is yet another method used to increase the availability of natural water sources.

Demand management and conservation are methods that target efficiency. Conservation begins by reducing high losses from water supply distribution systems. Demand

management has gone largely unaddressed since most water utilities still focus on infrastructure development rather than on conservation.

It is worth noting that industry's approach in recent years has been to reduce wastewater and minimize the quantity of processed water needed as this method has proven to be technically feasible and economically advantageous. The demand reduction and efficiency approach should be an integral part of modern water resources management. Its applicability should be promoted while recognizing that it requires a distinct change in the behavioral patterns of institutions, utilities, and individuals—a change that will require education, awareness-raising, and political commitment to achieve effective implementation.

Institutional responses at different levels are also needed. Some nations have implemented new laws and regulations that point the way forward toward protecting and restoring our water sources. Many nations are adapting emerging technical practices to secure and protect their existing natural water resources and use local knowledge as part of sustainable resource development.

14.2.5 THE BENEFITS

There will be economic, social, and environmental benefits from carrying out regular water resources assessments (WRAs) in all basins and aquifers in individual nations as well as regionally, where transboundary shared water resources are present.

Modern approaches to WRA are rapidly emerging and now go well beyond the traditional hydraulic and supply-biased studies carried out during the last century. WRAs have been extended to take advantage of the recently recognized benefits that come from using an integrated approach (Integrated Water Resources Management (IWRM)) and including ecosystems' services (ecosystem approach). WRAs continue to fundamentally require well-documented hydrological cycle component data—without this data the evaluation results are unreliable. To be comprehensive and to assist in sustainable practices, WRAs should include well-documented user consumption and water quality requirements, accurate use data, estimates of the environmental flow volumes needed to maintain ecosystem resilience, characterization of both point and nonpoint sources of pollution and the quality of the receiving waters, and the extensive engagement of all water users and other pertinent stakeholders.

14.3 FLOW AND HYDROSTATIC FORCES

14.3.1 DEFINITION OF A FLUID

A fluid may be defined broadly as a substance which deforms continuously when subjected to shear stress. This fluid can be made to flow if it is acted upon by a source of energy. This can be made clear by assuming the fluid being consisted of layers parallel to each other and letting a force act upon one of the layers in a direction parallel to its plane. This force divided by the area of the layer is called shear stress. As long as this shear stress is applied the layer will continue to move relative to its neighboring layers.

If the neighboring layers offer no resistance to the movement of fluid, this fluid is said to be frictionless fluid or ideal fluid. Practically speaking, ideal fluids do not exist in nature, but in many practical problems the resistance is either small or is not important, therefore it can be ignored. A fluid is always a continuous medium and there cannot be voids in it. The properties of a fluid, e.g., density, may however, vary from place to place in the fluid.

In addition to shear force, fluid may also be subjected to compressive forces. These compressive forces tend to change the volume of the fluid and in turn its density. If the fluid yields to the effect of the compressive forces and changes its volume, it is compressible, otherwise it is incompressible.

14.3.2 BERNOULLI'S EQUATION

According to Newton's second law, the force F used to accelerate a body must be equal to the product of the mass m of the body and its acceleration a. If this law is applied to an ideal (frictionless), incompressible (constant–density) fluid—Euler's law of conservation of momentum is obtained.

Assuming a fluid moving steadily through an imaginary fixed volume element xyz, in the x direction (across the yz–plane), as shown in Figure 14.1. If pressure P_1 and P_2 act upon the two opposite faces, the resultant force on the fluid in the volume element considered is

$$F = P_1 yz - P_2 yz + Bwxyz \qquad (14.1)$$

where
  B=body or mass forces per unit mass
  w=constant fluid density

If x is very small distance, the acceleration along it is constant,

$$a = (V_2 - V_1)/t \qquad (14.2)$$

where t is the time for the fluid through distance x.

Since the mass of the fluid element is $m = wxyz$, with Newton's law $F = ma$

$$P_1 yz - P_2 yz + Bwxyz = wxyz \frac{(V_2 - V_1)}{t} \qquad (14.3)$$

or,

$$P_1 - P_2 + Bwx = wx \frac{(V_2 - V_1)}{t} \qquad (14.4)$$

The mean velocity over the distance x is

$$(V_2 + V_1)/2 = x/t \qquad (14.5)$$

Substituting

$$P_1 - P_2 + Bwx = w \frac{(V_2^2 - V_1^2)}{2} \qquad (14.6)$$

This is Euler's equation for the x-direction, also known as the momentum equation.

If the x-direction is taken as vertically upward as shown in Figure 14.2, such that P_1 and V_1 are at height Z_1 and P_2 and V_2 are at height Z_2, the distance is

$$x = Z_2 - Z_1 \qquad (14.7)$$

The force B per unit mass would now be due to gravitation, and $B = -g$

$$P_1 - P_2 - gw(Z_2 - Z_1) = w(V_2^2 - V_1^2)/2 \qquad (14.8)$$

or

$$P_1 + w\left(\frac{V_1^2}{2}\right) + gwZ_1 = P_2 + w\left(\frac{V_2^2}{2}\right) + gwZ_2 \qquad (14.9)$$

This is Bernoulli's equation for the ideal incompressible fluid in terms of pressure. The Second Law of Thermodynamics limits the directions in which energy can be transformed or transferred. It is well known that friction causes a degradation

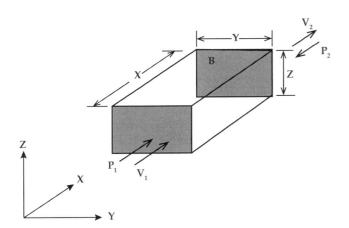

FIGURE 14.1   Illustrations of Euler's law.

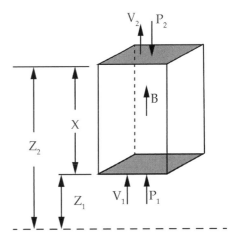

FIGURE 14.2   Illustrations of Euler's law for vertical control body.

of useful "pressure" energy. So, the heat will flow from a hot body to cold body; in terms of the Second Law, it is impossible for the reverse to occur.

or

$$\frac{P_1}{wg} + \frac{V_1^2}{2g} + Z_1 = \frac{P_2}{wg} + \frac{V_2^2}{2g} + Z_2 \qquad (14.10)$$

where P is in Pa (N/m^2).

The pressures P_1 and P_2 are referred to as static pressures; they act in all directions regardless of the direction of the flow. The terms $V_1^2/2$ and $V_2^2/2$ are referred as velocity or dynamic pressures. The summation of the two is referred as the total pressure of the oncoming fluid stream, also called the facing pressure. Each pressure term is expressed in length (m).

14.3.3 Hydrostatic Forces

Hydrostatic force is defined as the force due to the pressure of a fluid at rest. When a surface is submerged in a fluid, forces develop on the surface due to the fluid. The determination of these forces is important in the design of storage tanks, ships, dams, and other hydraulic structures. For fluids at rest we know that the force must be perpendicular to the surface since there are no shearing stresses present. The pressure varies linearly with depth if the fluid is incompressible. For a horizontal surface, such as the bottom of a liquid-filled tank as shown in Figure 14.3, the magnitude of the resultant force is simply $F_R = pA$ where p is the uniform pressure on the bottom and A is the area of the bottom.

Note that if atmospheric pressure acts on both sides of the bottom, the resultant force on the bottom is simply due to the weight of liquid in the tank. Since the pressure is constant and uniformly distributed over the bottom, the resultant force acts through the centroid of the area.

14.4 PRESSURIZED PIPE FLOW

14.4.1 General Characteristics of Pipe Flow

The general method of transporting fluid (liquid or gas) is the flow through a closed conduit. It is commonly called "pipe" if it is of round cross-section and "duct" if the cross-section is not round. The common examples include water pipes, hydraulic hoses, air distribution in a duct in an air conditioning plant, etc. The main driving force for the flow to occur is the pressure differential at both ends of the pipe and the walls of the pipe which is designed to withstand this pressure difference without any undue distortion of the shape.

The basic governing equations such as mass, momentum, and energy conservation can be applied for viscous, incompressible fluids in pipes and ducts. Following assumptions are made for the present analysis:

- Cross-section of the duct is circular unless otherwise specified.
- The pipe is completely filled with fluid being transported; otherwise the flow may be treated as open-channel flow where gravity alone is the driving force.
- The driving potential is the pressure difference across the pipe.

The flow of fluid in a pipe may be laminar, turbulent, or transitional depending upon the flow rates. Such a flow in the pipe is characterized by a dimensionless number called "Reynolds number" (R_e) as explained earlier.

14.4.2 Fully Developed Flow

The fluid typically enters to the pipe with nearly uniform velocity at some location. The region of the flow near which the fluid enters is known as "entrance region." Referring to Figure 14.4a, the velocity profile at section (1) is nearly uniform. As the fluid moves through the pipe, the velocity at the wall approaches to zero due to viscous effect and is commonly called "no-slip boundary condition." Thus a boundary layer is produced along the pipe wall such that the initial velocity profile changes with distance along the pipe. At the end of entrance length, i.e., beyond the section (2), the velocity profile does not vary with the distance. The boundary layer completely grows to fill the pipe where the viscous effects are predominant. For the fluid within the "inviscid core" and surrounding the centerline from (1) to (2), the viscous effects are negligible. The flow between the sections (2) and (3) is "fully developed" until there is any change in the character of the pipe.

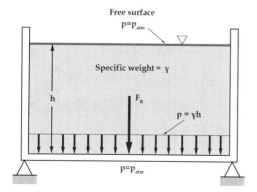

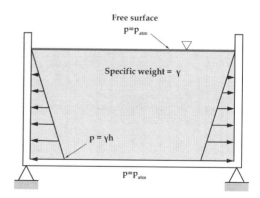

FIGURE 14.3 Hydrostatic force on a horizontal plane.

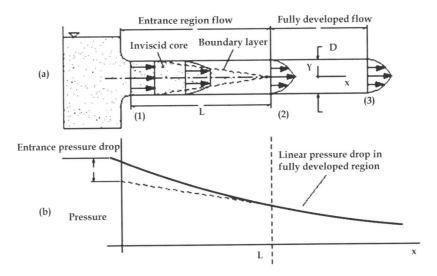

FIGURE 14.4 (a) Illustration of fully developed flow in a horizontal pipe; (b) pressure distribution along the horizontal pipe.

The shape of the velocity profile depends on whether the flow is laminar or turbulent which in turn affects the length of the entrance region (l_e). The dimensionless entrance lengths $\left(\dfrac{l_e}{D}\right)$ can be correlated with Reynolds number as

$$\frac{l_e}{D} = 0.06 R_e \quad \text{(for laminar flow)} \tag{14.11}$$

$$\frac{l_e}{D} = 4.4 R_e \quad \text{(for turbulent flow)} \tag{14.12}$$

Pressure differential is the driving potential for the flow through pipe. When the flow is fully developed, the pressure gradient is negative, i.e.,

$$\left(\frac{\partial p}{\partial x}\right) = -\left(\frac{\Delta p}{l}\right) \tag{14.13}$$

However, referring to Figure 14.4b, the pressure drop is more significant at the entrance region (outside the inviscid core) where the viscous effects are predominant.

14.4.3 MOODY CHART

The fundamental difference between laminar and turbulent flow is that the shear stress for laminar flow depends on the viscosity of the fluid, whereas in case of turbulent flow, it is the function of density of the fluid. In general, the pressure drop ΔP, for steady, incompressible turbulent flow in a horizontal round pipe of diameter D can be written in the functional form as

$$\Delta P_{\text{turb}} = f\left(V, D, l, \varepsilon, \mu, \rho\right) \tag{14.14}$$

where V is the average velocity, l is the length of the pipe, and ε is a measure of the roughness of the pipe wall. Similar expression can also be written for the case of laminar flow

in which the ε term will be absent because the pressure drop in laminar flow is found to be independent of pipe roughness, i.e.,

$$\Delta P_{\text{lam}} = f\left(V, D, l, \mu, \rho\right) \tag{14.15}$$

By dimensional analysis treatment,

$$\left(\frac{\Delta P_l}{\frac{1}{2}\rho V^2}\right)_{\text{lam}} = \Phi\left(\frac{\rho V D}{\mu}, \frac{1}{D}\right) \tag{14.16}$$

$$\left(\frac{\Delta P_l}{\frac{1}{2}\rho V^2}\right)_{\text{turb}} = \Phi\left(\frac{\rho V D}{\mu}, \frac{1}{D}, \frac{\varepsilon}{D}\right) \tag{14.17}$$

The only difference between the above two expressions is that the term (ε/D), which is known as the "relative roughness." In commercially available pipes, the roughness is not uniform, so it is correlated with pipe diameter and the contribution (ε/D) forms a significant value in friction factor calculation. From tests with commercial pipes, Moody gave the values for average pipe roughness listed in Table 14.1.

Now the equation can be simplified with reasonable assumption that the drop is proportional to pipe length. It can be done only when

$$\frac{\Delta P}{\frac{1}{2}\rho V^2} = \frac{l}{D}\bar{\phi}\left(R_e, \frac{\varepsilon}{D}\right) \tag{14.18}$$

It can be rewritten as

$$\Delta P = f\left(\frac{l}{D}\right)\left(\frac{\rho V^2}{2}\right) \tag{14.19}$$

where f is known as "friction factor" and is defined by

TABLE 14.1

Average Values of Roughness for Commercial Pipes

Material	ε (mm)
Riveted steel	0.9–9
Riveted steel	0.3–3
Wood	0.18–0.9
Cast iron	0.26
Galvanized iron	0.15
Commercial steel	0.045
Plastic and glass	0 (smooth pipes)

$$f = \phi\left(R_e, \frac{\varepsilon}{D}\right) \qquad (14.20)$$

Now, recalling the energy equation for a steady incompressible flow,

$$\frac{p_1}{\gamma} + \frac{V_1^2}{2g} + z_1 = \frac{p_2}{\gamma} + \frac{V_2^2}{2g} + z_2 + h_L \qquad (14.21)$$

where h_L is the head loss between two sections. With assumption of horizontal ($z_1 = z_2$) constant diameter pipe ($D_1 = D_2$ or $V_1 = V_2$) with fully developed flow,

$$\Delta p = p_1 - p_2 = \gamma h_L = \rho g h_L \qquad (14.22)$$

From Equations 14.21 and 14.22, the head loss can be determined

$$h_L = f\left(\frac{l}{D}\right)\left(\frac{V^2}{2g}\right) \qquad (14.23)$$

This is known as Darcy-Weisbach equation and is valid for fully developed, steady, incompressible horizontal pipe flow. If the flow is laminar, the friction factor will be independent on (ε/D) and simply,

$$f = \frac{64}{R_e} \qquad (14.24)$$

The functional dependence of friction factor on the Reynolds number and relative roughness is rather complex. It is found from exhaustive set of experiments and is usually presented in the form of curve-fitting formula or data. The most common graphical representation of friction factor dependence on surface roughness and Reynolds number is shown in "Moody Chart" that is shown in Figure 14.5. This chart is valid universally for all steady, fully developed, incompressible flows.

14.5 OPEN CHANNEL FLOW

Open channel flow involves the flow of a liquid in a channel or conduit that is not completely filled. There exists a free surface between the flowing fluid (usually water) and fluid above it (usually the atmosphere). The pressure on the surface is equal to the local atmospheric pressure. The main driving force is the fluid weight-gravity forces. Under steady, fully

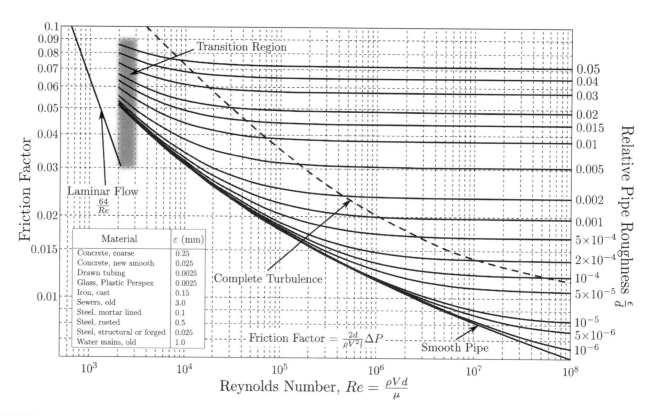

FIGURE 14.5 Moody chart.

developed flow conditions, the component of the weight force in the direction of flow is balanced by the equal and opposite shear force between the fluid and the channel surface.

14.5.1 TYPES OF OPEN CHANNEL

Open channel is natural or manmade conveyance structure which has a free surface at atmospheric pressure. For example, flow in rivers, streams, flow in sanitary and storm sewers flowing partially full. Figure 14.6 shows a trapezoidal shaped open channel with its different parts.

- Flume is the channel made of wood, metal, concrete, or masonry usually supported on or above to carry out water across a depression.
- A chute is a channel having steep slopes.
- A drop is similar to chute but the change in the elevation is affected in a short distance.
- A culvert when partially full is a covered channel installed to drain water through highways or railways embankment.

14.5.1.1 Prismatic and Nonprismatic Channels

A channel in which the cross-sectional shape, size, and the bottom slope are constant is termed as prismatic channel. All natural channels generally have varying cross-section and consequently are nonprismatic. Most of the man-made channels are prismatic channels over long stretches. The rectangle, trapezoid, triangle, and circle are commonly used shapes in man-made channels.

14.5.1.2 Rigid and Mobile Boundary Channels

Rigid channels are those in which the boundary is not deformable. The shape and roughness magnitudes are not functions of flow parameters; for example, lined canals and nonerodible unlined canals. In rigid channels, the flow velocity and shear stress distribution will be such that no major scouring, erosion, or deposition will take place in the channel and the channel geometry and roughness are essentially constant with respect to time.

When the boundary of the channel is mobile and flow carries considerable amounts of sediment through suspension and is in contact with the bed, such channels are classified as mobile channels. In the mobile channel, not only depth of flow but also bed width, longitudinal slope of channel may undergo changes with space and time depending on type of flow. The resistance to flow, quantity of sediment transported, and channel geometry all depend on interaction of flow with channel boundaries. A general mobile boundary channel can be considered to have four degrees of freedom. In rigid channel, we have one degree of freedom.

14.5.2 FLOW REGIMES

14.5.2.1 Steady and Unsteady Flows

A steady flow occurs when the flow properties, such as depth or discharge, at a section do not change with time. If the depth or discharge changes with time, the flow is termed as unsteady. Flood flows in rivers and rapidly varying surges in canals are some examples of unsteady flow.

14.5.2.2 Uniform and Nonuniform Flows

If the flow properties, say the depth of flow, in an open channel remain constant along the length of the channel, the flow is said to be uniform. Figure 14.7 shows a uniform flow through open channel. A flow in which the flow properties vary along the channel is termed as nonuniform flow. A prismatic channel carrying a certain discharge with a constant velocity is an example of uniform flow. In uniform flow, the gravity force

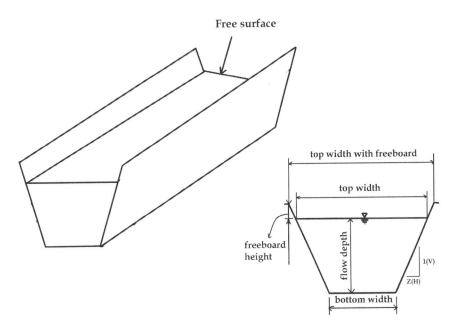

FIGURE 14.6 A trapezoidal-shaped open channel.

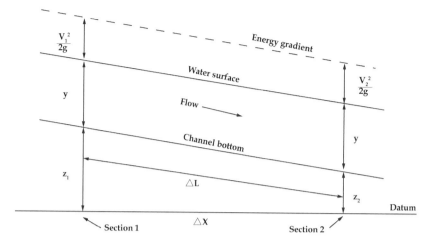

FIGURE 14.7 Uniform flow through open channel.

on the flowing liquid balances the frictional force between the flowing fluid and inside surface of the channel, which is in contact with the fluid. In case of nonuniform flow, the friction and gravity force are not in balance.

14.5.2.3 Gradually Varied and Rapidly Varied Flow

The nonuniform flow can be classified as gradually varied flow (GVF) and rapidly varied flow (RVF). Varied flow assumes that no flow is externally added to or taken out of channel system. The volume of water in a known time interval is conserved in the channel system. If the change of depth in a varied flow is gradual so that the curvature of streamlines is not excessive, such a flow is said to be GVF. Figure 14.8 shows water surface profile of a GVF. Here y_1 and y_2 are the depth at section 1 and 2, respectively. In GVF, the loss of energy is essentially due to boundary friction. Therefore, the distribution of pressure in the vertical direction may be taken as hydrostatic. If the curvature in a varied flow is large and the depth changes appreciably over short lengths, such a phenomenon is termed as RVF.

14.5.2.4 Spatially Varied Flow

If some flow is added to or subtracted from the system, the resulting varied flow is a spatially varied flow (SVF). SVF can be steady or unsteady. In steady SVF, the discharge while being steady varies along the channel length. The flow over a side weir is an example of steady flow. SVF through open channel is shown in Figure 14.9.

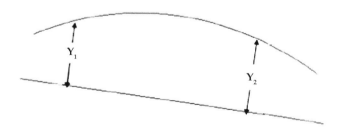

FIGURE 14.8 GVF through open channel.

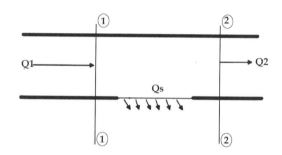

FIGURE 14.9 SVF through open channel.

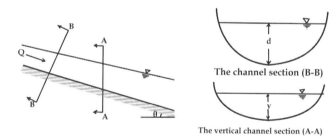

FIGURE 14.10 Channel section.

14.5.3 CHANNEL GEOMETRY

A channel built with constant cross-section and constant bottom slope is called a prismatic channel; otherwise the channel is nonprismatic. The term channel section means cross-section of channel taken normal to the direction of the flow. A vertical channel section is the vertical section passing through the lowest or bottom section. Two channel sections are shown in Figure 14.10.

Natural channel sections are generally very irregular in shape, whereas artificial channels are usually designed with sections of regular geometry shapes. The trapezoid is a commonly used shape; the rectangle and triangle are special case of trapezoid. Since the rectangle has vertical sides, it is commonly used for channels built of stable materials, such as lined masonry, rocks, metal, or timber. The depth of flow y is the vertical distance from the water surface to the lowest

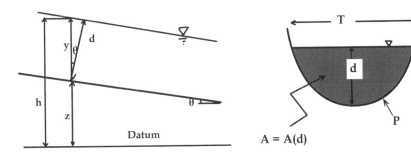

y- vertical distance from the lowest point of the channel section to the free surface

d- depth of flow normal to the direction of flow

θ- channel bottom slope

$d = y\cos\theta$ for mild slope $y = d$

T- top width   A- cross-sectional area of flow   p - wetted perimeter

FIGURE 14.11   Geometric elements of channel section.

point of the channel section. Stage is the elevation or vertical distance of the free surface above a datum. Top width T is the width of the natural section at the water surface. Water area A is the cross-sectional area of the flow measured normal to the direction of the flow. The wetted perimeter P is the length of the line of intersection of the channel wetted surface with the cross-sectional plane normal to the direction of the flow. Figure 14.11 shows the different geometric elements of channel section.

Hydraulic radius R is the ratio of the water area to its wetted perimeter

$$R = A/P \tag{14.25}$$

Hydraulic depth D is the ratio of the water area to the top width

$$D = A/T \tag{14.26}$$

Section factor Z for critical flow computation Z is the product of the water area and the square root of hydraulic depth

$$Z = A\sqrt{D} \tag{14.27}$$

In case of uniform flow, section factor is given by

$$Z = AR^{2/3} \tag{14.28}$$

Table 14.2 gives the equations for some of the common types of channel sections.

14.5.4   Velocity Distribution in Open Channel

Owing to the presence of a free surface and to the friction along the channel wall, the velocities in channel are not uniformly disturbed in channel section. The measured maximum velocity in ordinary channels usually appears to occur below the free surface at a distance of 0.05–0.25 of the depth. The velocity distribution in a channel section depends also on

factors as the unusual shape of the section, the roughness of the channel, and the presence of bends. The roughness of the channel will cause the curvature of the vertical velocity distribution curve to increase. On a bend, the velocity increases greatly at a convex side, owing to centrifugal action of the flow. Surface wind has little effect on velocity distribution. Velocity component in transverse direction is usually small and insignificant compared with the longitudinal velocity components. In a long and uniform reach, remote from the entrance, a double spiral motion will occur in order to permit equalization of shear stress on both sides of the channel. The pattern will include one spiral on each side of centerline, where the water level is highest. Surface velocity v is related to average velocity $\bar{v}$ as $\bar{v} = kv$, where k is the coefficient whose value lies between 0.8 and 0.95. The velocity distributions in vertical section of different types of open channels are shown in Figure 14.12.

14.5.5   Wide-Open Channels

In very wide-open channels, the velocity distribution in the central region of the section is essentially the same as it would be in a rectangular channel of infinite width. The sides of the channel have practically no influence on the velocity distribution in the central region. A wide-open channel can safely be defined as a rectangular channel whose width is greater than 10 times the depth of flow. For either experimental or analytical purposes, the flow in the central region of a wide-open channel may be considered the same as the flow in a rectangular channel of infinite width. Hydraulic radius for a very wide open-channel flow is just about equal to the flow depth.

Consider a rectangular channel, for a rectangular channel, we know that

$$R = \frac{A}{P} = \frac{By}{B + 2y} \tag{14.29}$$

For a wide rectangular channel, the denominator $P = B + 2y \approx B$ as y is very small.

TABLE 14.2

Equations for Different Types of Channel Sections

	Rectangular	Trapezoidal	Triangular	Circular	Parabolic
Flow area A	bh	$(b+mh)h$	mh^2	$\dfrac{1}{8}(\theta-\sin\theta)D^2$	$\dfrac{2}{3}Bh$
Wetted perimeter P	$b+2h$	$b+2h\sqrt{1+m^2}$	$2h\sqrt{1+m^2}$	$\dfrac{1}{2}\theta D$	$B+\dfrac{8}{3}\dfrac{h^2}{B}$
Hydraulic radius R_h	$\dfrac{bh}{b+2h}$	$\dfrac{(b+mh)h}{b+2h\sqrt{1+m^2}}$	$\dfrac{mh}{2\sqrt{1+m^2}}$	$\dfrac{1}{4}\left[1-\dfrac{\sin\theta}{\theta}\right]D$	$\dfrac{2B^2h}{3B^2+8h^2}$
Top width B	b	$b+2mh$	$2mh$	$(\sin\theta/2)D$ or $2\sqrt{h(D-h)}$	$\dfrac{3}{2}Ah$
Hydraulic depth D_h	h	$\dfrac{(b+mh)h}{b+2mh}$	$\dfrac{1}{2}h$	$\left[\dfrac{\theta-\sin\theta}{\sin\theta/2}\right]\dfrac{D}{8}$	$\dfrac{2}{3}h$

Valid for $0<\xi\le 1$, where $\xi=4h/B$

If $\xi>1$, then $P=(B/2)\left[\sqrt{1+\xi^2}+(1/\xi)\ln\left(\xi+\sqrt{1+\xi^2}\right)\right]$.

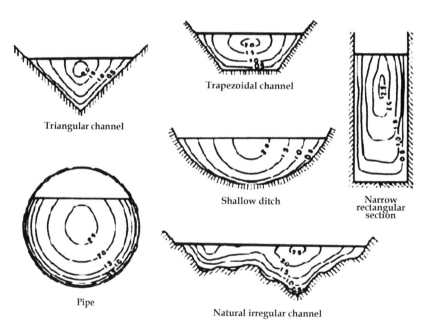

FIGURE 14.12 Velocity distribution in vertical section of open channel.

14.5.6 SPECIFIC ENERGY

Specific energy is the energy head relative to the channel bottom, expressed in terms of flow depth and velocity head. An illustration of specific energy is shown in Figure 14.13. When the channel slope is small, specific energy is given by

$$\text{Therefore } R = \frac{By}{B} = y \qquad (14.30)$$

$$E = y + \frac{v^2}{2g} \qquad (14.31)$$

In case of channels having large bottom slope, the specific energy is given by

$$E = y\cos\theta + \frac{v^2}{2g} \qquad (14.32)$$

where θ is the bottom slope of the channel.

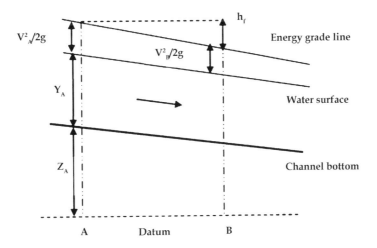

FIGURE 14.13 Definition sketch for energy equation.

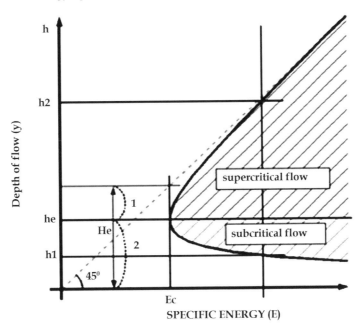

FIGURE 14.14 Specific energy diagram.

When the depth of flow is plotted against the specific energy for a given channel section and discharge, a specific curve is obtained as shown in Figure 14.14. For a given specific energy, there are two possible depths h_1 and h_2; these are called the alternate depth. Minimum specific energy corresponds to the critical state of flow. Thus, at the critical state the two alternate depths apparently become one, which is known as critical depth. When the depth of flow is greater than the critical depth, the velocity of flow is less than the critical velocity for the given discharge and hence the flow is subcritical. When the depth of flow is less than the critical depth, the flow is supercritical. For nonprismatic channel, however, the channel section varies along length of the channel, and hence, the specific energy curve differs from section to section.

The relationship between q and y is given by

$$q^2 = 2gy^2(E - y) \qquad (14.33)$$

For a given specific energy, maximum discharge occurs when

$$y = \frac{2}{3}E \qquad (14.34)$$

Flow beyond q_{max} is not possible. The maximum discharge occurs at the critical depth, for any other discharge, two alternate depths are possible. The relationship between discharge and specific energy is shown in Figure 14.15.

14.5.7 CRITICAL FLOW

The critical state of flow has been defined as the condition for which the Froude number is equal to unity. A more common definition is that it is the state of flow at which the specific energy is a minimum for a given discharge. When the depth of flow of water over a certain reach of a given channel is equal to the critical depth y_c, the flow is called critical flow. Critical slope is a slope such that normal flow occurs

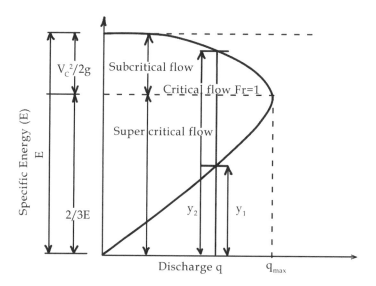

FIGURE 14.15 Discharge-specific energy diagram.

with Froude number, $F = 1$. The smallest critical slope for a specified channel shape, discharge, and roughness is termed as limiting slope. Furthermore, by adjusting the slope and discharge, critical uniform flow may be obtained at the given normal depth S_{cn}.

The equation for specific energy in channel of small slope with $\alpha = 1$ may be written as

$$E = y + \frac{Q^2}{2gA^2} \tag{14.35}$$

Differentiating with respect to y and noting that Q is constant,

$$\frac{dE}{dy} = 1 - \frac{Q^2}{gA^3}\frac{dA}{dy} = 1 - \frac{V^2}{gA}\frac{dA}{dy} \tag{14.36}$$

dA can be written as $T\,dy$. Therefore, $dA/dy = T$ and the hydraulic depth $D = A/T$ so the above equation becomes

$$\frac{dE}{dy} = 1 - \frac{V^2 T}{gA} = 1 - \frac{V^2}{gD} \tag{14.37}$$

At the critical state of flow, the specific energy is a minimum or $dE/dy = 0$. The above equation, therefore, gives

$$\frac{Q^2 T}{gA^3} = 1 \tag{14.38}$$

or

$$\frac{Q^2}{g} = \frac{A^3}{T} \tag{14.39}$$

At critical state of flow, velocity head is equal to half hydraulic depth. A flow at or near the critical state is unstable. This is because a minor change in specific energy at or close to critical state will cause a major change in depth. As the curve is almost vertical near the critical depth, a slight change in energy

would change depth to a much smaller or much greater alternate depth corresponding to specific energy after the change. It can be observed also that, when the flow is near the critical state, the water surface appears unstable and wavy. Such phenomena are generally caused by the minor changes in energy due to variations in channel roughness, cross-section, slope, or deposits of sediment. In the design of a channel, if the depth is found at or near the critical depth for a great length of channel, the shape or slope of channel should be altered in order to secure greater stability.

For uniform flow computation, conveyance (K) is given by

$$K^2 = Cy^n \tag{14.40}$$

where C is hydraulic parameter and n is hydraulic exponent for uniform flow computation.

For computation of critical flow, conveyance is given by

$$Z^2 = Cy^m \tag{14.41}$$

where the section factor Z is a function of depth of flow y given by

$$Z = \frac{A^{3/2}}{\sqrt{T}} \tag{14.42}$$

The exponential factor m is

$$m = \frac{y}{A}\left\{3T - \frac{A}{T}\frac{dT}{dy}\right\} \tag{14.43}$$

14.5.8 MOMENTUM IN OPEN CHANNEL

The momentum of the flow passing a channel section per unit time is expressed by

$$\beta w Q V / g \tag{14.44}$$

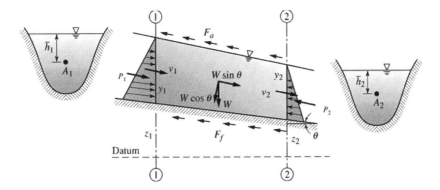

FIGURE 14.16 Momentum principle in open channels.

where β is the momentum coefficient, w is the unit weight of water, Q is the discharge, and V is the mean velocity. According to Newton's second law of motion, the change of momentum per unit of time in the body is equal to the resultant of all the external forces that are acting on the body. The momentum principle in open channels is shown in Figure 14.16.

Expression for momentum change per unit time in the body of water enclosed between two sections may be written as:

$$\frac{Qw}{g}(\beta_2 V_2 - \beta_1 V_1) = P_1 - P_2 + W\sin\theta - F_f \qquad (14.45)$$

where subscripts refers to sections 1 and 2; P is the resultant of pressure acting on the two sections; W is the weight of water enclosed between two section; and F_f is the total external force of friction an resistance acting along the surface of contact between the water and the channel.

When the slope of the channel is relatively small, therefore in a rectangular channel with a small slope

$$P_1 = \frac{1}{2}\gamma b y_1^2 \qquad (14.46)$$

$$P_2 = \frac{1}{2}\gamma b y_2^2 \qquad (14.47)$$

$$F_f = \gamma b h_f \bar{y} \qquad (14.48)$$

where $\bar{y}$ is the average depth
 and

$$Q = \frac{1}{2}(V_1 + V_2)b\bar{y} \qquad (14.49)$$

The weight of the water is

$$W = \gamma b L \bar{y} \qquad (14.50)$$

$$\sin\vartheta = \left(\frac{z_1 - z_2}{L}\right) \qquad (14.51)$$

Substituting them in the momentum equation, we get

$$z_1 + y_1 + \beta_1 \frac{v_1}{2g} = z_2 + y_2 + \beta_2 \frac{v_2}{2g} + h_f \qquad (14.52)$$

h_f is the losses due to external forces exerted on water by the wall of the channel.

For parallel or GVF, the values of P in the momentum equation may be computed by assuming a hydrostatic distribution of pressure and value of $\beta = 1$. For curvilinear or RVF, the value of P must be corrected for curvature effect of the streamlines of the flow. The momentum equation has certain advantages in application to problem involving high internal energy changes such as problem of hydraulic jump. If energy equation is applied to such problems, the unknown internal energy loss is indeterminate. Therefore, momentum equation is applied to these problems since it deals only with external forces.

14.5.9 SPECIFIC FORCE

Specific force is the sum of the pressure force and momentum flux per unit weight of the fluid at a section. Specific force is constant in a horizontal frictionless channel. By plotting the depth against the specific force for a given channel section and discharge, a specific force curve is obtained as shown in Figure 14.17.

For a short horizontal reach of a prismatic channel $\theta = 0$ and $F_f = 0$, also assuming $\beta_1 = \beta_2 = 1$

$$\frac{Q\gamma}{g}(V_2 - V_1) = P_1 - P_2 \qquad (14.53)$$

$$P_1 = \frac{1}{2}\gamma A_1 \bar{z}_1 \qquad (14.54)$$

$$P_2 = \frac{1}{2}\gamma A_2 \bar{z}_2 \qquad (14.55)$$

$\bar{z}_1$ and $\bar{z}_2$ are the distance of centroids below the water surface and A_1 and A_2 are the water area and

$$V = Q/A \qquad (14.56)$$

Substituting the above expression in the momentum equation, we get

$$A_1 \bar{z}_1 + \frac{Q^2}{gA_1} = A_2 \bar{z}_2 + \frac{Q^2}{gA_2} \qquad (14.57)$$

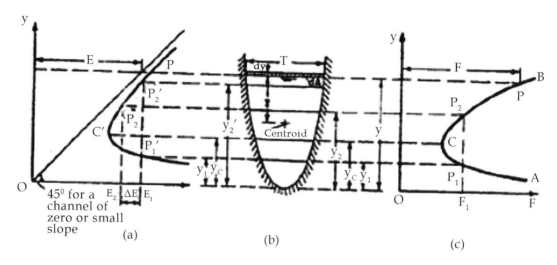

FIGURE 14.17   (a) Specific energy curve; (b) typical channel section; (c) specific force curve for an open channel.

Specific force F is a function of depth of flow, channel geometry, and discharge and is given by

$$F = A\bar{z} + \frac{Q^2}{gA} \qquad (14.58)$$

For a minimum value of specific force, the first derivative of F with respect to y should be zero

$$\frac{dF}{dy} = -\frac{Q^2}{gA^2}\frac{dA}{dy} + A = 0 \qquad (14.59)$$

This is also a criterion for critical state of flow as derived earlier. Therefore, depth at the minimum value of specific force is critical depth. From F vs. y curve, it can be seen that for a given value of specific force there are two possible depths.

14.6   GROUNDWATER FLOW

14.6.1   GROUNDWATER FLOW

In hydrogeology, groundwater flow is defined as the "part of stream flow that has infiltrated the ground, has entered the phreatic zone, and has been discharged into a stream channel, via springs or seepage water." It is governed by the groundwater flow equation. Groundwater is water that is found underground in cracks and spaces in the soil, sand, and rocks. An area where water fills these spaces is called a phreatic zone or saturated zone. Groundwater is stored in and moves slowly through the layers of soil, sand, and rocks called aquifers. The rate of groundwater flow depends on the permeability (the size of the spaces in the soil or rocks and how well the spaces are connected) and the hydraulic head (water pressure).

The direction of groundwater flow follows a curved path through an aquifer from areas of high water levels to areas where water levels are low; that is, from below high ground, which are recharge areas, to groundwater discharge points in valleys or the sea. The direction of flow is indicated by the slope of the water table which is called the hydraulic gradient.

An aquifer that is exposed at the surface is said to be unconfined. Because of earth movements in the past, many aquifers dip below younger strata of impermeable clay. As the thickness of the clay increases the aquifer becomes saturated throughout its entire thickness and the pressure of the water it contains increases. The water rises above the top of the aquifer and may overflow at the surface from a borehole that penetrates into the aquifer; it is said to be under artesian pressure. Such aquifers are called confined aquifers. The level to which the water rises in boreholes that penetrate into confined aquifers is known as the potentiometric surface. Figure 14.18 shows unconfined and confined aquifers.

Water flows through confined aquifers to discharge points some distance down-gradient at a spring or possibly offshore into the sea. The isolated oases in deserts exist because groundwater is issuing from a confined aquifer which may have locally intersected the ground surface, or where the water is rising up a fracture in the overlying confining rocks. Where such outlets do not exist, the water discharges from confined aquifers by slow upward seepage through the overlying clays. The velocity of flow under confined conditions is much slower than that in unconfined aquifers.

When groundwater is pumped from a borehole, the water level is lowered in the surrounding area. A hydraulic gradient is created in the aquifer which allows water to flow toward the borehole. The difference between the original water level and the pumping level is the drawdown, which is equivalent to the head of water necessary to produce a flow through the aquifer to the borehole—the greater the yield from the borehole, the greater the drawdown. The drawdown decreases with increasing distance from the borehole until a point is reached where the water level is unaffected. The surface of the pumping level is in the form of an inverted cone and is referred to as a cone of depression. Water flows into a borehole from all directions in response to pumping and, as it is flowing through a decreasing cylindrical area, the velocity increases as it converges toward the borehole. Figure 14.19 shows the drawdown of the water table around a pumping borehole to form a cone of depression. The shape and extent of the cone of depression depend

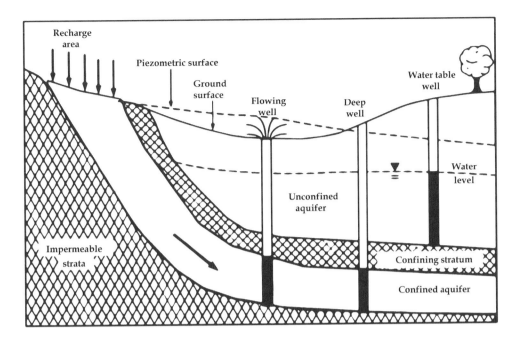

FIGURE 14.18 Unconfined and confined aquifers.

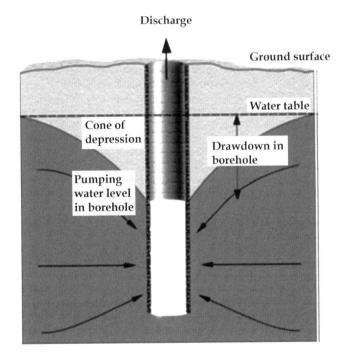

FIGURE 14.19 Drawdown of water table around pumping borehole.

upon the rate of abstraction, the duration of the abstraction, and the hydraulic properties of the aquifer.

14.6.2 GROUNDWATER FLOW EQUATION

The groundwater flow equation is the mathematical relationship which is used to describe the flow of groundwater through an aquifer. The transient flow of groundwater is described by a form of the diffusion equation, similar to that used in heat transfer to describe the flow of heat in a solid

(heat conduction). The steady-state flow of groundwater is described by a form of the Laplace equation, which is a form of potential flow and has analogs in numerous fields.

The groundwater flow equation is often derived for a small Representative Elemental Volume, where the properties of the medium are assumed to be effectively constant. A mass balance is done on the water flowing in and out of this small volume, the flux terms in the relationship being expressed in terms of head by using the constitutive equation called Darcy's law, which requires that the flow is slow.

14.6.2.1 Mass Balance

A mass balance must be performed, and used along with Darcy's law, to arrive at the transient groundwater flow equation. This balance is analogous to the energy balance used in heat transfer to arrive at the heat equation. It is simply a statement of accounting, that for a given control volume, aside from sources or sinks, mass cannot be created or destroyed. The conservation of mass states that, for a given increment of time (Δt), the difference between the mass flowing in across the boundaries, the mass flowing out across the boundaries, and the sources within the volume is the change in storage.

$$\frac{\Delta M_{stor}}{\Delta t} = \frac{M_{in}}{\Delta t} - \frac{\Delta M_{out}}{\Delta t} - \frac{\Delta M_{gen}}{\Delta t} \qquad (14.60)$$

14.6.2.2 Diffusion Equation (Transient Flow)

Mass can be represented as density times volume, and under most conditions, water can be considered incompressible (density does not depend on pressure). The mass fluxes across the boundaries then become volume fluxes (as are found in Darcy's law). Using Taylor series to represent the in and out flux terms across the boundaries of the control volume, and using the divergence theorem to turn the flux across the boundary into a flux over the entire volume, the final form of the groundwater flow equation (in differential form) is

$$S_s \frac{\partial h}{\partial t} = -\nabla \cdot q - G \qquad (14.61)$$

This is known in other fields as the diffusion equation or heat equation, it is a parabolic partial differential equation (PDE). This mathematical statement indicates that the change in hydraulic head with time (left-hand side) equals the negative divergence of the flux (q) and the source terms (G). This equation has both head and flux as unknowns, but Darcy's law relates flux to hydraulic heads, so substituting it in for the flux (q) leads to

$$S_s \frac{\partial h}{\partial t} = -\nabla \cdot (-K\nabla h) - G \qquad (14.62)$$

Now if hydraulic conductivity (K) is spatially uniform and isotropic (rather than a tensor), it can be taken out of the spatial derivative, simplifying them to the Laplacian, this makes the equation

$$S_s \frac{\partial h}{\partial t} = K\nabla^2 h - G \qquad (14.63)$$

Dividing through by the specific storage (S_s), puts hydraulic diffusivity ($\alpha = K/S_s$ or equivalently, $\alpha = T/S$) on the right-hand side. The hydraulic diffusivity is proportional to the speed at which a finite pressure pulse will propagate through the system (large values of α lead to fast propagation of signals). The groundwater flow equation then becomes

$$\frac{\partial h}{\partial t} = \alpha\nabla^2 h - G \qquad (14.64)$$

where the sink/source term, G, now has the same units but is divided by the appropriate storage term (as defined by the hydraulic diffusivity substitution).

14.6.2.3 Laplace Equation (Steady-State Flow)

If the aquifer has recharging boundary conditions a steady-state may be reached (or it may be used as an approximation in many cases), and the diffusion equation simplifies to the Laplace equation.

$$0 = \alpha\nabla^2 h \qquad (14.65)$$

This equation states that hydraulic head is a harmonic function and has many analogs in other fields. The Laplace equation can be solved using techniques but with the additional requirements of a steady-state flow field.

A common method for solution of this equation in civil engineering and soil mechanics is to use the graphical technique of drawing flow nets, where contour lines of hydraulic head and the stream function make a curvilinear grid, allowing complex geometries to be solved approximately.

Steady-state flow to a pumping well (which never truly occurs, but is sometimes a useful approximation) is commonly called the Thiem solution.

14.6.2.4 Two-Dimensional Groundwater Flow

The above groundwater flow equations are valid for three-dimensional flow. In unconfined aquifers, the solution to the 3D form of the equation is complicated by the presence of a free surface water table boundary condition: in addition to solving for the spatial distribution of heads, the location of this surface is also an unknown. This is a nonlinear problem, even though the governing equation is linear.

An alternative formulation of the groundwater flow equation may be obtained by invoking the Dupuit–Forchheimer assumption, where it is assumed that heads do not vary in the vertical direction (i.e., $\partial h/\partial z = 0$). A horizontal water balance is applied to a long vertical column with area $\delta x \delta y$ extending from the aquifer base to the unsaturated surface. This distance is referred to as the saturated thickness, b. In a confined aquifer, the saturated thickness is determined by the height of the aquifer, H, and the pressure head is nonzero everywhere. In an unconfined aquifer, the saturated thickness is defined as the vertical distance between the water table surface and the aquifer base. If $\partial h/\partial z = 0$, and the aquifer base is at the zero datum, then the unconfined saturated thickness is equal to the head, i.e., $b = h$.

Assuming both the hydraulic conductivity and the horizontal components of flow are uniform along the entire saturated thickness of the aquifer (i.e., $\partial q_x/\partial z = 0$ and $\partial K/\partial z = 0$), we can express Darcy's law in terms of integrated discharges, Q_x and Q_y:

$$Q_x = \int_0^b q_x dz = -Kb\frac{\partial h}{\partial x} \qquad (14.66)$$

$$Q_y = \int_0^b q_y dz = -Kb\frac{\partial h}{\partial y} \qquad (14.67)$$

Inserting these into our mass balance expression, we obtain the general 2D governing equation for incompressible saturated groundwater flow:

$$\frac{\partial nb}{\partial t} = \nabla \cdot (Kb\nabla h) + N \qquad (14.68)$$

where n is the aquifer porosity. The source term, N (length per time), represents the addition of water in the vertical direction (e.g., recharge). By incorporating the correct definitions for saturated thickness, specific storage, and specific yield, we can transform this into two unique governing equations for confined and unconfined conditions:

$$S\frac{\partial h}{\partial t} = \nabla \cdot (KH\nabla h) + N \qquad (14.69)$$

(confined), where $S = S_s b$ is the aquifer storage coefficient and

$$S_y\frac{\partial h}{\partial t} = \nabla \cdot (Kh\nabla h) + N \qquad (14.70)$$

(unconfined), where S_y is the specific yield of the aquifer.

Note that the PDE in the unconfined case is nonlinear, whereas it is linear in the confined case. For unconfined steady-state flow, this nonlinearity may be removed by expressing the PDE in terms of the head squared:

$$\nabla \cdot \left(K \nabla h^2\right) = -2N \qquad (14.71)$$

or, for homogeneous aquifers,

$$\nabla^2 h^2 = \frac{-2N}{K} \qquad (14.72)$$

This formulation allows us to apply standard methods for solving linear PDEs in the case of unconfined flow. For heterogeneous aquifers with no recharge, potential flow methods may be applied for mixed confined/unconfined cases.

14.6.3 Summary of Differential Equations for Groundwater Flow

All equations are derived by combining Darcy's Law with an equation of continuity (mass balance).

Symbols used:

b = aquifer thickness
h = piezometric head
K = hydraulic conductivity
S = storage coefficient
S_s = specific storage
S_{ya} = apparent specific yield
T = transmissibility
t = time
x, y, z = orthogonal coordinate directions
∇^2 = Laplacian operator

$$\nabla^2 = \frac{\partial^2}{\partial x^2} + \frac{\partial^2}{\partial y^2} + \frac{\partial^2}{\partial z^2} \qquad (14.73)$$

Relations:

$S = bS_s$
$T = bK$

14.6.3.1 Confined Flow

a. **Inhomogeneous anisotropic confined aquifer**

$$\frac{\partial}{\partial x}\left(K_x \frac{\partial h}{\partial x}\right) + \frac{\partial}{\partial y}\left(K_y \frac{\partial h}{\partial y}\right) + \frac{\partial}{\partial z}\left(K_z \frac{\partial h}{\partial z}\right) = S_s \frac{\partial h}{\partial t} \qquad (14.74)$$

Coordinate system must be selected so that axes are collinear with principal axes of hydraulic conductivity.

If the medium is homogeneous, the above equation reduces to:

b. **Homogeneous anisotropic confined aquifer**

$$K_x \frac{\partial^2 h}{\partial x^2} + K_y \frac{\partial^2 h}{\partial y^2} + K_z \frac{\partial^2 h}{\partial z^2} = S_s \frac{\partial h}{\partial t} \text{ or } K\nabla^2 h = S_s \frac{\partial h}{\partial t} \qquad (14.75)$$

Again, the coordinate axes must be chosen to coincide with the hydraulic conductivity axes.

c. **Homogeneous isotropic confined aquifer (diffusion equation)**

$$\frac{\partial^2 h}{\partial x^2} + \frac{\partial^2 h}{\partial y^2} + \frac{\partial^2 h}{\partial z^2} = \frac{S_s}{K} \frac{\partial h}{\partial t} \text{ or } \nabla^2 h = \frac{S_s}{K} \frac{\partial h}{\partial t} \qquad (14.76)$$

For an aquifer of constant thickness, b, this can be written as:

$$\frac{\partial^2 h}{\partial x^2} + \frac{\partial^2 h}{\partial y^2} + \frac{\partial^2 h}{\partial z^2} = \frac{S}{T} \frac{\partial h}{\partial t} \text{ or } \nabla^2 h = \frac{S}{T} \frac{\partial h}{\partial t} \qquad (14.77)$$

Note that because the medium is isotropic, the selection of coordinate axes is arbitrary—they may be placed in any convenient orientation.

d. **Horizontal flow in homogeneous isotropic confined aquifer of constant thickness**

If flow is horizontal, $\frac{\partial h}{\partial z} = 0$, and thus

$$\frac{\partial^2 h}{\partial x^2} + \frac{\partial^2 h}{\partial y^2} = \frac{S}{T} \frac{\partial h}{\partial t} \qquad (14.78)$$

In this equation, piezometric head does not vary with elevation. Note that if a flow has significant vertical components of velocity, there does not exist a single piezometric surface indicative of flow conditions in the confined aquifer as a whole.

e. **Steady flow in homogeneous isotropic confined aquifer**

Steady flow means that the flow rate, piezometric head, and amount of fluid in storage do not change with time. Thus the right-hand side of the equation given in (14.57) becomes zero and we get:

$$\frac{\partial^2 h}{\partial x^2} + \frac{\partial^2 h}{\partial y^2} + \frac{\partial^2 h}{\partial z^2} = 0 \text{ or } \nabla^2 h = 0 \qquad (14.79)$$

This is **Laplace's equation**, the subject of much study in other fields of science. Many powerful and elegant methods are available for its solution, especially in two dimensions.

14.6.3.2 Unconfined Flow (Water-Table Aquifers)

a. **Inhomogeneous anisotropic unconfined aquifer**

$$\frac{\partial}{\partial x}\left(K_x \frac{\partial h}{\partial x}\right) + \frac{\partial}{\partial y}\left(K_y \frac{\partial h}{\partial y}\right) + \frac{\partial}{\partial z}\left(K_z \frac{\partial h}{\partial z}\right) = 0 \qquad (14.80)$$

The right-hand side equals zero not because $\frac{\partial h}{\partial t} = 0$ (it does not), but because in unconfined aquifers $S_s = 0$. The flow domain for which solutions of this equation are sought is not constant because the water table position changes with time.

b. **Homogeneous anisotropic unconfined aquifer**

$$K_x \frac{\partial^2 h}{\partial x^2} + K_y \frac{\partial^2 h}{\partial y^2} + K_z \frac{\partial^2 h}{\partial z^2} = 0 \text{ or } K\nabla^2 h = 0 \qquad (14.81)$$

Coordinate axes must coincide with conductivity axes.

c. **Homogeneous isotropic unconfined aquifer**

$$\frac{\partial^2 h}{\partial x^2} + \frac{\partial^2 h}{\partial y^2} + \frac{\partial^2 h}{\partial z^2} = 0 \text{ or } \nabla^2 h = 0 \qquad (14.82)$$

Note that this is Laplace's equation again.

d. **Dupuit assumptions for unconfined flow**

Solution of the equations for unconfined groundwater flow is complicated by the fact that the aquifer thickness changes as groundwater is withdrawn, i.e., removal of water from the aquifer lowers the water table.

If vertical components of flow are negligible or small, we can use the Dupuit assumptions to simplify the solution of the equations. These assumptions are as follows:

1. The flow is horizontal in any vertical profile.
2. The Darcy velocity q_x is constant over the depth of flow z_f. z_f is a function of x.
3. The Darcy velocity at the free surface (water table) can be expressed as $q = -K \frac{\partial h}{\partial x}$, rather than $q = -K \frac{\partial h}{\partial l}$. This is reasonable for small water-table slopes $\frac{\partial h}{\partial l}$.
4. These assumptions require a hydrostatic pressure distribution along any vertical line.
5. The consequence of these assumptions is that $Q = -Kh \frac{\partial h}{\partial x}$. In this relation, h is the piezometric head at water table and thickness of flow as shown in Figure 14.20.

e. **Boussinesq equation: Homogeneous Isotropic Unconfined aquifer, Dupuit Assumptions**

1. Nonlinear equation for general case

$$\frac{\partial}{\partial x}\left(h\frac{\partial h}{\partial x}\right) + \frac{\partial}{\partial y}\left(h\frac{\partial h}{\partial y}\right) = \frac{S_{ya}}{K}\frac{\partial h}{\partial t} \qquad (14.83)$$

2. Linearized equation for situations where spatial variation of h is small with respect to magnitude of h

$$\frac{\partial^2 h}{\partial x^2} + \frac{\partial^2 h}{\partial y^2} = \frac{S_{ya}}{bK}\frac{\partial h}{\partial t} \qquad (14.84)$$

14.6.3.3 Aquifers with Vertical Accretion to Flow

a. **Homogeneous isotropic confined aquifer overlain by leaky confining layer (aquitard)**

$$\frac{\partial^2 h}{\partial x^2} + \frac{\partial^2 h}{\partial y^2} + \frac{h_0 - h}{B^2} = \frac{S}{bK}\frac{\partial h}{\partial t} \qquad (14.85)$$

where

$$B = \sqrt{\frac{Kbb_a}{K_a}} = \text{leakage factor}$$

K_a=hydraulic conductivity of overlying aquitard
b_a=thickness of overlying aquitard
$h_0 - h$ = difference in piezometric head between that of confined aquifer (h) and that of above aquitard (h_0)

b. **Homogeneous isotropic recharging unconfined aquifer**

$$\frac{\partial}{\partial x}\left(h\frac{\partial h}{\partial x}\right) + \frac{\partial}{\partial y}\left(h\frac{\partial h}{\partial y}\right) + \frac{W}{K} = \frac{S_{ya}}{K}\frac{\partial h}{\partial t} \qquad (14.86)$$

where
W=recharge or accretion rate (rate of percolation to water table)

14.7 WATER DISTRIBUTION

14.7.1 WATER DISTRIBUTION ON EARTH

Water covers approximately 71% of our planet area, which means that close to three quarters of its surface is composed of H_2O. The 96.5% of volume of water on planet earth is in salt-water form in our seas and oceans. Only 3% of the water on our planet is fresh water, of which 1.74% is actually in ice form in polar ice caps and glaciers, which are currently diminishing as a result of climate change. The rest of this freshwater is found in various forms, including rivers, lakes, and the atmosphere. Table 14.3 illustrates the various different types of water and their distribution around the earth.

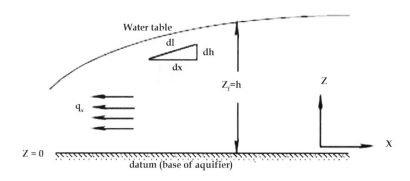

FIGURE 14.20 Illustration of Dupuit assumption for unconfined flow.

TABLE 14.3
Water Distribution on Earth

Location	Percentage of Total Water	Type of Water
Seas and oceans	96.5%	Salty
Polar ice caps and glaciers	1.74%	Fresh
Salted groundwater	0.94%	Salty
Fresh groundwater	0.76%	Fresh
Continental glaciers and permafrost	0.022%	Fresh
Freshwater lakes	0.007%	Fresh
Salt-water lakes	0.006%	Salty
Ground humidity	0.001%	Fresh
Atmosphere	0.001%	Fresh
Reservoirs	0.0008%	Fresh
Rivers	0.0002%	Fresh
Biological water	0.0001%	Fresh

Water represents between 50% and 90% of the mass of living beings, with it being 75% in human beings and reaching up to 90% in algae. This means that water is indispensable for the continued existence of living species, and it is also essential for the world to continue developing economically.

Since almost all the water on the planet is in salt-water form, desalination represents an efficient and very real option to obtain clean water for drinking, for our hygiene and to obtain our food; furthermore, water provided in this way is well suited for energy production, which is central to our economic activity.

14.7.2 Water Distribution System

Distribution system is a network of pipelines that distribute water to the consumers with appropriate quality, quantity, and pressure. They are designed to adequately satisfy the water requirement for a combination of Domestic, Commercial, Industrial, and Firefighting purposes.

A good distribution system should satisfy the following conditions:

- Water quality should not get deteriorated in the distribution pipes.
- It should be capable of supplying water at all the intended places with sufficient pressure head.
- It should be capable of supplying the requisite amount of water during firefighting.
- The layout should be such that no consumer would be without water supply, during the repair of any section of the system.
- All the distribution pipes should be preferably laid one meter away or above the sewer lines.
- It should be fairly water-tight as to keep losses due to leakage to the minimum.

The main elements of the distribution system are as follows:

- Pipe systems
- Pumping stations

- Storage facilities
- Fire hydrants
- House service connections
- Meters
- Other appurtenances

14.7.3 Components of a Network

The features of a minimum size community water distribution system are shown in Figure 14.21.

Water source: At the beginning of every water distribution network, there must be a raw water source (No. 1 in Figure 14.21) such as a lake, river, or groundwater source. To provide enough water for the network the water can be stored in a reservoir (No. 2 in Figure 14.21).

Raw water pumping: These pumps (No. 3 in Figure 14.21) transport water from the source to the raw water storage facility. The water is filtered before being injected into the network in order to prevent corrosion.

Treatment (Purification): At one stage of the distribution, the water is purified in order to reach the standard quality. In the illustration above (No. 4 in Figure 14.21) this is done in a centralized treatment plant by screening, coagulation-flocculation, sedimentation, and filtration (e.g., membrane filtration). As a last step it is disinfected (e.g., chlorination or ozonation).

High-service pumping: Pumping stations (No. 5 in Figure 14.21) may be needed to pump water up to service areas that have higher elevations than other areas of a community. Fill gravity tanks that float on the water supply distribution system. When pumping stations are used to distribute water, and no water storage is provided, the pumps force water directly into the network.

Elevated storage: An extremely important element in a water distribution system is water storage (No. 6 in Figure 14.21). System storage facilities have a far-reaching effect on a system's ability to provide adequate consumer consumption during periods of high demand while meeting fire protection requirements. The two common storage methods are ground-level storage and elevated storage.

Piping/distribution system: At the end of the system, water is distributed via pipes to the consumer, who has to pay for the used amount (No. 7 and No. 8 in Figure 14.21).

14.7.4 Layouts of Distribution Network

The distribution pipes are generally laid below the road pavements, and as such, their layouts generally follow the layouts of roads. There are, in general, four different types of pipe networks; any one of which either singly or in combinations, can be used for a particular place. They are Dead-End System, Grid Iron System, Ring System, and Radial System.

14.7.4.1 Dead-End System

In the dead-end system (also called tree system), one main pipeline runs through the center of the populated area and submains branch off from both sides. The submains divide into

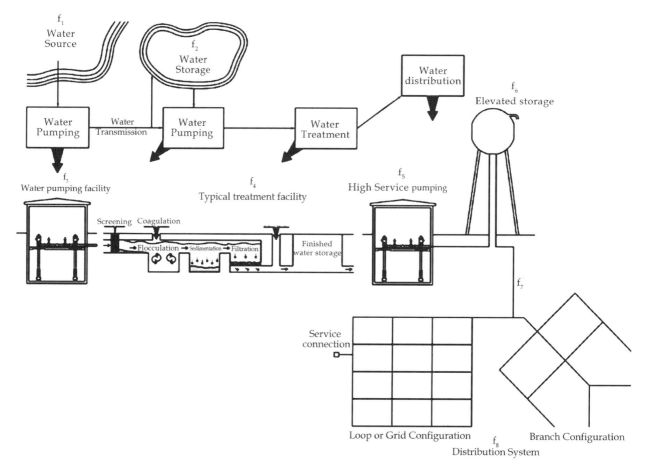

FIGURE 14.21   Features of a minimum size community water distribution system.

several branch lines from which service connections are provided. It is suitable for old towns and cities having no definite pattern of roads. Dead-End system is shown in Figure 14.22.

Advantages
- The design calculation is simple and easy.
- A smaller number of cut-off valves are required and the *operation and maintenance* cost is low.
- Pipe laying is simple

Disadvantages
- The system is less successful in maintaining satisfactory pressure in remote areas and is therefore not favored in modern waterworks practice.
- One main pipeline provides the entire city, which is quite risky.
- The head loss is relatively high, requiring larger pipe diameter, and/or larger capacities for pumping units. Dead ends at line terminals might affect the quality of water by allowing sedimentation and encouraging bacterial growth due to stagnation. Water hammer could also cause burst of lines. A large number of scour valves are required at the dead ends, which need to be opened periodically for the removal of stale water and sediment.

- The discharge available for firefighting in the streets is limited due to high head loss in areas with weak pressure

14.7.4.2   Grid Iron System

In this system, the main supply line runs through the center of the area and submains branch off in perpendicular directions. The branch lines interconnect the submains. This system is ideal for cities laid out on a rectangular plan resembling a gridiron. The distinguishing feature of this system is that all of the pipes are interconnected and there are no dead ends. Water can reach a given point of withdrawal from several directions, which permits more flexible operation, particularly when repairs are required. Figure 14.23 shows Grid Iron System.

Advantages of the Gridiron Distribution System
- The free circulation of water, without any stagnation or sediment deposit, minimizes the chances of pollution due to stagnation.
- Because of the interconnections water is available at every point with minimum loss of head.
- Enough water is available at street fire hydrants, as the hydrant draws water from the various branch lines.

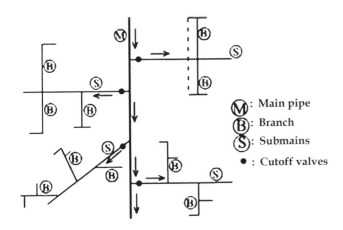

FIGURE 14.22   Dead-end system.

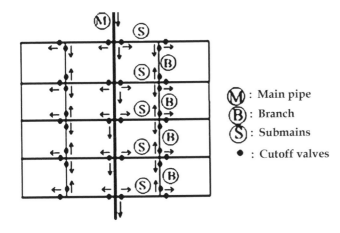

FIGURE 14.23   Grid iron system.

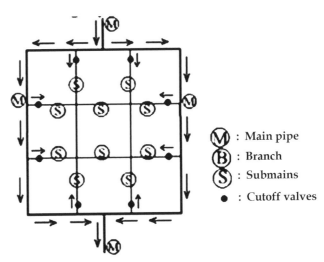

FIGURE 14.24   Ring system.

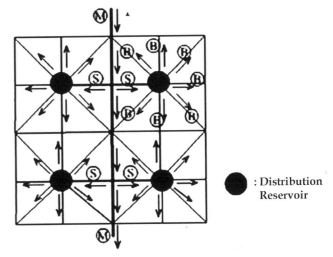

FIGURE 14.25   Radial system.

* During repairs, only a small area of distribution is affected.

Disadvantages of the Gridiron Distribution System
* A large number of cut-off valves are required.
* The system requires longer pipe lengths with larger diameters.
* The analysis of discharge, pressure, and velocities in the pipes is difficult and cumbersome.
* The cost of pipe laying is higher.

14.7.4.3   Circular or Ring System

In a circular or ring system, the supply main forms a ring around the distribution area. The branches are connected cross-wise to the mains and also to each other. This system is most reliable for a town with well-planned streets and roads. The advantages and disadvantages of this system are the same as those of the gridiron system. However, in case of fire, a larger quantity of water is available, and the length of the distribution main is much higher. Ring distribution system is shown in Figure 14.24.

Advantages
1. Water can be supplied to any point from at least two directions.

14.7.4.4   Radial System

In this system, the whole area is divided into a number of distribution districts. Each district has a centrally located distribution reservoir (elevated) from where distribution pipes run radially toward the periphery of the distribution district. This system provides swift service, without much loss of head. The design calculations are much simpler. Radial system is shown in Figure 14.25.

Advantages
1. It gives quick service.
2. Calculation of pipe sizes is easy.

14.7.5   Hydraulic Design

Most common methods used for the hydraulic analysis of distribution systems are as follows:

1. Dead-End method
2. Hardy-Cross method
3. Equivalent pipe method

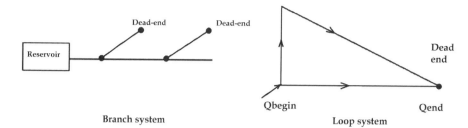

FIGURE 14.26   Dead-end method.

14.7.5.1   Dead-End Method

- Determine the locations of "dead-ends," as shown in Figure 14.26 providing that water will be distributed in the shortest way. At the dead-end points there will be no flow distribution.
- To apply dead-end method for loop systems, convert it to branch system. To do this, a dead-end point is identified for each loop. The location of dead-end point is chosen such that distance travelled to reach dead-end point from two different directions will almost equal to each other. Because in a closed loop

$$\sum \mathrm{HL}_{\text{from one direction}} \cong \sum \mathrm{HL}_{\text{from other direction}} \quad (14.87)$$

- Start calculations from dead-ends to service reservoir.
- Calculate the total flow rate to be distributed $Q_{\max,h} + Q_{\text{fire}}$
- To calculate design flow rate of each pipe,
  - Q distributed,
  - Q begin, and
  - Q end
    should be calculated.

To calculate Q distributed:

- Population density coefficients (k) are calculated from the areas to where water to be distributed. Population density in each area is determined according to the number of stories and their values are as given in Table 14.4.
  Unit of k=population/m length of pipe
- Equivalent pipe lengths are calculated for each pipe:

$$\left(L_{\text{eq}}\right)_i = kL_i \quad (14.88)$$

- Distributed flow in unit pipe length:

$$q = \frac{Q_{\text{total}}}{\sum L_{\text{eq}}} \quad (14.89)$$

TABLE 14.4

Population Density

Number of Story	1	2	3	4	5
One-sided buildings	0.5	1	1.5	1.75	2
Two-sided buildings	1	2	3	3.5	4

- Distributed flow (Q_{dist}) in each pipe:

$$\left(Q_{\text{dist}}\right)_i = q\left(L_{\text{eq}}\right)_i \quad (14.90)$$

To determine Design Flow

A. For the pipes having dead-end:

$$Q_{\text{design}} = 0.577Q_{\text{dist}} + Q_{\text{end}} + Q_{\text{fire}} \quad (14.91)$$

B. For the pipes having no dead-end:

$$Q_{\text{design}} = 0.55Q_{\text{dist}} + Q_{\text{end}} + Q_{\text{fire}} \quad (14.92)$$

- Diameter of each pipe is selected providing that velocity should be in the range.
- Head losses through each pipe are calculated by using Darcy-Weisbach or Hazen-Williams equation.
  - HL calculation according to Darcy-Weisbach:

$$\mathrm{HL} = kQ^2L \quad (14.93)$$

where $k = \dfrac{f}{DA^2 2g}$

  - HL calculation according to Hazen-Williams:

$$\mathrm{HL} = kQ^{1.85}L \quad (14.94)$$

where $k = \left(\dfrac{1}{0.278CD^{2.63}}\right)^{\frac{1}{0.54}}$

- Piezometric elevations and pressures are calculated. To do this, water level in the reservoir and diameter and length of the main line have to be known.

14.7.5.2   Hardy-Cross Method

- This method is applicable to closed-loop pipe networks.
- The outflows from the system are assumed to occur at the nodes (NODE: end of each pipe section). This assumption results in uniform flow in the pipelines.
- The Hardy-Cross analysis is based on the principles that

1. At each junction, the total inflow must be equal to total outflow.

$$\sum Q_{inflow} = \sum Q_{outflow} \text{ (flow continuity criterion)} \quad (14.95)$$

2. Head balance criterion: algebraic sum of the head losses around any closed-loop is zero.

$$\sum HL_{clockwise\ direction} = \sum HL_{counter\ clockwise\ direction} \quad (14.96)$$

- For a given pipe system, with known junction out-flows, the Hardy-Cross method is an iterative pro-cedure based on initially estimated flows in pipes. Estimated pipe flows are corrected with iteration until head losses in the clockwise direction and in the counter clockwise direction are equal within each loop.

Procedure:

1. Outflows from each node are decided.
2. Flows and direction of flows in pipes are estimated by considering the flow continuity condition.

$$\text{At each node: } \sum Q_{inflow} = \sum Q_{outflow} \quad (14.97)$$

3. Decide the sign of flow direction. Usually clockwise direction (+) and counter clockwise direction (−). Use the same sign for all loops.
4. Diameters are estimated for the initially assumed flow rates knowing the diameter, length and rough-ness of a pipe; headloss in the pipe is a function of the flowrate Q.

   Applying Darcy-Weisbach

$$HL = KQ^2 \quad (14.98)$$

where $K = \dfrac{fL}{D}\dfrac{1}{2gA^2}$

   Applying Hazen-Williams

$$HL = KQ^{1.85} \quad (14.99)$$

where $K = \dfrac{L}{\left(0.278CD^{2.63}\right)^{1.85}}$ for SI units

- Formula for flow correction, ΔQ

$$\Delta Q = \dfrac{-\sum HL}{2\sum\left(\dfrac{HL}{Q}\right)} \text{ for Darcy-Weisbach} \quad (14.100)$$

$$\Delta Q = \dfrac{-\sum HL}{1.85\sum\left(\dfrac{HL}{Q}\right)} \text{ for Hazen-Williams} \quad (14.101)$$

TABLE 14.5
Flow Values

Q Initial	ΔQ	Q New
0.1	+0.001	0.1+0.001
−0.2		−0.2+0.001
−0.3		−0.3+0.001
0.4		0.4+0.001

TABLE 14.6
Double Correction Applied to Pipes in Two Loops

	1st Loop	ΔQ₁	2nd Loop	ΔQ₂
Initially	+1	−x	−1	+y
After correction	+1−x−y		−1+y+x	

5. By using ΔQ value, new estimated flows are calcu-lated as per Table 14.5.

   For pipes common in two loops are subjected to double correction as per Table 14.6.
6. Computational procedure is repeated until each loop in the entire network has negligibly small correc-tions (ΔQ).

14.7.5.3 Equivalent Pipe Method

Equivalent pipe is a method of reducing a combination of pipes into a simple pipe system for easier analysis of a pipe network, such as a water distribution system. An equivalent pipe is an imaginary pipe in which the head loss and discharge are equivalent to the head loss and discharge for the real pipe system. There are three main properties of a pipe: diameter, length, and roughness. As the coefficient of roughness, C, decreases the roughness of the pipe decreases. For example, a new smooth pipe has a roughness factor of $C=140$, while a rough pipe is usually at $C=100$. To determine an equiva-lent pipe, any of the above two properties must be assumed. Therefore, for a system of pipes with different diameters, lengths, and roughness factors, one could assume a specific roughness factor and diameter. The most common formula for computing equivalent pipe is the Hazen-Williams formula.

14.8 WATER FOR HYDROELECTRIC GENERATION

14.8.1 HYDROELECTRIC POWER

Hydroelectric power, using the potential energy of rivers, now supplies 17.5% of the world's electricity. Apart from a few countries with an abundance of it, hydro capacity is normally applied to peak-load demand, because it is so readily stopped and started. It is not a major option for the future in the devel-oped countries because most major sites in these countries having potential for harnessing gravity in this way are either being exploited already or are unavailable for other reasons such as environmental considerations.

Hydroenergy is available in many forms, potential energy from high heads of water retained in dams, kinetic energy from current flow in rivers and tidal barrages, and kinetic energy also from the movement of waves on relatively static water masses. Many ingenious ways have been developed for harnessing this energy but most involve directing the water flow through a turbine to generate electricity. Those that do not usually involve using the movement of the water to drive use some other form of hydraulic or pneumatic mechanism to perform the same task.

14.8.2 Types of Hydroelectric Projects

Hydroelectric plants are classified commonly by their hydraulic characteristics, that is, with respect to the water flowing through the turbines that run the generators. Water for hydroelectric plants is generally received from streams, rivers, seas, oceans, etc. Broadly, the following classifications are made as shown in Figure 14.27.

14.8.2.1 Run-of-River Schemes

These are hydropower plants that utilize the stream flow as it comes, without any storage being provided as shown in Figure 14.27a. Generally, these plants would be feasible only on such streams which have a minimum dry weather flow of such magnitude which makes it possible to generate electricity throughout the year. Since the flow would vary throughout the year, they would run during the monsoon flows and would otherwise remain shut during low flows. Of course, the economic feasibility of providing the extra units apart from the regular units has to be worked out. Further, the monsoon tailwater (TW) in rivers with flat slopes becomes higher, causing the plants to become inoperative. Run-of-river plants may also be provided with some storage as shown in Figure 14.27b to take care of the variation of flow in the river as for snow-melt rivers. During off-peak hours of electricity demand, as in the night, some of the units may be closed and the water conserved in the storage space, which is again released during peak hours for power generation. A schematic cross-sectional view of a typical run-of-river scheme is shown in Figure 14.28.

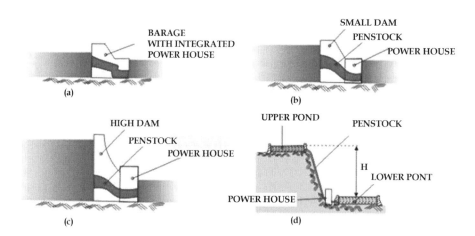

FIGURE 14.27 Types of hydroelectric schemes: (a) run-of-river without pondage; (b) run-of river with pondage; (c) storage schemes; (d) pump-storage schemes.

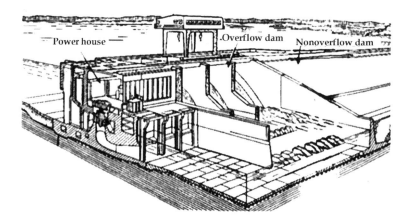

FIGURE 14.28 A typical run-of-river hydroelectric station using a dam and an in-stream power house.

14.8.2.2 Storage Schemes

Hydropower plants with storage are supplied with water from large storage reservoir as shown in Figure 14.27c that has been developed by constructing dams across rivers. Generally, the excess flow of the river during monsoon would be stored in the reservoir to be released gradually during periods of lean flow. Naturally, the assured flow for hydropower generation is more certain for the storage schemes than the run-of-river schemes. A typical schematic cross-sectional view of a storage scheme power plant is shown in Figure 14.29.

14.8.2.3 Pumped-Storage Schemes

Hydropower schemes of the pumped-storage type are those which utilize the flow of water from a reservoir at higher potential to one at lower potential as shown in Figure 14.27d. The upper reservoir (also called the head-water pond) and the lower reservoir (called the tail-water pond) may both be constructed by providing suitable structure across a river as shown in Figure 14.30. During times of peak load, water is drawn from the head-water pond to run the reversible turbine-pump units in the turbine mode. The water released gets collected in the tail-water pond. During off-peak hours, the reversible units are supplied with the excess electricity available in the power grid which then pumps part of the water of the tail-water pond back into the head-water reservoir. The excess electricity in the grid is usually used for the generation of the thermal power plants which are in continuous running mode. However, during night, since the demand of electricity becomes drastically low and the thermal power plants cannot switch off or start immediately, therefore a large amount of excess power is available at that time.

14.8.2.4 Tidal Power Development Schemes

These are hydropower plants which utilize the rise in water level of the sea due to a tide, as shown in Figure 14.31. During high tide, the water from the sea-side starts rising, and the turbines start generating power as the water flows into the bay. As the sea water starts falling during low tide the water from the basin flows back to the sea which can also be used to generate power provided another set of turbines in the opposite direction are installed. Turbines which generate electricity for either direction of flow may be installed to take advantage of the flows in both directions.

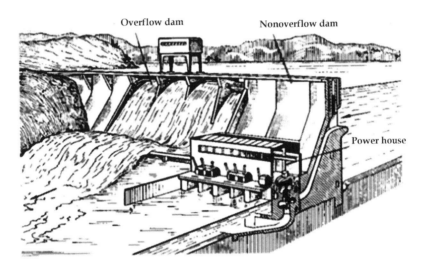

FIGURE 14.29   A typical storage-type hydroelectric station with a power house built at the toe of the dam.

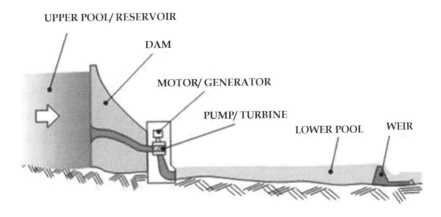

FIGURE 14.30   Pump-storage scheme development with upper and lower pools in the same river.

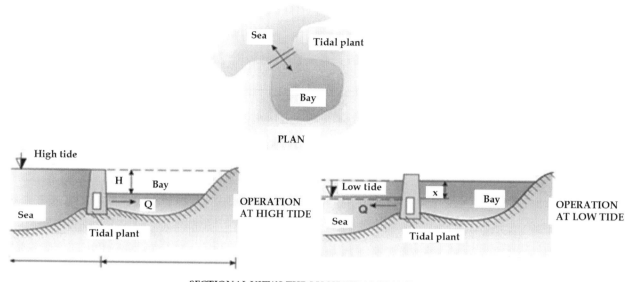

FIGURE 14.31   Tidal power plant scheme.

14.8.3   HYDROPOWER POTENTIAL

Electricity from water is usually referred to as hydropower, where the term "hydro" is the Greek word for water and hydropower is the energy contained in water. It can be converted in the form of electricity through hydroelectric power plants. All that is required is a continuous inflow of water and a difference of height between the water level of the upstream intake of the power plant and its downstream outlet. In order to evaluate the power of flowing water, we may assume a uniform steady flow between two cross-sections of a river, with H (meters) of difference in water surface elevation between two sections for a flow of Q (m³/s), the power (P) can be expressed as

$$P = \gamma Q \left(H + \frac{v_1^2 - v_2^2}{2g} \right) [\text{Nm/s}] \qquad (14.102)$$

where v_1 and v_2 are the mean velocities in the two sections. Neglecting the usually slight difference in the kinetic energy and assuming a value of γ as 9810 N/m², one obtains the expression of power as

$$P = 9810QH \,[\text{Nm/s}] \qquad (14.103)$$

Since energy of 1000 Nm/s can be represented as 1 kW (1 kilo-Watt), the above equation can be written as:

$$P = 9.81QH \,[\text{kW}] \qquad (14.104)$$

The above expression gives the theoretical power of the selected river stretch at a specified discharge.

In order to evaluate the potential of power that may be generated by harnessing the drop in water levels in a river between two points, it is necessary to have knowledge of the hydrology or stream flow of the site, since that would be varying every day. Even the average monthly discharges over a year would vary. Similarly, these monthly averages would not be the same for consecutive years. Hence, in order to evaluate the hydropower potential of a site, the following criteria are considered:

1. Minimum potential power is based on the smallest runoff available in the stream at all times, days, months and years having duration of 100%. This value is usually of small interest.
2. Small potential power is calculated from the 95% duration discharge.
3. Medium or average potential power is gained from the 50% duration discharge.
4. Mean potential power results by evaluating the annual mean runoff.

Since it is not economically feasible to harness the entire runoff of a river during flood (as that would require a huge storage), there is no reason for including the entire magnitude of peak flows while calculating potential power or potential annual energy.

Hence, a discharge-duration curve may be prepared as shown in Figure 14.32 which plots the daily discharges at a location in the decreasing order of magnitude starting from the largest daily discharge observed during the year and going up to the minimum daily discharge.

From this annual discharge curve, a truncation is made at a discharge Q_t which is the discharge corresponding to a time of "t" days, where t can be the median (say, 182 days or 50% duration), denoted by (Q_{182} or $Q_{50\%}$), or a higher Q_t (t less than 182 days) can be selected by specialists who are familiar with the local conditions and future plans for power supply. Accordingly, the annual magnitude of potential (theoretical) energy can be computed in KWh as below and referring to Figure 14.32.

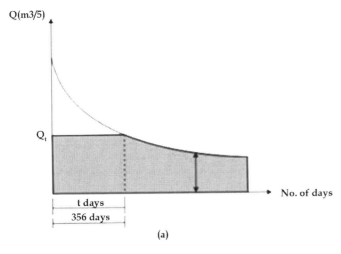 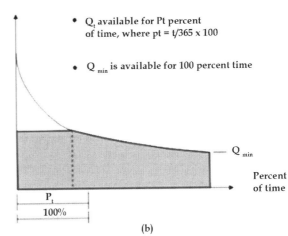

FIGURE 14.32 Flow curve for one year: (a) expressed in time and (b) expressed in percentage of time.

$$E_p = 24 \times 9.81 H \left(Q_t + \sum_{i}^{365} Q_i \right) \qquad (14.105)$$

$$\approx 235 H \cdot A \text{ (in kWh)} \qquad (14.106)$$

where Q_i denotes the daily mean flow during the period 365-t days and A, the hatched area cut by Q_t, where the area under the curve has a unit m³×day/s. The massive influx of water in the hydrologic cycle has an estimated potential for generating, on a continuous basis, 40,000 billion units (TWh) of power annually for the whole world (CBIP, 1992). Hydropower potential is commonly divided into three categories:

a. **Theoretical**: 40,000 TWh
b. **Technical**: 20,000 TWh
c. **Economical**: 9800 TWh

The terms used above are explained below:

a. **Theoretical**

The gross theoretical potential is the sum of the potential of all natural flows from the largest rivers to the smallest rivulets, regardless of the inevitable losses and unfeasible sites.

b. **Technical**

From technical point of view, extremely low heads (less than around 0.5 m), head losses in water ways, efficiency losses in the hydraulic and electrical machines are considered as infeasible. Hence, the technically usable hydropotential is substantially less than the theoretical value.

c. **Economic**

Economic potential is only that part of the potential of more favorable sites which can be regarded as economic compared to alternative sources of power like oil and coal. Economically feasible potential, therefore, would change with time, being dependent upon the cost of alternate power sources. This potential is constantly updated and shows an increasing trend with the exhausting stock of fossil fuel.

Table 14.7 taken from CBIP (1992) shows a continental breakup of world's economical hydropower potential.

Asia is seen to be endowed with the maximum hydropower potential.

Some nations have enough hydropower to become exporters of electricity. Switzerland, for example, exports electricity to neighboring France and Italy. Nepal, Bhutan, Peru, and Laos are similarly blessed with abundant hydro resources. Within India, Meghalay is probably the only state generating hydropower more than its requirements and exports power to the neighboring state of Assam.

14.9 FLOOD CONTROL

A flood is defined as "overflow of inland and tidal waters causing rapid and usual accumulation or runoff of surface water, from any source or a mud flow in a considerable area of land." It is a natural phenomenon associated with hydrological cycle and not an accumulation of large quantities. Flood control refers to all methods used to reduce or prevent the worst effects of flood water. In other words, flood control is the prevention or reduction of the flood damage.

TABLE 14.7

World's Economical Hydropower Potential

Region	Available Potential (Billions Units)
Asia (except CIS and Russia)	2700
CIS and Russia	1100
Africa	1590
North America	1580
South America	1910
Europe (except CIS and Russia)	720
Oceania	200
Total	9800

14.9.1 Causes of Floods

Floods are caused under the influence of both natural and man-made factors which vary from place to place.

14.9.1.1 Natural Causes

The important natural causes of floods are:

- Prolonged rainfall
- High-intensity rain fall
- Meandering passage of rivers
- Landslides and volcanic eruptions
- Blockage of river passage
- High tides, storm surge, or tsunamic coastal area

14.9.1.2 Man-Made Causes

The important man-made causes are:

- Construction work
- Urbanization
- Alteration of river passage
- Construction of clam, bridges, and storage bodies
- Deforestation
- Farming

Human activity in the coastal low lands and the river valleys can intensify the hazard of flooding, such activities include:

- Drained declaimed wetland for agriculture
- Removal of sand from beach
- Ground water extraction and mining of gas and oil
- Destroying natural vegetation in muddy coast lines
- Destroying or damage of coral reels and
- Deforestation and soil erosion

14.9.2 Effects of Floods

14.9.2.1 Primary Effects

The primary effects include physical damage to structures, including bridges, buildings, and sewerage systems. It can endanger the lives of humans and other species.

14.9.2.2 Secondary Effects

i. **Water supplies**: Floods can cause contamination of water and therefore clean drinking water will become scarce,
ii. **Diseases**: It can lead to unhygienic conditions and spread of water borne diseases,
iii. **Crops and food supplies**: Shortage of food crop can be caused due to loss of entire harvest,
iv. **Trees**: Nontolerant species may die from suffocation from floods.
v. **Transport**: Transport links may be destroyed from floods. As a result it will be so hard to get emergency aid to those who need it.

14.9.2.3 Tertiary and Long-Term Effects

Economic: Economic hardship due to temporary decline in tourism, rebuilding costs, food shortage leading to price increase, etc., may be caused due to floods.

Psychological: Flooding can be highly traumatic for individuals, in particular where deaths, serious injuries, and loss of property occur.

14.9.2.4 Benefits of Floods

Though there are many disruptive effects of flooding on human settlements and economic activities, it can sometimes bring benefits also. It can make soil more fertile and provide nutrients in which it is deficient. The viability for hydrologically based renewable sources of energy is higher in flood-prone regions.

14.9.3 Methods of Flood Control

14.9.3.1 Dams

Many dams and their associated reservoirs are designed completely or partially to aid in flood protection and control. Reservoirs are one of the most direct methods of flood control through storing surface runoff; thus, attenuating flood waves and storing flood water to be redistributed without exceed in downstream flood conditions. Many large dams have flood-control arrangements in which the level of reservoir must be kept below a certain elevation before the onset of the rainy/summer melt season to allow a certain amount of space in which floodwaters can fill. The term dry dam refers to a dam that serves purely for flood control without any conservation storage. For flood control, it is ideal to maintain the reservoir at the lowest level possible for storage. Keeping the reservoir at a high level provides the ability to maintain low flows and hydropower production in droughts.

14.9.3.2 Water-Gate

The Water-Gate Flood barrier is a rapid response barrier which can be rolled out in a matter of minutes. It is unique in the way that it self deploys using the weight of water to hold it back. The product has been FM Approved following testing from the US Army. It is used in 30 countries around the world, and notably by the Environment Agency in the UK.

14.9.3.3 Diversion Canals

Floods can be controlled by redirecting excess water to purpose-built canals or floodway, which in turn divert the water to temporary holding ponds or other bodies of water where there is a lower risk or impact to flooding.

14.9.3.4 Self-Closing Flood Barrier

The self-closing flood barrier is a flood defense system designed to protect people and property from inland waterway floods caused by heavy rainfall, gales, or rapid melting snow. It can be built to protect residential properties and whole communities, as well as industrial or other strategic areas. The barrier system is constantly ready to deploy in a flood situation, it can be installed in any length and uses the rising flood water to deploy.

14.9.3.5 Coastal Defenses

Coastal flooding has been addressed with coastal defenses, such as sea walls, beach nourishment, barrier islands, etc. Tides gates are used along with culverts and dykes which are placed at the mouth of streams or small rivers, where an estuary begins or where tributary streams or drainage ditches connect to sloughs. Tide gates close during incoming tides to prevent tidal waters from moving upland, and open during outgoing tides to allow waters to drain out via the culvert and into the estuary side of the dike. The opening and closing of the gates is driven by a difference in water level on either side of the gate.

14.9.3.6 River Defenses

In many countries, rivers are prone to floods and are often carefully managed. Defenses such as levees, reservoirs, and weirs are used to prevent rivers from bursting their banks. When these defenses fail, emergency measures such as sandbags, hydrosacks, or portable inflatable tubes are used.

14.9.3.7 Levees

Levees are one of the oldest forms of river flood control used to protect people and their property from damaging flood waters. They consist of earthen embankments built between the river and the area to be protected. Levees restrict the flood water's flow to the river side of the levee. This will increase the stage in the river. It should be built well outside the meander width of rivers prone to lateral migration.

The elevation of the levee crest is determined through careful consideration of many factors:

- Cost-risk analysis of the cost to build the larger levee versus the risk of damage brought on by a larger flood.
- Political, social, military, and environmental reasons will all play a part in determining a levee's design height.
- The cost to double the size of a levee is most often correlated to the square of the material needed. This means the cost can be quite high and often prohibit increased levee design elevations.

14.10   STORM SEWERS AND DETENTION

14.10.1   Storm Sewer

The storm sewer or storm drain or surface water drain/sewer is a system designed to carry rainfall runoff and other drainage. It is not designed to carry sewage or accept hazardous wastes. The runoff is carried in underground pipes or open ditches and discharges untreated into local streams, rivers and other surface water bodies. Storm sewer inlets are typically found in curbs and low-lying outdoor areas. Some older buildings have basement floor drains that connect to the storm sewer system.

Disposal of chemicals or hazardous substances to the storm sewer system damages the environment. Motor oil, cleaners, paints, and other common household items that get into storm drains can poison fish, birds, and other wildlife, and can find their way into drinking water supplies. In addition, grass clippings, leaves, litter, and organic matter can clog storm drains and cause flooding.

Storm sewer systems may be small residential or large municipal systems that allow the flow of water runoff away from developed areas. The components of storm sewer and their functions are as follows.

14.10.1.1   Inlet

There are mainly two types of storm sewer inlets: side inlets and grated inlets.

Side inlets are located adjacent to the curb and rely on the ability of the opening under the backstone or lintel to capture flow. They are usually depressed at the invert of the channel to improve capture capacity.

Many inlets have gratings or grids to prevent people, vehicles, large objects, or debris from falling into the storm drain. Grate bars are spaced so that the flow of water is not impeded, but sediment and many small objects can also fall through. However, if grate bars are too far apart, the openings may present a risk to pedestrians, bicyclists, and others in the vicinity. Storm drains in streets and parking areas must be strong enough to support the weight of vehicles, and are often made of cast iron or reinforced concrete.

14.10.1.2   Catch Basin

Some of the heavier sediment and small objects may settle in a catch basin, or sump, which lies immediately below the outlet, where water from the top of the catch basin reservoir overflows into the sewer proper. The catch basin serves much the same function as the "trap" in household waste water plumbing in trapping objects. The performance of catch basins at removing sediment and other pollutants depends on the design of the catch basin (for example, the size of the sump), and on routine maintenance to retain the storage available in the sump to capture sediment. Catch basins act as a first-line pretreatment for other treatment practices, such as retention basins, by capturing large sediments and street litter from urban runoff before it enters the storm drainage pipes.

14.10.1.3   Piping

Pipes can come in many different cross-sectional shapes (rectangular, square, bread-loaf-shaped, oval, inverted pear-shaped, egg shaped, and most commonly, circular). Pipes made of different materials can also be used, such as brick, concrete, high-density polyethylene or galvanized steel. Fiber reinforced plastic is being used more commonly for drain pipes and fittings.

14.10.1.4   Outlet

Most drains have a single large exit at their point of discharge (often covered by a grating) into a canal, river, lake, reservoir, sea, or ocean. Small storm sewers may discharge into individual dry wells. Storm sewers may be interconnected using slotted pipe, to make a larger dry well system. Storm sewers may also discharge into man-made excavations known as recharge basins or retention ponds.

14.10.2   Environmental Impacts of Storm Sewers

14.10.2.1   Water Quantity

Storm sewers are often unable to manage the quantity of rain that falls during heavy rains or storms. When storm sewers are inundated, basement and street flooding can occur. Unlike catastrophic flooding events, this type of urban flooding occurs in built-up areas where man-made drainage systems are prevalent. Urban flooding is the primary cause of sewer backups and basement flooding which can affect properties year after year.

14.10.2.2 Water Quality

The first flush from urban runoff can be extremely dirty. Storm water may become contaminated while running down the road or other impervious surface or from lawn chemical runoff, before entering the drain. They tend to pick up oils, fertilizers, pesticides, heavy metals, etc. Separation of undesired runoff can be achieved by installing devices within the storm sewer system. They are referred to as oil-grit separators or oil-sediment separators. They consist of a specialized manhole chamber, and use the water flow and/or gravity to separate oil and grit.

14.10.2.3 Reducing Stormwater Flows

Runoff into storm sewers can be minimized by including sustainable urban drainage system or low impact development or green infrastructure practices into municipal plans. To reduce stormwater from rooftops, flows from eaves troughs (rain gutters and downspouts) may be infiltrated into adjacent soil, rather than discharged into the storm sewer system. Storm water runoff from paved surfaces can be directed to unlined ditches before flowing into the storm sewers, again to allow the runoff to soak into the ground. Permeable paving materials can be used in building sidewalks, driveways, and, in some cases, parking lots, to infiltrate a portion of the stormwater volume.

In many areas, detention tanks are required to be installed inside a property and are used to temporarily hold rainwater runoff during heavy rains and restrict the outlet flow to the public sewer. This reduces the risk of the public sewer being overburdened during heavy rain. An overflow outlet may also be utilized which connects higher on the outlet side of the detention tank. This overflow would prevent the detention tank from completely filling up. By restricting the flow of water in this way and temporarily holding the water in a detention tank public sewers are far less likely to become surcharged.

14.10.2.4 Mosquito Breeding

Catch basins are commonly designed with a sump area below the outlet pipe level which is a reservoir for water and debris to help prevent the pipe from clogging. Unless they are constructed with permeable bottoms to allow water to infiltrate into the underlying soil, this subterranean basin can become a perfect mosquito breeding area because it is cool, dark, and retains stagnant water for long periods of time. Combined with standard grates which have holes large enough for mosquitoes to enter and leave the basin, this is a major problem in mosquito control.

Basins can be filled with concrete up to the pipe level to prevent this reservoir from forming. Without proper maintenance, the functionality of the basin is questionable, as these catch basins are most commonly not cleaned annually as is needed to make them perform as designed. The trapping of debris serves no purpose because once filled they operate as if no basins were present, but continue to allow a shallow area of water retention for the breeding of mosquito. Moreover, even if cleaned and maintained, the water reservoir remains filled, accommodating the breeding of mosquitoes.

14.10.3 Hydraulic Design of Storm Sewers

Storm water is collected from streets into the link drains, which in turn discharge into main drains of open type. The main drain finally discharges the water into open water body. As far as possible gravity discharge is preferred, but when it is not possible, pumping can be employed. While designing, the alignment of link drains, major drains, and sources of disposal are properly planned on contour maps. The maximum discharge expected in the drains is worked out. The longitudinal sections of the drains are prepared keeping in view the full supply level (FSL) so that at no place it should go above the natural surface level along the length. After deciding the FSL line, the bed line is fixed (i.e., depth of drain) based on following consideration.

a. The bed level should not go below the bed level of source into which storm water is discharged.
b. The depth in open drain should preferably be kept less than man height.
c. The depth is sometimes also decided based on available width.
d. The drain section should be economical and velocities generated should be nonsilting and nonscouring in nature.

The drain section is finally designed using Manning's formula.

$$v = \frac{1}{n} r^{2/3} s^{1/2} \qquad (14.107)$$

where

v = velocity of flow in the sewer, m/s
r = hydraulic mean depth of flow, m

$$r = a/p$$

a = cross-sectional area of flow, m^2
p = wetted perimeter, m
n = rugosity coefficient; depends upon the type of the channel surface, i.e., material, and lies between 0.011 and 0.015. For brick sewer, it could be 0.017 and 0.03 for stone facing sewers.
s = hydraulic gradient, equal to invert slope for uniform flows.

Adequate free board is provided over the design water depth at maximum discharge. Minimum of 0.3 m free board is generally provided in storm water drains.

14.10.4 STORMWATER DETENTION

Stormwater runoff overwhelms city sewers and can damage nearby streams and rivers through erosion. Handling the stormwater near its source can avoid costly repairs that would otherwise be directed at correcting erosion or controlling flooding. A common method for managing stormwater is to build a basin. Basins are meant to collect stormwater and slowly release it at a controlled rate so that downstream areas are not flooded or eroded. While effective for flood control, these practices have significant limitations for water quality treatment and for preventing impacts to stream systems. There are two main types of basins—detention and retention.

The main difference between a detention basin and retention basin is the presence or absence of a permanent pool of water, or pond. The water level is controlled by a low flow orifice. In most cases, the orifice is part of a metal or concrete structure called a riser. A detention, or dry, pond has an orifice level at the bottom of the basin and does not have a permanent pool of water. All the water runs out between storms and it usually remains dry. A retention basin or pond has a riser and orifice at a higher point and therefore retains a permanent pool of water. A retention pond looks like a regular pond but plays an important role in controlling stormwater runoff.

The basins are important for storing and slowing stormwater runoff from nearby areas, especially areas with asphalt or concrete development. Stormwater runoff flows much faster from these surfaces than naturally occurring areas and needs to be diverted to ensure the runoff occurs at the desired rate. The amount of cleaning and treatment of the water is limited. Dry basins, or detention basins, only control flood flows. A retention pond can also provide some water quality benefits by reducing pollutants and sediments.

14.10.4.1 Dry Detention Basins

Dry detention ponds are best used in areas where there are ten or more acres of land. On smaller sites, it is difficult to control water quality and other options may be more appropriate.

Dry detention ponds generally use a very small slope to divert water. The inlet needs to be not more than 15% higher than the outlet to ensure the correct amount of water flow through the system. The system works by allowing a large collection area, or basin, for the water. The water then slowly drains out through the outlet at the bottom of the structure. Sometimes concrete blocks and other structures act as a deterrent to slow the water flow and collect extra debris.

Advantages
i. Surrounding areas have vegetative buffer that can withstand dry or wet conditions.
ii. May cost less to implement than a wet retention pond because the size is generally smaller.

Disadvantages
i. Requires a large amount of space.
ii. Does not improve water quality.

iii. Can become a mosquito breeding ground.
iv. Can detract from property value, whereas retention ponds may add value.

14.10.4.2 Wet Retention Ponds

Wet retention ponds are a stormwater control structure that provides retention and treatment of contaminated stormwater runoff. By capturing and retaining stormwater runoff, wet retention ponds control stormwater quantity and quality. The ponds' natural processes then work to remove pollutants. Retention ponds should be surrounded by natural vegetation to improve bank stability and improve aesthetic benefits.

Water is diverted to a wet retention pond by a network of underground pipes connecting storm drains to the pond. The system allows for large amounts of water to enter the pond, and the outlet lets out small amounts of water as needed to maintain the desired water level.

From a health standpoint, there is always a concern with standing water. This can be a drowning hazard, particularly with children. Ponds can also draw mosquitoes, which may contribute to the transmission of some diseases.

Advantages
i. Retention ponds are simple if space is provided.
ii. Collects and improves water quality.
iii. Naturally processes water without additional equipment.
iv. Improved stormwater collection and flood control.
v. New habitats are created.
vi. Can be used for recreational purposes.

Disadvantages
i. Can be a drowning hazard.
ii. Large areas of land are needed.
iii. Negative water quality impacts if not properly designed.

14.10.4.3 Maintenance Considerations

One of the most important maintenance needed for either of these basins is to ensure that the orifice does not become blocked or clogged. Keeping the pipes clear of debris will ensure the ponds and basins are functioning properly. Keeping up with maintenance can reduce costly repairs in the future. Other maintenance includes:

i. **Identifying and repairing areas of erosion**: A few times a year and after major storms, check for gullies and other disturbances on the bank.
ii. **Removing sediment and debris**: Keeping pipes clear of debris and removing sediment ensures proper function. Remove debris around and in ponds before it reaches the outlets to prevent problems.
iii. **Maintaining vegetation**: The amount of maintenance depends on the type of vegetation surrounding the basin. Some grasses need weekly mowing, and others can be maintained a couple of times a year.

14.11 STREET AND HIGHWAY DRAINAGE AND CULVERTS

14.11.1 STREET AND HIGHWAY DRAINAGE

Urban streets not only carry traffic, but stormwater runoff as well. However, water on a street can create hydroplaning effects, and can severely impact the traffic flow and the safety of travelers. It can also adversely affect the engineering properties of the materials with which it was constructed. Provision for adequate drainage is of paramount importance in road design and cannot be overemphasized. Cut or fill failures, road surface erosion, and weakened subgrades followed by a mass failure are all products of inadequate or poorly designed drainage.

For these reasons, a street drainage system must be properly designed to quickly remove stormwater from the traffic lanes. Drainage design is most appropriately included in alignment and gradient planning. Street and highway drainage design mainly includes collecting, transporting, and disposing of surface/subsurface water originating on or near the highway right of way or flowing in streams crossing bordering that right of way. The water which are dangerous for highways are as follows:

- **Rainwater**: Cause erosion on surface or may seep downward and damage pavement (surface drains)
- **Groundwater**: May rise by capillary action and damage pavement (subsurface damage)
- **Water body**: May cross a road (river/stream) and may damage road (cross drainage works)

It is more appropriate to take care of drainage at the time of location survey. Ideal location for a drainage stand point would lie along the divides between large drainage areas. Then all streams flow away from the highway, and the drainage problem is reduced to caring for the water that falls on the roadway and back slope. In contrast, location paralleling large streams is far less desirable as they cross every tributary where it is largest. Ideal locations avoid steep grades and heavy cuts and fills as they create difficult problems in erosion control.

Street drainage includes both minor and major drainage systems. A minor system consists of street inlets and storm sewers, which can handle minor storm events. During a major storm event, street gutters and roadside ditches operate like wide and shallow channels to carry the flooding water away.

14.11.1.1 Draining the Roadway and Road Side

The highway engineer should ensure that the precipitation is removed from the pavement as soon as possible and that highway drainage is done efficiently. Water that falls on the roadway follows laterally or obliquely from it, under the influence of cross slope or super elevation in pavement and shoulder. A suitable value of cross fall for paved roads is about 3%, and for carriage way with a slope of 4%–6% for shoulders. An increased cross-fall for the carriage way, e.g., 4% is desirable if the quantity of the final shape of the road surface is likely to be low for any reason.

14.11.1.2 Drainage within Pavement Layers

It is an essential element of structural design because the strength of the subgrade used for design purposes depends on the moisture content during the most likely adverse conditions. It is evident that benefits are derived from applying steeper cross falls to layers at successive depths in the pavement.

The top of the subbase should have a cross-fall of 3%–4% and the top of the subgrade should be 4%–5%. These cross-falls not only improve the drainage performance of the various layers but also provide a slightly greater thickness of material at the edge of pavement where the structure is more vulnerable to damage. The design thickness should be maximum at the centerline of the pavement.

14.11.1.3 Road Way Drainage in Fill

The most common practice is to let the flow continue on the shoulder and down the slope to the natural ground. Little erosion would occur, if the slopes are protected by turf or if the water flows across the roadway and down the slope as a sheet. The unprotected slopes wash away badly and irregularities in shoulder or pavement concentrate water into small streams causing erosion. One way of preventing washing of side slope is to retain the water at the outer edge of shoulder.

14.11.1.4 Highway Drainage of Runoff in Cut

Water from road way and back slope is collected in road side channel, which may be trapezoidal or triangular, the design of which is based on how much water is to be accommodated. Intercepting channel (sometimes called crown ditch) may be employed at the top of cut slope.

The advantages of having highway drainage in cut are that it prevents erosion of the back slope by runoff from the hill above and also intercepts water, not allowing it to enter side drain which may cause greater discharge in side drains.

14.11.1.5 Road-Way Drainage in Urban Areas

Water falling on the road surface generally flows along the gutter to curbs or gutter inlets and then to underground storm drains. They are expensive as compared to rural area drainage works. But considering the large volume of traffic, pedestrian's property, etc., they are inevitable. They are designed to limit the spread of water over the traveled lanes to some arbitrary maximum. The inlets at low points should be designed for longer return period.

14.11.1.6 Highway Runoff Drainage in Rural Areas

Generally, open unlined drains with suitable cross-section and longitudinal slope are provided parallel to road alignment, called longitudinal drains. In embankment, they are provided on one or both side beyond the toe. In cutting area, it is installed on either side of formation.

Construction of deep open drains may be undesirable due to restriction of space. In such cases, covered drains or drainage trenches properly filled with layers of coarse sand and gravel may be used.

14.11.1.7 Cross-Drainage Structures and Works

When a low laying area or a stream or a river crosses the alignment of road, arrangements should be made to allow the water of stream or river to pass onto the other side of road. The water is passed by structures known as cross drainage works. These include road culverts, bridge, and cause ways.

i. **Culverts**:

It encompasses practically all closed conduits employed for highway drainage with the exception of storm drains. It can be defined as

a. A drain sewer that is totally enclosed and usually of a size through which a man can pass.

b. An opening through an embankment for the conveyance of water by means of pipe or an enclosed channel.

Culverts are usually installed in the original stream bed with their grades confirming to those of natural channel. In this way, distribution to stream flow and the erosion problem it creates are held to minimum.

Common culvert types are:
- Pipe culvert
- Arch pipe culvert
- Box culvert
- Bridge culvert
- Arch culvert

ii. **Bridges**:

Bridges are used in runoff drainage systems where the stream span is large, for which special designs are made almost in every case greater than 6 m.

iii. **Cut-off walls**:

They extend below the level of expected scour to reduce seepage beneath and adjacent to the dam.

iv. **Dips or cause ways in Highway drainage**:

A dip is formed by lowering the roadway grade to the level of the stream from the bank to bank of the stream. Vertical curves at each end transition back to the regular grade line. Washing of the roadway surface is prevented by curtain wall of concrete rubble masonry. With proper design it is damaged little by flood water, so that the maintenance cost is low. With long transition at ends they ride smoothly. One disadvantage of dips is that interruptions and hazard may occur to traffic when the dip is flowing.

v. **A dip culvert combined**:

High-level causeway is sometimes employed for better advantage. Partially lowered pipe culverts under the road surface at stream bed level carry small flows without inconvenience to traffic. The larger waterway capacity of the dip comes into play during major floods.

14.11.2 CULVERTS

A culvert is a transverse and fully enclosed drainage structure that runs under a road or portion of land. The size and type of culvert depends on the amount of water flowing, the area that is discharging to it and how deep the culvert is being installed. Some culverts can also serve as roadway surfaces but they will always serve to convey water through a pipe or channel. Generally selection and type of material depends on the comparative cost, location of structure, availability of skilled labor, time limitations, and design being proposed.

14.11.2.1 Materials Used for Culvert Construction

Culverts are like pipes but very large in size. They are made of many materials like

- Concrete (reinforced and nonreinforced)
- Steel (smooth and corrugated)
- Corrugated aluminum
- Vitrified clay
- Plastic
- Bituminous fiber
- Cast iron
- Wood
- Stainless steel

In most cases concrete culverts are preferred. Concrete culverts may be reinforced or nonreinforced. In some cases culverts are constructed in site called cast in situ culverts. Precast culverts are also available. By the combination of above materials, composite culverts are also made.

14.11.2.2 Location of Culverts

The location of culverts should be based on economy and usage. Generally, it is recommended that the provision of culverts under roadway or railway is economical. There is no need to construct separate embankment or anything for providing culverts. The culverts should be perpendicular to the roadway. The culverts should be of greater dimensions to allow maximum water level (MWL). The culvert should be located in such a way that flow should be easily done. It is possible by providing required gradient.

14.11.2.3 Types of Culverts

Following are the types of culverts generally used in construction:

- Pipe culvert (single or multiple)
- Pipe arch (single or multiple)
- Box culvert (single or multiple)
- Arch culvert
- Bridge culvert

14.11.2.4 Hydraulic Design Considerations for Culverts

1. Design Flood Discharge
   - Watershed characteristics
   - Design flood frequency or return interval
   - All designs should be evaluated for flood discharges greater than the design flood
2. Headwater (HW) Elevation: check upstream water surface elevation

3. TW: check that outlet will not be submerged, and
4. Outlet Velocity: usually controlled by barrel slope and roughness

14.11.2.5 Terminology

i. **HW**: Depth from the culvert inlet invert to the energy grade line. If the approach velocity head is small then HW is approximately the same as the upstream water depth above the invert.
ii. **TW**: Depth of water on the downstream side of the culvert. The TW depends on the flow rate and hydraulic conditions downstream of the culvert.

14.11.2.6 Types of Flow Control

1. **Inlet control**: Flow capacity is controlled at the entrance by the depth of HW and entrance geometry, including the barrel shape, cross-sectional area and the inlet edge.
2. **Outlet control**: Hydraulic performance controlled by all factors included with Inlet Control and additionally include culvert length, roughness, and TW depth.

14.11.2.6.1 Inlet Control

There are two possible conditions:

i. **Unsubmerged**: Steep culvert invert and HW not sufficient to submerge inlet. Culvert inlet acts effectively like a weir.

$$Q = C_w B (\mathrm{HW})^{3/2} \qquad (14.108)$$

where
B=width of weir crest
C_w=weir coefficient=3.0 may be assumed for initial calculations.

ii. **Submerged**: HW submerges top of culvert inlet but the barrel does not necessarily flow full. Culvert inlet acts like an orifice or sluice gate.

$$Q = C_d A \sqrt{2g(\mathrm{HW} - b/2)} \qquad (14.109)$$

where
b=culvert height
HW−b/2=head on culvert measured from barrel centreline
    Orifice discharge coefficient, C_d, varies with head on the culvert, culvert type, and entrance geometry. Nomographs and computer programs are usually. For initial calculations, a value C_d=0.60 may be used.

14.11.2.6.2 Outlet Control

Outlet control will govern if the HW is deep enough, the culvert slope is sufficiently flat, and the culvert is sufficiently long.

There are three possible flow conditions:

1. Both inlet and outlet submerged, with culvert flowing full.
2. Inlet is submerged but the TW does not submerge the outlet. In this case the barrel is full over only part of its length.
3. Neither the HW nor TW depths are sufficient for submergence.

Culvert capacity determined from energy equation:

$$\mathrm{HW} + S_0 L = \mathrm{TW} + h_e + h_f + h_v \qquad (14.110)$$

where
    HW−TW=headwater − tailwater=total energy head loss (ft)
    h_e = entrance head loss (ft)
    h_f = friction losses (ft)
    h_v=velocity head (ft)

Entrance Head Loss, h_e

$$h_e = K_e \frac{V^2}{2g} \qquad (14.111)$$

where
    K_e=Culvert Entrance Loss Coefficients

Pipe with projecting square-edged entrance −0.5
    Pipe mitered to conform to fill slope −0.7
    Box with wingwalls at 30°–75° to barrel −0.4

Friction Losses, h_f
    Manning's Equation

$$V = \frac{1.49}{n} R_h^{2/3} S^{1/2} \qquad (14.112)$$

or

$$h_f = 29 \frac{n^2 L}{R_h^{4/3}} \left(\frac{V^2}{2g} \right) \qquad (14.113)$$

Design equation for Case 1:

$$\mathrm{HW} = \mathrm{TW} + \left[K_e + 29 \frac{n^2 L}{R_h^{4/3}} + 1 \right] \left(\frac{V^2}{2g} \right) - S_0 L \quad (14.114)$$

Other cases are more difficult, and all require iteration.

Roadway Overtopping
    During roadway overtopping, the roadway acts as a weir:

$$Q = C_w L (\mathrm{HW}_r)^{3/2} \qquad (14.115)$$

HW_r = upstream depth, measured above roadway crest (ft).

14.11.2.7 Culvert Design Procedure

1. Establish design data – Q, L, S_0, HW_{max}, V_{max}, culvert material, cross-section, and entrance type.
2. Determine first trial size culvert (arbitrary, $A = Q/10$, etc.).
3. Assuming INLET CONTROL, determine HW depth HW. For unsubmerged and submerged conditions, use the respective equations.
4. Assuming OUTLET CONTROL, determine HW depth HW using the energy equation.
   i. Determine TW depth TW for downstream control (for uniform flow use Manning's equation).
   ii. If TW depth $> b$ (height of culvert) set $h_0 = $ TW.
   iii. If TW $< b$, set $h_0 = \dfrac{b + y_c}{2}$ or TW, whichever is greater.
   iv. Calculate the energy loss through the culvert, $H = h_e + h_f + h_v$.
      a. For full culvert flow use
      $$H = \left[K_e + 29\frac{n^2 L}{R_h^{4/3}} + 1 \right]\left(\frac{V^2}{2g} \right)$$
      b. For partially full culvert use the direct step method to determine the water level and energy losses along the culvert length:
      $$\Delta x = \frac{E_2 - E_1}{S_0 - S_f}, \text{ where } E = y + \frac{Q^2}{2gA^2} \text{ which}$$
      is equal to $y + \dfrac{q^2}{2gy^2}$ for box culverts. If y becomes $> b$, use the full culvert formula for the remaining distance, but neglect the velocity head term. Calculate the HW HW from $HW = E + H - S_0\Delta L$, where E is the specific energy at the location where the culvert becomes full and ΔL is the length of full-flowing culvert.
5. Compare HW values from steps 3 and 4. The higher HW governs and indicates the flow control existing under the specified conditions for the trial calculation.
6. Try alternate sizes and characteristics and repeat steps 3 and 4 until design specifications are met.
7. Compute outlet velocity assuming area based on TW, y_c, or y_n, as appropriate.

14.12 DESIGN OF SPILLWAYS AND ENERGY DISSIPATION FOR FLOOD CONTROL STORAGE

14.12.1 Spillways

A spillway is a structure designed to "spill" flood waters under controlled conditions. It is a provision for storage and detention dams to release surplus water or floodwater that cannot be contained in the allotted storage space. This action prevents overtopping, which can be particularly destructive in the case of earth fill and rock fill dams, both of which can fail completely when overtopped. The Spillways can be uncontrolled (normally) or controlled. Concrete dams normally incorporate an over-fall or crest spillway, but embankment dams generally require a separate side-channel or shaft spillway structure located adjacent to the dam.

14.12.2 Spillway Components

Spillways are generally made up of four components: a control structure, discharge channel, terminal structure, and entrance/outlet channels.

Control structures regulate the flows from the reservoir into the spillway, ensuring that flow will not enter the spillway until the water in the reservoir reaches the designed level, and moderating flow into the spillway once the design level has been reached. Control structures can be sills, weirs, orifices, or pipes.

Discharge channels, also known as waterways, convey flow that passes through the control structure down to the streambed below the dam. The conveyance structures may not always be present in a spillway design; at times, discharge may fall freely after passing through the control structure.

Terminal structures ensure that the flow, which oftentimes acquires a high velocity while traveling down a spillway, will not cause excessive erosion to the toe of the dam, or any other nearby structures. Plunge basins, flip buckets, and deflectors are all examples of terminal structures.

Entrance channels convey water from a reservoir to the control structure. Outlet channels convey flow that has reached the terminal structure to the river channel that resides below the dam. Entrance and outlet channels are not necessarily a component of all spillways; it is possible for the spillway to transport flow directly from the reservoir to the river channel.

14.12.3 Classification of Spillways

I. **According to the most prominent feature**
   - Ogee spillway
   - Chute spillway
   - Side channel spillway
   - Shaft spillway
   - Siphon spillway
   - Straight drop or overfall spillway
   - Tunnel spillway/Culvert spillway
   - Labyrinth spillway
   - Stepped spillway

II. **According to Function**
   - Service spillway
   - Auxiliary spillway
   - Fuse plug or emergency spillway

III. **According to Control Structure**
   - Gated spillway
   - Ungated spillway
   - Orifice of sluice spillway

14.12.4 Various Aspects Involved in a Spillway Design

The following aspects are involved in the design of spillways:

1. Hydrology
   - Estimation of inflow design flood
   - Selection of spillway design flood
   - Determination of spillway outflow discharge
   - Determination of frequency of spillway use
2. Topography and geology
   - Type and location of spillway
3. Utility and operational aspects
   - Serviceability
4. Constructional and structural aspects
   - Cost-effectiveness

14.12.5 Spillway Design Flood

Probable Maximum Flood

This is the flood that may be expected from the most severe combination of critical meteorological and hydrological conditions that are reasonably possible in the region. This is computed by using the Probable Maximum Storm.

Standard Project Flood

This is the flood that may be expected from the most severe combination of hydrological and meteorological factors that are considered reasonably characteristic of the region and is computed by using the Standard Project Storm.

14.12.6 Estimation of Spillway Design Flood

The estimation of spillway design flood or the inflow design flood is an exercise involving diverse disciplines of hydrology, meteorology, statistics and probability. There is a great variety of methods used around the world to determine exceptional floods and their characteristics. The methods can be under the two main categories:

1. Methods based mainly on flow data.
2. Methods based mainly on rainfall data.

14.12.7 Spillway Design

The ogee or overflow spillway is the most common type of spillway. It has a control weir that is Ogee or S-shaped. It is a gravity structure requiring sound foundation and is preferably located in the main river channel.

The basic shape of the overfall (ogee) spillway is derived from the lower envelope of the overall nappe flowing over a high vertical rectangular notch with an approach velocity, $V_0=0$, and a fully aerated space beneath the nappe ($p=p_0$) as shown in Figure 14.33.

14.12.7.1 Discharge Characteristics

Similar to the crest profile, the discharge characteristics of the standard spillway can also be derived from the characteristics of the sharp crested weir. The weir equation in the form:

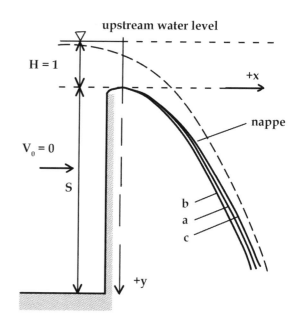

FIGURE 14.33 Overfall spilling shape derivation.

$$Q = C\sqrt{2g}LH_e^{3/2} \qquad (14.116)$$

where

Q = flow rate (m³/s)
C = the dimensionless coefficient of discharge
L = the crest length of perpendicular to the flow (m)
H_e = the total head on the crest, including the approach velocity head, h_a

If the discharge, Q, is used as the design discharge in above equation, then the term H_e will be the corresponding design head (H_d) plus the velocity head (H_a). i.e., $H_e=H_d+H_a$. For high ogee spillways, the velocity head is very small, and $H_e \cong H_d$. Figure 14.34 shows a schematic of flow and related quantities as they pertain to an ogee spillway.

However, if the design head and the actual head acting at the crest are the same, and the ogee spillway under consideration has a vertical face upstream, then the ogee crest coefficient, C_D can replace the C in the equation, leading to the following equation:

$$Q = C_D\sqrt{2g}LH_D^{3/2} \qquad (14.117)$$

Overflow spillways are named as high-overflow, and low-overflow depending upon to the relative upstream depth P/H_D. In high-overflow spillways, this ratio is ($P/H_D > 1.33$) and the approach velocity is generally negligible. Low spillways have appreciable approach velocity, which affects both the shape of the crest and the discharge coefficients. Figure 14.35 gives the variation of discharge coefficient C_D with P/H_D.

Figure 14.36 gives variation of C_D, the value of C when H equals the design head H_D, with the relative upstream depth P/H_D. Here P is the height of the spillway crest with respect to the channel bed.

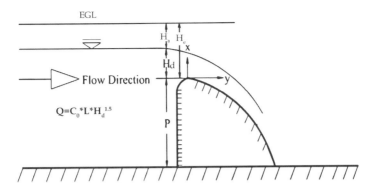

FIGURE 14.34 Schematic of flow and related quantities as they pertain to an ogee spillway.

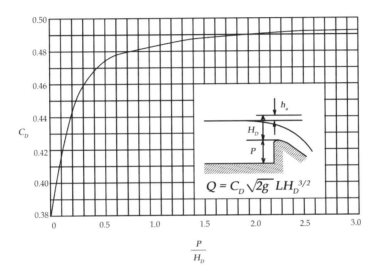

FIGURE 14.35 Discharge coefficient C_D versus P/H_D for design head.

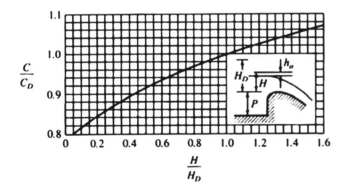

FIGURE 14.36 variation of C/C_D versus H/H_D.

Overflow spillways frequently use undershot radial gates for releases over the dam. The governing equation for gated flows is

$$Q = \frac{2}{3}\sqrt{2g}CL\left(H_1^{3/2} - H_2^{3/2}\right) \qquad (14.118)$$

where C is a coefficient of discharge, and H_1 and H_2 are total heads to the bottom and top of the gate opening. The coefficient C is a function of geometry and the ratio d/H_1, where d

is the gate aperture. Figure 14.37 shows the variation of coefficient of discharge for flow under gates with d/H_1.

14.12.7.2 The Spillway Crest Profile

On the crest shape based on a design head, H_D, when the actual head is less than H_D, the trajectory of the nappe falls below the crest profile, creating positive pressures on the crest, thereby reducing the discharge. On the other hand, with a higher than design head, the nappe-trajectory is higher than crest, which creates negative pressure pockets and results in increased discharge. Figure 14.34 shows the derivation of overfall spillway shape.

In the figure,

a: $p=p_0$, $H_D=H_{max}$; $H=H_D$
b: $p<p_0$, $H_D<H_{max}$; $H>H_D$
c: $p>p_0$, $H_D>H_{max}$; $H<H_D$

For $H_D=H_{max}$, the pressure is atmospheric and $C_D=0.745$. For $H_D>H_{max}$, the pressure on the spillway is greater than atmospheric and the coefficient of discharge will be $0.578<C_D<0.745$. The lower limit applies for broad-crested weirs with $C_D=1/\sqrt{3}$ and is attained at very small values of H_{max}/H_D. For $H_D<H_{max}$, negative pressures result, reaching

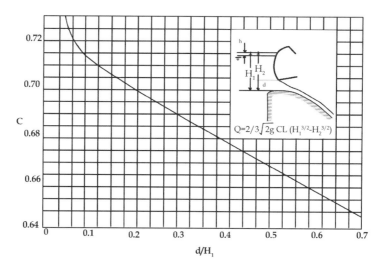

FIGURE 14.37   Variation of coefficient of discharge for flow under gates.

cavitation level for $H = 2H_D$ with $C_D = 0.825$. For safety, it is recommended not to exceed the value $H_{max} \approx 1.65H_D$ with $C_D \approx 0.81$, in which case the intrusion of air on the spillway surface must be avoided, as otherwise the overfall jet may start to vibrate.

Accordingly, it is considered desirable to under-design the crest shape of a high overflow spillway for a design head, H_D, less than the head on the crest corresponding to the maximum reservoir level, H_e ($\sim H_{max}$). However, with too much negative pressure, cavitation may occur.

The coefficient of discharge (or simply discharge) is influenced by a number of factors such as

1. the relation of the actual crest shape to the ideal nappe shape,
2. the depth of approach,
3. the inclination of the upstream face,
4. the contraction caused by the crest piers and abutment,
5. the interference due to downstream apron, and
6. the submergence of the crest due to downstream water level.

It is possible that due to the placement of the piers and abutments of the dam, the flow over the crest of the spillway could be contracted, in which case, the actual length of the crest is not the effective length, L, required for the discharge calculation. In such a situation, effective crest length can be computed using the following equation:

$$L = L' - 2\left(NK_p + K_a\right)H \qquad (14.119)$$

where
L = effective length of crest
L' = net length of crest
N = number of piers
K_p = pier contraction coefficient
K_a = abutment contraction coefficient
H = head on crest

Pier contraction coefficients, K_p, are functions of several factors, including shape and location of the pier's nose, thickness of the pier design head of the spillway, and the velocity of the approaching flow. Average K_p's vary between 0.0 and 0.02 depending on the shape a pier's nose. Average K_a, the abutment contraction coefficient, varies between 0.0 and 0.2, also depending on the shape of the abutments.

14.12.7.3   Spillway Toe

The spillway toe is the junction between the discharge channel and the energy dissipator. Its function is to guide the flow passing down the spillway and smoothly in the energy dissipator. A minimum radius of 3 times the depth of flow entering the toe is recommended. Figure 14.38 shows toe curve at the base of the spillway.

14.12.8   ENERGY DISSIPATION

Dissipation of the kinetic energy generated at the base of a spillway is essential for bringing the flow into the downstream river to the normal condition in as short of a distance as possible. This is necessary, not only to protect the riverbed and banks from erosion but also to ensure that the dam itself and adjoining structures like powerhouse, canal, etc., are not undermined by the high-velocity turbulent flow.

14.12.9   CLASSIFICATION OF ENERGY DISSIPATION

1. **Based on hydraulic action**: Turbulence and internal friction as in hydraulic jump stilling basins, roller buckets, and impact and pool diffusion as with ski-jump buckets and plunge pools.
2. **Based on the mode of dissipation**: Horizontal as in the hydraulic jump, vertical as with ski-jump buckets or free jets, and oblique as with spatial and cross flows. The vertical dissipation may be in the downward direction as with free jets and plunge pools and in upward direction as with roller buckets.

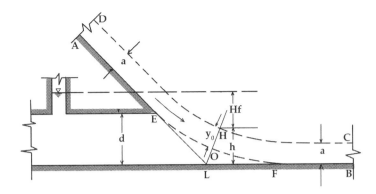

FIGURE 14.38   Toe curve at the base of the spillway.

3. **Based on geometry or form of the main flow**:
   Situations involving sudden expansion, contraction,
   counter acting flows, impact, etc.
4. **Based on the geometry or form of the structure**:
   Stilling basin employs hydraulic jump with or without appurtenances like chute blocks, baffle piers, etc.
   Buckets (ski-jump or flip buckets) include special
   shapes like serrated, dentated buckets, and roller
   buckets that are either solid roller bucket or slotted
   buckets.

14.12.10   Principal Types of Energy Dissipators

The energy dissipators for spillways can be grouped under the
following five categories:

1. Hydraulic jump stilling basins
2. Free jets and trajectory buckets
3. Roller buckets
4. Dissipation by spatial hydraulic jump
5. Impact type energy dissipators

14.12.11   Ananlysis of Parameters

Figure 14.39 shows the key parameters in energy dissipation.
They are as follows:

i. HW rating curve
ii. Energy line for head water
iii. TW rating curve
iv. Energy line for TW

Hence the energy equation can be written as:

$$y_0 + \frac{v_0^2}{2g} = y_1 + \frac{v_1^2}{2g} = y_2 + \frac{v_2^2}{2g} + \Delta E \qquad (14.120)$$

where ΔE=energy dissipation between upstream and
downstream.

The mass conservation equation can be written as:

$$Q_1 = Q_2 = Q_3 \qquad (14.121)$$

In case of hydraulic jump at the downstream
   From Figure 14.40, for a given discharge intensity over a
spillway, the depth y_1 is equal to q/V_1, and V_1 is determined by
drop H_1, being equal to $\sqrt{2gH_1}$.
   Hence

$$V_1 = (2gH_1)^{0.5} \qquad (14.122)$$

$$y_1 = q/V_1 \qquad (14.123)$$

Thus

$$q/y_1 = (2gH_1)^{0.5} \qquad (14.124)$$

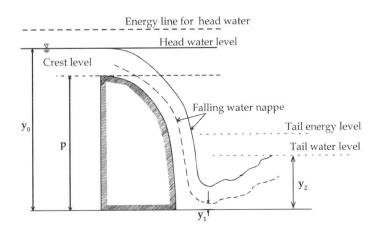

FIGURE 14.39   Key parameters in energy dissipation.

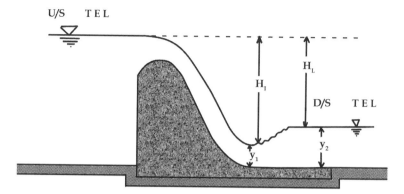

FIGURE 14.40 Spillway discharge.

Therefore,

$$y_2 = -\frac{y_1}{2} + \sqrt{\frac{y_1^2}{4} + \frac{2q^2}{gy_1}}$$ (14.125)

Assuming the bed to be horizontal, the energy dissipation equation can be written as:

$$\Delta E = \left(y_2 + \frac{v_2^2}{2g} \right) - \left(y_1 + \frac{v_1^2}{2g} \right)$$ (14.126)

Hence, for a given discharge intensity and given height of spillway, y_1 is fixed and thus y_2 (required for the formation of hydraulic jump) is also fixed. But the availability of a depth equal to y_2 in the channel on the downstream cannot be guaranteed as it depends upon the TW level, which depends upon the hydraulic dimensions and slope of the river channel at downstream. The problem should, therefore, be analyzed before any solution can be found by plotting the following curves:

i. **TW Curve**: A graph plotted between q and TW depth.
ii. **Jump Height Curve (JH Curve) also called y_2 curve**: A curve plotted on the same graph, between q and y_2.

Figure 14.41 shows the ideal condition for jump formation. In this case, the hydraulic jump will form at the toe of the spillway at all discharges. A simple concrete apron of length equivalent to length of jump is generally sufficient to provide here. Figure 14.42 shows the different situations of TW and JH curves.

Table 14.8 shows the suggested arrangements for the different situations of JH and TW curves.

14.12.12 STILLING BASIN

A stilling basin may be defined, as a structure in which the energy dissipating action is confined. If the phenomenon of hydraulic jump is basically used for dissipating this energy, it may be called a hydraulic jump type of stilling basin.

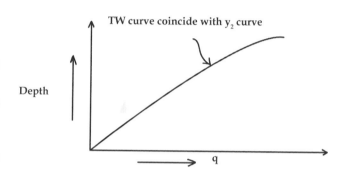

FIGURE 14.41 Ideal condition of jump formation.

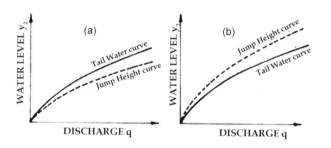

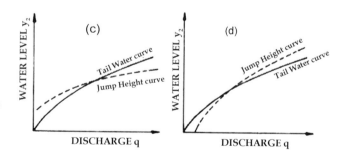

FIGURE 14.42 (a–d) Jump height and TW rating curves.

The auxiliary devices may be used as additional measures for controlling the jump, etc. Stilling basins are placed at the ends of dam spillways and at the ends of steep-sloped canal sections where elevation change has generated high kinetic energy. Stilling basin comes in a variety of types and can either contain a straight drop to a lower elevation or an inclined chute. Inclined chutes are the most common design for stilling basins.

TABLE 14.8

Choice of Energy Dissipator Based on Analysis of JH and TW Rating Curves

Case	Situation	Suggested Arrangement
A	TW curve is above the JH curve for all the discharges	Horizontal apron if $1 < y_1/y_2 < 1.1$ and sloping apron or roller bucket if $y_1/y_2 > 1.1$
B	Tw curve is below the JH curve for all the discharges	Flip bucket
C	TW curve is below the JH curve for lower discharges and above the JH curve for higher discharges	Sloping-cum-horizontal apron such that the jump forms on the horizontal portion for low discharges and on the slopping portion for high discharges
D	TW curve is above the JH curve for lower discharges and below the JH curve for higher discharges	A combination energy dissipator performing as a hydraulic jump apron for low discharges and flip bucket for high discharges.

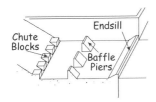

FIGURE 14.43   Standard stilling basin.

14.12.12.1   Elements of Stilling Basin

Figure 14.43 shows a standard stilling basin with all its elements. They are

i. **Chute blocks**: They are concrete blocks built into the inclined sections of the spillway. These features are commonly placed at the head of the stilling basin to create turbulence prior to the hydraulic jump.

ii. **Baffle blocks**: They are freestanding concrete blocks built in the main basin. These blocks are only used for flows less than 20m/s due to the high force they are subjected to and the potential for cavitation.

iii. **End sills or dentated sills**: It is a built-up lip at the tail of the basin, with or without blocks. The sill height has the most significant impact on energy dissipation and taller sills are used to reduce the overall length of the stilling basin.

14.12.13   DESIGN OF HYDRAULIC JUMP STILLING BASIN TYPE ENERGY DISSIPATORS

A hydraulic jump is the sudden turbulent transition of supercritical flow to subcritical. This phenomenon, which involves a loss of energy, is utilized at the bottom of a spillway as an energy dissipator by providing a floor for the hydraulic jump to take place as shown in Figure 14.44. The amount of energy dissipated in a jump increases with the rise in Froude number of the supercritical flow.

The two depths, one before (y_1) and one after (y_2) the jump, are related by the following expression:

$$\frac{y_1}{y_2} = \frac{1}{2}\left(-1 + \sqrt{1 + 8F_1^2}\right) \qquad (14.127)$$

where F_1 is the incoming Froude number $= \dfrac{v_1}{\sqrt{gy_1}}$ (14.128)

Alternatively, the expression may be written in terms of the outgoing Froude number $F_2\left(= \dfrac{v_2}{\sqrt{gy_2}}\right)$ as

$$\frac{y_2}{y_1} = \frac{1}{2}\left(-1 + \sqrt{1 + 8F_2^2}\right) \qquad (14.129)$$

where v_1 and v_2 are the incoming and outgoing velocities and g is the acceleration due to gravity.

The energy lost in the hydraulic jump (E_L) is given as:

$$E_L = \frac{(y_2 - y_1)^3}{4 y_1 y_2} \qquad (14.130)$$

In most cases, it is possible to find out the prejump depth (y_1) and velocity (v_1) from the given value of discharge per unit width (q) through the spillway. This is done by assuming the total energy is nearly constant right from the spillway entrance up to the beginning of the jump formation, as shown in Figure 14.45. v_1 may be assumed to be equal to $\sqrt{2gH_1}$, where H_1 is the total energy upstream of the spillway, and neglecting friction losses in the spillway. The appropriate expressions may be solved to find out the postjump depth (y_2) and velocity (v_2).

14.12.14   DEFLECTOR BUCKETS

Sometimes it is convenient to direct spillway into the river without passing through a stilling basin. This is accomplished with a deflector bucket designed so that the jet strikes the riverbed a safe distance from the spillway and dam. This type of spillway is often called a flip bucket or ski-jump spillway. The deflector bucket and the plunge pool are shown in Figure 14.45.

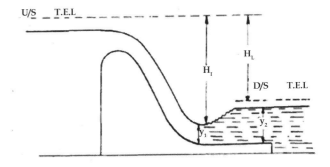

FIGURE 14.44   Hydraulic jump and associated parameters.

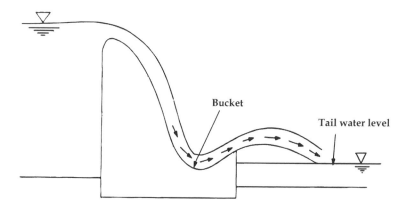

FIGURE 14.45 Definition sketch for deflector bucket and plunge pool.

The trajectory of the jump

$$y = x \tan\theta - \frac{x^2}{4\cos^2\theta(h_v + d)} \qquad (14.131)$$

where

h_v = velocity head
d = thickness of the jump

When the free jet discharging from the deflection bucket falls into an erodible riverbed, a plunge pool is eroded to a depth, D, given by:

$$D = 1.90 H^{0.225} q^{0.54} \qquad (14.132)$$

where

D = depth of scour measured from the TW (m)
q = discharge per unit width (m³/s/m)
H = drop in elevation from reservoir water surface to TW (m)

14.13 SEDIMENTATION AND EROSION HYDRAULICS

14.13.1 SEDIMENTATION

Sedimentation is the act or process of forming or accumulating sediment in layers, including such processes as the separation of rock particles from the material from which the sediment is derived, the transportation of these particles to the site of deposition, the actual deposition or settling of the particles, the chemical and other changes occurring in the sediment, and the ultimate consolidation of the sediment into solid rock. It is the terminal end of sediment transport. It includes the termination of transport by saltation or true bedload transport.

Sediments are material of varying size of mineral and organic origin. They range in particle distribution from micron-sized clay particles through silt, sand, gravel, rock, and boulders. Sediments originate from bed load transport, beach and bank erosion, and land runoff. They are naturally sorted by size through prevalent hydrodynamic conditions.

In general, fast-moving water will contain coarse-grained sediments and quiescent water will contain fine-grained sediments. Mineralogical characteristics of sediments vary widely and reflect watershed characteristics. Organic material in sediments is derived from the decomposed tissues of plants and animals, from aquatic and terrestrial sources, and from various point and nonpoint wastewater discharges. The content of organic matter increases in concentration as the size of sediment mineral particles decreases. Dissolved chemicals in the overlying and sediment pore waters are a product of inorganic and organic sedimentary materials, as well as runoff and ground water that range from fresh to marine in salinity. This sediment/water environment varies significantly over space and time and its characteristics are driven by complex biogeochemical interaction between the inorganic, living, and nonliving organic components. The sediment biotic community includes micro-, meso-, and macrofauna, and -flora that are independent of each other and their host sediment's biogeochemical characteristics.

Sedimentation is the direct result of the loss (erosion) of sediments from other aquatic areas or land-based areas. Sedimentation can be detrimental or beneficial to aquatic environments. Moreover, sediment impoverishment (erosion or lack of replenishment) in an area can be as bad as too much sedimentation. Sedimentation in one area is linked to erosion or impoverishment in another area and is a natural process of all water bodies (i.e., lakes, rivers, estuaries, coastal zones, and even the deep ocean). As an example, detrimental effects can be related to the burial of bottom-dwelling organisms and beneficial effects can be related to the building of new substrates for the development of marshes. These natural physical processes will continue whether or not they are influenced by the activities of humankind.

Human activities, however, have significantly enhanced sedimentation as well as sediment loss. Sedimentation activities can be land-based (i.e., agriculture, forestry, construction, urbanization, recreation) and water-based (i.e., dams, navigation, port activities, drag fishing, channelization, water diversions, wetlands loss, other large-scale hydrological modifications). Sediment impoverishment or loss is generally due to retention behind dams, bank or beach protection activities, water diversions, and many of the aquatic activities cited here.

Morphological changes (physical changes over a large area) to large aquatic systems can also result in major changes in natural sediment erosion and sedimentation patterns. As an example, the change in the size and shape of a water body will result in new water flow patterns leading to erosion or sediment removal from sensitive areas.

14.13.2 Effects of Sedimentation

The environmental impacts of sedimentation include loss of important or sensitive aquatic habitat, decrease in fishery resources, loss of recreation attributes, loss of coral reef communities, human health concerns, changes in fish migration, increases in erosion, loss of wetlands, nutrient balance changes, circulation changes, increase in turbidity, loss of submerged vegetation, and coastline alteration.

The effects of changes to sedimentation patterns will depend on whether the change results in an increase or decrease in sediment availability. Both effects have various physical and chemical consequences for water quality and aquatic ecosystem health. Sedimentation effects are usually local, but transboundary impacts may occur where major river systems form a common border and where littoral currents carry inputs across international boundaries

14.13.2.1 From Increased Sedimentation

- Smothering of marine communities and, in severe cases, complete burial leading to suffocation of corals, mangrove stands, and seagrass beds.
- Decreases in the amount of available sunlight which may in turn limit the production of algae and macrophytes, increase water temperatures, and reduce growth of natural vegetation.
- Damage fish by irritating or scouring their gills and degrade fish habitats as gravel containing buried eggs becomes filled with fine particles, thus reducing available oxygen.
- Reduce the success of visual predators and may also harm some benthic macroinvertebrates.
- Fill watercourses, storm drains, and reservoirs leading to costly dredging and an increased risk for flooding.
- Many toxic organic chemicals, heavy metals, and nutrients are physically and/or chemically adsorbed by sediments, so that an increase of sediment loading to the marine environment can also lead to increased deposition of these toxic substances that result in further negative impacts such as eutrophication.

14.13.2.2 Decreased Sedimentation

- If a decrease of sediment occurs, it can lead to degradation of an ecosystem by starving it of the elements needed to sustain production since sediments often carry a variety of minerals, nutrients, and organic matter.

- Increased velocity of the water which may cause erosion downstream and in coastal zones and cause damage to coastal and marine ecosystems and human settlements;

Overall the degradation of coastal ecosystems and coral reefs from increases or decreases in sediment loads may lead to important losses in revenue caused by impacts on the tourism and fishing industries and on coastal development.

14.13.3 Control of Sedimentation

Control of sedimentation can be successful if implemented on a broad land area or watershed scale and is directly related to improvement in land-use practices. Agriculture and forestry (logging) improvements where soil loss is minimized are not only technically feasible: They can be carried out at a moderate cost and with net benefits. The United Nations Environmental Program also has global programs, their Regional Seas activities, to guide countries in the management of land-based activities negatively impacting the coastal zone. Improved land-use practices are the primary measures to control sediment sources: terracing, low tillage, modified cropping, reduced agricultural intensity (e.g., no-till buffer zones), and wetlands construction as sediment interceptors. Forestry practices such as clear-cutting to the water's edge without replacement tree planting must be seriously curtailed because base soil in exposed areas will erode and import sediment to sensitive aqueous areas. Wetlands that separate upland areas from aquatic areas serve as natural filters for the runoff from the adjacent land. Wetlands thus serve to trap soil particles and associated agricultural contaminants. The construction of natural buffer zones and wetlands replenishment adjacent to logging areas are effective techniques. Watershed construction activities such as port expansion, water diversions, channel deepening, and new channel construction must undergo a complete environmental assessment, coupled with predictive sediment resuspension and transport modeling, so alternative courses of action and activities to minimize the negative impacts of sedimentation may be chosen.

Sediments are absolutely necessary for aquatic plant and animal life. Managed properly, sediments are a resource; improper sediment management results in the destruction of aquatic habitat that would have otherwise depended on their presence. The United Nations Group of Experts on the Scientific Aspects of Marine Environmental Protection recently recognized that on a global basis, changes in sediment flows are one of the five most serious problems affecting the quality and uses of the marine and coastal environment.

14.13.4 Erosion

Erosion is the detachment and movement of soil particles by natural forces, primarily water, and wind. More broadly, erosion is the process of wearing away rocks, geologic, and soil material via water, wind, or ice (e.g., glaciers). Erosion will transport materials after mechanical weathering has broken rock and geologic materials down into smaller, moveable pieces.

Sedimentation is the process in which particulate matter carried from its point of origin by either natural or human-enhanced processes is deposited elsewhere on land surfaces or in water bodies. Sediment is a natural product of stream erosion; however, the sediment load may be increased by human practices. Such enhanced sources of sediment in a watershed include unvegetated stream banks and uncovered soil regions, including construction sites, deforested areas, and croplands.

14.13.5 Erosive Forces

Through different types of erosion, the Earth's surface is continually being reshaped. Landmasses are altered as waves and tides erode old lands while silt and other sediments deposited in water bodies build up new lands. Over geologic time, some gullies become ravines and eventually valleys.

14.13.5.1 Ice

Glaciers are important means of erosion over long periods of time. Although a glacier moves slowly, it eventually removes material from the surface over which it travels through processes such as plucking, abrasion, crushing, and fracturing. Rock fragments, located at the bottom and sides of moving ice masses, will grind and scour bedrock to eventually form the walls and floors of mountain valleys.

14.13.5.2 Wind

Wind can move sediment grains over long distances when they are carried through the air. Sediments also can be blown along expanses of land, such as beaches, mudflats, unvegetated cropland, or construction areas. Obstructions help reduce the wind's erosive capacity; hence, windblown sediment often is deposited at these locations.

14.13.5.3 Water

Flowing water plays a major role in erosion by carrying away soils and other materials on the land surface. In general, there is a potential for water erosion where the land slope is at least 2%. The four types of water erosion of soil are sheet, rill, gully, and tunnel.

Sheet erosion is the uniform removal of soil without the development of visible water channels. It is the least apparent of the four erosion types. Rill erosion is soil removal through the cutting of many small, but conspicuous, channels. Gully erosion is the consequence of water that cuts down into the soil along the line of flow. Gullies develop more quickly in places like animal trails, plow furrows, and vehicle ruts. Tunnel erosion may occur in soils with sublayers that have a greater tendency to transport flowing water than does the surface layer.

Along the seacoast, erosion of rocky coastal cliffs and sandy beaches results from the strong, unceasing action of currents and waves. Severe storms exacerbate the problem of sandy beach erosion. In many world regions, land and economic losses due to coastal erosion represents a serious problem.

14.13.5.4 Human Activities

Erosion is a natural process, but human activity can make it happen more quickly. Trees and plants hold soil in place. When people cut down forests or plow up grasses for agriculture or development, the soil washes away or blows away more easily. Landslides become more common. Water also rushes over exposed soil rather than soaking into it, causing flooding.

14.13.6 Ecological and Economic Impacts

Excessive erosion can reduce the soil's inherent productivity, whereas the associated sedimentation can damage young plants and fill drainage ditches, lakes, and streams. Erosive processes can reduce farm income by decreasing crop yields and increasing maintenance costs for drainage systems. Additional erosion damages in both rural and urban areas include reduced property values, deteriorated water quality, and increased costs of removing sediment from roadways, roadside ditches, and surface-water supplies.

One of the principal causes of degraded water quality and aquatic habitat is the depositing of eroded soil sediment in waterbodies. Excessive amounts of sediment resulting from natural or human-induced causes can result in the destruction of aquatic habitat and a reduction in the diversity and abundance of aquatic life. Diversity and population size of fish species, mussels, and benthic (bottom-dwelling) macro invertebrates associated with coarse substrates can be greatly reduced if the substrates are covered with sand and silt. Where sand or silt substrates have historically predominated; however, increased deposition may have little detrimental impact on benthic aquatic life.

Suspended sediment causes the water to be cloudy (turbid). Increased turbidity reduces light transmission (and hence photosynthesis), thereby reducing the growth of algae and aquatic plants, which can adversely affect the entire aquatic ecosystem. Moreover, increased turbidity decreases the water's aesthetic appeal and the human enjoyment of recreational activities. If the river cross-section is sufficiently reduced by sediment buildup, sedimentation can increase downstream flooding. In addition, some metal ions, pesticides, and nutrients may adhere to sediment particles and be transported downstream.

Increased sediment in surface-water bodies (e.g., rivers, lakes, and reservoirs) may have an economic impact on public water systems that use them as a source of drinking water. High turbidity not only is aesthetically displeasing, but also interferes with disinfection of the water prior to it being pumped to customers. Communities whose water-supply source has become more turbid often invest millions of dollars to upgrade their treatment facilities in order to remove the increased sediment load.

14.13.7 PREVENTIVE MEASURES AND EROSION CONTROL

Erosion control is the process of reducing erosion by wind and water. During the last 40 years of the twentieth century, nearly one-third of the world's arable land was lost to erosion, and topsoils continue to be lost at a nonsustainable rate, i.e., faster than it can be naturally restored.

Farmers and engineers must regularly practice erosion control. Sometimes, engineers simply install structures to physically prevent soil from being transported. Gabions are huge wire frames that hold boulders in place, for instance. Gabions are often placed near cliffs. These cliffs, often near the coast, have homes, businesses, and highways near them. When erosion by water or wind threatens to tumble the boulders toward buildings and cars, gabions protect landowners and drivers by holding the rocks in place.

Erosion control can also be done by physically changing the landscape. Living shorelines, for example, are a form of erosion control for wetland areas. Living shorelines are constructed by placing native plants, stone, sand, and even living organisms such as oysters along wetland coasts. These plants help anchor the soil to the area, preventing erosion. By securing the land, living shorelines establish a natural habitat. They protect coastlines from powerful storm surges as well as erosion.

Global warming, the latest increase in temperature around the world, is speeding erosion. The change in climate has been linked to more frequent and more severe storms. Storm surges following hurricanes and typhoons threaten to erode miles of coastline and coastal habitat. These coastal areas have homes, businesses, and economically important industries, such as fisheries. The rise in temperature is also quickly melting glaciers. This is causing the sea level to rise faster than organisms can adapt to it. The rising sea erodes beaches more quickly.

Water quality can be protected by controlling and minimizing erosion. State and federal laws provide some consideration of erosion and sedimentation control along streams and lakes during flood control, drainage, highway, bridge, and other stream-related construction projects. A variety of programs administered by public and private agencies encompass research projects, education programs, technical assistance, regulatory measures, and cost-share financial assistance for erosion control.

Coordinated efforts by federal and state agricultural programs have been helping reduce the amount of soil erosion and sedimentation. Farmers now utilize a variety of conservation practices to reduce erosion and limit sedimentation, such as conservation tillage methods, use of conservation buffers and riparian habitat, and establishment of conservation easements. The Natural Resource Conservation Service, state and local conservation districts, and other organizations play critical roles in preserving cropland productivity and limiting adverse water quality impacts.

Preventive measures for reducing excessive sediment load in streams include:

- proper repair and maintenance of drainage ditches and levees;

- minimal disturbance of the streambanks;
- avoidance of structural disturbance of the river;
- reduction of sediment excesses arising from construction activities;
- application of artificial and natural means for preventing erosion; and
- use of proper land and water management practices on the water-shed.

These preventive measures are preferred over remedial measures, which include:

- construction of detention reservoirs, sedimentation ponds, or settling basins;
- development of side-channel flood-retention basins; and
- removal of deposited sediment by dredging.

14.14 WATER RESOURCES MANAGEMENT FOR SUSTAINABILITY

Water is a valuable resource, sustaining human life, production processes, and ecosystems. It supports the sustainability of national economy, society, and the environment. Factors, such as population growth, rainfall variability and climate change, water pollution, land use changes, institutional arrangements, and water demand for food production, affect future water availability and security. Sustainable water management is increasingly becoming important and we need innovation in technology, policy, and institutional arrangements to cope with future water scarcity and drive increased water use efficiency and productivity as well as environmental and social outcomes in the longer term.

Water resource management is the activity of planning, developing, distributing and managing the optimum use of water resources. It is a subset of water cycle management. Ideally, water resource management planning has regard to all the competing demands for water and seeks to allocate water on an equitable basis to satisfy all uses and demands. As with other resource management, this is rarely possible in practice.

The management of water is not only a technical issue; it requires a mix of measures including changes in policies, prices and other incentives, as well as infrastructure and physical installations. IWRM focuses on the necessary integration of water management across sectors, policies, and institutions. It is the process which promoted the coordinated development and management of water, land and related resources, in order to maximize the resultant economic and social welfare in an equitable manner without compromising the sustainability of vital ecosystems.

There are essentially five "C"s that need to be taken into account in developing water management policies:

- **Commitment**: The political will to achieve effective implementation is indispensable on the part of those taking part both in public and private sectors.

- **Content**: Policies must be meaningful especially the poor and marginalized in developing countries.
- **Cooperation**: Water management need to be prepared with the full involvement of the stakeholders concerned.
- **Checking**: Monitoring of implementation and of results is essential with all stakeholders involved.
- **Communication**: Successful communication includes reporting to the public on results, as well as listening to feedback.

Contemporary water managers have to deal with an increasingly complex picture. Their responsibilities entail managing variable and uncertain supplies to meet rapidly changing and uncertain demands; balancing ever-changing ecological, economic, and social values; facing high risks and increasing unknowns; and sometimes needing to adapt to events and trends as they unfold. Moreover, effective water management demands transboundary coordination in a context where a total of 276 international river basins cover almost half the earth's surface, and some 273 identified transboundary aquifers underpin various national economies.

The productivity of the water sector has declined over the past decade, along with other sectors of the economy. Improvements to productivity in the water sector will be underpinned by better resource management, more efficient use of labor and advances in technology as well as integration with other services such as electricity and waste disposal. Adaptive planning and using real options for investment decisions minimize the risk of unnecessary, high-cost investments. Efficient water markets ensure that water is most effectively allocated between competing uses to where it has highest value. Water pricing should reflect the value of water, but there is a need to improve the technical and economic evaluation of water externalities so that they can be incorporated into policy decisions.

Demand-side measures such as water efficiency programs are often the cheapest cost to implement, but when they extend to water restrictions, the external social and economic costs are born by the broader community. Policy barriers to rural-urban trading and potable reuse of recycled water should be removed. A holistic approach taking broad economic, environmental, and social issues into account is essential for sustainable water management.

14.14.1 Water Resource Management for Agriculture

Agriculture is the major user of water in most countries. It also faces the enormous challenge of producing almost 50% more food by 2030 and doubling production by 2050. This will probably have to be achieved with less water, mainly because of pressures from growing urbanization, industrialization, and climate change. Consequently, it will be important in future that farmers face the right signals to increase water use efficiency and improve water management, especially as agriculture is the major user of water. The scope of sustainable management of water resources in agriculture concerns the responsibility of water managers and users to ensure that water resources are allocated efficiently and equitably and used to achieve socially, environmentally and economically beneficial outcomes. It includes irrigation to smooth water supply across the production seasons; water management in rain-fed agriculture; management of floods, droughts, and drainage; and conservation of ecosystems and associated cultural and recreational values.

Agricultural water resource management covers a wide range of agricultural systems and climatic conditions, drawing on varying water sources, including surface water; groundwater; rainwater harvesting; recycled wastewater; and desalinated water. It also operates in a highly diverse set of political, cultural, legal, and institutional contexts, encompassing a range of areas of public policy: agriculture, water, environment, energy, fiscal, economic, social, and regional. Future policies to address the sustainable management of water resources in agriculture will be greatly influenced by climate change and climate variability, including seasonality problems, such as changes in the timing of annual rainfall patterns or periods of snow pack melt. In some regions, projections suggest that crop yields could improve. For other localities, climate change will lead to increased stress on already scarce water resources, while some areas are expected to see the growing incidence and severity of flood and drought events, imposing greater economic costs on farming and the wider economy. Irrigated agriculture, which accounts for most water used by agriculture, will continue to play a key role in agricultural production growth. To avoid a global water crisis, farmers will have to strive to increase productivity to meet growing demands for food, while industry and cities find ways to use water more efficiently.

14.14.2 Urban Water Management

As the carrying capacity of the Earth increases greatly due to technological advances, urbanization in modern times occurs because of economic opportunity. More people live in cities than in rural areas. In places, rapid population influx, inadequate public services, and out-of-date urban planning models have marginalized vast numbers of new arrivals into informal settlements or slums, exacerbating inequality and urban poverty, and compromising efforts to achieve and sustain water security. Urban water management is now on the verge of a revolution in response to rapidly escalating urban demands for water as well as the need to make urban water systems more resilient to climate change.

The goals of urban water management are to ensure access to water and sanitation infrastructure and services; manage rainwater, wastewater, stormwater drainage, and runoff pollution; control waterborne diseases and epidemics; and reduce the risk of water-related hazards, including floods, droughts, and landslides. All the while, water management practices must prevent resource degradation.

Conventional urban water management strategies, however, have strained to meet demand for drinking water, sanitation, wastewater treatment, and other water-related services. Some cities already face acute water shortages and deteriorating water quality.

Integrated Urban Water Management (IUWM) promises a better approach than the current system, in which water supply, sanitation, stormwater, and wastewater are managed by isolated entities, and all four are separated from land-use planning and economic development. IUWM calls for the alignment of urban development and basin management to achieve sustainable economic, social, and environmental goals. It offers a set of principles that underpin better coordinated, responsive, and sustainable resource management practice:

- It recognizes alternative water sources.
- It differentiates the qualities and potential uses of water sources.
- It views water storage, distribution, treatment, recycling, and disposal as part of the same resource management cycle.
- It seeks to protect, conserve, and exploit water at its source.
- It accounts for nonurban users that are dependent on the same water source.
- It aligns formal institutions (organizations, legislation, and policies) and informal practices (norms and conventions) that govern water in and for cities.
- It recognizes the relationships among water resources, land use, and energy.
- It simultaneously pursues economic efficiency, social equity, and environmental sustainability.
- It encourages participation by all stakeholders.

14.14.3 FUTURE OF WATER RESOURCES

One of the biggest concerns for our water-based resources in the future is the sustainability of the current and even future water resource allocation. As water scarcity increases, the importance of how it is managed grows vastly. Finding a balance between what is needed by humans and what is needed in the environment is an important step in the sustainability of water resources. Attempts to create sustainable freshwater systems have been seen on a national level in countries such as Australia, and such commitment to the environment could set a model for the rest of the world.

The field of water resources management will have to continue to adapt to the current and future issues facing the allocation of water. With the growing uncertainties of global climate change and the long term impacts of management actions, the decision-making will be even more difficult. It is likely that ongoing climate change will lead to situations that have not been encountered. As a result, alternative management strategies are sought for in order to avoid setbacks in the allocation of water resources.

14.15 RESERVOIR AND STREAM FLOW ROUTING

14.15.1 RESERVOIR

Reservoirs are those water bodies formed or modified by human activity for specific purposes, in order to provide a reliable and controllable resource. They are usually found in areas of water scarcity or excess, or where there are agricultural or technological reasons to have a controlled water facility. Where water is scarce, for example, reservoirs are mainly used to conserve available water for use during those periods in which it is most needed for irrigation or drinking water supply. When excess water may be the problem, then a reservoir can be used for flood control to prevent downstream areas from being inundated during periods of upstream rainfall or snow-melt.

The main uses of reservoirs include Drinking and municipal water supply, Industrial and cooling water supply, Power generation, Agricultural irrigation, River regulation and flood control, Commercial and recreational fisheries, Body contact recreation, boating, and other aesthetic recreational uses, Navigation, Canalization, and Waste disposal (in some situations).

14.15.2 TYPES OF RESERVOIRS

There are three main types of reservoirs: valley-dammed reservoirs, bank-side reservoirs, and service reservoirs.

14.15.2.1 Valley Dammed Reservoir

A dam constructed in a valley relies on the natural topography to provide most of the basin of the reservoir. Dams are typically located at a narrow part of a valley downstream of a natural basin. The valley sides act as natural walls with the dam located at the narrowest practical point, as shown in Figure 14.46 to provide strength and the lowest practical cost of construction. In many reservoir construction projects, people have to be moved and re-housed; historical artifacts are

FIGURE 14.46 Valley dammed reservoir

moved or relocated. Where the topography is poorly suited to a single large reservoir, a number of smaller reservoirs may be constructed in a chain.

14.15.2.2 Bank-Side Reservoirs

These reservoirs are made by diverting water from local rivers or streams to an existing reservoir. Such reservoirs are usually built partly by excavation and partly by the construction of a complete encircling bund or embankment which may exceed 6 km in circumference. Both the floor of the reservoir and the bund must have an impermeable lining or core. Although this can be applied to many different geographical areas, unlike the valley-dammed reservoir, which requires a valley, diverting water from a river can create problems.

14.15.2.3 Service Reservoirs

Service reservoirs are entirely man made. They are usually stored in concrete basins above or below ground. They are used to store fully treated potable water close to the point of distribution. Many service reservoirs are constructed as water towers, often as elevated structures on concrete pillars where the landscape is relatively flat. Other service reservoirs can be almost entirely underground, especially in more hilly or mountainous country. Service reservoirs perform several functions including ensuring sufficient head of water in the water distribution system and providing hydraulic capacitance in the system to even out peak demand from consumers enabling the treatment plant to run at optimum efficiency. Large service reservoirs can also be managed to so that energy costs in pumping are reduced by concentrating refilling activity at times of day when power costs are low.

14.15.3 Classification of Reservoirs

Depending on the purposes, the reservoir may be classified into the following:

14.15.3.1 Storage or Conservation Reservoirs

They are constructed to store water received when there is excess rainfall and is released gradually when it is needed for purposes like irrigation, hydroelectricity, domestic, industrial, etc., or when the river flow is low.

14.15.3.2 Flood Control Reservoirs

A flood control reservoir is constructed for the purpose of flood control. It protects the areas lying on its downstream side from the damages due to flood.

14.15.3.3 Retarding Reservoirs

A retarding reservoir is provided with spillways and sluiceways which are ungated. The retarding reservoir stores a portion of the flood when the flood is rising and releases it later when the flood is receding.

14.15.3.4 Detention Reservoirs

A detention reservoir stores excess water during floods and releases it after the flood. It is similar to a storage reservoir but is provided with large gated spillways and sluiceways to permit flexibility of operation.

14.15.3.5 Distribution Reservoir

It is a small storage reservoir used for water supply in a city. It accounts for varying rate of water supply during the day. Such distribution permits pumping plants and water treatment works, etc., to operate at constant rate.

14.15.3.6 Balancing Reservoirs

A reservoir downstream of the main reservoir for holding water let down from the main reservoir in excess of that required for irrigation, power generation or other purposes.

14.15.3.7 Conservation Reservoir

A reservoir impounding water for useful purposes, such as irrigation, power generation, recreation, domestic, industrial, and municipal supply, etc.

14.15.3.8 Multipurpose Reservoir

These reservoirs are meant to serve more than one purpose.

14.15.4 Zones of Storage

The different zones of storage of a reservoir are shown in Figure 14.47.

Some of the common terms used in reservoir storage are as follows:

Full reservoir level (FRL): The FRL is the highest water level to which the water surface will rise during normal operating conditions.

MWL: The MWL is the maximum level to which the water surface will rise when the design flood passes over the spillway.

Minimum pool level: The minimum pool level is the lowest level up to which the water is withdrawn from the reservoir under ordinary conditions.

The different zones are:

i. **Dead storage**: The volume of water held below the minimum pool level is called the dead storage. It is provided to cater for the sediment deposition by the impounding sediment laid in water. Normally it is equivalent to volume of sediment expected to be deposited in the reservoir during the design life reservoir.

ii. **Live or useful storage**: The volume of water stored between the FRL and the minimum pool level is called the useful storage. It assures the supply of water for specific period to meet the demand.

iii. **Bank storage**: It is developed in the voids of soil cover in the reservoir area and becomes available as seepage of water when water levels drops down. It increases the reservoir capacity over and above that given by elevation storage curves.

iv. **Valley storage**: The volume of water held by the natural river channel in its valley up to the top of its

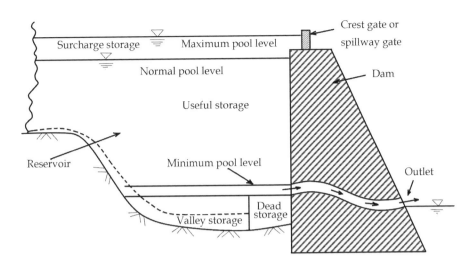

FIGURE 14.47 Zones of reservoir storage.

banks before the construction of a reservoir is called the valley storage. The valley storage depends upon the cross-section of the river.

v. **Flood/Surcharge storage**: It is the storage contained between maximum reservoir level and FRLs. It varies with spillway capacity of dam for given design flood.

14.15.5 Reservoir Capacity

The reservoir capacity depends upon the inflow available and demand. The inflow in the river is always greater than the demand, so no storage is required. If the inflow in the river is small but the demand is high, a large reservoir capacity is required.

The required capacity for a reservoir can be determined by the following methods:

1. Graphical method, using mass curves.
2. Analytical method

14.15.5.1 Graphical Method

i. Prepare a mass inflow curve as shown in Figure 14.48 from the flow hydrograph of the site for a number of consecutive years including the most critical years (or the driest years) when the discharge is low.

ii. Prepare the mass demand curve corresponding to the given rate of demand. If the rate of demand is constant, the mass demand curve is a straight line. The scale of the mass demand curve should be the same as that of the mass inflow curve.

iii. Draw the lines AB, FG, etc., such that they are parallel to the mass demand curve and are tangential to the crests A, F, etc., of the mass curve.

iv. Determine the vertical intercepts CD, HJ, etc., between the tangential lines and the mass inflow

curve. These intercepts indicate the volumes by which the inflow volumes fall short of demand.

Assuming that the reservoir is full at point A, the inflow volume during the period AE is equal to ordinate DE and the demand is equal to ordinate CE. Thus the storage required is equal to the volume indicated by the intercept CD.

v. Determine the largest of the vertical intercepts found in Step (iv). The largest vertical intercept represents the storage capacity required.

The following points should be noted while using graphical method:

- The capacity obtained in the net storage capacity which must be available to meet the demand. The gross capacity of the reservoir will be more than the net storage capacity. It is obtained by adding the evaporation and seepage losses to the net storage capacity.

- The tangential lines AB, FG, etc., when extended forward must intersect the curve. This is necessary for the reservoir to become full again. If these lines do not intersect the mass curve, the reservoir will not be filled again. However, very large reservoirs sometimes do not get refilled every year. In that case, they may become full after 2–3 years.

- The vertical distance such as FL between the successive tangents represents the volume of water spilled over the spillway of the dam.

14.15.5.2 Analytical Method

In this method, the capacity of the reservoir is determined from the net inflow and demand. Storage is required only when the demand exceeds the net inflow. The total storage required is equal to the sum of the storage required during the various periods.

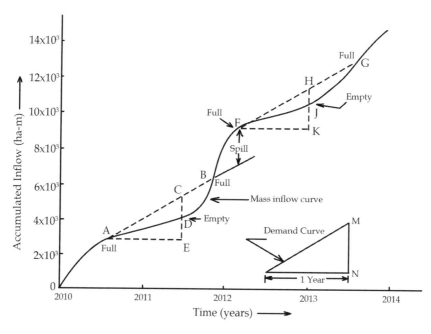

FIGURE 14.48   Graphical method: mass inflow curve.

i. Collect the stream flow data at the reservoir site during the critical dry period. Generally, the monthly inflow rates are required. However, for very large reservoirs, the annual inflow rates may be used.

ii. Ascertain the discharge to be released downstream to satisfy water rights or to honor the agreement between the states or the cities.

iii. Determine the direct precipitation volume falling on the reservoir during the month.

iv. Estimate the evaporation losses which would occur from the reservoir. The pan evaporation data are normally used for the estimation of evaporation losses during the month.

v. Ascertain the demand during various months.

vi. Determine the adjusted inflow during different months as follows:

Adjusted inflow = Stream inflow + Precipitation

$-$Evaporation $-$ Downstream Discharge   (14.133)

vii. Compute the storage capacity for each month.

Storage required = Adjusted inflow $-$ Demand   (14.134)

viii. Determine the total storage capacity of the reservoir by adding the storages required found in Step (vii).

14.15.6 Reservoir Yield

Yield is the volume of water which can be withdrawn from a reservoir in a specified period of time.

Safe yield is the maximum quantity of water which can be supplied from a reservoir in a specified period of time during a critical dry year.

Secondary yield: It is the quantity of water which is available during the period of high flow in the rivers when the yield is more than the safe yield.

Average yield: The average yield is the arithmetic average of the firm yield and the secondary yield over a long period of time.

Design yield: The design yield is the yield adopted in the design of a reservoir. The design yield is usually fixed after considering the urgency of the water needs and the amount of risk involved.

14.15.7 Determination of Yield of a Reservoir

The yield from a reservoir of a given capacity can be determined by the use of the mass inflow curve.

1. Prepare the mass inflow curve from the flow hydrograph of the river as shown in Figure 14.49.

2. Draw tangents AB, FG, etc., at the crests A, F, etc., of the mass inflow curve in such a way that the maximum departure (intercept) of these tangents from the mass inflow curve is equal to the given reservoir capacity.

3. Measure the slopes of all the tangents drawn in Step 2.

4. Determine the slope of the flattest tangent.

5. Draw the mass demand curve from the slope of the flattest tangent. The yield is equal to the slope of this line.

14.15.8 Stream Flow Routing

Stream flow increases as the variable source area extends into the drainage basin. The variable source area is the area of the watershed that is actually contributing flow to the stream at any point. The variable source area expands during rainfall and contracts thereafter. Flow routing is a technique of determining the flow hydrograph at a watercourse from known or

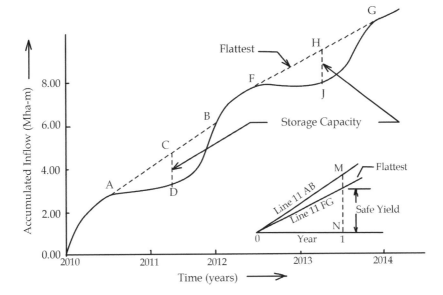

FIGURE 14.49   Yield determination: mass inflow curve.

assumed hydrographs at one or more points upstream. If the flow is a flood, the procedure is specially known as flood routing. It is used for forecasting, design of spillways, reservoir, and flood protection works, etc.

14.15.9   CLASSIFICATION OF FLOOD ROUTING

14.15.9.1   Reservoir Routing

It considers modulation effects on a flood wave when it passes through a water reservoir. It results in outflow hydrographs with attenuated peaks and enlarged time bases. Variations in reservoir elevation and outflow can be predicted with time when relationships between elevation and volume are known.

14.15.9.2   Channel Routing

It considers changes in the shape of input hydrograph while flood waves pass through a channel downstream. Flood hydrographs at various sections predicted when input hydrographs and channel characteristics are known.

14.15.10   TYPES OF FLOW ROUTING

14.15.10.1   Lumped or Hydrological Routing

Routing by lumped system methods is called hydrologic (lumped) routing. In lumped flow routing, the flow is a function of time at particular location. It employs essentially the continuity equation and flow/storage relationship.

14.15.10.2   Distributed or Hydraulic Routing

Routing by distributed system methods is called hydraulic (distributed) routing. In distributed flow routing, the flow is a function of space and time throughout the system. It uses continuity and momentum equations along with the equation of motion of unsteady flow (St. Venant equations).

The different methods of lumped flow routing based on the storage function are:

i. **Level pool reservoir routing**: Storage is a nonlinear function of the outflow only.

$$S = f(Q) \qquad (14.135)$$

ii. **Muskingum method for flow routing in channels**: Storage is a linearly related to inflow and outflow.

iii. **Linear reservoir models**: Storage is a linear function of outflow and its time derivatives.

Effect of reservoir storage is to redistribute the hydrograph by shifting the centroid of the inflow hydrograph to the position of that of the outflow hydrograph in time.

For hydrologic routing, input $I(t)$, output $Q(t)$, and storage $S(t)$ as functions of time are related by the continuity equation:

$$\frac{dS}{dt} = I(t) - Q(t) \qquad (14.136)$$

Even if an inflow hydrograph $I(t)$ is known, equation cannot be solved directly to obtain the outflow hydrograph $Q(t)$ because both Q and S are unknown. A second relationship or storage function is required to relate S, I, and Q; coupling the storage function with the continuity equations provides a solvable combination of two equations and two unknowns.

The specific form of the storage function depends on the nature of the system being analyzed. In reservoir routing by level pool method, storage is a nonlinear function of Q, $S=f(Q)$, and the function $f(Q)$ is determined by relating reservoir storage and outflow to reservoir water level. In the Muskingum method for flow routing in channels, storage is linearly related to I and Q. The effect of storage is to redistribute the hydrograph by shifting the centroid of the inflow hydrograph to the position of the outflow hydrograph in a time of redistribution. In very long channels, the entire flood wave also travels considerable distance, and the centroid of its hydrograph may then be shifted by a time period longer than the time of redistribution. This additional time may be considered the time of translation. The total time of flood movement between the centroids of the inflow and outflow hydrographs is equal to the sum of the time of redistribution and time of translation. The process of redistribution modifies the shape of the hydrograph, while translation changes its position.

14.15.11   RESERVOIR ROUTING

It is a procedure for calculating the outflow hydrograph from a reservoir with a horizontal water surface. Flow of flood waves from rivers/streams keeps on changing the head of water in the reservoir, $h = h(t)$. It is required to find variations of storage (S), outflow (Q), and the head of water in the reservoir with time (h), for given inflow with time.

In a small interval of time, the continuity equation can be written as:

$$\bar{I}\Delta t - \bar{Q}\Delta t = \Delta S \qquad (14.137)$$

Considering average inflow in time t, average outflow in time t, and change in storage in t, the above equation can be written as:

$$\left(\frac{I_1 + I_2}{2}\right)\Delta t - \left(\frac{Q_1 + Q_2}{2}\right)\Delta t = S_2 - S_1 \qquad (14.138)$$

For reservoir routing, the following data should be known in order to predict the outflow conditions and storage variations.

  i. Elevation vs storage
 ii. Elevation vs outflow discharge and hence storage vs outflow discharge
iii. Inflow hydrograph, and
 iv. Initial values of inflow, outflow Q, and storage S at time $t=0$.

Note: Δt must be shorter than the time of transit of the flood wave through the reach.

A variety of methods are available for reservoir routing, the important ones being Pul's method and Goodrich's method.

14.15.11.1   Pul's Method

On rearrangement of Equation 14.138 we get

$$\left(\frac{I_1 + I_2}{2}\right)\Delta t + \left(S_1 - \frac{Q_1 \Delta t}{2}\right)\Delta t = \left(S_2 + \frac{Q_2 \Delta t}{2}\right) \qquad (14.139)$$

All terms on the LHS are known at the starting of the routing and RHS is a function of elevation h for a chosen time interval Δt.

Graphs can be prepared for h Vs Q, h Vs S, and h Vs $\left(S + \frac{Q\Delta t}{2}\right)$ in order to identify the possible outflow and storage with respect to the given time.

Procedure is then repeated for full inflow hydrograph for complete analysis.

14.15.11.2   Goodrich Method

On rearrangement of Equation 14.138 we get

$$(I_1 + I_2) + \left(\frac{2S_1}{\Delta t} - Q_1\right) = \left(\frac{2S_2}{\Delta t} + Q_2\right) \qquad (14.140)$$

Graphs can be prepared for h Vs Q, h Vs S, and h Vs $\left(\frac{2S}{\Delta t} + Q\right)$.

Flow routing through time interval Δt, all terms on the LHS and hence RHS are known t. Value of outflow Q for $\left(\frac{2S}{\Delta t} + Q\right)$ can be read from the graph and value of $\left(\frac{2S}{\Delta t} - Q\right)$ is calculated from $\left(\frac{2S}{\Delta t} + Q\right) - 2Q$ for the next time interval. Computations can be repeated for subsequent routing periods.

14.15.12   CHANNEL ROUTING: MUSKINGUM METHOD

It is a hydrologic routing method used for handling variable discharge–storage relationship. Here storage is a function of both outflow and inflow discharges. Water surface in a channel reach is not only parallel to the channel bottom but also varies with time. The model storage in channel routing is a combination of wedge and prism as shown in Figure 14.50. Prism storage is the volume that would exist when uniform flow occurs at the downstream depth. Wedge storage is the volume formed between actual water surface profile and top surface of prism storage.

During the advance of flood wave, inflow exceeds outflow and a positive wedge is formed. Whereas during recession, outflow exceeds inflow and a negative wedge is formed.

Let us assume that cross-sectional area of the flood flow section is directly proportional to the discharge at the section as shown in Figure 14.51. Then,

- Volume of the prism storage $= KO$
- Volume of the wedge storage $= KX(I - O)$
where
  K – proportionality coefficient, and
  X – weighting factor having the range $0 < X < 0.5$

Total storage can be written as:

$$S = K(XI + (1 - X)O) \qquad (14.141)$$

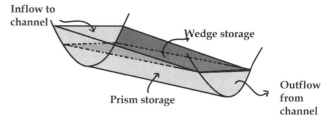

FIGURE 14.50   Model storage using Muskingum method.

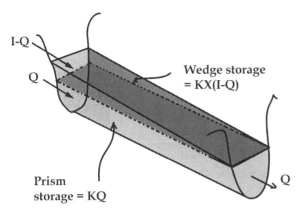

I-Q

Q

Wedge storage
= KX(I-Q)

Prism
storage = KQ

Q

FIGURE 14.51 Flood flow section.

The above equation is known as Muskingum storage equation.

Here a linear model is considered for routing flow in streams; therefore, the value of X depends on shape of modeled wedge storage. So $X=0$ for level pool storage, $(S = KO)$; and $X=0.5$ for a full wedge. K is the time of travel of flood wave through channel reaches.

Values of storage at time j and $j+1$ can be written as

$$S_j = K\left(XI_j + (1-X)O_j\right) \quad (14.142)$$

and

$$S_{j+1} = K\left(XI_{j+1} + (1-X)O_{j+1}\right) \quad (14.143)$$

Change in storage over time interval t is given by

$$S_{j+1} - S_j = K\left(X\left(I_{j+1} - I_j\right) + (1-X)\left(O_{j+1} - O_j\right)\right) \quad (14.144)$$

From the continuity equation

$$\left(\frac{I_j + I_{j+1}}{2}\right)\Delta t - \left(\frac{O_j + O_{j+1}}{2}\right)\Delta t = S_{j+1} - S_j \quad (14.145)$$

Equating the above two equations,

$$K\left(X\left(I_{j+1} - I_j\right) + (1-X)\left(O_{j+1} - O_j\right)\right)$$
$$= \left(\frac{I_j + I_{j+1}}{2}\right)\Delta t - \left(\frac{O_j + O_{j+1}}{2}\right)\Delta t_j \quad (14.146)$$

On simplification, we get

$$O_{j+1} = C_1 I_{j+1} + C_2 I_j + C_3 O_j \quad (14.147)$$

This equation is known as Muskingum's routing equation.
where

$$C_1 = \frac{0.5\Delta t - KX}{K(1-X) + 0.5\Delta t} \quad (14.148)$$

$$C_2 = \frac{0.5\Delta t + KX}{K(1-X) + 0.5\Delta t} \quad (14.149)$$

$$C_3 = \frac{K(1-X) - 0.5\Delta t}{K(1-X) + 0.5\Delta t} \quad (14.150)$$

$$C_1 + C_2 + C_3 = 1 \quad (14.151)$$

Δt should be so chosen such that $K>t>2KX$ for best results. If $\Delta t<2KX$ then the coefficient C_1 will be negative.

14.15.12.1 Required Input for Muskingum Routing

- Inflow hydrograph through a channel reach.
- Values of K and X for the reach.
- Value of the outflow O_j from the reach at the start.
- For a given channel reach, K and X are taken as constant.
- K is determined empirically (e.g., Clark's method: $K = cL/s^{0.5}$, where c is a constant; L is the length of stream; s is the mean slope of channel) or graphically.
- X is determined by trial and error procedure.

14.15.12.2 Routing Procedure Using Muskingum Method

- Knowing K and X, select an appropriate value of Δt.
- Calculate C_1, C_2, and C_3.
- Starting from the initial conditions known inflow, outflow, calculate the outflow for the next time step.
- Repeat the calculations for the entire inflow hydrograph.

14.15.13 Flood Routing by Saint Venant Equations

It is a physically based theory of flood propagation derived from Saint Venant equations for gradually varying flow in open channels. It is a hydraulic routing method. The flow is considered as one-dimensional. The conservation of mass-continuity equation and momentum-dynamic wave equation takes place.

The governing equations are:
Equation of continuity:

$$\frac{\partial Q}{\partial x} + \frac{\partial A}{\partial t} - q = 0 \quad (14.152)$$

Momentum equation:

$$\frac{\partial Q}{\partial t} + \frac{\partial}{\partial x}\left(\frac{Q^2}{A}\right) = gA\left(S_0 - S_f\right) - gA\frac{\partial h}{\partial x} \quad (14.153)$$

where q=lateral inflow, Q=discharge in the channel, A=area of flow in the channel, S_0=bed shape, and S_f=friction slope of channel

Diffusion approach : $\frac{\partial Q}{\partial x} + \frac{\partial A}{\partial t} - q = 0 \quad (14.154)$

Initial conditions: $\frac{\partial h}{\partial x} + \frac{\partial A}{\partial t} - q = 0 \quad (14.155)$

Boundary conditions: $Q = \frac{1}{n} R_h^{2/3} S_f^{1/2} A \quad (14.156)$

Kinematic approach : For $S_0 = S_f, \quad \frac{\partial Q}{\partial x} + \frac{\partial A}{\partial t} - q = 0 \quad (14.157)$

14.16 WATER WITHDRAWALS AND USES

14.16.1 WATER WITHDRAWALS

Water is one of the most vital natural resources for all life on Earth. The availability and quality of water always have played an important part in determining not only where people can live but also their quality of life. Even though there always has been plenty of fresh water on Earth, water has not always been available when and where it is needed, nor it is always of suitable quality for all uses. Water must be considered as a finite resource that has limits and boundaries to its availability and suitability for use.

Water withdrawals, or water abstractions, are defined as freshwater taken from ground or surface water sources, either permanently or temporarily, and conveyed to a place of use. If the water is returned to a surface water source, abstraction of the same water by the downstream user is counted again in compiling total abstractions and it may lead to double counting. The data include abstractions for public water supply, irrigation, industrial processes, and cooling of electric power plants. Mine water and drainage water are included, whereas water used for hydroelectricity generation is normally excluded. This indicator is measured in m^3 per capita.

Figure 14.52 shows the different ways in which water is withdrawn for different uses. The offstream uses (depicted on the left) are those in which water is removed from its source, either by pumping or diversion. Instream uses (depicted on the right) are those in which water remains in place and typically refers to stream (rather than groundwater). Where water supply is limited, conflicts may result between and among the various uses.

In the figure, the demands on water are also represented as a tug-of-war among the various offstream and instream uses. The balance between supply and demand for water is a delicate one. The availability of usable water has and will continue to dictate where and to what extent development will occur. Water must be in sufficient supply for an area to develop, and an area cannot continue to develop if water demand far outstrips available supply. Further, a water supply will be called upon to meet an array of offstream uses (in which the water is withdrawn from the source) in addition to instream uses (in which the water remains in place).

14.16.2 THE WATER-USE CYCLE

Water is constantly in motion by way of the hydrological cycle. Water evaporates as vapor from oceans, lakes, and rivers; is transpired from plants; condenses in the air and falls as precipitation; and then moves over and through the ground into water bodies, where the cycle begins again.

The water-use cycle is composed of the water cycle with the added influence of human activity. Dams, reservoirs, canals, aqueducts, withdrawal pipes in rivers, and groundwater wells all reveal that humans have a major impact on the water cycle. In the water-use cycle, water moves from a source to a point of use, and then to a point of disposition. The sources of water are either surface water or groundwater. Water is withdrawn and moved from a source to a point of use, such as an industry, restaurant, home, or farm. After water is used, it must be disposed of (or sometimes, reused). Used water is either directly returned to the environment or passes through a treatment processing plant before being returned.

14.16.3 CATEGORIES OF WATER USE

Water use can be categorized into the following for analyzing current patterns and predicting future trends.

i. **Commercial water use**: It includes fresh water for motels, hotels, restaurants, office buildings, other commercial facilities, and civilian and military institutions. Domestic water use is probably the most important daily use of water for most people.

ii. **Domestic water use**: It is the most important daily use of water for most people. It includes water that is used in the home every day, including water for normal household purposes, such as drinking, food preparation, bathing, washing clothes and dishes, flushing toilets, and watering lawns and gardens.

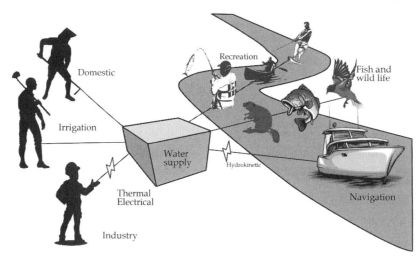

FIGURE 14.52 Different types of water use.

iii. **Industrial water use**: It is a valuable resource to the nation's industries for such purposes as processing, cleaning, transportation, dilution, and cooling in manufacturing facilities. Major water-using industries include steel, chemical, paper, and petroleum refining. Industries often reuse the same water over and over for more than one purpose.

iv. **Irrigation water use**: It includes water artificially applied to farm, orchard, pasture, and horticultural crops, as well as water used to irrigate pastures, for frost and freeze protection, chemical application, crop cooling, harvesting, and for the leaching of salts from the crop root zone. Nonagricultural activities include self-supplied water to irrigate public and private golf courses, parks, nurseries, turf farms, cemeteries, and other landscape irrigation uses.

v. **Livestock water use**: It includes water for stock animals, feed lots, dairies, fish farms, and other nonfarm needs. Water is needed for the production of red meat, poultry, eggs, milk, and wool, and for horses, rabbits, and pets. Livestock water use only includes fresh water.

vi. **Mining water use**: It includes water for the extraction of naturally occurring minerals; solids, such as coal and ores; liquids, such as crude petroleum; and gases, such as natural gas. The category includes quarrying, milling (such as crushing, screening, washing, and flotation), and other operations as part of mining activity. A significant portion of the water used for mining, about 32%, is saline.

vii. **Public supply water use**: It refers to water withdrawn by public and private water suppliers, such as municipal water works, and delivered to users for domestic, commercial, and industrial purposes.

viii. **Thermoelectric power water use**: It is the amount of water used in the production of electric power generated with heat. The source of the heat may be from fossil fuels, nuclear fission, or geothermal. Fossil fuel power plants typically reuse water. They generate electricity by turning a turbine using steam power. After the steam is used to turn the turbines, it is condensed back to water by cooling it. The condensed water is then routed back to the boiler, where the cycle begins again.

14.16.4 FUTURE WATER USE

The intensive water use by humanity is rapidly changing the dynamics within water bodies such as rivers, lakes, groundwater, and seas. The water withdrawals for public supply and domestic uses will continue to increase as population increases. Higher water prices and active water conservation programs, however, may reduce the per capita use rates. With increased competition for water for instream uses, such as river-based recreation, aesthetic enjoyment, fish and wildlife habitat, and hydroelectric power, along with higher municipal uses, irrigators will have increasing difficulty competing

economically for available water supplies. As a result, the science of sustainable water use has become a central challenge of our time. Technologies which minimize the effect of human activities on water systems are both expensive and challenging to develop; to that end, water science requires interdisciplinary and cross-sector approaches, including from the fields of engineering, chemistry, economics, as well as the environmental, medical, and social sciences.

Constraints must be imposed to conserve and limit the use of available water in the study area. By reducing the demand for water, the recommended conservation measures will have a positive effect on water quality and the environment. Voluntary domestic water conservation measures include the following:

- Limiting toilet flushing.
- Adopting water-saving plumbing fixtures, such as toilets and shower heads.
- Adopting water-efficient appliances (notably washing machines).
- Limiting outdoor uses of water, as by watering lawns and gardens during the evening and early morning, and washing cars on lawns and without using a hose.
- Adopting water-saving practices in commerce, such as providing water on request only in restaurants and encouraging multiday use of towels and linens in hotels.
- Repairing household leaks.
- Limiting use of garbage disposal units.

In conclusion, various known methods can lead to significant savings in both indoor and outdoor water use. To implement these methods, government agencies in the study area should consider encouraging their adoption through education, incentives, pricing, taxation, and regulation, and to this end will be involved in setting priorities at various times for the support of needed measures, taking into account the uncertainties attached to the available evidence.

Through rationing, research, and possibly economic pricing policies, agricultural water use can become more efficient. A number of useful practices are already used to some degree in the study area, and these practices should be expanded to help conserve agricultural water use:

- Harvesting local water runoff and floodwater to increase water supplies for dry land agriculture.
- Reducing evaporative water loss by cropping within closed environments (desert greenhouses). This method is economic with land and water use, avoids soil salinization, and produces high yields of exportable crops, such as ornamentals, fruits, vegetables, and herbs.
- Using computer-controlled drip "fertigation" (fertilizer applied with irrigation water) and soilless substrates in greenhouses, which economizes on water and fertilizer use and helps prevent groundwater pollution.

- Considering the use of brackish water for irrigation of salinity-tolerant crops.
- Saving more freshwater by switching to irrigation with treated wastewater or with brackish water if possible.
- Changing production from crops with high water requirements to crops with lower water requirements.

The use of rooftop cisterns can help in water harvesting for individual domestic supplies. Catchment systems and storage ponds should also be expanded for agricultural water use. Even where conventional sources of water are available, cisterns can provide supplemental water inexpensively and relieve the demand on the water distribution system.

Where brackish waters can be desalted, this approach offers a clear promise of augmenting the available water supply. Such desalination is technologically feasible and will not usually have adverse environmental impacts. Desalination technologies can be really expensive. Economic feasibility depends on the quality of the feedwater, the technology used, and the relative attractiveness of other alternatives.

Reclamation of wastewater will theoretically double the amount of increased supply brought about by new sources of freshwater as well. Thus, widespread reclamation would decrease the amounts of water needed to meet the probably increased regional demand. Urban reuse of wastewater requires dual municipal distribution systems—one for potable, the other for reclaimed water. The prospect of major urban expansion in the area provides the incentive to plan communities with an initial dual-water system.

15 Soil Mechanics

15.1 INTRODUCTION

The word "soil" is derived from the Latin word soilum which means the upper layer of the earth that may be dug or plowed, especially the loose surface material of the earth in which plants grow. The term "soil" in soil engineering is defined as an unconsolidated material, composed of soil particles produced by the disintegration of rocks.

The term "soil mechanics" was coined by Dr. Karl Terzaghi in 1925 when his book Erdbaumechanics on the subject was published in Germany. According to Terzaghi, "Soil mechanics is the application of laws of mechanics and hydraulics to engineering problems dealing with sediments and other unconsolidated accumulations of solid particles produced by mechanical and chemical disintegration of rocks, regardless of whether or not contain an admixture of organic constituents." Soil mechanics is, therefore, a branch of mechanics which deals with the action of forces on soil and with the flow of water in soil.

Soil engineering in an applied science dealing with applications of principles of soil mechanics to practical problems. It includes site investigations, design and construction of foundations, earth retaining structures, and earth structures. Geotechnical engineering is a broader term which includes soil engineering, rock mechanics, and geology.

15.2 FORMATION OF SOIL

Soils are formed by the weathering of rocks due to mechanical disintegration or chemical decomposition. When rock surface gets exposed to atmosphere for an appreciable time, it disintegrates or decomposes into small particles and thus soils are formed.

Soil may be considered as an incident material obtained from the geologic cycle which goes on continuously in nature. The geologic cycle consists of erosion, transportation, deposition, and upheaval of soil. Exposed rocks are eroded and degraded by various physical and chemical processes. The products of erosion are picked up by agencies of transportation such as water and wind and are carried to new locations where they are deposited. This shifting of the material disturbs the equilibrium of forces on the earth and causes large-scale earth movements and upheavals. This process results in further exposure of rocks and the geologic cycle gets repeated.

If the soil stays at the place of its formation just above the parent rock, it is known as residual soil or sedentary soil. When the soil has been deposited at a place away from the place of its origin, it is called a transported soil.

The various agencies of transporting soil and redepositing soils are water, ice, wind, and gravity. Water-formed transported soils are termed as alluvial, marine, or lacustrine. All the material, picked up, mixed, disintegrated, transported, and redeposited by glaciers either by ice or by water issuing from melting of glaciers, is termed glacial drift or simple drift. The glacial deposits in general consist of a heterogeneous mixture of rock fragments and soils of varying sizes and proportions and except the stratified drift deposited by glacial streams are without any normal stratification. Dune sand and loess are the wind-blown (Aeolian) deposits. Loess is the windblown silt or silty clay having little or no stratification. Soils transported by gravitational forces are termed colluvial soils such as talus. The accumulation of decaying and chemically deposited vegetable matter under conditions of excessive moisture results in the formation of cumulose soils, such as peak and muck.

Soils are formed by either physical disintegration or chemical decomposition of rocks.

15.3 SOIL STRUCTURE AND CLAY MINERALS

Soil structure is usually defined as the arrangement and state of aggregation of soil particles in a soil mass. Soil structure is an important factor which influences many soil properties, such as permeability, compressibility, and shear strength, etc.

15.3.1 TYPES OF SOIL STRUCTURE

The following are the types of soil structures.

i. **Single-grained structures** (Figure 15.1): Cohesionless soil such as gravel and sand are composed of grains in which gravitational forces are more predominant than surface forces. When deposition of these soils occurs, the particles settle under gravitational forces and take an equilibrium position. Each particle is in contact with those surrounding it. The soil structure so formed is known as single-grained structure. Depending upon the relative position of soil particles, the soil may be loose structure or a dense structure.

It can be proved that for the loosest condition, the void ratio is 0.90 and for the densest state, it is 0.35. Figure 15.2 shows sphere in loosest and densest condition.

ii. **Honeycomb structures** (Figure 15.3): Such a structure exists in grains of silts or rock fur smaller than 0.02 mm diameter and larger than 0.0002 mm. When such grains settle under gravity, the surface forces also play an equally important role. The grains coming in contact are held until miniature arches are formed, bridging over relatively large void spacing and forming a honeycomb structure.

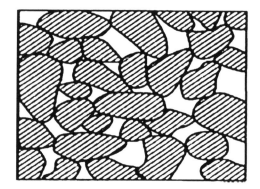

FIGURE 15.1 Single-grained structure.

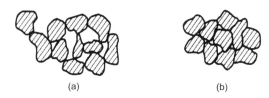

(a)   (b)

FIGURE 15.2 Particles in (a) loosest and (b) densest condition.

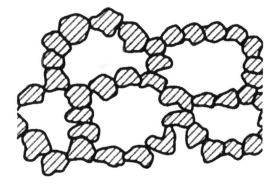

FIGURE 15.3 Honeycomb structure.

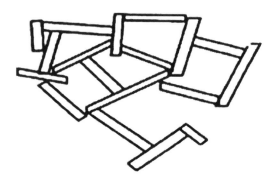

FIGURE 15.4 Flocculated structure.

iii. **Flocculated structures** (Figure 15.4): Flocculated structure occurs in clay. The clay particles have large surface area, and therefore, the electrical forces are important in such soils. The clay particles have a negative charge on the surface and a positive charge on the edges. Interparticle contact develops between

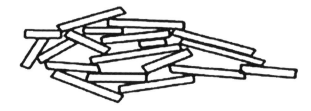

FIGURE 15.5 Dispersed structure.

the positively charged edges and the negatively charged faces. This results in flocculated structure. The particles are oriented "edge-to–edge" or "edge-to-face" with respect to one another.

Flocculated structure is formed when there is net attractive force between particles. When clay particles settle in water, deposits formed have a flocculated structure. The degree of flocculation of a clay deposit depends on the type and concentration of clay.

In general, the soils in a flocculated structure have a low compressibility, a high permeability, and high shear strength.

iv. **Dispersed structures** (Figure 15.5): Dispersed structure develops in clays that have been reworked or remoulded. The particles develop more or less a parallel orientation. Clay deposits with flocculent structure when transported to other places by nature or man get remoulded. Remoulding converts the edge-to-face orientation to face-to-face orientation. The dispersed structure is formed in nature when there is a net repulsive force between particles.

The soils in dispersed structure generally have low shear strength, high compressibility, and low permeability.

15.3.2 BASIC STRUCTURAL UNITS OF CLAY MINERALS

Clay minerals are hydrous aluminum silicate with other metallic ions in sheet-like structure. Their particles are very small in size, very flaky in shape, and thus have considerable surface area. These clay minerals are evolved mainly from the chemical weathering of certain rock minerals. Clay minerals are composed of two fundamental structural units.

1. Tetrahedral unit
2. Octahedral unit

1. Tetrahedral unit:

A tetrahedral unit comprises a central silicon atom surrounded by four oxygen atoms positioned at the vertices of the tetrahedron as shown in Figure 15.6.

The tetrahedral units are combined with each other such that each oxygen atom at the base of tetrahedron lies in same plane and is being shared

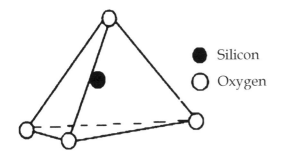

FIGURE 15.6   Silica tetrahedron.

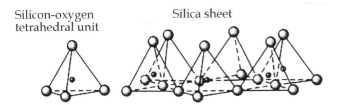

FIGURE 15.7   Silica sheet.

between two tetrahedron units. This combination of tetrahedral units is called silica sheet as shown in Figure 15.7. In the silica sheet, each of the oxygen ions at the base is common to two adjacent units. The sharing of charges leaves three negative charges at the base per tetrahedral unit. This, along with the two negative charges at the apex, makes a total of five negative charges to four positive charges of silicon ion. Thus there is a net charge of −1 per unit.

2. Octahedral unit:

An octahedral unit comprises a central ion of either aluminum, magnesium, or iron surrounded by six hydroxyl ions as shown in Figure 15.8.

The octahedral sheet is a combination of octahedral units. If the atom at the center is aluminum, the resulting sheet is called gibbsite sheet. If the atom at the center is magnesium, the sheet is called brucite sheet. If the atom at the center is iron, the sheet is called ferrite sheet.

Each hydroxyl ion in gibbsite sheet is being shared between 3 octahedral units. Therefore, net charge present over gibbsite is −1.

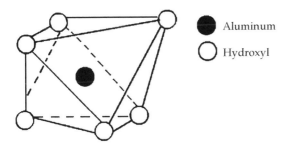

FIGURE 15.8   Octahedral unit.

15.3.3   TYPES OF CLAY MINERALS

Clayey soils are made of three basic minerals. They are as follows.

1. **Kaolinite mineral**: Kaolinite is the most common mineral of the kaolin group. The kaolinite structural unit is made up of gibbsite sheets joined to silica sheets through the unbalanced oxygen atoms at the apexes of the silicas as shown in Figure 15.9. This structural unit is symbolized by which is about 7 Å thick.
2. **Montmorillonite mineral**: Montmorillonite mineral is the most common mineral of the montmorillonite group of minerals. The basic structural unit consists of an alumina sheet sandwiched between two silica sheets as shown in Figure 15.10. Successive structural units are stacked one over another. The thickness of each structural unit is about 10 Å.
3. **Illite mineral**: Illite is the main mineral of the Illite group. The basic structural unit is similar to that of montmorillonite mineral. However, the mineral has properties different from montmorillonite due to the following reasons.
   a.   There is always a substantial amount of isomorphous substitution if silicon by aluminum in silica sheet. Consequently, the mineral has a larger negative charge than in montmorillonite.

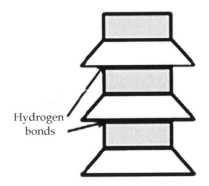

FIGURE 15.9   Kaolinite structural unit.

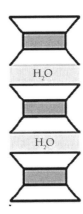

FIGURE 15.10   Montmorillonite structural unit.

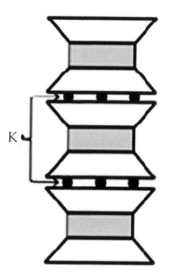

FIGURE 15.11   Illite structural unit.

b. The link between different structural units is through nonexchangeable potassium and not through water as shown in Figure 15.11. This bonds the units, more firmly than in montmorillonite.
c. The lattice of the illite is stronger than that of montmorillonite and is therefore less susceptible to cleavage.
d. Illite swells less than montmorillonite. However, swelling is more than in kaolinite.
e. The space between different structural units is much smaller than in montmorillonite, as the potassium ions just fit in-between the silica sheet surfaces.
f. The properties of illite minerals are somewhat intermediate between that of kaolinite and montmorillonite. The bond between the nonexchangeable potassium ions, though stronger than that in montmorillonite, is considerably weaker than hydrogen bond of kaolinite. The swelling of illite is more than that of kaolinite but less than montmorillonite. The specific surface is about $80 \, m^2/kg$.

15.4   BASIC DEFINITIONS AND RELATIONSHIPS

A soil mass consists of solid particles which form a porous structure. The voids in the soil mass may be filled with air, with water or partly with air and partly with water. If the voids are filled with air only, then the soil will be in dry state. If the voids are filled with water, then the soil will be in saturated condition. Thus soil can be considered as either two-phase system or three-phase system. The diagrammatic representation of the different phases in a soil mass is called phase diagram. A soil mass is a three-phase system consisting of solid particles, water, and air. A three-phase diagram is applicable for a partially saturated soil.

The phase diagram of two-phase system (Figure 15.12) and three-phase system (Figure 15.13) is shown below.

In the above figures,

V = total volume of the soil mass
V_a = volume of air
V_w = volume of water
V_v = volume of voids = $V_a + V_w$
V_s = volume of solids
W_a = weight of air (considered to negligible)
W_w = weight of water
W_s = weight of solids
W = total weight of soil mass = $W_s + W_w$

15.4.1   Basic Definitions

1. **Water content** [w]: Water content is also called moisture content. It is the ratio of weight of water to the weight of soil solids.

$$w = W_w/W_s \; ; \; w \geq 0 \tag{15.1}$$

This is represented as percentage. The water content of an oven dry soil is zero but natural water content for moist soil is around 60%. There is no upper limit for water content. It can be greater than 100%.

2. **Degree of saturation** [S]: Degree of saturation of a soil is defined as the volume of water to the volume of voids in the soil mass.

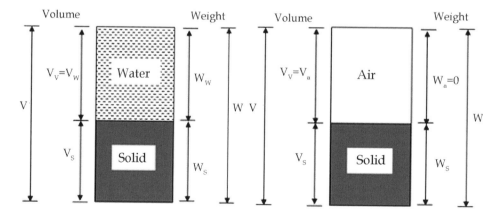

FIGURE 15.12   Phase diagram for two phase system.

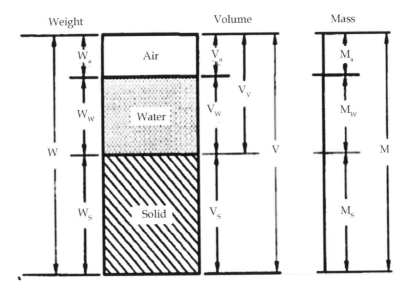

FIGURE 15.13   Phase diagram for three phase system.

$$S = \left(V_w/V_v\right) \times 100 \tag{15.2}$$

It is represented as percentage. For dry soil $S=0\%$ and for fully saturated soil $S=100\%$, whereas partially saturated soil have $0<S<100\%$.

3. **Void ratio** [e]: The void ratio of a soil is defined as the ratio of the total volume of voids to the volume of solids.

$$e = V_v/V_s; \quad e>0 \tag{15.3}$$

It is expressed in decimal. Void ratios of fine grained soil are generally higher than those of coarse grained soils. In general $e>0$, thus there is no upper limit.

4. **Porosity** [n]: The porosity of a soil is defined as the ratio of volume of voids to the total volume of soil.

$$n = \left(V_v/V\right) \times 100\% \tag{15.4}$$

It is expressed as percentage. In porosity, the total volume of soil is used which includes volume of voids. Hence porosity of soil cannot exceed 100%. The range of porosity is $0<n<100\%$. It is also called percentage voids.

5. **Air content** [a_c]: The air content of soil is defined as the ratio of the volume of air to the total volume of soil voids. It is expressed as percentage.

$$a_c = V_a/V_v \tag{15.5}$$

6. **Percentage air voids** [n_a]: The percentage air voids is defined as the ratio of volume of air to the total volume of soil mass. It is expressed as percentage.

$$n_a = \left(V_a/V\right) \times 100 \tag{15.6}$$

7. **Unit weight of soil**: The unit weight of soil is defined as its weight per unit volume.
   a. Bulk unit weight [γ]: It is the ratio of total weight of soil to the total volume of soil mass. It is expressed as kN/m³.

$$\gamma = W/V = \left(W_s + W_w\right)/\left(V_a + V_w + V_s\right) \tag{15.7}$$

   b. Dry unit weight [γ_d]: It is the ratio of total dry weight of soil to the total volume of soil mass. Dry unit weight is used as a measure of denseness of soil. More dry unit weight means the more dense or compacted is the soil.

$$\gamma_d = W_s/V \tag{15.8}$$

   c. Saturated unit weight [γ_{sat}]: It is defined as the ratio of total saturated weight of soil to the total volume of soil mass.

$$\gamma_{sat} = W_{sat}/V \tag{15.9}$$

   d. Submerged unit weight [γ_{sub}]: It is the ratio of buoyant weight of soil to the total volume of soil mass. When soil is below water, that is, in submerged condition, a buoyant force acts on the soil solids which are equal in magnitude to the weight of water displaced by the soil solids. Hence the net weight of soil is reduced and the reduced weight is known as buoyant weight or submerged weight. Submerged unit is roughly equal to (1/2) of saturated unit weight.

$$\gamma_{sub} = W_{sub}/V = \gamma_{sat} - \gamma_w \tag{15.10}$$

where γ_w=unit weight of water.

c. Unit weight of water [γ_w]: Unit weight of water is the ratio of weight of water to the volume occupied by the water. It depends on its temperature. However, the unit weight of water is taken to be constant as 9.81 kN/m³ or 1 g/cc. It is expressed as kN/m³.

$$\gamma_w = W_w/V_w \qquad (15.11)$$

f. Unit weight of solids [γ_s]: It is the ratio of weight of soil solids to the volume occupied by the soil solids. It is expressed as kN/m³.

$$\gamma_s = W_s/V_s \qquad (15.12)$$

8. **Density of soil**: Density of soil is defined as the mass of the soil per unit volume.
   a. Bulk density [ρ]: Bulk density or moist density is defined as the ratio of total soil mass to the total volume. It is expressed as kg/m³.

$$\rho = M/V \qquad (15.13)$$

   b. Dry density [ρ_d]: Dry density is defined as the ratio of total dry mass to the total volume.

$$\rho_d = M_s/V \qquad (15.14)$$

   c. Saturated density [ρ_{sat}]: It is defined as the ratio of total saturated soil mass to the total volume of soil mass.

$$\rho_{sat} = M_{sat}/V \qquad (15.15)$$

   d. Submerge density [ρ_{sub}]: It is the ratio of submerged mass of soil solid to the total volume of soil mass.

$$\rho_{sub} = M_{sub}/V \qquad (15.16)$$

9. **Specific gravity** (G): Specific gravity of soil solids is the ratio of the weight of a given volume of solids to the weight of an equivalent volume of water at 4°C.

$$G = \gamma_s/\gamma_w \qquad (15.17)$$

The specific gravity of most of the organic soil lies in the range of 2.65–2.80. For organic soils, it lies in the range of 1.2–1.40.

10. **Apparent or mass specific gravity** [G_m]: Mass specific gravity is defined as the ratio of the total weight of a given volume of soil to an equivalent volume of water. Mass specific gravity can be defined as the ratio of bulk unit weight of soil to unit weight of water.

$$G_m = \gamma_t/\gamma_w$$

If soil is in saturated state, $G_m = \gamma_{sat}/\gamma_w$ (15.18)

If soil is in dry state, $G_m = \gamma_d/\gamma_w$ (15.19)

11. **Relative density or density index** [I_D]: It is also known as degree of density. The term is used to express the relative compactness of a natural cohesionless soil deposit. The density index is defined as the ratio of the difference between the void ratio of the soil in its loosest state e_{max} and its natural voids e to the difference between the void ratios in the loosest and densest state.

$$I_D = (e_{max} - e)/(e_{max} - e_{min}) \qquad (15.20)$$

where
e_{max} = void ratio in loosest state
e_{min} = void ratio in densest state
e = void ratio in natural state

15.4.2 Some Important Relationships

1. **Relation between** W_s, W, **and** w

$$W_s = W/(1+w) \qquad (15.21)$$

2. **Relation between** e **and** n

$$e = n/(1-n) \ \& \ n = e(1+e) \qquad (15.22)$$

3. **Relation between** e, S, w, **and** G

$$e\,S = w\,G \qquad (15.23)$$

4. **Relation between** e, S, **and** n_a

$$n_a = e(1-S)/(1+e) \qquad (15.24)$$

5. **Relation between** n, a_c, **and** n_a

$$n_a = n\,a_c \qquad (15.25)$$

6. **Relation between** G, γ_d, **and** e

$$\gamma_d = G\gamma_w/(1+e) \qquad (15.26)$$

7. **Relation between** G, γ_{sat}, **and** e

$$\gamma_{sat} = (G+e)\gamma_w/(1+e) \qquad (15.27)$$

8. **Relation between** G, γ_{sub}, **and** e

$$\gamma_{sub} = (G-1)\gamma_w/(1+e) \qquad (15.28)$$

9. **Relation between** G, γ_t, S, **and** e

$$\gamma_t = (G+e\,S)/(1+e) \qquad (15.29)$$

10. **Relation between γ_d, γ_t, and w**

$$\gamma_d = \gamma_t/(1+w) \qquad (15.30)$$

11. **Relation between γ_d, G, w, and n_a**

$$\gamma_d = (1-n_a)G\gamma_w/(1+wG) \qquad (15.31)$$

15.5 INDEX PROPERTIES OF SOILS

Index properties are those properties which are used for the identification and classification of soils. Index properties can be divided in to two categories.

  i. Soil grain properties
 ii. Soil aggregate properties

The soil grain properties depend on the individual grains of soil mass, whereas soil aggregate properties depend on the soil mass as a whole, that is, soil history, soil formation, and soil structure. Hence soil aggregate properties are of great engineering importance.

  i. Soil grain properties
       The most important soil grain properties of soil are as follows:
       a. Grain size distribution: by sieve and sedimentation analysis
       b. Grain shape: Bulky, flaky, and needle shaped, etc.
 ii. Soil aggregate properties
       The various soil aggregate properties are as follows:
       a. Unconfined compressive strength (q_u)
       b. Consistency and Atterberg's limits
       c. Sensitivity
       d. Thixotrophy and soil activity
       e. Relative density

15.5.1 PARTICLE SIZE ANALYSIS

15.5.1.1 Grain Size Distribution

Particle size analysis is done in two stages: (1) sieve analysis and (2) sedimentation analysis. Grain size analysis of coarse grained soil is carried out by sieve analysis, whereas analysis of fine grained soils is by sedimentation method (either by hydrometer or pipette method). Generally, most of the soil contains both coarse as well as fine grain constituents. Hence a combined analysis is usually carried out. In combined analysis, dry soil fraction retained on sieve size 4.75 mm is called gravel fraction which is subjected to coarse sieving analysis and soil fraction passing through 4.75 mm sieve is further subjected to fine sieve analysis. Fraction passing through 0.075 mm sieve is analyzed by hydrometer or pipette method.

Grain size distribution curve

A graph is plotted between % finer and sieve size in semi-log paper (Figure 15.14). Sieve size is taken on log scale on x-axis and % finer in arithmetic scale in y-axis. From the grain distribution, curve sizes corresponding to 60% finer, 30% finer, and 10% finer are computed. They are represented as D_{60}, D_{30}, and D_{10}, respectively. This curve gives us idea about the type and gradation of the soil. Grading of soil means the distribution of particles of different sizes in a soil mass. D_{60} represents a size in mm such that 60% of the particles are finer than this size. D_{10} is called effective size or effective diameter.

A soil sample may be either well graded or poorly graded (uniformly graded). A soil is said to be well graded when it has good representation of particles of all sizes. A soil is said to be poorly graded if it has an excess of certain particles and deficiency of other or if it has most of the particles of about the same size; in the latter case, it is known as uniformly graded soil. A curve with a flat portion represents a soil in which some intermediate size particles are missing. Such a soil is known as gap-graded or skip-graded. The grading of soil is shown in Figure 15.15.

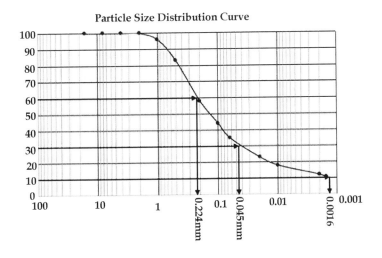

Particle Size Distribution Curve

Particle size D,mm

FIGURE 15.14 Particle size distribution curve.

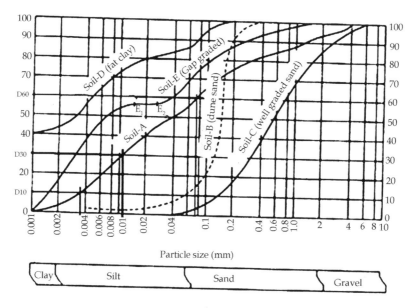

FIGURE 15.15   Grading of soil.

Uniformity Coefficient (Cu)

Uniformity coefficient is measure of particle size range and is given as:

$$C_u = D_{60}/D_{10} \tag{15.32}$$

For uniformly graded soil, C_u is nearly unity.

For a well graded gravel, $C_u > 4$.

For a well graded sand, $C_u > 6$.

Coefficient of Curvature (C_c)

The shape of the particle size curve is represented by coefficient of curvature.

$$C_c = (D_{30})^2/(D_{10} \times D_{60}) \tag{15.33}$$

For a well graded gravel, $1 \leq C_c \leq 3$.

For a well graded sand, $1 \leq C_c \leq 3$.

1. **Fine sieving**:

    Soil fraction passing through 4.75 mm sieve is subjected to fine sieve analysis. It can be performed either in dry state or wet state. Wet sieving is preferred when clay content is present in the sand. In fine sieving, following sieves are arranged in decreasing order as 2 mm, 1 mm, 600 μm, 425 μm, 150 μm, and 75 μm. The procedure of analysis is same as that of coarse analysis.

    Sedimentation analysis

    Sedimentation analysis is used to determine grain size distribution of the soil fraction passing through 0.075 mm size sieve. It is based on "Stoke's Law," which gives the terminal velocity of a small sphere settling in a fluid of infinite medium. When a small sphere settles in a fluid, its velocity first increases under the action of gravity, but the drag force comes into action and retards the velocity. After an initial

adjustment period, steady conditions are attained and the velocity becomes constant. The velocity attained is known as terminal velocity. Terminal velocity is given by

$$V = (\gamma_s - \gamma_w)D^2/18\mu \tag{15.34}$$

where

V = terminal velocity

γ_s = unit weight of spherical particle

γ_w = unit weight of water /liquid

D = diameter of the spherical particle

μ = dynamic viscosity (Ns/m²)

2. **Pipette method**

    In this method, 500 ml of soil suspension is required. All the quantities required for 1000 ml of suspension are halved to get a 500 ml of a suspension. This suspension is taken in a sedimentation tube. The pipette is fitted with a suction inlet.

    If m is the mass of dispersing agent per ml of suspension, the weight of soil solids per ml is given by

$$m_D = m'_D - m$$

where

m_D = actual mass of solid/ml

m'_D = mass of solids/ml as obtained from the sample.

    Merits and demerits of pipette method

    The pipette method (Figure 15.16) is a standard laboratory method for the particle size analysis of fine grained soil. It is very accurate method. However, the apparatus is quite delicate and expensive; it requires a very sensitive weighing balance. For quick particle size analysis, the hydrometer method is more convenient.

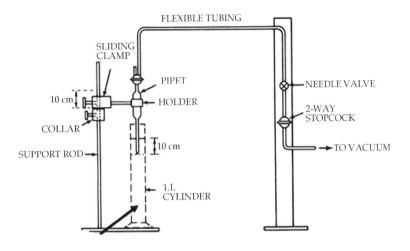

FIGURE 15.16 Pipette.

3. **Hydrometer method:**

A hydrometer is an instrument used for the determination of the specific gravity of liquids. A special type of hydrometer with a long stem is used as shown in Figure 15.17. The stem is marked from top to bottom, generally in the range of 0.995–1.030. The lower layers of the suspension have specific gravity greater than that of the upper layers.

Casagrande has shown that hydrometer measures the specific gravity of suspension at a point indicated by the center of immersed volume. If the volume of the stem is neglected, the center of the immersed volume of hydrometer is same as the center of the bulb. Thus, hydrometer gives the specific gravity of suspension at the center of the bulb.

15.5.2 CONSISTENCY OF SOIL

Consistency is meant for the relative ease with which soil can be deformed. This term is mostly used for fine grained soils

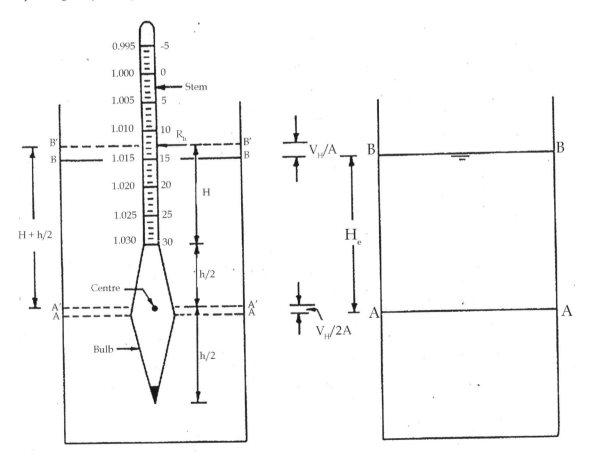

FIGURE 15.17 Hydrometer.

for which the consistency is related to a large extent to water content. Consistency denotes degree of firmness of the soil which may be termed as soft, firm, stiff, or hard. In 1911, the Swedish agriculturalist Atterberg divided the entire range from liquid to solid state into four stages: (1) the liquid state, (2) the plastic state, (3) semi-solid state, (4) solid state. Thus, the consistency limits are the water contents at which the soil mass passes from one state to the other. This is shown in Figure 15.18. These limits are expressed as per cent water content.

1. **Liquid limit [L.L] or** $[w_L]$: Liquid limit is the water content corresponding to the arbitrary limit between liquid and plastic state of consistency of a soil. It is defined as the minimum water content at which soil is still in the liquid state, but has a small shearing strength against flowing which can be measured by standard available means. With reference to standard liquid limit device, it is defined as the minimum water content at which a part of soil cut by groove of standard dimensions will flow together for a distance of 12 mm under an impact of 25 blows in the device.

2. **Plastic limit [P.L] or** $[w_P]$: Plastic limit is the water content corresponding to an arbitrary limit between the plastic and the semi-solid state states of consistency of a soil. It is defined as the minimum water content at which a soil will just begin to crumble when rolled into a thread of 3 mm in diameter.

3. **Shrinkage limit** $[w_s]$: Shrinkage limit is defined as the maximum water content at which a reduction in water content will not cause a decrease in the volume of a soil mass. It is the lowest water content at which a soil can still be completely saturated.

The shrinkage limit $[w_s]$ can be determined from:

$$w_s = \left[(W_1 - W_s) - (V_1 - V_d)\gamma_w\right]/W_d$$
$$= w_1 - \left[(V_1 - V_d)\gamma_w\right]/W_d \quad (15.35)$$

where
W_1 = original weight of soil
V_1 = original volume of soil
W_s = dry weight of soil
W_d = dry weight of soil
V_d = dry volume of soil
w_l = natural water content

15.5.2.1 Shrinkage Parameters

a. **Shrinkage index**: The shrinkage index is defined as the numerical difference between the liquid limit and the shrinkage limit.

$$I_s = w_l - w_s \quad (15.36)$$

b. **Shrinkage ratio (SR)**: SR is defined as the ratio of a given volume change expressed as a percentage of dry volume to the corresponding change in water content.

$$SR = \left[(V_1 - V_2)/V_d\right]/\left[w_1 - w_2\right] \times 100 \quad (15.37)$$

where
V_1 = volume of soil mass at water content w_1
V_2 = volume of soil mass at water content w_2
V_d = volume of dry soil mass
At shrinkage limit, $w_2 = w_s$ and $V_2 = V_d$

$$SR = \left[(V_1 - V_d)/V_d\right]/\left[w_1 - w_s\right] \times 100$$

SR can also be defined as

$$SR = \gamma_d/\gamma_w$$

c. **Volumetric shrinkage (VS)**: VS is defined as the change in volume expressed as a percentage of the dry volume when the water content is reduced from a given value to the shrinkage limit.

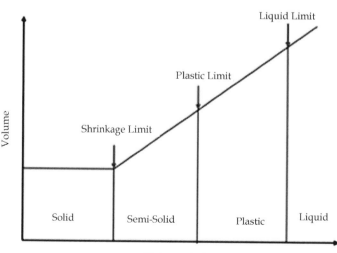

FIGURE 15.18 Atterberg limits.

$$VS = (V_1 - V_d)/V_d \times 100 \quad (15.38)$$

$$= SR(w_1 - w_s)$$

d. **Degree of shrinkage**: It is the percentage loss in volume of soil on drying corresponding to initial volume.

$$DOS = (V_1 - V_d)/V_1 \times 100 \quad (15.39)$$

where
V_1=initial volume of soil sample
V_d=dry volume of soil sample

e. **Linear shrinkage**: It is defined as the change in length divided by the initial length when the water content is reduced to the shrinkage limit.

15.5.3 Basic Definitions

a. **Plasticity index**: Plasticity index [I_P or PI] is the range of water content over which the soil remains in the plastic state. It is equal to the difference between the liquid limit and the plastic limit.

$$I_P = w_l - w_p \quad (15.40)$$

When either liquid limit or plastic limit cannot be determined, the soil is nonplastic. When the plastic limit is equal to or greater than the liquid limit, the plasticity index is reported as zero.

b. **Liquidity index** [I_L or LI]: Liquidity index or water-plasticity ratio is the ratio expressed as a percentage of the natural water content of a soil minus its plastic limit to its plasticity index.

$$I_L = (w - w_l)/I_P \quad (15.41)$$

where w=natural water content of the soil.

The liquidity index of a soil indicates the nearness of its water content to its liquid limit. When the soil is at liquid limit, its liquidity index is 100% and it behaves as a liquid. When the soil is at the plastic limit, its liquidity index is zero. Negative values of the liquidity index indicate water content smaller than the plastic limit. The soil is then in hard state.

c. **Consistency index** [I_C or CI]: Consistency index or relative consistency is defined as the ratio of the liquid limit minus the natural water content to the plasticity index.

$$I_C = (w_l - w)/I_P \quad (15.42)$$

The consistency index indicates the firmness of soil. It shows the nearness of the water content of soil to its plastic limit.

d. **Flow index** [I_f]: Flow index is defined as the slope of the flow curve obtained between the number of blows and the water content in Casagrande's method of determination of the liquid limit.

$$I_f = (w_1 - w_2)/\log(N_2/N_1) \quad (15.43)$$

where
N_1=number of blows required at water content w_1
N_2=number of blows required at water content w_2

e. **Toughness index** [I_t]: Toughness index of a soil is defined as the ratio of the plasticity index to the flow index.

$$I_t = I_P/I_f \quad (15.44)$$

Toughness index of a soil is a measure of the shearing strength of the soil at the plastic limit. The value of the toughness index of most soils lies between 0 and 3.

f. **Sensitivity**: When the structure is disturbed, the soil becomes remoulded and its engineering properties change considerably. Sensitivity [S_t] of a soil indicates its weakening due to remoulding. It is defined as the ratio of the undisturbed strength to the remoulded strength at the same water content. Soil type based on sensitivity is given in Table 15.1

$$S_t = (q_u)_u/(q_u)_r \quad (15.45)$$

where
$(q_u)_u$=unconfined compressive strength of the undisturbed clay
$(q_u)_r$=unconfined compressive strength of the remoulded clay

g. **Thixotropy**: The word Thixotropy is derived from two words: thixis meaning touch and tropo meaning to change. Therefore, thixotropy means any change that occurs by touch.

The loss of strength of a soil due to remoulding is partly due to change in the soil structure and partly due to disturbance caused to water molecules in the absorbed layer. Some of these changes are reversible. If a remoulded soil is allowed to stand, without loss

TABLE 15.1
Classification of Soils Based on Sensitivity

Serial No:	Sensitivity	Type of Soil
1	<1.00	Insensitive
2	1.0–2.0	Little sensitive
3	2.0–4.0	Moderately sensitive
4	4.0–8.0	Sensitive
5	8.0–16.0	Extra sensitive
6	>16.0	Quick

of water, it may regain some of its lost strength. In soil engineering, this gain in strength of the soil with passage of time after it has been remoulded is called thixotropy. It is mainly due to a gradual reorientation of molecules of water in the absorbed water layer and due to re-establishment of chemical equilibrium.

h. **Activity of soils**: Activity (A) of a soil is the ratio of the plasticity index and the percentage of clay fraction.

$$A = I_P/F \qquad (15.46)$$

where
I_P = plasticity index
F = percentage clay fraction (finer than 2 μm size)

The amount of water in a soil mass depends upon the type of clay mineral present. Activity is a measure of the water-holding capacity of clayey soils. The changes in the volume of clayey soil during swelling and shrinkage depend up on the activity. Soil classification based on activity is given in Table 15.2.

Activity gives information about the type and effect of clay mineral in a soil. The following two points are worth noting:

15.6 SOIL CLASSIFICATION

Soil classification is the arrangement of soils into different groups such that the soils in a particular group have similar behavior. As there is wide variety of soils, it is desirable to systematize or classify the soils into broad groups of similar behavior. For general engineering purposes, soil may be classified by the following systems.

1. **Particle size distribution**: In this system, soils are arranged according to the grain size. Terms such as gravel, sand silt, and clay are used to indicate grain sizes. These terms are used only as designation of particles size and do not signify the naturally occurring soil types which are mixtures of particles of different sizes and exhibit definite characteristics. It is preferable to use the word "silt size" and "clay size" in place of silt and clay in this system. There are various grain size classifications in use, but the more commonly used systems are as follows:
   a. U.S. Bureau of Soil and Public Road Administration System of United States
   b. International soil classification, proposed at the International Soil Congress at Washington, D.C., 1927
   c. The M.I.T. classification (Figure 15.19) proposed by Prof. Gilboy of Massachusetts Institute of Technology, a simplification of the Bureau of Soils Classification
   d. Indian Standard Classification (Figure 15.19) based on the M.I.T. system
2. **Textural classification**: Texture means visual appearance of the surface of a material such as fabric or cloth. The visual appearance of a soil is called its texture. The texture depends upon the particle size, shape of particles, and gradation of particles.

According to the textural classification system, the percentage of sand (size 0.05–2.0 mm), silt (size 0.005–0.05 mm), and clay (size less than 0.005 mm) are plotted along the three sides of an equilateral triangle. The equilateral triangle is divided into 10 zones. Each zone indicates a type of soil. The soil

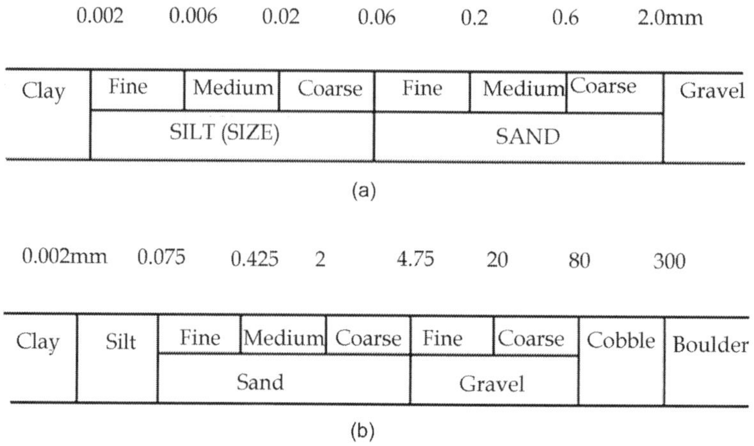

TABLE 15.2
Classification of Soils Based on Activity

Serial No:	Activity	Soil Type
1	$A < 0.75$	Inactive
2	$A = 0.75$ to 1.25	Normal
3	$A > 1.25$	Active

	0.002	0.006	0.02	0.06	0.2	0.6	2.0mm

Clay	Fine	Medium	Coarse	Fine	Medium	Coarse	Gravel
	SILT (SIZE)			SAND			

(a)

	0.002mm	0.075	0.425	2	4.75	20	80	300

Clay	Silt	Fine	Medium	Coarse	Fine	Coarse	Cobble	Boulder
		Sand			Gravel			

(b)

FIGURE 15.19 (a) MIT and (b) IS soil classification.

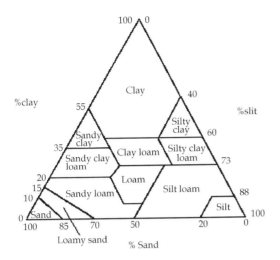

FIGURE 15.20 Textural classification.

can be classified by determining the zone in which it lies. A key is given that indicates the directions in which the lines are to be drawn to locate the point.

The textural classification system (Figure 15.20) is useful classifying soils consisting of different constituents. The system assumes that the soil does not contain particles larger than 2.0 mm size. However, if the soil contains a certain percentage of soil particles larger than 2.0 mm, a correction is required in which the sum of the percentages of sand, silt, and clay is increased to 100%.

3. **Highway research board classification or AASHTO classification system**: American Association of State Highway and Transportation Official (AASHTO) classification system is useful for classifying soils for highways. The particle size analysis and the plasticity characteristics are required to classify a soil. The classification system is a complete system which classifies both coarse-grained and fine-grained soils. In this system, the soils are divided into 7 types, designated as A-1 to A-7. The soils A-1 and A-7 are further subdivided into two categories and the soil A-2 into four categories.

To classify a soil, its particle size analysis is done and the plasticity index and liquid limit are determined. A characteristics group index is used to describe the performance of the soils when used for pavement construction. Group index is not used to place a soil in particular group, it is actually a means of rating the value of the index, the poorer is the quality of the material.

$$G.I = 0.2a + 0.005ac + 0.01bd \qquad (15.47)$$

where

a = that portion of percentage passing 75 μm sieve greater than 35 and not exceeding 75 expressed as a whole number (0–40)

b = that portion of percentage passing 75 μm sieve greater than 15 and not exceeding 55 expressed as a whole number (0–40)

c = that portion of the numerical liquid limit greater than 40 and not exceeding 60 expressed as positive whole number (0–20)

d = that portion of the numerical liquid limit greater than 10 and not exceeding 30 expressed as positive whole number (0–20)

The group index of a soil depends upon (1) the amount of material passing the 75 μm IS sieve, (2) the liquid limit, and (3) the plastic limit. The smaller the value of group index, the better is the soil. A group index of zero indicates a good subgrade whereas a group index of 20 or greater shows a very poor subgrade.

4. **Unified soil classification system**: This system was first developed by Casagrande in 1948 and later in 1952, and was modified by the Bureau of Reclamation and the Corps of Engineers of the United States of America. The system has also been adopted by American Society of Testing Machine. The system is the most popular system for use in all types of engineering problems involving soils.

The system uses both the particle size analysis and plasticity characteristics of soils like AASHTO system. In this system the soils are classified into 15 groups. The soil is first classified into two categories.

Coarse-grained soils: If more than 50% of the soil is retained in IS sieve of size 75 μm (No. 200 sieve), it is designated as coarse-grained soil. There are 8 groups of coarse-grained soil.

Fine-grained soils: If more than 50% of the soil passes in IS sieve of size 75 μm (No. 200 sieve), it is designated as fine-grained soil. There are 6 groups of coarse-grained soil. Symbols used in Universal Soil Classification (USC) system are presented in Table 15.3.

a. Coarse-grained soils: The coarse-grained soils are designated as gravel (G) if 50% or more

TABLE 15.3

Symbols Used in USC System

Symbols	Descriptions
G	Gravel
S	Sand
M	Silt
C	Clay
O	Organic
W	Well-graded
P	Poorly graded
M	Nonplastic fines
C	Plastic fines
L	Low plasticity
H	High plasticity

of coarse fraction (plus 0.075 mm) is retained on 4.75 mm sieve (No. 4 sieve); otherwise it is termed as sand (S).

If the coarse-grained soils contain less than 5% fines and are well graded (W), they are given the symbols GW and SW, and if poorly graded (P), symbols GP and SP. If the coarse-grained soils contain more than 12% fines, these are designated as GM, GC, SM, or SC. If the percentage of fines is between 5% and 12% dual symbols such as GW-GM and SW-SM are used.

Figure 15.21 represents the plasticity chart and Figures 15.22 and 15.23 represent the US classification of coarse-grained soils and fine-grained soil, respectively.

Plasticity index, I_P
Liquid limit, W_L
CH
$I_P = 0.73 (W_L = 20)$
A-Line
$W_L = 50$
$W_L = 35$
CI
MH or OH
C_L-M_L
CL
MI or OI
ML
ML
0 10 20 30 40 50 60 70 80 90 100
10 4 7 20 30 40 50 60

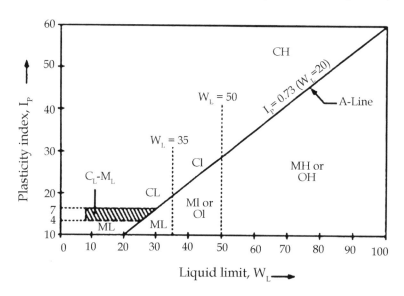

FIGURE 15.21 Plasticity chart.

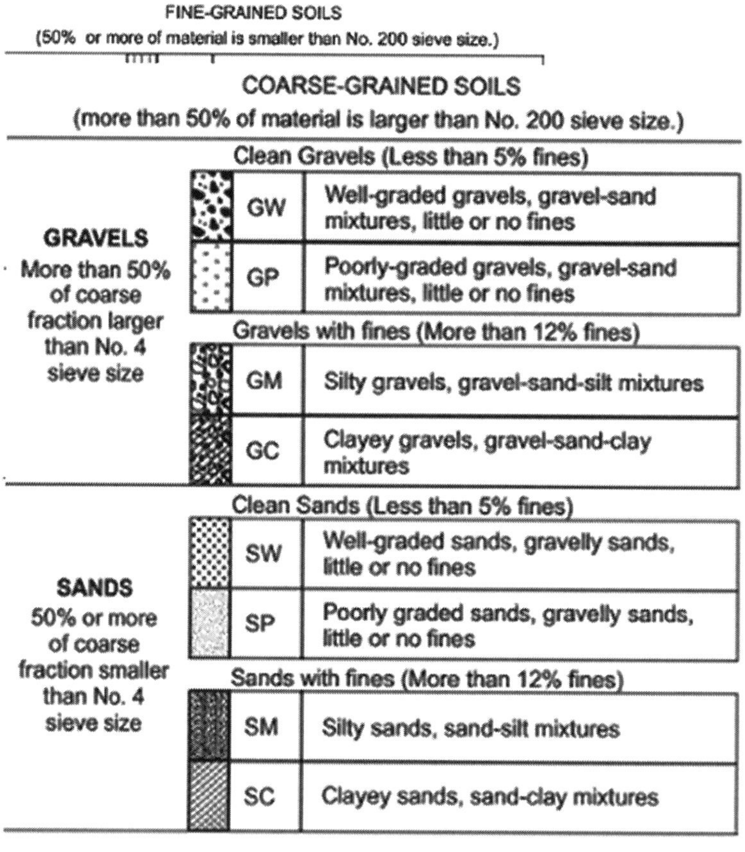

FINE-GRAINED SOILS
(50% or more of material is smaller than No. 200 sieve size.)
COARSE-GRAINED SOILS
(more than 50% of material is larger than No. 200 sieve size.)
Clean Gravels (Less than 5% fines)
GRAVELS
More than 50% of coarse fraction larger than No. 4 sieve size
GW
Well-graded gravels, gravel-sand mixtures, little or no fines
GP
Poorly-graded gravels, gravel-sand mixtures, little or no fines
Gravels with fines (More than 12% fines)
GM
Silty gravels, gravel-sand-silt mixtures
GC
Clayey gravels, gravel-sand-clay mixtures
Clean Sands (Less than 5% fines)
SANDS
50% or more of coarse fraction smaller than No. 4 sieve size
SW
Well-graded sands, gravelly sands, little or no fines
SP
Poorly graded sands, gravelly sands, little or no fines
Sands with fines (More than 12% fines)
SM
Silty sands, sand-silt mixtures
SC
Clayey sands, sand-clay mixtures

FIGURE 15.22 US classification of coarse-grained soil.

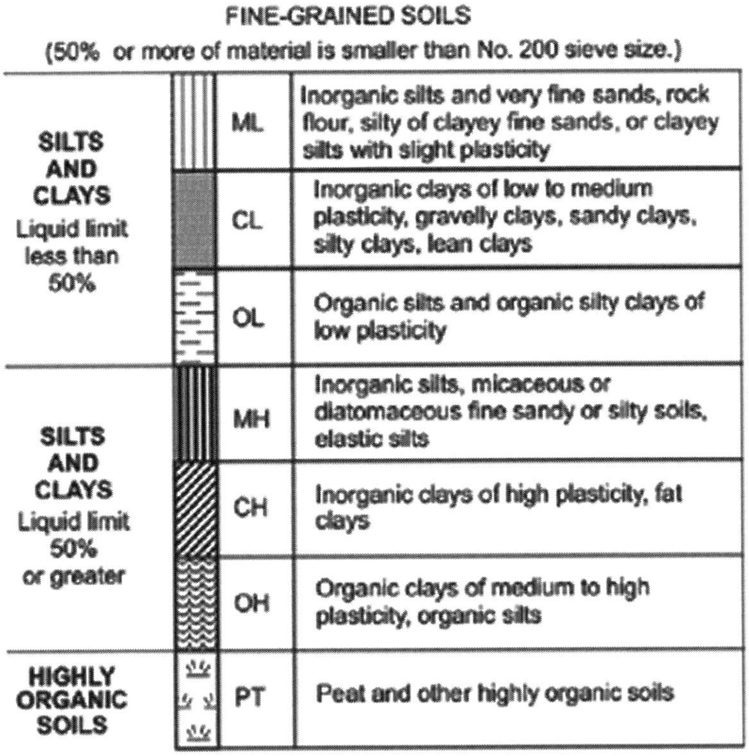

FINE-GRAINED SOILS

(50% or more of material is smaller than No. 200 sieve size.)

SILTS AND CLAYS Liquid limit less than 50%		ML	Inorganic silts and very fine sands, rock flour, silty of clayey fine sands, or clayey silts with slight plasticity
		CL	Inorganic clays of low to medium plasticity, gravelly clays, sandy clays, silty clays, lean clays
		OL	Organic silts and organic silty clays of low plasticity
SILTS AND CLAYS Liquid limit 50% or greater		MH	Inorganic silts, micaceous or diatomaceous fine sandy or silty soils, elastic silts
		CH	Inorganic clays of high plasticity, fat clays
		OH	Organic clays of medium to high plasticity, organic silts
HIGHLY ORGANIC SOILS		PT	Peat and other highly organic soils

FIGURE 15.23 US classification of fine-grained soil.

b. Fine-grained soils: Fine-grained soils are further divided into two types:

   1. Soils of low compressibility (L): if the liquid limit is 50% or less. These are given the symbols ML, CL, and OL.
   2. Soils of high compressibility (H): if the liquid limit is more than 50%. These are given the symbols MH, CH, and OH.

   The exact type of the soil is determined from the plasticity chart. The A-line has the equation $I_p = 0.73(w_1 - 20)$. It separates clays from silt. When the plasticity index and the liquid limit plot in the hatched portion of the plasticity chart, the soil is given double symbol CL-ML.

   The inorganic soil ML and MH and the organic soil OL and OH plot in the same zones of the plasticity chart. The distinction between the inorganic and organic soils is made by oven-drying. If oven-drying decreases the liquid limit by 30% or more, the soil is classified organic (OL or OH) and inorganic (ML or MH).

   Highly organic soils are identified by visual inspection. These soils are termed as peat.

5. **Indian standard classification system**: Indian standard classification system adopted by Bureau of Indian standards is in many respects similar to the Unified Soil classification system. However, there is one basic difference in the classification of fine grained soil. The fine-grained soil are subdivided into three categories of low, medium, and high compressibility instead of two categories of low and high compressibility in the US classification system.

   Soils are divided into three broad divisions.

a. Coarse-grained soils, when 50% or more of the total material by weight is retained on 75 μm IS sieve.

b. Fine-grained soils, when more than 50% of the total material passes 75 μm IS sieve.

c. If the soil is highly organic and contains a large percentage of organic matter and particles decomposed vegetation, it is kept in a separate category marked as peat (P).

   In all, there are 18 groups: 8 groups of coarse-grained 9 groups of fine-grained soil and one of peat.

a. Coarse-grained soils:

   Coarse-grained soils are subdivided into gravel and sand. The soil is termed as gravel (G) when more than 50% of coarse fraction (plus 75 μm) is retained on 4.75 mm IS sieve, and termed as sand (S) if more than 50% of the coarse fraction is smaller than 4.75 mm IS sieve.

b. Fine-grained soils:

   Fine-grained soils are further divided into three subdivisions, depending upon the values of the liquid limits:

   i. Silt and clays of low compressibility: These soils have liquid limit less than 35 (represented by L).

TABLE 15.4

Basic Soil Components in IS Classification System

Soil	Soil Components	Symbols	Particle Size Range and Description
1. Coarse-grained components	Boulders	None	Rounded to angular, bulky, hard rock, particle; average diameter more than 300 mm
	Cobble	None	Rounded to angular, bulky, hard rock, particle; average diameter smaller than 300 mm but retained on 80 mm IS sieve
	Gravel	G	Rounded to angular, bulky, hard rock, particle: passing 80 mm IS sieve but retained on 4.75 mm IS sieve. Coarse: 80 mm to 20 mm IS sieve; Fine: 20 mm to 4.75 mm IS sieve
	Sand	S	Rounded to angular, bulky, hard rock, particle: passing 4.75 mm IS sieve but retained on 75 μm IS sieve. Coarse: 4.75 mm to 2.0 mm IS sieve; Medium: 2.0 mm to 425 μm IS sieve; Fine: 425 μm to 65 μm IS sieve
2. Fine-grained components	Silt	M	Particles smaller than 75 μm IS sieve; identified by behavior, that is, slightly plastic or nonplastic regardless of moisture and exhibits little or no strength when air-added.
	Clay	C	Particles smaller than 75 μm IS sieve; identified by behavior, that is, it can be made to exhibit plastic properties within a certain considerable strength when air-dried.
	Organic matter	O	Organic matter in various sizes and stages of decomposition.

ii. Silt and clays of medium compressibility: These soils have liquid limit greater than 35 but less than 50 (represented by I).

iii. Silt and clays of high compressibility: These soils have liquid limit greater than 50 (represented by H).

Basic soil components in IS classification system are presented Table 15.4.

15.7 PERMEABILITY

The property of the soil which permits flow of water through it is called the permeability. The permeability is the ease with which water can flow through it. Permeability is a very important engineering property of soils. A knowledge of permeability is essential in a number of soil engineering problems such as settlement of buildings, yield of wells, seepage through and below the earth structure. It controls the hydraulic stability of soil masses. The permeability of soils is also required in the design of filters used to prevent piping in hydraulic structures.

15.7.1 HYDRAULIC HEAD

The total head at any point in a flowing fluid is equal to the sum of the elevation (datum) head, the pressure head, and the velocity head. The elevation head (z) is equal to the vertical distance of the point above the datum. The pressure (p/γ_w) is equal to the head indicated by a piezometer with its tip at that point.

The velocity head is equal to $v^2/2g$. However, for flow of water through soils, as the velocity (v) is extremely small, the velocity head is neglected. Therefore, the total head of water in soil engineering problems is equal to the sum of the elevation head and the pressure head. For flow problems in soils, the downstream water level is generally taken as datum. The piezometric level is the water level shown by a piezometer

inserted at that point. The line joining the piezometric levels at various points is called piezometric surface. The piezometric surface also represents the hydraulic gradient line. The sum of the pressure head and the elevation head is known as the piezometric head.

The loss of head per unit length of flow through the soil is equal to the hydraulic gradient (i).

$$i = h/L \tag{15.48}$$

where

h=hydraulic head
L=length of the specimen

15.7.2 DARCY'S LAW

The flow of free water through soil is governed by Darcy's law. In 1856, Darcy demonstrated experimentally that for laminar flow in a homogenous soil, the velocity of flow (v) is given by

$$v = k\,i \tag{15.49}$$

where

k=coefficient of permeability
i=hydraulic gradient

The velocity of flow is also known as the discharge velocity or the superficial velocity. The discharge q is obtained by multiplying the velocity of flow (v) by the total cross sectional area of soil (A) normal to the direction of flow.

$$q = v\,A = k\,i\,A \tag{15.50}$$

where k=coefficient of permeability

Coefficient of permeability is defined as the velocity of flow which would occur under unit hydraulic gradient. The coefficient of permeability has the dimensions of velocity (L/T). It is measured in mm/s, cm/s, m/s, m/day, or other velocity units.

15.7.2.1 Seepage Velocity

The discharge velocity is based on the gross cross sectional area while the seepage velocity is due to the voids present in the soil. So the seepage velocity is also called actual velocity of the soil.

If the quantity of water flowing through the soil in unit time is q, then

$$q = vA = v_s A_v \qquad (15.51)$$

where

v_s = seepage velocity
A_v = area of voids in the soil

From the equation we can get

$$v = \frac{A_v}{A} v_s$$

$$v = n v_s \qquad (15.52)$$

15.7.2.2 Value of Hydraulic Conductivity (k)

Typical value of hydraulic conductivity or permeability of saturated soil is given in Table 15.5.

15.7.2.3 Empirical Relation for k

Various equations have been proposed in the past. These are given below.

For uniform sand

$$k = c D^2_{10} \qquad (15.53)$$

where

c = constant of value 100
D = effective size in cm

For dense compacted sand

$$k = 0.35 D^2_{15} \qquad (15.54)$$

TABLE 15.5

Typical Value of Hydraulic Conductivity or Permeability of Saturated Soil

Soil Type	K, cm/s
Clean gravel	100–1
Coarse sand	1–0.01
Fine sand	0.01–0.001
Silty clay	0.001–0.00001
Clay	<0.00001

For medium-to-fine sand

$$k = 1.4 e^2 k_{0.85} \qquad (15.55)$$

where

e = void ratio
$k_{0.85}$ = hydraulic conductivity at void ratio 0.85

15.7.3 Determination of Coefficient of Permeability

The coefficient of permeability of a soil can be determined using following methods.

a. **Laboratory methods**: The coefficient of permeability of a soil sample can be determined by the following methods.
   i. Constant head permeability test
   ii. Variable head permeability test
   The former test is suitable for relatively more pervious soils and latter for less pervious soils.
b. **Field methods**: The coefficient of permeability of a soil deposit in-situ conditions can be determined by the following field methods.
   i. Pumping-out test
   ii. Pumping-in test
   The pumping-out tests influence a large area around the pumping well and give an overall value of the coefficient of permeability of the soil deposit. The pumping-in test influences a small area around the hole and therefore gives a value of the coefficient of permeability of the soil surrounding the hole.
c. **Indirect methods**: The coefficient of permeability of the soil can also be determined indirectly from the soil parameters by
   i. computation from the particle size or its specific surface
   ii. computation from the consolidation test data
   The first method is used if the particle size is known. The second method is used when the coefficient of volume change has been determined from the consolidation test on the soil.
d. **Capillarity–permeability test**: The coefficient of permeability of an unsaturated soil can be determined by the capillarity–permeability test.

15.7.3.1 Constant Head Permeability Test

Water flows from the overhead tank consisting of the three tubes: the inlet tube, the overflow tube, and the outlet tube. The constant hydraulic gradient i causing the flow is the head h (difference in the water levels of the overhead and bottom tanks) divided by the length L of the sample. If the length of the sample is large, the head lost over a length L of the specimen is measured by inserting piezometric tubes. Figure 15.24 represents the constant head permeability test.

If Q is the total quantity of flow in a time interval t, we have from Darcy's law,

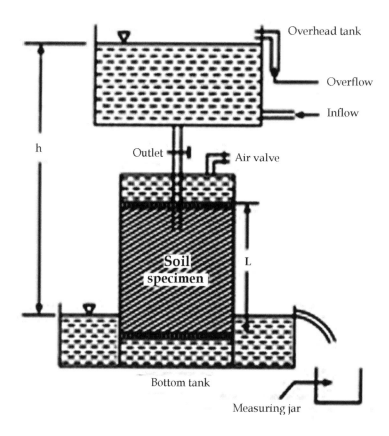

FIGURE 15.24 Constant head permeability test setup.

$$q = Q/t$$

$$= k\,i\,A$$

Therefore, $\quad k = (Q/t)(1/i\,A)$

$$= (QL)/(t\,h\,A) \qquad (15.56)$$

where A = total cross sectional area of sample.

When steady state of flow is reached, the total quantity of water Q in time t is collected in a measuring jar. It is based on the assumption that flow is laminar where K is independent of i.

15.7.3.2 Falling Head Permeability Test

The constant head permeability test is used for coarse-grained soil only where a reasonable discharge can be collected in a given time. However, the falling head permeability test is used for relatively less permeable soils where the discharge is small. Figure 15.25 shows the diagrammatic representation of a falling head test arrangement.

A stand pipe of known cross-sectional area a is fitted over the permeameter and water is allowed to run down. The water level in the stand pipe constantly falls as water flows. Observations are started after steady state of flow has reached. The head at any time instant t is equal to the difference in the water level in the stand pipe and the bottom tank. Let h_1 and h_2 be heads at time intervals t_1 and t_2 $(t_2 > t_1)$, respectively.

Let h be the head at any intermediate time intervals t, and $-dh$ be the change in the head in a smaller time interval dt (minus sign has been used since h decreases as t increases). Hence from Darcy's law, the rate of flow q is given by

$$q = (-dh\,a)/dt$$

$$= k\,i\,A$$

where I = hydraulic gradient at time $t = h/L$

$$(k\,h/L)A = -(dh/dt)a$$

$$(Ak/aL)dt = -(dh/dt)$$

Integrating between two time limits (t_1 and t_2), we get

$$(Ak/aL)(t_2 - t_1) = \log_e\left(h_1/h_2\right)$$

Denoting $t_2 - t_1 = t$, we get

$$k = (aL/At)\log_e\left(h_1/h_2\right)$$

$$k = 2.303(aL/At)\log_{10}\left(h_1/h_2\right) \qquad (15.57)$$

15.7.3.3 Pumping Tests

The main principle behind pumping tests is that if we pump water from a well and calculate the discharge of the well and the drawdown in the well and in piezometers at known

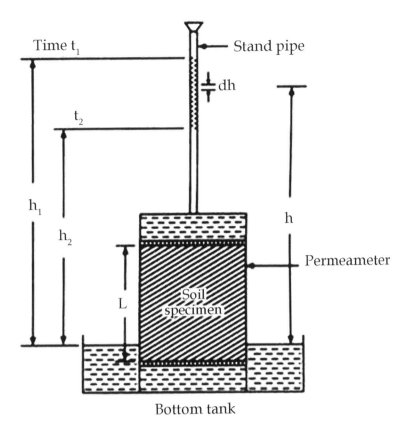

FIGURE 15.25   Falling head permeability test setup.

distance from the well, we can substitute these measurements into well-defined flow equations and find the hydraulic characteristics.

15.7.3.4   Capillary Rise in Soil

Above the water table, when the soil is saturated, the pore pressure will be negative (less than atmospheric). The height that is above the water table to which the soil is saturated is called capillary rise, this depends on the grain size and the size of the pores. In case of coarse soils, the capillary rise is very small.

The continuous void spaces in the soil can behave as bundles of capillary tubes of various cross section. Because of the surface tension force produced, the water may rise to the phreatic surface. The height of rise of the water in the capillary tube can be given by summing the forces in the vertical direction.

$$h_c = \frac{4T_s \cos\alpha}{d\gamma_w} \qquad (15.58)$$

The surface tension T_s of water at 20°C can be equal to 75×10^{-8} kN/cm.

15.7.4   Factors Affecting Permeability of Soils

The factors affecting permeability are given by an equation

$$k = \left[D^2\gamma_w e^3 C\right]/\left[\mu(1+e)\right] \qquad (15.59)$$

1. **Effect of size of particle**: From Equation 15.59, it is clear that the coefficient of permeability of a soil is proportional to the square of the particle size. The permeability of coarse-grained soils is very large as compared to that of fine grained soils. Allen Hazen based on his experiments on filter sands of particles size between 0.1 and 3 mm found that the permeability can be expressed as

$$k = C D^2_{10} \qquad (15.60)$$

where
k = coefficient of permeability (cm/s)
D_{10} = effective diameter (cm)
C = constant approximately equal to 100

2. **Structure of soil mass**: The coefficient C in Equation 15.59 takes into account the shape of the flow passage. For the same void ratio, the permeability is more in the case of flocculated structure as compared to that in the dispersed structure.

   Stratified soil deposits have greater permeability parallel to the plane of stratification than that perpendicular to this plane. Loess deposits have greater permeability in the vertical direction than in the horizontal direction.

3. **Shape of particles**: The permeability of a soil depends upon the shape of particles. Angular particles have greater specific surface area as compared with rounded particles. For the same void ratio, the

soils with angular particles are less permeable than those with rounded particles, as the permeability is inversely proportional to the specific surface area.

4. **Void ratio**: Equation 15.59 indicates that the coefficient of permeability varies as $e^3/(1+e)$. Increase in void ratio increases the area available for flow. So for a given soil, the greater the void ratio, the higher is the value of the coefficient of permeability.

5. **Properties of pore fluid**: Pore fluids are fluids that occupy the void spaces. As Equation 15.59 indicates, the coefficient of permeability is directly proportional to the unit weight of water (γ_w) and is inversely proportional to its viscosity (μ). The coefficient of permeability increases with an increase in temperature due to reduction in the viscosity.

6. **Degree of saturation**: If the soil is not fully saturated, it contains air pockets formed due to entrapped air or due to air liberated from percolating water. Whatever may be the cause of the presence of air in soils, the permeability is reduced due to presence of air which causes blockage of passage. Consequently, the permeability of a partially saturated soil is considerably smaller than that of a fully saturated soil.

7. **Adsorbed water**: Adsorbed water is thin microscopic film of water surrounding the individual soil grains. The fine-grained soils have a layer of adsorbed water strongly attached to their surface. This adsorbed water layer is not free to move under gravity. It causes an obstruction to flow of water in the pores and hence reduces the permeability of soils.

8. **Entrapped air and impurities in water**: Any foreign matter in water has a tendency to plug the flow passage and reduce the effective voids and hence reduce the permeability of soils.

15.7.5 Permeability of Stratified Soil Deposits

A stratified soil deposit consists of a number of soil layers having different permeabilities. Each layer, assumed to be homogeneous and isotropic, has its own value of coefficient of permeability. The average permeability of the deposit as a whole parallel to the planes of stratification and that normal of the planes of stratification can be determined.

1. Flow parallel to planes of stratification:
   Let us consider a deposit consisting of two horizontal layers of soil of thickness H_1 and H_2.
   For flow parallel to the planes of stratification, the loss of head (h) over a length L is same for both the layers. Therefore, the hydraulic gradient (i) for each layer is equal to the hydraulic gradient of the entire deposit. From the continuity equation, the total discharge (q) per unit width is equal to the sum of the discharges in the individual layers.

$$q = q_1 + q_2$$

Let k_1 and k_2 be the permeability of the layers 1 and 2, respectively, parallel to the plane of stratification and k_h be the average permeability of the soil deposit parallel to the planes of stratification.

$$q = k_1 i H_1 + k_2 i H_2$$

$$k_h = (k_1 H_1 + k_2 H_2)/(H_1 + H_2)$$

If there are n number of layers, then

$$k_h = (k_1 H_1 + k_2 H_2 + \cdots + k_n H_n)/(H_1 + H_2 + \cdots H_n) \quad (15.61)$$

2. Flow perpendicular to the plane of stratification:
   Let us consider a soil deposit consisting of two layers of thickness H_1 and H_2 in which flow occurs perpendicular to the plane of stratification.
   Let k_1 and k_2 be the permeability of the layers 1 and 2, respectively, in the direction perpendicular to the plane of stratification and k_v be the average permeability of the soil deposit perpendicular to the plane of stratification. In this case the discharge per unit width is the same for each layer and is equal to the discharge in the entire deposit. However, the hydraulic gradient and hence the head loss through each layer will be different.

$$q = q_1 = q_2$$

$$h = h_1 + h_2$$

$$h = i_1 H_1 + i_2 H_2$$

We have $v = k_v i = k_v (h/H)$

$$h = v H / k_v$$

Also $i_1 = v/k_1$ and $i_2 = v/k_2$
Substituting these values, we will get,

$$k_v = (H_1 + H_2)/\left[(H_1/k_1) + (H_2/k_2)\right]$$

If there are n number of layers, then

$$k_v = (H_1 + H_2 + \cdots + H_n)/\left[(H_1/k_1) + (H_2/k_2) + \cdots + (H_n/k_n)\right]$$

$$(15.62)$$

15.8 SEEPAGE ANALYSIS

Seepage is the flow of water under gravitational forces in a permeable medium. Flow of water takes place from a point of high head to a point of low head. The flow is generally laminar.

The path taken by a water particle is represented by a flow line. Although an infinite number of flow lines can be drawn, for convenience, only a few are drawn. At certain points on different flow lines, the total head will be the same. The lines connecting points of equal total head can be drawn. These lines are known as equipotential lines. As flow always takes place along the steepest hydraulic gradient, the equipotential lines cross flow lines at right angles. The flow lines and equipotential lines together form a flow net. The flow net gives a pictorial representation of the path taken by water particles and the head variation along that path.

15.8.1 SEEPAGE PRESSURE

By virtue of the viscous friction exerted on water flowing through soil pores, an energy transfer is affected between the water and the soil. The force corresponding to this energy transfer is called the seepage force or seepage pressure. Thus, seepage pressure is the pressure exerted by water on the soil through which it percolates. It is this seepage pressure that is responsible for the phenomenon known as quick sand and is of vital importance in the stability analysis of earth structures subjected to the action of seepage.

If h is the hydraulic head or the head lost due to frictional drag of water flowing through a soil mass of thickness z, the seepage pressure p_s is given by

$$p_s = h\gamma_w$$

$$p_s = (h/z)z\gamma_w$$

$$= iz\gamma_w$$

where

z=length over which the head h is lost
i=hydraulic gradient

The seepage force J transmitted to the soil mass of total cross sectional area A is

$$J = p_s A$$

$$J = iz\gamma_w A$$

The seepage force per unit volume is given by $j=i\,z\,\gamma_w\,A/z$ $A=i\gamma_w$

The seepage pressure always acts in the direction of flow. The vertical effective pressure may be decreased or increased due to the seepage pressure depending up on the direction of flow. Thus the effective pressure in a soil mass subjected to seepage pressure is given by

$$\sigma' = z\gamma' \pm p_s = z\gamma' \pm iz\gamma_w$$

If the flow occurs in the downward direction, the effective pressure is increased and hence+sign is used. If, however, flow occurs in upward direction, the effective pressure is decreased.

15.8.2 QUICK SAND CONDITION

When flow takes place in an upward direction, the seepage pressure also acts in the upward direction and the effective pressure is reduced. If the seepage pressure becomes equal to the pressure due to submerged weight of the soil, the effective pressure is reduced to zero. In such a case, a cohesionless soil loses all its shear strength and the soil particles have a tendency to move up in the direction of flow. This phenomenon of lifting of soil particles is called quick sand condition, boiling condition, or quick sand. Thus, during the quick condition

$$\sigma' = z\gamma_{sub} - p_s$$

or

$$p_s = z\gamma_{sub} \text{ or } iz\gamma_w = z\gamma_{sub}$$

From which

$$i = i_c = \gamma_{sub}/\gamma_w$$

$$i_c = (G-1)/(1+e) \qquad (15.63)$$

where i_c=critical hydraulic gradient.

The hydraulic gradient at such a state is called critical hydraulic gradient. For loose deposits of sand or silt, if void ratio e is taken as 0.67 and G as 2.67, the critical hydraulic gradient works out to unity. It should be noted that quick sand is not a type of sand but a flow condition occurring within a cohesionless soil when its effective pressure is reduced to zero due to upward flow of water.

15.9 STRESS DISTRIBUTION IN SOIL

Stresses are induced in as soil mass due to weight of overlying soil and due to the applied loads. These stresses are required for the stability analysis of the soil mass, the settlement analysis of foundations, and the determination of the earth pressures.

The stresses induced in the soil due to applied loads depend upon its stress–strain characteristic and also depends up on a large number of factors such as drainage condition, water content, void ratio, rate of loading, the load level, and the stress path.

15.9.1 GEOSTATIC STRESSES

The stresses due to self-weight of soils are generally large as compared with those induced due to imposed loads. In soil engineering problems, the stresses due to self-weight are significant. In many cases the stresses due to self-weight are a large proportion of the total stresses and may govern the design.

When the ground surface is horizontal and the properties of the soil do not change along a horizontal plane, the stresses due to self-weight are known as geostatic stresses. Such a condition generally exists in sedimentary soil deposits. In such a case, the stresses are normal to the horizontal and vertical planes and there are no shearing stresses on these planes. In other words, these planes are principal planes. The vertical stress and horizontal stress can be determined as below.

1. **Vertical stresses**: The vertical stresses are determined by considering the horizontal plane A-A at a depth z below the ground surface.

   Let the area of cross section of the prism be A. If the unit weight of the soil (γ) is constant, the vertical stress (σ_z) is equal to the weight of the soil in prism divided by the area of the base. Thus,

   $$\sigma_z = \text{weight of the soil in prism/area of the base.}$$

   $$= \gamma(z \times A)/A$$

   $$\sigma_z = \gamma\, z \qquad (15.64)$$

2. **Horizontal stresses**: The horizontal stresses (σ_x and σ_y) act on vertical planes. The horizontal stresses at a point in a soil mass are highly variable. These depend not only upon the vertical stresses but also on the type of the soil and on the conditions whether the soil is stretched or compressed laterally. In the treatment that follows, it would be assumed that $\sigma_x = \sigma_y$.

   The ratio of the horizontal stress to the vertical stress is known as the coefficient of lateral stress or lateral stress ratio (K). Thus

   $$K = \sigma_x/\sigma_z \qquad (15.65)$$

   $$\sigma_x = K\,\sigma_z$$

   In natural deposits generally there is no lateral strain. The lateral stress coefficient for this case is known as the coefficient of lateral pressure at rest (K_0). The value of K_0 can be obtained from the theory of elasticity and is given by

   $$K_0 = \mu\,/\,1-\mu \qquad (15.66)$$

   The value of K_0 can be obtained if the Poisson's ratio μ is known or estimated. But in practical use as the soil is not a purely elastic material and it is difficult to estimate Poisson's ratio.

   The value of K_0 is determined from actual measurement of soil pressure or from experience. For a sedimentary sand deposit, its value varies from 0.30 to 0.6, and for a normally consolidated clay, its value varies from 0.5 to 1.0.

   Jacky's formula is commonly used to find K_0.

   $$K_0 = 1 - \sin\varphi \qquad (15.67)$$

where φ = angle of shearing resistance.

15.9.2 Vertical Stress due to Concentrated Load

Boussinesq gave the theoretical solutions for the stress distribution in an elastic medium subjected to a concentrated load on its surface. The solutions are commonly used to obtain the stresses in a soil mass due to externally applied loads.

15.9.2.1 Assumptions

1. The soil mass is an elastic continuum, having a constant value of modulus of elasticity (E). That is, the ratio between the stress and strain is constant.
2. The soil is homogenous, that is, it has identical properties at different points.
3. The soil is isotropic, that is, it has identical properties in all directions.
4. The soil mass is semi-infinite, that is, it extends to infinity in the downward direction and lateral directions. In other words, it is limited on its top by a horizontal plane and extends to infinity in all other directions.
5. The soil is weightless and is free from residual stresses before the application of the load.

Let a point load P act on the ground surface at a point O which may be taken as the origin of the x, y, and z axes as shown in Figure 15.26. Let us find the stress components at a point A in the soil mass, having coordinates x, y, and z or having a radial horizontal distance r and vertical distance z from the point O; thus the vertical stress at a point situated at a radial distance r and depth z below the loaded point is given by

$$\sigma_z = (3/2\pi)\left(P/z^2\right)\left[1/1+(r/z)^2\right]^{5/2} \qquad (15.68)$$

$$\sigma_z = I_B\left(P/z^2\right) \qquad (15.69)$$

where I_B = Boussinesq's influence factor

The intensity of vertical stress just below load point ($r=0$) is given by

$$\sigma_z = 0.4775\left(P/z^2\right) \qquad (15.70)$$

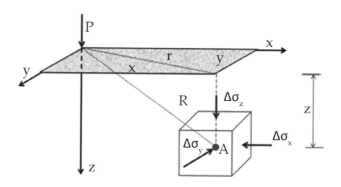

FIGURE 15.26 Boussinesq stresses.

15.9.2.2 Limitations of Boussinesq's Solution

The solution was initially obtained for determination of stresses in elastic solids. Its application to soils may be questioned as the soils are far purely elastic solids. However, experience indicates that the results obtained are satisfactory.

The application of Boussinesq's solution can be justified when the stress changes are such that only a stress increase occurs in the soil. The real requirement for use of the solution is not that the soil is elastic, but it should have a constant ratio between stress and strain. When the stress decrease occurs, the relation between stress and strain is not linear and therefore the solution is not strictly applicable.

For practical cases, the Boussinesq's solution can be safely used for homogenous deposits of clay, man-made fills and for limited thickness of uniform sand deposits. In deep sand deposits, the modulus of elasticity increases with an increase in depth, and therefore, the Boussinesq's solution will not give satisfactory results. The point loads applied below ground surface cease somewhat smaller stresses than are caused by surface loads, and therefore, the Boussinesq's solution is not strictly applicable. However, the solution is frequently used for shallow footings, in which z is measured below the base of the footing.

15.9.3 Pressure Distribution Diagrams

By means of Boussinesq's stress distribution theory, the following vertical pressure distribution diagrams can be prepared.

1. **Stress isobar or isobar diagram**: An isobar is a curve or contour connecting all points below the ground surface of equal vertical pressure. An isobar is a spatial, curved surface of the shape of a bulb, because the vertical pressure on a given horizontal plane is the same in all directions at points located at equal radial distances around the axis of loading. The zone in a loaded soil mass bounded by an isobar of given vertical pressure intensity is called a pressure bulb. The vertical pressure at every point on the surface of pressure bulb is the same.

2. **Vertical pressure distribution on a horizontal plane** (Figure 15.27): Vertical pressure distribution diagram on the horizontal plane at a constant depth z can be analyzed by using the vertical pressure given by Boussinesq theory.

$$\sigma_z = I_B\left(Q/z^2\right)$$

σ_z is maximum occurs just below the load ($r=0$), which is equal to

$$\sigma_z = 0.4775\left(Q/z^2\right)$$

The diagram is symmetrical about the vertical axis. The maximum stress occurs just below

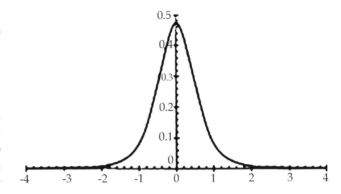

FIGURE 15.27 Vertical pressure distribution on a horizonal plane.

the load and it decreases rapidly as the distance r increases.

3. **Vertical pressure distribution on a vertical plane**: The vertical stress distribution diagram on a vertical plane at a radial distance of r can be obtained by

$$\sigma_z = I_B\left(Q/z^2\right)$$

In this case value, radial distance r is constant and depth changes. The vertical is plotted horizontally along r-axis and depth parallel to the z-axis as shown in Figure 15.28. The vertical stress on the vertical line first increases with increase in depth and reaches up to its maximum value and starts decreasing beyond it with further increase in depth.

Maximum vertical pressure on vertical line occurs when the angle made by the polar ray attains a value of 39°13′53.3″ from the point load.

15.9.4 Vertical Stress due to a Uniform Line Load

The vertical stress in a soil mass due to a vertical line load can be obtained using Boussinesq's solution. Let the vertical line load of intensity q' per unit length, along y – axis, acting on the surface of a semi-infinite soil mass.

The vertical stress at a point P in the soil mass with coordinates (x, z) is given by

$$\sigma_z = 2q'\big/\pi z\left[1+\left(1+(x/z)^2\right)\right]^2 \tag{15.71}$$

If the point P is just below the line load at a depth $z(x=0)$, the

$$\sigma_z = 2q'/\pi z \tag{15.72}$$

15.9.5 Vertical Stress due to Strip Load

The vertical stress at a point due to uniform load of intensity q on a strip of width B can be obtained from Boussinesq's solution. The expression will depend on whether the point lies below the center of strip load or not.

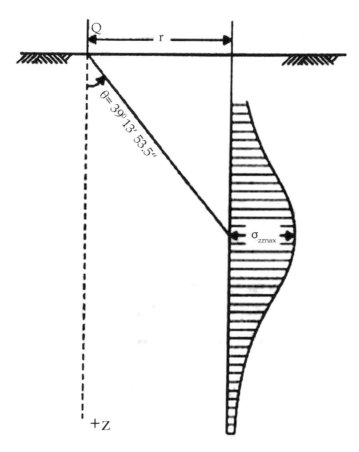

FIGURE 15.28 Vertical pressure distribution on a vertical plane.

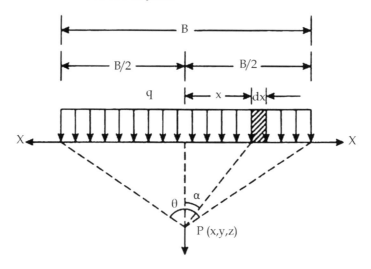

FIGURE 15.29 Vertical stress due to point load.

a. Point P below the center of strip load (Figure 15.29)
   The vertical stress at a point P is given by

$$\sigma_z = q/\pi[2\theta + \sin 2\theta \ ] \qquad (15.73)$$

b. Point P not below the center of strip load (Figure 15.30)
   The vertical stress at a point P is given by

$$\sigma_z = q/\pi[2\theta + \sin 2\theta \cos 2\varphi] \qquad (15.74)$$

where $2\theta = \beta_2 - \beta_1$ and $2\varphi = \beta_2 + \beta_1$

15.9.6 Vertical Stress Distribution below Uniformly Loaded Circular Area

The loads applied to soil surface by footings are not concentrated loads. These are usually spread over a finite area of footing. It is generally assumed the footing is flexible and the contact pressure is uniform. In other words, the load is assumed to be uniformly distributed over area of the base of footings.

The vertical stress at the point A at a depth z below the center of a uniformly loaded circular area (Figure 15.31) can

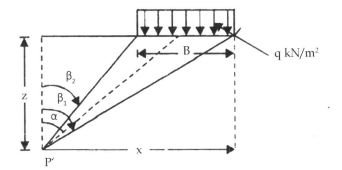

FIGURE 15.30   Vertical stress due to strip load.

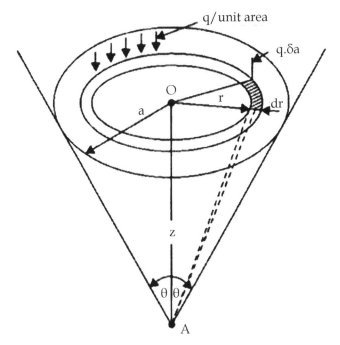

FIGURE 15.31   Vertical stress due to load on circular area.

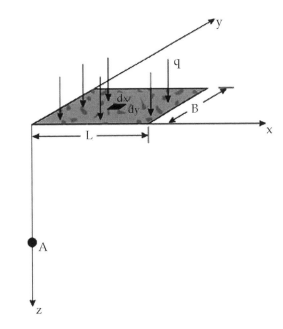

FIGURE 15.32   Vertical stress under corner of uniformly loaded rectangular area.

15.9.7   VERTICAL STRESS UNDER A CORNER OF A RECTANGULAR AREA

The vertical stress under a corner of a rectangular area with uniformly distributed load of intensity q as shown in Figure 15.32 can be obtained from Newmark's equation.

$$\sigma_z = q/2\pi\left[\left\{mn/\left(m^2+n^2+1\right)^{1/2}\right\}\right.$$

$$\times\left\{\left(m^2+n^2+2\right)/\left(m^2+n^2+m^2n^2+1\right)\right\}$$

$$\left.+\sin^{-1}\left\{mn/\left(m^2+n^2+m^2n^2+1\right)^{1/2}\right\}\right] \quad (15.77)$$

$$\sigma_z = I_N q$$

where

$m=B/z$ and $n=L/z$
I_N=Newmark's influence coefficient

The values of m and n can be interchanged without any effect on the values of σ_z.

Fadum's charts (Figure 15.33) can be used to determination of influence factor (I_N). These charts can be used in a design office. The chart can be also be used for determination of vertical stress under a strip load in which case the length tends to infinity and the curve n=infinity can be used.

15.9.8   VERTICAL STRESS AT ANY POINT ON A RECTANGULAR AREA

The equations developed in the preceding section can also be used for finding vertical stress at a point which is not located below the corner. The rectangular area is subdivided into

be determined by Boussinesq's solution. Let the intensity of the load q per unit area and R be the radius of the loaded area.

The vertical stress is given by

$$\sigma_z = q\left[1-\left\{1/1+(R/z)^2\right\}^{3/2}\right] \quad (15.75)$$

$$\sigma_z = I_c q$$

where I_c=influence coefficient for circular area = $[1-\{1/(1+(R/z)^2)\}^{3/2}]$

Let $\tan\theta=R/z$

$$I_c = \left[1-\left\{1/1+(\tan\theta)^2\right\}^{3/2}\right]$$

$$I_c = 1-\cos^3\theta$$

Therefore, $\sigma_z = q\left[1-\cos^3\theta\right] \quad (15.76)$

where 2θ=angle subtended at point P by load.

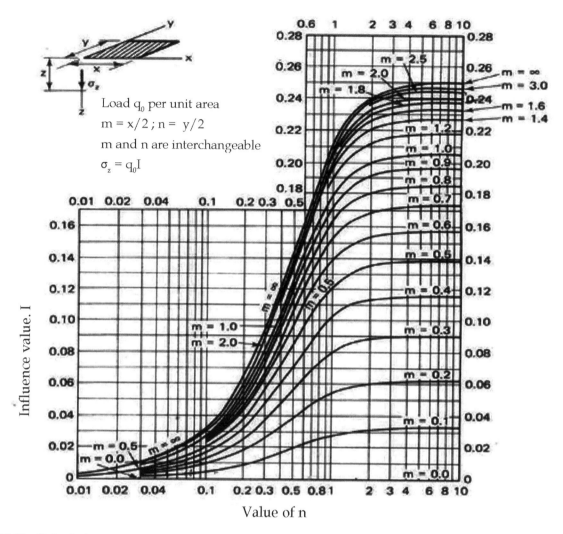

FIGURE 15.33 Fadum's chart.

rectangles such that each rectangle has a corner at the point where the vertical stress is required. The vertical stress is determined using the principle of superposition.

a. Point anywhere below the rectangular area (Figure 15.34)

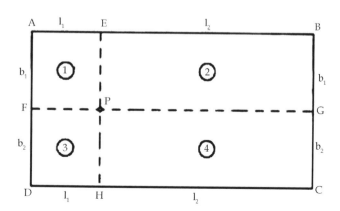

FIGURE 15.34 Rectangular loaded area.

The given rectangle is subdivided into 4 small rectangles, AEPH, EBFP, PFCG, and HPGD, each having one corner at P. The vertical stress at P due to given rectangular load is equal to that from the four small rectangles. Therefore,

$$\sigma_z = q\left[(I_N)_1 + (I_N)_2 + (I_N)_3 + (I_N)_4\right] \qquad (15.78)$$

where $(I_N)_1$, $(I_N)_2$, $(I_N)_3$, and $(I_N)_4$ are Newmark's influence factors obtained for four rectangles marked as (1), (2), (3), and (4).

For the special case, when the point P is at the center of the rectangle ABCD, all the four small rectangles are equal. Therefore,

$$\sigma_z = 4I_N q$$

b. Point outside the loaded area (Figure 15.35)

In this case, a large rectangle AEPF is drawn with its one corner P.

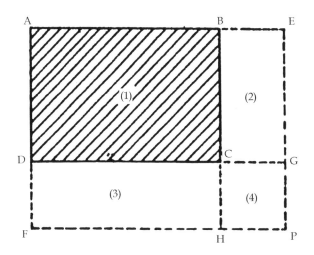

FIGURE 15.35 Stress outside point of rectangular loaded area.

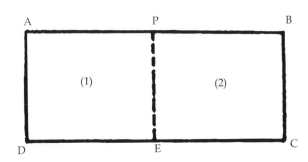

FIGURE 15.36 Stress below edge of the loaded rectangular area.

Now,

Rectangle ABCD=rectangle AEPF−rectangle BEPH−rectangle DGPF+rectangle CGPH

The last rectangle CGPH is given plus sign because this area has been deducted twice, once in rectangle BEPH and once in rectangle DGPF.

Therefore, the stress at P is given by

$$\sigma_z = q\left[(I_N)_1 - (I_N)_2 - (I_N)_3 + (I_N)_4\right] \qquad (15.79)$$

c. Point below the edge of the loaded area (Figure 15.36)

If the point P is below the edge of the loaded area ABCD, the given rectangle is divided into two small rectangles APED and PBCE. In this case,

$$\sigma_z = q\left[(I_N)_1 + (I_N)_2\right] \qquad (15.80)$$

15.9.9 Westergaard's Solution

Boussinesq's solution assumes that the soil deposit is isotropic. Actual sedimentary deposits are generally anisotropic. There are generally thin layers of sand embedded in homogenous clay strata. Westergaard's solution assumes that there are thin sheets of rigid materials sandwiched in a homogenous soil mass. These thin sheets are closely packed and are of infinite rigidity and are, therefore, incompressible. These permit only downward displacement of the soil mass as a whole without any lateral displacement. Therefore, Westergaard's solution represents more closely the actual sedimentary deposits.

According to Westergaard, the vertical stress at a point P at a depth z below the concentrated load is given by

$$\sigma_z = (c/2\pi)\left(Q/z^2\right)/\left[c^2 + (r/cz)^2\right]^{3/2} \qquad (15.81)$$

where c depends on the Poisson ratio (μ) given by

$$c = \left[(1-2\mu)/(2-2\mu)\right]^{1/2} \qquad (15.82)$$

For an elastic material, the value of μ varies between 0.0 and 0.50. For the case when $\mu=0$, then $c=1/\sqrt{2}$.

$$\sigma_z = \left(Q/\pi z^2\right)/\left[1+2(r/z)^2\right]^{3/2}$$

$$\sigma_z = I_w\left(Q/z^2\right) \qquad (15.83)$$

where I_w=Westergaard's influence coefficients

15.9.10 Approximate Methods

1. Equivalent point load method

The vertical stress at a point under a loaded area of any shape can be determined by dividing the loaded area into small areas as shown in Figure 15.37 and replacing the distributed load on each small area by an equivalent point load acting at the centroid of the area. The total load is thus converted into number of point loads. The vertical stress at any point below or outside the loaded area is equal to the sum of the vertical stresses due to these equivalent point loads.

$$\sigma_z = \left[\begin{matrix} Q_1(I_B)_1 + Q_2(I_B)_2 + Q_3(I_B)_3 \\ +Q_4(I_B)_4 + \cdots + Q_n(I_B)_n \end{matrix}\right]\Bigg/ z^2 \qquad (15.84)$$

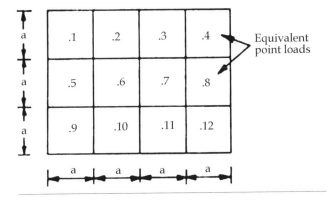

FIGURE 15.37 Load distribution.

2. Two-to-one load distribution method:

The actual load distribution with the depth is complex. However, it can be assumed to spread approximately at a slope of 2V:1H. Thus the vertical pressure at any depth z below the soil surface can be determined approximately by constructing a frustum of pyramid or cone of depth z and side slopes (2:1). The pressure distribution is assumed to be uniform on a horizontal plane at that depth as shown in Figure 15.38.

The average vertical stress σ_z depends upon the shape of the loaded area.

a. Square area $(B \times B)$, $\sigma_z = q\, B^2/(B+z)^2$
b. Rectangular area $(B \times L)$ $\sigma_z = q\, (B \times L)/(B+z)(L+z)$

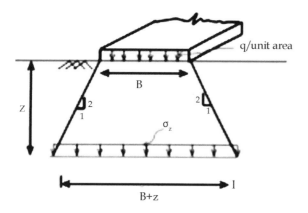

FIGURE 15.38   Two-to-one method of load distribution.

c. Strip area $(B$, unit length) $\sigma_z = q\, (B \times 1)/(B+z) \times 1$
d. Circular area (diameter D) $\sigma_z = q\, D^2/(D+z)^2$

15.10 COMPACTION

Compaction means pressing the soil particles close to each other by mechanical methods. Air, during compaction, is expelled from the void space in the soil mass, and therefore, the mass density is increased. Compaction of a soil mass is done to improve its engineering properties. Compaction generally increases the shear strength of the soil and, hence, the stability and bearing capacity. It is also useful in reducing the compressibility and permeability of the soil.

15.10.1 STANDARD PROCTOR TEST

To assess the amount of compaction and the water content required in the field, compaction tests done on the same soil in the laboratory. The tests provide a relationship between the water content and the dry density. The water content at which the maximum dry density achieved can be obtained from the relationships provided by the tests.

IS:2720 (part VII) recommends essentially the same specifications as in Standard Proctor test with some minor modifications and metrification.

The mould recommended is of 100 mm diameter, 127.3 mm height, and 1000 ml capacity. The rammer recommended is of 2.6 kg mass with a free drop of 310 mm and a face diameter of 50 mm. The soil is compacted in three layers. The mould is fixed to a detachable base plate. The collar is of 60 mm height. Standard Proctor test is shown in Figure 15.39.

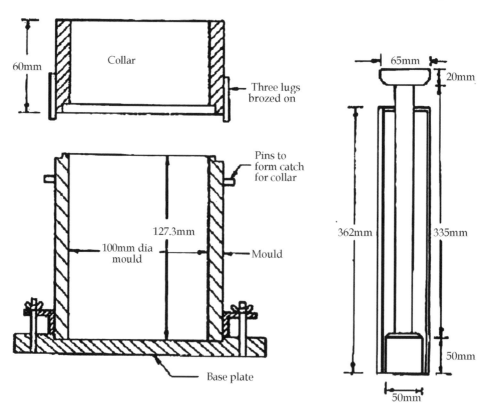

FIGURE 15.39   Compaction mould and rammer of Standard Proctor test.

If the percentage of soil retained on 4.75 mm sieve is more than 20%, a larger mould of internal diameter 150 mm, effective height of 127.3 mm, and capacity 2250 ml is recommended.

$$\text{Bulk mass density } \rho = M/V \text{ gm/ml}$$

where

M = mass of compacted soil (gm)
V = volume of the mould (ml)

$$\text{Dry density } \rho_d = \rho/(1+w)$$

where w = water content

15.10.1.1 Compaction Curve

A compaction curve is plotted between the water content as abscissa and the corresponding dry density as ordinate as shown in Figure 15.40. It is observed that the dry density initially increases with an increase in water content till the maximum density (ρ_{max}) is attained. With further increase in water content, the dry density decreases. The water content corresponding to the maximum dry density is known as the optimum water content (OWC) or the optimum moisture content.

At water content lower than the optimum, the soil is rather stiff and has lot of void spaces, and therefore, the dry density is low. As the water content is increased, the soil particles get lubricated and slip over each other, and move into densely packed positions and the dry density is increased. However, at water content more than the optimum, the additional water reduces the dry density, as it occupies the space that might have been occupied by solid particles.

For a given water content theoretical maximum density, $(\rho_d)_{theomax}$, is obtained corresponding to the condition when there are no air voids (i.e., degree of saturation is equal to 100%). The theoretical maximum dry density is also known as saturated dry density $(\rho_d)_{sat}$. In this condition, the soil becomes saturated by reduction in voids to zero but no change

in water content. The soil could become saturated by increasing the water content such that all air voids are filled.

$$\text{Dry density } \rho_d = G\rho_w/(1+e)$$

$$\text{At } eS = wG, \rho_d = G\rho_w/1+(wG/S)$$

The theoretical maximum dry density occurs when S = 100%.

$$\left(\rho_d\right)_{theomax} = G\rho_w/1 + wG \tag{15.85}$$

It may be mentioned that compaction methods cannot remove all the air voids, and therefore, the soil never becomes fully saturated. Thus the theoretical maximum dry density is only hypothetical. It can be calculated from Equation 15.85. For any value of w, if the value of G is known. The line indicating the theoretical maximum dry density can be plotted along with the compaction curve. It is also known as zero air void line or 100% saturation line.

15.10.2 MODIFIED PROCTOR TEST

The Modified Proctor test was developed to represent heavier compaction than that in the Standard Proctor test.

In the Modified Proctor test, the mould used is the same as in the Standard Proctor test. However, the rammer used is much heavier and has a greater drop than that in the Standard Proctor test. Its mass is 4.89 kg and the free drop is 450 mm. The face diameter is 50 mm as in the Standard Proctor test. The soil is compacted in five equal layers; each layer is given 25 blows. The compacted effort in the Modified Proctor test measured in kJ/m³ of soil is about 4.56 times that in the Standard Proctor test. Thus a much heavier compaction is attained.

The dry densities are obtained for different water contents and the compaction curve is drawn. Figure 15.41 shows the compaction curve for Modified Proctor test (curve No: 2).

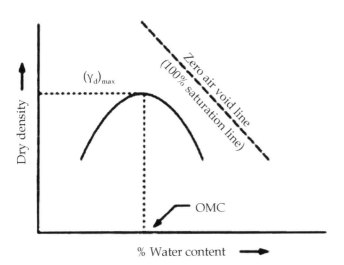

FIGURE 15.40 Compaction curve.

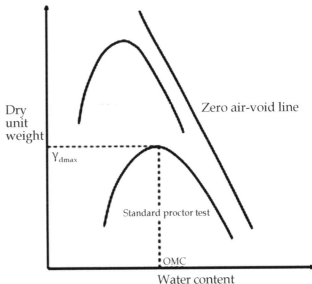

FIGURE 15.41 Compaction curve for Modified Proctor test.

The curve is higher than and to the left of that obtained from a Standard Proctor test (curve No: 1). The heavier compaction increases the maximum dry density but decreases the OWC.

15.10.3 Factors Affecting Compaction

The following are factors affecting compaction.

1. **Water content**: At low water content, the soil is stiff and offers more resistance to compaction. The dry density of the soil increases with an increase in the water content till the OWC is reached. At that stage, the air voids attain approximately a constant volume. With further increase in water content, the air voids do not decrease, but the total voids increase and the dry density decreases.

2. **Amount of compaction**: The effect of increasing the amount of compactive effort is to increase the maximum dry density and to decrease the OWC. Compactive effort is the measure of mechanical energy applied to the soil mass. It may be mentioned that the maximum dry density does not go on increasing with an increase in compactive effort. For a certain increase in the compactive effort, the increase in dry density becomes smaller and smaller. Finally, a stage is reached beyond which there is no further increase in the dry density with an increase in the compactive effort.

3. **Type of soil**: The dry density achieved depends upon the type of soil. The maximum dry density and the OWC for different soils are shown in Figure 15.42.

   In general, coarse-grained soils can be compacted to higher dry density than fine-grained soils. With the addition of even a small quantity of fines to a coarse-grained soil, the soils attain a much higher dry density for the same compactive effort. A well-graded sand attains a much higher dry density than a poorly graded soil.

   Cohesive soils have high air voids. These soils attain a relatively lower maximum dry density as compared to cohesionless soils. Such soils require more water than cohesionless soils, and therefore, the OWC is high. Heavy clays of very high plasticity have very low dry density and very high OWC.

4. **Method of compaction**: The dry density achieved depends not only on the amount of compaction effort but also on the method of compaction. For the same amount of compaction effort, the dry density will depend upon whether the method of compaction utilizes kneading action, dynamic action, or static action. In kneading compaction, compaction is done by applying a given pressure for a fraction of a second, whereas in static compaction, soil is simply pressed under a constant pressure.

5. **Admixtures**: The compaction characteristics of the soils are improved by adding other materials known as admixtures. The most commonly used admixtures are lime, cement, and bitumen. The dry density achieved depends upon type and amount of admixtures.

15.10.4 Relative Compaction and Compaction Control

Relative compaction is defined as the ratio of the field dry unit weight to the laboratory maximum dry unit weight as per specified standard test, e.g., Standard Proctor test, Indian standard (light/heavy) compaction test, etc.

Compaction control is done by measuring dry density and water content of the compacted soil in the field. For the quick determination of water content in field, sand bath method, calcium carbide method, or alcohol method are used. Proctors compaction needle is a useful tool for the rapid determination of compaction in the field.

15.11 CONSOLIDATION OF SOILS

When a soil mass is subjected to compressive force like all other materials, its volume decreases. The property of the soil due to which a decrease in volume occurs under compressive

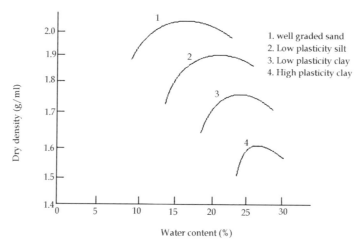

FIGURE 15.42 Typical standard compaction curve for different soils.

forces is known as the compressibility of soil. The compression of soils can occur due to one or more of the following causes.

i. Compression of solid particles and water in the voids
ii. Compression and expulsion of air in the voids
iii. Expulsion of water in the voids

When the soil is fully saturated, compression of soil occurs mainly due to the third cause, namely, expulsion of water. The compression of a saturated soil under a steady static pressure is known as consolidation. It is entirely due to expulsion of water from the voids. As the consolidation of soils occurs, the water escapes.

15.11.1 TYPES OF CONSOLIDATION

The consolidation of a soil deposit can be divided into 3 stages.

1. **Initial consolidation**: When a load is applied to a partially saturated soil, a decrease in volume occurs due to expulsion and compression of air in the voids. A small decrease in volume also occurs due to compression of solid particles. The reduction in volume of the soil just after the application of the load is known as initial consolidation or initial compression. For saturated soils, the initial consolidation is mainly due to compression of solid particles.
2. **Primary consolidation**: After initial consolidation, further reduction in volume occurs due to expulsion of water from voids. When a saturated soil is subjected to a pressure, initially all the applied pressures are taken up by water as an excess pore water pressure, as water is almost incompressible as compared with solid particles. A hydraulic gradient develops and the water starts flowing out and a decrease in volume occurs. The decrease depends upon the permeability of the soil and is therefore time-dependent. This reduction in volume is called primary consolidation.
3. **Secondary consolidation**: The reduction in volume continues at a very slow rate even after the excess hydrostatic pressure developed by the applied pressure is fully dissipated and the primary consolidation is complete. This additional reduction in the volume is called secondary consolidation. The causes for secondary consolidation are not fully established. It is attribute to the plastic readjustment of the solid particles and the adsorbed water to the new stress system. In most inorganic soils, it is generally small.

15.11.2 CONSOLIDATION OF UNDISTURBED SPECIMEN

Soil deposits may be divided into three classes as regards to consolidation theory.

1. **Preconsolidated clay or overconsolidated clay**: A clay is said to be precompressed or preconsolidated or overconsolidated if it has ever been subjected to a pressure in excess of its present overburden pressure. The temporary overburden pressure to which a soil has been subjected and under which it go consolidated is known as preconsolidation pressure.
2. **Normally consolidated clay**: A normally consolidated soil is one which has never been subjected to an effective pressure greater than the existing overburden pressure and which is also completely consolidated by the existing overburden.
3. **Under consolidated clay**: A soil which is not fully consolidated under the existing overburden pressure is called under consolidated soil.

15.11.2.1 Overconsolidation Ratio

Over consolidation ratio (OCR) is defined as the maximum applied effective stress in the past to the present applied effective stress.

$$OCR = \frac{\text{maximum applied effective stress in the past}(\sigma_0')}{\text{present applied effective stress}(\sigma')}$$

For normally consolidated soils, $OCR \leq 1$.
    For over consolidated soils, $OCR > 1$.
    If $\sigma' < \sigma_0'$, then soil is called over consolidated soils.
    If $\sigma' > \sigma_0'$, then soil is called normally consolidated soils.

15.11.3 CONSOLIDATION TEST

The compressibility and consolidation characteristics of soil are determined in the laboratory by the oedometer or consolidometer test. It consists of a loading device and a cylindrical container called consolidation cell. The soil specimen is placed in the cell between the top and the bottom porous stones. The consolidation cells are of two types:

a. Floating or free ring cell (Figure 15.43) in which both the top and bottom porous stones are free to move. The top porous stone can move downward, and the bottom porous stone can move upward as the sample consolidates.
b. Fixed ring cell (Figure 15.43) in which the bottom porous stone cannot move. Only the top porous stone can move downward as the specimen consolidates. The fixed ring cell can also be used as variable head permeability test apparatus. For this purpose, a piezometer is attached to the base of the cell.

The test involves the measurement of one-dimensional compression of saturated soil under several increments of vertical stress. Under each increment of loading, soil is allowed to consolidate till there is no or little further compression. Each increment is maintained for at least 24 hours. The compression of the specimen is noted at several intervals of time for each increment. During consolidation under the

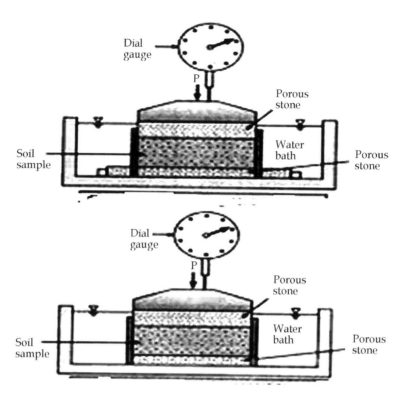

FIGURE 15.43 Free and fixed ring cell.

maximum load, the excess pore water has dissipated completely. Now the specimen is unloaded in two or three stages and the soil is allowed to swell; during the unloading stage, final swell readings are taken. After complete unloading, the wet and dry weights of sample are determined. The results of oedometer test are plotted on graph between the void ratio at the end of each increment period and the corresponding effective stress.

15.11.3.1 Computation of Void Ratio

The void ratio at the end of each increment period is known as increment void ratio. The increment void ratio may be found out by the following two methods.

1. Height of solids method
2. Change in void ratio method

1. **Height of solids method**: During compression of a soil specimen, the volume of soil solids remains constant, while the volume of voids decreases. For a constant height of solids over the area of cross section A of sample, knowing the dry weight of sample and specific gravity, the height of the soil solids H_s can be computed as

$$H_s = W_s/G\gamma_w A \qquad (15.86)$$

The equilibrium void ratio can be calculated as

$$e = H - H_s/H_s \qquad (15.87)$$

where
H_s=height of solid
W_s=dry weight of soil
G=specific gravity of soil
A=cross-sectional area of soil specimen
H=height of specimen at the end of the particular stress increment=$H_1 + \Delta H$
H_1=height of specimen at the beginning of the stress increment
ΔH=change in thickness under the stress increment

2. **Change in void ratio method**: In this method, it is assumed that the specimen is fully saturated at the end of the test. Now void ratio of the saturated soil at the end of the test is calculated from the equation

$$eS = wG$$

$$\text{Hence } e_f = w_f G_s \quad (S = 1)$$

where w_f=final water content at the end of the test.

If the change in volume is a consequence of a decrease in void ratio Δe, then

$$\Delta H/H = \text{change in volume/original volume} = \Delta e/1 + e_0$$

where e_0=initial void ratio of sample before any compression.

Now knowing ΔH, H, and e_0, we can calculate change in void ratio Δe.

Knowing Δe, increment void ratio at the end of every effective stress can be calculated as

$$e = e_0 - \Delta e$$

The change in void ratio can also be found directly by

$$\Delta e = \Delta H (1 + e_f) / H_f \qquad (15.88)$$

Change in void ratio Δe under each stress increment is calculated from above formula working from backward; from the known value of e_f, the equilibrium void ratio at the end of each stress increment can be deducted.

15.11.3.2 Consolidation Test Results

1. Dial gauge reading–time plot:

   The plot between dial gauge reading and time is required for determining the coefficient of consolidation, which is useful for obtaining the rate of consolidation in the field.

2. Final void ratio–effective stress plot:

   The thickness of the specimen after 24 hours of application of load increment is taken as the final thickness for that increment. The final void ratio e_f corresponding to the final thickness for each increment is determined using change in void ratio method.

   As the sand is relatively less compressible, the change in void ratio is small. The plot between final void ratio and effective stress (Figure 15.44) is required for the determination of the magnitude of the consolidation settlement in the field.

3. Final void ratio–log σ plot:

   Figure 15.45 shows the plot between final void ratio and effective stress. The curve has convexity

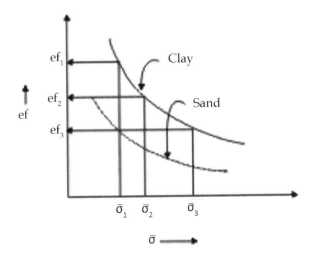

FIGURE 15.44   Void ratio-effective stress plot.

upward. The slope decreases with an increase in effective stress.

It is more convenient to plot the results in a semi-log graph, in which final void ratio is plotted on the natural scale and the effective stress as abscissa on the log scale. The plot is practically a straight line for normally consolidated clay within the range of pressure usually encountered in practice.

15.11.4 Basic Definitions

1. **Coefficient of compressibility** (a_v): Coefficient of compressibility is defined as the decrease in void ratio per unit increase in effective stress. It is equal to the slope of the void ratio–effective stress curve on arithmetic scale.

   $$a_v = -\left(\Delta e / \Delta \sigma'\right) \qquad (15.89)$$

   The coefficient of compressibility has dimensions of $[L^2/F]$. The unit is m^2/kN.

2. **Coefficient of volume change** (m_v): The coefficient of volume change or volume compressibility is defined as the ratio of unit volume change or volumetric strain per unit increase in effective stress.

   $$m_v = -\left(\Delta V / V_0\right) / \Delta \sigma' \qquad (15.90)$$

   where
   $\Delta V =$ change in volume
   $V_0 =$ initial volume
   $\Delta \sigma' =$ change in effective stress

The volumetric strain ($\Delta V / V_0$) can be expressed in terms of either void ratio or the thickness of the specimen.

1. Let e_0 be the initial void ratio. Let the volume of solids be unity. Therefore, the initial volume V_0 is equal to $1 + e_0$. If Δe is change in void ratio due to change in volume ΔV, we have $\Delta V = \Delta e$. Thus

   $$\Delta V / V_0 = \Delta e / 1 + e_0 \qquad (15.91)$$

2. As the area of cross section of the sample in the consolidometer remains constant, the change in volume is proportional to the change in height. Thus $\Delta V = \Delta H$

   $$\Delta V / V_0 = \Delta H / H_0 \qquad (15.92)$$

   Therefore, $m_v = -\left(\Delta H / H_0\right) / \Delta \sigma'$

   $$= -\left(\Delta e / 1 + e_0\right) / \Delta \sigma' = a_v / 1 + e_0 \quad (15.93)$$

Thus m_v depends on the effective stress. Its value also decreases with an increase in effective stress. The unit is m^2/kN.

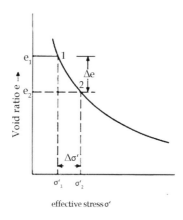 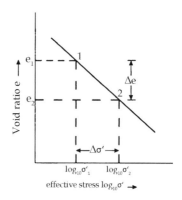

FIGURE 15.45   Void ratio-log effective stress plot.

3. **Compression index** (C_c): The compression index is equal to the slope of the linear portion of void ratio versus $\log \sigma'$ plot.

$$C_c = -\Delta e / \log_{10}\left(\sigma'/\sigma'_0\right)$$

$$C_c = -\Delta e / \log_{10}\left[(\sigma'_0 + \Delta\sigma')/\sigma'_0\right] \qquad (15.94)$$

The compression index is extremely used for determination of the settlement in the field. The compression index of clay is related to its index properties, especially liquid limit (w_l) and is given by

For undistributed soils, $C_c = 0.009\left(w_l - 10\right).$ (15.95)

For remoulded soils, $C_c = 0.007\left(w_l - 10\right).$ (15.96)

The compression index is also related to the in-situ void ratio (e_0) and water content (w_0), and is given by

$$C_c = 0.54\left(e_0 - 0.35\right) \qquad (15.97)$$

$$C_c = 0.0054\left(2.6w_0 - 0.35\right) \qquad (15.98)$$

4. **Expansion index** (C_e): The expansion index or swelling index is the slope of the $e - \log \sigma'$ plot obtained during unloading.

$$C_e = \Delta e / \log_{10}\left[(\sigma' + \Delta\sigma')/\sigma'\right] \qquad (15.99)$$

Expansion index is much smaller than compression index.

5. **Recompression index** (C_r): The recompression index is the slope of the recompression curve obtained during reloading when void ratio is plotted against effective stress on semi-log scale.

$$C_r = -\Delta e / \log_{10}\left[(\sigma' + \Delta\sigma')/\sigma'\right] \qquad (15.100)$$

The recompression index is appreciably smaller than compression index. It is usually in the range of 1/10–1/5 of the compression index.

15.11.4.1 Determination of Coefficient of Consolidation

The following two methods are commonly used.

1. **Square root of time fitting method**: The method developed by Taylor utilizes the similarity in shape between theoretical curve of U vs $\sqrt{T_v}$ and dial gauges versus $\sqrt{t}$. The theoretical curve is characterized by a straight line portion at least up to $U=60\%$. It has been further established that at $U=90\%$, the value of $\sqrt{T_v}$ as 1.15 times the value obtained by extension of the initial straight line portion. Taylor used this characteristic feature to locate the 90% consolidation point on the experimental plot. For a given load increment, the dial gauge readings are taken for different time intervals. A curve is plotted between the dial gauge reading (R) as ordinate and the $\sqrt{t}$ as abscissa.

   As load increment is applied, there is an initial compression. It is obtained by producing back the initial linear part of the curve to intersect the dial gauge reading axis. This corresponds to the corrected zero reading (R_c). The consolidation between the dial gauge reading R_0 and R_c is the initial compression. The Terzaghi theory of consolidation is not applicable in this range. Starting from R_0, a second straight line is drawn such that its abscissa is 1.15 times the abscissa of the first line. The intersection of this line with the experimental curve identifies R_{90} and the time required for 90% consolidation t_{90} is read off. With t_{90} thus determined C_v can be computed using the equation,

$$T_v = C_v t / d^2 \qquad (15.101)$$

   An advantage of this method is that plotting of curve can be done as the time progresses.

2. **Logarithm of time fitting method**: The method given by Casagrande uses the theoretical curve between U and $\log T_v$. Since the time corresponding to $U=100\%$ approaches infinity, Casagrande suggested

that if two tangents are drawn to the experimental curve their intersection point will give $U=100\%$. The dial reading at $U=100\%$ is R_{100}. Casagrande method aims at finding R_{50} so that t_{50} the time corresponding to $U=50\%$ can be determined. Since R_{100} is already defined it remains to locate R_0, the dial reading corresponding to $U=0$ percent before R_{50} can be located midway between R_0 and R_{100}.

To determine R_0 two values of time t_1 and t_2 are chosen on the initial part of the curve such that $t_2=4t_1$. The points corresponding to these values of time are marked on the curve and the vertical distance between them measured say z. Now a horizontal line is drawn above the first point at the same vertical distance z. This line cuts the ordinate at R_0 called the corrected zero reading. The compression that occurs between these two points and the time corresponding to R_{50} is read off as t_{50}.

15.11.5 Settlement Analysis

The total settlement of a loaded soil can be grouped into two categories.

1. **Immediate settlement** (S_i): Immediate settlement occurs almost immediately after the load is imposed, as a result distortion of the solid without any volume change may occur. It is due to compression, expulsion of air, elastic deformation of solids, and squeezing of water.

   For cohesionless soils, immediate settlement can be found out by

   $$S_i = H_0 \Big/ C_s \left\{ \log_{10}\left[(\sigma_0' + \Delta\sigma')/\sigma_0' \right] \right\} \quad (15.102)$$

   where H_0=total thickness of soil layer initially

   $$C_s = 1.5\big(C_r/\sigma_0'\big);$$

   where
   C_r=static cone resistance in kN/m²
   σ_0' =initial effective stress due to overburden pressure at the center of layer
   $\Delta\sigma'$=increase in effective stress at the center of layer due to application of load.

   In case of saturated clays (cohesive soils), immediate settlement is insignificant. However, small elastic settlement below the corners may occur due to elastic deformation to molecules and squeezing of water.

   The immediate elastic deformation below the corner of a rectangular base foundation is given by

   $$S_i = q \cdot B\big(1 - \mu^2\big) I_t / E_s \quad (15.103)$$

   where
   q=pressure at the base of foundation
   B=width of foundation

μ=Poisson's ratio=0.3–0.45
E_s=Young's modulus of soil
I_t=influence factor or shape factor of foundation which depends on (B/L) ratio

2. **Consolidation settlement**: The total consolidation settlement of soil can be further divided into two parts.
   a. Settlement due to primary consolidation: Settlement due to primary consolidation (S_c) occurs due to expulsion of pore water from a loaded saturated soil mass. In primary consolidation, the rate of flow is controlled by pore pressure, the permeability, and the compressibility of the soil. It can be computed by four methods.
      i. **Method (1): Using voids ratio**
         If Δe is change in void ratio due to increase in effective stress on soil layer, as per change in void ratio method,

         $$\Delta H/H_0 = \Delta e/1 + e_0 \quad (15.104)$$

         $$S_c = \Delta H = H_0\big(\Delta e/1 + e_0\big) \quad (15.105)$$

         where
         e_0=initial void ratio at beginning of consolidation
         H_0=initial thickness of compression layer
            This method is suitable for both over-consolidated and normally consolidated soils.
      ii. **Method (2): Using coefficient of compressibility** (a_v)
         We know $a_v=\Delta e/\Delta\sigma'$
         Hence $\Delta e=a_v \cdot \Delta\sigma'$
         On substituting Δe in Equation 15.105, we get

         $$S_c = H_0\big(\Delta e/1 + e_0\big) = \big(H_0 \cdot a_v \cdot \Delta\sigma'\big)/(1 + e_0)$$

         $$S_c = \big[a_v/(1 + e_0)\big]H_0\Delta\sigma' \quad (15.106)$$

         where $\Delta\sigma'$=increase in effective stress at mid-height of the layer
      iii. **Method (3): Using coefficient of volume change** (m_v)
         We know $m_v=a_v/1 + e_0$
         Hence $a_v=m_v(1 + e_0)$
         On substituting a_v in Equation 15.106, we get

         $$S_c = m_v\big(1 + e_0\big)\big(H_0\Delta\sigma'\big)/(1 + e_0)$$

         $$S_c = H_0 m_v\Delta\sigma' \quad (15.107)$$

         Since m_v and $\Delta\sigma'$ vary with the depth and variation of m_v with the depth is not linear. Hence computation of m_v and $\Delta\sigma'$ at center does not give accurate results. For more precise computation, the total thickness should be divided into more number of layers and settlement of each layer should be computed separately by making computation at the center of each layer for m_v and $\Delta\sigma'$.

iv. **Method (4): Using compression index (C_c)**

We know $C_c = -\Delta e/\log_{10}\left[(\sigma'_0 + \Delta\sigma')/\sigma'_0\right]$

Hence $\Delta e = C_c \log_{10}\left[(\sigma'_0 + \Delta\sigma')/\sigma'_0\right]$

On substituting Δe in Equation 15.105, we get

$$S_c = \left\{H_0 C_c \log_{10}\left[(\sigma'_0 + \Delta\sigma')/\sigma'_0\right]\right\}/(1+e_0) \quad (15.108)$$

b. Settlement due to secondary consolidation: Settlement due to secondary consolidation (S_s) is due to plastic readjustment of solids. It occurs at very low rate and is negligible in granular soils.

The secondary consolidation after time t from the completion of primary settlement is given by

$$S_s = C_s H_{100} \log_{10}\left(t/t_{100}\right)/(1+e_{100}) \quad (15.109)$$

where

C_s = secondary compression index

H_{100} = thickness of compressible layer after primary consolidation

In absence of data $H_{100} \approx H_0$

e_{100} = void ratio at the center of layer after primary consolidation

In absence of data $e_{100} \approx e_0$

t_{100} = time required to complete primary consolidation

t = time at which secondary settlement is computed after the completion of primary consolidation

Thus total settlement of soil is given by

$$S_t = S_i + S_c + S_s \quad (15.110)$$

15.12 SHEAR STRENGTH

When the soil is loaded, shearing stresses are induced in it. When the shearing stresses reach a limiting value, shear deformation takes place, leading to the failure of the soil mass. The shear strength of soil is the resistance to deformation by continuous shear displacement of soil particles or on masses upon the action of a shear stress. The failure conditions for a soil may be expressed in terms of limiting shear stress called shear strength or as a function of the principal stresses. All stability analyses in soil mechanics involve a basic knowledge of the shearing properties and shearing resistance of the soil. The shear strength is most difficult to comprehend and one of the most important of the soil characteristics.

The shearing resistance of soil is constituted basically of the following components.

1. The structural resistance to displacement of the soil because of the interlocking of the particles.
2. The frictional resistance to translocation between the individual soil particles at their contact points.
3. Cohesion or adhesion between the surface of the soil particles.

The shear strength in cohesionless soil results from intergranular friction alone, while in all other soils, it results both from internal friction as well as cohesion. However, plastic undrained clay does not possesses internal friction.

15.12.1 STRESS AT A POINT: MOHR'S CIRCLE OF STRESS

In a stressed soil mass, shear failure can occur along any plane. At any stressed point, there exist three mutually perpendicular planes on which there are no shearing stresses acting. These are known as principal planes. The normal stresses that act on these planes are called principal stresses; the largest of these is called the major principal stress (σ_1), the smallest is called the minor principal stress (σ_3), and the third one is called the intermediate principal stress (σ_2). The corresponding planes are, respectively, designated as the major, minor, and intermediate principal planes. However, the critical stress conditions occur only at σ_1 and σ_3. In two-dimensional stress systems, the major and minor principal planes occur on horizontal and vertical directions.

If σ_1 and σ_3 are known, it can be shown that on any plane AB inclined at angle θ to the direction of major principal plane, the normal stress σ and the shear stress τ are given by

$$\sigma = \left[(\sigma_1 + \sigma_3)/2\right] + \left[(\sigma_1 - \sigma_3)/2\right]\cos 2\theta \quad (15.111)$$

$$\tau = \left[(\sigma_1 - \sigma_3)/2\right]\sin 2\theta \quad (15.112)$$

Mohr demonstrated that these equations tend themselves to graphical representation. It can be shown that "the locus of stress coordinates (σ, τ) for all planes through a point is a circle called the Mohr's circle of stress" (Figure 15.46).

15.12.2 MOHR: COULOMB THEORY

According to Mohr, the failure is caused by a critical combination of the normal and shear stresses. The soil fails when the shear stress (τ_f) on the failure plane at failure is a unique function of the normal stress (σ) acting on that plane.

$$\tau_f = f(\sigma) \quad (15.113)$$

Since the shear stress on the failure plane at failure is defined as the shear strength (s), the above equation can be written as

$$s = f(\sigma) \quad (15.114)$$

The Mohr's theory is concerned with the shear stress at failure plane at failure. A plot can be made between shear stress τ and the normal stress σ at failure. The curve defined by Equation

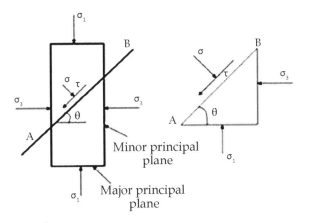

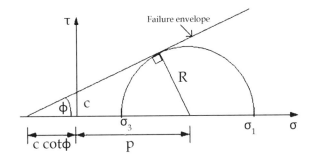

FIGURE 15.46 Mohr's stress circle.

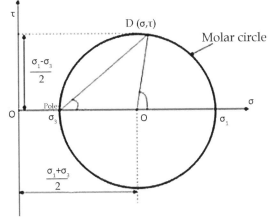

FIGURE 15.47 Mohr's envelope.

15.114 is known as Mohr's envelope (Figure 15.47). There is a unique failure envelope for each material.

Failure of the material occurs when the Mohr's circle of stresses touches the Mohr's envelope. The Mohr's circle represents all possible combinations of shear and normal stresses at the stressed point. At the point of contact (D) of the failure envelope and the Mohr's circle, the critical combination of shear and normal stresses is reached and the failure occurs. The plane indicated by the line PD is therefore the failure plane. Any Mohr's circle which does not cross the failure envelope and lies below the envelope represents a stable condition.

The shear strength (s) of a soil at a point on a particular plane was expressed by Coulomb as a linear function of the normal stress on that plane as,

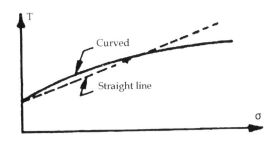

FIGURE 15.48 Straight line representation of Mohr's envelope.

$$s = c + \sigma \tan \varphi \qquad (15.115)$$

In other words, the Mohr's envelope is replaced by a straight line by Coulomb as shown in Figure 15.48.

In the above equation, c is equal to the intercept on τ axis and φ is the angle which the envelope makes with σ-axis. The component c of shear strength is known as cohesion. Cohesion holds the particles of the soil together in a soil mass and is independent of the normal stress. The angle φ is called the angle of internal friction. It represents the frictional resistance between the particles, which is directly proportional to the normal stress.

15.12.3 Modified Mohr: Coulomb Theory

Later it was found that the parameters c and φ depend upon a number of factors such as the water content, drainage conditions, and conditions of testing. The current practice is to consider c and φ as mathematical parameters which represent the failure conditions for a particular soil under given condition. These indicate the intercept and slope of the failure envelope, respectively.

Terzaghi established that the normal stresses which control the shear strength of a soil are the effective stresses and not the total stresses. In terms of effective stresses, Equation 15.115 is written as

$$s = c' + \sigma' \tan \varphi' \qquad (15.116)$$

where c' and φ' are the cohesion intercept and the angle of shearing resistance in terms of the effective stresses.

This equation is known as the Modified Mohr–Coulomb equation for the shear strength of the soil. It is one of the most important equations of soil engineering.

The Mohr–Coulomb theory shows a reasonably good agreement with the observed failures in the field and in the laboratory. The theory is ideally suited for studying the behavior of soils at failure. The theory is used for estimation of the shear strength of soils.

15.12.4 Different Types of Tests and Drainage Conditions

The following tests are used to measure the shear strength of a soil.

1. Direct shear test
2. Triaxial compression test
3. Unconfined compression test
4. Shear vane test

The shear test must be conducted under approximate drainage conditions that simulate the actual field problem. In shear tests, there are two stages:

1. Consolidation stage in which the normal stress or confining pressure is applied to the specimen and it is allowed to consolidate.
2. Shear stage in which shear stress or deviator stress is applied to the specimen to shear it.

Depending upon the drainage conditions, there are three types of tests. They are as follows:

a. **Unconsolidated–undrained condition**: In this type of test, no drainage is permitted during the consolidation stage. The drainage is also not permitted in the shear stage.

As no time is allowed for consolidation or dissipation of excess pore water pressure, the test can be conducted quickly in a few minutes. The test is known as unconsolidated–undrained test (UU) test or quick test (Q test).

b. **Consolidated–undrained condition**: In a consolidated–undrained test, the specimen is allowed to consolidate in the first stage. The drainage is permitted until the consolidation is complete.

In the second stage when the specimen is sheared, no drainage is permitted. The test is known as consolidated–undrained test (CU) test. It is also known as R test, as the alphabet R falls between the alphabet Q used for quick test and the alphabet S used for slow test.

The pore water pressure can be measured in the second stage if the facilities for its measurement are available. In this case, the test is known as CU test.

c. **Consolidated–drained condition:** In a consolidated–drained test, the drainage of the specimen is permitted in both the stages. The sample is allowed

to consolidate in the first stage. When the consolidation is complete, it is sheared at a very slow rate to ensure that fully drained conditions exist and the excess pore water is zero.

The test is known as a consolidated–drained (CD) test or drained test. It is also known as the slow test or S test.

15.12.4.1 Direct Shear Test (Figure 15.49)

It is the oldest shear test, still in use, quite simple to perform. The soil specimen that is to be tested is confined in a metal box of square cross section that is split into two halves horizontally, a small clearance being maintained between the two halves of the box. This test is also called shear box test. There is no control over drainage conditions and no mechanism to measure pore pressure. Hence this test is preferred for drained condition (CD). Also a constant shear strain rate of 1.25 mm/minute is applied.

15.12.4.1.1 Presentation of Results of Direct Shear Test

Stress–strain curve: A stress–strain curve is a plot between shear stress and the shear displacement as shown in Figure 15.50. In case of dense sand (over consolidated clays), the shear stress attains a peak value at a small strain. With further increase in strain, the shear stress decreases slightly and becomes more or less constant, known as ultimate stress. In case of loose sands or normally consolidated clays, the shear stress increases gradually and finally attains a constant value, known as the ultimate stress or residual strength. It has been observed that the ultimate shear stress attained by both dense and loose sands tested under similar conditions is approximately the same. Figure 15.51 is a plot between shear strain and void ratio.

It may be observed that the void ratio of an initial loose sand decreases with an increase in shear strain, whereas that for initial dense sand increases with an increase in strain. The void ratio at which there is no change in it with an increase in strain is known as critical void ratio. If the sand initially is at the critical void ratio, there would be practically no change in volume with an increase in shear strain.

i. **Failure envelope**: For obtaining a failure envelope, a number of identical specimens are tested under

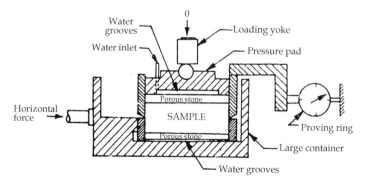

FIGURE 15.49 Direct shear test.

Direct shear tests on sand

Stress-strain relationship

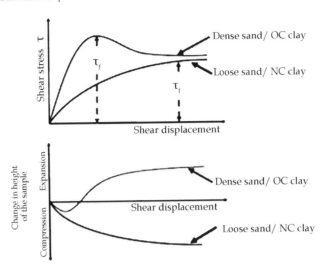

FIGURE 15.50 Shear stress-shear strain plot.

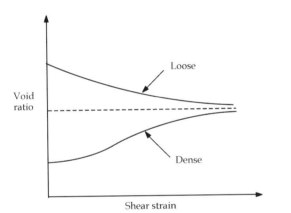

FIGURE 15.51 Plot between shear strain and void ratio.

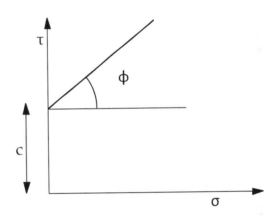

FIGURE 15.52 Graphical representation of Coulomb equation.

different normal stresses. The shear stress required to cause failure is determined for each normal stress. The failure envelope is obtained by plotting the points corresponding to shear strength at different normal stresses and joining them by a straight line. The inclination of the failure envelope to the horizontal gives the angle of shearing resistance φ and its intercept on the vertical axis is equal to the cohesion intercept c as shown in Figure 15.52.

For dense sands, the failure envelope can be drawn either for peak stress or for ultimate stress. The values of the parameters φ and c for the two envelopes will be different. For loose sands, the failure envelope is drawn for ultimate stress, which is usually taken as the shear stress at 20% shear strain.

ii. **Mohr's circle**: In a direct shear test, the stresses on planes other than the horizontal plane are not known. It is therefore not possible to draw Mohr's stress circle at different shear loads. However, the Mohr's circle can be drawn at the failure condition assuming that the failure plane is horizontal.

In Figure 15.53, the point B represents the failure condition for particular normal stress. The Mohr's circle at failure is drawn such that it is tangential to the failure envelope at B. The horizontal line BP gives the direction of the failure plane. The point P is the pole. The lines PD and PA give the directions of the major and minor principal planes, respectively.

15.12.4.2 Triaxial Compression Test

The triaxial compression test or triaxial test (Figure 15.54) is used for the determination of shear characteristics of all types of soils under different drainage conditions. The tests are performed on cylindrical soil samples. The samples have normally a height-to-diameter ratio of 2:1.

The test equipment specially consists of a high-pressure cylindrical cell, made of perspex, fitted between the base and the top cap. The sample is fitted between rigid end caps and covered with a rubber membrane. It is then placed in perspex cell, which is filled with water.

The soil specimen is subjected to three compressive stresses in mutually perpendicular directions, one of the three

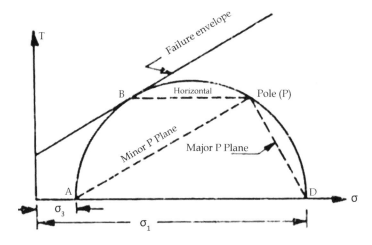

FIGURE 15.53 Angle of failure plane and angle of shearing resistance.

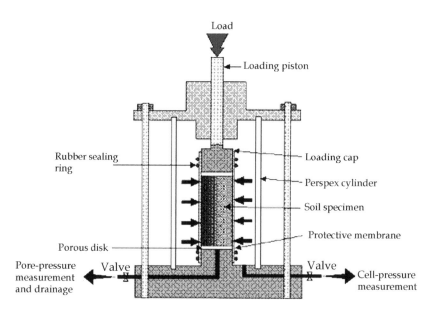

FIGURE 15.54 Triaxial test apparatus.

stresses being increased until the specimen fails in shear. The desired three-dimensional stress system is achieved by an initial application of all-around fluid pressure or confining pressure through water. While this confining pressure is kept constant throughout the test, only vertical loading is increased gradually at uniform rate. The axial stress thus constitutes the major principal stress and the confining pressure acts in the other two principal stresses being equal to the confining pressure. Basically this test is conducted in two stages.

i. **Cell pressure or consolidation stage**: In the first stage of confining pressure, all-around constant pressure σ_c is applied. This confining pressure σ_c acts as minor principal stress, that is, $\sigma_3 = \sigma_c$. For drain condition, expulsion of pore water is permitted by opening the valve and for undrained condition, valve is kept close. In drain condition, consolidation is completed when expulsion of pore water stop and

change in volume of soil is equal to the volume of water collected.

ii. **Shear stage or Deviator stage**: After the completion of consolidation stage, lateral pressure σ_3 is kept constant whereas axial stress is increased by σ_d, which is called deviator stress. Deviator stress is increased till soil sample fails in shear. At failure, deviator stress (σ_d) is called confined compressive strength.

Let $(\sigma_d)_f$ be the deviator stress at failure and is given by

$$(\sigma_d)_f = P/A_f \qquad (15.117)$$

where
$P =$ axial force applied
$A_f =$ area at failure

The total axial stress t failure will be a major principal stress at failure and is equal to

$$(\sigma_1)_f = \sigma_3 + (\sigma_d)_f \qquad (15.118)$$

This test is repeated with identical samples using different values of confining pressure (σ_c).

15.12.4.2.1   Area Correction

If A_0, h_0, and V_0 are the initial area of cross section, height, and volume of soil sample, respectively. Let at any stage of test, these values become A, h, and V, respectively. Also the corresponding changes in values being designated as ΔA, Δh, and ΔV, then
    at any stage,

$$V = V_0 \pm \Delta V = A(h_0 + \Delta h)$$

$$A = (V_0 \pm \Delta V)/(h_0 + \Delta h)$$

But for axial compression, Δh is known to be negative
    Therefore,

$$A = (V_0 \pm \Delta V)/(h_0 - \Delta h)$$

$$A = V_0\left[1 \pm (\Delta V/V_0)\right]/h_0\left[1 - (\Delta h/h_0)\right]$$

$$A = A_0\left[1 \pm \varepsilon_v\right]/\left[1 - \varepsilon_L\right]$$

where

ε_v = volumetric strain
ε_L = linear strain

If drainage is not permitted, there will be no volume change. Hence $\varepsilon_v = 0$

$$\text{Therefore} \quad A = A_0/\left[1 - \varepsilon_L\right] \qquad (15.119)$$

where $1/[1 - \varepsilon_L]$ = correction factor

15.12.4.2.2   Presentation of Results of Triaxial Tests

1. Stress–strain curves:
   a. Drained test: In Figure 15.55, the y-axis shows the deviator stress ($\sigma_1 - \sigma_3$) and the x-axis, the axial stain (ε_1). For dense and overconsolidated clay, the deviator stress reaches a peak value and then it decreases and becomes almost constant, equal to the ultimate stress, at large strains. For loose sand and normally consolidated clay, the deviator stress increases gradually till the ultimate stress is reached.
       In dense sand and overconsolidated clay, there is decrease in the volume at low strains, but at large strains, there is an increase in the volume. In the loose sand and normally consolidated clay, the volume decreases at all strains.
   b. Consolidated undrained test: The shape of the curves is similar to that obtained in a consolidated drained test. In consolidated undrained test, there is an increase in the pore water pressure throughout for loose sand and normally consolidated clay as shown in Figure 15.56.
   c. However, in the case of dense sand and overconsolidated clay, the pore water pressure increases at low strains but at large strains it becomes negative.

2. Mohr's envelope
   For drawing the failure envelopes, it is necessary to test at least three samples at three different cell pressures in the stress range of interest. For dense sands and overconsolidated clays, the failure envelope can be drawn either for the peak stress or for the ultimate stress, which is usually taken at 20% strain. Further, the failure envelope can be drawn either in terms of effective stresses or in terms of total stress. Of course, the two envelopes will give different values of strength parameters (c and φ).
   a. Effective stress: Figure 15.57 shows the failure envelope for normally consolidated clay in terms of effective stresses obtained from consolidated drained test. The failure envelope has an angle of shearing resistance of φ and passes through origin. First, the Mohr's circles for the three tests are drawn in terms of effective stress corresponding to failure conditions. Then the best common tangent is drawn to the three circles. The common tangent is the failure envelope.
       Thus, for normally consolidated clays, shear strength is

$$s = \sigma' \tan \varphi' \qquad (15.120)$$

The failure envelope for overconsolidated clay in terms of effective stresses is slightly

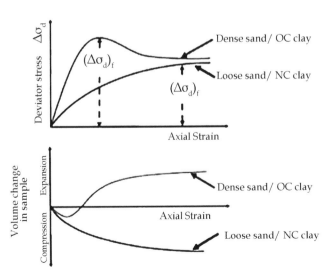

FIGURE 15.55   Plot between deviator stress and axial strain for drained test.

Consolidated- Undrained test (CU test)

Stress-strain relationship curve during shearing

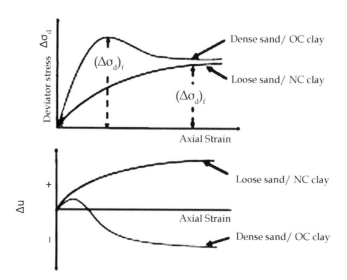

FIGURE 15.56   Variation of deviator stress and pore pressure with axial strain.

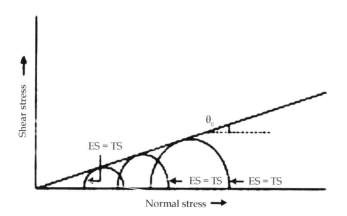

FIGURE 15.57   Failure envelope for normally consolidated clays.

curved in the initial portion, but for convenience, it is approximated as a straight line. The failure envelope has an intercept c' on the τ axis. The angle of shearing resistance is φ'. In the case of overconsolidated clays, shear strength is

$$s = c' + \sigma' \tan \varphi' \qquad (15.121)$$

The failure envelopes in terms of effective stresses can also be drawn from the results of a consolidated undrained test when the pore water pressure measurements are also taken. The shear strength parameters c' and φ' obtained from the consolidated undrained tests and that from consolidated drained tests are approximately equal.

Drained tests on dense sands and overconsolidated clays give slightly higher values of φ' due to extra work required during dilation (increase

in volume), but the difference is small and therefore usually neglected.

b. Total stresses: The failure envelope in terms of total stresses can be drawn from the test results of a consolidated undrained test. The failure envelopes are similar in shape to that in terms of effective stresses, but the values of the strength parameters are quite different. Figure 15.58 shows the failure envelopes for effective stresses and also total stresses for normally consolidated clay. The angle of shearing resistance in terms of total stress (φ_{cu}) is much smaller than that for the effective stresses (φ').

In case of normally consolidated clays, shear strength is

$$s = \sigma \tan \varphi_{cu} \qquad (15.122)$$

Figure 15.59 shows failure envelope for overconsolidated clay in terms of total stress. The angle of shearing resistance (φ_{cu}) is much smaller than the angle φ' obtained in terms of effective stresses. In the case of over consolidated clays, shear strength is

$$s = c_{cu} + \sigma \tan \varphi_{cu} \qquad (15.123)$$

Figure 15.60 shows the failure envelope in terms of total stress from an unconsolidated undrained test on normally consolidated clay. The failure envelope is horizontal ($\varphi=0$) and has a cohesion intercept c_u. In this case shear strength is $s=c_u$. The failure envelope for an overconsolidated clay is also horizontal, but the value of

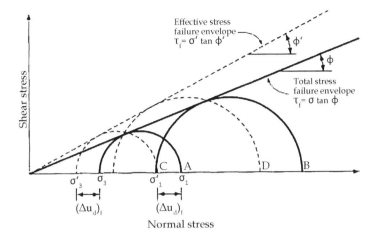

FIGURE 15.58 Failure envelope in terms of effective stress and normal stress.

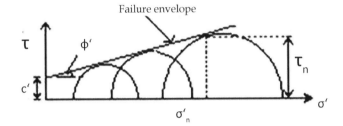

FIGURE 15.59 Failure envelope for overconsolidated clays.

c_u will be more, depending upon the degree of overconsolidation.

For an unconsolidated undrained test, the failure envelope cannot be drawn in terms of effective stresses. In all the tests conducted at different confining pressures, the effective stress remains the same. This is due to the fact that an increase in confining pressure results in an equal increase in pore water pressure for a saturated soil under undrained conditions.

15.12.4.3 Unconfined Compression Test

The unconfined compression test is a special form of the triaxial test in which the confining pressure is zero. The test can be conducted only on clayey soils which can stand without confinement. The test is generally performed on intact saturated clay specimens. Although the test can be conducted in a triaxial test apparatus as UU test, it is more convenient to perform it in an unconfined compression testing machine. There are two types of machines.

a. Machine with a spring
b. Machine with a proving ring

15.12.4.3.1 Presentation of Results

In an unconfined compression test, the minor principal stress (σ_3) is zero. The major principal stress (σ_1) is equal to the deviator stress and is found out by

$$\sigma_1 = P/A \qquad (15.124)$$

where

P = axial load
A = area of cross section

The axial stress at which the specimen fails is known as the unconfined compressive strength (q_u). The stress–strain curve

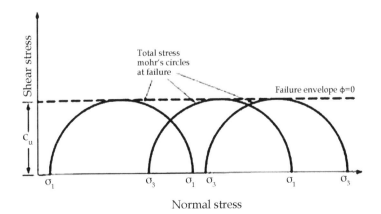

FIGURE 15.60 Failure envelope for unconsolidated undrained test on normally consolidated clay.

can be plotted between the axial stress and the axial strain at different stages before failure.

While calculating the axial stress, the area of cross section of the specimen at that axial strain should be used. The corrected area can be obtained from

$$A = A_0/[1-\varepsilon] \qquad (15.125)$$

The Mohr's circle can be drawn for stress conditions at failure. Figure 15.61 shows the Mohr's circle for unconfined compression test. As the minor principal stress is zero, the Mohr's circle passes through the origin. The failure envelope is horizontal ($\varphi_u = 0$). The cohesion intercept is equal to the radius of the circle.

$$s = c_u = \sigma_1/2 = q_u/2 \qquad (15.126)$$

15.12.4.4 Vane Shear Test

The undrained shear strength of soft clays can be determined in a laboratory by a vane shear test. The test can also be conducted in the field on the soil at the bottom of a bore hole. The field test can be performed even without drilling a bore hole by direct penetration of the vane from the ground surface if it is provided with a strong shoe to protect it. The vane shear test is suitable under the following conditions

a. The clay is normally consolidated and sensitive in nature

b. Only the undrained shear strength is required

The soil mass should be in a saturated condition for the vane test to be applicable.

The apparatus consists of a vertical steel rod having four thin stainless steel blades or vanes fixed at its bottom end as shown in Figure 15.62. IS: 2720-XXX-1980 recommends that the height H of the vane should be equal to twice the overall diameter D. The diameter and the length of the rod are recommended as 2.5 mm and 60 mm, respectively.

The shear strength is given by

$$T = S \pi \left[\left(D^2 H/2 \right) + D^3/6 \right]$$

$$S = T/\pi \left[\left(D^2 H/2 \right) + \left(D^3/6 \right) \right] \qquad (15.127)$$

where

T = torque applied
S = shear strength
D = diameter
H = height of the vane inside the sample

If the test is carried out such that the top end of the vane does not shear the soil, then shear strength is given by

$$S = T/\pi \left[\left(D^2 H/2 \right) + \left(D^3/12 \right) \right] \qquad (15.128)$$

The vane shear test can be used to determine the sensitivity of the soil. After the initial test, the vane is rotated rapidly through several revolutions such that the soil becomes remoulded. The test is repeated on the remoulded soils and the shear strength in remoulded state is determined.

15.13 MECHANICALLY STABILIZED EARTH WALL

Mechanically stabilized earth (MSE) is soil constructed with artificial reinforcing. It can be used for retaining walls, bridge abutments, sea walls, and dikes. The reinforcing elements used can vary but include steel and geosynthetics. MSE is a term usually used in the USA to distinguish it from the name "Reinforced Earth." A trade name of the Reinforced Earth Company, but elsewhere Reinforced soil is the generally accepted term. MSE wall stabilizes unstable slopes and retains the soil on steep slopes and under crest loads. The wall face is often precast, segmental blocks, panels, or geocells that can tolerate some differential movements. The walls are in-filled with granular soil with or without reinforcement, while retaining the backfill soil. Reinforced walls utilize horizontal layers typically of geogrids.

15.13.1 TERMINOLOGY

The generic cross section of an MSE structure is shown in Figure 15.63.

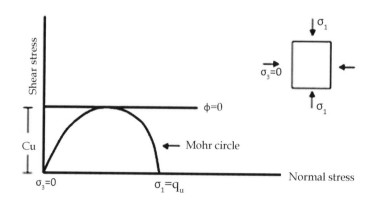

FIGURE 15.61   Mohr's circle for unconfined compression test.

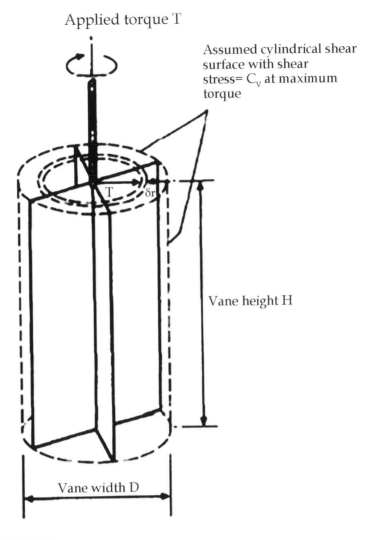

Applied torque T

Assumed cylindrical shear surface with shear stress= C_v at maximum torque

Vane height H

Vane width D

FIGURE 15.62   Vanes shear test setup.

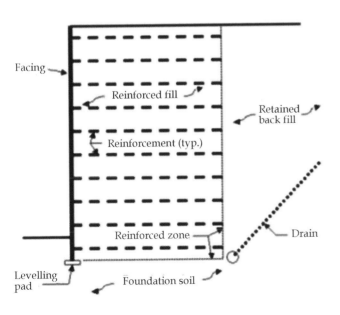

Facing

Reinforced fill

Retained back fill

Reinforcement (typ.)

Reinforced zone

Drain

Levelling pad

Foundation soil

FIGURE 15.63   Cross section of MSE.

1. **Inclusion**: Inclusion is a generic term that encompasses all man-made elements incorporated in the soil to improve its behavior. Examples of inclusions are steel strips, geotextile sheets, steel or polymeric grids, steel nails, and steel tendons between anchorage elements. The term reinforcement is used only for those inclusions where soil-inclusion stress transfer occurs continuously along the inclusion.

2. **MSE Wall or MSEW**: MSE wall is a generic term that includes reinforced soil (a term used when multiple layers of inclusions act as reinforcement in soils placed as fill). Reinforced Earth is a trademark for a specific reinforced soil system.

3. **Reinforced Soil Slopes (RSSs)**: RSSs are a form of reinforced soil that incorporate planar reinforcing elements in constructed earth-sloped structures with face inclinations of less than 70°.

4. **Geosynthetics**: Geosynthetics is a generic term that encompasses flexible polymeric materials used in geotechnical engineering such as geotextiles, geomembranes, geonets, and geogrids.

5. **Facing**: Facing is a component of the reinforced soil system used to prevent the soil from raveling out between the rows of reinforcement. Common facings include precast concrete panels, dry cast modular blocks, gabions, welded wire mesh (WWM), shotcrete, timber lagging and panels, polymeric cellular confinement systems, and wrapped sheets of geosynthetics. The facing also plays a minor structural role in the stability of the structure.
6. **Retained backfill**: Retained backfill is the fill material located behind the mechanically stabilized soil zone.
7. **Reinforced fill**: Reinforced fill is the fill material in which the reinforcements are placed.

15.13.2 Applications of MSE Walls

MSEW structures are cost-effective alternatives for most applications where reinforced concrete or gravity type walls have traditionally been used to retain soil. These include bridge abutments and wing walls, as well as areas where the right-of-way is restricted, such that an embankment or excavation with stable side slopes cannot be constructed. They are particularly suited to economical construction in steepsided terrain, in ground subject to slope instability, or in areas where foundation soils are poor.

MSE walls offer significant technical and cost advantages over conventional reinforced concrete-retaining structures at sites with poor foundation conditions. In such cases, the elimination of costs for foundation improvements such as piles and pile caps, that may be required for support of conventional structures, has resulted in cost savings of greater than 50% on completed projects.

Temporary MSE wall structures have been especially cost-effective for temporary detours necessary for highway reconstruction projects. Temporary MSE walls are used to support temporary roadway embankments and temporary bridge abutments.

MSE walls are also used as temporary support of permanent roadway embankments for phased construction.

15.13.3 Types of Systems

A system for either MSEW or RSS structures is defined as a complete supplied package that includes design, specifications, and all prefabricated materials of construction necessary for the complete construction of a reinforced soil structure. MSE/RSS systems can be described by the reinforcement geometry, stress transfer mechanism, reinforcement material, extensibility of the reinforcement material, and the type of facing and connections.

a. **Reinforcement Geometry**:
   Three types of reinforcement geometry can be considered.
   1. Linear unidirectional: Strips, including smooth or ribbed steel strips, or coated geosynthetic strips over a load-carrying fiber.

2. Composite unidirectional: Grids or bar mats characterized by grid spacing greater than 6 in. (150 mm).
3. Planar bidirectional: Continuous sheets of geosynthetics, WWM, and woven wire mesh. The mesh is characterized by element spacing of less than 6 in. (150 mm).

b. **Reinforcement Material**:
   Distinction can be made between the characteristics of metallic and nonmetallic reinforcements:
   1. Metallic reinforcements: Typically of mild steel. The steel is usually galvanized.
   2. Nonmetallic reinforcements: Generally polymeric materials consisting of polyester or polyethylene (PET).

c. **Reinforcement Extensibility**:
   There are two classes of extensibility relative to the soil's extensibility.
   1. Inextensible: The deformation of the reinforcement at failure is much less than the deformability of the soil. Steel strip and bar mat reinforcements are inextensible.
   2. Extensible: The deformation of the reinforcement at failure is comparable to or even greater than the deformability of the soil. Geogrid, geotextile, and woven steel wire mesh reinforcements are extensible

15.13.4 Facing Systems Used in MSE Walls

The types of facing elements used in the different MSE systems control their aesthetics because they are the only visible parts of the completed structure. In addition, the facing provides protection against backfill sloughing and erosion, and provides, in certain cases, drainage paths. The type of facing influences settlement tolerances. Major facing types are as follows.

a. **Segmental precast concrete panels**: The precast concrete panels have a minimum thickness of 5-½ in. (140 mm) and are of square, rectangular, cruciform, diamond, or hexagonal geometry. Typical nominal panel dimensions are 5-foot (1.5 m) high and 5- or 10-foot (1.5 or 3 m) wide. Temperature and tensile reinforcement of the concrete are required and should be designed in accordance with AASHTO LRFD Specifications for Highway Bridges (2007).
b. **Dry cast modular block wall (MBW) units**: These are relatively small, squat concrete units that have been specifically designed and manufactured for retaining wall applications. The weight of these units commonly ranges from 30 to 110 lbs (15 to 50 kg), with units of 75 to 110 lbs (35 to 50 kg) routinely used for highway projects. Unit heights typically range from 4 to 12 in. (100 to 300 mm) for the various manufacturers, with 8-in. (200 mm) typical.

Exposed face length usually varies from 8 to 18 in. (200 to 450 mm). Nominal front to back width (dimension perpendicular to the wall face) of units typically ranges between 8 and 24 in. (200 and 600 mm). Units may be manufactured solid or with cores. Full height cores are filled with aggregate during erection.

c. **WWM**: Wire grid can be bent up at the front of the wall to form the wall face. This type of facing is used for example in the Hilfiker, Tensar, and Reinforced Earth wire faced retaining wall systems. This type of facing is commonly used for RSS with face angles of about 45° and steeper.

d. **Gabion Facing**: Gabions (rock-filled wire baskets) can be used as MSE wall or RSS facing with reinforcing elements consisting of WWM, welded bar-mats, geogrids, geotextiles, or the double-twisted woven mesh placed between or integrally manufactured with the gabion baskets.

e. **Geosynthetic Facing**: Geosynthetic reinforcements are looped around at the facing to form the exposed face of the MSEW or RSS. These faces are susceptible to ultraviolet light degradation, vandalism, and damage due to fire. Geogrid used for soil reinforcement can be looped around to form the face of the completed retaining structure in a similar manner to WWM and fabric facing. Vegetation can grow through the grid structure and can provide both ultraviolet light protection for the geogrid and a pleasing appearance.

f. **Post-construction Facing**: For wrapped faced walls, the facing—whether geotextile, geogrid, or wire mesh—can be attached after construction of the wall by shotcreting, guniting, cast-in-place concrete, or by attaching prefabricated facing panels made of concrete, wood, or other materials. This multistaging facing approach adds cost but is advantageous where significant settlement is anticipated.

15.13.5 TYPES OF REINFORCEMENTS USED IN MSE WALLS

Most, although not all, MSE wall systems with precast concrete panels use steel reinforcements that are typically galvanized. The two types of steel reinforcements currently in use with segmental panel faced MSE walls are:

a. **Steel strips**: The currently commercially available strips are ribbed top and bottom, 2 in. (50 mm) wide and 5/32-in. (4 mm) thick. Smooth strips 2- to 4¾-in. (60 to 120 mm) wide, 1/8 to 5/32-in. (3 to 4 mm) thick have been used.

b. **Steel grids**: Welded wire grid using two to six W7.5 to W24 longitudinal wire spaced at either 6 or 8 in. (150 or 200 mm). The transverse wire may vary from W11 to W20 and are spaced based on design requirements from 9 to 24 in. (230 to 600 mm). Welded steel wire mesh spaced at 2 by 2-in. (50 by 50 mm) of thinner wire has been used in conjunction with a welded wire facing. Some MBW systems use steel grids with two longitudinal wires.

Most MBW systems use geosynthetic reinforcement, predominantly geogrids. The following soil reinforcement types are widely used and available:

a. **High-density polyethylene geogrid**: These are of uniaxial manufacture and are available in up to 6 grades of strength. This type of reinforcement is also used with segmental panel facing.

b. Polyvinyl chloride **(PVC)-coated PET geogrid**: Available from a number of manufacturers. They are characterized by bundled high-tenacity PET fibers in the longitudinal load carrying direction. For longevity the PET is supplied as a high molecular weight fiber and is further characterized by a low carboxyl end group number.

Other types of soil reinforcements, and their applications, include:

a. **Geotextiles**: High-strength geotextiles can be used principally in connection with RSS construction. Both PET and polypropylene geotextiles have been used.

b. **Double-twisted steel mesh**: The Terramesh® system by Maccaferri, Inc., uses a metallic, soft-temper, double-twisted mesh soil reinforcement that is galvanized and then coated with PVC. This reinforcement is used for RSS and gabion faced MSE wall construction. Note that this reinforcement is classified as an extensible type of reinforcement due to its manufacturing geometry even though it is metallic.

c. **Geosynthetic strap**: Although not (currently) widely used, geosynthetic strap type reinforcement has been used with segmental panel faced MSE walls. The strap consists of PET fibers encased in a polyethylene sheath.

15.14 SOIL NAILING

Soil nailing is the technique used in slope stabilization and excavation with the use of passive inclusions, usually steel bars, termed as soil nail. It consists of passive reinforcement which is encased in grout to provide corrosion protection and improved load transfer to ground.

Soil nailing system, the mechanism of interaction between the soil and nail in the passive zone, is quite important. The effect of different parameters like nail roughness, rigidity, flexibility, nail inclination and soil properties such as angle of internal friction and dilation have also been studied by various researchers. The application of Finite Element Method has been proposed for the estimation of maximum shear stress generation at the soil–nail interface and comprehending the concept of dilation and stress release after drilling of the nail bore.

15.14.1 Concept of Soil Nailing

The function of soil nailing is to strengthen or stabilize the existing steep slopes and excavations as construction proceeds from the top to bottom. Soil nails develops their reinforcing action through soil–nail interaction due to the ground deformation which results in development of tensile forces in soil nail. The major part of resistances comes from the development of axial force which is basically a tension force. Conventionally, shear and bending have been assumed to provide little contribution in providing resistance.

The effect of soil nailing is to improve the stability of slope or excavation through

a. increasing the normal force on shear plane and hence increase the shear resistance along slip plane in friction soil;
b. reducing the driving force along slip plane both in friction and cohesive soil.

In soil nailing, the reinforcement is installed horizontally or gently inclined parallel to the direction of tensile strain so that it develops maximum tensile force.

15.14.1.1 Ground Condition Best Suited for Soil Nailing

The ground conditions which are best suited for soil nailing can be enlisted as:

a. Residual soil and weathered rocks.
b. Stiff cohesive soils such as clayey silts and other soils that are not prone to creep deformation.
c. Dense sand and gravel with some cohesive properties.
d. Ground conditions located above the ground water table.

15.14.1.2 Ground Condition Not Suited for Soil Nailing

The ground conditions which render the application of soil nailing technique to be unsuitable are as follows:

a. Loose clean granular sand as they do not have adequate stand-up time prior to nail installation.
b. Soils having excessive moisture or wet pockets as they will create stability problems.
c. Soils having frost susceptibility and expansive properties result into excessive facing load.
d. Highly fractured rocks with open joints or voids due to problem in grouting.

15.14.2 Various Types of Soil Nailing

Various types of soil nails which are commonly encountered in practice are as follows:

a. **Grouted Soil nail**: First the nails are centrally placed in the drill hole and then the grout is inserted in the hole with the help of pressure or gravity.
b. **Driven nails**: These nails are directly driven after each excavation step.
c. **Jet grouted nails**: These are composite inclusion made of grouted soil with central steel rod. It uses high-pressure grout for vibro-percussion driving and then the nails are installed using high frequency.
d. **Launched nails**: In this process, nails are launched into the ground with very high speed, using compressed air launcher. This method is very rapid, flexible, and economical. Nails are installed at speed of around 320 km/h.

15.14.3 Elements of Soil Nail

The various elements of a soil nail (Figure 15.64) can be enlisted as follows:

a. **Steel bar**: This is the main component of soil nail system. It may be solid or hollow with necessary required strength. It acts as a tension member.
b. **Centralizers**: It is fixed with steel bar so that nail can be placed centrally in drill hole.
c. **Grout**: It is used to fill the space between ground and installed nails. Proper bonding is achieved through grouting and the loads are transfer first to the grout and then to the nail.
d. **Nail head**: It works as a reaction pad for generation of tensile force in the nails and it also prevents local failure between the nails.
e. **Hex nut, washer and bearing plate**: These are used to connect nail to the facing and form the integral part of nail head.
f. **Temporary and permanent facing**: It provides support to the exposed surface of soil nail and act as bearing surface for bearing plate. After that, permanent facing is installed over temporary facing.
g. **Drainage system**: A prefabricated synthetic drainage is placed vertically against the excavation face to prevent any seepage against the excavation face.

15.15 LATERAL STRESS AND RETAINING WALL

A retaining structure is required to provide lateral support to the soil mass. Generally the soil masses are vertical or nearly vertical behind the retaining structure. Thus retaining wall maintains the soil at different elevations on its either side. In the absence of a retaining wall, the soil on the higher side would have a tendency to slide and may not remain stable.

The design of the retaining structure requires the determination of the magnitude and the line of action of the lateral earth pressure. The magnitude of the lateral earth pressure depends up on a number of factors such as the mode of movement of the wall, the flexibility of the wall, the properties of the soil, and the drainage conditions.

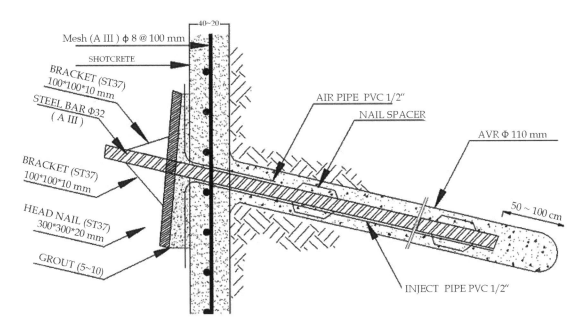

FIGURE 15.64 Parts of soil nail.

15.15.1 Different Types of Lateral Earth Pressure

Lateral earth pressures can be grouped into three categories depending up on the movement of the retaining wall with respect to the soil retained. The soil retained is also known as the backfill.

1. **At-rest earth pressure**: The lateral earth pressure is called at-rest pressure when the soil mass is not subjected to any lateral yielding or movement. This case occurs when the retaining wall is firmly fixed at its top and is not allowed to rotate or move laterally. The at-rest condition is also known as the elastic equilibrium, as no part of soil mass has failed and attained the plastic equilibrium.

2. **Active pressure**: A state of active pressure occurs when the soil mass yields in such a way that it tends to stretch horizontally. It is a state of plastic equilibrium as the entire soil mass is on the verge of failure. A retaining wall when moves away from the backfill, there is a stretching of the soil mass and the active state of earth pressure exists.

3. **Passive pressure**: A state of passive pressure exists when the movement of the wall is such that the soil tends to compress horizontally. It is another extreme of the limiting equilibrium condition. The passive pressure develops on the left side of the wall below the ground level, as the soil on this zone is compressed when the movement of the wall is toward left.

15.15.1.1 Variation of Pressure

Figure 15.65 shows the variation of earth pressure with the wall movement. Point B represents the case when there is no movement of the wall. It indicates the at-rest pressure.

Point A indicates active pressure. When the wall moves away from the backfill, some portion of the backfill located immediately behind the wall tries to break away from the rest of the soil mass. This wedge-shaped portion, known as the failure wedge or the sliding wedge, moves downward and outward. The lateral earth pressure exerted on the wall is minimum in this case. The soil is at the verge of failure due to a decrease in the lateral stress. The horizontal strain required to reach the active state of plastic equilibrium.

Point C indicates passive earth pressure. When the wall moves toward the backfill, the earth pressure increases. The failure wedge moves upward and inward. The maximum value of the earth pressure is the passive earth pressure. The soil is at the verge of failure due to an increase in the lateral stress. Figure 15.66 represents failure wedges in case of active and passive earth pressure conditions.

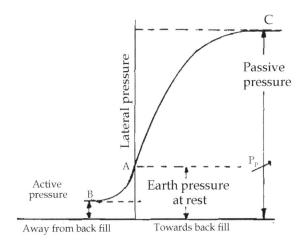

FIGURE 15.65 Variation of earth pressure.

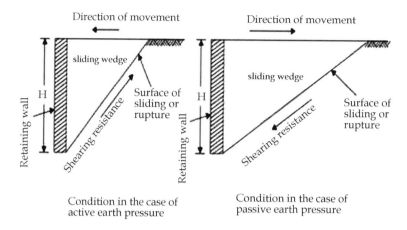

FIGURE 15.66 Failure wedge in case of active and passive earth pressure.

15.15.2 Earth Pressure at Rest

Consider a vertical wall of height H, retaining a soil having a unit weight of γ as shown in Figure 15.67. A uniformly distributed load, q/unit area, is also applied at the ground surface. The shear strength s of the soil is

$$s = c + \sigma' \tan \varphi \qquad (15.129)$$

where

  C=cohesion
  φ=angle of friction
  σ'=effective normal stress

At any depth z below the ground surface, the vertical subsurface stress is

$$\sigma_z = q + \gamma z \qquad (15.130)$$

If the wall is at rest and is not allowed to move at all either away from the soil mass or into the soil mass (e.g., zero horizontal strain), the lateral pressure at a depth z is

$$\sigma_h = K_0 \sigma_z + u$$

$$\sigma_h = K_0 \left(q + \gamma z \right) + u \qquad (15.131)$$

where K_0=coefficient of lateral earth pressure at rest which is equal to the ratio of horizontal stress to the vertical stress.

$$K_0 = 1 - \sin \varphi \,(\text{Jacky's formula}) \qquad (15.132)$$

where u=pore water pressure
  If the surcharge q=0 and the pore water pressure u=0, the pressure diagram will be a triangle. The total force, P_0, per unit length of the wall can now be obtained from the area of the pressure diagram.

$$P_0 = P_1 + P_2$$

$$P_0 = K_0 q H + \tfrac{1}{2} \left(K_0 \gamma H^2 \right) \qquad (15.133)$$

where P_1 and P_2=area of pressure diagram.
  The point application of the resultant pressure P_0 is determined from the pressure distribution diagram. For triangular pressure distribution, it acts at height $H/3$ from the base.
  If the water table is located at depth $z < H$, then the effective unit weight of soil below the water table equals $\gamma' \,(\gamma_{sat} - \gamma_w)$

$$\text{At } z = 0, \quad \sigma'_h = K_0 q$$

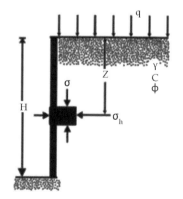 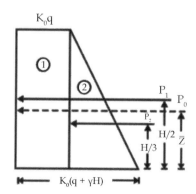

FIGURE 15.67 Earth pressure distribution with uniform surcharge load.

$$\text{At } z = H_1, \quad \sigma_h' = K_0\left(q + \gamma H_1\right)$$

$$\text{At } z = H_2, \quad \sigma_h' = K_0\left(q + \gamma H_1 + \gamma' H_2\right)$$

Hence the total force per unit length of the wall can be determined from the area of the pressure diagram. Thus

$$P_0 = K_0 q + \tfrac{1}{2} K_0 \gamma H_1^2 + K_0\left(q + \gamma H_1\right)H_2$$

$$+ \tfrac{1}{2} K_0 \gamma' H_2^2 + \tfrac{1}{2} \gamma_w H_2^2 \qquad (15.134)$$

15.15.3 RANKINE'S EARTH PRESSURE THEORY

Rankine considered the equilibrium of a soil element within a soil mass bounded by a plane surface.

15.15.3.1 Active Earth Pressure

Let us consider an element of dry soil at a depth z below the level soil surface as shown in Figure 15.68. If a wall tends to move away from the soil a distance Δx, the soil pressure on the wall at any depth will decreases as the soil expands outward.

In active state, minor principal stress

$$\sigma_3 = \sigma_h = p_a \qquad (15.135)$$

Major principal stress

$$\sigma_1 = \sigma_v = \gamma z \qquad (15.136)$$

From the stress relationship between σ_1 and σ_3 at failure,

$$\sigma_1 = \sigma_3 \tan^2 \alpha + 2c \tan \alpha \qquad (15.137)$$

where $\alpha = 45° - \varphi/2$.

For cohesionless soil $c = 0$

For active state within cohesionless soil,

$$\sigma_v = \sigma_h \tan^2 \alpha = \sigma_h \tan^2\left(45° - \varphi/2\right) \qquad (15.138)$$

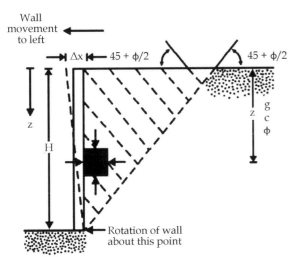

Wall movement to left

Δx 45 + φ/2   45 + φ/2

z

H

Rotation of wall about this point

z g c φ

FIGURE 15.68 Active Rankine state.

$$\sigma_h = \cot^2\left(45° - \varphi/2\right)\sigma_v \qquad (15.139)$$

$$p_a = K_a \sigma_v \qquad (15.140)$$

where K_a = coefficient of active earth pressure.

$$K_a = \cot^2\left(45° - \varphi/2\right) = 1 - \sin\varphi / 1 + \sin\varphi \qquad (15.141)$$

When the wall moves away from the backfill, the failure wedge moves downward, and the resisting force due to the shearing strength of the soil is developed in the upward direction along the failure plane. The resisting force causes a decrease in the earth pressure acting on the wall. The decrease in earth pressure continues till the maximum resistance has been mobilized. The earth pressure does not decrease beyond this point and the active state is reached and the soil has attained plastic equilibrium.

15.15.3.2 Passive Earth Pressure

The passive Rankine state of plastic equilibrium can be explained by considering the element of soil at a depth z below the soil surface as shown in Figure 15.69. As the soil is compressed laterally, the horizontal stress is increased, whereas the vertical stress remains constant.

In passive state, minor principal stress

$$\sigma_3 = \sigma_h = p_p \qquad (15.142)$$

Major principal stress

$$\sigma_1 = \sigma_v = \gamma z \qquad (15.143)$$

From the stress relationship between σ_1 and σ_3 at failure,

$$\sigma_1 = \sigma_3 \tan^2 \alpha + 2c \tan \alpha \qquad (15.144)$$

where $\alpha = 45° + \varphi/2$

For cohesionless soil $c = 0$

For active state within cohesionless soil,

$$\sigma_v = \sigma_h \tan^2 \alpha = \sigma_h \tan^2\left(45° + \varphi/2\right) \qquad (15.145)$$

$$p_p = \tan^2\left(45° + \varphi/2\right)\sigma_v$$

$$p_p = K_p \sigma_v \qquad (15.146)$$

where K_p = coefficient of passive earth pressure.

$$K_p = \tan^2\left(45° + \varphi/2\right) = 1 + \sin\varphi / 1 - \sin\varphi \qquad (15.147)$$

When the wall moves toward the backfill, the lateral earth pressure increases because the resistance builds up in the direction toward the wall. The pressure reaches a maximum value when the full shearing resistance has been mobilized. Further movement of the wall does not increase the pressure and the passive state is reached and the soil has attained plastic equilibrium.

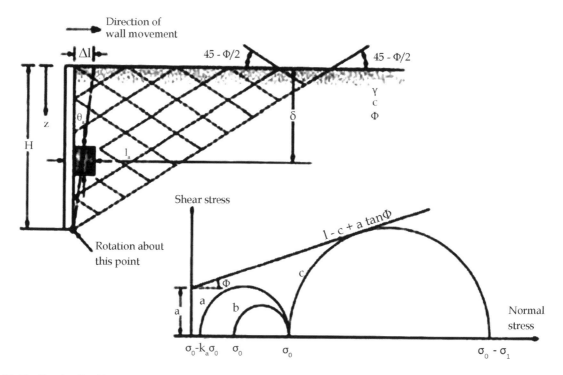

FIGURE 15.69 Passive Rankine state.

15.15.3.3 Rankine's Earth Pressure when the Surcharge is Inclined

Two stresses are called conjugate stresses when the direction of one stress is parallel to the plane on which the other stress acts. Rankine assumed that the vertical stress on an element of the soil within the inclined backfill and the lateral stress on the vertical plane of the element are conjugate stresses. In other words, he assumed that the lateral stress is parallel to the inclined backfill. If the backfill is rises at an angle α with respect to the horizontal, then the coefficients of active and passive earth pressure (Figure 15.70) are given by:

$$K_a = \cos\alpha \left\{ \frac{\left[\cos\alpha - \left(\cos^2\alpha - \cos^2\varphi \right)^{1/2} \right]}{\left[\cos\alpha + \left(\cos^2\alpha - \cos^2\varphi \right)^{1/2} \right]} \right\} \quad (15.148)$$

$$K_p = \cos\alpha \left\{ \frac{\left[\cos\alpha + \left(\cos^2\alpha - \cos^2\varphi \right)^{1/2} \right]}{\left[\cos\alpha - \left(\cos^2\alpha - \cos^2\varphi \right)^{1/2} \right]} \right\} \quad (15.149)$$

where φ = angle of friction of soil.

At any depth z, Rankine active and passive pressure can be expressed as

$$p_a = K_a \gamma z \quad (15.150)$$

$$p_p = K_p \gamma z \quad (15.151)$$

Total force per unit length of the wall is

$$P_a = \tfrac{1}{2} K_a \gamma H^2 \text{ and } P_P = \tfrac{1}{2} K_p \gamma H^2 \quad (15.152)$$

15.15.3.4 Rankine's Earth Pressure in Cohesive Soil

Rankine's original theory was for cohesionless soils. It was extended by Resal and Bell for cohesive soils. The basics is similar to that of cohesionless soil with one basic difference is that the failure envelope has a cohesion intercept c, whereas for cohesionless soil it is zero.

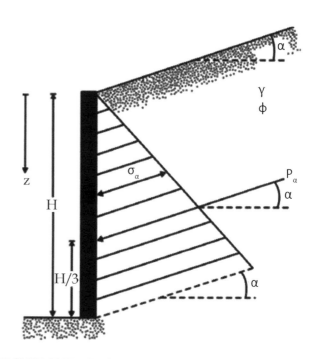

FIGURE 15.70 Earth pressure distribution on an inclined back wall with sloping surface.

15.15.3.5 Active State

In active state (Figure 15.71), lateral stress reduces to its minimum value p_a while the vertical stress remains unchanged.

From stress relationship at failure, now we have

$$\sigma_1 = \sigma_3\left(1 + \sin\varphi/1 - \sin\varphi\right) + 2c\left(1 + \sin\varphi/1 - \sin\varphi\right)^{1/2} \quad (15.153)$$

Hence

$$\sigma_3 = \sigma_h = p_a$$

$$\sigma_1 = \sigma_v = \gamma z$$

Substituting σ_1 and σ_3 in above equation, we get

$$\gamma z = p_a\left(1 + \sin\varphi/1 - \sin\varphi\right) + 2c\left(1 + \sin\varphi/1 - \sin\varphi\right)^{1/2}$$

$$p_a = \gamma z\left(1 - \sin\varphi/1 + \sin\varphi\right) - 2c\left(1 - \sin\varphi/1 + \sin\varphi\right)^{1/2}$$

$$p_a = K_a\gamma z - 2c\left(K_a\right)^{1/2} \quad (15.154)$$

$$\text{At } z = 0, \ p_a = -2c\left(K_a\right)^{1/2}$$

$$\text{At } z = H, \ p_a = K_a\gamma H - 2c\left(K_a\right)^{1/2}$$

Negative pressure in cohesive soils in the active state at top indicates that these soils are in the state of tension and these tensile stresses gradually reduce to zero at a depth Z_c and it is given by

$$0 = K_a\gamma Z_c - 2c\left(K_a\right)^{1/2}$$

$$Z_c = 2c\big/\gamma\left(K_a\right)^{1/2} \quad (15.155)$$

The depth Z_c is known as the depth of tensile crack. The tensile stress eventually causes a crack to from along the soil–wall interface.

The pressure is negative at top region and it becomes zero at a depth Z_c. If the wall has a height of $2Z_c$, the total earth pressure is zero. This height is known as the critical height H_C.

$$H_C = 2Z_c \quad (15.156)$$

If the height of an unsupported vertical cut is smaller than H_C, it should be able to stand. However, the conditions in unsupported vertical cut are different from those near a retaining wall. In the vertical cut, the lateral stress is everywhere zero, whereas in the retaining wall, it varies from $-2 c (K_a)^{1/2}$ to $+2 c (K_a)^{1/2}$. Because of this difference in the stress condition, the safe height of the vertical cut is given by

$$H_C = 2 \times 2 \ c\big/\gamma\left(K_a\right)^{1/2}$$

$$H_C = 4c/\gamma; \quad \text{for } \varphi = 0 \quad (15.157)$$

15.15.3.6 Passive State

An expression for passive pressure in a cohesive soil can be determined by referring equation obtained for active pressure.

$$\text{Therefore,} \quad p_p = K_p\gamma z - 2c\left(K_p\right)^{1/2}$$

$$\text{At} \quad z = 0, \ p_p = 2c\left(K_p\right)^{1/2}$$

$$\text{At} \quad z = H, \ p_p = K_p\gamma H - 2c\left(K_p\right)^{1/2}$$

The total pressure on the retaining wall per unit length is given by

$$P_p = \tfrac{1}{2}H\left(K_p\gamma H\right) - 2c\left(K_p\right)^{1/2} H \quad (15.158)$$

15.15.4 Retaining Wall

Retaining walls are structures used to retain earth or other loose material which would not be able to stand vertically itself. The retained material exerts a push on structure and this tends to overturn and slide it. The weight of the retaining

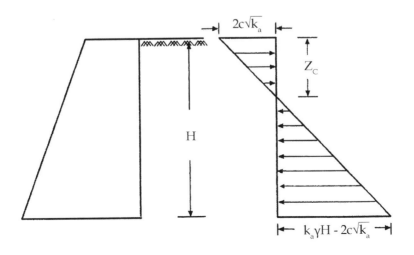

FIGURE 15.71 Active earth pressure distribution c-φ soil.

walls is of considerable significance in achieving and maintaining stability of the entire system. The different types of retaining walls (Figure 15.72) are as follows:

1. **Gravity wall**: A gravity wall is of plain concrete and provides stability by its weight.
2. **Cantilever wall**: This is the most commonly used retaining wall. It consists of three components: vertical wall, heel slab, and toe slab. Each of the three components acts as a cantilever beam. Stability is provided by weight of the earth on the base and weight of the retaining wall.
3. **Counterfort wall**: In a counterfort retaining wall the vertical slab and the horizontal slabs, that is, heel and toe are tied together by counterforts. Counterforts are transverse walls spaced at certain intervals and act as tension ties to support the vertical wall. Stability is provided by weight of the earth on the base slab and weight of the retaining wall.
4. **Buttress wall**: A buttress wall is similar to a counterfort wall except that the transverse walls are located on the side of the vertical wall opposite to the retained material and act as compression struts. Although, a buttress as a compression member is more economical than the tension counterfort, still the latter is more widely used than a buttress because the counterfort is hidden beneath the retained material, whereas the buttress is exposed. Moreover, buttress occupies more space in front of the wall which otherwise could be utilized more efficiently.

5. **Bridge abutments**: A wall-type bridge abutment acts similar to a cantilever-retaining wall except that the bridge deck provides an additional horizontal restraint at top of the vertical slab. This type of abutment is designed as a beam fixed at the bottom and simply supported or partially restrained at the top.
6. **Box culvert**: A box culvert is a box-like structure having either single cell or multiple cells. It acts as a closed rigid frame that not only resists lateral earth pressure but also vertical load from soil above it or from both soil and highway vehicles.

15.15.4.1 Forces on Retaining Walls

The main force that acts on a retaining wall (Figure 15.73) is pressure due to the retained material. The magnitude and direction of the earth pressure that tends to overturn and slides a retaining wall may be determined by applying the principles of soil mechanics. The pressure exerted by the retained material is proportional to its density and to the distance below the earth surface, that is,

$$p = K \gamma h \tag{15.159}$$

where

p = earth pressure
γ = density of retained material
h = depth of the section below earth surface
K = a coefficient that depends on the physical properties of the soil

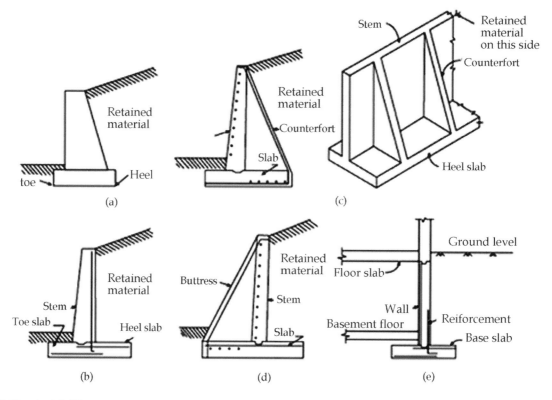

FIGURE 15.72 (a–e) Different types of retaining wall.

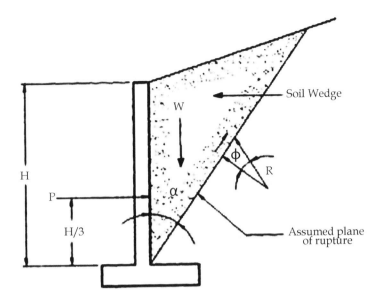

FIGURE 15.73   Forces acting on retaining wall.

The value of factor C could be obtained by using either Coulomb's theory or Rankine's theory for cohesionless materials such as sand gravel. Both theories give equally conservative results for vertical wall with horizontal fill. Coulomb's theory gives better results for other cases. Earth pressure behind a retaining wall is generally calculated by Rankine's theory.

W.J.M. Rankine presented a theory of earth pressure for homogeneous incompressible cohesionless soil assuming a hydrostatic pressure direction along the path. Rankine's equation for active earth pressure is given as:

$$p = K_a \gamma h \qquad (15.160)$$

where K_a=coefficient of active pressure.

$$K_a = \cos\delta \left\{ \frac{\left[\cos\delta - \left(\cos^2\delta - \cos^2\varphi\right)^{1/2}\right]}{\left[\cos\delta + \left(\cos^2\delta - \cos^2\varphi\right)^{1/2}\right]} \right\}$$

where

   δ=angle of surcharge, that is, slope of the retained earth with horizontal
   φ=angle of repose of soil

The pressure p acts parallel to top surface of the material retained. If the top of the earth retained is horizontal, that is, $\delta=0$.

$$K_a = \left(1 - \sin\varphi\right) / \left(1 + \sin\varphi\right) \qquad (15.161)$$

The total force behind the wall is given by area of the pressure triangle, that is,

$$P_a = \tfrac{1}{2} K_a \gamma h^2 \qquad (15.162)$$

per unit length of wall and acts at a depth of $2/3\,h$ below the top.
   Rankine's equation for passive pressure is given by

$$P_p = \tfrac{1}{2} K_p \gamma h^2 \qquad (15.163)$$

where K_p=coefficient of passive earth pressure

$$K_p = \cos\delta \left\{ \frac{\left[\cos\delta + \left(\cos^2\delta - \cos^2\varphi\right)^{1/2}\right]}{\left[\cos\delta - \left(\cos^2\delta - \cos^2\varphi\right)^{1/2}\right]} \right\}$$

If top of the earth retained is horizontal, that is, $\delta=0$, the coefficient of passive earth pressure becomes,

$$K_p = 1 + \sin\varphi / 1 - \sin\varphi \qquad (15.164)$$

The passive pressure is quite high as compared with the active pressure. The computation of passive pressure is generally more in error than is the calculation of active pressure. Passive pressure contribution in the design of retaining wall is neglected which is on the conservative side.

Other forces acting on a retaining wall may be due to the construction of buildings and moving loads on the surface of the backfill. When buildings are constructed near top of a retaining wall, the lateral pressure applied to the wall is increased. Such a surcharge can be converted into an additional height of backfill soil material which will approximate the lateral earth pressure effect. If the surcharge is uniform over the sliding area behind the wall, equivalent height of backfill is given as total surcharge intensity per unit length divided by the density of backfill.

15.16   STABILITY OF SLOPES AND LANDSLIDES

An earth slope is an unsupported, inclined surface of a soil mass. Earth embankments are commonly required for railways, roadways earth dams, leeves, and river training works. The stability of these embankments or slopes, as they are commonly called, should be very thoroughly analyzed since their failure may lead to loss of human life as well as colossal economic loss.

The failure of a mass of soil located beneath a slope is called a slide. It involves a downward and outward movement of the entire mass of soil that participates in the failure. The failures of slopes take place mainly due to (1) the action of gravitational forces and (2) seepage forces within soil. They may also fail due to excavation or undercutting of its foot or due to gradual disintegration of the structure of the soil. Slides may occur in almost every conceivable manner, slowly or suddenly and with or without any apparent provocation.

The analysis of stability of slope consists of two parts: (1) the determination of the most severely stressed internal surface and the magnitude of the shearing stress to which it is subjected; (2) the determination of the shearing strength along this surface. The shearing stress to which any slope can be subjected depends up on the unit weights of the material and the geometry of the slope, while shearing strength which can be mobilized to resist the shearing stress depends on the character of the soil, its density, and drainage conditions.

Slopes may be of two types. They are infinite slope and finite slope. If a slope represents the boundary surface of a semi-infinite soil mass and the soil properties for all identical depths below the surface are constant, it is called an infinite slope. If the slope is of limited extent, it is called a finite slope. Slopes extending to infinity do not exist in nature.

15.16.1 Different Definitions of Factor of Safety

Three different definitions of the factor of safety are used.

1. **Factor of safety with respect to shear strength**: In common usage, the factor of safety is defined as the ratio of the shear strength to the shear stress along the surface of failure. The factor of safety as defined above is known as the factor of safety with respect to shear strength.

$$\text{Thus} \quad F_s = s/\tau_m \qquad (15.165)$$

where
F_s=factor of safety with respect to shear strength
s=shear strength
τ_m=mobilized shear strength (equal to applied shear stress)
The equation can be written in terms of cohesion intercept and the angle of shearing resistance as

$$F_s = c + \sigma' \tan\varphi / c_m + \sigma' \tan\varphi_m \qquad (15.166)$$

where
c_m=mobilized cohesion
φ_m=mobilized angle of shear resistance
σ'=effective pressure

Rearranging Equation 15.166

$$c/F_s + \sigma' \tan\varphi / F_s = c_m + \sigma' \tan\varphi_m$$

$$\text{Therefore,} \quad c_m = c/c/F_s \qquad (15.167)$$

$$\tan\varphi_m = \tan\varphi / F_s \qquad (15.168)$$

Equations 15.165 and 15.166 indicate the factor of safety with respect to cohesion intercept and that with respect to the angle of shearing resistance are equal to the factor of safety with respect to shear strength.

2. **Factor of safety with respect to cohesion**: The factor of safety with respect to cohesion (F_c) is the ratio of the available cohesion intercept (c) and the mobilized cohesion intercept.

$$\text{Thus} \quad F_c = c/c_m \qquad (15.169)$$

where
c=cohesion intercept
c_m=mobilised cohesion intercept
F_c=factor of safety with respect to cohesion

3. **Factor of safety with respect to friction**: The factor of safety with respect to friction is the ratio of the available frictional strength to the mobilized frictional strength.

$$\text{Thus} \quad F_\varphi = \sigma' \tan\varphi / \sigma' \tan\varphi_m \qquad (15.170)$$

where
F_φ=factor of safety with respect to friction
φ=angle of shearing resistance
φ_m=angle of mobilized shearing resistance
In the analysis of stability of slopes, generally the three factors of safety are taken equal ($F_s=F_c=F_\varphi$).

15.16.2 Stability of an Infinite Slope of Cohesionless Soils

The stability criteria of an infinite slope of cohesionless soils will depend on whether the soil is dry or submerged or has steady seepage.

a. Dry soil
The factor of safety against shear failure is given by

$$F_s = s/\tau$$

$$F_s = \left(\gamma H \cos^2 i\right) \tan\varphi' / \gamma H \cos i \sin i$$

$$F_s = \tan\varphi' / \tan i \qquad (15.171)$$

The above equation indicates that the slope is just stable when $\varphi'=i$. The factor of safety is greater than unity when i is less than φ'. For the slope angle i greater than φ', the slope is not stable.

It is also noticed that the factor of safety of an infinite slope of a cohesionless soil is independent of the height H of the assumed failure prism.

b. Submerged soil

If the slope is submerged under water, the normal effective stress and the shear stress are calculated using the submerged unit weight and not the bulk unit weight as was used for dry soil.

Thus, the factor of safety is given by

$$F_s = s/\tau$$

$$F_s = \left(\gamma' H \cos^2 i\right)\tan\varphi'/\gamma' H \cos i \sin i$$

$$F_s = \tan\varphi'/\tan i$$

c. Steady seepage along the slope

Factor of safety is given by

$$F_s = s/\tau = \gamma' H \cos^2 i \tan\varphi'/\gamma_{sat} H \cos i \sin i$$

$$F_s = \gamma' \tan\varphi'/\gamma_{sat} \tan i$$

As the submerged unit weight is about one-half of the saturated unit weight, the factor of safety of the slope is reduced to about one-half of that corresponding to the condition when there is no seepage. The angle φ' in the wet condition of a cohesionless soil is approximately the same as in dry condition.

15.16.3 STABILITY OF AN INFINITE SLOPE OF COHESIVE SOILS

The stability analysis of an infinite slope of cohesive soils is similar to that in the case of cohesionless soils, with one basic difference that the shear strength of a cohesive $(c-\varphi)$ soil is given by

$$s = c' + \sigma' \tan\varphi'$$

a. Dry soil

Factor of safety is given by

$$F_s = c' + \left(\gamma H \cos^2 i\right)\tan\varphi'/\gamma H \cos i \sin i \quad (15.172)$$

Thus, the factor of safety of an infinite slope in cohesive soil soils depends not only on φ' and i but also on γ, H, and c'.

The height at which the slope is just stable is known as the critical height (H_C). Thus

$$H_C = c'/\gamma \left(\tan i - \tan\varphi'\right)\cos^2 i \quad (15.173)$$

For heights less than critical height, the factor of safety is given by

$$F_s = c' + \left(\gamma H \cos^2 i\right)\tan\varphi'/\gamma H \cos i \sin i \quad (15.174)$$

b. Submerged slope

As in the case of cohesionless soils, the normal and tangential components of the weight are taken for submerged unit weights and not for bulk unit weights. Thus,

$$F_s = c' + \left(\gamma' H \cos^2 i\right)\tan\varphi'/\gamma' H \cos i \sin i \quad (15.175)$$

c. Steady seepage along the slope

The case is similar to that for a cohesionless soil. In this case, the factor of safety is given by

$$F_s = c' + \left(\gamma' H \cos^2 i\right)\tan\varphi'/\gamma_{sat} H \cos i \sin i \quad (15.176)$$

The critical height (H_C) is obtained corresponding to a factor of safety of unity. Thus

$$c' + \left(\gamma' H_C \cos^2 i\right)\tan\varphi' = \gamma_{sat} H_C \cos i \sin i \quad (15.177)$$

$$H_C = c'/\left(\gamma_{sat} \tan i - \gamma' \tan\varphi'\right)\cos^2 i \quad (15.178)$$

15.16.4 STABILITY CHARTS

The stability number (S_n) is given by

$$S_n = c_m/\gamma H = c/F_c \gamma H \quad (15.179)$$

The reciprocal of the stability number is known as stability factor. The stability number is a dimensionless quantity.

For slope angle i greater than 53°, the toe failure occurs. For $i \le 53°$ and small values of φ_m, a more critical surface may pass below the toe.

The chart is applicable for $\varphi_m = 0$. In soils with $\varphi_m = 0$ and the slope angle less than 53°, the failure surface extends below the toe as deep as possible. The stability number also depends up on parameter D_f.

15.16.4.1 Uses of Stability Chart

1. The stability number can be used to determine the factor of safety of a given slope. For the known values if i and φ_m, the value of stability number S_n is determined from the chart (Figure 15.74) and the factor of safety is determined as

$$F_c = c/c_m = c/S_n \gamma H \quad (15.180)$$

If $\varphi_m = 0$, Figure 15.75 is used to determine the stability number S_n for the given values of i and D_f. The chart can also be used to determine the distance nH from the toe where the slip circle cuts the horizontal line.

2. The stability charts can also be used to determine the steepest slope for a given factor of safety. In this case, the stability number is computed from the relation

$$S_n = c/F\gamma H \quad (15.181)$$

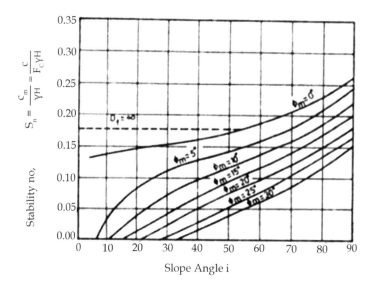

FIGURE 15.74   Taylor's stability number charts for $\varphi > 0$.

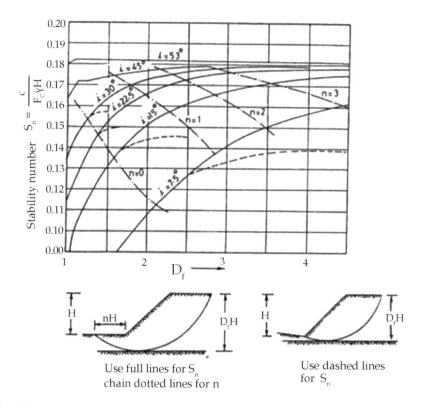

FIGURE 15.75   Taylor's stability number charts for $\varphi = 0$.

For the computed value of S_n, the value of i is read from the stability chart for a given value of φ_m.

15.16.5   Improving Stability of Slopes

The slopes which are susceptible to failure by sliding can be improved and made usable and safe. Various methods are used to stabilize the slopes. The methods generally involve one or more of the following measures which either reduce the mass which may cause sliding or improve the shear strength of the soil in the failure plane.

a. Slope flattening reduces the weight of the mass tending to slide. It can be used wherever possible.
b. Providing a berm below the toe of the slope increases the resistance to movement. It is especially useful when there is a possibility of base failure.
c. Drainage helps in reducing the seepage forces and hence increases the stability. The zone of subsurface water is lowered and infiltration of the surface water is prevented.
d. Densification by use of explosives, vibroflotation or terra probe helps in increasing the shear

strength of cohesionless soils and thus increasing the stability.

e. Consolidation by surcharging, electro-osmosis or other methods help in increasing the stability of slopes in cohesive soils.

f. Grouting and injection of cement or other compounds into specific zones help in increasing the stability of slopes.

g. Sheet piles and retaining walls can be installed to provide lateral support and to increase the stability. However, the method is quite expensive.

h. Stabilization of the soil helps in increasing the stability of slopes.

15.16.6 LANDSLIDES

The term landslide is self-explanatory and refers to the downward sliding of huge quantities of land masses. Generally, such sliding occurs along steep slopes of hills or mountains. It may be sudden or slow in its occurrence. Also, in magnitude, it may be major or minor. Often, loose and unconsolidated surficial material undergoes sliding. But sometimes, huge blocks of consolidated rocks may also be involved. In general, landslides are viewed as slope failures resulting in sudden slipping of enormous volumes of material along downward slopes.

15.16.6.1 Classification of Earth Movement

Generally, all movements of land masses are referred to as landslides. But they differ in many aspects and hence it would be appropriate to group all of them under "earth movement." All the types of earth movements may be classified as

1. Earth flow
2. Landslides
3. Subsidence

This classification is mainly based on the type of the movement which the displaced mass has suffered, that is, whether flowage or failure on a definite shearing surface. In earth flows, the movement is distributed throughout the displaced mass. In landslides, the movement is confined to a definite shearing plane or zone. In subsidence, the movement is vertically downward.

1. **Earth flows**: There are three types of earth flow, namely, solifluction, creep, and rapid flows.

   Solifluction refers to the downward movement of wet soil along the slopes under the influence of gravity.

   Creep refers to the extremely slow downward movement of dry surficial matter. It is always limited to the surface or the area just below it. This is important from the civil engineering point of view because the rate of movement is so slow that it may not be detected until its defects on engineering structures call attention to it. Thus, railway tracks, highways,

retaining walls, or tunnels built in or on creeping slopes may be thrown out of line or destroyed. On careful examination, bending of strata along a down slope, dislodgment if fence posts or telephone poles, curvature of tree trunks, buldged or broken retaining walls, etc., offer clues to recognize creep.

Rapid flows are similar to creep but differ with respect to the speed and depth of the material involved. These are rapid earth flows and involve considerable depth. These generally accompany heavy rains. Mud flows are similar to rapid flows.

2. **Landslides**: If a mass of earth or rock moves along a definite zone or surface, the failure is called a landslide. Debris slides, rock slides, and rock falls are the important types of landslides.

   a. Debris slides: Debris slides are the failures of unconsolidated material on a surface of rupture. In a majority of the cases, debris slides represent readjustments of the slope of the ground. They are common along the steep sides of rivers, lakes, etc. They may occur on any slope where internal resistance to shear is reduced below a safe limit. Debris slides of small magnitude are called slumps. Slump is often accompanied by complementary bulges at the toe. Debris slides are generally graded into earth flows.

   b. Rock slides: Rock slides are the movements of essentially consolidated material which mainly consists of recently detached bedrock. The causes of occurrence of rock slides are given separately. These cause considerable loss of life and property.

   c. Rock falls: Rock falls refer to the blocks of rocks of varying sizes suddenly crashing downwards (from cliffs) along steep slopes. These are common along steep shore lines and in the higher mountain regions during rainy season.

3. **Subsidence**: This essentially represents the downward movement of the surface. It may be due to both water and artificial causes. Subsidence may take place due to plastic outflow of underlying strata or due to compaction of underlying material or due to collapse.

   a. Subsidence due to plastic outflow: Beneath heavy loads, plastic layers which become plastic due to disturbance may be squeezed outward, allowing surface settlement or subsidence. Thus, for example, clay may be extruded from beneath a structure.

   b. Subsidence due to compaction: Sediments often become compact because of load. Excessive pumping out of water and the withdrawal of oil from the ground also cause subsidence, locally.

   c. Subsidence due to collapse: In regions where extensively underground mining has removed a large volume of material, the weight of the overlying rock may cause collapse and subsidence.

It may also happen when underground formations are leached by surface water.

15.16.6.2 Causes of Landslides

Landslides occur due to various causes. Broadly, they may be grouped in to two types. They are inherent or internal causes and immediate causes. Of these, the internal causes are responsible to the extent of creating favorable or suitable conditions for landslide occurrence. Their role is just short of the actual slip of the land mass which is because of frictional resistance movement and inertia. The other set of causes, that is, immediate causes, plays the role of overcoming this frictional resistance or inertia by providing necessary energy in the form of a sudden jerk, for the actual occurrence of landslide.

1. Internal causes

    The causes which are inherent in the land mass concerned are again of various types such as influence of slope, associated water, constituent lithology, associated geological structures, human factors, etc.
    a. Effect of slope:

        This is a very important factor which provides favorable conditions for landslide occurrence. It is both directly and indirectly responsible for land slips of loose overburdens due to greater gravity influence, whereas gentle slopes are not prone to such land slips because, in such cases, loose overburden encounters greater frictional resistance; hence any possible slip is stalled.
    b. Effect of water:

        This is the most important factor which is mainly responsible for landslide occurrence. This is so because it adversely affects the stability of the loose ground in different ways as follows.
        i. The presence of water greatly reduces the intergranular cohesion of the particles of loose ground. This weakens the ground inherently and therefore makes it prone to landslide occurrence. Frequent landslide occurrences soon after rainfall support this observation.
        ii. On hill slopes, water on percolation through the overlying soil zone may flow down as a film or thin sheet of water above the underlying hard rocks. This not only acts as a lubricating medium between the two but also induces the (downward) movement in the overlying loose material along its own direction of flow. Thus the water creates yet another condition favorable for landslide occurrence.
        iii. Along hill slopes, rain water, while percolating down, carries with it fine clay and silty material which may form a thin band at the interface of loose overlying material and underlying hard work. In the presence of water the clayey matter becomes very plastic and may form a slippery base. This further enhances the chances of loose overburden to slip downward.
        iv. Along steep slopes the presence of water in the intergranular spaces in loose, unconsolidated, or weathered material adds to its weight, thereby increasing the influence of gravity, which, in turn, promotes landslide occurrence.
        v. Water, being the most powerful solvent, not only causes decomposition of minerals but also leaches out the soluble matter of rocks. This reduces the compaction or cohesion of the rock body and makes it a weak mass.
        vi. In the regions near the snowline, water frequently changes its state of occurrence, that is, either into solid condition or into the liquid condition in response to slight changes in climatic conditions there. This phenomenon leads to the disintegration of rocks in the form of frost wedging and frost heaving. These factors not only add to the weakening of rocks but also enhance the instability of the region and lead to an increase in gravity effect.
    c. Effect of lithology:

        The nature of rock types also influences landslide occurrence. For example:
        i. Rocks which are highly fractured, porous, and permeable are prone to landslide occurrence because they give scope for the water to play an effective role.
        ii. Rocks which are rich in certain constituents like clay (montmorillonite, bentonite), mica, calcite, glauconite, gypsum, rock salt, and calcareous cementing material are more prone to landslide occurrence because they are easily leached out, causing porosity and permeability. The clay content contributes to the slippery effect.
        iii. Thinner strata are more susceptible to sliding than thicker strata.
        iv. A sequence of strata having thin, soft, and weak beds lying in-between hard and thick beds provide a congenial setup for landslide occurrence.
    d. Effect of associated structures:

        The geological structures which increase the chances of landslide occurrences are inclined bedding planes, joints, faults, or shear zones. All these are planes of weakness. When their dip coincides with that of the surface slope they create conditions of instability.

In case of joints and faults, they not only create the aforementioned unstable conditions but also, being open fractures, provide scope for easy percolation of rain water. This naturally enhances the instability. Further, if thin clay beds happen to be underlying in such cases, the percolated water reaches the same and changes the clay bed into a slippery medium. This means that the overburden becomes highly susceptible to sliding.

e. Effect of the human factors:

Sometimes, human being interfere with nature by virtue of their activities and cause landslides. Thus, for example, this may happen when undercuttings are made along the hill slopes for laying roads or railways tracks. Such an activity eliminates lateral support which means gravity will become more effective, leading to landslide occurrence. Of course, this is not always so as inferred.

When construction works are carried out on hilltops, they will act as heavy loads on the loose zone of overburden and create scope for sliding of the underlying loose ground.

Of course, the aforementioned situation may develop even when heavy snowfall occurs on the top of mountain. It leads to massive piling up of snow mass. Such a snow body will have an effect similar to that of a construction made on a hill top.

2. Immediate causes

The different causes listed earlier simply create favorable conditions for the occurrence of landslides. But they themselves do not bring about the actual occurrence of landslides. Otherwise, landslides could have occurred anywhere and at any time just if the causes were present. But the fact that the landslides occur suddenly at certain times only indicates that these causes only prepare the ground but because of factors such as frictional resistance the overlying mass will remain in the same plane in a delicate, critical, or unstable condition. In such contexts if conditions to overcome this situation are created by some sudden impulse, landslides will occur and may cause destruction. Hence such an impulse, which is a sudden jerk or vibration of the ground, acts as the immediate cause for the occurrence of landslides. This sudden jolting phenomenon of the ground may, in turn, be due to different natural and artificial reasons like an avalanche, violent volcanic eruption, fall of a meteorite, occurrence of an earthquake, tsunamis or blasting of explosives in quarrying, tunneling, road cutting or mining.

15.16.7 Effects of Landslides

From the civil engineering point of views, if landslides occur at vulnerable, they may cause:

a. disruption of transport or blocking of communications by damaging roads and railways and telegraph poles
b. obstruction to the river flow in valleys, leading to their overflow and floods
c. damage to sewer and other pipelines
d. burial or destruction of buildings and other constructions, and
e. earthquakes.

15.16.7.1 Preventive Measures for Landslides

To prevent the occurrence of landslides, it would be logical to take such steps which would counter the effects of those factors responsible for landslide occurrence. The main factors which contribute to landslide occurrence are slope, water content, structural defects, unconsolidated or loose character of the overburden, lithology, and human interference. These may be tackled as follows.

1. To counter the effect of slope:
   Retaining walls may be constructed against the slopes, so that the material which rolls down is not only prevented from further fall but also reduces the slope. Terracing of the slope is another effective measure in this regard.
2. To counter the effect of water:
   A proper drainage system is the suitable measure. This involves the quick removal of percolated moisture by means of surface drainage and subsurface drainage. Construction of suitable ditches and waterways along slopes and provision of trenches at the bottom and drainage tunnels helps in draining off the water from the loose overburden.
3. To counter the structural defects:
   The different structural defects such as weak planes and zones may be either covered or grouted suitably so that they are effectively sealed off. These measures not only prevent the avenues for percolation of water but also increase the compaction or cohesion of the material concerned.
4. Not to resort to reduce the stability of existing slopes:
   This is done by not undertaking any undercuttings on the surface slope and by not undertaking any construction at the top of the hills.
5. To counter the loose nature of overburden:
   Growing vegetation, plants, and shrubs on loose ground help in keeping the loose soil together.
6. Avoiding heavy traffic and blasting operations near the vulnerable places naturally help in preventing the occurrence of landslides.

16 Foundation Engineering

16.1 GEOTECHNICAL INVESTIGATION AND REPORT

Site investigations or subsurface explorations are done for obtaining the information about subsurface conditions at the site of proposed constructions. Site investigations in one form or the other is generally required for every big engineering project. Information about the surface and subsurface features is essential for the design of structures and for planning construction techniques.

Site investigations consist of determining the profile of the natural soil deposits at the site, taking the soil samples and determining the engineering properties of the soils. It also includes in-situ testing of the soils.

Site investigations are generally done to obtain the information that is useful for one or more of the following purposes.

1. To select the type and depth of foundation for a given structure.
2. To determine the bearing capacity of the soil.
3. To estimate the probable maximum and differential settlements.
4. To establish the ground water level and to determine the properties of water.
5. To predict the lateral earth pressure against retaining walls and abutments.
6. To select suitable constructions techniques.
7. To predict and to solve potential foundation problems.
8. To ascertain the suitability of the soil as a construction material.
9. To investigate the safety of the existing structures and to suggest the remedial measures.

The relevant information is obtained by drilling holes, taking the soil samples, and determining the index and engineering properties of the soil. In-situ tests are also conducted to determine the properties of the soils in natural conditions.

16.1.1 STAGES IN SUBSURFACE EXPLORATIONS

Subsurface explorations are generally carried out in three stages:

1. **Reconnaissance**: Site reconnaissance is the first step in a surface subsurface exploration program. It includes a visit to the site and to study the maps and other relevant records. It helps in deciding future program of site investigations, scope of work, methods of exploration to be adopted, types of samples to be taken, and the laboratory testing and in-situ testing.

2. **Preliminary explorations**: The aim of preliminary explorations is to determine the depth, thickness, extent, and composition of each soil stratum at the site. The depth of the bed rock and the ground water table is also determined.

   The preliminary explorations are generally in the form of a few borings or test pits. Tests are conducted with cone penetrometers and sounding rods to obtain information about the strength and compressibility of soils.

   Geophysical methods are also used in preliminary explorations for locating the boundaries of different strata.

3. **Detailed explorations**: The purpose of the detailed explorations is to determine the engineering properties of the soils in different strata. It includes an extensive boring program, sampling, and testing of the samples in a laboratory.

   Field tests, such as vane shear tests, plate load tests, and permeability tests, are conducted to determine the properties of the soils in natural state. The tests for the determination of dynamic properties are also carried out, if required.

   For complex projects involving heavy structures, such as bridges, dams, multistorey buildings, it is essential to have detailed explorations.

16.1.2 DEPTH OF EXPLORATION

The depth of exploration required at a particular site depends up on the degree of variation of the subsurface data in the horizontal and vertical directions.

The depth of exploration is governed by the depth of the influence zone. The depth of the influence zone depends up on the structure, intensity of loading, shape and disposition of the loaded area, the soil profile, and the physical characteristics of the soil. The depth up to which the stress increment due to superimposed loads can produce significant settlement and shear stress is known as the significant depth. The depth of exploration should be at least equal to the significant depth.

The significant depth is generally taken as the depth at which the vertical stress is 20% of the load intensity. According to the above criterion, the depth of exploration should be about 1.5 times the width of the square footing and about 3.0 times the width of the strip footing. However, if the footings are closely spaced, the whole of the load area acts as a raft foundation. In that case, the depth of boring should be at least 1.5 times the width of the entire loaded area as shown in Figure 16.1.

In the case of the foundation, the depth of exploration below the tip of bearing piles is kept at least 1.5 times the width of the pile group as shown in Figure 16.2. However, in the case of friction piles, the depth of exploration is taken 1.5 times the width of the pile group measured from the lower third point as shown in Figure 16.3.

It is more logical to relate the increase in stress to the in-situ stress. The depth of exploration is usually taken up to the level at which the increase in stress is 1/20th of the in-situ stress before the application of the load.

When the foundations are taken up to rock, it should be ensured that large boulders are not mistaken as bed rock. The minimum depth of core boring into the bed rock should be 3 m to establish it as a rock.

In case of multistoreyed buildings, the depth of exploration can be taken from following formula:

$$D = C(S)^{0.7} \qquad (16.1)$$

where

D = depth of exploration (m),

C = constant, equal to 3 for light steel buildings and narrow concrete buildings. It is equal to 6 for heavy steel buildings and wide concrete buildings.

S = number of storeys.

If loose soils or recently deposited soil or a week stratum is encountered, it should be explored thoroughly. Exploration should be carried to be a depth at which the net increase in the vertical stress is less than the allowable bearing pressure of the soil.

For two adjacent footings, each of size $B \times L$, spaced at a clear spacing A, IS: 1892–1972 suggests that the minimum depth of boring should be $1.5B$ when $A \geq 4B$, and it should be $1.5L$ when $A < 2B$. For adjacent rows of such footings, the minimum recommended depth of exploration is $4.5B$ when $A < 2B$; it is $3.5B$ when $A > 2B$; and it is $1.5B$ when $A \geq 4B$.

For explorations of deep excavations, the depth of exploration below the proposed excavation level should be at least 1.5 times the depth of excavation. In case of road cuts, it is taken at least equal to the width of the cut.

In case of road fills, the minimum depth of boring is 2 m below the ground surface or equal to the height of the fill, whichever is greater.

In case of gravity dams, the minimum depth of boring is twice the height of the dam.

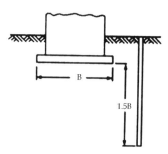

FIGURE 16.1 Depth of boring should be at least 1.5 times the width of the entire loaded area.

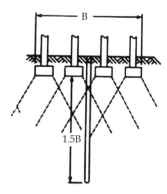

FIGURE 16.2 Pile group.

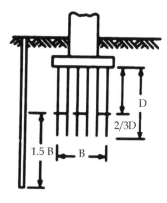

FIGURE 16.3 Friction piles.

16.1.3 BORINGS FOR EXPLORATION

When the depth of exploration is large, borings are used for exploration. A vertical bore hole is drilled in the ground to get the information about the subsoil strata. Samples are taken from the bore hole and tested in a laboratory. The bore hole may be used for conducting in-situ tests and for locating the water table. Extensometers or pressure meter may also be installed in the bore hole for the measurement of deformation in the substrata.

Depending upon the type of soil and the purpose of boring, the following methods are used for drilling the holes.

1. Auger Boring
2. Washing Boring
3. Rotary Boring
4. Percussion Boring
5. Core Boring

A few holes are drilled during the preliminary investigation. In the detailed investigations, a large number of holes are drilled to thoroughly investigate the subsoil strata.

The results of boring are presented in the form of boring-log and subsurface profiles.

16.1.3.1   Auger Boring

An auger is a boring tool similar to the one used by a carpenter for boring holes in wood. It consists of a shank with a cross-wise handle for turning and having central tapered feed screw as shown in Figure 16.4. The augers can be operated manually or mechanically.

The hand augers used in boring are about 15–20 cm in diameter. These are suitable for advancing holes up to a depth of 3–6 m in soft soils. The hand auger is attached to the lower end of a pipe of about 18 mm diameter. The pipe is provided with a cross-arm at its top. The hole is advanced by turning the cross-arm manually and at the same time applying thrust in the downward direction. When the auger is filled with soil, it is taken out. If the hole is already driven, another type of auger, known as post-hole auger, is used for taking soil samples.

Auger boring is generally used in soils which can stay open without casing or drilling mud. Auger borings are particularly useful for subsurface investigations of highways, railways, and air fields, where the depth of exploration is small. The investigations are done quite rapidly and economically by auger boring.

The main disadvantage of the auger boring is that the soil samples are highly disturbed. Further, it becomes difficult to locate the exact changes in the soil strata.

16.1.3.2   Washing Boring

In washing boring (Figure 16.5), the hole is drilled by first driving a casing, about 2–3 m long, and then inserting it into a hollow drill rod with a chisel-shaped chopping bit at its lower end. Water is pumped down the hollow drill rod, which is known as wash pipe. Water emerges as a strong jet through a small opening of the chopping bit. The hole is advanced by

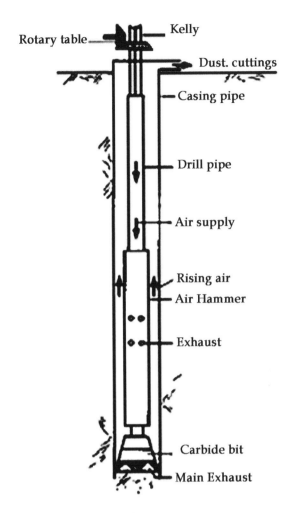

FIGURE 16.5   Washing boring.

a combination of chopping action and the jetting action, as the drilling bit and the accompanying water jet disintegrate the soil. The water and the chopped soil particles rise upward through the annual space between the drill rod and the casing. The return water, also known as wash water, is laden with the soil cuttings. It is collected in a tub through a T-shaped pipe fixed at the top of the casing.

The hole is further advanced by alternately raising and dropping the chopping bit by a winch. The swivel joint provided at the top of the drill rod facilities the turning and twisting of the rod. The process is continued even below the casing till the hole begins to cave in. At that stage, the bottom of the casing can be extended by providing additional pieces at the top.

The wash boring is mainly used for advancing a hole in the ground. Once the hole has been drilled, a sampler is inserted to obtain soil samples for testing in a laboratory.

The equipment used in washing borings is relatively light and inexpensive. The main disadvantage of the method is that it is slow in stiff soils and coarse grained soils. It cannot be used efficiently in hard soils, rock, and the soils containing boulders. The method is not suitable for taking good quality undisturbed samples above ground water table, as the wash water enters the strata below the bottom of the hole and causes an increase in its water content.

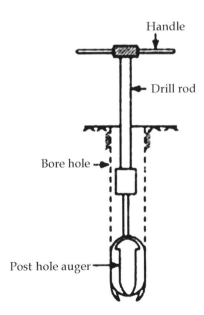

FIGURE 16.4   Auger boring.

16.1.3.3 Rotatory Drilling

In the rotatory drilling method, the bore hole is advanced by rotating a hollow drill rod which has a cutting bit at its lower end. A drill head is provided at the top of the drill rod. It consists of a rotary mechanism and an arrangement for applying downward pressure.

As the drilling rod is rotated, the cutting bit shears off chips of the material penetrated. A drilling fluid under pressure is introduced through the drilling rod to the bottom of the hole. The fluid carries the cuttings of the material penetrated from the bottom of the hole to the ground surface through the annular space between the drilling rod and the walls of the hole. The drilling fluid also cools the drilling bit. In case of an uncased hole, the drilling fluid also supports the walls of the hole.

When the soil sample is required to be taken, the drilling rod is raised and the drilling bit is replaced by a sampler.

Rotary drilling can be used in clay, sand, and rocks. Bore holes of diameter 50–200 mm can be easily made by this method.

16.1.3.4 Percussion Drilling

The percussion drilling method is used for making holes in rocks, boulders, and other hard strata. In this method, a heavy chisel is alternately lifted and dropped in a vertical hole. The material gets pulverized. If the point where the chisel strikes is above the water table, water is added to the hole. The water forms a slurry with the pulverized material, which is removed by a sand pump or a bailer at intervals. Percussion drilling may require a casing. Percussion drilling is also used for drilling of tube wells.

The main advantage of Percussion drilling method is that it can be used for all types of materials. It is particularly used for drilling holes in glacial tills containing boulders. One of the major disadvantages is that the material at the bottom of the hole is disturbed by heavy blows of the chisel. It is not possible to get good quality undisturbed samples. Further, the method is generally more expensive than other methods. Moreover, it becomes difficult to detect minor changes in the properties of the strata penetrated.

16.1.3.5 Core Drilling

The core drilling method is used for drilling holes and for obtaining rock cores. In this method, a core barrel fitted with drilling bit is fixed to a hollow drilling rod. As the drilling rod is rotated, the bit advances and cuts an annular hole around an intact core. The core is then removed from its bottom and it is retained by a core lifter and brought to the ground surface. Water is pumped continuously in to the drilling rod to keep the drilling bit cool and to carry the disintegrated material to the ground surface.

The core drilling may be done using either a diamond studded bit or a cutting edge consisting of chilled shot. The diamond drilling is superior to the other type of drilling, but is costlier. The core barrel consists of a single tube or a double tube. A double-tube barrel gives a good quality sample of the rock.

16.1.4 Subsoil Investigation Report

A subsoil investigation report should contain the data obtained from bore holes, the site observations, and laboratory results. It should also give the recommendations about the suitable type of foundations, allowable soil pressure, and expected settlements. Figure 16.6 represents a bore log.

It is essential to give a complete and accurate record of data collected. Each bore hole should be identified by a code number. The location of each bore hole should be fixed by measurement of its distance or angles from some permanent feature. All relevant data for bore hole are recorded in a boring log. A boring log gives the description or classification of various strata encountered at different depths. Any additional information obtained at the field, such as soil consistency, unconfined compression strength, standard penetration test (SPT), cone penetration test, is also indicated on the boring log. It should also show the water table. If the laboratory tests have been conducted, the information about index properties, compressibility, shear strength, permeability, etc., should also be provided.

The data obtained from a series of bore holes are presented in the form of a subsurface profile. A subsurface profile is a vertical section through the ground along the line of exploration. It indicates the boundaries of different strata, along with their classification. It is important to remember that conditions between bore holes are estimated by interpolation, which may not be correct. Obviously, the larger the number of holes, the more accurate is the subsurface profile.

The site investigation report should contain the discussion of the results. The discussion should be clear and concise. The recommendations about the type and depth of foundation, allowable soil pressure, and expected settlements should be specific. The main findings of the report are in conclusions.

A soil exploration report generally consists of the following.

1. Introduction, which gives the scope of investigation.
2. Description of the proposed structure, the location, and the geological conditions at the site.
3. Details of the field exploration program, including the number of borings, their location, and depths.
4. Details of methods of exploration.
5. General description of the subsoil conditions as obtained from in-situ tests, such asSPT, cone test.
6. Details of the laboratory test conducted on the soil samples and the results are obtained.
7. Depth of the ground water table and the changes in water levels.
8. Discussion of the results.
9. Recommendation about the allowable bearing pressure, the type of foundation or structure.
10. Conclusions. The main findings of investigations should be clearly stated. It should be brief but should mention the salient points.

Limitations of the investigations should also be briefly stated.

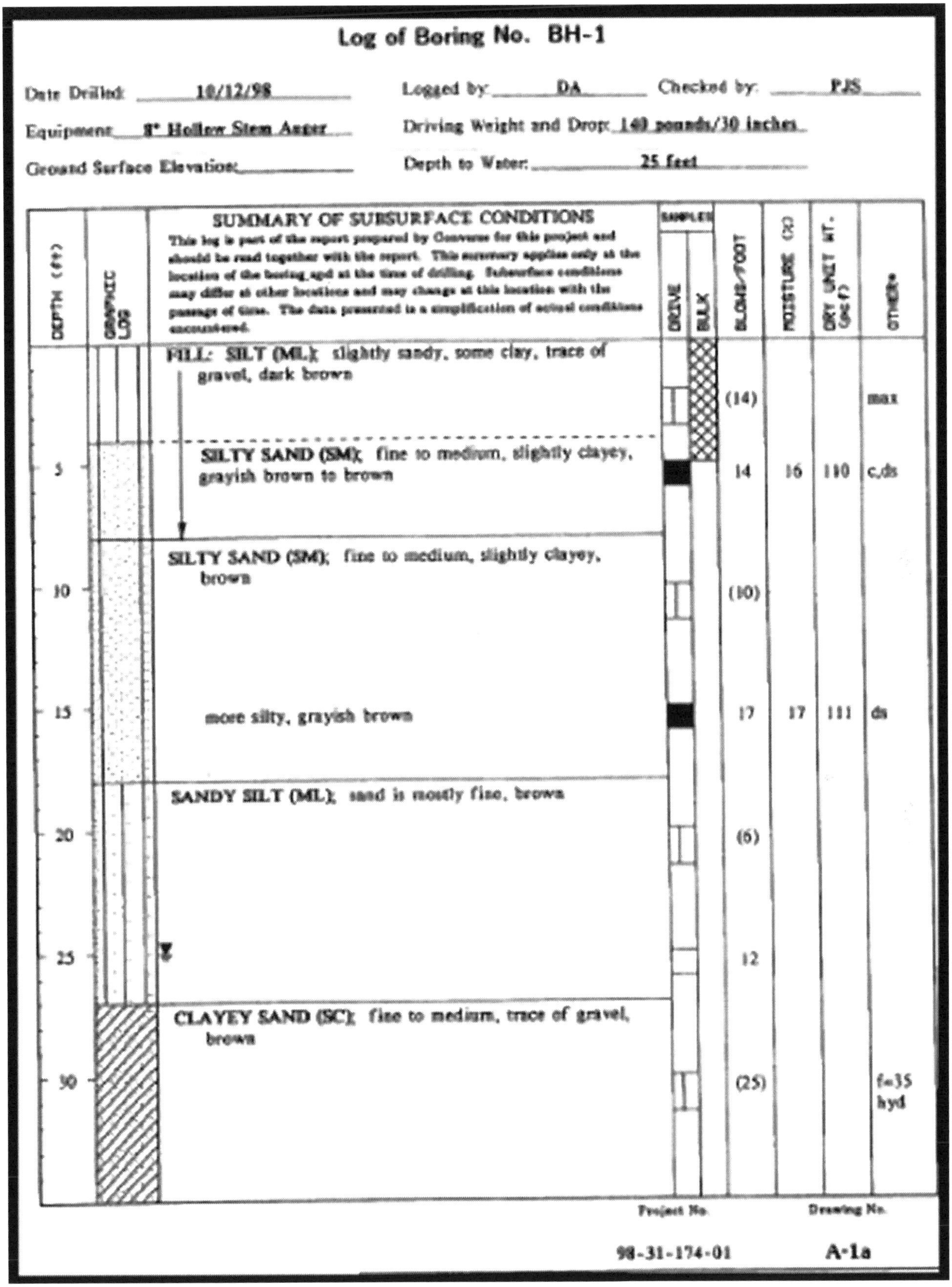

Log of Boring No. BH-1

Date Drilled: _____10/12/98_____   Logged by: _____DA_____   Checked by: _____PJS_____

Equipment: _____8" Hollow Stem Auger_____   Driving Weight and Drop: _____140 pounds/30 inches_____

Ground Surface Elevation: _____   Depth to Water: _____25 feet_____

DEPTH (ft)	GRAPHIC LOG	SUMMARY OF SUBSURFACE CONDITIONS	SAMPLES		BLOWS/FOOT	MOISTURE (%)	DRY UNIT WT. (pcf)	OTHER
			DRIVE	BULK				
		FILL: SILT (ML); slightly sandy, some clay, trace of gravel, dark brown			(14)			max
5		SILTY SAND (SM); fine to medium, slightly clayey, grayish brown to brown			14	16	110	c,ds
10		SILTY SAND (SM); fine to medium, slightly clayey, brown			(10)			
15		more silty, grayish brown			17	17	111	ds
20		SANDY SILT (ML); sand is mostly fine, brown			(6)			
25					12			
30		CLAYEY SAND (SC); fine to medium, trace of gravel, brown			(25)			f=35 hyd

This log is part of the report prepared by Converse for this project and should be read together with the report. This summary applies only at the location of the boring and at the time of drilling. Subsurface conditions may differ at other locations and may change at this location with the passage of time. The data presented is a simplification of actual conditions encountered.

Project No.   98-31-174-01

Drawing No.   A-1a

FIGURE 16.6   Represent a bore log.

16.2 IN-SITU SOIL TESTS

16.2.1 Standard Penetration Test

The SPT (Figure 16.7) is most commonly used in in-situ test, especially for cohesionless soils which cannot be easily sampled. The test is extremely useful for determining the relative density and the angle of shearing resistance of cohesionless soils. It can also be used determine the unconfined compressive strength of cohesive soils.

The SPT is conducted in a bore hole using a standard split-spoon sampler. When the bore hole has been drilled to the desired depth, the drilling tools are removed and the sampler is lowered to the bottom of the hole. The sampler is driven into the soil by a drop hammer of 63.5 kg mass falling through a height of 750 mm at the rate of 30 blows per minute. The number of hammer blows required to drive 150 mm of the sample is counted. The sampler is further driven by 150 mm and the number of blows recorded. Likewise, the sampler is once again further driven by 150 mm and the number of blows recorded. The number of blows recorded for the first 150 mm is disregarded. The number of blows recorded for the last two 150 mm intervals is added to give the standard penetration number (N). In other words, the standard penetration number is equal to the number of blows required for 300 mm of penetration beyond a seating drive of 150 mm.

If the number of blows for 150 mm drive exceeds 50, it is taken as refusal and the test is discontinued. The standard penetration number is corrected for dilatancy correction and overburden correction as below.

1. Dilatancy correction

Silty fine sands and fine sands below the water table develop pore pressure which is not easily dissipated. The pore water pressure increases the resistance of the soil and hence the penetration number (N).

Terzaghi and Peck recommend the following correction in the case of silty fine sands when the observed value of N exceeds 15.

The corrected penetration number, $N_C = 15 + \frac{1}{2}(N_R - 15)$

(16.2)

where
N_R = recorded value
N_C = corrected value

$$\text{If } N_R \leq 15, \ N_C = N_R \qquad (16.3)$$

2. Overburden pressure correction

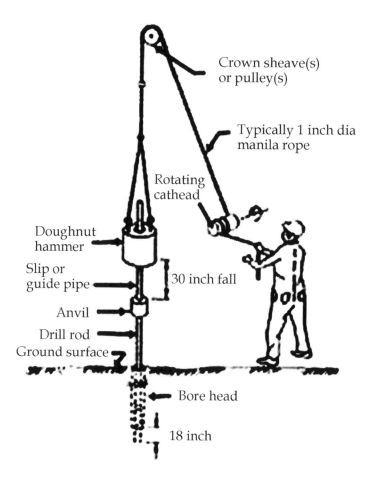

FIGURE 16.7 The standard penetration test.

In granular soils, the overburden pressure affects the penetration resistance. If the two soils having same relative density but different confining pressures are tested, the one with a higher confining pressure gives a higher penetration number. As the confining pressure in cohesionless soils increases with the depth, the penetration number for soils at shallow depths is underestimated and that at greater depths is overestimated. For uniformity, the N values are obtained from field tests under different effective overburden pressures.

Gibbs and Holtz recommend the use of the following equation for dry or moist clean sand.

$$N_C = N_R \times 350/(\sigma'_0 + 70) \tag{16.4}$$

where
N_R = observed N value
N_C = corrected N value
σ'_0 = effective overburden pressure (kN/m²)

The above equation is applicable for $\sigma'_0 \leq 280$ kN/m².

The ratio of (N_C/N_R) should lie between 0.45 and 2.0. If (N_C/N_R) ratio is greater than 2.0, N_C should be divided by 2.0 to obtain the design value used in finding the bearing capacity of soil.

The correction may be extended to saturated silty sand and fine sand after modifying the N_R according to above equation. N_C obtained from Equation 16.4 would be taken as N_R in Equation 16.2.

Thus the overburden correction is applied first and then the dilatancy correction is applied.

Peck, Hansen, and Thornburn give the chart for correction of N values to an effective overburden pressure of 96 kN/m². According to them,

$$N = 0.77\, N_R \log(1905/\sigma'_0) \text{ for } \sigma'_0 \geq 24 \text{ kN/m}^2 \tag{16.5}$$

Figure 16.8 shows the correction diagram.
At $\sigma'_0 = 0$, the value of N/N_R is 2.0.

The correction given by Bazaraa and also by Peck and Bazaraa is one of the commonly used corrections. According to them,

$$N = 4N_R/(1 + 0.0418\,\sigma'_0) \text{ if } \sigma'_0 < 71.8 \text{ kN/m}^2 \tag{16.6}$$

$$N = 4N_R/(3.25 + 0.0104\,\sigma'_0) \text{ if } \sigma'_0 < 71.8 \text{ kN/m}^2 \tag{16.7}$$

$$N = N_R \text{ if } \sigma'_0 = 71.8 \text{ kN/m}^2 \tag{16.8}$$

Correlation of N with Engineering Properties

The value of the standard penetration number N depends upon the relative density of the cohesionless soil and the unconfined compressive strength of the cohesive soil. If the soil is compact or stiff, the penetration number is high.

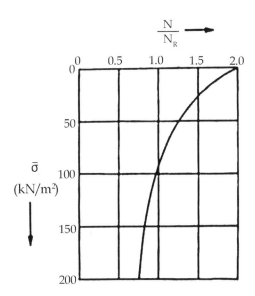

FIGURE 16.8 Shows the correction diagram.

The angle of shearing resistance (φ) of the cohesionless soil depends upon the number N. Figure 16.9 shows the variations of N and φ. In general, the greater the N value, the greater is the angle of the shearing resistance.

The consistency and the unconfined shear strength of the cohesive soils can be approximately determined from the SPT number N. As the correlation is not dependable, it is advisable to determine the shear strength of the cohesive soils by conducting shear tests on the undisturbed samples or by conducting in situ vane shear test.

The unconfined compressive strength can also be determined from the following relation:

$$q_u = 12.5N \tag{16.9}$$

where q_u = unconfined compressive strength (kN/m²).
Correlation between N and φ is given by Table 16.1
Correlation between N and q_u is given by Table 16.2.

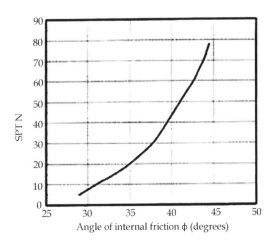

FIGURE 16.9 Shows the variation of N and φ.

TABLE 16.1

Correlation between N and φ

Value of N	Denseness	Value of φ
0–4	Very loose	25°–32°
4–10	Loose	27°–35°
10–30	Medium	30°–40°
30–50	Dense	35°–45°
>50	Very Dense	>45°

TABLE 16.2

Correlation between N and q_u

Value of N	Consistency	Value of q_u in kN/m²
0–2	Very Soft	< 25
2–4	Soft	25–50
4–8	Medium	50–100
8–15	Stiff	100–200
15–30	Very Stiff	200–400
>30	Hard	>400

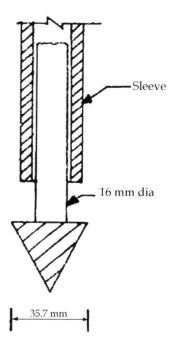

FIGURE 16.10 Shows a Dutch cone.

16.2.2 CONE PENETRATION TEST

Sounding methods are frequently used to determine the penetration resistance and the engineering properties of the soil. The sounding methods mainly consist of the cone test and the SPT.

The cone test was developed by the Dutch Government, Soil Mechanics Laboratory at Delft, and is, therefore, also known as Dutch Cone Test. The test is conducted either by the static method or by dynamic method.

a. Static cone penetration test

The Dutch cone has an apex angle of 60° and an overall diameter of 35.7 mm, giving an end area of 10 cm². Figure 16.10 shows a Dutch cone.

For obtaining the cone resistance, the cone is pushed downward at a steady rate of 10 mm/s through a depth of 35 mm each time. The cone is pushed by applying thrust and not by driving.

After the cone resistance has been determined, the cone is withdrawn. The sleeve is pushed onto the cone and both are driven together into the soil and the combined resistance is also determined. The resistance of the sleeve alone is obtained by subtracting the cone resistance from the combined resistance.

For effective use of the cone penetration test, some reliable calibration is required. This consists of comparing the results with those obtained from conventional tests conducted on undisturbed samples in a laboratory. It is also convenient to compare the cone test results with the SPT results. As the SPTs have been more commonly conducted in the past, good correlation studies are available between the SPT number (N) and the engineering properties of

the soil. If the cone penetration results are related to the SPT number N, indirect correlations are obtained between the cone test results and the engineering properties of soil.

The following relations hold approximately good between the point resistance of the cone (q_c) and the standard penetration number (N).

i.   Gravels $q_c = 800N$ to $1000N$
ii. Sands $q_c = 500N$ to $600N$
iii. Silty sands $q_c = 300N$ to $400N$
iv. Silts and clay $q_c = 200N$

where q_c is in kN/m².

b. Dynamic cone test

The test is conducted by driving the cone by blows of a hammer. The number of blows for driving the cone through a specified distance is a measure of the dynamic cone resistance.

Dynamic cone tests are performed either by using a 50 mm cone without bentonite slurry or by using a 65 mm cone with bentonite slurry (IS: 4968- part I and II -1976). The driving energy is given by a 65 kg hammer falling through a height of 75 cm. The number of blows for every 10 cm penetration is recorded. The number of blows required for 30 cm of penetration is taken as the dynamic cone resistance (N_{cbr}). If the skin friction is to be eliminated, the test is conducted in a cased bore hole. Figure 16.11 shows a Dynamic Cone.

When a 65 mm cone with bentonite slurry is used, the setup should have arrangements for circulating slurry so that the friction on the driving rod is eliminated.

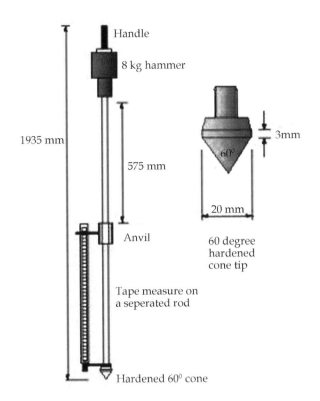

FIGURE 16.11   Shows a dynamic cone.

The dynamic cone resistance (N_{cbr}) is correlated with the SPT number N. The following approximate relations may be used when a 50 mm diameter cone is used.

$N_{cbr} = 1.5 N$ for depths up to 3 m
$N_{cbr} = 1.75 N$ for depths between 3 and 6 m
$N_{cbr} = 2.0 N$ for depths greater than 6 m

The Central Building Research Institute, Roorkee, has developed the following correlation between the dynamic cone resistance N_{cbr} of 65 mm diameter cone bentonite slurry and the SPT number (N).

$N_{cbr} = 1.5 N$ for depths up to 4 m
$N_{cbr} = 1.75 N$ for depths between 4 and 9 m
$N_{cbr} = 2.0 N$ for depths greater than 9 m

The above relations are applicable for medium to fine sand.

16.3   BEARING CAPACITY OF SHALLOW FOUNDATION

A foundation is that part of a structure which transmits the weight of the structure to the ground. All structures constructed on land are supported on foundations. A foundation is, therefore, a connecting link between the structure proper and the ground which supports it. The word "foundation" is derived from latin word fundare, meaning to set or ground on something solid. In other words, a foundation is an artificially laid base on which a structure is set or built up.

Foundations may be broadly classified into two categories: (1) shallow foundations and (2) deep foundations. A shallow foundation transmits the loads to the strata at a shallow depth. A deep foundation transmits the loads at considerable depth below the ground surface. The distinction between a shallow foundation and a deep foundation is generally made according to Terzaghi's criterion. According to which, a foundation is termed shallow, if it is laid at a depth equal to or less than its width. A foundation is termed deep, if it is laid at a depth more than its width.

16.3.1   Basic Definitions

1. **Gross pressure intensity (q)**: It is the total pressure at the base of foundation due to the weight of super structure, self-weight of footing, and the weight of earth fill.
2. **Net pressure intensity (q_n)**: It is defined as the excess pressure or the difference in intensities of the gross pressure after the construction of the structure and the original overburden pressure.

$$q_n = q - \gamma D_f \qquad (16.10)$$

3. **Ultimate bearing capacity (q_u)**: It is defined as the minimum gross pressure at the base of the foundation at which the soil just fails in shear.
4. **Net ultimate bearing capacity (q_{nu})**: It is the minimum net pressure at which soil just fails in shear.

$$q_{nu} = q_u - \gamma D_f \qquad (16.11)$$

5. **Net safe bearing capacity (q_{ns})**: It is that net pressure which can be applied safely at the base of footing without risk of shear failure.

$$q_{ns} = q_{nu} / \text{fos} \qquad (16.12)$$

where fos is between 2 and 3

6. **Safe bearing capacity (q_s)**: The maximum pressure which the soil can carry safely without risk of shear failure is called safe bearing capacity.

$$q_s = q_{ns} + \gamma D_f \qquad (16.13)$$

7. **Allowable bearing capacity or pressure (q_a)**: It is the net loading intensity at which neither the soil fails nor there is excessive settlement detrimental to the structure.
8. Net safe settlement pressure (q_{np})
   It is the net pressure which soil can carry without exceeding the allowable settlement. The maximum allowable settlement is 25–40 mm for individual footing. It is also known as unit soil pressure or safe bearing pressure or safe bearing capacity
9. Net allowable bearing pressure (q_{na})

It is the net pressure which can be used for the design of foundation

$$q_{na} = q_{ns} \text{ if } q_{np} > q_{ns} \qquad (16.14)$$

$$q_{na} = q_{np} \text{ if } q_{ns} > q_{np} \qquad (16.15)$$

It is also known as allowable soil pressure or allowable bearing capacity or pressure

16.3.2 Mode of Shear Failure

Three principal modes of shear failures are as follows:

1. **General shear failure** (Figure 16.12a): In this case of failure, continuous failure surfaces develop between the edges of the footing and the ground surface. The failure is accompanied by tilting of footing and considerable bulging of sheared soil mass. Such failure occurs in soils of low compressibility.
2. **Local shear failure** (Figure 16.12b): In this failure, there is a significant compression of soil under the footing and only partial development of state of plastic equilibrium. Failure is defined by large settlement. There is only slight bulging of soil around the footing and no titling of footing. Such failures occur in soils with high compressibility.
3. **Punching shear failure** (Figure 16.12c): This failure occurs when there is relatively high compression of soil under footing, accompanied by shearing in the vertical direction around the edges of the footing. Failure is defined by very large settlement. It may occur in relatively loose sand.

16.3.3 Terzaghi's Method of Analysis

Terzaghi made the following assumptions for determination of bearing capacity of soil (Figure 16.13).

a. The footing is shallow (depth of footing is equal to or less than the width of the footing).
b. Base of foundation is assumed to be rough.
c. Footing is continuous (it is a strip footing).
d. The angle of roughness between base of footing and soil is considered equal to the friction angle of the soil.
e. Failure is assumed to be general shear failure and at the time of failure soil reaches into plastic state.
f. The soil above the base of the footing is replaced by an equivalent surcharge.
g. The stress zone of the soil extends up to foundation level and not up to ground level.

16.3.3.1 Terzaghi's Failure Criterion

As the base of the footing is rough, the soil in the wedge ABC immediately beneath the footing is prevented from undergoing any of the lateral yield, so the soil in wedge 1 remains in a state of equilibrium. That is the first zone. The second zone

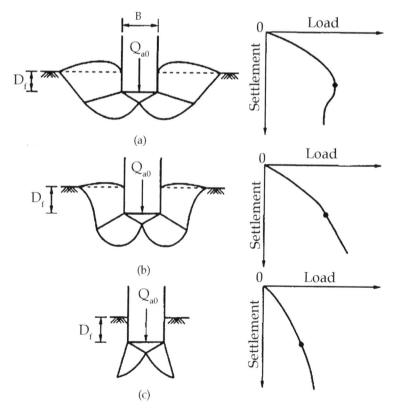

FIGURE 16.12 (a) General shear failure, (b) local shear failure, (c) punching shear failure.

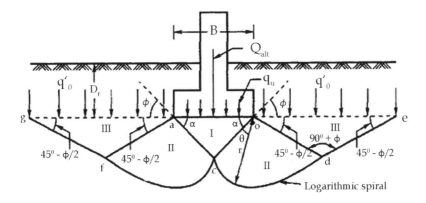

FIGURE 16.13 Terzaghi's method of analysis.

is the radial shear zone and the third zone is the Rankine's passive pressure zone.

The failure does not extend above the horizontal planes, that is, the shearing resistance of soil above the base of footing is neglected. The loading conditions are similar to that of a retaining wall under the passive pressure. Failure occurs when the downward pressure exerted on the soil adjoining the inclined surfaces of the soil wedges is equal to the upward pressure.

This assumption can be used to develop an equation for bearing capacity. He also suggested the following empirical relation to the actual cohesion and angle of shearing resistance in the case of local shear failure.

Mobilized cohesion,

$$c_m = 2/3 \, c' \quad (16.16)$$

Mobilized angle of shearing resistance,

$$\tan \varphi'_m = 2/3 \tan \varphi' \quad (16.17)$$

16.3.3.1.1 Types of Failure

1. **General shear failure**: Terzaghi assumes general shear failure while developing bearing capacity equation for strip footing and is given by

$$q_u = c \, N_c + \gamma \, D_f \, N_q + 0.5 \, B \, \gamma \, N_\gamma \quad (16.18)$$

where N_c, N_q, and N_γ are bearing capacity factors based on friction angle φ.

It mainly occurs to dense sand and stiff clay. After a certain load intensity equal to q_u, the settlement increases suddenly. The failure surface extends to the ground surface and heave is observed on both sides.

2. **Local shear failure**: For local shear failure, Terzaghi has recommended modified parameters c_m and φ_m to calculate ultimate bearing capacity for strip footings, where

$$c_m = 2/3 \, c \quad (16.19)$$

$$\tan \varphi_m = 2/3 \tan \varphi \quad (16.20)$$

The ultimate bearing capacity for local shear failure is given by the equation

$$q_u = 2/3 \, c \, N'_c + \gamma \, D_f \, N'_q + 0.5 \, B \, \gamma \, N'_\gamma \quad (16.21)$$

where N'_c, N'_q, and N'_γ are bearing capacity factors based on friction angle φ_m.

It mainly occurs to medium dense sand and clay of medium consistency. At a load of q_u, sudden jerks occur and the failure surface gradually extends outward. Beyond q_u, there is a large increase in settlement. Heave is observed only when there is a substantial vertical settlement.

3. **Punching shear failure**: It is observed mainly in loose and soft clay. In this there is no heave observed and the failure surface does not extend over ground surface.

16.3.3.2 Terzaghi's Bearing Capacity Equation

16.3.3.2.1 Ultimate Bearing Capacity Equations

1. Strip footing

$$q_u = c \, N_c + \gamma \, D_f \, N_q + 0.5 \, B \, \gamma \, N_\gamma \quad (16.22)$$

where

B = width of footing
D_f = depth of footing below ground level
$\gamma \, D_f$ = surcharge at foundation level
N_c, N_q, and N_γ = bearing capacity factors

2. Square footing

$$q_u = 1.3 \, c \, N_c + \gamma \, D_f \, N_q + 0.4 \, B \, \gamma \, N_\gamma \quad (16.23)$$

3. Circular footing

$$q_u = 1.3 \, c \, N_c + \gamma \, D_f \, N_q + 0.3 \, B \, \gamma \, N_\gamma \quad (16.24)$$

4. Rectangular footing

$$q_u = \left[1 + 0.3(B/L)\right] c \, N_c + \gamma \, D_f \, N_q + 0.5 \left[1 - 0.2(B/L)\right] B \, \gamma \, N_\gamma \quad (16.25)$$

In case pure cohesive soil, $N_c = 5.7$, $N_q = 1$, and $N_\gamma = 0$.

Terzaghi's bearing capacity factors are given in Table 16.3.

16.3.4 Factors Affecting Ultimate Bearing Capacity

1. **Relative density and angle of shearing resistance**: Higher the relative density, greater the angle of shearing resistance φ' of cohesionless soil deposit. The bearing capacity factors N_q and N_γ increase with increase in φ' and hence the bearing capacity increases with an increase in relative density. However, bearing capacity increases appreciably even with a moderate increase in φ provided the soil deposit possesses medium-to-high relative density. In a loose deposit, the gain in bearing capacity with same increase in φ value is not significant.

2. **Width of the footing**: The ultimate bearing capacity of footing in sand increases with width of footing. The increase is substantial when N_γ is large. In case of loose granular soil, an increase in width alone will not bring about any significant increase in bearing capacity.

3. **Depth of footing**: An increase in the depth of foundation will result in an increase in surcharge at the base level of the foundation, leading to an increase in bearing capacity.

4. **Position of water table**: It was observed that the rise of water table from below the foundation result in a decrease in the bearing capacity in a granular soil deposit because of the decrease in effective unit weight of soil consequent to the soil being submerged. If the water table reaches the ground level, rising from a depth equal to or greater than width of the footing measured from the base of footing, the bearing capacity could be reduced by 50%.

5. **Unit weight of soil**: The unit weight of granular soil usually varies from 14 to 21 kN/m³. A soil with higher density will lead to higher bearing capacity.

16.3.5 Effect of Water Table in Bearing Capacity

The general equation of bearing capacity is based on the assumption that the water table is located well below the foundation. It has been found that the effect of water table is negligible, if the depth of water table d_w is sufficiently below the base of foundation.

In the general equation of bearing capacity, there are two terms which are affected by water table movement.

    a. The surcharge term, $\gamma\, D_f\, N_q$
    b. The soil weight component, $0.5\, B\, \gamma_s\, N_\gamma$

The general bearing capacity (Equation 16.18) is

$$q_u = c\, N_c + \gamma\, D_f\, N_q + 0.5\, B\, \gamma_s\, N_\gamma$$

where

γ = effective unit weight of soil in surcharge zone
γ_s = effective unit weight of soil in shear failure zone

Case I: when water table is in the zone I $[d_w \geq B]$ (Figure 16.14)

When the water table is well below the foundation $[d_w \geq B]$, neither the surcharge zone nor shear zone will be affected. Hence there will be no effect of water table on bearing capacity.

Case II: when water table is in zone II $[0 < d_w < B]$

In this case, surcharge zone will not be affected by water table. Hence $\gamma = \gamma_{bulk}$. Here, water is within shear zone. Hence unit weight of soil below water table will be submerged and above water table will remain γ_{bulk}.

$$\text{Therefore, } \gamma_s = \left\{ \gamma_{bulk}\, z_\gamma + \gamma'\left[B - z_\gamma \right] \right\}/B \quad (16.26)$$

Thus the equation of bearing capacity becomes

$$q_u = c\, N_c + \gamma_{bulk}\, D_f\, N_q + 0.5\, B\, \gamma_s\, N_\gamma \quad (16.27)$$

Case III: when water table is in zone I

In this case zone II will be in fully submerged state but zone I will be partially affected.

$$\gamma_s = \gamma' \text{ and } \gamma = \left\{ \gamma_{bulk}\, z_q + \gamma'\left[D_f - z_q \right] \right\}/D_f \quad (16.28)$$

Thus the equation of bearing capacity becomes

$$q_u = c\, N_c + \gamma\, D_f\, N_q + 0.5\, B\, \gamma'\, N_\gamma \quad (16.29)$$

16.3.5.1 Teng's Reduction Factor

The general equation for bearing capacity using Teng's reduction factor with the influence of water table is given below

$$q_u = c\, N_c + \gamma\, D_f\, N_q\, R_{wq} + 0.5\, B\, \gamma\, N_\gamma\, R_{w\gamma} \quad (16.30)$$

where R_{wq} and $R_{w\gamma}$ are Teng's reduction factors for N_q and N_γ, respectively.

TABLE 16.3
Terzaghi's Bearing Capacity Factors

Φ'	General Shear Failure		
	N_c	N_q	N_γ
0	5.7	1	0
5	7.3	1.6	0.5
10	9.6	2.7	1.2
15	12.9	4.4	2.5
20	17.7	7.4	5
25	25.1	12.7	9.7
30	37.2	22.5	19.7
35	57.8	41.4	42.4
40	95.7	81.3	100.4
45	172.3	173.3	297.5
50	347.5	415.1	1153.2

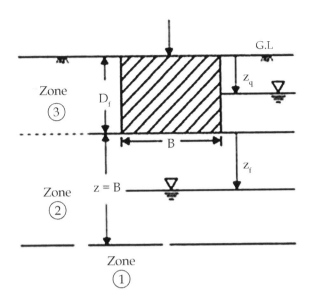

FIGURE 16.14   Effect of water table on bearing capacity.

$$R_{w\gamma} = 0.5 + \frac{0.5b}{B} \leq 1 \qquad (16.31)$$

$$R_{wq} = 1 - \frac{0.5a}{D_f} \leq 1 \qquad (16.32)$$

Case 1:

The water table is between base footing and depth B. In this case, $R_{wq} = 1$.

Case 2:

The water table is at the base of footing. In this case, $R_{wq} = 1$ and $R_{w\gamma} = 0.5$.

Case 3:

The water table is in between the base of footing and the ground surface. In this case, $R_{w\gamma} = 0.5$.

Case 4:

In this case, the water table is at the ground surface.

So both the reduction factors become 0.5. i.e., $R_{wq} = 0.5$ and $R_{w\gamma} = 0.5$.

16.4   DEEP FOUNDATION

A foundation is said to be deep when the depth of foundation is greater than width of foundation. Pile foundation, pier foundation, well foundation, and caisson foundation are the types of deep foundation. Pile foundation is generally used when simple spread footing at a suitable depth is not possible either because the strata of required bearing capacity is available at a greater depth or steep slopes are encountered. Pier foundation is used in situation where the loads to be transmitted are heavier than loads that can be accomplished by piles. To make a pier foundation, an opening is drilled to the desired depth and concrete is poured. A well is large hollow open-ended structure in which excavation and casting continue till it sinks to prescribed depth. Wells are more commonly provided under bridge piers and abutments. A caisson

is a structural box or chamber that is sunk into place or built in place by systematic excavation below the bottom. Caissons are classified as open caissons, pneumatic caissons, and box or floating caisson.

16.4.1   CLASSIFICATION OF PILES

a. **Based on load transfer**

Figure 16.15 represents different types of piles.

1. End-bearing piles: If piles rests over hard or stiff strata then load supporting power is essentially from base, such piles are called end-bearing piles.
2. Friction piles: If piles are driven through soft strata then load supporting power is due to skin friction along the surface area of pile.
3. Bearing and friction piles: In the piles, the load supporting power is due to bearing and skin friction.

b. **Based on function**

1. Anchor piles: These piles are used for providing anchorage against horizontal pull from sheet piles or retaining wall.
2. Sheet piles: Generally, these are used to retain backfill of soil and support soil in open excavation. These are also provided below hydraulic structures to minimize uplift force or to prevent piping failure.
3. Tension or uplift piles: These piles are used to resist uplift force. A pile may be under tension if it resists overturning of wall structurally connected to the pile cap, and due to the tension developed, these piles are known tension piles.
4. Compaction piles: The piles driven in granular soils in order to improve density and bearing capacity are known as compaction piles.
5. Fender piles: These piles are used to protect concrete deck or other water front structure from impact which may be caused by the waves generated by ships and floating objects.
6. Batter piles: These are driven in inclined direction to resist the large horizontal or inclined force.

c. **Based on soil displacement**

1. Displacement piles: The piles whose installation causes displacement and disturbance of soil are known as displacement piles.
2. Nondisplacement piles: The piles which don not cause any displacement and disturbance of soil are known as nondisplacement piles.

d. **Based on method of installation**

1. Driven piles: These piles are driven through dynamic action and load-carrying capacity is due to end-bearing and skin friction resistance.
2. Driven and cast in-situ piles: Piles formed by driving a tube fitted with a driving shoe into the soil and filling the tube with sand or concrete.

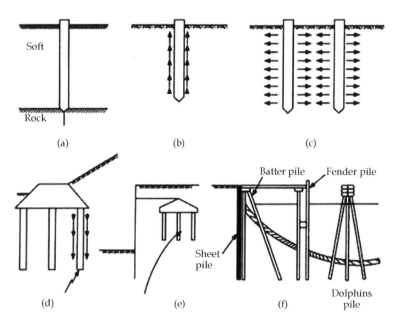

FIGURE 16.15 Represents different types of piles: (a) end-bearing pile, (b) friction pile, (c) compression pile, (d) tension pile, (e) anchor pile, (f) miscellaneous.

3. Bored and cast in-situ piles: Piles formed by making a bore hole into the soil and filling it with concrete are known as bored and cast in-situ piles.

e. **Based on material used**

1. Timber piles: They are made from tree trunks with its branches chopped off. They are provided with a driving point or shoe at the lower end and iron cap at the top to prevent damage during driving. Timber piles are easy to handle, but in situations of fluctuating water table, timber piles should be suitably treated to enhance their life as untreated piles above water table are likely to be eaten by insects, termites, etc. Usually available length will be 4–6 m. Timber piles are used where good bearing stratum is available at a relatively shallow depth.

2. Steel piles: Steel piles are usually of rolled H-sections or thick pipe sections. These piles are used to withstand large impact stresses and where fewer disturbances from driving are desired. These piles are also used to support open excavations and to provide seepage barrier. They can be used to carry heavy loads. However they are likely to be affected by corrosive action and hence need painting or encasement in concrete to resist corrosion.

3. Concrete piles: Concrete piles are either precast or cast in-situ. Precast piles are cast and cured at the casting yard and then transported to the site for installation. These piles are adequately reinforced to withstand handling stresses along with working stress. Precast piles are generally used for short lengths. Cast in-situ piles are constructed by drilling hole in the ground and then filling that hole with freshly prepared concrete after placing the reinforcement.

4. Composite piles: A pile made up of two different materials like concrete and timber or concrete and steel is called composite pile. Composite piles are mainly used where a part of the pile is permanently under water. The part of the pile which will be under water can be made of untreated timber and the other part can be of concrete.

16.4.2 METHOD OF DETERMINING BEARING CAPACITY OF PILES

The ultimate bearing capacity of a single pile is generally estimated by the following methods.

1. **Static method**: This is an analytical method which is suitable for friction piles in cohesive soils. This method is based on the assumption that the ultimate bearing capacity of a pile is sum of the total ultimate skin friction (Q_f) and total ultimate end-bearing resistance (Q_{pu}).

$$Q_u = Q_{pu} + Q_f \quad (16.33)$$

The ultimate end bearing resistance is given by

$$Q_{pu} = A_p \left[c\,N_c + \sigma' N_q + 0.5\,B\,\gamma\,N_\gamma \right] (\text{for } c - \varphi \text{ soil}) \quad (16.34)$$

$$Q_{pu} = c\,N_c\,A_p (\text{for cohesive soil}) \quad (16.35)$$

where
c = unit cohesion at the base of the pile

N_c, N_q, and N_γ = bearing capacity factors
σ' = effective overburden pressure at the tip of the pile.
B = width or diameter of the pile.
γ = effective unit weight of soil
A_p = sectional area of the pile at its base

The ultimate skin friction is given by

$$Q_f = c_u\, \alpha\, A_s \text{(for cohesive soil)} \qquad (16.36)$$

where

A_s = circumferential area of the pile
α = adhesion factor
c_u = cohesion in the embedded length of the pile

$$Q_f = K\sigma'_{av} \tan \delta A_s \text{(for cohesionless soil)} \qquad (16.37)$$

where

K = coefficient of lateral earth pressure
σ'_{av} = average effective overburden pressure over embedded length of pile
δ = angle of friction between the pile and the soil
A_s = circumferential area of the pile

2. **Dynamic method**: It is based on hammer test and is suitable for friction pile in cohesive soil. This analysis is based on the assumption that the dynamic resistance to drive the pile is equal to ultimate load-carrying capacity of the pile under static loading.

   a. Engineers News Record formulae:

   $$Q_u = W H/(S + C) \qquad (16.38)$$

   where

   Q_u = ultimate pile load capacity
   W = weight of hammer (kN)
   H = height of fall (cm)
   S = penetration or set per blow (cm)
   C = 2.5 cm for drop hammer
   C = 0.25 cm for steam hammer
   Safe bearing capacity = Q_u / FOS

   where FOS = 6

   b. Modified Hiley's formula
      This formula accounts various energy losses during impact of pile and hammer and elastic compression of pile, pile cap, and soil.
      The ultimate load-carrying capacity of a pile is given by

   $$Q_u = \eta_h \eta_b W\, H/(S + 0.5C) \qquad (16.39)$$

   where

   W = weight of hammer (kN)
   H = height of fall (cm)

S = set per blow in cm
η_h = efficiency of hammer
η_h = 1 (drop hammer)
η_h = 0.75 – 0.85 (single-acting steam hammer)
η_h = 0.70 – 0.80 (double-acting steam hammer)
η_b = efficiency of hammer blow

$$\eta_b = \left(W + e^2 P\right)\big/(W + P) \quad [W > eP]$$
$$\eta_b = \left[\left(W + e^2 P\right)\big/(W + P)\right] - \left[(W - eP)/(W + P)\right]$$
$$[W < e\,P]$$

P = weight of pile
e = coefficient of restitution between pile and hammer
$C = B_1 + C_2 + C_3$ = total elastic compression
C_1 = temporary elastic compression of dolly and packing
C_1 = 1.77 R/A, where driving with dolly or helmet and cushion of 2.5 cm thick
C_1 = 9.05 R/A, where driving is with short dolly up 60 cm long, helmet, and cushion up to 7.5 cm thick
C_2 = temporary elastic compression of pile = 0.657 RL/A
C_3 = temporary elastic compression of soil = 3.55 RL/A
A = area of pile in cm^2
L = length of pile in m

3. **Pile load test**: The pile load test is the only direct method to determine allowable loads on piles. Two categories of tests on piles, namely, initial test and routine test are usually carried out. Initial test should be carried out on test piles to estimate the allowable load or to predict the settlement at a working load. Routine test is carried out as a check on working piles and to assess the displacement corresponding to the working load.
   A test pile is a pile which is used only in a load test and does not carry load of superstructure. The minimum test load on such piles should be twice the safe load or the load at which the total settlement attains a value of 10% of pile diameter in case of a single pile and 40 mm in case of a pile group.
   A working pile is a pile which is driven or cast in situ along with other piles to carry load from the superstructure. The test load on such piles should be up to one and half times the safe load or up to the load at which settlement attains a value of 12 mm for a single pile and 40 mm for a group of piles whichever is earlier.
   According to the code, the test shall be carried out by applying a series of vertical downward loads on a reinforced cement concrete (RCC) cap over the pile. The load shall preferably be applied by means of a remote controlled hydraulic jack taking reaction

against a loaded platform. The test load shall be applied in increments of about 20% of the assumed safe load. Settlement shall be recorded with at least three dial gauges of sensitivity 0.02 mm, suitably mounted on independent datum bars at least five times the pile diameter away with a minimum distance of 1.5 m. Each stage of loading shall be maintained till the rate of movement of pile top is not more than 0.1 mm/hours whichever is later.

The allowable load on a single pile shall be the lesser of the following:

a. Two-thirds of the final load at which the total settlement attains a value of 12 mm.
b. Fifty percent of the final load at which the total settlement equals 10% of the pile diameter in case of uniform diameter piles and 7.5% of bulb diameter in case of under reamed piles.

The allowable load on a group of piles shall be lesser of the following:

a. Final load at which the total settlement attains a value of 25 mm.
b. Two-thirds of the final load at which the total settlement attains a value of 40 mm.

4. **Correlation with penetration test data**: Static cone penetration test data and SPT data are often used to determine the pile load capacity.

a. SPT: The point resistance of driven piles in sand including H piles can also be determined using N values as per the below equation.

$$q_{pu} = 40N(L/D) \text{ kN/m}^2 \qquad (16.40)$$

where N is the standard penetration resistance as observed in the field for bearing stratum without the overburden corrections, L is the length of the pile, and D is the diameter of the pile. The value of q_{pu} is limited to $400 N$.

For bored piles in sand

$$q_{pu} = 14 \, N\left(D_b/B\right) \text{ kN/m}^2 \qquad (16.41)$$

where D_b is the actual penetration into the granular soil.

b. Dutch cone test: Skin friction resistance for the driven piles can also be obtained with the help of penetration test data by using Meyerhof correlations.

$$f_s = q_c/200 \text{ kg/cm}^2 \qquad (16.42)$$

Point resistance $q_p = \left(q_c/10\right) \times (D/B) \qquad (16.43)$

16.4.3 GROUP ACTION OF PILES

Piles are always used in a group to ensure that the structural loads from a member like a column or a well lies within zone of influence of the foundation. If a single driven pile is used as the foundation, one cannot be certain that pile would be located centrally below the foundation element because quite often the pile moves laterally during driving. The resultant eccentricity of loading may result in the development of bending stresses in the pile, and consequently, the pile may fail structurally. Hence, a minimum number of three piles are used under a column in a triangular pattern, even if load does not warrant the use of three piles. When the number of piles required is more than three, the piles are so arranged that they are symmetrical with respect to the load. Piles under a wall are arranged on either side of central line of the wall in a staggered formation. Figure 16.16 represents the group action of pile.

The load is transferred to the piles in the group through a reinforced slab or beam called the pile cap. The pile tops are connected together to the pile cap which helps the pile group act as an integral unit.

The ratio of the ultimate load capacity of the pile group Q_{ug} to the sum of the individual load capacities of the piles in the group is called the group efficiency.

$$\eta = Q_{ug}/n \times Q_u \qquad (16.44)$$

where n is the number of piles in the group and Q_u is the load capacity of one pile.

Experimental evidence has indicated that a group of piles may fail in one of the following two ways.

i. by block failure
ii. by individual pile failure

A block failure normally occurs when piles are spaced less than 2–3 pile diameters while individual pile failure occurs for wider spacing. In block failure, the soil bound by the perimeter of the pile group and the embedded length of the piles act as one unit or a block.

The ultimate load capacity of the pile group by block failure Q_{ug} is given by

$$Q_{ug} = c_{ub} \, N_c \, A_b + P_b \, L \, c_u \qquad (16.45)$$

where

c_{ub} = undrained strength of clay at the base of the pile group
c_u = average undrained strength of clay along the length of block
N_c = bearing capacity factor taken as 9
A_b = cross-sectional area of the block
P_b = perimeter of the block
L = embedded length of the pile

Q_{ug} for individual pile failure is given by:

$$Q_{ug} = n \, Q_u \qquad (16.46)$$

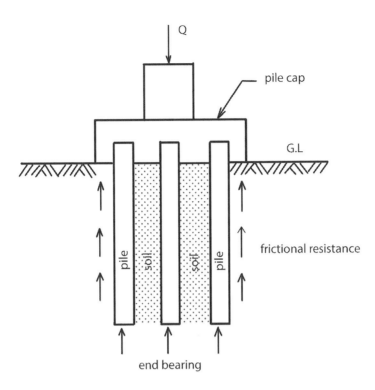

FIGURE 16.16   Represents the group action of pile.

16.4.4   SETTLEMENT OF PILE GROUP

The vertical movement that occurs at the level of the pile cap is largely due to the settlement of the soil supporting the pile. This has to be restricted to a value within the permissible settlement for the structure in question. The settlement of a group of pile is more than the settlement of a single pile even when the load on the single pile and the load on each pile of the pile group are the same. This is because the zone of influence of pile group is much deeper than that of a single pile.

  a. Clayey soil

     The widely used approach to calculate the settlement of pile group in clay is the equivalent raft approach. In this method, the pile group is assumed to act as a single, large raft assumed to be placed at some arbitrary depth inside the soil. For displacement piles, the common practice is to assume the equivalent raft at two-thirds the pile length over the area enclosed by the piles at that depth. For bored piles the equivalent raft is assumed at the base of the piles over the area enclosed by the piles at this depth.

     Generally a 2:1 load distribution is assumed from the level at which the load acts. Sometimes the load is assumed to spread outward from the edge of the block at an angle of 30° to the vertical. For 2:1 distributing the stress increase at the middle of each layer is calculated as

$$\Delta\sigma_z = Q_g/(B + Z_i)(L + Z_i) \qquad (16.47)$$

where Z_i is the distance from the level of the application of the load to the middle of the clay layer i.

$$\Delta s = C_c H_i/(1 + e_0(i))\log_{10}\left[(\sigma' + \Delta\sigma')/\sigma'\right]. \qquad (16.48)$$

  b. For estimating the settlement of a pile group in sand, the common practice is to extrapolate this from the settlement of an individual test pile measured in a load test. Skempton expressed the settlement of a pile group S_g as a ratio of the settlement of an individual pile S_i. S_g/S_i is called the settlement ratio.

$$S_g/S_i = \left[(4B + 2.7)/(B + 3.6)\right]^2 \qquad (16.49)$$

where B is the width of the pile group.

     Meyerhoff expressed the settlement ratio for square pile group driven in sand as

$$S_g/S_i = s(5 - s/3)/4(1 + 1/r) \qquad (16.50)$$

where s is the ratio of pile spacing to diameter and r is the number of rows in pile group.

16.4.5   NEGATIVE SKIN FRICTION

Piles installed in freshly placed fills of soft compressible deposits are subjected to a downward drag, a consequence of the consolidation of the strata after the piles are installed. This downward drag on the pile surface, when the soil moves down relative to the pile, adds to the structural loads and is called negative skin friction. Thus negative skin friction has an effect of reducing the allowable load on pile. Negative skin friction may also develop if the fill material is a loose sand deposit. It can occur due to the lowering of the ground water table which increases the effective stress, thus causing consolidation of the soil with the resultant down-drag on piles.

The magnitude of negative skin friction F_n for a single pile in a filled-up soil deposit is given by

$$F_n = P\,L\,c_a \text{(cohesive soil)} \tag{16.51}$$

where

P = perimeter of pile
L = length of the pile in the compressible stratum
c_a = unit adhesion = $\alpha\,c_u$
α = adhesion factor
c_u = undrained cohesion of the compressible layer

$$F_n = 0.5\,P\,L^2\gamma\,K\tan\delta \tag{16.52}$$

where

K = lateral earth pressure coefficient
δ = angle of friction between pile and soil

The magnitude of negative skin friction in pile group is the higher of the values obtained from the equations given below.

$$F_{ng} = n\,F_n \tag{16.53}$$

$$F_{ng} = P_g\,L\,c_u + L\,A_g\,\gamma \tag{16.54}$$

where

n = number of piles in a group
P_g = perimeter of the group
γ = unit weight of the soil within the pile group up to depth L
A_g = area of pile group within the perimeter

16.4.6 Types of Caissons

Caissons are of three types, namely, open caissons (well), box caissons, and pneumatic caissons.

1. Open caisson or well:
    The top and bottom of the caisson are open during construction; it may have any shape in plan (circular, rectangle, etc.). It has a cutting edge which is fabricated at the site and the first segment of shaft is built on it. The soil inside the shaft is dredged by suitable means and another segment added to it. The advantage is that it can be constructed up to any depth and the cost of construction is relatively low.
2. Box caisson:
    A box caisson is open at top but is closed at the bottom. This type of caisson is first cast on land and then towed to the site where it is sunk onto a previously levelled foundation base. The box caisson is also called floating caisson. This caisson is used where loads are not very heavy and a bearing stratum is available at a shallow depth.
3. Pneumatic caisson:
    A pneumatic caisson has a working chamber at the bottom of the caisson which is kept dry by forcing out water under pressure, thus permitting excavation under dry conditions. Air locks are provided at the top. The caisson gradually sinks as excavation is made. On reaching the final depth, the working chamber is filled with concrete. Such caisson has the advantage of better control in sinking and supervision.

17 Traffic and Transportation Engineering

17.1 PROPERTIES OF TRAFFIC ENGINEERING ELEMENTS

The accomplishment of traffic engineering relies on the coordination among the stated three essential components, i.e., the vehicles, the highways, and the street clients. This chapter expounded factors like human, vehicle, and road that influence transportation.

17.1.1 Human Variables Influencing Transportation

Street clients can be described as drivers, travelers, individuals by walking, and so forth who utilize the streets and expressways. The transportation engineer ought to deal with a collection of street client qualities. For instance, a traffic signal composed to enable a typical passerby to cross the street securely may make a serious risk to an elderly individual. Consequently, the design contemplations ought to securely and proficiently oblige the elderly individuals, the kids, the weakened, the moderate and brisk, and the extraordinary and terrible drivers.

17.1.1.1 Variability

The human qualities like capability to respond to a circumstance, vision and hearing, and other physical and mental variables change from individual to individual and rely upon age, fatigue, nature of improvement, nearness of medications/liquor, and so forth. The effect of every single of these components and their corresponding changeability cannot be accounted when a transportation facility is structured. So an institutionalized esteem, i.e., the 85th percentile estimation of various qualities is taken as design esteem.

85th percentile esteem: denotes a trademark that can be met or outperformed by 85% of the population.

Model: If we state that the 85th percentile estimation of speed of walking is around 2 m/s, it implies that 85% of individuals have a speed of walking quicker than 2 m/s. The inconstancy is in this manner settled by choosing appropriate 85th percentile estimations of the attributes.

17.1.1.2 Basic Qualities

The street client qualities may be of two fundamental sorts, among which some of them are quantifiable like response time, visual keenness, and so forth while some others are less quantifiable like the mental variables, physical quality, fatigue, and expertise.

17.1.1.3 Response Time

The street client is exposed to a progression of upgrades both expected and unforeseen. The time made to play out a move as indicated by the improvement includes a progression of stages like:

Perception: Perception is the way toward seeing the sensations got through the sense organs, nerves, and minds. It is really the acknowledgments that a boost on which a response is to happen exists.

For instance, if a driver approaches a crossing point where the signal is red, the driver initially observes the signal (i.e., he perceives it first), he analyzes that it is a red/STOP signal, he chooses to stop, and lastly applies the brake (volition). This arrangement is known as the perception-identification-emotion-volition (PIEV) time or perception-response time.

Aside from the above time, the vehicle itself going at beginning pace would require some more opportunity to stop. That is, the vehicle going at starting rate u will go for a separation, $d = ut$, where t is the PIEV time. The vehicle will travel some more distance after the brake is applied.

17.1.1.4 Visual Keenness and Driving

The perception–response time depends enormously on the adequacy of driver's vision in seeing the objects and traffic control measures. The PIEV time will be diminished if the vision is clear and precise. Visual sharpness identifies with the field of clearest vision.

Most intense vision: inside a cone of 3°–5°
Genuinely clear vision: inside 10°–12°
Peripheral vision: inside 120°–180°

Different components to be considered for accurate design:

- Dynamic visual sharpness
- Depth perception
- **Glare vision**: Glare recuperation time is the time needed to recoup from the impact of glare after the light source passes. This will be greater for aged individuals.
- **Color vision**: Very imperative as it can come into picture in the event of sign and signal acknowledgment.

17.1.1.5 Walking

- The most common of the street clients are the people on foot. People on foot traffic along the pathways, walkways, cross-walks, security zones, islands, and over and under passes ought to be considered.
- On a normal, walking speed of pedestrians is taken as 1.5–2 m/s.
- Parking spaces and facilities like signs, transport stops, and over and under passes are to be found and planned by the greatest separation to which a customer will walk.
- It was found that in residential communities 90% park inside 185 m of their goals while only 66% park so close in huge city.

17.1.2 OTHER CHARACTERISTICS

Hearing is adequate for recognizing sounds, yet absence of perceiving sounds sharpness can be repaid by utilization of listening devices. The variability of demeanor of drivers concerning age, sex, knowledge, and aptitude in driving and so on is additionally vital.

17.1.3 VEHICLE FACTORS

The street ought to be with the end goal that it ought to take into account the requirements of existing and foreseen vehicles. A portion of the vehicle factors that influence transportation is talked about beneath.

17.1.3.1 Design Vehicles

Roadway frameworks oblige a wide grouping of sizes and sorts of vehicles, from smallest compact traveler cars to the largest double and triple tractor-trailer blends. As indicated by the distinctive geometric highlights of highways like the path width, path widening on curves, minimum curb and corner sweep, clearance statures, and so forth, some standard physical estimations for the vehicles have been suggested. Road authorities are constrained beyond what many would consider possible on vehicular characteristics primarily:

- to give down as far as possible for street designers to work to,
- to make sure that the street space and its geometry is accessible to typical vehicles,
- to actualize control over traffic successfully and proficiently, and
- protect and secure other street clients moreover.

Thinking about the above points, the vehicles can be categorized into:

- motorized two wheelers
- motorized three wheelers
- passenger vehicle
- bus
- single-pivot trucks
- multi-pivot trucks
- truck trailer mixes
- slow on mechanized automobiles

17.1.3.2 Vehicle Measurements

The vehicular measurements that can influence the street and traffic configurations are chiefly:

Width: Vehicular width influences the width of paths, shoulders, and the vehicle parking facility. The capacity of the street will likewise diminish if the width surpasses the calculated esteems.

Height: Affects the freedom tallness of structures like over-bridges, under-bridges, and electric and other administration lines and furthermore setting of signs and signals.

Length: Affects the additional width of pavement, slightest turning radius, safe surpassing separation, street capacity, and the parking facility.

Rear overhang: Mainly vital when the vehicle takes a right/left turn about a stationary point.

Ground clearance: Comes into scene while planning ramps and property get to and as bottoming out on a peak can prevent a vehicle from moving under its own pulling power.

17.1.3.3 Weight, Axle Configuration

The heaviness of the vehicle is transmitted to the pavement through the axles thus the plan parameters are settled based on number of pivots.

Power to weight proportion:

- Measure of the straight forwardness with which a vehicle can run.
- It decides the working productivity of vehicles out and about and is increasingly vital for substantially substantial vehicles.
- Important principle which decides the length to which a positive inclination can be allowed mulling over the instance of heavy vehicles.

17.1.3.4 Turning Radius and Turning Path

- The least turning radius is subject to the design and class of vehicle. The compelling width of the automobile is expanded on a turning.
- Imperative at crossing points, circular junctions, terminals, and parking areas.

17.1.3.5 Visibility

The perceivability of the driver is impacted by the vehicular measurements. To the extent forward perceivability is concerned, the measurements of the vehicle and the incline and curvature of wind screens, windscreen wipers, entry way columns, and so on ought to be with the end goal that:

- vision is clear even at awful climate conditions like haze, ice, and rain;
- it ought not veil the people on foot, cyclists, or different vehicles during intersection maneuver.

Equally imperative is the side-and-rear perceivability while moving particularly at convergences when the driver changes his speed so as to union or cross a traffic stream. Rear vision proficiency may be accomplished by legitimately positioning the inner or outer mirrors.

17.1.3.6 Acceleration Characteristics

The speeding up limit of an automobile is reliant on its mass, the opposition from movement and accessible power. The increasing speed rates are most elevated at low speeds and diminish as speed increments. Substantial vehicles have brought down rates of speeding up than traveler cars. The distinction in increasing speed rates winds up noteworthy in blended rush hour gridlock streams. For instance, substantial vehicles like trucks will postpone all travelers at a crossing point. Once more, the holes created can be involved by other smaller automobiles just on the off chance that they are allowed the chances to pass. The nearness of overhauls makes the issue progressively severe. Trucks back off on grades because their power is inadequate to keep up their ideal speed. As trucks backs off on grades, long gaps will be shaped in the rush hour gridlock stream which cannot be productively filled by typical passing moves.

17.1.3.7 Braking Performance

The time and separation taken to cease the automobile are extremely critical to an extent the design of different traffic facilities is concerned. The components on which the braking separation relies are on the sort of the street and its condition, the type and state of tyre, and the kind of the stopping mechanism. The separation required to decelerate from one speed to next one is demonstrated by

$$d = \frac{v^2 - u^2}{f + g} \qquad (17.1)$$

where

d = separation required for braking,
v = initial speed of vehicle,
u = final speed of vehicle,
f = forward rolling and skidding friction coefficient, and
g = grade (in decimals).

The fundamental qualities of a traffic framework affected by braking and deceleration execution are as follows:

Safe ceasing sight distance: The least ceasing sight separation incorporates both the response time required and the length canvassed in ceasing.

Clearance and change interval: All automobiles at a separation further away than one ceasing sight distance from the signal when the yellow is flashed is thought to have the capacity to cease safely and securely. Such an automobile which is at a separation equivalent or more noteworthy than the ceasing sight distance should head out a distance equivalent to the ceasing sight distance in addition to the width of the road, in addition to the length of the automobile. Consequently, the yellow and every single red time ought to be determined to oblige the sheltered leeway of those vehicles.

Sign placement: The driver ought to see the sign board from a separation in any event equivalent to or more prominent than the stopping sight distance.

17.1.4 ROAD FACTORS

17.1.4.1 Road Surface

The factors relating to road surface, which are to be given extraordinary consideration in the plan, development, and support and repair of highways for their safe and economical practical activity.

- Road unpleasantness
- Wear and tear of tire
- Tractive opposition
- Noise
- Light reflection
- Electrostatic properties

For substantial traffic quantities, a smooth riding surface with great all-climate hostile to slide properties is alluring.

17.1.4.2 Lighting

Brightening is utilized to enlighten the physical highlights of the street path and to help in the driving assignment. It is vital that pathway lighting be moved toward the premise of many traffic data, for example, night vehicular traffic, person on-foot volumes, and mishap encounter.

A luminaire is a total light-producing gadget which disseminates light into various patterns like a garden hose spout circulates water.

17.1.4.3 Roughness

Roughness is one of the fundamental variables that an engineer ought to give significance amid the structural design, development, and upkeep of a roadway framework. Drivers will in general look for smoother surface when they are given a choice. On four-path highways, where the surface of the inward pathway is rougher than the outside path, passing automobiles will in general come back to the outside path after execution of the passing move. Shoulders or even speed-switch to another paths might intentionally be roughened as a method of outline.

17.1.4.4 Pavement Colors

- **Light-shaded pavements (cement concrete pavements)**: better perceivability amid day time.
- **Dark-shaded pavements (bituminous pavements)**: better perceivability amid night.
  Contrasting pavements might be utilized to demonstrate preferential utilization of traffic paths.

17.1.4.5 Night Visibility

Since most mishaps happen during late evenings on account of diminished perceivability, the traffic designer must endeavor to enhance night time perceivability all around he can. A vital factor is the proportion of light which is reflected by the street surface to the drivers' eyes. Glare caused by the light of approaching automobile on reflection is irrelevant on a dry pavement; however, it is a vital factor when the pavement is wet.

17.1.4.6 Geometric Aspects

- Central segments of the asphalt are marginally elevated and are slanted to either side in order to keep the unwanted pooling of water out and about the surface. This will break down the riding quality since the asphalt will be exposed to numerous failures like potholes and so forth.
- Least path width ought to be given to decrease the chances of mishaps. Additionally, the speed of the

automobile will be decreased and time expended to achieve the goals will likewise be more.

- Right of way (ROW) width ought to be appropriately given. On the off chance that the option to proceed width turns out to be less, future extension will wind up troublesome and the development of that area will be antagonistically influenced.
- Gradient diminishes the tractive exertion of large vehicles bringing about more fuel utilization.
- **Curves**: Near bends, odds of mishaps are more. Speed of automobiles is likewise influenced.

17.2 HIGHWAY GEOMETRIC DESIGN

The goal of geometric plan is to give perfect capability in track activity and most extreme security at sensible expense. The geometric structure of pathway manages the measurements and format of noticeable highlights of the roadway, for example, alignment, cross-section components, locate distances, horizontal curvature, horizontal gradients, and convergences. The emphasis of the geometric structure is to address the necessity of the driver and the vehicles, for example, well-being, solace, capability, and so forth. Geometric plan of highways manages the accompanying components:

1. Cross-section components
2. Sight distance contemplations
3. Details of horizontal alignment
4. Details of vertical alignment
5. Intersection components

Cross-section components: It incorporates cross-incline, different widths of street (i.e., pavement width, formation width, and street land width), surface attributes, and highlights in street edges.

Sight distance contemplations: It is the noticeable land in front of the driver at flat and vertical bends and at crossing points for the protected developments of vehicles.

Horizontal alignment: Horizontal bends are acquainted to alter the bearing of street. It incorporates highlights like cant, radius of curvature, transition curve, additional augmenting, and setback distance.

Vertical alignment: Its parts like gradients, vertical bends sight distance, and structure of length of curves.

Intersection components: Proper structure of crossing point is especially fundamental for the safe and proficient traffic development. Its highlights are format, limit, and so forth.

17.2.1 Factors Influencing the Geometric Designs

17.2.1.1 Design Speed

- Defined as the most vital ceaseless speed at which a particular vehicle can go with safety on the expressway if climatic conditions are favorable.

- Different from the legitimate speed limit and ideal speed.
- **Legal speed limit**: Speed limit compelled to control a typical propensity of drivers to go past a recognized safe speed.
- **Desired speed**: Maximum speed at with which a driver travels when unconstrained by either track or nearby geometry.

17.2.1.2 Topography

It is less requesting to develop streets with necessary rues for a plain scene. However, for a given design speed, the construction cost increments multiform with the angle and the terrain. Geometric structure standard is described by sharper bends and more extreme inclinations.

17.2.1.3 Traffic Factors

One of the urgent elements to be considered in expressway configuration is the gridlock information, in both present and future appraisals. Gridlock system volume demonstrates the level of services (LOS) for which the roadway is being arranged and straightforwardly influences the geometric highlights such as width of road, its alignment, grades, and so on, without information on the traffic is exceptionally hard to design any interstate roadway.

17.2.1.4 Design Hourly Volume and Capacity

The general unit for estimating traffic on interstate roadway is the Annual Average Daily Traffic volume, condensed as AADT. It is uneconomical to structure the roadway facilities for the pinnacle traffic stream. Hence a sensible estimation of gridlock system volume is chosen for the design and this is known as Design Hourly Volume which is resolved from broad traffic volume considers. The proportion of volume to capacity influences the level of service of the street.

17.2.1.5 Environmental and Other Factors

The ecological elements like air contamination, noise contamination, landscaping, aesthetics, and other overall conditions ought to be given due contemplations in the geometric structure of streets.

17.2.2 Passenger Car Unit

- The stream of traffic with unhindered blending of various vehicle classes frames the "Mixed Traffic Flow." To gauge the traffic volume and capacity of roadway facilities under blended traffic stream, the different vehicle classes are changed over to one regular standard vehicle unit. It is a usual practice to consider the traveler car as the standard vehicle unit to change over the other vehicle classes and this unit is called Passenger Car Unit (PCU).
- Thus in a mixed rush hour gridlock stream, traffic volume and capacity are commonly communicated as pcu/hour (or) pcu/lane/hour and traffic density as pcu/km length of lane.

17.2.2.1 Components Influencing PCU Values

- Vehicles qualities, for example, measurements, power, speed, acceleration, and braking attributes.
- Transverse and longitudinal gaps (or) clearances between moving vehicles which relies on speed and driver qualities.
- Traffic stream qualities, for example, composition of various vehicle classes, mean speed and speed approximation of mixed traffic stream, volume-to-capacity proportion, and so on.
- Roadway attributes, for example, street geometrics incorporates angles, curve and so on, provincial or urban street, nearness of convergences, and the kinds of crossing points.
- Regulation and control of traffic such as speed limit, one-way traffic, presence of different traffic control devices, etc.
- Ecological and climatic conditions.

17.2.3 Pavement Surface Characteristics

Four highlights of the pavement surface for safe and convenient driving:

- the grating between the wheels and the asphaltic surface;
- smoothness of the street surface;
- light reflection qualities of the highest point of pavement surface; and
- drainage of water.

17.2.3.1 Friction

- Crucial factor in the structure of even bends and consequently the protected working speed.
- Affects the increasing speed and deceleration capacity of automobiles.
- Lack of sufficient rubbing can cause sliding or slipping of automobiles.

Sliding happens when the way went along the street surface is greater than the circumferential wheel movement because of friction.

Sliding happens when wheel rotates greater than the comparing longitudinal development along the street.

Different components which influence friction are as follows:

i. Pavement type
ii. Pavement condition
iii. Condition of tyre
iv. Speed and load of vehicle
   - The frictional power that is made between the wheel and the asphalt is the load acting multiplied by a factor called the frictional coefficient "f."

- The coefficient of longitudinal frictional contact as 0.35–0.4 relying upon the speed, and the coefficient of lateral friction as 0.15. The former is useful for insight distance estimation and the latter in curve design.

17.2.3.2 Unevenness

Unevenness influences the vehicle working cost, speed, riding comfort, safety, fuel utilization, and mileage of tyres.

Unevenness index: proportion of unevenness which is the aggregate proportion of vertical undulations of the pavement surface recorded per unit horizontal length of the street (Table 17.1).

17.2.3.3 Drainage

- The surface of pavement ought to be totally impermeable to counteract drainage of water into the layers of pavement.
- Both texture of pavement and its geometry should aid in emptying water out of the surface in less time.

17.2.3.4 Camber

Camber or cant is the cross-inclination given to elevate widely the central portion of the road surface in the transverse direction to empty the storm water out from the street surface. The goals of giving camber are as follows:

- Surface security particularly for gravel and bituminous streets
- Subgrade security by providing appropriate drainage
- Quick drying of pavement thereby builds safety

Camber is estimated in 1 out of n or $n\%$ (e.g., 1 out of 50 or 2%) and the esteem relies upon the kind of pavement surface. The normal types of camber are parabolic, straight, or blend of them.

17.2.3.5 Right of Way

ROW or width of land (Figure 17.1) is the land width obtained for the street, along its alignment. It ought to be satisfactory to suit all the cross-sectional components of the interstate roadway and may sensibly accommodate for future enhancement. To anticipate strip development along expressways, control lines and building lines might be given.

TABLE 17.1
Unevenness Index

Unevenness Index	Remarks
Less than 1500 mm/km	Good
Under 2500 mm/km	Satisfactory (up to speed of 100 km/hour)
More noteworthy than 3200 mm/km	Uncomfortable (even for 55 km/hour)

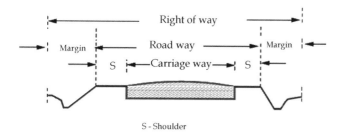

FIGURE 17.1 Right of way.

Control line: Line which speaks to the nearest furthest limits of future uncontrolled building action in connection to a street.

Building line: Line on either side of the street; among which and the street, no building action is allowed by any stretch of the imagination.

The ROW width is administrated by the following:

- **Width of formation**: relies upon the class of the interstate roadway and width of roadway and street edges.
- **Height of embankment or profundity of cutting**: administrated by the geology and vertical alignment.
- **Side slopes of embankment**: relies upon the tallness of slope, type of soil, and so on.
- **Drainage framework and their size**: relies upon precipitation, geography, and so on.
- **Sight distance considerations**: On bends and so on there is limitation to the perceivability on the inward side of the bend because of the nearness of few obstacles like structures and so forth.
- **Reserve space for future augmenting**: Some area must be obtained ahead of time foreseeing future advancements like widening of the street.

17.2.3.6 Width of Carriage Way

Width of carriage way or the pavement width relies upon the width of the gridlock path and the number of paths. Width of a gridlock path relies upon the width of the automobile and the leeway. Side freedom enhances working speed and security.

The greatest reasonable width of an automobile is 2.44 and the attractive side leeway for single path gridlock is 0.68 m. This requires least of path width of 3.75 m for a single path street. However, 0.53 m is required for side leeway, on both sides and 1.06 m at the center. Along these lines, a two-path street requires a least of 3.5 m for every path.

17.2.3.7 Kerbs

Kerbs demonstrate the limit between the carriage way and the pathways.

- **Low kerbs**: This kind of kerb is given to such an extent that they urge the traffic to stay in through rush hour gridlock paths and furthermore enable the driver to enter the shoulder area with little trouble. The tallness of the kerb is around 10 cm over the

pavement edge with an inclination which enables the vehicle to climb effectively. This is generally given at medians and channelization plans and furthermore helps in longitudinal seepage.

- **Semi-barrier-type kerbs**: When the person on foot traffic is high, these kerbs are given. Their stature is 15 cm over the pavement edge. This kind of kerb counteracts infringement of parking vehicles, yet at intense crisis, it is conceivable to roll over this kerb with some trouble.
- **Barrier-type kerbs**: They are intended to dishearten vehicles from leaving the pavement. They are given when there is impressive amount of person on foot traffic. They are put at a stature of 20 cm over the pavement edge with a steep batter.
- **Submerged kerbs**: They are utilized in rural streets. The kerbs are given at pavement edges between the pavement edge and shoulders. They give horizontal repression and strength to the pavement.

17.2.3.8 Roadway Width

Roadway width or formation width includes both the widths of separators as well as shoulders. This does not include the additional space in embankment/cutting.

17.2.4 Sight Distance

Sight distance open from a point is the authentic partition along the road surface, over which a driver from a predefined stature over the carriage way has detectable quality of stationary or moving things.

Mainly three-sight distance circumstances are considered for plan:

- Stopping sight distance (SSD) or the absolute least sight distance
- Intermediate sight distance (ISD) is characterized as twice the SSD
- Overtaking sight distance (OSD) for safe surpassing task
- Head light sight distance is the separation noticeable to a driver amid night driving under the illumination of head lights
- Safe sight distance to go into a crossing point

The most basic idea in all these is that reliably the driver going at the design speed of the pathway must have adequate carriageway distance inside his line of vision to enable him to stop his vehicle before crashing with a bit-by-bit moving or stationary article appearing out of the blue in his own rush hour gridlock path. The estimation of sight distance relies upon:

i. Response time of driver:
   Response time is the time taken by a driver from the instant he perceives the object to the moment when he applies the brakes. The total response time

may be split up into four segments based on PIEV hypothesis. A considerable lot of the investigations demonstrated that drivers need about 1.5–2 seconds during ordinary circumstances. The response time is usually taken as 2.5 seconds.

ii. Speed of the vehicle:
- As the speed increases, the time needed for stopping the vehicle increases. Consequently, it is clear that speed increments with sight distance increments.

iii. Brake efficiency:
- Brake efficiency relies on the age of the vehicle, vehicle quality, and so forth. On the off chance that the brake proficiency is 100%, the vehicle will stop the minute the brakes are applied. However, essentially, it is absurd to expect to accomplish 100% brake productivity. Consequently, the sight distance required will be more when the proficiency of brake is less.
- For safe geometric plan, we expect that the vehicles have just 50% brake proficiency.

iv. Frictional resistance between the tyre and the street:
- When the frictional obstruction is more, the vehicles stop quickly. As such sight required will be less. No different arrangement for brake effectiveness is given while figuring the sight distance. This is considered alongside variable of longitudinal frictional contact.
- IRC has determined the estimation of longitudinal friction in the middle of 0.35–0.4

v. Angle of street:
- While ascending a slope, the automobile can stop quickly. Along these lines sight distance required is less.
- While sliding an inclination, gravitational force additionally comes without hesitation and additional time is adequate to stop the vehicle. Sight distance needed will be more for this situation.

17.2.4.1 Stopping Sight Distance

SSD is the least sight distance accessible on a roadway at any spot having adequate length to empower the driver to stop a vehicle going at configuration speed, securely without impact with some other obstacles.

Safe stopping distance is the separation an automobile goes from the point at which a circumstance is first seen to the time the decreasing speed rate is finished. In interstate roadways configuration, sight distance at any rate equivalent to the safe stopping distance ought to be provided.

Stopping sight distance = Lag distance + Braking separation
$$(17.2)$$

- **Lag distance**: Separation gone by the vehicle amid the response time and is denoted by $v*t$, where v is the velocity of vehicle in m/s^2.

- **Braking distance**: Distance gone by the vehicle amid braking activity.

For a level street this is acquired by comparing the work done in ceasing the automobile and the dynamic vitality of the automobile. If F is the most extreme force of friction created and the braking separation is l, at that point work done against frictional force in ceasing the automobile is

$$F_1 = fWl \qquad (17.3)$$

where
W: total weight of vehicle.

The kinetic vitality at design speed is

$$\frac{1}{2}mv^2 = \frac{1}{2}\frac{Wv^2}{g} \qquad (17.4)$$

$$l = \frac{v^2}{2gf} \qquad (17.5)$$

Therefore, the SSD is given by:

$$SSD = vt + \frac{v^2}{2gf} \qquad (17.6)$$

where
v: design velocity in m/s
t: reaction time in seconds,
g: acceleration due to gravity; and
f: coefficient of friction.

At the point when there is a rising inclination, the segment of gravity aids to braking action and consequently the braking separation is diminished.

$$l = \frac{v^2}{2g\left(f + \dfrac{n}{100}\right)} \qquad (17.7)$$

Likewise, the braking separation can be determined for a descending inclination. In this way the general equation for SSD is given by

$$SSD = vt + \frac{v^2}{2g\left(f \pm 0.01n\right)} \qquad (17.8)$$

17.2.4.2 Overtaking Sight Distance/ Passing Sight Distance

- Least separation open to the visibility of the driver of an automobile meaning to overwhelm the moderate automobile ahead securely against the rush hour gridlock in the other way.

- Calculated at the middle line of the street over which the driver with eye level above 1.2 m from the street surface can see the highest point of an item 1.2 m over the street surface.

Elements influencing the OSD:

i. Velocity of the overwhelming automobile, surpassed automobile, and of the automobile coming in the other way.
ii. Spacing between automobiles, which thus relies upon the speed.
iii. Professionalism and response time of the driver.
iv. Rate of increasing speed of overwhelming vehicle.
v. Slope of street.

The overwhelming sight distance comprises three sections:

- d_1 is the distance gone by overwhelming vehicle A amid the response time $t = t_1 - t_0$
- d_2 is the distance gone by the vehicle during the actual surpassing activity

$$T = t_3 - t_1$$

- d_3 is the distance gone by on-coming vehicle C amid the overwhelming task (T).

$$OSD = d_1 + d_2 + d_3 \qquad (17.9)$$

$$OSD = v_b t + 2s + v_b\sqrt{\frac{4s}{a}} + vT \qquad (17.10)$$

where
v_b = velocity of the moderate moving automobile in m/s,

v = velocity of quick moving automobile in m/s,
t = response time of driver in seconds,
s = spacing between the two vehicles in m,
a = overtaking vehicle's acceleration in m/s².

On the off chance that the speed of the overwhelmed automobile is not given, it tends to be exceptionally all around expected that it moves 16 km for every hour slower than the design speed.

Overtaking zones:

Overtaking zones (Figure 17.2) are given when OSD cannot be given all through the length of the roadway. These are zones committed for surpassing activity, set apart with wide streets. The attractive length of overwhelming zones is five times the OSD and the least is three times the OSD.

17.2.5 Geometric Design Elements

i. **Horizontal alignment**:
   - minimum curve radius (maximum degree of curvature);
   - minimum length of tangent between compound or reverse curves;
   - transition curve parameters;
   - minimum passing sight distance and stopping sight distance on horizontal bends.

ii. **Vertical alignment**:
   - maximum gradient;
   - length of maximum gradient;
   - minimum passing sight distance or stopping sight distance on summit (crest) curves;
   - length of sag curves.

iii. **Cross-section**:
   - carriageway width;
   - carriageway cross-fall;

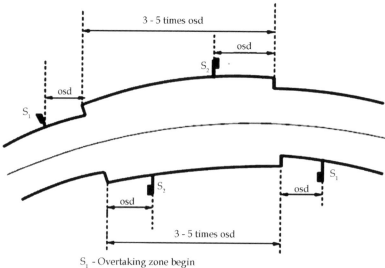

S₁ - Overtaking zone begin
S₂ - End of overtaking zone

FIGURE 17.2 Overtaking zones.

- rate of super elevation of road;
- widening of curves;
- width of shoulder;
- shoulder cross-fall;
- width of structure;
- width of right-of-way;
- sight distance;
- cut and fill slants and dump cross-area.

17.2.5.1 Horizontal/Level Alignment

The horizontal/level alignment comprises a progression of converging tangents and round curves, with or without transition bends.

The nearness of even bend confers centrifugal force which is responsive force acting outward on an automobile negotiating it. This outward force relies upon speed and span of the horizontal curve and is neutralized to a specific degree by transverse rubbing between the tyre and asphalt surface. On a bended road, this power will in general reason the vehicle to overwhelm or to slide outward from the focal point of street curvature. For appropriate structure of the curve, an understanding of the forces following up on a vehicle taking a horizontal curve is important. Various forces following up on the automobile are shown in Figure 17.3.

1. Centrifugal force (P in outward direction)
2. Weight of vehicle (W in downward direction)
3. Reaction of ground on wheels (R_A and R_B)
   - Centre of gravity is at h distance over the ground level
   - Wheel base: b units

The centrifugal force is given by

$$P = \frac{Wv^2}{gR} \tag{17.11}$$

where
v is the speed of the vehicle in m/s,
g is the acceleration due to gravity in m/s^2, and
R is the radius of the curve in m.

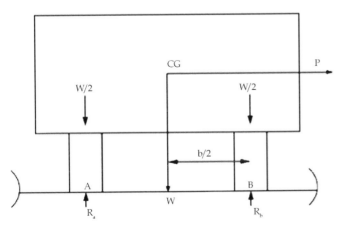

FIGURE 17.3   Effect of horizontal curve.

The impact factor denoted as P/W is given by

$$\frac{P}{W} = \frac{v^2}{gR} \tag{17.12}$$

The centrifugal force has the following impacts: An inclination to upset the automobile about the external wheels and propensity transverse slipping.

At equilibrium, overturning is possible only when

$$\frac{v^2}{gR} = \frac{b}{2h} \tag{17.13}$$

and for well-being the accompanying condition must be fulfilled:

$$\frac{b}{2h} > \frac{v^2}{gR}$$

When the centrifugal force P is more prominent than the greatest conceivable transverse skid resistance, transverse skidding occurs because of the rubbing between the asphalt surface and tyre. The transverse slip resistance (F) is evaluated by using the equation:

$$F = F_A + F_B = fW \tag{17.14}$$

where
F_A and F_B are the fractional force at tyre A and B,
R_A and R_B are the reaction at tyre A and B,
f is the lateral coefficient of friction, and
W is the weight of the vehicle.

At equilibrium, when skidding takes place

$$\frac{P}{W} = f = \frac{v^2}{gR} \tag{17.15}$$

And for safety the following condition must be satisfied:

$$f > \frac{v^2}{gR}$$

Analysis of super-elevation:

Super-elevation or cant is the transverse inclination given at horizontal bend to neutralize the centrifugal force, by raising the external edge of the asphalt regarding the inner edge, all through the length of the horizontal curve. So as to discover how much this raising ought to be, the accompanying analysis might be finished. The forces following up on a vehicle while taking a horizontal curve with cant are demonstrated in Figure 17.4.

Forces following up on a vehicle on horizontal curve of radius R m with a speed of v m/s^2 are as follows:

- P is the centrifugal force acting on the horizontal direction outward through the focal point of gravity,

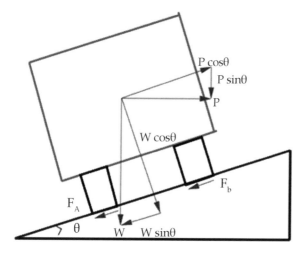

FIGURE 17.4 Forces acting on vehicle moving along a horizontal curve with super elevation.

- W is the weight of vehicle acting downward through the focal point of gravity, and
- F is the friction force between wheels and pavement, along the internal surface.

For harmony condition,

$$P \cos \theta = W \sin \theta + F_A + F_B \qquad (17.16)$$

$$e + f = \frac{v^2}{gR} \qquad (17.17)$$

where
e is the rate of super elevation,
f is the coefficient of lateral friction usually taken as 0.15,
v is the speed of vehicle in m/s,
R is the radius of curve in m, and
$g = 9.8\text{m/s}^2$.

Three explicit cases that may emerge from condition above are as per the following:

- If there is no contact because of some viable reasons, at that point, $f = 0$ and equation moves toward becoming

$$e = \frac{v^2}{gR} \qquad (17.18)$$

This results in the circumstance where the weight on the external and inward wheels is same, requiring high cant e.
- If there is no super-rise given because of some pragmatic reasons, then at that point $e = 0$ and condition progresses toward becoming

$$f = \frac{v^2}{gR} \qquad (17.19)$$

This results in a high frictional coefficient.

- When $e = 0$ and $f = 0.15$, for safe voyaging, speed from condition is given by

$$v_b = \sqrt{fgR} \qquad (17.20)$$

where v_b is the confined speed.

Guidelines on cant:
While planning the different components of the street like super elevation, if it is structured for a specific vehicle it is called design vehicle which has some standard weight and estimations. In the real case, the street needs to cater for mixed traffic rather than a design vehicle. Taking into down to earth contemplations of all such unsafe circumstances, IRC has given a few rules about the most extreme and least cant, and so forth.

Design of super-elevation:
IRC recommends accompanying design methodology:
Step 1: Find e for 75% of design speed, disregarding f, i.e.,

$$e_1 = \frac{(0.75v)^2}{gR} \qquad (17.21)$$

Step 2: If $e_1 \leq 0.07$, then at that point,

$$e = e_1 = \frac{(0.75v)^2}{gR} \qquad (17.22)$$

Else if $e_1 > 0.07$ go to step 3.
Step 3: Find f_1 for the design speed and max e, i.e.,

$$f_1 = \frac{v^2}{gR} - e = \frac{v^2}{gR} - 0.07 \qquad (17.23)$$

In the vent that $f_1 < 0.15$, then the most extreme, $e = 0.07$ is okay for the design speed, else go to step 4.
Step 4: Find the reasonable speed v_a for the greatest $e = 0.07$ and $f = 0.15$,

$$v_a = \sqrt{0.22gR} \qquad (17.24)$$

In the off chance $v_a \geq v$ design is sufficient; generally use speeds embrace control measures for speed control measures.
Table 17.2 demonstrates the IRC determinations for the most extreme and least cant.

TABLE 17.2
IRC Specifications for Super Elevation

Maximum Super Elevation	Minimum Super Elevation
7%—Plain and rolling terrain	2%–4% For drainage and for large radius of the horizontal curve
10%—Hilly terrain	
4%—Urban road	

- Depends on (1) moderate moving automobile and (2) substantially stacked trucks with high center of gravity (CG).

Attainment of cant:

1. Removal of the crown of the cambered segment by:
   a. Turning the external edge about the crown: The external portion of the cross-inclinations is pivoted about the crown at an ideal rate with the end goal that this surface falls on the indistinguishable plane from the internal half.
   b. Shifting the location of the crown: This strategy is otherwise called as diagonal crown method. Here the location of the crown is logically moved outward, along these lines thus in this manner expanding the width of the internal portion of cross-segment dynamically.
2. Rotation of the asphalt cross-segment to accomplish full cant by: there are two techniques for achieving cant by pivoting the asphalt.
   a. Rotation about the middle line: The asphalt is turned to an extent that the inward edge is discouraged and the external edge is raised both considerably the aggregate sum of cant, i.e., by $E = 2$ as for the inside.
   b. Rotation about the internal edge: Here the asphalt is turned raising the external edge and also the inside to such an extent that the external edge is raised by everything of super-rise as for the inward edge.

Radius of horizontal curve:

$$R_{\text{ruling}} = \frac{v^2}{g(e+f)} \qquad (17.25)$$

Extra-widening:

Extra-widening alludes to the extra-width of carriageway that is required on a bend section of a street over and above that required on a straight alignment. This augmenting is done because of two reasons: (1) the extra-width required for a vehicle taking a horizontal curve; (2) propensity of the drivers to employ from the edge of the carriageway as they drive on a bend. The first is alluded as the mechanical widening and the second is known the psychological widening.

i. Mechanical widening:

The reasons for the mechanical widening are as follows: when a vehicle negotiates a horizontal curve, the rear wheels follow a path of shorter radius than the front wheels as shown in Figure 17.5. This phenomenon is called off-tracking and has the effect of increasing the effective width of a road space required by the vehicle.

$$W_m = \frac{nl^2}{2R} \qquad (17.26)$$

ii. Psychological widening:

There is a propensity for drivers to drive near to the edges of pavement on the bends. Some additional space needs to be accommodated for more freedom for the intersection and overwhelming activities on bends. IRC proposed an observational connection for the psychological broadening at the horizontal bends

$$W_{ps} = \frac{v}{2.64\sqrt{R}} \qquad (17.27)$$

Consequently, the aggregate widening required at a horizontal curve W_e is

$$W_e = W_m + W_{ps} = \frac{nl^2}{2R} + \frac{v}{2.64\sqrt{R}} \qquad (17.28)$$

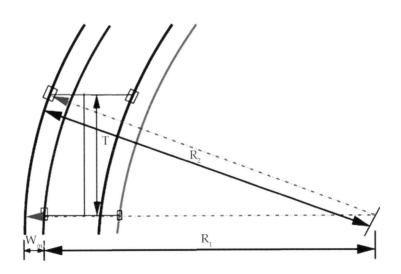

FIGURE 17.5 Extra-widening at a horizontal curve.

Horizontal transition curves:

Transition curve is given to alter the horizontal arrangement from straight to round curve continuously and has a radius which diminishes from boundlessness at the straight end to the ideal radius of the roundabout curve at the opposite end.

Five goals for giving transition curve:

- to present step-by-step the centrifugal force between the tangent point and the start of the circular bend, maintaining sudden yank on the vehicle. This builds the solace of travelers;
- to empower the driver, turn the steering step-by-step for his own solace and security;
- to give steady presentation of super elevation;
- to give steady presentation of additional widening;
- to upgrade the stylish appearance of the street.

Kinds of transition curve:

Spiral, cubic parabola, and lemniscate.

IRC prescribes spiral as the transition curve as it satisfies the necessity of a perfect transition curve, that is,

- rate of variation or centrifugal acceleration is steady (smooth) and
- radius of the transition curve is ∞ at the straight edge and it changes to R at the curve point $L_s \infty 1/R$ and estimation and field execution is simple.

Length of transition curve:

Most extreme of the accompanying three criteria: rate of change of outward increasing speed; rate of change of cant; and an experimental formula given by IRC.

- **Rate of variation of increasing outward speed (centrifugal acceleration):** If c is the rate of change of centrifugal acceleration, it tends to be composed as:

$$c = \frac{v^3}{L_s R} \tag{17.29}$$

Length of the transition bend L_{s1} in m is expressed as

$$L_{s1} = \frac{v^3}{cR} \tag{17.30}$$

where c represents the rate of variation of centrifugal acceleration given by an observational equation proposed by IRC as beneath:

$$c = \frac{80}{75 + 3.6v} \tag{17.31}$$

where
$c_{min} = 0.5$
$c_{max} = 0.8$

- **Rate of change of cant**: Raise (E) of the external edge regarding inward edge is given by

$$E = eB = e(W + W_e) \tag{17.32}$$

The rate of change of this raise from 0 to E is accomplished bit by bit with an angle of 1 in N over the length of the transition curve (typical range of N is 60–150). Along these lines, the length of the transition curve L_{s2} is:

$$L_{s2} = Ne(W + W_e) \tag{17.33}$$

- **By observational formula**: IRC recommends that the length of the transition curve is least for a plain and moving terrain;

$$L_{s3} = \frac{35v^2}{R} \tag{17.34}$$

For steep and bumpy landscape,

$$L_{s3} = \frac{12.96v^2}{R} \tag{17.35}$$

And the shift s as

$$s = \frac{L_s^2}{24R} \tag{17.36}$$

Setback Distance

The separation required from the middle line of a horizontal curve to an impediment on the inward side of the bend to give sufficient sight distance at a horizontal bend is referred to as the setback distance or the clearance distance. It relies upon

- sight distances (OSD, ISD, and OSD),
- curve radius,
- curve length.

17.2.5.2 Vertical Alignment

The vertical alignment of a street comprises angles (straight lines in a vertical plane) and vertical bends. The vertical alignment is normally drawn as a profile, which is a plot with rise as vertical axis and the horizontal distance along the middle line of the street as the horizontal axis.

When two vertical curves connecting two gradients meet, they form either convex (summit curve) or concave (valley curve).

Gradient:

Gradient is the rate of rise or fall along the street length as for the horizontal.

Portrayal of gradient:

The positive inclination or the ascending slope is signified as +n and the negative inclination as −n. The deviation angle N is when two grades meet, the angle which estimates alter of direction and is given by the arithmetical contrast between the two grades

$$\left(n_1 - \left(-n_2 \right) \right) = n_1 + n_2 = \alpha_1 + \alpha_2 \qquad (17.37)$$

Types of gradient:

- **Ruling gradient**: The most extreme inclination with which the originator endeavors to structure the vertical profile of the street is referred to as the ruling gradient or the design gradient. Table 17.3 demonstrates the IRC determinations for the equivalent.
- **Limiting gradient**: This is taken when the ruling inclination results in huge increment in expense of development. On rolling surface and bumpy surfaces it might be every now and again important to achieve limiting gradient. The permissible limits for the limiting gradient values according to IRC are as represented by Table 17.4.
- **Exceptional gradient**: Exceptional gradients are extremely more extreme gradients given at unavoidable circumstances. They ought to be restricted for short stretches not surpassing around 100m at a stretch. In rocky and steep surfaces, progressive excellent inclinations must be isolated by minimum 100m long gentle slope. At hairpin curves, the inclination is confined to 2.5%. Table 17.5 gives

TABLE 17.3
IRC Recommendation for Ruling Gradients for Various Streets

Terrain	Ruling
Plain/rolling	3.3
Hilly	5.0
Steep	6.0

TABLE 17.4
Recommendation for Limiting Gradients for Different Roads

Terrain	Limits
Plain/rolling	5.0
Hilly	6.0
Steep	7.0

TABLE 17.5
Recommendation for Exceptional Inclinations for Various Roads

Terrain	Exceptional
Plain/rolling	6.7
Hilly	7.0
Steep	8.0

the suggested qualities for exceptional gradients for various streets.

- **Critical length of grade**: The most extreme distance of the rising inclination which a stacked truck can work without undue decrease in speed is called critical length of the grade. A speed of 25 km/hour is a sensible esteem.
- **Minimum gradient**: Minimum inclination is accommodated for seepage reason and it relies on the precipitation, soil type, and other site circumstances. At least 1 of every 500 might be adequate for concrete drain and 1 out of 200 for open soil channels are found to give attractive execution.

Grade compensation:

While a vehicle is negotiating a horizontal curve, in the event that there is an angle in addition, there occurs enhanced protection for traction due to both bend and the inclination. In such cases, the aggregate opposition ought not surpass the obstruction because of inclination determined. The grade compensation can be characterized as the decrease in slope at the horizontal curve due to the extra tractive force required because of curve resistance which is expected to set the additional tractive force required at the bend.

specifications:

i. For grades flatter than 4%, grade compensation is not adequate since the loss of tractive force is immaterial.
ii. Grade compensation is given by $(30+R)/R\%$, where R is the horizontal curve radius in meters.
iii. The most extreme grade compensation is restricted to $75/R\%$.

17.2.6 SUMMIT CURVE

Summit curves are vertical bends with upward gradients. They are framed when two angles meet as delineated in Figure 17.6 in any of the accompanying four different ways:

a. when a positive slope meets another positive inclination
b. when positive slope meets a flat inclination
c. when a rising gradient meets a descending angle
d. when a descending slope meets another descending angle

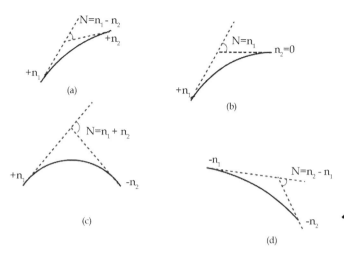

FIGURE 17.6 (a–d) Types of summit curve.

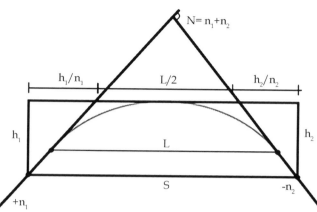

FIGURE 17.7 Length of summit curve ($L < S$).

17.2.6.1 Design Ceasing

- The stopping sight distance or absolute least sight distance ought to be given on these bends and where surpassing is not disallowed, overtaking sight distance or intermediate sight distance ought to be given beyond what many would consider conceivable.
- On the occasion that a quick moving vehicle goes along a summit curve, the centrifugal force will act upward, which is against the gravitational force and a piece of the tyre pressure is alleviated making less distress to the travelers. On the off chance that the curve is furnished with satisfactory sight distance, the length would be adequate to facilitate the vibration because of the progress in slope.
- Circular summit curves are indistinguishable since the radius stays same all through and subsequently the sight distance.
- Transition curves are not attractive since it has fluctuating range of radius thus the sight distance will likewise change.
- The deviation angle gave on summit curves for roadways is extensive, thus the simple parabola is relatively compatible to a circular segment, between similar tangent points. Parabolic curves are simple for calculation and furthermore it had been discovered that it gives great riding solace to the drivers. Simple parabolic curve is favored as summit curve.

17.2.6.2 Length of Summit Curve

The length of the curve is guided by the sight distance thought. A driver ought to be able to stop his vehicle securely if there is a check on the opposite side of the street.

Condition of the parabola is given by $y = ax^2$,

where $a = N/2L$,

where
N is the deviation angle;
L is the inferring length of the bend.

Two circumstances can emerge contingent upon the upward and downward inclinations when the length of the bend is more noteworthy than the sight distance.

Case a: Length of summit curve more noteworthy than sight distance ($L > S$)

$$L = \frac{NS^2}{2\left(\sqrt{h_1} + \sqrt{h_2}\right)^2} \tag{17.38}$$

Case b: Length of the summit curve less than sight distance is given by (Figure 17.7)

$$L = 2S - \frac{\left(\sqrt{2h_1} + \sqrt{2h_2}\right)^2}{N} \tag{17.39}$$

If stopping sight distance is considered,

- height of the driver's eye above the road surface (h_1) is adopted as 1.2 m;
- height of the obstruction over the pavement surface (h_2) is adopted as 0.15 m.

If overtaking sight distance is considered,

- the height of the driver's eye (h_1) as well as the height of the object (h_2) are taken as 1.2 m.

17.2.7 VALLEY CURVE

Valley curve or sag curves are vertical bends with convexity downward. They are framed when two gradients meet as outlined in Figure 17.8 in any of the accompanying four different ways:

a. When a diving slope meets another diving slope.
b. When a diving slope meets a flat gradient.
c. When a diving slope meets a climbing slope.
d. When a climbing slope meets another climbing slope.

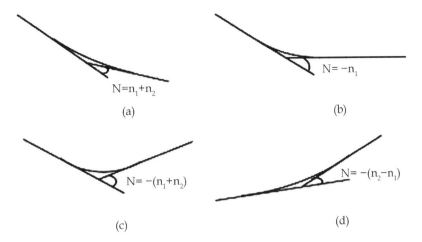

FIGURE 17.8 (a–d) Valley curves.

17.2.7.1 Design Contemplations

There are no restrictions to sight distance at valley curves amid day time. In the absence or insufficiency of road light, the main hotspot for perceivability is with the assistance of headlights amid night. Henceforth valley curves are structured considering the headlight distance.

In valley curves, the outward power will be forced downward alongside the heaviness of the automobile, and henceforth, effect to the automobile will be more. This will bring about snapping of the automobile and cause distress to the travelers. In this manner the most imperative design factors considered in valley curves are as follows:

- **Impact-free movement of vehicles at configuration speed**: For slowing introducing and incrementing the outward radial power acting downward, the best shape that could be given for a valley curve is a transition curve. Cubic parabola is commonly favored in vertical valley bends.
- **Accessibility of ceasing sight distance under headlight of automobiles for late evening drives**: During night, under front light driving condition, sight distance decreases and accessibility of stopping sight distance under head light is critical. The head light sight distance ought to be at any rate equivalent to the ceasing sight distance. There is no issue of overtaking sight distance during the night since alternate vehicles with headlights could be seen from a significant distance.

17.2.7.2 Length of the Valley Curve

The valley curve is made completely transitional by giving two comparative transition curves of equivalent length. The transitional curve is set out by a cubic parabola expressed as $y = bx^3$, where $b = 2N/3L_2$.

The length of the valley transition curve is designed on the basis of two criteria:

i. Comfort criteria:

The length of the valley curve is dependent on the rate of progress of outward increasing speed that will guarantee comfort: Let c is the rate of change of acceleration, R is the least radius of the curve, v is the configuration speed, and t is the time, at that point c is given as

$$c = \frac{v^3}{LR} \qquad (17.40)$$

For a cubic parabola, the estimation of R for length L_s is given by

$$R = \frac{L}{N} \qquad (17.41)$$

Passable rate of change of centrifugal acceleration is restricted to an agreeable dimension of about $0.6\,\text{m/s}^3$.

ii. Safety criteria:

The driver ought to have satisfactory headlight sight distance at any point of the alignment.

Length of the valley curve for front light distance might be resolved for two conditions:

1. Length of valley curve more prominent than the ceasing sight distance.
2. Length of valley curve not exactly the same as the stopping sight distance.

17.2.8 INTERSECTION

It is the most intricate area on any expressway. Crossing point is an area shared by at least two streets. It has some zone assigned for vehicles to swing to various directions to achieve their ideal goals. Its principle work is to give channelization of route direction.

Requirement for study of intersection/crossing point:

Drivers need to settle on split second choice at a crossing point by thinking about this course, convergence geometry, different vehicles, their speed, direction, and henceforth. A little mistake in judgment can cause severe mishaps. It additionally causes postponement and it relies upon type, geometry, and kind of control. In general, overall traffic flow relies upon the execution of convergence. It likewise influences the capacity of the street.

17.2.8.1 Interchange

Interchange is where the traffic between at least two pathways at various levels in the grade-isolated intersections. Diverse sorts are as follows:

i. Diamond interchange:
   - Most prominent type of four-leg interchange found in the urban areas where major and minor streets cross.
   - It can be structured regardless of whether the major road is moderately restricted.
ii. Clover leaf interchange:
   - A four-leg interchange, utilized when two expressways of high volume and speed cross one another.
   - Advantage: it gives finish detachment of traffic. Likewise fast speed at crossing points can be accomplished.
   - Disadvantage: vast region of land is needed.
iii. Trumpet interchange
   - It is a three-leg interchange. On the off chance that one of the legs of the interchange meets a roadway at some edge however does not cross it; at that point the interchange is known trumpet interchange.

17.2.8.2 Channelized Intersection

The vehicles moving toward the crossing point are coordinated to positive ways by islands, stamping, and so on, and this process is known as channelization. There can be channelized or un-channelized crossing points but channelized crossing points give more well-being and effectiveness. On the off chance that no channelizing is given, the driver will have less inclination to decrease the speed while entering the crossing point from the carriageway. The nearness of traffic islands, markings, and so on makes the driver to decrease the speed and turns out to be more cautious while moving the crossing point. A channelizing island fills in as an asylum for people on foot and makes walker crossing more secure.

17.2.8.3 Traffic Rotaries

Rotary crossing points or the roundabouts are extraordinary types of at-grade convergences that spread out for the development of traffic in one way around a central traffic island.

Advantages:

i. Traffic stream is managed to just a single direction of movement, along these lines taking out serious clashes between intersection movements.
ii. The automobiles which are entering the rotary are tenderly compelled to decrease the speed and keep on moving at a slower speed. Accordingly, a greater amount of the vehicles should be halted.
iii. Because of lower speed of negotiation and disposal of extreme clashes, mishaps and their seriousness are significantly less in rotaries.
iv. Rotaries are self-governing and do not require for all intents and purposes any control by police or traffic signals.
v. These are best suited in areas of moderate traffic, particularly with unpredictable geometry, or crossing points with more than three or four approaches.

Disadvantages:

- All the vehicles are compelled to back off and negotiate the convergence. In this manner the combined delay might be a lot greater than the channelized convergence.
- In situations of low traffic also, the vehicles are compelled to decrease their speed.
- These require vast area of land making them expensive at urban regions.
- Since the automobiles are not ceasing, and they accelerate as rotary exits, they are not reasonable when pedestrians are more.

Traffic tasks in a rotary:

The traffic tasks at a rotary are as follows:

- **Diverging operation**: It is a traffic task when the automobile moving one way is isolated into various streams as indicated by their goals.
- **Merging operation**: Merging operation is the inverse of diverging operation. At the point when traffic streams originating from different places and going to a common location are consolidated into a solitary stream then the operation is alluded to as merging.
- **Weaving operation**: Weaving operation is the joined development of both the merging and diverging operations in similar ways.

17.3 TRAFFIC FLOW

Traffic flow is the investigation of the movement of individual drivers and vehicles between two points and the connections they make with each other. Shockingly, contemplating traffic stream is troublesome in light of the fact that driver conduct is something that cannot be anticipated with 100% assurance.

To more readily represent traffic stream, relationships have been built up between the three main fundamental attributes: (1) stream, (2) density, and (3) velocity. These connections help in arranging, design, and activities of roadway facilities.

17.3.1 TRAFFIC FLOW VARIABLES

The traffic stream incorporates a blend of driver and automobile conduct. The driver or human conduct being nonuniform, traffic stream is additionally nonuniform in nature. It is affected not just by the individual attributes of both automobile and human yet; in addition by the manner in which a gathering of such units connects with one another.

The traffic engineer, however to plan and design, accepts that these progressions are inside sure ranges. For instance, if the most extreme reasonable speed of an expressway is 60 km/hour, the entire traffic flow can be accepted to proceed onward at 40 km/hour instead of 100 or 20 km/hour.

The traffic flow parameters can be predominantly named as:

- **Macroscopic characteristics**: Measurement of amount or quality, i.e., stream, density, and speed.
- **Microscopic characteristics**: Measures of partition, i.e., the headway or separation between the automobiles which can be either time or space progress.

17.3.1.1 Speed

Speed is considered as a quality estimation of movement as the drivers and travelers will be more concerned about the speed of the voyage than the design aspects of the traffic. It is characterized as the rate of movement in distance per unit of time. Scientifically velocity v is expressed as

$$v = \frac{d}{t} \tag{17.42}$$

where d is the distance traveled in m and t is the time in seconds.

Speed of various vehicles will shift as for existence. To represent these variations, following sorts of speed can be characterized.

Spot speed:

Spot speed is referred to as the instantaneous speed of an automobile at a predetermined area.

- Used to structure the layout of street like horizontal and vertical bends, cant, and so forth.
- Required for deciding the area and size of signs, structure of signs, and safe speed, and speed zone.
- Used as the fundamental information given by a traffic designer for accident analysis, street maintenance, and blockage.

Spot speed is estimated by utilizing an endoscope or direct timing system or by time-lapse photographic strategies. It very well may be dictated by speeds extricated from video pictures by recording the separation traveled by all automobiles between a specific match of casings.

Running speed:

Running speed is referred to as the normal speed kept over a specific course while the automobile is running and is found by partitioning the length of the course by the time duration the automobile was in movement. The running speed will dependably be greater than or equivalent to the voyage speed, as postponements are not taken into account during computation of running speed.

Journey speed:

Journey speed is referred to as the viable speed of the automobile to traverse between two points divided by the aggregate time taken for the automobile to finish the journey including any ceased time. On the off chance that the voyage speed is not exactly running speed, it demonstrates that the voyage pursues a stop-run condition with authorized acceleration and deceleration. The spot speed here may differ from zero to some most extreme in overabundance of the running speed.

Time mean speed and space mean speed:

Time mean speed is characterized as the normal speed of the considerable number of vehicles passing a point on a roadway over some predetermined time period. Space mean speed is characterized as the normal speed of the considerable number of vehicles occupying a given area of a roadway over some predetermined time period. Time mean speed is point estimation whereas space mean speed is a measure identifying with length of expressway or lane, i.e., the mean speed of vehicles over undefined time frame at a point in space is the time mean speed and the mean speed over a space at a given moment is the space mean speed.

17.3.1.2 Flow

Flow or volume is characterized as the quantity of vehicles which passes across a point on a roadway or a provided path or guidance of a parkway amid an explicit time interim. The estimation is done by checking the quantity of vehicles, n_t, passing a specific point in one path in a characterized period t. At that point the volume of traffic, q (vehicles/hour), is given by

$$q = \frac{n_t}{t} \tag{17.43}$$

Flow is communicated in planning and configuration field accepting a day as the estimation of time.

Variations of volume:

The assortments of volume with respect to time, i.e., month to month, day to day, and hour to hour, and inside an hour are additionally as essential as volume figuring.

- Volume will be better than expected in a lovely monitoring month of summer, yet will be more articulated in rustic than in urban zone.
- The most huge variety is from hour to hour. The pinnacle hour saw amid mornings and night times of Monday to Friday, which is typically 8%–10% of aggregate everyday flow or two to three times the normal hourly volume.

Types of volume estimation:

Since there is extensive variation in the volume of traffic, a few sorts of estimation of volume are ordinarily taken which will average these varieties into a solitary volume count which is utilized in many configuration purposes.

i. **Average annual daily traffic**: The normal 24-hour traffic volume at a specified area over a year, i.e., the aggregate number of vehicles passing the site in a year divided by 365.
ii. **Average annual weekday traffic (AAWT)**: The normal 24-hour traffic volume happening on Monday to Friday over an entire year. It is figured by partitioning the yearly total weekday traffic volume by 260.
iii. **Average daily traffic (ADT)**: A normal 24-hour traffic volume at a predefined area for some time frame less than 365 days. It might be estimated for a half year, a season, a month, 7 days, or as meager as 2 days. An ADT is a legitimate number just for the period over which it was estimated.
iv. **Average weekday traffic (AWT)**: A normal 24-hour traffic volume happening on weekdays for some time frame short of 1 year, like for a month or a season.

The connection between AAWT and AWT is closely resembling to that among AADT and ADT. Volume is estimated utilizing distinctive ways like manual checking, detector/sensor counting, and so on. Volume is treated as the most essential of the considerable number of variables of traffic flow.

17.3.1.3   Density

Density is characterized as the quantity of automobiles possessing a given distance of parkway or path and is commonly communicated as automobiles per km. One can photograph a length of street x, tally the quantity of the number of vehicles, n_x, in one path of the street by then of time and determine density as

$$k = \frac{n_x}{x} \tag{17.44}$$

Using the basic traffic flow qualities like flow, density, and speed, a couple of different variables of traffic volume can be determined. They are:

i. Time headway:

The minuscule character identified with volume is the time headway or simple headway. It is characterized as the time distinction among any two progressive automobiles when they cross a given point. Basically, it includes the estimation of time between the section of one back guard and the following past a given point. On the off chance that all types of headways h in time period, t, over which volume has been estimated are included at that point

$$\sum_{1}^{n_t} h_i = t \tag{17.45}$$

Be that as it may, the volume is characterized as the quantity of automobiles n_t estimated in time interim t, i.e.,

$$q = \frac{n_t}{t} = \frac{1}{h_{av}} \tag{17.46}$$

where h_{av} is the normal headway. In this manner, normal headway is expressed as the converse of volume. Time headway is frequently alluded to as headway.

ii. Distance headway:

It is characterized as the separation among points of two progressive automobiles at some random time. It includes the estimation from a photo, the separation from back guard of lead automobile to back guard of following automobile at a point of time. In the event that all the space headways in distance x over which the density has been estimated are included,

$$\sum_{1}^{n_x} s_i = x \tag{17.47}$$

Be that as it may, the density (k) is the quantity of automobile n_x at a separation of x, i.e.,

$$k = \frac{n_x}{x} = \frac{1}{s_{av}} \tag{17.48}$$

where s_{av} is the average distance headway which is the reverse of density and is some of the time referred as spacing.

iii. Travel time:

Travel time is characterized as the time taken to finish a voyage. As the speed builds, travel time required to achieve the destination also diminishes and the other way round. In this way travel time is contrarily corresponding to the speed. Be that as it may, practically speaking, the speed of an automobile varies after some time and thus the travel time speaks to an average esteem.

Time-space chart:

Time space chart is a helpful apparatus in analyzing the movement of automobiles. It demonstrates the direction of automobiles as a 2D plot. Time space chart can be plotted for a solitary automobile and additionally as various automobiles.

Single vehicle:

Taking one automobile once, investigation can be completed on the situation of the automobile regarding time. This will create a plot which facilitates the connection of its situation on a street extend with respect to the time. This plot therefore will be between separation x and time t. x will be a function of position of the vehicle for each time t along the street stretch. This graphical portrayal of $x(t)$ in a (t, x) plane is a bend which is known as a trajectory or direction. The trajectory provides an instinctive, clear, and finish synopsis of automobile movement in one measurement.

In Figure 17.9a, the separation x goes continues expanding regarding the starting place as the time advances. The automobile is moving in a smooth situation along the street way. In Figure 17.9b, the automobile initially moves with smooth pace, in the pace of achieving position, inverts its direction of movement. In Figure 17.9c, the automobile in-between becomes stationary and keeps the same position.

From the figure, steeply expanding segment of $x(t)$ means a quickly propelling automobile and horizontal portions of $x(t)$ denote a ceased automobile whereas the shallow sections demonstrate a moderate-moving automobile. A straight line signifies constant speed movement and bending sections denote accelerated motion; if the bend is curved downward it indicates increasing speed. However, a bend that is curved upward means deceleration.

Multiple vehicles:

Time-space plot can also be utilized to decide the key variables of traffic stream like speed, density, and volume. It can

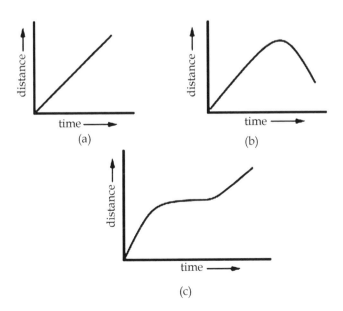

FIGURE 17.9 (a–c) Time space diagram for a single vehicle.

additionally be utilized to locate the determined attributes like space headway and time headway. Density, by definition, is the quantity of automobiles per unit length. An onlooker investigating into the stream can check four automobiles passing the street stretch somewhere at x_1 and x_2 at time t. Henceforth, the density is expressed as

$$k = \frac{4 \text{ vehicles}}{x_2 - x_1} \qquad (17.49)$$

According to the definition, volume is the quantity of automobiles calculated for a specific interim of time. For 6 numbers of automobiles available between the time t_1 and t_2, the volume q is given as

$$q = \frac{3 \text{ vehicles}}{t_2 - t_1} \qquad (17.50)$$

The average of these volumes in an explicit area (i.e. time extending over an interim) is called time means and those taken at a moment over a space interim are named as space means. Also, vertical gap between two back-to-back lines represents space headway. The reciprocal of density generally provides the space headway between automobiles around then.

The horizontal gap between the automobiles denoted by lines gives the time headway. The reciprocal of flow gives the normal time headway between automobiles by then.

17.4 DESIGN OF TRAFFIC FACILITIES

17.4.1 PARKING SYSTEM

17.4.1.1 On-Street Parking

On-road parking implies the automobiles are parked on road sides which are controlled by the government organizations. Regular sorts of on-road parking are as recorded beneath. This characterization depends on the edge in which the vehicles are parked with respect to the street alignment. According to IRC the standard component of an automobile is taken as 5m × 2.5 m and that for a truck it is 3.75 m × 7.5 m.

i. **Parallel parking**: The automobiles are left along the street length. No regression movement is included while parking or unparking the automobile. Consequently, it is the safest parking considering the accident point of view. Be that as it may, it devours the greatest curb length and therefore just a base number of automobiles can be left for a given kerb length. This technique of parking gives slightest check to the on-going traffic out on the road as least street width is utilized. Parallel parking of automobiles is shown in Figure 17.10.

The length available to park N number of vehicles,

$$L = \frac{N}{5.9} \qquad (17.51)$$

FIGURE 17.10 Illustration of parallel parking.

ii. **30° parking**: In 30° parking, the automobiles are left at 30° regarding the street alignment. For this situation, more automobiles can be left contrasted with parallel parking. Additionally, there is better mobility. Deferral caused to the traffic is likewise least.

iii. **45° parking**: As the angle of parking expands, huge quantity of automobiles can be parked. Henceforth contrasted with parallel parking and 30° parking, huge quantity of automobiles can be suited in this type.

iv. **60° parking**: The automobiles are parked at 60° to the heading of street. Progressive number of automobiles can be obliged in this type.

v. **Right angle parking**: In right angle parking or 90° parking, the automobiles are left at right angles to the heading of the street. Even though it requires maximum width kerb length required is practically nothing. The automobiles require complex moving and this may cause extreme accidents which creates hindrance to the street traffic especially when the street width is less. Nonetheless, it can oblige greatest number of automobiles for a given kerb length.

17.4.1.2 Off-Street Parking

In numerous urban focuses, a few regions are solely dispensed for parking which will be at some separation far from the main stream of traffic. Such a parking is alluded to as off-street parking. They might be worked by either public offices or private firms.

Parking prerequisites:

- Residential zone with area less than 300 m² requires just community parking spot.
- For residential plot zone with area between 500 and 1000 m², at least one-fourth of the open area ought to be held for parking.
- Offices require no less than one space for every 70 m² as parking space.

- One parking spot is adequate for 10 situates in an eatery whereas for theatres and film lobbies 1 parking spot for 20 seats is adequate.

17.4.1.3 Effects of Parking

- **Congestion**: Parking consumes impressive road space prompting to the bringing down of the road limit.
- **Accidents**: Careless moving of parking and unparking prompts to accidents which are alluded to as parking accidents. Normal type of parking accidents happens while driving out a car from the parking area, indiscreet opening of the doors of parked cars, and while bringing in the vehicle to the parking lot for parking.
- **Environmental contamination**: They additionally cause contamination to the earth since ceasing and starting of vehicles while parking and unparking result in clamor and fumes. They additionally influence the aesthetic magnificence of the structures since vehicles parked at every accessible space make a feeling that the structure ascends from a plinth of cars.
- **Obstruction to fire-fighting operations**: Parked automobiles might hinder the movement of firefighting automobiles. Now and again they block access to hydrants and access to structures.

17.4.1.4 Parking Statistics

Before taking any measures for the improvement of conditions and to appraise the parking charges additionally, information regarding accessibility of parking spots, degree of its use, and parking request is fundamental. Parking surveys are planned to give all these data. Since the duration of parking changes with various vehicles, the accompanying parking insights are typically vital.

- **Parking accumulation**: It is characterized as the quantity of automobiles left at a given moment of time. Usually this is represented by accumulation curve. Accumulation curve is the chart acquired by plotting the quantity of bays possessed as for time.
- **Parking volume**: Parking volume is the aggregate number of vehicles parked at a given duration of time. This does not give consideration for repetition of vehicles. The real volume of vehicles entered in the parking zone is recorded.
- **Parking load**: Parking load represents the area under the accumulation curve. It can likewise be gotten by basically duplicating quantity of vehicles possessing the parking area at each time interim with the time interim. It is represented as vehicle hours.
- **Average parking duration**: It is the proportion of aggregate vehicle hours to the quantity of vehicles parked.

$$\text{parking duration} = \frac{\text{parking load}}{\text{parking volume}} \quad (17.52)$$

- **Parking turnover**: It is the proportion of quantity of vehicles parked in duration to the number of parking bays accessible. This can be represented as quantity of vehicles per bay per time duration.

$$\text{parking turnover} = \frac{\text{parking volume}}{\text{no. of bays available}} \quad (17.53)$$

- **Parking index**: Parking index is likewise called occupancy or proficiency. It is characterized as the proportion of number of bays possessed in time duration to the aggregate space accessible. It gives a total measure of how successfully the parking spot is used.

$$\text{parking index} = \frac{\text{parking load}}{\text{parking capacity}} \times 100 \quad (17.54)$$

17.4.2 Traffic Signs

Traffic control gadget is moderately utilized for imparting connection among traffic engineer and street clients. The original traffic control gadgets utilized are traffic signals, street markings, traffic signals, and parking control. Distinctive types of traffic signs are regulatory signals, warning signals, and informatory signals.

17.4.2.1 Prerequisites

- Each control gadget must have an explicit reason for the safe and effective operation of traffic stream. The unnecessary gadgets ought not to be utilized.
- It should order consideration from the street clients. For directing consideration, the sign ought to be distinctive and clear. The sign ought to be set so that the driver does not need any additional push to see the signal.
- It ought to pass on an unmistakable, basic meaning. Clarity and simplicity of message is fundamental for the driver to appropriately comprehend the importance in brief time. The utilization of shading, shape, and legend as codes ends up vital in such manner. The legend ought to be kept short and simple so that even a less instructed driver could comprehend the message in less duration.
- Road clients must regard the signals. Regard is told just when the drivers are molded to expect that all gadgets convey significant and vital messages. Overuse, misuse, and confusing messages from the gadgets tend the drivers to disregard them.
- The control gadgets ought to give satisfactory time to legitimate reaction from the street clients. The sign boards ought to be set at a separation to such

an extent that the driver could see it and gets enough time to react to the circumstance. For instance, the STOP sign which is constantly set at the stop line of the convergence ought to be visible for something like one safe stopping sight distance far from the stop line.

17.4.2.2 Communication Tools

Various components are utilized by the traffic specialist to connect with the street clients. Messages are passed on through the accompanying elements.

- **Color**: Usage of various hues for different signs is vital. The most usually utilized hues are red, green, yellow, black, blue, and brown. These are utilized to code certain gadgets and to reinforce explicit messages. Consistent utilization of hues helps the driver in recognizing the nearness of sign board ahead.
- **Shape**: The classifications of shapes typically utilized are circular, triangular, rectangular, and diamond shape. Two extraordinary shapes utilized in rush hour gridlock signs are octagonal shape for STOP sign and utilization of inverted triangle for GIVE WAY (YIELD) signal. Diamond shaped signals are not commonly utilized in India.
- **Legend**: For the simple comprehension by the driver, the legend ought to be short, simple, and explicit with the goal that it does not redirect the consideration of the driver. Symbols are typically utilized as legends so that even an individual unfit to peruse the language will have the capacity to comprehend that. There is no need of it on account of traffic signals and street markings.
- **Pattern**: Two-fold solid and dotted lines are utilized. Each pattern passes on various kinds of significance. The regular and reliable utilization of pattern to pass on data is prescribed with the goal that the drivers get acclimated with the distinctive kinds of markings and can in a split second remember them.

17.4.2.3 Varieties of Traffic Signals

There are a few several traffic signals accessible covering wide assortments of traffic circumstances. They can be arranged into three principle classifications.

- **Regulatory signals**: This signal needs the driver to comply with the signal for the security of other street clients.
- **Warning signals**: These signals are for the security of drivers and guide them to comply with these signals.
- **Informative signals**: These signals provide data to the driver about the offices accessible ahead, and the route and separation to achieve the explicit goals.

Furthermore, uncommon type of traffic signals in particular *work zone signals* is additionally accessible. These kinds of signals are utilized to offer cautioning to the street clients when some construction work is going out and about. They are put just for brief term and will be expelled not long after the work is finished and when the street is taken back to its ordinary condition.

i. Regulatory signs:

These signals are additionally called mandatory signals since it is required that the drivers must comply with these signals. In the event that the driver neglects to obey them, the control office has the privilege to make legitimate move against the driver. These signals have commonly black legend on white foundation. They are circular with red outskirts. The administrative signals might be additionally grouped as follows:

• ROW series: These incorporate two special signals that allot the ROW to the chose methodologies of a crossing point. They are the STOP sign and GIVE WAY sign. For instance, when one minor street and major street meet at a convergence, inclination ought to be given to the vehicles going through the major street. Thus the give way sign board will be put on the minor street to advice the driver on the minor street that he should give path for the vehicles on the significant street. In the event that the two noteworthy streets are meeting, at that point the traffic engineer decides based on the traffic on which approach the sign board must be set up. Stop sign is another case of administrative signs that comes in ROW series which requires the driver to stop the vehicle at the stop line.

• Speed series: The number of speed signals that might be utilized to control the speed of the automobile on street. They include typical speed limit signals, truck speed, minimum speed signals, etc. Speed limit signals are placed to limit the speed of the automobile to a particular speed for a number of reasons. Separate truck speed limits are applied on high-speed roadways where heavy commercial vehicles must be limited to slower speeds than passenger cars for safety reasons. Minimum speed limits are applied on high-speed roads like expressways, freeways, etc., where safety is again a predominant reason.

• Movement series: They contain various signals that influence explicit automobile moves. These incorporate turn signals, alignment signals, exclusion signals, one-way signals, and so on. Turn signals incorporate turn forbiddances and path use control signals. Path use signals make utilization of arrows to determine the movements which all automobiles in the path must take. Turn signals are utilized to securely suit turns in unsignalized crossing points.

• Parking series: They incorporate parking signals which demonstrate not only parking restrictions or confinements but also show areas where parking is allowed, the sort of automobile to be parked, parking duration, and so forth.

• Pedestrian series: They incorporate both legend and image signals. These signals are intended for the security of walkers and incorporate signals showing people on foot-only streets, people on foot-crossing sites, and so forth.

• Miscellaneous: Wide assortment of signals that are incorporated into this class is as follows: a "KEEP OF MEDIAN" signal, signals indicating street terminations, signals confining automobiles carrying hazardous cargo or items, signals demonstrating automobile weight limitations, and so forth.

A few instances of the regulatory signs are shown in Figure 17.11. They incorporate a stop signal, give way signal, signals for no passage, signal demonstrating preclusion for right turn, automobile width limit signal, speed limit signal, and so on.

ii. Warning signals:

Warning signals or cautionary signals offer data to the driver regarding the looming street condition. They encourage the driver to comply with the guidelines. The shading tradition utilized for this type of signals is that the legend will be black in color with white foundation. Shapes like upward triangular or diamond shape with red outskirts are utilized. A part of the precedents for this type of signals are given in Figure 17.12 and incorporate right-hand bend signal board, signals for narrow street, signal showing railway track ahead, and so on.

iii. Informative signal:

Informative signals likewise called guide signals are given to help the drivers to achieve their ideal

FIGURE 17.11 Regulatory signs.

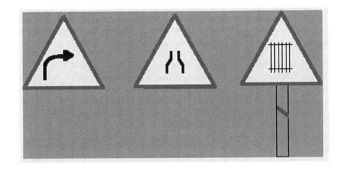

FIGURE 17.12 Warning signs.

goals. These are transcendently implied for the drivers who are new to the area. The guide signals are excess for the users who are accustomed with the area. A few examples for these types of signals are route markers, destination signals, mile posts, service information, and recreational and cultural interest zone signing, and so on.

- Route markers are utilized to distinguish numbered interstate roadways. They have designs that are particular and one of a sort. They are represented as black letters on yellow foundation.
- Destination signals are utilized to demonstrate the direction to the critical destination points, and to stamp critical crossing points. Distance in kilometers is once in a while set to the right side of the destination. They are, usually, rectangular shaped with the long dimension in the horizontal direction and are color coded as white letters in green foundation.
- Mile posts are given to illuminate the driver about the advancement along a route to achieve his goal.
- Service control signals offer data to the driver with respect to different services such as sustenance, fuel, medical help, and so on. They are composed with white letters on blue foundation. Data on historic, recreational, and other cultural zone are given on white letters in brown background.

17.5 PAVEMENT ENGINEERING

Pavement engineering is a branch of civil engineering that utilizes designing methods to design and maintain flexible (asphalt) and rigid (concrete) pavements. This incorporates roads and highways and includes learning of soils, hydraulics, and material properties. Pavement engineering includes new construction as well as rehabilitation and maintenance of existing pavements. Maintenance frequently includes utilizing designing judgment to make upkeep repairs with the highest long-term advantage and least expense.

Sorts of pavements:

1. Flexible pavement
2. Rigid pavement
3. Semi-rigid pavement

Components which influence choice of these pavements:

- Initial cost
- Good material availability
- Maintenance cost
- Environmental conditions
- Industrial waste availability
- Intensity of traffic

Distinction among flexible and rigid pavements is dependent on the manner in which the loads are distributed to the subgrade. Figure 17.13 shows the pavement components.

17.5.1 FLEXIBLE PAVEMENTS

Flexible pavement can be characterized as a blend of asphaltic or bituminous material and aggregates set on a bed of compacted granular material of proper quality in layers over the subgrade. Flexible pavements include water bound macadam (WBM) roads and road stabilized with or without asphaltic toppings.

The structure of flexible pavement is dependent on the rule that for a load of any magnitude, the intensity of a load lessens as the load is transmitted downward from the surface by virtue of spreading over an increasingly larger area, by conveying it deep enough into the ground through progressive layers

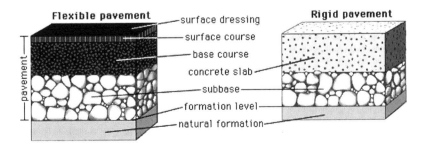

FIGURE 17.13 Rigid pavement and flexible pavement.

of granular material. The strength of subgrade essentially affects the thickness of the flexible pavement.

17.5.1.1 Design Parameters

Thickness is the vertical compressive strain that goes to the subgrade because of the standard axle load to extent of 8.17 kN (8170 kg), if more than this causes perpetual deformation in the type of rutting. The most extreme rutting can be acknowledged in town streets as 50 mm before upkeep and the diagnostic assessment should be possible as indicated by IRC: 37. For rigid and semi-rigid asphalt tensile stress is taken as the configuration criteria.

Traffic:

According to IRC: 37, design traffic ought to be 0.1–2 msa (million standard axles). Weight of commercial automobile (loaded) is considered as 3 tons or more. For design traffic existing traffic and rate of traffic grown are considered. Traffic study ought to be done according to IRC: 9.

Design life:

For provincial streets life span of 10 years is considered. In low-volume streets for the thin bituminous surfacing life span of 5 years is considered.

Computation of design traffic:

$$a = p(1 + r)^{(n+x)} \qquad (17.55)$$

where
- a = number of commercial automobiles per day for design
- p = number of commercial automobiles per day at last count
- r = annual growth rate of commercial traffic
- n = number of years between last count and year of completion of construction
- x = life span in years

17.5.1.2 Parts of Pavement
Subgrade:

i. Act as a foundation and thus provide support to the pavements.
ii. In rural streets, usually the top 30 cm of the cutting or embankment at formation level is considered as subgrade.
iii. Standard proctor compaction ought to attain 100% in subgrade.
iv. In the case of clayey soils, 95% standard proctor compaction and water content of 2% in excess of optimum are recommended.
v. Soil underneath subgrade ought to be compacted to 97% of standard proctor compaction.

NB: If CBR (California Bearing Ratio) is under 2% for 100 mm thickness, at that point least CBR of 10% is to be given to the sub-base for CBR of 2%. On the off chance that CBR is over 15%, sub-base is not required.

Sub-base course:

Chosen materials set on subgrade which is compacted to 98% of heavy compaction. The purpose of sub-base is to circulate the stresses over a wide zone of the subgrade induced by the traffic.

Materials:

- CBR of 15%
- Material passing through 425 μm Indian Standards (IS) sieve
- Liquid limit should be less than 25%
- Plasticity index should be less than 6%

Waste material, for example, fly ash, Iron and steel slag, recycled concrete and municipal waste is additionally utilized. A drainage layer of 100 mm formation width should be provided in case the subgrade is silty or clayey soil and yearly precipitation of zone exceeds 1000 mm.

Base course:

In order to withstand high stress intensities which develop because of traffic under the wearing surface. Various sorts of base course utilized are as follows:

1. Water-bound macadam
2. Crusher-run macadam
3. Dry lean concrete
4. Soft aggregate base course
5. Lime-fly ash concrete

Surface course:

Surface course thickness relies on the volume of traffic and the type of material utilized for it. For gravel roads additional thickness ought to be given because of lost in thickness due to the traffic activity. Bituminous wearing courses must consist of good quality aggregate with aggregate impact esteem not surpassing 30% so as to reduce degradation of aggregates by crushing.

17.5.2 Rigid Pavements

A rigid pavement is built from cement concrete or reinforced concrete slabs. Grouted concrete roads are in the class of semi-rigid pavements.

The design of rigid pavement depends on giving a structural cement concrete slab of adequate strength to resist the loads from traffic. The rigid pavement has rigidity and high modulus of elasticity to distribute the load over a relatively wide area of soil.

In the design of a rigid pavement, the flexural strength of concrete is the central factor and not the strength of subgrade. Because of this property of pavement, when the subgrade deflects underneath the rigid pavement, the concrete slab is able to bridge over the localized failures and regions of inadequate support from subgrade because of slab action.

17.5.2.1 Elements Affecting Pavement Design

There are significant number of variables which influence the pavement design. The variables may be of loading applied, environment conditions, materials used, and so forth which are as stated.

1. Wheel load
2. Axle configuration
3. Contact pressure
4. Speed of vehicle
5. Repetition of loads
6. Type of subgrade
7. Temperature
8. Rainfall

Wheel Load Influence on Pavements:

Wheel load on asphalt is a vital factor to decide the asphalt thickness to be attained. By giving sufficient thickness, the load coming from wheels does not influence the subgrade soil. The wheel load acts at specific point on asphalt and causes deformations. In the event that the automobile contains dual wheels on one side of pivot, then convert it into equivalent single wheel load (ESWL). Dual-wheeled axle vehicles control the contact pressure within the limits.

Axle Configuration:

Axles are the vital part of the automobiles which empower the wheels to turn while moving. By giving multiple axles, vehicle can carry more loads. So, the axle load also influences the pavement configuration. In the layer theory of flexible pavement design wheels on one side of axles are considered to structure the pavement. Also in the plate hypothesis of rigid pavement design, wheels on both the sides are considered.

Tire Contact Pressure on Pavement:

When the automobile is moving on asphalt, a pressure is created between the tyre and pavement. On the off chance that the tyre is low-pressure tyre, then contact pressure will be more prominent than tyre pressure. If it is high-pressure tyre, then contact pressure will be not as much as tyre pressure. Even though the contact area is elliptical, for calculation, easiness circular shape is considered.

Speed of vehicle:

On the off chance that the automobile is moving at creep speed then also damage occurs to the asphalt. In the event that automobile speed is gradually increased then it will cause little strains on the asphalt.

Repetition of Loads:

The wheel loads are repeated all the time because of this some deformation happens on the asphalt. Total deformation occurring is the total of all wheel loads acting on it. So, frequency of load is also considered in the pavement design. Also, single axle with dual wheels conveying 80 kN load is considered as standard axle.

Type of subgrade:

Different tests like CBR, triaxial, and so on will help to evaluate the nature of subgrade. From this we can adopt the expected thickness of pavement. If subgrade soil is poor then the pavement gets damaged easily.

Temperature Effects on Pavements Design:

- In the instance of asphalt roads, temperature influences the resilient modulus of surface course. In extremely hot condition, asphalt layers lose their stiffness. At low temperature, asphalt layers become brittle and splits are formed.
- In case of rigid pavement, temperature stresses are produced. Curling of concrete is likewise possible because of the temperature variation at top and bottom layers of pavement.

Rainfall:

Dampness variations or precipitation from rain influences the profundity of groundwater table. Good drainage facilities ought to be accommodated for good strength and support. The ground water table in any event should be below 1 m from the pavement surface.

17.5.2.2 Flexible Pavement Design

A flexible pavement structure is normally made out of a few layers of materials. The load is distributed from one layer to the layer beneath it and is spread uniformly which is then passed to the following layer underneath. Thus, decreasing the stresses which are most extreme at the top layers and least at the top of subgrade. Hence, layers are normally organized in the order of decreasing load-bearing capacity with the most noteworthy load-bearing capacity material (and most costly) on the top and the least load-bearing capacity material (and least costly) on the bottom.

Design procedures:

For flexible pavements, the primary consideration of structural design is the determination of thickness and composition of different layers. The fundamental design elements are stresses because of traffic load and variations in temperature. Two strategies of flexible pavement structural design are common nowadays, namely, empirical design and mechanistic empirical design.

1. Empirical design:
   An empirical methodology depends on the aftereffects of experience. Some of them are either founded on physical properties or strength parameters of soil subgrade. An empirical analysis approach for flexible pavement configuration can be done with or without conducting a soil strength test.
2. Mechanistic-empirical design:
   Empirical-mechanistic method of design is based on the mechanics of materials that relate input, such as wheel load, to an output or pavement response.

In the pavement design, the responses are the stresses, strains, and deflections within a pavement structure and the physical causes are the loads and material properties of the pavement structure. The relationship between these phenomena and their physical causes are typically described using some mathematical models. Along with this mechanistic approach, empirical elements are used when defining what values of the calculated stresses, strains, and deflections result in pavement failure. The relationship between physical phenomena and pavement failure is described by empirically derived equations that compute the number of loading cycles to failure.

17.6 PAVEMENT ANALYSIS

17.6.1 TRAFFIC AND LOADING

There are three distinct methodologies for considering vehicular and traffic attributes, which influence the pavement design.

- **Fixed traffic**: In this methodology pavement thickness is expressed by considering the single load and not by considering the number of load repetitions. The heaviest wheel load foreseen is utilized for configuration reason. Since the technique is old it is rarely used nowadays.
- **Fixed automobile**: In this methodology the pavement thickness is administrated by the number of repetitions of a standard axle load. In the event that the axle load is certifiably not a standard one, it must be converted to an equivalent axle load by the number of repetitions of given axle load and its equivalent axle load factor (EALF).
- **Variable traffic and automobile**: Both traffic and automobile are considered exclusively, so there is no compelling reason to assign an equivalent factor for each axle load. The loads can be divided into a number of groups and the stresses, strains, and deflections under each load group can be determined separately, and utilized for design purposes. The traffic and loading components to be considered incorporate axle loads, load repetitions, and tyre contact area.

17.6.2 EQUIVALENT SINGLE-WHEEL LOAD

Automobiles can have numerous axles which will distribute the load into various axles, and thus to the asphalt through the wheels. ESWL is the single-wheel load having a similar contact pressure, which produces same estimation of most extreme stress, deflection, tensile stress, or contact pressure at the desired profundity. The strategy of finding the ESWL for equivalent stress criteria is given underneath. This is a semi-rational technique, known as Boyd and Foster strategy, in view of the accompanying assumptions:

- equivalency concept depends on equivalent stress;
- contact area is circular;
- influence angle is 45°; and
- soil medium is elastic, homogeneous, and isotropic half space.

The ESWL is given by

$$\log_{10} \text{ESWL} = \log_{10} P + \frac{0.301 \log_{10}\left(\dfrac{z}{d/2}\right)}{\log_{10}\left(\dfrac{2S}{d/2}\right)} \quad (17.56)$$

where

P is the wheel load,
S is the middle-to-focus distance between the two wheels,
d is the clear distance between two wheels, and
z is the ideal depth.

Since the design of flexible asphalts is by layered hypothesis, just the wheels on one side should have been considered. The design of rigid asphalt is by plate hypothesis and henceforth the wheel loads on both sides of axle should be taken into consideration.

- **Legal axle load**: The greatest permitted axle load on the roads is called legal axle load. For interstate roadways the maximum legal axle load in India, determined by IRC, is 10 tons.
- **Standard axle load**: It is a single axle load with dual wheel conveying 80 kN load and the structure of asphalt depends on the standard axle load.
- **Repetition of axle loads**: The deformation of asphalt because of a single application of axle load might be little yet because of repeated application of load there would be accumulation of unrecovered or permanent deformation which results in failure of asphalt. If the asphalt structure fails with N_1 number of repetition of load W_1 and for the similar failure criteria on the off chance that it requires N_2 number of repetition of load W_2, then $W_1 N_1$ and $W_2 N_2$ are viewed as equivalent.
- **Equivalent axle load factor**: An EALF defines the damage per pass to asphalt by the ith type of axle relative to the damage per pass of a standard axle load. While finding the EALF, the failure criterion is important. Two types of failure criteria are commonly adopted: fatigue cracking and rutting.

17.6.3 STRENGTH CHARACTERISTICS OF PAVEMENT MATERIALS

Pavements are an aggregation of materials. These materials, their related properties, and their interactions decide the properties of the resultant asphalt.

Subgrade soil:

Subgrade soil is the basic piece of the road pavement structure as it gives the support to the pavement from underneath. Undistorted soil underneath the pavement is called natural subgrade. Compacted subgrade is the soil compacted by controlled development of heavy compactors. The subgrade soil and its properties are vital in the design of pavement structure. The fundamental function of the subgrade is to give satisfactory support to the pavement, and for this, the subgrade should have adequate stability under worse climatic and loading conditions.

Desirable properties:

The desirable properties of subgrade soil as a roadway material are as follows:

  i. Stability
 ii. Incompressibility
iii. Permanency of strength
 iv. Minimum changes in volume and stability under adverse conditions of weather and ground water
  v. Good drainage
 vi. Ease of compaction

17.6.3.1 Tests on Soil

The tests conducted to assess the quality properties of soils may be extensively separated into three categories:

* Shear tests
* Bearing tests
* Penetration tests

  i. CBR test:
    CBR test was created by the California Division of Highway as a technique for classifying and assessing soil subgrade and base course materials for flexible pavements. It is a penetration test wherein a standard piston, having a cross-sectional area of 3 in.^{2} (or 50 mm diameter), is utilized to penetrate the soil at a standard rate of 1.25 mm/minute. The pressure up to a penetration of 12.5 mm and its proportion to the bearing estimation of a standard crushed rock is named as the CBR.
 ii. Plate-bearing test:
    Plate-bearing test is utilized to assess the support capability of subgrades, bases, and, in some cases, complete pavement. Information from the tests is appropriate for the design of both flexible and rigid pavements. In plate-bearing test, a compressive stress is applied to the soil or pavement layer through rigid plates relatively large size and the deflections are measured for different stress values. The deflection level is commonly limited to a low value, in the order of 1.25–5 mm and thus the deformation caused might be halfway elastic and halfway plastic because of compaction of the stressed mass with negligible plastic deformation.

The plate-bearing test has been devised to assess the supporting power of subgrades or any other pavement layer by utilizing plates of bigger diameter. The plate-bearing test was initially intended to discover the modulus of subgrade reaction in the Westergaard's analysis for wheel load stresses in cement concrete pavements.

17.7 PAVEMENT DESIGN

17.7.1 IRC Method of Design of Flexible Pavements

17.7.1.1 Design Criteria

The flexible pavements have been demonstrated as a three-layer structure, and stresses and strains at critical locations have been registered utilizing the linear elastic model. To give legitimate thought to the aspects of execution, the accompanying three sorts of pavement distress resulting from cyclic application of traffic loads are considered:

1. Vertical compressive strain at the highest point of the subgrade which can cause subgrade deformation bringing about permanent deformation at the asphalt surface.
2. Horizontal elastic strain or stress at the base of the bituminous layer which can cause crack of the bituminous layer.
3. Pavement deformation inside the bituminous layer.

While the perpetual deformation inside the bituminous layer can be controlled by meeting the mix design necessities, thickness of granular and bituminous layers is chosen utilizing the scientific structure approach so that strains at the critical points are inside the allowable limits. For calculating tensile strains at the base of the bituminous layer, the stiffness of dense bituminous macadam (DBM) layer with 60/70 bitumen has been utilized in the analysis.

17.7.1.2 Design Methodology

In view of the execution of existing designs and utilization of scientific methodology, basic design charts and a list of asphalt designs are included into the code. The asphalt designs are given for subgrade CBR esteems going from 2% to 10% and design traffic extending from 1 to 150 msa for a normal yearly asphalt temperature of 35°C. Utilizing the accompanying simple input variables, proper designs could be decided for the given traffic and soil quality.

* Design traffic regarding aggregate number of standard axles; and
* CBR estimation of the subgrade.

  i. Design traffic:
    The technique takes into account the traffic as far as the aggregate number of standard axles (8160 kg) to be conveyed by the asphalt during the life span. This requires the accompanying data:

1. Initial traffic as far as commercial vehicles per day (CVPD)
2. Rate of increase in traffic amid the design life
3. Life span in years
4. Vehicle damage factor (VDF)
5. Distribution of business traffic over the carriage way.

Initial traffic:

Initial traffic is characterized as far as business vehicles per day (CVPD). For the basic structure of the pavement, just business vehicles are viewed as accepting laden load of 3 tons or greater and their pivot loading will be considered. Gauge of the initial everyday normal traffic stream for any street should ordinarily be based on 7-day 24-hour classified traffic counts (ADT- Average Daily Traffic). In the event of new streets, traffic assessments can be made based on potential land use and traffic on existing courses in the zone.

Traffic growth rate:

Traffic growth rates can be evaluated

i. by considering the previous trends of traffic development, and
ii. by building up econometric models. In the event that sufficient information is not accessible, it is recommended that an average yearly development rate of 7.5% might be adopted.

Design life:

It is suggested that asphalt for arterial streets like NH, SH ought to be proposed for an actual existence of 15 years, EH and urban streets for 20 years, and different other classes of streets for 10–15 years.

Vehicle Damage Factor:

The VDF is a multiplier for changing the quantity of business vehicles of various axle loads and axle configurations to the quantity of standard axle-load repetitions. It is characterized as a proportionate number of standard axles per business vehicle. The axle load equivalency factors are utilized to change over various axle load repetitions into equivalent standard axle load repetitions. For these equivalency factors, refer IRC:37 2001.

Vehicle distribution:

A realistic evaluation of distribution of commercial traffic by direction and by path is important as it specifically influences the total equivalent standard axle load application utilized in the design.

i. **Single lane streets**: Traffic will be more channelized on single roads than two path roads, and to take into account this concentration of wheel load repetitions, the design ought to be founded on aggregate number of business vehicles in both directions.

ii. **Two-lane single carriageway streets**: The structure ought to be based on 75% of the commercial vehicles in both directions.
iii. **Four-lane single carriageway streets**: The structure ought to be based on 40% of the aggregate number of commercial vehicles in both directions.
iv. **Dual carriageway roads**: The design of dual two-lane carriageway roads ought to be based on 75% of the quantity of commercial vehicles in each direction. For dual three-lane carriageway and dual four-lane carriageway the distribution factor will be 60% and 45%, respectively.

Design traffic:

The design traffic is considered as far as the total number of standard axles in the path conveying most extreme traffic during the structure life of the street.

$$N = \frac{365 \times \left[(1+r)^n - 1 \right]}{r} \times A \times D \times F \qquad (17.57)$$

where

N is the cumulative number of standard axles to be catered for the design in terms of million standards axle (msa),
A is the initial traffic in the year of completion of construction in terms of the number of commercial vehicles per day,
D is the path distribution factors,
F is the VDF, n is the design life in years, and
r is the yearly growth rate of commercial vehicles ($r = -0.075$ if development rate is 7.5% per annum).

The traffic in the time of finishing is assessed utilizing the accompanying equation:

$$A = P(1+r)^x \qquad (17.58)$$

where

P is the quantity of business vehicles according to last tally, and
x is the quantity of cars between the last tally and the time of finishing between the last tally and the time of completion of the project.

17.7.1.3 Pavement Thickness Design Charts

For the design of pavements to convey traffic in the scope of 1–10 msa, chart 1 is used and for traffic in the scope of 10–150 msa, chart 2 of IRC:37 2001 is used. The design curves relate asphalt thickness to the combined number of standard axles to be persisted the design life for different subgrade CBR esteems extending from 2% to 10%. The design charts will represent the aggregate thickness of the asphalt for the above inputs. The aggregate thickness comprises granular sub-base, granular base, and bituminous surfacing.

17.7.1.4 Composition of Pavement

- Sub-base:

    Sub-base materials consist of natural sand, gravel, laterite, brick metal, crushed stone, or blends thereof meeting the recommended grading and physical necessities. The sub-base material ought to have a minimum CBR value of 20% and 30% for traffic up to 2 msa and traffic exceeding 2 msa separately. Sub-base as a rule comprises granular or WBM and the thickness should not be under 150 mm for configuration traffic under 10 msa and 200 mm for configuration traffic of 1:0 msa or more.

- Base:

    The suggested designs are for unbounded granular bases which contain ordinary WBM or wet mix macadam or equal confirming to MOST determinations. The materials ought to be of good quality with least thickness of 225 mm for traffic up to 2 msa and 150 mm for traffic exceeding 2 msa.

- Bituminous surfacing:

    The surfacing comprises a wearing course or a binder course in addition to wearing course. The most generally utilized wearing courses are surface dressing, open-graded premix carpet, mix seal surfacing, semi-dense bituminous concrete, and bituminous concrete. For binder course, MOST determines, it is attractive to utilize bituminous macadam for traffic up to 5 msa and DBM for traffic more than 5 msa.

17.7.1.5 Group Index Technique

Group Index technique of flexible pavement configuration is an experimental strategy which relies on physical properties of the soil sub-grade.

By sieve analysis test we can decide the Group index value of soil subgrade from underneath stated equation

$$GI = 0.2a + 0.005ac + 0.01bd \qquad (17.59)$$

where

a = percentage of soil passing 0.074 mm sieve in excess of 35%, not exceeding 75.

b = percentage of soil passing 0.074 mm sieve in excess of 15%, not exceeding 55.

c = Liquid limit greater than 40.

d = Plasticity index greater than 10.

17.7.1.6 Data Required for Flexible Pavement Design

1. Group index of soil subgrade:

    Group index value range of various soils is given below:

    i. For good soil: 0–1

    ii. For fair soil: 2–4

    iii. For poor soil: 5–9

    iv. For exceptionally poor soil: 10–20

2. Traffic volume:

    It is the proportion of yearly average everyday traffic, peak-hour traffic. It is named by commercial vehicles/day or CVPD. It is characterized in three classifications based on number of vehicles per day.

    Assuming number of vehicles per days is

    i. <50: light traffic

    ii. 50–300: medium traffic

    iii. >300: heavy traffic

17.7.2 Rigid Pavement Design

Rigid pavements do not flex a lot under loading-like flexible pavements. They are developed utilizing cement concrete. For this situation, the load-carrying capacity is primarily because of rigidity and high modulus of elasticity of the slab. H. M. Westergaard is viewed as the pioneer in giving the normal treatment of the rigid pavement experimental investigations.

17.7.2.1 Modulus of Subgrade Reaction

Westergaard considered the rigid pavement slab as a thin elastic plate laying on soil subgrade, which is accepted as a thick liquid. The upward response is thought to be corresponding to the deformation. Based on this assumption, Westergaard characterized a *modulus of subgrade response K* in kg/cm^3 given by $K = \dfrac{p}{\Delta}$, where Δ is the displacement level taken as 0.125 cm and p is the pressure sustained by the rigid plate of 75 cm diameter at a deflection of 0.125 cm.

17.7.2.2 Relative Stiffness of Slab to Subgrade

A specific degree of resistance to slab deflection is offered by the subgrade. The slab deflection is an immediate estimation of the magnitude of the subgrade pressure. This pressure deformation qualities of rigid asphalt lead Westergaard to characterize the term *radius of relative stiffness l* in cm which is given by the equation

$$l = \sqrt[4]{\frac{Eh^3}{12K\left(1-\mu^2\right)}} \qquad (17.60)$$

where

E = modulus of elasticity of cement concrete in kg/cm^2 (3.0×10^5),

μ = Poisson's ratio of concrete (0.15),

h = thickness of slab in cm,

K = modulus of subgrade response.

17.7.2.3 Critical Load Positions

There are three typical areas to be specific: the *interior, edge* and *corner*, where varying conditions of slab continuity exist. These areas are named as critical load positions.

17.7.2.3.1 Equivalent Radius of Resisting Section

At the point where the inside point is loaded, just a little area of the pavement is opposing the bending moment of the plate.

Westergaard's gave a connection relation for equivalent radius of the resisting section in cm in the equation

$$b = \sqrt{1.6a^2 + h^2} - 0.675h, \quad \text{if } a < 1.724h \quad (17.61)$$

$$b = a, \quad \text{otherwise}$$

where
   a = radius of wheel load distribution in cm,
   h = thickness of slab in cm.

17.7.3 WHEEL LOAD STRESSES—WESTERGAARD'S STRESS EQUATION

The cement concrete slab is thought to be homogeneous and to have uniform flexible properties with vertical subgrade response being corresponding to the deflection. Westergaard created connections for the stress at interior, edge, and corner regions, signified as $\sigma_i, \sigma_e, \sigma_c$ in kg/cm² separately and given by the condition

$$\sigma_i = \frac{0.316P}{h^2}\left[4\log_{10}\left(\frac{l}{b}\right) + 1.069\right] \quad (17.62)$$

$$\sigma_e = \frac{0.572P}{h^2}\left[4\log_{10}\left(\frac{l}{b}\right) + 0.359\right] \quad (17.63)$$

$$\sigma_c = \frac{3P}{h^2}\left[1 - \left(\frac{a\sqrt{2}}{l}\right)^{0.6}\right] \quad (17.64)$$

where
   h = thickness of slab in cm,
   P = load on wheel in kg,
   a = radius of the wheel load distribution in cm,
   l = radius of the relative stiffness in cm,
   b = radius of the resisting section in cm.

17.7.3.1 Temperature Stresses
Temperature stresses are produced in cement concrete asphalt because of variety in slab temperature. This is caused by

   i. *Daily variation* bringing about a temperature gradient across the thickness of the slab;
   ii. *Seasonal variation* bringing about an overall change in the slab temperature.

The previous cause results in warping stresses and the later in frictional stresses.

   i. Warping stress:
      The warping stress occurring at the interior, edge, and corner regions, denoted as $\sigma_{t_i}, \sigma_{t_e}, \sigma_{t_c}$ in kg/cm² separately and given by the equation

$$\sigma_{t_i} = \frac{E\grave{o}t}{2}\left(\frac{C_x + \mu C_y}{1 - \mu^2}\right) \quad (17.65)$$

$$\sigma_{t_e} = \text{Max}\left(\frac{C_x E\grave{o}t}{2}, \frac{C_y E\grave{o}t}{2}\right) \quad (17.66)$$

$$\sigma_{t_c} = \frac{E\grave{o}t}{3(1-\mu)}\sqrt{\frac{a}{l}} \quad (17.67)$$

where
   E = modulus of elasticity of concrete in kg/cm² (3×10^5),
   $\grave{o}$ = thermal coefficient of concrete per °C (1×10^{-7}),
   t = temperature difference between the top and base of the slab,
   C_x and C_y = coefficient based on L_x/l in the desired direction,
   L_y/l = right angle to the desired direction,
   μ = Poisson's ration (0.15),
   a = radius of the contact area,
   l = radius of the relative stiffness.

   ii. Frictional stresses:
      The frictional stress σ_f in kg/cm² is expressed as

$$\sigma_f = \frac{WLf}{2 \times 10^4} \quad (17.68)$$

where
   W = unit weight of concrete in kg/cm² (2400),
   f = coefficient of sub grade friction (1.5),
   L = slab length in m.

   iii. Combination of stresses:
      The combined impact of the various stresses gives rise to the accompanying three basic cases:
      • Summer, mid-day: The basic stress for edge region is expressed as

$$\sigma_{\text{critical}} = \sigma_e + \sigma_{t_e} - \sigma_f \quad (17.69)$$

      • Winter, mid-day: The critical combination of stress for the edge region is expressed as

$$\sigma_{\text{critical}} = \sigma_e + \sigma_{t_e} + \sigma_f \quad (17.70)$$

      • Mid-nights: The critical combination of stress for the corner region is expressed as

$$\sigma_{\text{critical}} = \sigma_c + \sigma_{t_c} \quad (17.71)$$

17.7.4 DESIGN OF JOINTS

Expansion joints:
   The reason for the expansion joint is to permit the expansion of the pavement because of increase in temperature

regarding construction temperature. The structural contemplations are as follows:

- Provided along the longitudinal course.
- Design includes finding the joint spacing for a given expansion joint thickness (say 2.5 cm determined by IRC) subjected to some most extreme spacing (say 140 according to IRC).

Contraction joints:

Contraction joints are provided to allow the contraction of slab on account of fall in slab temperature underneath the construction temperature. The structural contemplations are as follows:

- Movement is confined by subgrade friction
- Design includes slab length which is expressed as

$$L_c = \frac{2 \times 10^4 S_c}{W \cdot f} \tag{17.72}$$

where
S_c = allowable stress in cement concrete in tension (0.8 kg/cm²),
W = unit weight of the concrete (2400 kg/cm³),
f = coefficient of subgrade friction (1.5).

- Steel reinforcements with a maximum spacing of 4.5 m as per IRC can be used.

17.7.4.1 Dowel Bars

Dowel bar is used to successfully exchange the load between two concrete slabs and to keep the two slabs in same level. The dowel bars are given toward the direction of traffic (longitudinal). The structural contemplations are as follows:

- Mild steel rounded bars,
- Bonded on one side and free on other side.

Bradbury's analysis:

Bradbury's analysis gives load-exchange capacity of single dowel bar in shear, bending, and bearing as pursues:

$$P_s = 0.785 d^2 F_s \tag{17.73}$$

$$P_f = \frac{2 d^3 F_f}{L_d + 8.8\delta} \tag{17.74}$$

$$P_b = \frac{F_b L_d^2 d}{12.5 (L_d + 1.5\delta)} \tag{17.75}$$

where
P = load transfer capacity of a single dowel bar in shear s, bending f, and bearing b,
d = diameter of the bar in cm,
L_d = length of the embedment of dowel bar in cm,
δ = joint width in cm,

F_s, F_f, F_b = permissible stress in shear, bending, and bearing for the dowel bar in kg/cm².

Design methodology:

Step 1: Find the length of dowel bar inserted in slab L_d.

$$L_d = 5d \sqrt{\frac{F_f (L_d + 1.5\delta)}{F_b (L_d + 8.8\delta)}} \tag{17.76}$$

Step 2: Find the load exchange capacities P_s, P_f, and P_b of single dowel bar with the L_d.

Step 3: Assume load capacity of dowel bar is 40% wheel load, find the load capacity factor f as

$$\max \left\{ \frac{0.4P}{P_s}, \frac{0.4P}{P_f}, \frac{0.4P}{P_b} \right\} \tag{17.77}$$

Step 4: Dowel bar spacing.

- Effective separation up to which effective load transfer occur is given by $1.8l$, where l is the radius of relative stiffness.
- Assume a direct variation of capacity factor of 1.0 under load to 0 at $1.8l$.
- Assume dowel spacing and find the capacity factor for the above spacing.
- Actual capacity factor ought to be more noteworthy than the required capacity factor.
- If not, do one more trial with the new spacing.

17.7.4.2 Tie Bars

As opposed to dowel bars, tie bars are not load transfer gadgets but rather serve as a means to tie two sections of slab. Consequently, tie bars should be deformed or hooked and ought to be solidly anchored into the concrete for proper functioning. These are little than dowel bars and set at greater interims. These are given crosswise over longitudinal joints.

Design methodology:

Step 1: Diameter and spacing:
The diameter and spacing are first discovered by equating the aggregate subgrade friction to the aggregate tensile stress for a unit length (1 m). Thus the area of steel per meter in cm² is expressed as

$$A_s \times S_s = b \times h \times W \times f \tag{17.78}$$

$$A_s = \frac{bhWf}{100 S_s} \tag{17.79}$$

where
b = width of pavement panel in m,
h = profundity of the pavement in cm,
W = unit weight of the concrete (assume 2400 kg/cm²),
f = frictional coefficient (assume 1.5),

S_s = permissible working tensile stress in steel (assume 1750 kg/cm^2).

Assume 0.8–1.5 cm diameter bars for the design.

Step 2: Length of tie bar:

Length of the tie bar is double the length expected to create bond stress equivalent to the working tensile stress and is expressed as

$$L_t = \frac{dS_s}{2S_b} \qquad (17.80)$$

where

d = diameter of the bar,
S_s = allowable tensile stress in kg/cm^2,
S_b = allowable bond stress (for plain bars: 17.5kg/cm^2 and for deformed bars 24.6 kg/cm^2).

17.8 HIGHWAY MAINTENANCE

Protecting and keeping each kind of roadway, roadside, structures as nearly as possible in its original condition as built or as subsequently enhanced and the operation of interstate roadway facilities and administrations to give agreeable and safe transportation is called maintenance of highways.

The various maintenance functions include

1. Surface maintenance
2. Roadside and drainage maintenance
3. Shoulder and approaches maintenance
4. Snow and ice control
5. Bridges maintenance
6. Traffic service
   • Highway maintenance is closely related to the nature of construction of original street.
   • Insufficient pavement or base thickness or inappropriate construction of these components soon results in costly patching or surface repair.
   • Shoulder care turns into a difficult issue where limited path force heavy vehicle to go with one set of wheels off the pavement.
   • Improperly planned drainage facilities, mean erosion, or deposition of material and costly cleaning operation or other restorative measures.
   • Sharp ditches and steep slopes require manual maintenance as compared to cheap maintenance of flatter ditch and soil by machine.
   • In snowy country, inappropriate area greatly low fills and narrow cuts leave no room for snow storage, creating great degree troublesome snow evacuation issues.

17.8.1 FAILURES OF FLEXIBLE PAVEMENTS

Distinctive kinds of failures experienced in flexible pavements are stated below.

1. Alligator cracking or map cracking (Fatigue)
2. Consolidation of pavement layers (Rutting)
3. Shear failure cracking
4. Longitudinal cracking
5. Frost heaving
6. Lack of binding to the lower course
7. Reflection cracking
8. Formation of waves and corrugation
9. Bleeding
10. Pumping

Alligator Cracking:

These are caused due to the overall movement of material or failure of material of the pavement payer on the pavement surface. This might be also caused by the repeated use of the heavy wheel loads bringing about weakness failure because of the moisture variations bringing about swelling and shrinkage of subgrade and other pavement materials. Cracking of the surface course in this pattern also occurs due to localized weakness of the underlying base course.

Shear Failure:

These are caused due to the weakness of the pavement mixtures or the shearing resistance being low due to insufficient stability of excessive heavy loading. The shear failure produces upheaval of pavement material resulting in fracture or cracking.

Frost Heaving:

Frost heaving is most prone on cold climatic zones. At the point when the water present in the pores of the layers transforms into ice, it results in swelling of the ice, and hence upheaval of the zone affected by the frost occurs. There will be no depression formed in frost heaving.

Longitudinal Cracking:

Longitudinal cracking may happen due to the differential settlement of the pavement coming about because of the differential volume change. Zones very close to the pavement edge are more prone to dampness and consequently it might swell more as contrast with the inside region of the pavement subgrade. This will create differential volume change of the pavement and therefore may prompt to the longitudinal cracking of the pavement. Usually, these longitudinal cracks navigate through the full asphalt thickness.

Consolidation Failure:

Consolidation failure occurs because of the continuous action of wheel load along the wheel path resulting in the formation of the ruts along the wheel path.

Wearing of the Surface:

Generally wearing of the surface might be caused because of the utilization of second rate material or because of the absence of the interlocking of the surface layer with the base layers. Absence of the interlocking might be a consequence of the nonuse of the prime and tack coat. Explicitly if there should be an occurrence of the overlays over the current cement concrete pavements or in case if the soil cement roads have poor interlocking.

Reflection cracks:

Reflection cracks are shaped in the overlays laid over the current cement concrete pavements. In such overlays, if any cracks are there in the current cement concrete pavements, this will be reflected in the surface layer also. These cracks are referred to as reflection cracks.

17.8.2 FAILURES IN RIGID PAVEMENTS

Failures happen because of two factors:

1. Deficiency of asphalt materials.
2. Structural insufficiency of the pavement framework.

17.8.2.1 Deficiency of Pavement Materials

Soft aggregates:

- Poor workmanship in joint development
- Poor joint filler or sealer material
- Poor surface wrap-up
- Improper and inadequate curing

Defects caused by above causes are as follows:

- Disintegration of cement concrete
- Formation of fracture
- Spalling of joints
- Poor riding surface
- Slippery surface
- Formation of shrinkage cracks
- Ingress of surface water and further dynamic failures.

17.8.2.2 Structural Inadequacy of Pavement System

Deficient subgrade support and reduced pavement thickness would be the real reason for the development of the structural cracking in pavements. Causes and sorts of failures are listed below:

- Insufficient pavement thickness
- Insufficient subgrade support and poor subgrade soil
- Inappropriate dispersing of joints

The above would offer ascent to the failures of the accompanying sorts:

- Cracking of slab corners
- Cracking of pavement longitudinally
- Settlement of slabs
- Widening of joints
- Mud Pumping

Typical pavement failures occurring in rigid pavements are as follows:

- Scaling of cement concrete
- Shrinkage cracks
- Spalling of joints
- Warping cracks
- Mud Pumping

17.8.2.3 Structural Cracking

Scaling of Cement Concrete:

At whatever points there are deficiencies in the concrete mix or nearness of some chemical undesirable materials may result into the scaling of cement concrete. Likewise due to excessive vibrations on the cement concrete mix, cement mortar comes to the top amid the construction, and in this manner, with use, the cement mortar gets abraded exposing the aggregates in the mix. This makes the pavement surface rough and pitiful in appearance.

Shrinkage Cracks:

Shrinkage crack occurs in cement concrete pavement during the curing stage after the construction stage. It may occur in longitudinal as well in transverse direction.

Spalling of Joints:

In some cases, when preformed filler materials are put during the casting of pavement slabs, the placement is by one way or another dislocated and filler is thus set at an angle. The concreting is finished without observing this faulty arrangement of the filler material. Consequently this results in an overhang of a concrete layer on the top and the joint later on shows more cracking and subsidence.

Warping Cracks:

In the event that the joints are not well intended to oblige the warping of slabs at edges, this results in the development of excessive stresses because of warping and the slab forms cracking at the edges in an irregular pattern. Hinge joints are usually given for relieving the slabs of warping stresses. There is no structural deformity because of warping cracks if appropriate support is given at the longitudinal and transverse joints as it manages the structural insufficiency.

17.9 TRANSPORT ECONOMICS

17.9.1 SCOPE OF TRANSPORTATION ECONOMICS

The study of economics or the financial matter is distinguished into macro-economics and micro-economics. Micro-economics is related with the wealth of society on a regional scale and deals with the behavior of aggregate concepts. On the other hand, micro-economics involves the behavior of relatively smaller entities such as firms and individuals. Transportation economics, while considered a branch of applied micro-economics, is associated with certain unique issues such as

- the demand for transportation is not direct, but is derived;
- the consumption of each transportation facility (i.e., each trip) is unique in time and space;
- technological differences among different modes and economies of scale;
- governmental interventionist policies and regulations in transportation.

Transportation economics specifically addresses the demand of transportation services, supply of transportation facilities, elasticities of demand and supply, price mechanisms, and transportation cost analysis.

In addition to providing benefits to their users, transport networks impose both positive and negative externalities on nonusers.

- **Positive externalities of transport networks**: include the ability to provide emergency services, increases in land value, and agglomeration benefits.
- **Negative externalities**: include local air pollution, noise pollution, light pollution, safety hazards, community severance, and congestion.

17.10   INTELLIGENT TRANSPORT SYSTEMS

Intelligent Transportation Systems (ITS) is the application of computer, electronics, and communication technologies and management strategies in an integrated manner to provide traveller information to increase the safety and efficiency of the surface transportation systems. These systems involve vehicles, drivers, passengers, road operators, and managers all interacting with each other and the environment, and linking with the complex infrastructure systems to improve the safety and capacity of road systems. Conventional driver training, infrastructure, and safety improvements may add to certain degree to decrease the number of accidents but not enough to combat this menace. ITSs are the best solution to the problem. Safety is one of the principal driving forces behind the evolution, development, standardization, and implementation of ITSs.

ITSs include an expansive range of remote and wire line correspondence based data and electronics advancements to all the more likely manage traffic and boost the use of the current transportation foundation. It enhances driving experience, security, and capacity of street framework, reduces risks in transportation, mitigates traffic blockage, enhances transportation effectiveness, and decreases contamination.

17.10.1   ITS User Services

So as to send ITS, a system is produced featuring different administrations the ITS can offer to the clients. A rundown of 33 client administrations has been given in the National ITS

Program Plan. All the above services are divided into eight groups. The division of these services is based on the perspective of the organization and sharing of common technical functions. The eight groups are described as follows:

   i. Travel and traffic management
  ii. Public transportation operations
 iii. Electronic payment
  iv. Commercial vehicle operations
   v. Advance vehicle control and safety systems
  vi. Emergency management
 vii. Information management
viii. Maintenance and construction management

i. **Travel and traffic management**:
   The principle goal of this gathering of administrations is to utilize real-time data on the status of the transportation framework to enhance its effectiveness and efficiency and to alleviate the antagonistic ecological effects of the framework. This gathering of client administrations is further divided in ten client administrations.
Pretrip data:
   This client administration provides data to the voyagers regarding the transportation framework before they start their trips with the goal that they can make more informed decisions with regard to their season of flight, the mode to utilize, and route to take to their destinations. The voyagers can get this data through PC or telephone frameworks at home or work and at significant public places. The data incorporate ongoing stream condition, genuine incidents and proposed alternate routes, scheduled street construction and maintenance assignments, travel routes, schedules, charges, exchanges, and parking facilities.
En-route driver information:
   This client benefit provides voyage-related details to the explorers in travel after they begin their trips through variable message signs, vehicle radio, or versatile communication gadgets. This encourages the voyagers to more readily utilize the existing facility by changing routes to avoid congestion. Likewise, this gives cautioning messages for roadway signs, for example, stop signs, sharp curves, decreased speed warnings, wet street condition flashed with in-vehicle showcases to the voyager to enhance the security of working a vehicle. The information can be presented as voice yield also.
Route guidance:
   This service provides information to the travelers with a suggested route to reach a specified destination, along with simple instructions on upcoming turns and other moves. This additionally provides travelers of all modes the real-time information about the transportation system,

including traffic conditions, street closures, and the status and schedule of transit systems. The benefits of this service are reduced delay and drivers stress levels particularly in an unfamiliar area.

Ride matching and reservation:

This user service gives real-time ride matching information to travellers in their homes, offices or other locations, and assists transportation providers with vehicle assignments and scheduling. Travelers give data to the service center and get numbers of ride-sharing options from which they can choose the best.

Traveller Services Information:

This service provides a business directory of data on travel-related services and facilities like the location, operating hours, and availability of food, lodging, parking, auto repair, hospitals, gas stations, and police facilities. This also makes reservations for many of these traveler services. The traveler services information is accessible in the home, office, or other public locations to help plan trips. These services are accessible en-route also.

Traffic Control:

This service gathers the real-time data from the transportation system, processes it into usable information, and uses it to determine the optimum assignment of right-of-way to vehicles and pedestrians. This helps in improving the flow of traffic by giving preference to transit and other high-occupancy vehicles or by adjusting the signal timing to current traffic conditions. The information collected by the Traffic Control service is also disseminated for use by many other user services.

Incident Management:

This service aims to improve the incident management and response capabilities of transportation and public safety officials, the towing and recovery industry, and others involved in incident response. Advanced sensors (close circuit TV cameras), data processors, and communication technologies are used to identify incidents quickly and accurately and to implement response which minimizes traffic congestion and the effects of these incidents on the environment and the movement of people and goods.

Travel Demand Management:

This client benefit creates and actualizes procedures to decrease the quantity of single occupancy vehicles while empowering the utilization of high occupancy vehicles and the utilization of progressively efficient travel mode. The techniques adopted are as follows:

1. Congestion pricing
2. Parking management and control
3. Mode change support
4. Telecommuting and alternate work schedule.

Emissions Testing and Mitigation:

The principle goal of this service is to monitor and actualize methodologies to divert traffic away from sensitive air quality zones, or control access to such regions utilizing advanced sensors. This likewise utilized to distinguish vehicles emanating pollutants surpassing the standard values and to advise drivers to empower them to make remedial move. This aids in encouraging implementation and assessment of different contamination control methodologies by specialists.

Roadway Rail Intersection:

This administration is to give enhanced control of roadway and train traffic to keep away or reduce the seriousness of crashes among trains and vehicles at expressway-rail intersections. This likewise screens the state of different highway-railway intersection (HRI) equipment.

ii. **Public transportation activities**:

This gathering of administrations is worried about enhancing the public transportation frameworks and empowering their utilization.

Public Transportation Management:

This client benefit gathers information through advanced correspondence and data frameworks to enhance the working of vehicles and facilities and to robotize the planning and management functions of public transit framework. This offers three errands:

1. To give ongoing computer examinations of vehicles and facilities to enhance transit tasks and support by checking the area of transit vehicles, by distinguishing deviations from the timetable, and offering potential answers to dispatchers and administrators.
2. To keep up transportation timetables and to guarantee transfer connections from vehicle to vehicle and between modes to encourage snappy reaction to service delays.
3. To improve security of travel, work force by giving access board of travel vehicles.

On the Way Transit Information:

This administration is proposed to give data on expected landing times of the vehicles, transfers, and connections to voyagers after they start their trips utilizing public transportation. This gives ongoing, exact travel benefit data on-board the vehicle, at travel stations, and transport stops to help voyagers in settling on choices and adjusting their trips in progress.

Customized Public Transit:

   The point of this administration is to offer public transport facility to explorers by allocating or planning vehicles by
   1. diverting adaptable routed travel vehicles;
   2. assigning secretly worked vehicles on interest which incorporate little buses, cabs, or other little, shared-ride vehicles.

Under this administration, voyagers give data of their outing inception and destination to service station. The center then allots the nearest vehicle to benefit the demand and to advise the voyagers with respect to entry of such vehicles well ahead of time to diminish their nervousness.

Public Travel Security:

   The client benefit makes a safe domain for open transportation administrators and care staff and screens the environment in travel facilities, travel stations, parking garages, transport stops, and on-board travel vehicles, and creates cautions (either naturally or physically) when important. It likewise gives security to the frameworks that screen key foundation of travel (rail track, bridges, tunnels, transport control ways, and so forth.).

iii. **Electronic payment**:

   This client benefit enables voyagers to pay for transportation administrations with a typical electronic installment mechanism for various transportation modes and capacities. Toll collection, travel fare payment, and parking payment are connected through a multimodular multiutility electronic framework. With a coordinated payment framework an explorer driving on a toll street, utilizing parking area would have the capacity to utilize the equivalent electronic gadget to pay toll, parking price, and the transit fare.

iv. **Commercial vehicle activities**:

   The goal is to improve the productivity and security of commercial vehicle activities. This incorporates going with administrations:
   1. CV electronic clearance
   2. Automated street side well-being inspection
   3. On-board well-being monitoring administrative process
   4. Hazardous material incident reaction
   5. Freight mobility

Commercial Vehicle Electronic Clearance:

   This administration enables authorization work force to electronically check security status, vehicle's accreditations, and size and weight information for the commercial vehicles before they achieve a review site. The experts send the illicit or possibly hazardous vehicles just for examination and bypass safe and legal transporters to go ceaselessly for consistence checks at checkpoints and other evaluation destinations.

Automated Roadside Safety Inspection:

   At review station the safety requirements are checked even more quickly and even more precisely in amid security assessment using computerized investigation capacities. Advanced equipment are used to check brake, steering and suspension execution and moreover the driver's execution relating to driver sharpness and readiness for duty.

On-board Safety Monitoring:

   This administration screens the driver, vehicle, and payload and advises rises driver, transporter, and also to the implementation work force, if a hazardous situation emerges amid activity of the vehicle. This client benefit also ensures cargo holder, trailer, and commercial vehicle integrity by seeing on-board sensors for a rupture or tamper event.

Commercial Vehicle Administrative Processes:

   This administration empowers transporters to buy credentials, for example, fuel use charges, trip licenses, overweight allow, or perilous material permits automatically. The mileage and fuel detailing and reviewing parts are given to the bearers normally which decline huge measure of time and printed material.

Hazardous Materials Incident Response:

   This client benefit furnishes incite data concerning the sorts and amounts of dangerous materials present at occurrence area to the crisis work force so as to encourage a fast and fitting reaction. The crisis faculty are educated with respect to shipment of any touchy risky materials so auspicious move could be made if there should be an occurrence of accidents.

Freight Mobility:

   This administration offers data to the drivers, dispatchers, and intermodular transportation suppliers, empowering carriers to exploit advantage of constant traffic data, and also vehicle and load area information, to build efficiency.

v. **Advanced vehicle control and safety systems**:

   This client service intends to upgrade the security of the transportation system by improving drivers' abilities to keep up watchfulness and control of the vehicle by enhancing the accident avoidance abilities of vehicles. Following client administrations are incorporated in this gathering:

Longitudinal Collision Avoidance:

   This client benefit offers direction to vehicle operators in evading longitudinal crashes to the front and additionally back of the vehicle. This is accomplished by executing rear-end collision warning and control, Adaptive Cruise Control, head-on collision cautioning and control, and backing collision cautioning to the drivers.

Lateral Collision Avoidance:

This helps drivers in warding off accidents that result when a vehicle leaves its own way of travel, by warning drivers and by accepting impermanent control of the vehicle. This administration provides for the drivers the way change/blind spot circumstance show, collision cautioning control, and path takeoff alerted and control.

Intersection Collision Avoidance:

This client benefit is explicitly aimed at giving vehicle operators help with maintaining a strategic distance from crashes at intersections. The framework tracks the situation of vehicles inside the intersection zone through the use of vehicle-to-vehicle communications or vehicle to infrastructure correspondence.

Vision Enhancement for Crash Avoidance:

This administration helps in decreasing the quantity of vehicle crashes that occur amid periods of poor detectable quality by in-vehicle sensors fit for of getting an image of driving condition and giving a graphical showcase of the image to the drivers.

Safety Readiness:

This helps to give drivers with warnings about their own driving execution, the state of the vehicle, and the state of the roadway as detected from the vehicle.

Precrash Restraint Deployment:

This administration helps in decreasing the number and seriousness of wounds caused by vehicle crashes by foreseeing an inevitable crash and by activating traveler safety framework before the real impact.

Automated Vehicle Operations:

This administration gives a totally computerized vehicle-highway framework in which instrumented vehicles chip away at instrumented roadways without operator intervention.

vi. **Emergency management**:

This administration has two functions:

1. Emergency warning and individual security: This is to enable voyagers to counsel appropriate crisis reaction work force with regard to the requirement for help because of crisis or noncrisis circumstances either by physically or consequently from the vehicle on the event of an accident.

2. Emergency vehicle management: This client benefit is to reduce the time from the receipt of a crisis warning to the landing of the crisis vehicles at event zone thereby reducing the severity of accident wounds.

vii. **Information management**:

This administration is planned to give the functionality expected to store and record the huge measures of information being gathered on a nonstop premise by various ITS advancements.

viii. **Maintenance and construction management**:

This client benefit is planned to give the functionality required for dealing the fleets of maintenance vehicles, dealing with the roadway concerning construction and support and safe roadway tasks.

17.10.2 ITS Architecture

The ITS architecture gives a typical structure for planning, characterizing, and coordinating intelligent transportation framework. It decides how the diverse ITS parts would communicate with one another to help tackling transportation issues. It recognizes and depicts different functions and allocates responsibilities to different partners of ITS. The ITS architecture ought to be common and of specified standards all through the state or region so it can deliver answer to several problems while interacting with different organizations.

1. **Interoperability**: The ITS architecture ought to be with the end goal that the data gathered, function implemented, or any equipment introduced be interoperable by various organizations in different state and regions.
2. **Capable of sharing and trading data**: The information by traffic tasks might be useful to the crisis administrations.
3. **Resource sharing**: Local correspondence towers built by various private organizations are required to be shared by ITS activities.

National ITS architecture:

This is produced by US Department of Transportation to give direction and co-ordinate all areas in sending ITS. It records all data accessible and keeps refreshing persistently. The national architecture comprises the accompanying segments.

User services and their prerequisites:

Various functions are expected to achieve the client services. These valuable explanations are called client services prerequisites. For all the client benefits, the necessities have been determined. In the event that any new function is included, new necessities are to be characterized.

18 Water Supply Engineering

18.1 INTRODUCTION

Water is absolutely essential not only for survival of human beings but also for animals, plants, and all other living beings. It is necessary that the water required for their needs must be good and it should not contain unwanted impurities or harmful chemical compounds or bacteria in it. Therefore, in order to ensure the availability of sufficient quantity of good quality water, it becomes necessary to plan and build suitable water supply schemes which may provide potable water to the various sections of community in accordance with their demands and requirements.

18.2 QUANTITY OF WATER

While designing the water supply scheme for a town or city, it is necessary to determine the total quantity of water required for various purposes by the city.

Following are the various types of water demands of a city or town:

1. **Domestic water demand**: The quantity of water required in the houses for drinking, bathing, cooking, washing, etc., is called domestic water demand and mainly depends upon the habits, social status, climatic conditions, and customs of the people. The domestic consumption of water in a developing country is nearly 100–150 liters/day/capita. But in developed countries, it may be 350 liters/day/capita because of the use of air coolers, air conditioners, maintenance of lawns, and automatic household appliances.

2. **Industrial demand**: The water required in the industries mainly depends on the type of industries which are existing in the city. The water required by factories, paper mills, cloth mills, cotton mills, breweries, sugar refineries, etc., comes under industrial use. The quantity of water demand for industrial purpose is around 20%–25% of the total demand of the city.

3. **Institution and commercial demand**: Universities, Institution, commercial buildings, and commercial centers including office buildings, warehouses, stores, hotels, shopping centers, health centers, schools, temple, cinema houses, railway and bus stations, etc., come under this category. On an average, a per capita demand of 20 liters/hour/day is usually considered to be enough to meet such commercial and institutional water requirements.

4. **Demand for public use**: Quantity of water required for public utility purposes such as for washing and sprinkling on roads, cleaning of sewers, watering of public parks, gardens, public fountains, etc., comes under public demand. To meet the water demand for public use, provision of 5% of the total consumption is made designing the water works for a city.

5. **Fire demands**: All the big cities have full firefighting squads. As during the fire breakdown, large quantity of water is required for throwing it over the fire to extinguish it. Generally, for a city with 50 lakh population, the fire demand is 1 liter/capita/day.

6. **Loses and wastes**: All the water, which goes in the distribution, pipes do not reach the consumers. The following are the reasons:
   a. Losses due to defective pipe joints, cracked and broken pipes, faulty valves and fittings.
   b. Losses due to consumers keep their taps open of public taps even when they are not using the water and allow the continuous wastage of water and losses due to unauthorized and illegal connections.

   While estimating the total quantity of water of a town, allowance of 15% of total quantity of water is made to compensate for losses, thefts, and wastage of water.

18.3 SOURCES OF WATER SUPPLY

Water is the most abundant compound in nature. It covers 75% of the earth surface. About 97.3% of water is contained in the great oceans that are saline and 2.14% is held in icecaps glaciers in the poles, which are also not useful. Barely the remaining 0.56% found on earth is in useful form for general livelihood.

The various sources of water available on the earth can be divided into the following two categories.

18.3.1 SURFACE SOURCE

The surface sources are further divided into streams, rivers, ponds, lakes, impounding reservoirs, etc.

a. **Natural ponds and lake**: In mountains at some places natural basins are formed with impervious bed by springs and streams are known as lakes. But lakes and ponds situated at higher altitudes contain almost pure water which can be used without any treatment. But ponds formed due to construction of houses, road, and railways contain large amount of impurities and therefore cannot be used for water supply purposes.

b. **Streams and rivers**: Rivers and streams are the main source of surface source of water. In rainy season the run-off water also carries with clay, sand, silt, etc., which make the water turbid. So river and stream water require special treatments.

c. **Impounding reservoirs**: In some rivers the flow becomes very small and cannot meet the requirements of hot weather. In such cases, the water can be stored by constructing a bund, a weir, or a dam across the river at such places where minimum area of land is submerged in the water and maximum quantity of water to be stored. This water should be used after purification.

18.3.2 SUBSURFACE SOURCE

These are further divided into infiltration galleries, infiltration wells, springs, and wells.

a. **Infiltration galleries**: A horizontal nearly horizontal tunnel which is constructed through water-bearing strata for tapping underground water near rivers, lakes, or streams are called infiltration galleries. For maximum yield the galleries may be placed at full depth of the aquifer.

b. **Infiltration wells**: In order to obtain large quantity of water, the infiltration wells are sunk in series in the banks of river. The wells are closed at top and open at bottom. For the purpose of inspection of well, the manholes are provided in the top cover. The water filtrates through the bottom of such wells, and as it has to pass through sand bed, it gets purified to some extent.

c. **Springs**: Sometimes ground water reappears at the ground surface in the form of springs. Springs generally supply small quantity of water and hence suitable for the hill towns. Some springs discharge hot water due to presence of sulfur and useful only for the cure of certain skin disease patients.

d. **Wells**: A well is defined as an artificial hole or pit made in the ground for the purpose of tapping water. In India 75%–85% of Indian population has to depend on wells for its water supply.

18.4 PUMPS FOR WATER SUPPLY PROJECT

The function of a pump is to lift the water or any fluid to higher elevation or at higher pressure. Pumps are driven by electricity, diesel, or steam power. They are helpful in pumping water from the sources, that is, from intake to the treatment plant and from treatment plant to the distribution system or service reservoir. In homes also pumps are used to pump water to upper floors or to store water in tanks over the buildings.

18.4.1 TYPES OF PUMPS AND THEIR SUITABILITY

Based on the mechanical principle of water, lifting pumps are classified as the follows:

a. **Displacement pumps**: This type of pumps is suitable for moderate heads and small discharges suitable for fire protection and water supply of individual houses. Reciprocating pumps, rotary, chain, gear wheel, pump, and wind mills comes under this category.

b. **Velocity pumps**: These types of pumps are used widely in water supply schemes containing sand, silt, etc. Centrifugal pumps, deep well, turbine pumps, jet pumps are the examples of velocity pump.

c. **Buoyancy pumps**: Airlifting pumps are generally adopted for pumping of water from deep wells to a lift of about 60 m containing mud, silt, debris, etc.

d. **Impulse pumps**: They are mainly used for small water supply projects to lift the water for a height of about 30 m or so. Hydraulic ram is an example of impulse pump.

18.4.2 CENTRIFUGAL PUMPS

Centrifugal force is made use of in lifting water. Electrical energy is converted to potential or pressure energy of water.

18.4.2.1 Component Parts of Centrifugal Pump

Centrifugal pump consists of the following parts:

a. **Casing**: The impellor is enclosed in the casing, which is so designed that kinetic energy of the liquid is converted into pressure energy before it leaves the casing.
b. Delivery pipe
c. Delivery valve
d. Impeller
e. Prime mover
f. Suction pipe
g. Strainer and foot valve

18.4.2.2 Description

The pumps shown in Figure 18.1 consist of an impeller that is enclosed in a water-tight casing. Water at lower level is sucked into the impellor through a suction pipe. Suction pipe should be air tight and bends in this pipe should be avoided. A strainer foot valve is connected at the bottom of the suction pipe to prevent entry of foreign matter and to hold water during pumping. Suction pipe is kept larger in diameter than delivery pipe to reduce cavitations and losses due to friction. An electric motor is coupled to the central shaft to impart energy.

18.4.2.3 Working Principle

When the impellor starts rotating it creates reduction of pressure at the eye of the impellor, which sucks in water through the suction pipe. Water on entering the eye is caught between the vanes of the impeller. Rapid rotation of the impellor sets up a centrifugal force and forces the water at high velocity outward against the casing convert

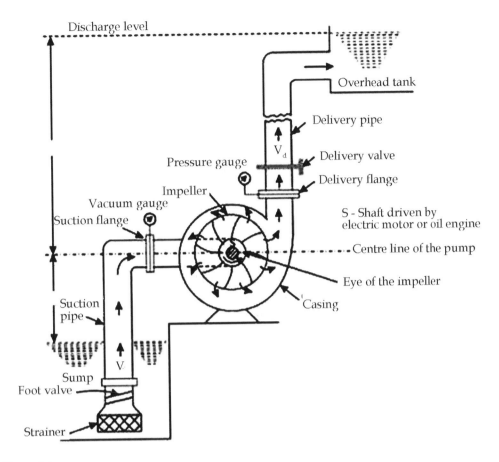

FIGURE 18.1 The pump.

the velocity energy into pressure energy which is utilized to overcome the delivery head.

18.4.3 Operation and Maintenance

Priming means filling up of the suction and casing completely with water.

Pressure and suction developed by the impellor is proportional to the density of the fluid and the speed of rotation. Impellor running in air will produce only negligible negative pressure on the head. Hence it is required, that is, the casing and impellor is filled with water through a funnel and cock. Trapped air is released through pet cock. Initially the delivery valve is closed and the pump started. The rotation impellor pushes the water in the casing into the delivery pipe and the water in the casing into the delivery pipe, and the resulting vacuum is filled by water rising through the suction pipe. The pass valve is opened while closing the bypass valve, while stopping the pump delivery valve is closed first and the pump switched off.

18.4.4 Selection of Pump Horse Power

Basic data regarding the water availability like diameter, depth of the well, depth of the water table, seasonal variations of water table, drawdown duration of pumping, and safe yield are to be collected accurately before selecting a pump.

18.4.4.1 Point to be Observed in Selecting a Pump

a. **Capacity and efficiency**: The pump should have the capacity required and optimum efficiency.
b. **Lift**: Suction head from the water level to the pump level
c. **Head**: It is also called delivery head. Generally the total heads (suction and delivery heads) should meet all possible situations with respect to the head.
d. **Reliability**: A reputed manufacturer or similar make pump already in use may give the failure rate and types of troubles.
e. **Initial cost**: The cost of the pump and its installation cost should be minimum.
f. **Power**: Power requirements should be less for operation.
g. **Maintenance**: Maintenance cost should be minimum. Availability of spares and cost of spares are to be ascertained.

18.4.4.2 Horse Power of Pump

The horse power of a pump can be determined by calculated the work done by a pump in raising the water up to H height.

Let the pump raise "W" kg of water to height "H" m.

Then work done by pump $= W \times H$ Kg m

$$= WQH \text{ mkg/s} \qquad (18.1)$$

where

W = density of water in kg/m^3.
Q = water discharge by pump in m^3/s

The water horse power = discharge × head/75

$$\text{W.H.P.} = WQH/75 \qquad (18.2)$$

Brake horse power = W.H.P./efficiency

$$= WQH/75 \times \eta \qquad (18.3)$$

18.5 QUALITY OF WATER

Absolutely pure water is never found in nature and contains number of impurities in varying amounts. The rainwater which is originally pure also absorbs various gases, dust, and other impurities while falling. This water when moves on the ground further carries salt, organic, and inorganic impurities. So this water before supplying to the public should be treated and purified for the safety of public health, economy, and protection of various industrial processes.

18.5.1 Characteristics of Water

For the purpose of classification, the impurities present in water may be divided into following three categories. They are physical, chemical, and biological characteristics.

18.5.1.1 Physical Characteristics of Water
The following are the physical characteristics.

1. **Turbidity**: Turbidity is caused due to the presence of suspended and colloidal matter in the water. Turbidity is a measure of resistance of water to the passage of light through it. Turbidity is expressed as NTU (Nephelometric Turbidity Units) or PPM (parts per million) or milligrams per liter (mg/l). Turbidity is measured by turbidity rod or tape, Jacksons Turbidimeter, or Bali's Turbidimeter. Drinking water should not have turbidity more than 10 NTU.

2. **Color and temperature**: Color in water is usually due to organic matter in colloidal condition but sometimes it is also due to mineral and dissolved organic impurities. The color produced by one milligram of platinum in a liter of water has been fixed as the unit of color. The permissible color for domestic water is 20 ppm on platinum cobalt scale. The color in water is not harmful but objectionable.

   The temperature of water is measured by means of ordinary thermometers. The temperature of surface water is generally at atmospheric temperature. The most desirable temperature for public supply is between 4.4°C and 10°C. The temperatures above 35°C are unfit for public supply, because it is not palatable.

3. **Taste and odor**: Taste and odor in water may be due to the presence of dead or live micro-organisms, dissolved gases such as hydrogen sulfide, methane, carbon dioxide, or oxygen combined with organic matter, mineral substances such as sodium chloride, iron compounds and carbonates, and sulfates of other substances. The water having bad smell and odor is objectionable and should not be supplied to the public. The intensities of the odors are measured in terms of threshold number. This number is numerically equal to the amount of sample of water in sewage required to be added to 1 l of fresh odorless water.

18.5.1.2 Chemical Characteristics
In the chemical analysis of water, these tests are done that will reveal the sanitary quality of the water. Chemical tests involve the determination of total solids, pH value, hardness of water, chloride content, etc.

1. **Total solids and suspended solids**: Total solids include the solids in suspension colloidal and in dissolved form. The quantity of suspended solids is determined by filtering the sample of water through fine filter, drying, and weighing. The quantity of dissolved and colloidal solids is determined by evaporating the filtered water obtained from the suspended solid test and weighing the residue.

2. pH value of water: pH value denotes the concentration of hydrogen ions in the water and it is a measure of acidity or alkalinity of a substance.

$$\text{pH} = -\log 10 \left[H^+ \right] \text{ or } 1/\log 10 \left[H^+ \right] \qquad (18.4)$$

   Depending upon the nature of dissolved salts and minerals, the pH value ranges from 0 to 14. For pure water, pH value is 7 and 0 to 7 acidic and 7 to 14 alkaline range. For public water supply pH value may be 6.5–8.5. The lower value may cause corrosion, whereas high value may produce incrustation, sediment deposits, and other bad effects.

   pH value of water is generally determined by pH papers or by using pH meter. pH can read directly on scale or by digital display using pH meter.

3. **Hardness of water**: It is a property of water, which prevents lathering of the soap. Hardness is of two types.
   a. Temporary hardness: It is caused due to the presence of carbonates and sulfates of calcium and magnesium. It is removed by boiling.
   b. Permanent hardness: It is caused due to the presence of chlorides and nitrates of calcium and magnesium. It is removed by zeolite method.

   Hardness is usually expressed in gm/l or ppm of calcium carbonate in water. Hardness of water is determined by ethylene diamine tetraacetic acid (EDTA) method. Generally, a hardness of 100–150 mg/l is desirable.

4. **Chloride content**: The natural waters near the mines and sea dissolve sodium chloride and also the presence of chlorides may be due to mixing of saline water and sewage in the water. Excess of chlorides is dangerous and unfit for use. The chlorides can be reduced by diluting the water. Chlorides above 250 ppm are not permissible in water.

5. **Nitrogen content**: The presence of nitrogen in the water indicates the presence of organic matters in the water. Presence of free nitrogen in water indicates its recent pollution. Free ammonia should not be more than 0.15 ppm. The presence of organic ammonia indicates the quantity of nitrogen before decomposition has started. It should be limited to 0.3 ppm. The presence of nitrite indicates partly decomposed condition and its permissible limit is zero. The presence of nitrate indicates fully oxidized condition. Excess presence of nitrogen will cause methemoglobinemia disease to the children.

6. **Metals and other chemical substances**: Water contains various minerals or metal substances such as iron, manganese, copper, lead, barium, cadmium, selenium, fluoride, arsenic, etc.

The concentration of iron and manganese should not allow more than 0.3 ppm. Excess will cause discoloration of clothes during washing and incrustation in water mains. Lead and barium are very toxic, and low ppm of these is allowed. Arsenic and selenium are poisonous and may cause totally; therefore, they must be removed totally. Human beings are affected by the presence of high-quality copper in the water. Fewer cavities in the teeth will be formed due to excessive presence of fluoride in water, more than 1 ppm. A laxative effect is caused in the human body due to excessive presence of sulfate in the water.

7. **Dissolved gases**: Oxygen and carbon dioxide are the gases mostly found in the natural water. The surface water contains large amount of dissolved oxygen because they absorb it from the atmosphere. Algae and other tiny plant life of water also give oxygen to the water. The presence of oxygen in the water in dissolved form keeps it fresh and sparkling. But more quantity of oxygen causes corrosion to the pipes material.

8. **Bio-chemical oxygen demand**: If the water is contaminated with sewage, the demand of oxygen by organic matter in sewage is known as biochemical oxygen demand. The aerobic action continues till the oxygen is present in sewage. As the oxygen exhausts the anerobic action begins due to which foul smell starts coming. Therefore, indirectly the decomposable matters require oxygen, which is used by the organisms.

18.5.1.3 Biological Characteristics

The examination of water for the presence of bacteria is important for the water supply engineer from the viewpoint of public health. The bacteria may be harmless to mankind or harmful to mankind. The former category is known as nonpathogenic bacteria and the later category is known as pathogenic bacteria. Many of the bacteria found in water are derived from air, soil, and vegetation. Some of these are able to multiply and continue their existence while the remaining die out in due course of time. The selective medium that promotes the growth of particular bacteria and inbuilt the growth of other organisms is used in the lab to detect the presence of the required bacteria, usually Coliform bacteria. For bacteriological analysis, the following tests are done:

1. **Count test**: In this method, the total number of bacteria present in a milliliter of water is counted. 1 ml of sample water is diluted in 99 ml of sterilized water and 1ml of dilute water is mixed with 10 ml of agar of gelatine. This mixture is then kept in incubator at 37°C for 24 hours or 20°C for 48 hours. After the sample is taken out from the incubator, colonies of bacteria are counted by means of microscope.

Drinking water should not have more than 10 coliforms/100 ml.

2. **MPN test (most probable number)**: The detection of bacteria by mixing different dilutions of a sample of water with fructose broth and keeping it in the incubator at 37°C for 48 hours. The presence of acid or carbon dioxide gas in the test tube will indicate the presence of B. coli. After this the standard statistical tables (Maccardy's) are referred and the "MOST PROBABLE NUMBER" (MPN) of B. coli per 100 ml of water is determined.

For drinking water, the MPN should not be more than 2.

18.6 TREATMENT OF WATER

Water available in various sources contains various types of impurities and cannot be directly used by the public for various purposes, before removing the impurities. For potability water should be free from unpleasant tastes, odors, and must have sparkling appearance. The water must be free from disease-spreading germs. The amount and type of treatment process will depend on the quality of raw water, the standards of quality of raw water, and the standards of quality to be required after treatment.

18.6.1 TYPES OF TREATMENT

Water treatment includes many operations like aeration, flocculation, sedimentation, filtration, softening, chlorination, and demineralization. Depending upon the quality of raw water and the quality of water desired, The treatment process of water can be divided into three main classes governed by their main principle of operation and the prevailing quality characteristics of the flow.

a. **Physical treatment process**: This process depends mainly on purely physical characteristics of the impurities to be removed. Characteristics such as size, density, viscosity, solubility are of importance in physical treatment operations. Treatments include screening and straining, sedimentation, flocculation, filtration, and gas transfer.

b. **Chemical treatment process**: These processes depend on the chemical properties of the impurities; chemical reagents are added to remove the impurities. They include adsorption, coagulation, ion exchange, and precipitation.

c. **Biological treatment process**: Biological treatment utilizes biological activity to stabilize or remove impurities and they are particularly useful for the removal of organic impurities. Biological processes may be aerobic, anaerobic, or facultative and are mostly done to purify waste waters. They include suspended growth system (activated sludge, oxidation pond) and attached growth system (biological filter, trickling filter).

18.6.2   Screening

Screens are fixed in the intake works or at the entrance of treatment plant so as to remove the floating matters as leaves, dead animals, etc.

18.6.3   Sedimentation

It is the process in which the suspended solids are made to settle by gravity under still water conditions is called plain sedimentation.

18.6.3.1   Plain Sedimentation

By plain sedimentation, the following are the advantages.

1. Plain sedimentation lightens the load on the subsequent process.
2. The operation of subsequent purification process can be controlled in better way.
3. The cost of cleaning the chemical coagulation basins is reduced.
4. No chemical is lost with sludge discharged from the plain settling basin.
5. Less quantity of chemicals is required in the subsequent treatment processes.

The amount of matter removed by sedimentation tank depends upon the factors such as velocity of flow, size, and shape of particles and viscosity of water. The particles which do not change in size, shape, or mass during settling are known as the discrete particles.

The velocity of discrete particles with diameter less than 0.1 mm is given by

$$V = 418(S - S_1)\, d^2 \left(3T + 70/100\right) \qquad (18.5)$$

where

V = velocity of settlement in mm/sec
S = specific gravity of the particle
S_1 = specific gravity of water
D = diameter of the particle in mm
T = temperature in °C

If the diameter of the particle is greater than 0.1 mm, then the velocity is measured by

$$V = 418(S - S_1)\, d \left(3T + 70/100\right) \qquad (18.6)$$

18.6.3.2   Sedimentation Aided with Coagulation

When water contains fine clay and colloidal impurities which are electrically charged are continually in motion and never settle down due to gravitational force. Certain chemicals are added to the water so as to remove such impurities which are not removed by plain sedimentation. The chemical form, insoluble, gelatinous, flocculent precipitate, absorbs and entangles very fine suspended matter and colloidal impurities during its formation and descent through water. These coagulants further have an advantage of removing color, odor, and taste from the water. Turbidity of water reduced up to 5–10 ppm and bacteria removes up to 65%.

Alum or aluminum sulfate is the normally used coagulant in all treatment plants because of the low cost and ease of storage as solid crystals over long periods.

The dosage of coagulants, which should be added to the water, depends upon the kind of coagulant, turbidity of water, color of water, pH of water, temperature of water, and mixing and flocculation time.

18.6.4   Filtration

The process of passing the water through beds of sand or other granular materials is known as filtration. For removing bacteria, color, taste, odors and producing clear and sparkling water, filters are used by sand filtration and 95%–98% of suspended impurities are removed.

The following are the mechanisms of filtration:

a. **Mechanical straining**: Mechanical straining of suspended particles in the sand pores.
b. **Sedimentation**: Absorption of colloidal and dissolved inorganic matter in the surface of sand grains in a thin film.
c. **Electrolytic action**: The electrolytic charges on the surface of the sand particles, which is opposite to that of charges of the impurities, are responsible for binding them to sand particles.

d. **Biological action**: Biological action due to the development of a film of micro-organism layer on the top of filter media, which absorb organic impurities.

18.6.5 Disinfection of Water

The process of killing the infective bacteria from the water and making it safe to the user is called disinfection. The water which comes out from the filter may contain some disease—causing bacteria in addition to the useful bacteria. Before the water is supplied to the public, it is almost necessary to kill all the disease-causing bacteria. The chemicals or substances which are used for killing the bacteria are known as disinfectants.

18.7 COAGULATION OF WATER

Very fine suspended mud particles and colloidal particles present in water cannot settle down in plain sedimentation tank of ordinary detention period. Such particles can be removed easily by increasing their size by changing them into flocculated particles. For this purpose, certain chemical compounds called coagulants are added to the water which on thorough mixing form a gelatinous precipitate called floc. The very fine colloidal particles present in water get attracted and absorbed in these flocs, forming the bigger sized flocculated particles. Coagulation is a chemical technique which is directed toward the destabilization of the charged colloidal particles. Flocculation is the slow mixing technique which promotes the agglomeration of the destabilized particles.

18.7.1 Chemicals Used for Coagulation

The important coagulants used are as follows. They are most effective when the water is slightly alkaline.

1. **Use alum as coagulant:** Alum is the name given to the aluminum sulfate with its chemical form as $(Al_2SO_4) \cdot 18H_2O$. The alum when added to raw water reacts with the bicarbonate alkalinities, which are generally present in raw water supplies, so as to form a gelatinous precipitate of colloids and thus grows in size and finally settles down to the bottom of the tank. The chemical reaction is

$$(Al_2SO_4) \cdot 18H_2O + 3Ca(HCO_3)_2 \rightarrow$$
$$3CaSO_4 + 2Al(OH)_3 \downarrow + 6CO_2 \uparrow$$

From the above equation, it is evident that the addition of alum to water imparts permanent hardness to it, in the form of calcium sulfate. The carbon dioxide gas which is evolved causes corrosiveness.

Alum has proved to be an effective coagulant and is now extensively used throughout the world. It is quite cheap, forms an excellent stable floc, and does not require any skilled supervision for handling.

It helps in reducing taste and color of raw water in addition to removing its turbidity. The only drawback of using alum is that the effective pH range for its use is less (6.5–8.5).

2. **Use of copperas as coagulant**: Copperas is the name given to ferrous sulfate with its chemical form as $FeSO_4 \cdot 7H_2O$. Copperas is generally added to raw water in conjunction with lime. Lime may be added either to copperas or vice versa. When lime is first added, the following reaction takes place:

$$FeSO_4 \cdot 7H_2O + Ca(OH)_2 \rightarrow CaSO_4 + Fe(OH)_2 + 7H_2O$$

Similarly, when copperas is added earlier to lime, the reaction takes place is

$$FeSO_4 \cdot 7H_2O + Ca(HCO_3)_2 \rightarrow Fe(HCO_3)_2 + CaSO_4 + 7H_2O$$

$$Fe(HCO_3)_2 + 2Ca(OH)_2 \rightarrow Fe(OH)_2 + 2CaCO_3 + 2H_2O$$

The ferrous hydroxide formed in either case, further gets oxidized, forming ferric hydroxide, as given below:

$$4Fe(OH)_2 + O_2 + 2H_2O \rightarrow 4Fe(OH)_3 \downarrow$$

The ferric hydroxide forms the floc and helps in sedimentation. Copperas is extensively used as a coagulant for raw waters that are not colored. It is generally cheaper than alum and functions effectively in the pH range of 8.5 and above. The quantity of copperas required is almost the same as that of alum.

3. **Use of chlorinated copperas as coagulant**: When chlorine is added to a solution of copperas (ferrous sulfate), the two react chemically, so as to form ferric sulfate and ferric chloride. The chemical equation is as follows.

$$6(FeSO_4 \cdot 7H_2O) + 3Cl_2 \rightarrow 2Fe_2(SO_4)_3 + 2FeCl_3 + 42H_2O$$

The resultant combination of ferric sulfate and ferric chloride is known as chlorinated copperas and is a valuable coagulant for removing colors, especially where the raw water has a low pH value.

Both the constituents of the chlorinated copperas along with lime are effective coagulants and their combination is often quite effective. The chemical reactions that take place are given below.

$$2Fe_2(SO_4)_3 + 3Ca(OH)_2 \rightarrow 3CaSO_4 + 2Fe(OH)_3 \downarrow$$

$$2FeCl_3 + 3Ca(OH)_2 \rightarrow 3CaCl_2 + 2Fe(OH)_3 \downarrow$$

The resulting ferric hydroxide forms the floc and helps in sedimentation. Ferric sulfate is quite effective in the pH range of 4–7 and above 9, whereas ferric chloride is quite effective in the pH range of 3.5–6.5 and above 8.5. The combination has therefore proved to be a very effective coagulant for treating low pH waters.

4. **Use of sodium aluminate as a coagulant**: Besides alum and iron salts, sodium aluminate ($Na_2Al_2O_4$) is also sometimes used as a coagulant. This chemical when dissolved and mixed with water reacts with the salts of calcium and magnesium present in raw water, resulting in the formation of precipitates of calcium or magnesium aluminate. The chemical reactions that are involved are as follows:

$$Na_2Al_2O_4 + Ca(HCO_3)_2 \rightarrow CaAl_2O_4 \downarrow + Na_2CO_3$$
$$+ CO_2 \uparrow + H_2O$$

$$Na_2Al_2O_4 + CaCl_2 \rightarrow CaAl_2O_4 \downarrow + 2NaCl$$

$$Na_2Al_2O_4 + CaSO_4 \rightarrow CaAl_2O_4 \downarrow + Na_2SO_4$$

The coagulant is about 1.50 times costlier than alum and is therefore generally avoided for treating ordinary public supplies; however, it is very useful for treating waters which do not have the natural desired alkalinity and thus cannot be treated with pure alum. This coagulant is widely used for treating boiler feed waters, which permit very low values of hardness.

18.8 FILTRATION OF WATER

The process of passing the water through beds of sand or other granular materials is known as filtration. For removing bacteria, color, taste, odors, and producing clear and sparkling water, filters are used by sand filtration and 95%–98% suspended impurities are removed.

The following are the mechanisms of filtration

a. **Mechanical straining**: Mechanical straining of suspended particles in the sand pores.

b. **Sedimentation**: Absorption of colloidal and dissolved inorganic matter in the surface of sand grains in a thin film.

c. **Electrolytic action**: The electrolytic charges on the surface of the sand particles, which opposite to that of charges of the impurities, are responsible for binding them to sand particles.

d. **Biological action**: Biological action due to the development of a film of micro-organism layer on the top of filter media, which absorb organic impurities.

18.8.1 TYPES OF FILTERS

Filtration is carried out in three types of filters.

18.8.1.1 Slow Sand Gravity Filter

Slow sand filters (Figure 18.2) are best suited on smaller plants at warm places. The sand used for the filtration is specified by the effective size and uniformity coefficient.

18.8.1.1.1 Construction

The filtering media consist of sand layers about 900–1100 mm in depth. The effective size D_{10} of sand varies from 0.2 to 0.4 mm and the uniformity coefficient (D_{60}/D_{10}) varies from 1.8 to 3.0. Below the fine sand layer, a layer of coarse sand of such size whose voids do not permit the fine sand to pass through it. The thickness of this layer may be 300 mm. The lowermost layer is a graded gravel of size 2–45 mm and thickness is about 200–300 mm. The gravel is laid in layers such that the smallest sizes are at the top. The gravel layer is retained for the coarse sand layer and is laid over the network of open jointed clay pipe or concrete pipes called under drainage. Water collected by the under drainage is passed into the outlet chamber.

18.8.1.1.2 Operation

The water from sedimentation tanks enters the slow sand filter through a submersible inlet. This water is uniformly spread over a sand bed without causing any disturbances. The water passes through the filter media at an average rate of 2.4–3.6 m³/m²/day. This rate of filtration is continued until

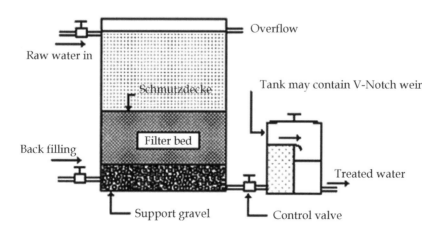

FIGURE 18.2 Slow sand filters.

the difference between the water level on the filter and in the inlet chamber is slightly less than the depth of water above the sand. The difference of water above the sand bed and in the outlet chamber is called the loss of head.

During filtration as the filter media gets clogged due to the impurities, which stay in the pores, the resistance to the passage of water and loss of head also increases. When the loss of head reaches 600 mm, filtration is stopped and about 20–30 mm from the top of bed is scrapped and replaced with clean sand before putting back into service to the filter. The scrapped sand is washed with the water, dried, and stored for return to the filter at the time of the next washing. The filter can run for 6 to 8 weeks before it becomes necessary to replace the sand layer.

18.8.1.1.3 Uses

The slow sand filters are effective in the removal of 98%–99% of bacteria of raw water. Slow sand filters also remove odors, tastes, and colors from the water but not pathogenic bacteria which requires disinfection to safeguard against water-borne diseases. The slow sand filter requires large area for their construction and high initial cost for establishment. The rate of filtration is also very slow.

18.8.1.1.4 Maintenance

The algae growth on the overflow weir should be stopped. Rate of filtration should be maintained constant and free from fluctuation. Filter head indicator should be in good working condition. Trees around the plant should be controlled to avoid bird droppings on the filter bed. No coagulant should be used before slow sand filtration since the floc will clog the bed quickly.

18.8.1.2 Rapid Sand Gravity Filter

Rapid sand filters are replacing the slow sand filters because of high rate of filtration ranging from 100 to 150 m³/m²/day and small area of filter is required. The main features of rapid sand filter are as follows:

a. **Effective size of sand**: 0.45–0.70 mm
b. **Uniformity coefficient of sand**: 1.3–1.7
c. **Depth of sand**: 600–750 mm
d. **Filter gravel**: 2–50 mm size (increase size toward bottom)
e. **Depth of gravel**: 450 mm
f. **Depth of water over sand during filtration**: 1–2 m
g. **Overall depth of filter including 0.5 m free board**: 2.6 m
h. **Area of single filter unit**: 100 m² in two parts of each 50 m²
i. **Loss of head**: Max 1.8–2.0 m
j. **Turbidity of filtered water**: 1 NTU

18.8.1.2.1 Operation

The water from coagulation sedimentation tank enters the filter unit through inlet pipe and uniformly distributed on the whole sand bed. Water after passing through the sand bed is collected through the under-drainage system in the filtered water well. The outlet chamber in this filter is also equipped with filter rate controller. In the beginning the loss of head is very small. But as the bed gets clogged, the loss of head increases and the rate of filtration becomes very low. Therefore, the filter bed requires its washing.

18.8.1.2.2 Washing of Filter

Washing of filter is done by the backflow of water through the sand bed. First inlet valve is closed and the water is drained out from the filter leaving a few centimeters depth of water on the top of sand bed. Keeping all valves closed the compressed air is passed through the separate pipe system for 2–3 minutes, which agitates the sand bed and stirs it well causing the loosening of dirt, clay, etc., inside the sand bed. Now open waste water valve to drain water from main drain and wash water supply valve are opened gradually, the wash water tank rises through the laterals, the strainers gravel, and sand bed. Due to backflow of water, the sand expands and all the impurities are carried away with the wash water to the drains through the channels, which are kept for this purpose.

18.8.1.2.3 Construction

Rapid sand filter consists of the following five parts.

a. **Enclosure tank**: A water tight tank is constructed either masonry or concrete.
b. **Under drainage system**: May be perforated pipe system or pipe and stretcher system.
c. **Base material**: Gravel should be free from clay, dust, silt, and vegetable matter. It should be durable, hard, round, and strong, and depth 400 mm.
d. **Filter media of sand**: The depth of sand, 600–750 mm.
e. **Appurtenances**: Air compressors useful for washing of filter and wash water troughs for collection of dirty water after washing of filter.

Washing process is continued till the sand bed appears clearly. The washing of filter is done generally after 24 hours and it takes 10 minutes, and during backwashing, the sand bed expands by about 50%.

Rapid sand filter (Figure 18.3) brings down the turbidity of water to 1 NTU. This filter needs constant and skilled supervision to maintain the filter gauge, expansion gauge, and rate of flow controller and periodical backwash.

18.8.1.3 Pressure Filter

Pressure filter is a type of rapid sand filter in closed water tight cylinder through which the water passes through the sand bed under pressure as shown in Figure 18.4. All the operations of the filter are similar to rapid gravity filter, except that the coagulated water is directly applied to the filter without mixing and flocculation. These filters are used for industrial plants but these are not economical on large scale.

Pressure filters may be vertical pressure filter and horizontal pressure filter. Backwash is carried by reversing the flow with values. The rate of flow is 120–300 m³/m²/day.

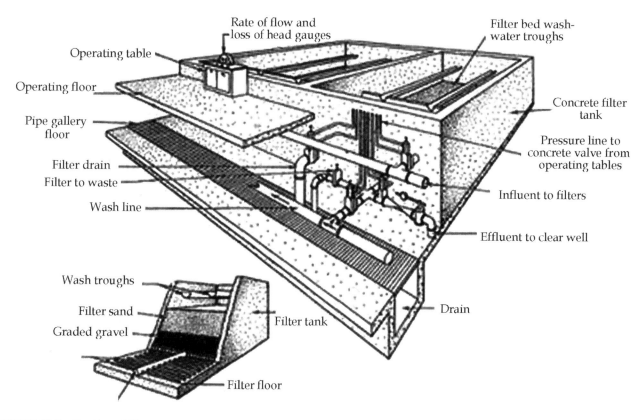

FIGURE 18.3   Rapid sand filter.

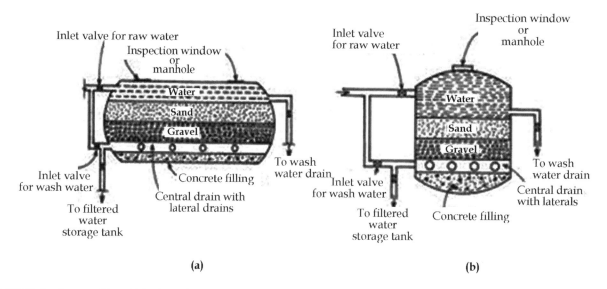

FIGURE 18.4   Pressure filter: (a) horizontal type and (b) vertical type.

18.9   DISINFECTION OF WATER

The process of killing the infective bacteria from the water and making it safe to the user is called disinfection. The water which comes out from the filter may contain some disease-causing bacteria in addition to the useful bacteria. Before the water is supplied to the public, it is almost necessary to kill all the disease-causing bacteria. The chemicals or substances which are used for killing the bacteria are known as disinfectants.

Requirements of good disinfectants

1. They should destroy all the harmful pathogens and make it safe for use.
2. They should not take more time in killing bacteria.
3. They should be economical and easily available.

4. They should not require high skill for their application.
5. After treatment the water should not become toxic and objectionable to the user.
6. The concentration should be determined simply and quickly.

18.9.1 Methods of Disinfection

Disinfection of water by different physical and chemical methods.

18.9.1.1 Physical Methods

1. **Boiling**: Boil the water for 15–20 minutes and kill the disease-causing bacteria. This process is applicable for individual homes.
2. **Ultraviolet rays**: Water is allowed to pass through 10 cm thickness by ultraviolet rays. This process is very costly and not used at water works, but suitable for institutions.
3. **Ultrasonic rays**: Suitable for institutions.

18.9.1.2 Chemical Methods

1. **Chlorination**: Using chlorine gas or chlorine compounds.
2. **Bromine and iodine**: It is expensive and leaves taste and odor.
3. **Potassium permanganate**: This method is used for disinfection of dug well water, pond water, or private source of water.
4. **Ozone**: Very expensive process leaves no taste, odor, or residue.
5. **Excess lime treatment**: Needs long detention time for time interval and large lime sludge to be treated.

18.9.2 Chlorination

Chlorination is the addition of chlorine to kill the bacteria. It is very widely adopted in all developing countries for the treatment of water for public supply. Chlorine is available in gas, liquid, or solid form (bleaching powder).

18.9.2.1 Residual Chlorine and Chlorine Demand

When chlorine is applied in water some of it is consumed in killing the pathogens, some react with organic and inorganic substances, and the balance is detected as "Residual Chlorine." The difference between the quantity applied per liter and the residual is called "Chlorine Demand." Polluted waters exert more chlorine demand. If the water is pretreated by sedimentation and aeration, chlorine demand may be reduced. Normally residual chlorine of 0.2 mg/l is required.

18.9.2.2 Behavior of Chlorine in Water

When chlorine is dissolved in water it forms hypochlorous acid and hydrochloric acid.

$$Cl_2 + H_2O \rightarrow HOCl + HCl$$

After some time hydochlorous acid further ionizes as follows:

$$HOCl \leftrightarrow H^+ + OCl^-$$

The two prevailing species (HOCl) and (OCl$^-$) are called free available chlorine and are responsible for the disinfection of water.

Chlorine reacts with ammonia in water to form and release monochloramine (NH$_2$Cl) dichloramine (NHCl$_2$), and trichloramine (NCl$_3$) and their distribution depends on the pH value of water.

18.9.2.3 Dosage of Chlorine

1. **Plain chlorination**: Plain chlorination is the process of addition of chlorine only when the surface water with no other treatment is required. The water of lakes and springs is pure and can be used after plain chlorination. A rate of 0.8 mg/l/hour at 15 N/cm^2 pressure is the normal dosage so as to maintain in resided chlorine of 0.2 mg/l.
2. **Super chlorination**: Super chlorination is defined as administration of a dose considerably in excess of that necessary for the adequate bacterial purification of water. About 10–15 mg/l is applied with a contact time of 10–30 minutes under the circumstances such as during epidemic breakout water is to be dechlorinated before supply to the distribution system.
3. **Break point chlorination (Figure 18.5)**: When chlorine is applied to water containing organics, micro-organisms, and ammonia, the residual chlorine levels fluctuate with increase in dosage. Up to portion AB, it is absorbed by reducing agents in water (like nitrates, iron, etc.) and further increases form chloramines with ammonia in water.

   Chloramines are effective as Cl$^-$ and OCl$^-$ formed. When the free chlorine content increases it reacts with the chloramines, reducing the available chlorine. At point "C," all the chloramines are converted to effective N$_2$, N$_2$O, and NCl$_3$.

   Beyond point "C" free residual chlorine appears again. This point "C" is called break-point chlorination. Dosage beyond this point is the same as super chlorination. In super chlorination no such rational measurement is made and the dosage is taken at random.
4. **Dechlorination**: Removal of excess chlorine resulting from super chlorination in part or completely is called "dechlorination." Excess chlorine in water gives pungent smell and corrodes the pipe lines. Hence excess chlorine is to be removed before supply. Physical methods like aeration, heating, and absorption on charcoal may be adopted. Chemical methods like sulfur dioxide (SO$_2$), sodium bi-sulfate (NaHSO$_3$), and sodium thiosulfate (Na$_2$S$_2$O$_8$) are used.

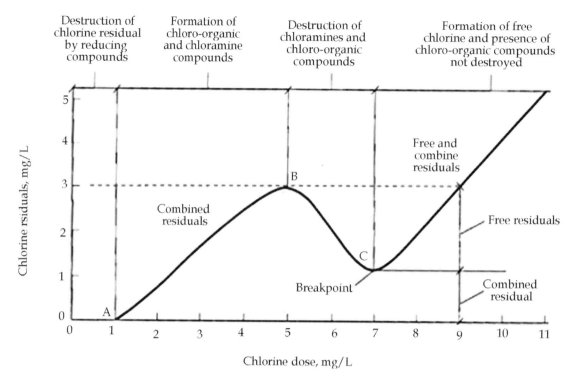

FIGURE 18.5 Break-point chlorination.

18.10 SOFTENING OF WATER

The reduction or removal of hardness from water is known as water softening. Hardness is of two types. They are temporary hardness and permanent hardness. Temporary hardness is caused due to the presence of carbonates and sulfates of calcium and magnesium. Permanent hardness is caused due to the presence of chlorides and nitrates of calcium and magnesium. Excess hardness leads to the following effects.

1. Large soap consumption in washing and bathing.
2. Fabrics when washed become rough and strained with precipitates.
3. Hard water is not fit for industrial use like textiles, paper making, dye, and ice cream manufactures.
4. The precipitates clog the pores on the skin and make the skin rough.
5. Precipitates can choke pipe lines and values.
6. It forms scales in the boilers tubes and reduces their efficiency and cause incrustations.
7. Very hard water is not palatable.

18.10.1 METHODS OF REMOVING TEMPORARY HARDNESS

Temporary hardness can be removed either by boiling or by adding lime to water.

18.10.1.1 By Boiling

Calcium carbonate, being soluble in water, will usually exist in water as calcium bicarbonate because it easily dissolves in natural water containing carbon dioxide. When such water is boiled, the carbon dioxide gas will get out, leading to the precipitation of $CaCO_3$, which can be sedimented out in settling tank.

$$Ca(HCO_3)_2 \rightarrow CaCO_3 \downarrow + CO_2 \uparrow + H_2O$$

Boiling cannot satisfactorily remove the temporary hardness due to magnesium. Hence boiling cannot be used be used for softening public water supplies.

18.10.1.2 Addition of Lime

Lime (CaO) generally hydrated lime $(Ca(OH)_2)$ is added to the water. The following reaction will takes place:

$$Mg\,CO_3 + Ca(OH)_2 \rightarrow Mg(OH)_2 \downarrow + CaCO_3 \downarrow$$

$$Mg(HCO_3)_2 + Ca(OH)_2 \rightarrow Ca(HCO_3)_2 + Mg(OH)_2 \downarrow$$

$$Ca(HCO_3)_2 + Ca(OH)_2 \rightarrow 2CaCO_3 \downarrow + 2H_2O$$

The calcium carbonate and magnesium hydroxide are precipitated and can be removed in the sedimentation tank. This method is generally adopted for softening waters which contain only temporary hardness.

18.10.2 METHODS OF REMOVING PERMANENT HARDNESS

The permanent hardness is more permanent and difficult to remove. It can be removed by certain special methods. The three methods which are commonly adopted for softening waters containing either permanent hardness or temporary hardness are as follows.

18.10.2.1 Lime Soda Process

In this process, lime and soda ash are added to the hard water, which react with the calcium and magnesium salts, so as to form insoluble precipitates of calcium carbonate and magnesium hydroxide. The lime soda process is shown in Figure 18.6. These precipitates are sedimented out in sedimentation tank. The following are the chemical reaction involved:

$$CO_2 + Ca(OH)_2 \rightarrow CaCO_3 + H_2O \quad (\text{removal of } CO_2)$$

$$Ca(HCO_3) + Ca(OH)_2 \rightarrow 2CaCO_3 + 2H_2O$$

$$(\text{removal of temporary hardness})$$

$$Mg(HCO_3) + Ca(OH)_2 \rightarrow CaCO_3 + Mg(CO_3) + 2H_2O$$

$$MgSO_4 + Ca(OH)_2 \rightarrow Mg(OH)_2 + CaSO_4$$

$$(\text{conversion of } MgSO_4 \text{ to } CaSO_4)$$

$$CaSO_4 + Na_2CO_3 \rightarrow CaCO_3 + Na_2SO_4$$

$$(\text{removal of sulphates})$$

$$CaCl_2 + Ca(OH)_2 \rightarrow Ca(OH)_2 + CaCl_2$$

$$MgCl_2 + Ca(OH)_2 \rightarrow Mg(OH)_2 + CaCl_2$$

$$(\text{removal of chlorides})$$

$$CaCl_2 + Na_2CO_3 \rightarrow CaCO_3 + 2NaCl$$

$$MgCl_2 + Ca(OH)_2 \rightarrow Mg(OH)_2 + CaCl_2$$

$$(\text{removal of chlorides})$$

$$CaCl_2 + Na_2CO_3 \rightarrow CaCO_3 + 2NaCl$$

$$MgCl_2 + Na_2CO_3 \rightarrow Mg\,CO_3 + 2NaCl$$

$$(\text{removal of chlorides})$$

18.10.2.2 Zeolite Process

The zeolite process is shown in Figure 18.7. Zeolites are compounds (silicates of aluminum and sodium) which replace sodium ions with calcium and magnesium ions when hard water passes through a bed of zeolites. The zeolite can be regenerated by passing a concentrated solution of sodium chloride through the bed. The chemical reactions involved are as follows:

$$2SiO_2Al_2O_3Na_2O + Ca(HCO_3)_2 \rightarrow$$

$$2SiO_2Al_2O_3CaO + 2NaHCO_3 \quad (\text{Zeolite})$$

$$2SiO_2Al_2O_3Na_2O + CaSO_4 \rightarrow 2SiO_2Al_2O_3CaO + Na_2SO_4$$

$$2SiO_2Al_2O_3Na_2O + CaCl_2 \rightarrow 2SiO_2Al_2O_3CaO + 2NaCl$$

Regeneration:

$$2SiO_2Al_2O_3Na_2O + 2NaCl \rightarrow 2SiO_2Al_2O_3Na_2O + CaCl_2$$

$$2SiO_2Al_2O_3MgO + 2NaCl \rightarrow 2SiO_2Al_2O_3Na_2O + MgCl_2$$

18.10.2.3 Demineralization

Demineralization (Figure 18.8) means removing the minerals from water. Since this process helps us in completely removing or reducing the mineral content by any desired extent, it is very suitable for producing water of any desired hardness or even mineral-free water. This demineralized water is sometime called as deionized (DI) water and is as pure as distilled water. This water is useful for industrial purposes especially for steam rising in high-pressure boilers. The complete removal of mineral is carried out by first passing through a bed of cation exchange resins and then through a bed of anion exchange resin.

Both cations and anions are removed by resins similar to zeolites in two columns by ion exchange method. Resins may be regenerated with sulfuric acid and sodium carbonate. This process is used in industries to get distilled water or quality water motion of water through the atmosphere, earth, plants, trees, rivers, and oceans in a cyclic motion through liquid, solid, and gaseous phases is called hydrological cycle.

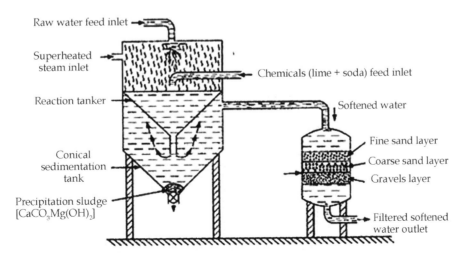

FIGURE 18.6　The lime soda process.

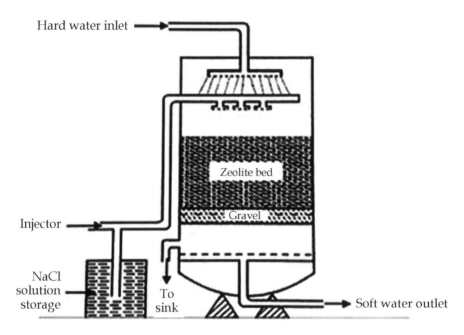

FIGURE 18.7 The zeolite process.

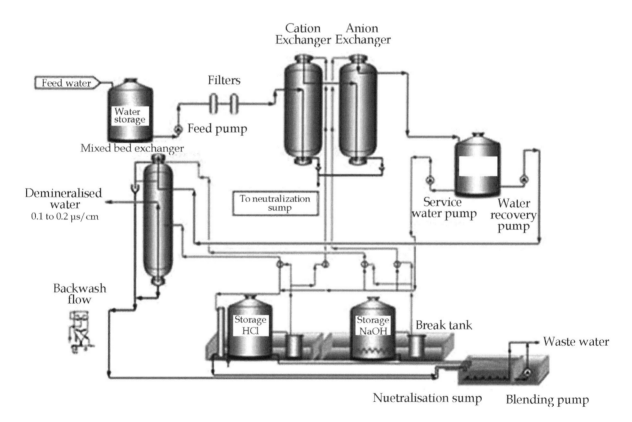

FIGURE 18.8 Demineralization.

18.11 MISCELLANEOUS METHOD FOR WATER TREATMENT

Besides the normal treatment processes such as coagulation, sedimentation, filtration, and disinfection, certain other special treatments are sometimes required in order to remove the special minerals, tastes, odors, and colors from water.

18.11.1 REMOVAL OF COLORS, ODORS, AND TASTES FROM WATERS

18.11.1.1 Aeration

Under the process of aeration, water is brought in intimate contact with air, so as to absorb oxygen and to remove carbon dioxide gas. It may also help in killing bacteria to a certain extent. It also helps in removing hydrogen sulfide gas and iron and manganese to a certain extent, from the treated water. The aeration of water can be carried out in one of the following ways:

1. by using spray nozzle
2. by permitting water to tickle over cascades
3. by air diffusion
4. by using trickling heads

Aeration also helps in removing iron and manganese and also in removing volatile gases such as carbon dioxide and hydrogen sulfide, yet cannot be relied up on to remove or even reduce the tastes and odors of all kinds. Aeration also helps in removing oils and decomposing products of algae and other aquatic vegetation and thus helps in removing the odors, tastes, and odor due to their presence.

18.11.1.2 Treatment with Activated Carbon

Activated carbon is a specially treated carbon which possesses the property of absorbing and attracting impurities such as gases, liquids, and finely divided solids. Because of its excellent property of absorbing impurities, it is widely used for removing tastes and odors from public supplies.

It can be manufactured by heating or charring wood or sawdust or some other similar carbonaceous materials at 500°C in a closed vessel and then slowly burning it under very closed controlled conditions at 800°C, thus removing hydrocarbons from it.

The activated carbon is mostly used in the powdered form and may be added to the water either before or after the coagulation, but before filtration. The most common method adopted is to add a portion in the mixing tank. This method of using activated carbon at two stages is called split method. The usual dosage of activated carbon varies from 5 to 20 mg/l and optimum dose may be determined first in the laboratory and then in the field.

18.11.1.3 Treatment with Copper Sulfate

Copper sulfate helps in removing colors, tastes, and odors from water. It may be added to treated water in the distribution pipes, but its main advantage is obtained by adding it to open reservoirs or lakes. The chief function served by copper sulfate in the reservoirs is to kill the algae or to rather check the growth of algae, even before its production. These algae, if permitted to complete their life cycle or of killed after attaining growth, give off oils and other decomposing products which impart highly disagreeable tastes and odors to the water. The use of copper sulfate at the proper time will help in preventing the growth of such algae and thereby keeping the water free from the bad tastes and odors likely to be caused by them.

18.11.1.4 Treatment with Oxidizing Agents

The colors, odors, and tastes from the water may also be removed by oxidizing the organic matter responsible for them. The oxidizing agents commonly used are potassium permanganate, chlorine, ozone, etc. The usual doses of potassium permanganate may vary between 0.05 and 0.1 mg/l. Chlorine also helps in removing the organic matter, provided sufficient doses are used.

Chlorine dioxide gas and ozone may also be used as oxidizing agents for obtaining good tasty water, but their use has not been found economical, as yet.

18.11.2 Removal of Salt and Dissolved Solids from Water

The water which contains common salt or sodium chloride dissolved in it has a peculiar salty or brackish taste, and therefore, it is named as salt water or brackish water. Since the salt waters have a brackish taste, they cannot be used for drinking, unless the salt content is removed or reduced. The process of removing this salt from water is known as desalination, and the resultant water which is free from salt is known as fresh water.

The various methods which are generally adopted for the conversion of salt water into fresh water are as follows.

18.11.2.1 Electrodialysis

Electrodialysis as shown in Figure 18.9 consists of applying a direct current across a body of water separated into vertical layers by membranes alternately permeable to cations and anions. Cations migrate toward the cathode and anions toward the anode. Cations and anions both enter one layer of water, and both leave the adjacent layer. Thus, layers of water enriched in salts alternate with those from which salts have been removed. The water in the brine-enriched layers is recirculated to a certain extent to prevent excessive accumulation of brine.

Fouling caused by various materials can cause problems with reverse osmosis treatment of water. Although the relatively small ions constituting the salts dissolved in wastewater readily pass through the membranes, large organic ions (proteins, for example) and charged colloids migrate to the membrane surfaces, often fouling or plugging the membranes and reducing efficiency. In addition, growth of microorganisms on the membranes can cause fouling.

Electrodialysis has the potential to be a practical and economical method to remove up to 50% of the dissolved inorganics from secondary sewage effluent after pretreatment to eliminate fouling substances. Such a level of efficiency would permit repeated recycling of water without dissolved inorganic materials reaching unacceptably high levels.

18.11.2.2 Ion Exchange Process

The ion exchange method is usually used for softening water. The ion exchange process used for removal of inorganics consists of passing the water successively over a solid cation exchanger and a solid anion exchanger, which replace cations and anions by hydrogen ion and hydroxide ion, respectively, so that each equivalent of salt is replaced by a mole of water. For the hypothetical ionic salt MX, the reactions are given

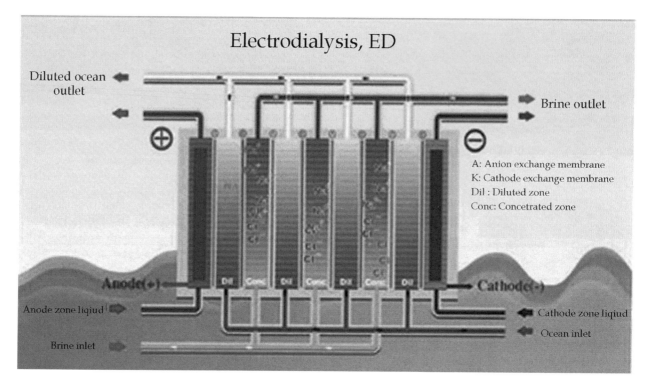

FIGURE 18.9 Electrodialysis.

in the following, where {Cat(s)} represents the solid cation exchanger and +{An(s)} represents the solid anion exchanger.

$$H^+ - \{Cat(s)\} + M^+ + X^- \rightarrow M^+ - \{Cat(s)\} + H^+ + X^-$$

$$OH^- + \{An(s)\} + H^+ + X^- \rightarrow X^- + \{An(s)\} + H_2Os$$

The cation exchanger is regenerated with strong acid and the anion exchanger with strong base.

Demineralization by ion exchange generally produces water of a very high quality. Unfortunately, some organic compounds in wastewater foul ion exchangers, and microbial growth on the exchangers can diminish their efficiency. In addition, regeneration of the resins is expensive, and the concentrated wastes from regeneration require disposal in a manner that will not damage the environment.

18.11.2.3 Reverse Osmosis

Reverse osmosis is a very useful and well-developed technique for the purification of water. Basically, it consists of forcing pure water through a semipermeable membrane that allows the passage of water but not of other material. This process, which is not simply sieve separation or ultrafiltration, depends on the preferential sorption of water on the surface of a porous cellulose acetate or polyamide membrane. Pure water from the absorbed layer is forced through pores in the membrane under pressure.

18.11.3 Removal of Iron and Manganese from Water

Iron and manganese salts are generally found dissolved together in well water or anaerobic reservoir waters, in invisible dissolved state. The iron and manganese may be present in water either in combination with organic matter; they can be easily removed by aeration, followed by coagulation, sedimentation, and filtration. During aeration, the soluble ferrous and manganese compounds present in water may get oxidized into insoluble ferric and manganic compounds which can be sedimented out easily.

On the other hand, when iron and manganese are present in combination with organic matter, it becomes difficult to break the bond between them and to cause their removal. However, when once this bond is broken, they can be removed as above. This bond may be removed either by adding lime and thereby increasing the pH value of water to about 8.5–9 or by adding chlorine or potassium permanganate.

Manganese zeolites, natural green sand coated with manganese dioxide, can also be used for removing soluble iron and manganese from solution. After the zeolite becomes saturated with metal ions, it can be regenerated by backwashing with potassium permanganate.

18.11.4 Removal of Fluorides from Water

Fluorides in drinking water must neither be totally absent nor should exceed an optimum value of about 1 ppm. To ensure this, fluorides are added to waters found deficient in fluoride concentrations, under a process known as fluoridation. When the fluoride concentration in given water exceeds the limiting value of 1–1.5 ppm, the fluorides are removed from water under a process known as defluoridation. Fluorides may enter the human body through drinking water 96%–99% of it combines with bones, since fluorides have affinity for calcium phosphate in the bones. Excess intake of fluoride can lead to dental fluorosis, skeletal fluorosis, or nonskeletal fluorosis.

18.11.4.1 Absorption by Activated Alumina

In this method, the raw water containing high contents of fluoride is passed through the insoluble granular beds of substance like activated alumina or activated carbon or serpentine or activated bauxite, which absorbs fluoride from the percolating water, giving out defluoridated water. Activated alumina is found to be an excellent medium for removal of excess fluoride. It is highly selective to fluoride in the presence of sulfates and chlorides, when compared to synthetic ion exchange resins. The activated alumina after becoming saturated with adsorbed fluoride can be cleaned and regenerated by backwashing with 1% caustic soda solution.

18.11.4.2 Ion Exchange Adsorption Method

The ion exchange process is usually used for removing hardness from water. The process here uses a strong base anion exchange resin in the chloride form. As the water passes through the bed of the resin contained in a pressure vessel, fluorides and other anions like arsenic, nitrates, etc., present in the water are exchanged with the chloride ions of the resin, thus releasing chlorides into water and adsorbing fluoride ions into resin. The arsenic and nitrate ions also get removed in the process. When the resins get saturated with anions like fluoride, nitrate, arsenic, etc., as indicated by their increased concentrations in the out flowing water, the same can be cleaned and regenerated with 5%–10% sodium chloride solution and the bed is returned to service.

This method ensures high efficiency in removal of resin and large amount of salt for regeneration of resin saturated with fluorides. The method is found to be very costly and good.

18.11.4.3 Reverse Osmosis

In this method the raw water is passed through a semipermeable membrane barrier, which permits the flow of clear water through itself and blocks the flow of salts including fluorides. This method is generally adopted for desalination for removing salt from water.

18.12 COLLECTION AND CONVEYANCE OF WATER

The main components of a water conveyance system consist of the following:

1. Intake structure
2. Water conducting system comprising of different structures
3. Outflow structure, which is usually a part of the turbine tail end

18.12.1 Intake Structure

The main function of the intakes' works is to collect water from the surface source and then discharge water so collected, by means of pumps or directly to the treatment water.

Intakes are a structure which essentially consists of opening, grating, or strainer through which the raw water from river, canal, or reservoir enters and carried to the sump well by means of conducts water from the sump well is pumped through the rising mains to the treatment plant.

The following points should be kept in mind while selecting a site for intake works.

a. Where the best quality of water available so that water is purified economically in less time.
b. At site there should not be heavy current of water, which may damage the intake structure.
c. The intake can draw sufficient quantity of water even in the worst condition, when the discharge of the source is minimum.
d. The site of the work should be easily approachable without any obstruction.
e. The site should not be located in navigation channels.
f. As per as possible the intake should be near the treatment plant so that conveyance cost is reduced from source to the water works
g. As per as possible the intake should not be located in the vicinity of the point of sewage disposal for avoiding the pollution of water.
h. At the site sufficient quantity should be available for the future expansion of the water works.

18.12.1.1 Types of Intake Structures

a. **Lake Intake**:

For obtaining water from lakes mostly submersible intakes are used. These intakes are constructed in the bed of the lake below the water level, so as to draw water in dry season also. These intakes have so many advantages such as no obstruction to the navigation, no danger from the floating bodies, and no trouble due to ice. As these intakes draw small quantity of water, these are not used in big water supply schemes or on rivers or reservoirs. The main reason being that they are not easily approachable for maintenance.

b. **River intake**:

Water from the rivers is always drawn from the upstream side, because it is free from the contamination caused by the disposal of sewage in it. The river intake process is shown in Figure 18.10. It is circular masonry tower of 4–7 m in diameter constructed along the bank of the river at such place from where required quantity of water can be obtained even in the dry period. The water enters in the lower portion of the intake known as sump well from penstocks.

c. **Reservoir intake**:

It consists of an intake well, which is placed near the dam and connected to the top of dam by Foot Bridge as shown in Figure 18.11.

The intake pipes are located at different levels with common vertical pipe. The valves of intake

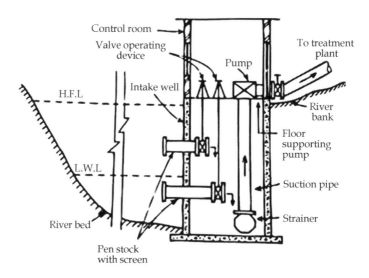

FIGURE 18.10   The river intake process.

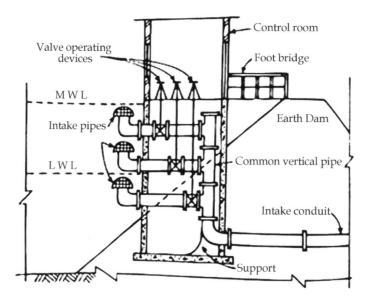

FIGURE 18.11   Reservoir intake.

pipes are operated from the top and they are installed in a valve room. Each intake pipe is provided with bell mouth entry with perforations of fine screen on its surface. The outlet pipe is taken out through the body of dam. The outlet pipe should be suitably supported. The location of intake pipes at different levels ensures supply of water from a level lower than the surface level of water.

When the valve of an intake pipe is opened the water is drawn off from the reservoir to the outlet pipe through the common vertical pipe. To reach up to the bottom of intake from the floor of valve room, the steps should be provided in Zigzag manner.

d. **Canal intake**: An intake chamber is constructed in the canal section (Figure 18.12). This results in the reduction of water way which increases the velocity

of flow. It therefore becomes necessary to provide pitching on the downstream and upstream portion of canal intake. The entry of water in the intake chamber takes through coarse screen and the top of outlet pipe is provided with fine screen. The inlet to outlet pipe is of bell-mouth shape with perforations of the fine screen on its surface. The outlet valve is operated from the top and it controls the entry of water into the outlet pipe from where it is taken to the treatment plant.

18.13   DISTRIBUTION SYSTEM OF WATER

After treatment, water is to be stored temporarily and supplied to the consumers through the network of pipelines called distribution system. The distribution system also includes pumps, reservoirs, pipe fittings, instruments for measurement

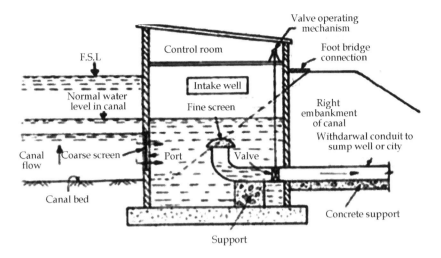

FIGURE 18.12 Canal intake.

of pressures, flow leak detectors, etc. The ultimate aim is to supply potable water to all the consumers whenever required in sufficient quantity with required pressure with least lost and without any leakage.

18.13.1 REQUIREMENT OF A DISTRIBUTION SYSTEM

a. They should convey the treated water up to consumers with the same degree of purity.
b. The system should be economical and easy to maintain and operate.
c. The diameter of pipes should be designed to meet the fire demand.
d. It should be safe against any future pollution. As per as possible it should not be laid below sewer lines.
e. Water should be supplied without interruption even when repairs are undertaken.
f. The system should be so designed that the supply should meet maximum hourly demand. A peak factor 2.5 is recommended for the towns of population 0.5–2 lakhs. For larger population a factor of 2.0 will be adequate.

18.13.2 LAYOUTS OF DISTRIBUTION SYSTEM

Generally in practice there are four different systems of distribution which are used. They are discussed in Section 14.7.4.

18.13.3 SYSTEM OF DISTRIBUTION

For efficient distribution it is required that the water should reach to every consumer with required rate of flow. Therefore, some pressure in pipeline is necessary, which should force the water to reach at every place. Depending upon the methods of distribution, the distribution system is classified as the follows.

18.13.3.1 Gravity System (Figure 18.13)

When some ground sufficiently high above the city area is available, this can be best utilized for distribution system in maintaining pressure in water mains. This method is also much suitable when the source of supply such as lake, river, or impounding reservoir is at sufficiently higher than city. The water flows in the mains due to gravitational forces. As no pumping is required, therefore, it is the most reliable system for the distribution of water.

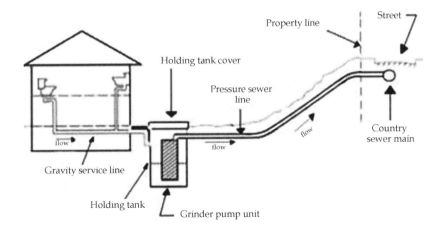

FIGURE 18.13 Gravity system.

18.13.3.2   Pumping System (Figure 18.14)

Constant pressure can be maintained in the system by direct pumping into mains. Rate of flow cannot be varied easily according to demand unless numbers of pumps are operated in addition to stand by ones. Supply can be affected during power failure and breakdown of pumps. Hence diesel pumps also in addition to electrical pumps as standby to be maintained. During fires, the water can be pumped in required quantity by the stand by units.

18.13.3.3   Combined Pumping and Gravity System (Figure 18.15)

This is also known as dual system. The pump is connected to the mains as well as elevated reservoir. In the beginning when demand is small the water is stored in the elevated reservoir, but when demand increases the rate of pumping, the flow in the distribution system comes from the both the pumping station as well as elevated reservoir. As in this system water comes from two sources: one from reservoir and the second from pumping station, it is called dual system. This system is more reliable and economical, because it requires

uniform rate of pumping but meets low as well as maximum demand. The water stored in the elevated reservoir meets the requirements of demand during breakdown of pumps and for fire-fighting.

18.14   PIPE APPURTENANCES

The various devices fixed along the water distribution system are known as appurtenances.

The necessity of the various appurtenances in distribution system is as follows:

a. To control the rate of flow of water.
b. To release or admit air into pipeline according to the situation.
c. To prevent or detect leakages.
d. To meet the demand during emergency.
e. To improve the efficiency of the distribution.

The following are the some of the fixtures used in the distribution system.

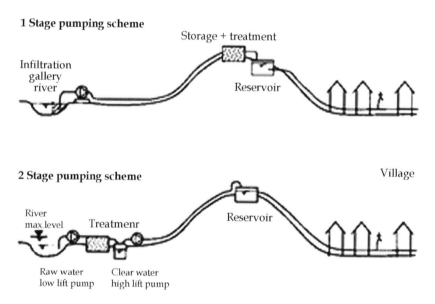

FIGURE 18.14   Pumping system.

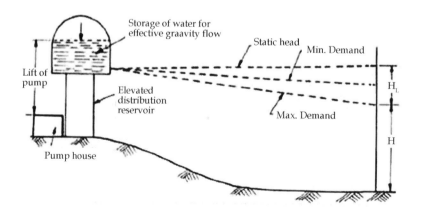

FIGURE 18.15   Combined pumping and gravity system.

a. Valves
b. Fire hydrants
c. Water meter

18.14.1 Types of Valves

In water works practice, to control the flow of water, to regulate pressure, to release or to admit air, prevent flow of water in opposite direction, valves are required.

1. **Sluice valves**: These are also known as gate valves or stop valves as shown in Figure 18.16. These valves control the flow of water through pipes. These valves are cheaper, offer less resistance to the flow of water than other valves. The entire distribution system is decided into blocks by providing these valves at appropriate places. They are provided in straight pipeline at 150–200 m intervals. When two pipelines interest, valves are fixed in both sides of intersection. When sluice valve is closed, it shuts off water in a pipeline to enable to undertake repairs in that particular block. The flow of water can be controlled by raising or lowering the handle or wheel.

2. **Check valve or reflux valve (Figure 18.17)**: These valves are also known as nonreturn valves. A reflux valve is an automatic device which allows water to go in one direction only. When the water moves in the direction of arrow, the valve swings or rotates around the pivot and it is kept in open position due to the pressure of water. When the flow of water in this direction ceases, the water tries to flow in a backward direction. But this valve prevents passage of water in the reverse direction.

Reflux valve is invariably placed in water pipe, which obtain water directly from pump. When pump fails or stops, the water will not run back to the pump and thus pumping equipment will be saved from damage.

3. **Air valves**: These are automatic valves and are of two types, namely, air inlet valves and air relief valves.
   a. Air inlet valves (Figure 18.18): These valves open automatically and allow air to enter into the pipeline so that the development of negative pressure can be avoided in the pipelines. The vacuum pressure created in the down stream-side in pipelines due to sudden closure of sluice valves. This situation can be avoided by using the air inlet valves.

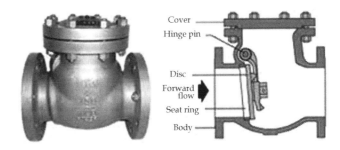

FIGURE 18.17 Check valve or reflux valve.

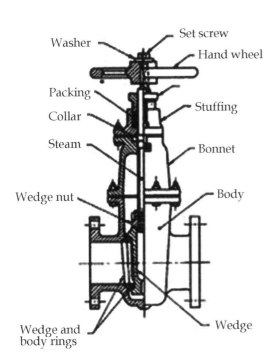

FIGURE 18.16 Sluice valves.

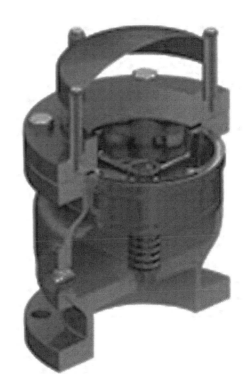

FIGURE 18.18 Air inlet valves.

FIGURE 18.19 Air relief valves.

4. **Drain valves or blow off valves (Figure 18.20):** These are also called wash-out valves they are provided at all dead ends and depression of pipelines to drain out the waste water. These are ordinary valves operated by hand.

5. **Scour valves (Figure 18.21):** These are similar to blow off valves. They are ordinary valves operated by hand. They are located at the depressions and dead ends to remove the accumulated silt and sand. After the complete removal of silt, the value is to be closed.

18.14.2 Water Meter

These are the devices which are installed on the pipes to measure the quantity of water flowing at a particular point along the pipe. The readings obtained from the meters help in working out the quantity of water supplied and thus the consumers can be charged accordingly. The water meters are usually installed to supply water to industries, hotels; big institutions, etc., metering prevents the wastage of purified water.

18.14.3 Fire Hydrants

A hydrant is an outlet provided in water pipe for tapping water mainly in case of fire. They are located at 100–150 m a part along the roads and also at junction roads. They are of two types.

i. **Flush hydrants:** The flush hydrants are kept in underground chamber flush with footpath covered by cast iron (CI) cover carrying a sign board "F-H."

ii. **Posthydrants:** The posthydrant remains projected 60–90 cm above ground level as shown in Figure 7.4. They have long stem with screw and nut to regulate the flow. In case of fire accident, the fire-fighting squad connects their hose to the hydrant and draws the water and sprays it on fire.

b. Air relief valves (Figure 18.19): Sometimes air is accumulated at the summit of pipelines and blocks the flow of water due to air lock. In such cases the accumulated air has to be removed from the pipe lines. This is done automatically by means of air relief valves.

This valve consists of a chamber in which one or two floats are placed and is connected to the pipe line. When there is flow under pressure in the pipeline water occupies the float chamber and makes the float to close the outlet. But where there is accumulation of air in the pipeline, air enters the chamber, makes the float to come down, thus opening the outlet. The accumulated air is driven out through the outlet.

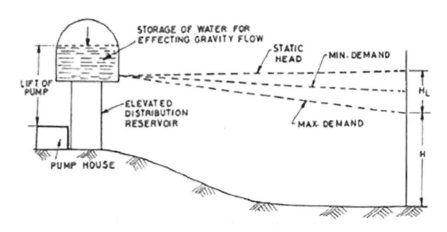

FIGURE 18.20 Drain valves or blow-off valves.

FIGURE 18.21   Scour valves.

A good fire hydrant

a. should be cheap
b. easy to connect with hose
c. easily detachable and reliable
d. should draw large quantity of water

18.15   WATER POLLUTION CONTROL AND WATER MANAGEMENT

Water pollution is the contamination of water bodies. This form of environmental degradation occurs when pollutants are directly or indirectly discharged into water bodies without adequate treatment to remove harmful compounds. Water pollution affects the entire biosphere—plants and organism living in these bodies of water. In almost all cases the effect is damaging not only to individual species and population but also the natural biological communities Water pollution is a major global problem which requires ongoing evaluation and revision of water resource policy. Water is typically referred to as polluted when it is impaired by anthropogenic contaminants and either does not support a human use, such as drinking water or undergoes a marked shift in its ability to support its constituent biotic communities such as fish. Natural phenomena such as volcanoes, algae, blooms, storms, and earthquakes also cause major changes in water quality and the ecological status of water.

18.15.1   Control of Water Pollution

Decisions on the type and degree of treatment and control of wastes and the disposal and use of adequately treated wastewater must be based on a consideration all the technical factors of each drainage basin, in order to prevent any further contamination or harm to the environment.

18.15.1.1   Sewage Treatment

In urban areas of developed countries domestic sewage is typically treated by centralized sewage treatment plants. Well-designed and operated systems (secondary treatment) can remove 90% or more of the pollutant load in sewage.

Some plants have additional systems to remove nutrients and pathogens.

Cities with sanitary sewer overflows or combined sewer overflows employ one or more engineering approaches to reduce discharges of untreated sewage including:

a. utilizing a green infrastructure approach to improve storm water management capacity throughout the system and reduce the hydraulic overloading of the treatment plant;
b. repair and replacement of leaking and malfunctioning equipment;
c. increasing overall hydraulic capacity of sewage collection system.

A household or business not served by a municipal treatment plant may have an individual septic tank, which pretreats the wastewater on site and infiltrates it into the soil.

18.15.1.2   Industrial Waste Water Treatment

Some industrial facilities generate ordinary domestic sewage that can be treated by municipal facilities. Industries that generate waste water with high concentrations of conventional pollutants (e.g., oil and grease) and conventional pollutants (e.g., heavy metals, volatile organic compounds) or other nonconventional pollutants such as ammonia need specialized treatment systems. Some of these facilities can install a pretreatment system to remove the toxic components and then send the partially treated waste water to the municipal system. Industries generating large volumes of waste water typically operate their own complete on-site treatment systems. Some industries have been successful at redesigning their manufacturing processes to reduce or eliminate pollutants, through a process called pollution prevention. The dissolved air floatation plant is shown in Figure 18.22.

Heated water generated by power plants or manufacturing plants may be controlled with:

a. cooling ponds, man-made bodies of water designed for cooling by evaporation, convection, and radiation.
b. Cooling towers, which transfer waste heat to the atmosphere through evaporation and/or heat transfer.
c. Cogeneration, a process where waste heat is recycled for domestic and/or industrial heating purposes.

18.15.1.3   Agricultural Waste Water Treatment

a. **Nonpoint source controls**: Sediment (loose soil) washed off fields are the largest source of agricultural pollution in the United States. Farmers may utilize erosion controls to reduce run-off flows and retain soil on their fields. Common techniques include contour plowing, crop mulching, crop rotation, planting perennial crops, and installing riparian buffers.

Nutrients (nitrogen and phosphorus) are typically applied to farm land as commercial fertilizer, animal

FIGURE 18.22 The dissolved air floatation plant.

manure, or spraying of municipal or industrial waste water (effluent) or sludge. Nutrients mat also enter runoff from crop residues, irrigation water, wildlife, and atmospheric deposition. Farmers can develop and implement nutrient management plans to reduce excess application of nutrients and reduce the potential for nutrient pollution.

To minimize pesticide impacts, farmers may use Integrated Pest Management techniques (which include biological pest control) to maintain control over pets, reduce reliance on chemical pesticides, and protect water quality.

b. **Point source waste water treatment**: Farms with large livestock and poultry operations such as factory farms are called concentrated animal feeding operations or feedlots in the United States and are being subject to increasing government regulation.

Animal slurries are usually treated by contaminant in anaerobic lagoons before disposal by spray or trickle application to grassland. Constructed wetlands are sometimes used to facilitate treatment of animal wastes. Some animal slurries are treated by mixing with straw and composted at high temperature to produce a bacteriologically sterile and friable manure for soil improvement.

18.15.1.4 Erosion and Sediment Control from Construction Sites

Sediment from construction sites is managed by installation of the following:

a. Erosion controls such as mulching and hydroseeding with natural fiber geotextiles.
b. Sediment controls such as sediment basins and silt fences.

Discharge of toxic chemicals such as motor fuels and concrete washout is prevented by the use of:

a. spill prevention and control plans and

b. specially designed containers (e.g., for concrete washout) and structures such as overflow controls and diversion berms.

18.15.1.5 Control of Urban Runoff (Storm Water)

Effective control of urban runoff involves reducing the velocity and flow of storm water as well as reducing pollutant discharges. Local governments use a variety of storm water management techniques, called best management practices, in the US that may focus on improving water quality and some perform both functions. Retention basin for controlling urban runoff is one of such activities.

Pollution prevention practices include low-impact development techniques, installation of green roofs, and improved chemical handling (e.g., management of motor fuels and oil, fertilizers, and pesticides). Runoff mitigation systems include infiltration basins, bioretention systems, constructed wetlands, retention basins, and similar devices.

Thermal pollution from runoff can be controlled by storm water management facilities that absorb the runoff or direct it into ground water such as bioretention systems and infiltration basins. Retention basins tend to be less effective at reducing temperature, as the water may be heated by the sun before being discharged to a receiving stream.

18.15.2 WATER MANAGEMENT

Water resource management is the activity of planning, developing, distributing, and managing the optimum use of water resources. It is a subset of water cycle management. Most effort in water resource management is directed at optimizing the use of water and in minimizing the environmental impact of water use on the natural environment.

Wastewater management should consider the sustainable management of wastewater from source to re-entry into the environment ("reuse/disposal" in the sanitation service chain) and not only concentrate on single or selected areas or segments of the service provision process. The aim of treatment is to reduce the level of pollutants in the wastewater before reuse or disposal into the environment; the standard of treatment required will be location and use-specific.

18.15.2.1   Water Management Plans

Water management plans help individual facilitates set long- and short-term water conservation goals. Following are the best water management practices for water conservation.

a. Meter/measure/manage:

Metering and measuring facility water use help to analyze saving opportunities. This also assures that the equipment is run correctly and maintained properly to help prevent water waste from leaks or from malfunctioning mechanical equipment.

b. Optimize cooling towers:

Cooling towers provide air-conditioning for laboratories and are large consumers of water. Cooling tower operations can be optimized by carefully controlling the ratio of evaporation to blow down is called the cycle of concentration. For maximum water efficiency, cooling towers should be operated at six or more cycles of concentration. Metering water put into and discharged from the cooling tower ensures the cooling tower is operating properly and can help identify leaks or other malfunctions.

c. Eliminate single-pass cooling:

Single-pass cooling circulates a continuous flow of water just once through the system for cooling purposes before it goes down the drain. In laboratories, single-pass cooling is eliminated. Instead, facilities have air-cooled or recirculating chilled water systems.

d. Use water smart landscaping and irrigation:

Planting native and drought tolerant plant species minimizes the need for supplemental irrigation. Landscape water use can also be reduced 10%–20% by having an irrigation water audit. Water Sense-labeled weather-based irrigation controllers or soil moisture sensors are used to water only when plants need it.

e. Control steam sterilizer water:

Steam sterilizer use cooling water to temper steam condensate discharge from the sterilizer to the laboratory drain. Many older sterilizers discharge a continuous flow of tempering water to the drain, even when it is not needed.

f. Reuse laboratory culture water:

Several laboratories require water for aquatic culture research. In some cases, culture water is pumped into laboratory specimen tanks from local bodies of water such as lakes or bays. It is then discharged into sewer or treated and returned to the body of water.

g. Control reverse osmosis:

Up to 10% of a laboratory's water consumption can be related to the multistep process of generating DI purified water through reverse osmosis. Water savings can be achieved by carefully regulating purified water generation rates to meet laboratory demand and making sure that systems are sized accordingly.

h. Recover rain water:

Recovery systems capture rainwater from the roof and redirect it to storage tank. This water is used for flushing toilets, supplying cooling towers and irrigating the landscape.

i. Recover air handler condensate:

Air-conditioning units produce condensate water from the cooling coils. Many laboratories are capturing this water for use as cooling tower makes up water.

18.16   RADIOACTIVITY AND WATER SUPPLIES

Radiological hazards may derive from ionizing radiation emitted by a number of radioactive substances (chemicals) in drinking water. Such hazards from drinking water are rarely of public health significance, and radiation exposure from drinking water must be assessed alongside exposure from other sources.

Environmental radiation originates from a number of naturally occurring and human-made sources. Radioactive materials occur naturally everywhere in the environment (e.g., uranium, thorium, and potassium-40). By far the largest proportion of human exposure to radiation comes from natural sources—from external sources of radiation, including cosmic and terrestrial radiation, and from inhalation or ingestion of radioactive materials.

18.16.1   Radiation Exposure through Drinking Water

Radioactive constituents of drinking-water can result from:

1. naturally occurring radioactive species (e.g., radionuclides of the thorium and uranium decay series in drinking water sources), in particular radium-226/228 and a few others;
2. technological processes involving naturally occurring radioactive materials (e.g., the mining and processing of mineral sands or phosphate fertilizer production);
3. radionuclides discharged from nuclear fuel cycle facilities;
4. manufactured radionuclides (produced and used in unsealed form), which might enter drinking water supplies as a result of regular discharges and, in particular, in case of improper medical or industrial use and disposal of radioactive materials; such incidents are different from emergencies, which are outside the scope of these guidelines; and
5. Past releases of radionuclides into the environment, including water sources.

18.16.2   Removal of Radioactivity in Water

Dangerous radioactive substances such as uranium, thorium, strontium, cesium, etc., from the wastes of the atomic reactors of from the atomic explosions may sometimes pollute the water sources. The radioactive pollutions of water sources must be controlled even with the greater use of radioactive materials, by properly controlling their waste discharge. The radioactivity present in water can be measured with the help of some very sensitive instruments and detectors. Radioactive metals if present in water can be removed only partly by coagulation and filtration. Only up to 80%–90% removals are expected. The radioactive substances such as strontium and cesium, which are held in the true solution in water, cannot be removed either by coagulation sedimentation or filtration.

The lime soda softening method is quite effective in bringing down the radioactivity present in water. With relatively high doses of lime stone such as 200 mg/l or so up to 99% of radioactivity can be removed. Similarly, the zeolite softeners are also very effective in removing radioactivity from water, especially if vermiculite zeolites are used.

Research is also being undertaken to find out other methods of removing radioactivity in water. Certain promising methods which are studied for removing radioactivity from water are

a. by phosphate coagulation method
b. by electrodialysis method
c. by addition of clay materials in water
d. by addition of metallic dusts to water

19 Sanitary Engineering

19.1 SANITARY ENGINEERING— AN INTRODUCTION

The branch of Public Health Engineering which deals with the perseveration and maintenance of health of the individual and the community by preventing communicable diseases is called Sanitary Engineering. It consists of scientific and methodical collection, conveyance, treatment and disposal of the waste matter by which public health can be protected from offensive and injurious substances.

If the wastewater created and given out by the human and animal life, and also by industries etc., is allowed to accumulate, it will lead to decomposition and will contaminate or pollute air, water and food. Hence sanitary disposal of the waste, either in the solid form or in the liquid form, is considered most essential. The sanitary sewage includes excreta (waste matter eliminated from the body), domestic sewage (used water from home or community which includes toilet, bath, laundry, lavatory, and kitchen sink wastes), and industrial wastes.

19.2 WASTEWATER

Sewage or wastewater is a dilute mixture of different wastes from residential, commercial, industrial, and other public places. This wastewater includes objectionable organic and inorganic compounds that may not be amenable to conventional treatment processes. The organic and inorganic matter can be in dissolved, suspended, and colloidal state. The inorganic matter consists of ash, cinder, sand, grit, mud, and other mineral salts. The organic matter is either nitrogenous or nitrogen-free. The chief sources of nitrogenous matter are urea and protein and nitrogen-free compounds include carbohydrates, fats, and soaps.

19.2.1 Important Terms and Definitions

1. **Refuse**: Refuse is used to indicate what is rejected or left out as worthless. It may be in liquid, semi-solid, or solid form and may be divided into six categories: Garbage, rubbish, sewage, sullage, subsoil waste, and storm water.
2. **Garbage**: Garbage is the dry refuse. It includes wastepaper, decayed fruits, and vegetables and leaves, and sweepings from streets, markets, and other public places. Thus, garbage contains large amounts of organic and purifying matter.
3. **Rubbish**: Rubbish indicates various solid wastes from offices, residences, and other buildings that are generally dry and combustible in nature. It also includes waste building materials, broken furniture, paper, rags, etc.
4. **Sullage**: The wastewater generated from bathrooms, kitchens, washing place, and wash basins, etc. Composition of this waste does not involve higher concentration of organic matter and it is found to be less polluted water as compared to sewage.
5. **Sewage**: The liquid waste originating from the domestic uses of water is called sewage. It contains sullage, discharge from toilets, urinals, wastewater generated from commercial establishments, institutions, industrial establishments, and also the groundwater and storm water that may enter into the sewers. It consists of disease-causing bacteria and when decomposed it forms foul gases.
6. **Subsoil water**: Groundwater entering into the sewers through leakages is called subsoil water.
7. **Storm water**: The rainwater of the locality is called storm water.
8. **Sanitary sewage**: It is also known as domestic sewage. It is mainly generated from the residential dwellings and industries. They produce foul smell. They are further classified into domestic sewage and industrial swage.
9. **Domestic sewage**: The sewage obtained from lavatory basins, urinals, and water closets of residential buildings, office buildings, theatres, and other institutions is called domestic sewage. The presence of human excreta and urine makes it extremely foul in nature.
10. **Industrial sewage**: The wastewater generated from the industrial and commercial areas is called Industrial sewage. Due to the presence of organic and inorganic compounds it may not be amenable to conventional treatment processes.
11. **Night soil**: It includes the human and animal excreta.
12. **Sewer**: Drain or an underground conduit through which sewage is carried to a point of discharge or disposal. Three types of sewer systems that are commonly used for sewage collection:
    a. *Separate sewers* carry the industrial and domestic wastes only.
    b. *Storm water drains* carry rainwater from the roofs and street surfaces.
    c. *Combine sewers* carry both storm water and sewage water together in the same conduit.
       - *House sewer* (or drain or individual house connection) discharges sewage from a building to a street sewer.
       - *Lateral sewer* collects sewage directly from the household buildings.
       - *Branch sewer* or *submain sewer* receives sewage from a relatively small area.

- *Main sewer* or *trunk sewer* receives sewage from many tributary branches and sewers, serving as an outlet for a large territory.
- *Depressed sewer* is a part of sewer constructed lower than adjacent sections to pass beneath an obstacle or obstruction. It fully runs under the force of gravity and at greater than atmospheric pressure. The sewage enters and leaves the depressed sewer at atmospheric pressure.
- *Intercepting sewer* is laid transversely to main sewer system to intercept the dry weather flow of sewage and additional surface and storm water as may be desirable. An intercepting sewer is usually a large sewer which is flowing parallel to a natural drainage channel, into which a number of main or outfall sewers discharge.
- *Outfall sewer* receives entire sewage from the collection system which is finally discharged to a common point.
- *Relief sewer* or *overflow sewer* carries the flow in excess of the capacity of an existing sewer.

13. **Sewerage**: Sewerage refers the infrastructure which includes device, equipment and appurtenances for the collection, transportation and pumping of sewage, but excluding works for the treatment of sewage.
14. **Wastewater**: Wastewater includes both organic and inorganic constituents, in soluble or suspended form, and mineral content of liquid waste carried through liquid media. The organic portion of the wastewater undergoes biological decompositions and the mineral matter may combine with water to form dissolved solids.

19.2.2 Sources of Sewage

The wastewater generated from the household activities contributes to the major part of the sewage. The wastewater created from recreational activities, public utilities, commercial complexes, and institutions is also discharged into sewers. The wastewater discharged from small- and medium-scale industries situated within the municipal limits and discharging partially treated or untreated wastewater into sewers also contributes to municipal wastewater.

19.3 COLLECTION AND CONVEYANCE OF REFUSE (WASTEWATER)

Depending upon the type of waste, two systems may be employed for its collection, conveyance, and disposal, mainly conservancy system and water carriage system.

19.3.1 Conservancy System or Dry System

This is an old system in which various types of wastes such as night soil, garbage, etc., are collected once in 24 hours.

Different methods of collection of various types of wastes in the system:

1. **Night soil**: Night soils or human excreta in latrines, privies, or cesspools, etc., are collected separately in pans or pails and carried on heads of sweepers to a central place from where it is transported in bullock carts or motor vans to a place away from the town for its final disposal. Normally, it is buried into ground, in trenches, to give excellent manure in 1 or 2 years.
2. **Garbage**: Garbage is collected separately, in dust bins and conveyed on hand carts or motor van once or twice in a day. It may consist of wastewater of both noncombustible as well as combustible type. Garbage disposal methods include the open dump, hog feeding, incineration, dumping into sanitary fill, fermentation, or biological digestion. Incineration if properly controlled is satisfactory for burning combustible refuse.
3. **Sullage and storm water**: Sullage and storm water are gathered and passed on independently in shut or open drains. The fluid and semi-fluid mass of filth which most of the time floods the repositories in privies is cleared away by the sweepers to deplete from the privies, which convey it to channels conveying sullage and storm water, along the general population paths or boulevards.

The disadvantages of this system are problems related to hygiene and sanitary aspect, involvement of man power, risk of epidemic, pollution problems, disposal, and cost issues, etc.

19.3.2 Water Carriage System

In this system, water is used as the medium to convey the waste from its point of production to the point of its treatment or final disposal. Sufficient quantity of water is required to be mixed with the wastes so that dilution ratio is so great that the mixture may flow just like water. In this system, specially designed latrines called water closets are used which are flushed with 5–10 liters of water after its use by every person. The human excreta are thus flushed away leading to sustainable designed and maintained sewers. The wastes from kitchens, baths, wash basins, etc., are also led to the sewers. The *sewers* are the underground closed pipes which are laid on suitable longitudinal gradient so that flow takes under gravity and proper flow velocity is maintained to keep the sewer clean. The sewers lead the sewerage so collected to a suitable site where it is treated suitably and then it is disposed off by irrigation or by dilution. It should be noted that the garbage is collected separately and conveyed in the same manner as done in the conveyance system.

The system requires large initial cost of installation and it requires large quantity of water also to create efficient flow conditions. This is the most efficient and hygienic system of sewage disposal and may be adopted in stages if sufficient funds are not available in the beginning.

19.3.2.1 Classification of Water Carriage System

1. Separate system

The separate system provides two separate systems of sewer—the one intended for the conveyance of foul sewage only such as fecal matter, domestic wastewaters, the washings, and draining of places such as slaughter houses, laundries, stables, and the wastewaters derived from the manufacturing processes and the other for the rainwater including the surface washing from certain streets, overflow from public baths, and foundations, etc. The sewage from the principal arrangement of sewers can be directed to the treatment works, while the spillout of the second arrangement of sewers can be released specifically to common streams and so forth with no treatment.

   Advantages:
   a. The cost of installation is low.
   b. The load on the treatment units will be lowered, since only the foul sewage carried by the separate sewers needs to be treated.
   c. There is no necessity of providing automatic flushing tanks, for use in dry weather, because the flow in a sewer of smaller section is much more efficient.
   d. Sewer with smaller section can be ventilated easily than those with larger section.
   e. Rainwater can be discharged in to streams or rivers without any treatment.

   Disadvantages:
   a. Since the sewers are of small sizes, it is difficult to clean them.
   b. They are likely to get choked.
   c. Two sets of sewers may ultimately prove to be costly.
   d. Because of lesser air contact in small size sewers, foul smell may be there due to the sewage gas formed.
   e. Storm water sewers or drains come in use only during the rainy season.

2. Combined system:

The joined system gives only a single sewer to pass on both the foul sewage and moreover the rainwater. The rainwater and sewage are brought to the sewage treatment plant, before its last exchange. The joined system is pushed on the ground that the street surface washings are as spoiled as the sewage itself and should in this manner be sensibly treated before being allowed to enter the regular stream.

   Advantages:
   a. The system requires only one set of sewers. Hence the maintenance costs are reduced.
   b. The sewers are of larger size and therefore the chances of their choking are rare and also it is easy to clean them.
   c. The strength of the sewage is reduced by dilution.

d. There is more air in the larger sewers and hence the chance of foul smell is less.

   Disadvantages:
   a. The cost of construction is very high because of its larger size.
   b. Their handling and transportation are also very difficult due to its larger size.
   c. The system is uneconomical in the circumstances when pumping is required for lifting sewage.
   d. During heavy rainfall, sewers may overflow and may thus create unhygienic conditions and cause pollution problems.

3. Partially combined system:

In this system just a single set of underground sewers is laid. These sewers concede the foul sewage and in addition the early washings by downpours. When the amount of storm water surpasses a specific limit, the storm water floods and is in this manner gathered and passed on in open channels to the common streams. The foul sewage, however, continues to flow in the sewers.

   Advantages:
   a. The sewers are of reasonable size; hence their cleaning is not very difficult.
   b. It has both the advantage of separate as well as combined systems.
   c. The storm water is permitted in the sewers which eliminates its chances of choking. The sewers are completely cleaned during the rainy season.
   d. The problem of disposing off storm water from homes is simplified.

   Disadvantages:
   a. During the dry weather, when there is no rainwater, the velocity of flow will be low. Thus self-cleansing velocity may not be achieved.
   b. The storm water increases the load on treatment units.
   c. The storm water also increases the cost of pumping.

19.4 QUANTITY OF SEWAGE

Preceding arranging the sewer, it is essential to know the discharge, i.e., measure of sewage, which will stream in it after finish of the endeavor. Correct estimation of sewage discharge is essential for water-driven structure of the sewers. Far lower estimation than reality will a little while later lead to lacking sewer measure resulting to designating of the arrangement or the sewers may not remain attractive for the entire structure time allotment. So likewise, high discharge evaluated will incite greater sewer gauge impacting economy of the sewerage plot, and the lower discharge extremely spilling in the sewer may not meet the criteria of self-cleaning speed and in this way prompting affidavit in the sewers. Genuine estimation of the discharge is beyond the realm of imagination if the sewers do not exist, and where the limit of the present sewers is lacking and ought to be extended, still real present release estimation may not be exact due to unaccounted surge and spillages that may occur in the present framework.

19.4.1 Sources of Sanitary Sewage

a. Water provided by the water specialist for domestic use, after desired use, it is released in to sewers as sewage.
b. Water was provided to different industries for different industrial procedures by local authority. Some amount of this water after use in various industrial applications is released as wastewater.
c. The water provided to the different public places, for example, schools, cinema theaters, inns, hospitals, and business edifices. Some portion of this water after wanted use joins the sewers as wastewater.
d. Water drawn from wells by people to satisfy domestic interest. After utilizations this water is released into sewers.
e. The water drawn for different purposes by businesses, from individual water sources, for example, wells, tube wells, lake, stream, and so on. Part of this water is changed over into wastewater in various industrial procedures or utilized for public utilities inside the industry creating wastewater. This is released into sewers. Infiltration of groundwater into sewers through leaky joints.
f. Entrance of rainwater in sewers during rainy season through faulty joints or cracks in sewers.

19.4.1.1 Dry Weather Flow

Dry weather flow is the stream that happens in sewers in independent sewerage system or the stream that happens amid dry seasons in consolidated system. This stream shows the stream of sanitary sewage. This relies on the rate of water supply, sort of territory served, monetary states of the general population, climate conditions, and penetration of groundwater in the sewers, if sewers are laid beneath the groundwater table.

19.4.2 Evaluation of Sewage Discharge

Besides accounted water given by water master that will be changed over to wastewater, following sums are considered while evaluating the sewage sum:

a. **Addition due to unaccounted private water supplies**: People utilizing water supply from private wells, tube wells, and so on add to the wastewater age more than the water provided by the civil specialist. Additionally, certain ventures use their very own well-spring of water. Some portion of this water, after wanted utilizations, is changed over into wastewater and at last released into sewers. This amount can be evaluated by genuine field perceptions.

b. **Addition because of infiltration**: This is extra amount due to groundwater leakage into sewers through defective joints or splits framed in the channels. The amount of the water relies on the stature of the water table over the sewer modify level. In the event that water table is well underneath the sewer modify level, the penetration can happen simply after rain when water is moving down through soil. Amount of the water entering in sewers relies on the permeability of the ground soil and it is exceptionally hard to evaluate. Storm water waste may likewise penetrate into sewers. For the most part, no additional arrangement is made for this amount. This additional amount can be dealt with by additional vacant space left at the best in the sewers, which are intended for running ¾ full at most maximum design discharge.
a. Subtraction because of water losses: The water loss, through spillage in water circulation system and house associations, does not achieve consumers and consequently not show up as sewage.
b. Subtraction because of water not entering the sewerage system: Certain measure of water is utilized for such purposes, which may not create sewage, e.g., boiler feed water, water sprinkled over the streets, lanes, yards, and greenery enclosures, water devoured in mechanical item, water utilized in air coolers, and so on.

Net amount of sewage=(Accounted amount of water provided from the water works+Addition due to unaccounted private water supplies+Addition because of infiltration−Subtraction because of water misfortunes−Subtraction because of water not entering the sewerage system)

For the most part, 75%–80% of accounted water provided is considered as amount of sewage delivered.

19.4.3 Variation in Sewage Flow

Variety happens in the stream of sewage over yearly normal day-by-day stream. Vacillation in stream happens from hour to hour and from season to season. The size of variety in the sewage amount fluctuates from place to place and it is extremely hard to anticipate. For littler township, this variety will be progressively articulated because of shorter length and travel time before sewage reaches to the primary sewer and for substantial urban areas this variety will be less. Figure 19.1 demonstrates the common place hourly varieties in sewage stream. For evaluating configuration release the following connection can be considered:

Most extreme every day stream = Multiple times the yearly

normal day by day stream   (speaking to regular varieties)

(19.1)

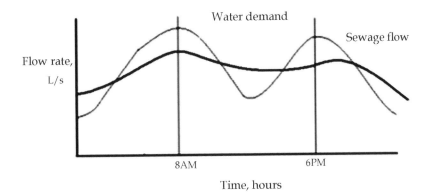

FIGURE 19.1 Typical hourly variations in sewage flow.

Most extreme hourly stream = 1.5 occasions the greatest every day stream (bookkeeping hourly varieties) = Multiple times the yearly normal day by day stream

$$(19.2)$$

As the tributary zone expands, top hourly stream will diminish. For littler populace served (under 50,000) the pinnacle factor can be 2.5, and as the populace served builds, its esteem diminishes. For extensive urban areas it tends to be considered about 1.5–2.0. Along these lines, for outfall sewer, the pinnacle stream can be considered as 1.5 occasions the yearly normal every day stream. Notwithstanding for plan of the treatment office, the pinnacle factor is considered as 1.5 occasions the yearly normal every day stream.

The base stream going through sewers is likewise critical to create self-purging speed to abstain from silting in sewers. Sewers must be checked for least speed as pursues:

Minimum daily flow = 2/3 Annual average daily flow   (19.3)

Minimum hourly flow = ½ minimum daily flow

= 1/3 Annual average daily flow (19.4)

The general variety between the greatest and least stream is more in the laterals and less in the primary or trunk sewers. This proportion might be more than six for laterals and around two to three if there should arise an occurrence of principle sewers.

19.5 CONSTRUCTION OF SEWERS

The development of sewers is costly, and in the event that they are not manufactured appropriately, therapeutic works can be troublesome, tedious, and exorbitant. In numerous occasions, a great deal of slip-ups can be stayed away from by considering the detail and necessities previously and amid development.

19.5.1 SHAPES OF SEWER PIPES

Sewers are ordinarily round channels that are laid subterranean dimension, slopping constantly toward the outfall and intended to stream under gravity. A portion of alternate shapes utilized for sewers are standard egg-shape, new egg-shape, horse shoe shape, allegorical shape, semi-curved, rectangular shape, U-shape, semi-roundabout shape, bin dealt with shape. They are appeared in Figure 19.2a–i.

- Standard egg-formed sewers are otherwise called ovoid molded sewer, and new or modified egg-formed sewers are utilized in consolidated sewers. These sewers can create self-purifying speed amid dry climate stream.
- Horse shoe molded sewers and semi-round areas are utilized for substantial sewers with overwhelming release, for example, trunk and outfall sewers.
- Rectangular or trapezoidal segment is regularly utilized as free secured storm water surface channels.
- U-molded segment is utilized for bigger sewers and particularly in open cuts.

19.5.2 SEWER MATERIALS

While selecting the material for manufacturing sewer pipes, the material must possess important factors such as resistance to corrosion, resistance to abrasion, sufficient strength and durability, less specific weight, imperviousness, hydraulic efficiency, and also less cost.

The different types of sewers according to material and their comparative utilities are as follows:

i. **Asbestos cement sewer**

Sewers which are manufactured from a mixture of cement, silica, and asbestos fiber converted under pressure to a dense homogeneous material possessing considerable strength is called an Asbestos cement (AC). They are suitable for carrying domestic sanitary sewage. The vertical pipes carry sullage from upper floors of multistory buildings in two-pipe system of plumbing. They are smooth and light weighted and can also be effectively cut, fitted, and bored. They are not much affected by soil corrosion. However, AC sewers are brittle and hence they fail under heavy load. They get easily broken during handling and transport.

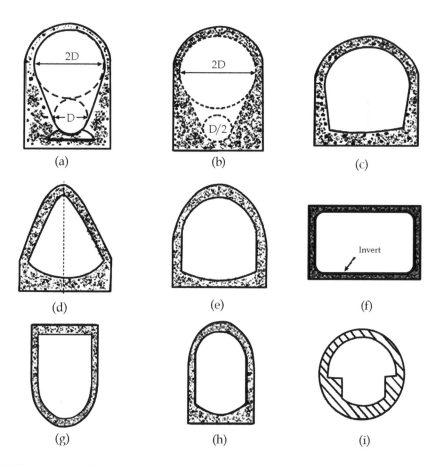

FIGURE 19.2 Different shapes of sewer sections: (a) standard egg-shaped sewer. (b) New/modified egg-shaped sewer. (c) Horse-shaped sewer section. (d) Parabolic section. (e) Semi-elliptical section. (f) Rectangular section. (g) U-shaped section. (h) Semi-circular section. (i) Basket handle section.

ii. **Brick sewers**

Block sewers are typically developed at the site and are utilized for building expansive size sewers. They are extremely helpful for the development of storm sewers or consolidated sewers. These days block sewers are replaced by concrete sewers. Brick sewers may get deformed and spillage may happen, because of the section of tree roots and groundwater through brick joints. To keep away from spillage they ought to be plastered from outside. They are additionally lined from inside with stone product or ceramic block to make them smooth and hydraulically efficient. Lining likewise makes the pipes resistant to corrosion. A lot of work is required.

iii. **Plain cement concrete and reinforced cement concrete sewers**

Plain cement concrete pipes of diameter up to 0.45 m and reinforcement cement concrete pipes of diameter up to 1.8 m diameter are available. These pipes can either be cast in situ or precast. Precast pipes have better quality than the cast in situ pipes. Both the pipes are strong in tension and compression and can also withstand erosion and abrasion. These pipes can be made of any desirable strength and are also economical for medium and large sizes.

However, these pipes get corroded and pitted when it is in contact with sulfuric acid. These pipes can be reinforced in different ways such as single-cage reinforced pipes, used for internal pressure less than 0.8 m; double-cage reinforced pipes used for both internal and external pressure greater than 0.8 m; elliptical cage reinforced pipes used for larger diameter sewers subjected to external pressure; and hume pipes with steel shells coated with concrete from inside and outside. An average longitudinal reinforcement of 0.25% is provided in these pipes.

iv. **Cast iron sewers**

This type of sewer is structurally strong and capable of withstanding greater tensile, compressive, and bending stresses compared to other sewers. These sewers can be utilized for outfall sewers, rising mains of pumping stations, modified siphons, and so forth, all running under strain. They are suitable when sewers are required under heavy traffic load, such as sewers below railways and highways, and also over piers in case of low-lying areas. They form 100% leak-proof sewer line to avoid groundwater contamination. However, CI sewers are costly and cannot withstand the corrosive action of gases and other acids present in sewage and hence they are usually lined from the inside with cement concrete or painted with coal tar.

v. **Vitrified clay or stoneware or salt-glazed sewers**

Vitrified clay sewer pipes are widely used as domestic connections as well as lateral sewers. They are available in size of 5–30 cm internal diameter and 0.9–1.2 m length. They are, however, rarely made in sizes bigger than 90 cm diameter. They are joined by bell and spigot flexible compression joints. They can withstand any effect caused by sulfide corrosion and, therefore, can be used for carrying polluted sewage and industrial wastes. They have very smooth interiors and are highly impervious. They are weak in tension, yet quite strong in compression. They are cheap, durable, and easily available and can be easily laid and jointed but these pipes are heavy, bulky, and brittle and therefore transportation is difficult.

vi. **Lead sewers**

These pipes are smooth, soft, and can take odd shapes. They can highly resist sulfide corrosion but these pipes are very costly. They are used in house connection. Lead sewers are sometimes used in smaller sizes and lengths in the toilets.

vii. **Plastic sewers**

These are ongoing materials utilized for non-weight sewer funnels. They are utilized for inward seepage works in house. These sewers are accessible in sizes 75–315 mm outer distance across and utilized in inward water supply and seepage fittings. They offer smooth internal surface and ease in fabrication and transport. They can withstand corrosion, light in weight, economical in laying, jointing, and maintenance. Theses pipes are tough and rigid and also these sewers have high co-efficient of thermal expansion and hence cannot be used in very hot regions.

19.5.3   Laying of the Sewer Pipes

Sewer pipes are usually laid starting from their outfall ends toward their starting point, and thus, they have the advantage of utilization of the tail length even during the initial periods of its construction is gained. Thus the functioning of the sewerage scheme need not wait till the completion of the entire scheme.

- **Locating the position of manholes**: First of all, the position of the manhole on the ground along the longitudinal section of the sewer line is located. It is a common practice, to lay the sewer pipe between the two manholes. The central line of the sewer is marked on the ground and an offset line is also marked parallel to the central line at a suitable fixed horizontal distance of 2–3 m away from it, about half the trench width plus 0.6 m. This line can be marked on the ground by fixing the pegs at 15 m intervals and can be used for finding out the center line of the sewer simply by offsetting the fixed distances from it.

- **Excavation of trench**: A sewer pipe is laid between the trench of suitable width which is excavated between the two manholes. Further excavations are carried out for laying the pipes between the next consecutive manholes. Thus the process is continued from the outfall end of the sewer toward the uphill until the entire sewers are laid out. The excavated material is deposited on one side of the trench, while the other side is used for the offset line and for lowering the sewer, pipes, and other construction materials into the trench. The trench width at the bottom is usually kept 15 cm more than the diameter of the sewer pipe, with a minimum width of 60–75 cm for smaller diameter pipes to facilitate laying and jointing of pipes.

- **Placing of bedding layer**: A bedding layer of concrete can be provided before laying the sewer pipe in case of ordinary or soft soil, while in case of rocky or hard soil, no bedding is required. The bottom half portion of the trench is excavated to confirm the shape of the pipe itself if the sewer pipes are not embedded in concrete. The trench is dug up to a level of the bottom embedding concrete or up to the invert level of the sewer pipes along with the pipe thickness if no embedding concrete is provided. Invert levels which is designed and desired slope as per the longitudinal section of the sewer should be precisely transferred to the trench bottom.

- **Lowering of sewer pipes into the excavated trench**: The sewer pipes are lowered down into the trench either manually or with the help of machines for bigger pipe diameters just after bedding concrete is laid in required alignment and levels. The length of sewer pipes is usually laid from the lowest point with their sockets facing up the gradient, on desired bedding and hence the spigot end of new pipe can be easily inserted on the socket end of the already laid pipe.

19.5.4   Testing of Sewers

The laid and jointed sewers are tested for watertight joints and straight alignment as described below:

1. **Test for leakage or water test**

The sewers are tested for no leakage just after giving sufficient time for the joints to set. These sewer pipe sections are tested both for the manholes to manhole under a test pressure of about 1.5 m water head for this purpose. For carrying out this, the lower end (i.e., downstream end) of the sewer is connected and water is filled in the manhole at the upper end. Water depth in manhole is maintained at about 1.5 m. Inspection of the sewer line is along the trench and the joints which leak or sweat are repaired.

2. **Test for straightness of alignment and obstruction**

The test for straightness of alignment can be performed by setting a mirror toward one side of the sewer line and a light at the contrary end. A full hover of light will be watched, if the pipe line is straight. Or something bad might happen, the mirror would show any impediments in the pipe barrel. The impediments in the pipe can be tried by embedding a smooth chunk of breadth 13 mm not exactly the inside distance across of the sewer pipe, at the upper end of the sewer. In the event that there is no deterrent, the ball will move down the alternate of the sewer pipe and develop at the lower end.

19.5.5 Backfilling the Trenches

After the sewer line has been laid and tried, the trenches are back filled. The earth ought to be laid similarly on either side of the sewer in layers of 15 cm thickness. Each layer ought to be appropriately watered, packed, and smashed. Over the crown of the sewer pipe, earth filling ought to be done cautiously by hand shoveling in layer utilizing chosen soils. Following a couple of long periods of exposure, when the top layer is completely settled, the street pavements might be built to avoid subsidence and splitting.

19.6 DESIGN OF SEWER

The hydraulic design of sewers and channels implies discovering their segments and slopes. It is usually carried out on the same lines as that of the water supply pipes but there are always two major differences between the characteristic flows in sewers and water supply pipes.

The water supply channels convey pure water with no solid particles under strain. The sewage contains particles in suspension, the heavier of which may settle down at the base of the sewers when the stream speed lessens, bringing about the obstructing of sewers. To abstain from silting of sewers, it is vital that the sewer channels be laid at such an angle, as to create self-purifying speeds at various possible discharges. The sewer channels convey sewage as gravity conduits and are in this way laid at a constant angle the descending way up to the outfall point, from where it will be lifted up, treated, and disposed.

Design of sewers is done assuming steady-state conditions. Steady-state means that the discharge or flow-rate at a point remains time-invariant.

The main objectives of the design of sewers are as follows:

1. Carry the peak flow rate for which the sewer is designed. It is directly connected with the maximum achievable velocity in the sewers. The sewage pipe materials should not get worn out. The wastewater manual recommends a maximum velocity of 3 m/s.
2. Transport suspended solids in such a manner that the siltation in a sewer is kept to a minimum. This

condition gives us an idea about the minimum velocity that has to be maintained inside a sewer during a low flow period.

Sewers are never designed to run full; there is always an empty space provided at the top. This is to carry the extra sewage in case of an unforeseen circumstance and also biodegradation causes generation of gases like methane, hydrogen sulfide, ammonia, etc., which can get dissolved if the sewers are running under pressure; they are unpressurised. At same slopes, the velocity and carrying capacity are more when it runs partially full.

The maximum velocity is achieved when the sewers are designed to run at 80% of the full depth. Generally, the sewer pipes of sizes less than 0.4 m diameter are designed as running half full at maximum discharge, and the sewer pipes greater than 0.4 m in diameter are designed as running two-third or three-fourth full at maximum discharge.

19.6.1 Hydraulic Formula for Determining Flow Velocities in Sewers and Drains

The various empirical formulas for determining the gradients necessary to obtain design velocities of flow in sewers are given below:

1. **Chezy's formula**

$$V = c\sqrt{rs} \tag{19.5}$$

where
V = velocity of flow in the channel (m/s)
c = Chezy's constant
r = hydraulic mean radius of the channel, i.e., hydraulic mean depth of channel (m)

$$r = \frac{a}{p} \tag{19.6}$$

where
a = area of the channel
p = wetted perimeter of the channel
s = hydraulic gradient equal to the ground slope for uniform flows (m/m)

The constant "c" depends on size, shape, and smoother roughness of the channel, the mean depth, etc. It can be obtained using either Kutter's formula or Bazin'formula.

a. Kutter's formula

$$c = \frac{\left(23 + \dfrac{0.00155}{s}\right) + \dfrac{1}{n}}{1 + \left(23 + \dfrac{0.00155}{s}\right)\dfrac{n}{\sqrt{r}}} \tag{19.7}$$

where
c is the Chezy's coefficient and can be used in Equation 19.6,

n=rugosity coefficient depending upon the type of channel surface,

s=bed slope of the sewer,

r=hydraulic mean depth.

b. Bazin's formula

$$c = \frac{157.6}{1.81 + \dfrac{K}{\sqrt{r}}} \qquad (19.8)$$

where

r=hydraulic mean depth of channel

K=Bazin's constant

2. **Manning's formula**

$$V = \frac{1}{n} r^{2/3} s^{1/2} \qquad (19.9)$$

where

n=rugosity coefficient depending upon the type of channel surface,

s=bed slope of the sewer,

r=hydraulic mean depth.

3. **Crimp and Burge's formula**

$$V = 83.5 \cdot r^{2/3} s^{1/2} \qquad (19.10)$$

This formula is comparable to Manning's formula having $\dfrac{1}{n} = 83.5$ or $n = 0.012$.

4. **Hazen–William's formula**

$$V = kCr^{0.63} s^{0.54} \qquad (19.11)$$

where

k=conversion factor for the unit system (k=0.849 for SI units),

C=roughness coefficient,

r=hydraulic mean depth,

S=slope of the energy line (head loss per length of pipe).

19.6.2 Minimum and Maximum Velocities to be Generated in Sewers

The flow velocity should be such that

i. it will not allow the particles to settle inside the sewer;

ii. the sewage pipe material does not get scoured out.

19.6.2.1 Minimum Velocity

The flow velocity in the sewers should be such that the suspended materials in sewage do not get silted up; i.e., the velocity should be such as to cause automatic self-cleansing effect. The generation of self-cleansing velocity should occur within the sewer for at least once in a day, because if certain

deposition takes place and is not removed, it will obstruct free flow, causing further deposition and finally leading to the complete blocking of the sewer. Since velocity depends upon the hydraulic mean depth and the slope on which the sewer is laid, values of slope which are required for generating self-cleansing velocity in different sized pipes can be easily found out.

Self-cleansing velocity is given by

$$V_S = c\sqrt{kd'(G-1)} \qquad (19.12)$$

where c=Chezy's constant. It can be equated to $\sqrt{\dfrac{8g}{f'}}$ by comparing Chezy's formula and Darcy-Weisbach formula.

$$V_S = \sqrt{\frac{8g}{f'} kd'(G-1)} \qquad (19.13)$$

The usual value of f' for sewer pipes is 0.03. Similarly by equating Chezy's formula with Manning's formula, we get

$$V_S = \frac{1}{n} r^{1/6} \sqrt{kd'(G-1)} \qquad (19.14)$$

where

n=rugosity coefficient (for sewer pipes, n=0.013);

r=hydraulic mean depth;

G=specific gravity of the particle;

k=dimensionless constant, 0.04 for granular particles, 0.8 for organic matters;

d'=diameter of the particle for which the sewer will be designed; this is the maximum particle size the sewer can safely carry;

19.6.2.2 Maximum Velocity

Interior surface of a sewer pipe which is smooth gets scoured due to continuous abrasion caused by the suspended solids present in sewage. The damages caused to the sewer pipes will affect their life spans and carrying capacities. Thus it becomes necessary to limit the maximum velocity in the sewer pipe. The material of the sewer will decide the limiting or nonscouring velocity. Table 19.1 shows the nonscouring limiting velocities in sewers and drains.

TABLE 19.1

Non-Scouring Limiting Velocities in Sewers and Drains

Sewer Material	Limiting Velocity in m/s
Vitrified tiles and glazed bricks	4.5–5.5
CI sewers	3.5–4.5
Stone ware sewers	3.0–4.0
Cement concrete sewers	2.5–3.0
Ordinary brick lined sewers	1.5–2.5
Earthen channels	0.6–1.2

19.6.3 Flow Variation Effects on Velocity in a Sewer

The variation in discharge affects the depth of flow varies, and hence, the hydraulic mean depth (r) varies. The change in the hydraulic mean depth affects the flow velocity from time to time. There is a necessity to check the sewer for maintaining a minimum velocity of about 0.45 m/s at the time of minimum flow (assumed to be 1/3rd of average flow).

A velocity of 0.9 m/s is caused at least at the time of maximum flow and mostly during the average flow periods also should be ensured by the designer. More than that, care should be taken to see that at the time of maximum flow, the velocity generated is not more than scouring value.

19.7 SEWER APPURTENANCES

Structures which are constructed at suitable intervals along the sewerage system to help its efficient operation and maintenance are called as sewer appurtenances. Manholes, Drop manholes, Lamp holes, Clean-outs, Street inlets called Gullies, Catch basins, Flushing Tanks, Grease and Oil traps, Inverted Siphons, and Storm Regulators are included in this category.

19.7.1 Manholes

A masonry or RCC chamber developed at reasonable interims along the sewer lines for giving access into them is known as a sewer vent. This helps sewer vent in examination, cleaning, and support of sewer. These are given at each twist, intersection, change of angle, or change of width of the sewer. The sewer line between the two sewer vents is laid straight with even slope. For straight sewer line sewer vents are given at normal interim relying on the distance across of the sewer. The dividing which ought to be accommodated in the sewer vent is suggested in IS 1742-1960. Minimal width of the sewer vent ought not to be in exactly interior distance across the sewer pipe in addition to 150 mm sidelining on both the sides. Tables 19.2 and 19.3 present the recommended spacing of manholes and minimum internal dimensions for manhole chambers, respectively.

19.7.1.1 Classification of Manholes

The manholes can be classified depending upon the depth as Shallow Manholes, Normal Manholes, and Deep Manholes.

a. **Shallow manholes (Figure 19.3)**: These are 0.7–0.9 m depth, built toward the beginning of the branch sewer or at a place not exposed to substantial traffic conditions. These are furnished with light cover at best and called as inspection chamber.

b. **Normal manholes (Figure 19.4)**: These manholes are 1.5 m profound with measurements 1.0 m × 1.0 m square or rectangular with 1.2 m × 1 m. These are furnished with heavy cover at its top to help the foreseen traffic load.

c. **Deep manholes**: The depth of these manholes is more than 1.5 m. The area of such sewer vent is not uniform all through. The size in upper bit is decreased by giving an offset. Steps are given in such sewer vents to slipping into the manhole. These are furnished with substantial cover at its top to help the traffic load. Figure 19.5 shows the rectangular and circular deep manhole.

19.7.1.2 Other Types of Manholes

a. **Straight-through manholes**: The simplest type of manhole, which is built on a straight run of sewer with no side junctions. Wherever there is change in the size of sewer, the soffit or crown level of the two sewers should be the same, except where special conditions require otherwise.

b. **Junction manholes**: These manholes are constructed at every junction of two or more sewers, and on the curved portion of the sewers, with curved portion situated within the manhole. These manholes can be constructed with the shape other than rectangular to suit the curve requirement and expected economy can be achieved. The smaller sewer's soffit at junction should not be lower than that of the larger sewer. The smaller sewer's gradient may be made steeper from the previous manhole to reduce the difference of invert at the point of junction to a convenient amount.

c. **Side entrance manholes**: When large sewers are used it is difficult to obtain direct vertical access to the sewer from the top ground level due to obstructions such as other pipe lines like water, gas, etc.,

TABLE 19.2
Spacing of Manholes

Pipe Diameter	Spacing
Small sewers	45 m
0.9–1.5 m	90–150 m
1.5–2.0 m	150–200 m
Greater than 2.0 m	300 m
Very large sewers	100 m/1 m diameter (general rule)

TABLE 19.3
The Minimum Internal Dimensions for Manhole Chambers

Depth of Sewer	Internal Dimensions
0.9 m or less depth	0.90 m × 0.80 m
For depth between 0.9 and 2.5 m	1.20 m × 0.90 m, 1.2 m diameter for circular
For depth above 2.5 m and up to 9.0 m	For circular chamber 1.5 m diameter
For depth above 9.0 m and up to 14.0 m	For circular chamber 1.8 m diameter

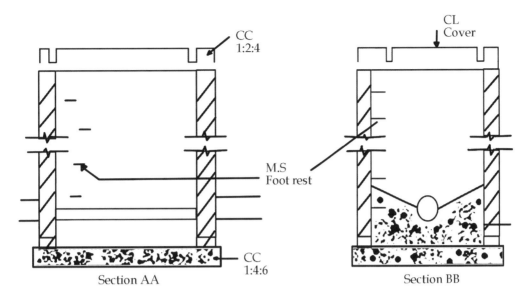

FIGURE 19.3 Shallow manhole.

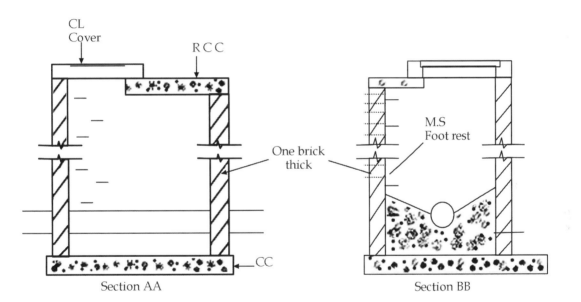

FIGURE 19.4 Normal manhole.

the access shaft should be constructed in the nearest convenient position off the line of sewer, and it should connected to the manhole chamber by a lateral passage. The floor of the side entrance passage which should fall at about 1 in 30 toward the sewer should enter the chamber not lower than the soffit level of the sewer. Large sewers should be provided with necessary steps or a ladder (with safety chain or removable handrail) should be provided to reach the benching from the side entrance above the soffit.

d. **Drop manholes (Figure 19.6)**: A sewer which connects with another sewer and where the difference in level between invert level of branch sewer and water line in the main sewer at maximum discharge is greater than 0.6 m; a manhole may be built either

with vertical or nearly vertical drop pipe from higher sewer to the lower one. The drop manhole is also necessary in the same sewer line in sloping ground if the drop more than 0.6 m is required to control the gradient and to satisfy the maximum velocity, i.e., nonscouring velocity.

e. **Scraper (service)-type manhole**: All sewers having diameter of 450 mm should have one manhole at intervals of 110–120 m of scraper type. This type of manhole should have a clear opening of 1.2 m×0.9 m at the top to facilitate lowering of buckets.

f. **Flushing manholes**: Branch sewers with flat ground, it is not possible to obtain self-cleansing velocity at all flows; due to very little flow, it is necessary to incorporate flushing device. This can be achieved by making grooves at intervals of 45–50 m in the

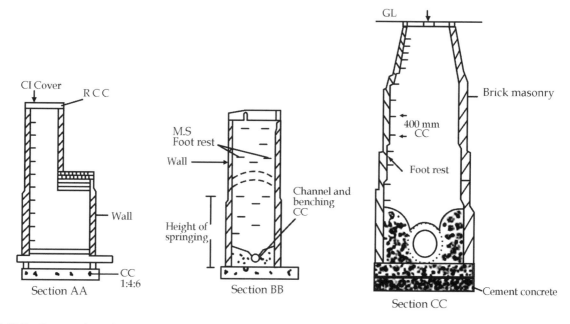

FIGURE 19.5   Rectangular and circular deep manhole.

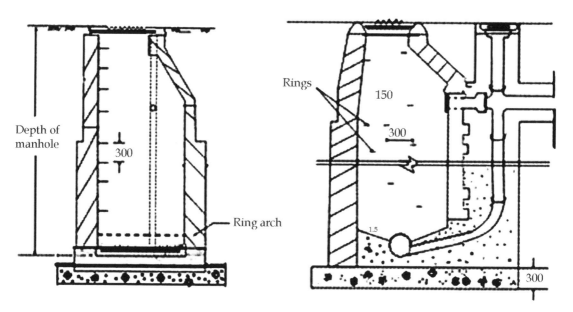

FIGURE 19.6   Drop manhole.

main drains in which wooden planks are inserted and water is allowed to head up. The planks when removed will allow the water to rush with high velocity facilitating cleaning of the sewers. Other method is by flushing by using water from overhead water tank through pipes and flushing hydrants or through fire hydrants or tankers and hose.

19.7.2 Inverted Siphons

The sewer that runs full under gravity flow at a pressure above atmosphere in the sewer is called an inverted siphon or depressed sewer (as shown in Figure 19.7). These are used to

pass under obstacles such as buried pipes, subways, etc. Since the inverted siphon requires considerable attention for maintenance, it can be used only where other means of passing an obstacle in line of the sewer are impracticable.

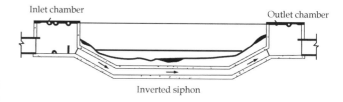

FIGURE 19.7   Inverted siphon.

19.7.3 Storm Water Inlets

a. Storm water deltas are utilized to concede the surface overflow to the sewers. These are isolated into three noteworthy gatherings, viz., control gulfs, drain channels, and joined deltas. They are utilized either as a discouraged or flush regarding the height of the asphalt surface. The reasonable opening ought not to be in excess of 25 mm. The pipe which is associating from the road channel to the sewer ought to be least of 200 mm distance across and laid with adequate incline. A dispersing of 30 m is prescribed a most extreme limit between the channels, which relies out and about surface, size, and sort of gulf and precipitation.

b. **Curb inlet (Figure 19.8)**: These are vertical openings in the road curbs through which storm water flow enters the storm water drains. These are preferred where heavy traffic is anticipated.

c. **Gutter inlets (Figure 19.9)**: Horizontal openings in the gutter which are covered by one or more gratings through which storm water is admitted are called as gutter inlets.

d. **Combined inlets**: In combined inlets, the curb and gutter inlet both are used to act as a single unit. They are normally placed right in front of the curb inlets.

19.7.4 Catch Basins

Catch basins (Figure 19.10) are utilized to stop the passage of overwhelming flotsam and jetsam present in the tempest water into the sewers; however, their utilization is debilitated as the disturbance because of mosquito rearing separated from presenting generous upkeep issues, gathering of contaminations. A space is accommodated with the gathering of polluting influences at the base. Punctured cover is given at the highest point of the bowl to permit rainwater to be gathered into the bowl. A hood is given to forestall getaway of sewer.

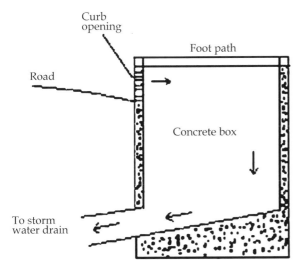

FIGURE 19.8 Curb inlet.

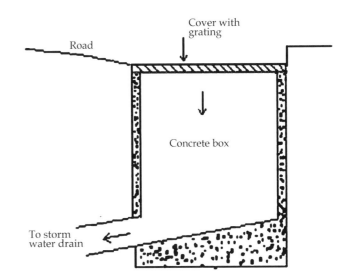

FIGURE 19.9 Gutter inlet.

19.7.5 Clean-Outs

Clean-out (Figure 19.11) is a pipe connected to the sewer placed under the ground. One end of the clean-out pipe is brought up to ground level and a cover is placed at ground level. A clean-out is usually provided at the upper end of lateral sewers as an alternate to manholes. When pipe blockage occurs, the cover is taken out and water is forced through the clean-out pipe to lateral sewers to remove obstacles in the sewer line. When met with large obstacles, flexible rod may be inserted through the clean-out pipe and moved forward and backward to remove such obstacles.

19.7.6 Regulator or Overflow Device

They avoid overloading of sewers, pumping stations, treatment plants, or disposal arrangement, by diverting the excess flow to relief sewer. The overflow device may be side flow or leaping weirs depending upon the position of the weir, siphon spillways, or float-actuated gates and valves.

a. **Side-flow weir**: It is constructed along one or both sides of the combined sewer and releases the excess flow during storm period to relief sewers or natural drainage courses. The crest of the weir is set at an elevation depending upon the desired depth of flow in the sewer. The weir length must be sufficient long for effective regulation of the flow.

b. **Leaping weir (Figure 19.12)**: A gap found in the invert of a sewer through which the dry weather flow falls and over which a portion of the entire storm leaps is used to form a leaping weir. It operates as a regulator without including moving parts but the disadvantage of this weir is that the grit material gets concentrated in the lower flow channel which is one of its advantages. From a practical viewpoint, it is desirable to use moving crests to make the opening adjustable. Whenever the discharge is

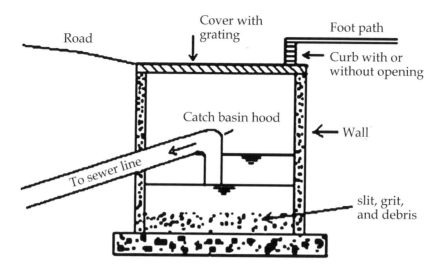

FIGURE 19.10   Catch basin.

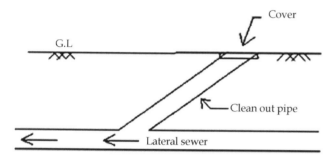

FIGURE 19.11   Clean-outs.

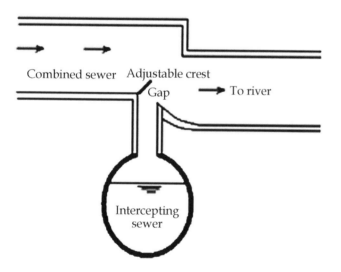

FIGURE 19.12   Leaping weir.

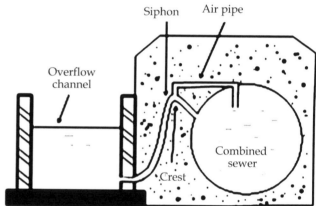

FIGURE 19.13   Siphon spillway.

automatic mechanical regulators. These are actuated by the float depending upon the water level in the sump interconnected to the sewers. As moving part is involved in this, regular maintenance of this regulator is essential.

d. **Siphon spillway (Figure 19.13)**: This arrangement of diverting excess sewage from the combined sewer is most effective as it works on the principle of siphon action and it operates automatically. The overflow channel is connected to the combined sewer through the siphon. An air pipe at the crest level of siphon activates the siphon when water reaches the combined sewer at the stipulated level.

19.7.7   Flap Gates and Flood Gates

Flap gates or backwater gates are installed at or near sewer outlets to prevent backflow of water during high tide, or at high stages in the receiving stream. These gates can be rectangular or circular in shape and made up of wooden planks or metal alloy sheets. Such gates should be designed such that the flap should get open at a very small head difference.

small, the sewage falls directly into the intercepting sewer through the opening. But when the discharge exceeds a certain limit, the excess sewage leaps or jumps across the weir and it is carried to natural stream or river.

c. **Float-actuated gates and valves**: The excess flow in the sewer can also be controlled by means of

Adequate storage in outfall sewer is also necessary to prevent back flow into the system due to the closure of these gates at the time of high tides, if pumping is to be avoided.

19.7.8 SEWER VENTILATORS

It is necessary to ventilate the sewer to make provision for the escape of air to take care of the exigencies of full flow and to keep the sewage as fresh as possible. When storm water is involved, this can be done by providing ventilating manhole covers. It is not necessary to provide provision of ventilators' elimination of intercepting traps in the house connections in case of modern sewerage system.

19.7.9 LAMP HOLE

Lamp hole is an opening or hole constructed in a sewer for the purpose of lowering a lamp inside it. It consists of stoneware or concrete pipe, which is connected to sewer line through a T-junction as shown in Figure 19.14. Concrete is made stable by covering it with pipe. Manhole cover of sufficient strength is provided at the ground level to take the load of traffic. An electric lamp is inserted in the lamp hole and the light of lamp is observed from manholes. When sewer length is unobstructed, the light of lamp will be seen. When construction of manhole is difficult, it should be constructed. The use of lamp hole is avoided as far as possible in present practices. This lamp hole can also be used for flushing the sewers. If the top cover is perforated it helps in ventilating the sewer, such lamp hole is known as fresh air inlet.

19.8 SEWAGE PUMPS

There are certain locations where it is possible to convey sewage by gravity to a central treatment facility or storm water is conveyed up to disposal point entirely by gravity. Whereas, in case of large area being served with flat ground, localities at lower elevation or widely undulating topography it may be

essential to employ pumping station for conveyance of sewage to central treatment plant. Sewage and storm water are required to be lifted up from a lower level to a higher level at various places in a sewerages system. Pumping of sewage is also generally required at the sewage treatment plant.

Pumping of sewage is different than water pumping due to polluted nature of the wastewater containing suspended solids and floating solids, which may clog the pumps. The dissolved organic and inorganic matter present in the sewage may chemically react with the pump and pipe material and can cause corrosion. The disease-causing bacteria present in the sewage may pose health hazard to the workers.

19.8.1 TYPES OF PUMPS

The following types of pumps are used in the sewerage system for pumping of sewage, sewage sludge, grit matter, etc., as per the suitability:

1. **Radial-flow centrifugal pumps**: These pumps have main two parts: the casing and the impeller. The impeller of the pump rotates at high speed inside the casing. The centrifugal force will cause the sewage to be drawn from the suction pipe into the pump and curved rotating vanes throw it up through the outlet pipe. Radial-flow pump throws the liquid entering the center of the impeller into a spiral volute or casing. The impellers of all centrifugal pumps can be shut, semi-open, or open according to the application. Open-impeller-type pumps are more apt as the suspended solids and floating matter present in the sewage can be easily pumped without clogging. These pumps have a horizontal or vertical design. These pumps are commonly used for any capacity and head. These pumps have low specific speed up to 4200.

2. **Axial-flow centrifugal pumps**: Axial-flow designs handle large capacities but only with reduced discharge heads. They are vertically constructed. The vertical pumps have positive submergence of the impeller. These are used for pumping large sewage flow, more than 2000 m³/hour and head up to 9.0 m. These pumps have relatively high specific speed of 8000–16,000. The water enters in this pump axially and the head is modified by the propelling action of the impeller vanes.

3. **Mixed flow pumps**: These pumps develop leads by a combination of centrifugal action and the lift of the impeller vane on the liquid. They have single impeller. The flow enters the pump axially and releases in an axial and radial direction into volute type casing. The specific speed of the pump varies from 4200 to 9000. These are applied for medium heads ranging from 8 to 15 m.

4. **Positive displacement pumps**: These pumps comprise reciprocating piston, plunger, and diaphragm pumps. Most of the reciprocating pumps that are

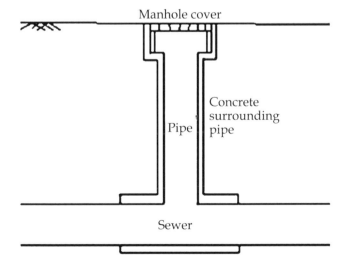

FIGURE 19.14 Lamp hole.

applied in environmental engineering are metering or power pumps. A piston or plunger is provided in a cylinder that draws forward and backward by a crankshaft connected to an outside driving unit. Altering metering pump flow involves merely changing the length and number of piston strokes. A diaphragm pump is almost alike to a reciprocating piston or plunger, but instead of a piston, it contains a flexible diaphragm which oscillates as the crankshaft rotates. Plunger and diaphragm pumps feed metered amounts of chemicals (acids or caustics for pH adjustment) to a water or wastewater stream. They are not suitable for sewage pumping because solids and rugs present in the sewage may clog them. They have high initial cost and very low efficiency.

5. **Rotary screw pumps**: A motor rotates a vane screw or a rubber stator on a shaft to lift or feed sludge or solid waste material to a higher level or the inlet of another pump for these types of pumps. These are applied in the square grit chamber for removal of grit.

6. **Air pumps**: These pumps include pneumatic ejectors and airlifts. In pneumatic ejector wastewater flows into a receiver pot and an air pressure system then blows the liquid to a treatment process at a higher elevation. The air system uses plant air (or steam), a pneumatic pressure tank, or an air compressor. This pumping system has no moving parts in contact with the waste; thus, no clogging of impeller is involved. Ejectors are normally maintenance-free and operate for longer time. Airlift pumps consist of an updraft tube, an airline, and an air compressor or blower. Airlifts blow air at the bottom of a submerged updraft tube. They expand reducing density and pressure within the tube as the air bubbles travel upward. In this way, higher flows can be lifted for short distances. Airlifts have their application in wastewater treatment to transfer mixed liquors or slurries from one process to another. These pumps have very low efficiency and they can lift the sewage up to small head.

19.8.2 Pumping Station

Pumping stations are often required for pumping of untreated domestic wastewater, storm water runoff, combined domestic wastewater and storm water runoff, sludge at a wastewater treatment plant, treated domestic wastewater, and recycling treated water or mixed liquor at treatment plants.

Pumping stations can be configured in a wide variety of arrangements, depending on size and application. The classifications for such pumping-station configurations are wet well/dry well, wet well only with submersible pumps, and wet well only with nonsubmersible pumps.

1. **Wet well and dry well**: In this configuration, two pits (wells) are required: one to hold the fluid, and one to house the pumps and appurtenances. This is required for fluids that cannot be primed or conveyed long distances in suction piping; this option is typically used to pump large volumes of raw wastewater, where uninterrupted flow is critical and wastewater solids could clog suction piping. While construction costs of this type may be higher and a heating, ventilation, and cooling system is necessary due to installation below ground. This configuration is best for operation and maintenance activities because operators can see and touch the equipment. The pumping station with horizontal pump and vertical pump in a dry well is shown in Figure 19.15.

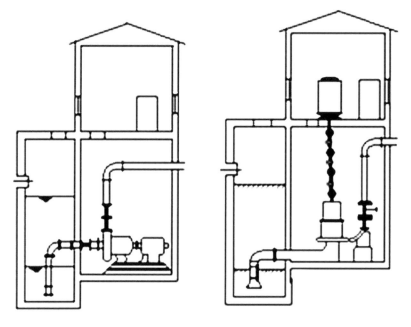

FIGURE 19.15 Pumping station with horizontal pump and vertical pump in a dry well.

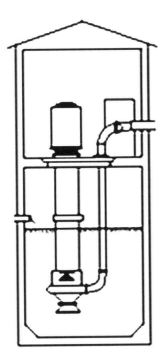
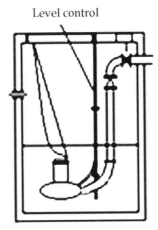

Level control

FIGURE 19.16   Pumping station with vertical pump in a wet well and wet well with submersible pump.

2. **Wet well with submersible pumps**: In this configuration, one well holds both the pumps and the wastewater being pumped. The pump impeller is submerged or nearly submerged in the wastewater. Additional piping is not required in this type to convey the wastewater to the impeller. This option is common worldwide, and the submersible centrifugal pumps can be installed and operated cost-effectively. When vertical pumps are installed the driving motor is mounted on the floor above the ceiling of the wet well.

3. **Wet well with nonsubmersible pumps**: In this configuration, one well holds the wastewater. The pumps are installed above the water level in wet well. This option is used in areas where the wastewater can be "pulled" through suction piping, e.g., treated or finished water or where shutdowns or failures would not be immediately critical, e.g., a package plant's raw wastewater lift stations, equalization of secondary treated wastewater, etc. Pumping station with vertical pump in a wet well and wet well with submersible pump is shown in Figure 19.16.

19.9   HOUSE DRAINAGE

The arrangement provided in a house or building for collecting and conveying wastewater through drain pipes, by gravity to join either a public sewer or a domestic septic tank that is termed as house drainage or building drainage.

For the proper design and construction of house drainage system, the following general principles are followed:

i. It is advisable to lay sewers by the side of building rather than below the building.

ii. The drains should be laid straight between inspection manholes.

iii. The house drain should be connected to the public sewer only if the level permits, i.e., only when public sewer is deeper than the house drain. Otherwise there will be reverse flow from the public sewer to the house drain.

iv. The entire system should be properly ventilated.

v. The house drainage should be containing enough number of traps at suitable points for efficient functioning of it.

vi. The joints should be watertight.

vii. The lateral sewers should be laid at proper gradient so that they will develop self-cleansing velocity.

viii. The layout house drainage system permits easy cleaning and removal of obstructions.

ix. The sewage formed should be conveyed as early as possible after its formation.

x. The materials of sewer should comply with the standard requirements.

xi. The rainwater from houses is collected from roofs and is allowed to flow freely on the road surface for catch basins to convey it to the storm water drains.

19.9.1   PIPES

In a house drainage system, a pipe may have the following designations depending up on the function it carries:

a. **Soil pipe**: A soil pipe is a pipe through which human excreta flows.

b. **Waste pipe**: It is a pipe which carries only the liquid waste. It does not carry human excreta.

c. **Vent pipe**: It is a pipe which is provided for the purpose of the ventilation of the system. A vent pipe is open at top and bottom to facilitate exit foul gases. It is carried at least 1 m higher than the roof level.

d. **Rainwater pipe**: It is a pipe which carries only the rainwater.

e. **Antisiphonage pipe**: It is a pipe which is installed in the house drainage to preserve the water seal of traps.

19.9.2 TRAPS

A trap is a depressed or bent fitting which when provided in a drainage system always remains full of water, thus maintaining a water seal. It prevents the passage of foul air or gas through it and allows the sewage or wastewater to flow through it.

Classification of traps

a. **Classification according to shape**:
   1. P-trap: This resembles the shape of letter P in which the legs are at right angles to each other.
   2. Q-trap or half- S-trap: This resembles the shape of letter Q in which the two legs meet at an angle other than right angle.
   3. S-trap: This resembles the shape of letter S in which both the legs are parallel to each other, discharging in the same direction.

b. **Classification according to use**:
   1. Floor trap or nahni trap: A floor trap is mainly used to collect wash water from floors, kitchens, and bathrooms. It forms the starting point of wastewater floor.
   2. Gully trap: These are special types of traps, provided at the external faces of the wall, which disconnect sullage drain from the main drainage system. Thus it receives wastewater from baths, kitchens, etc., and passes it onto house drain carrying excremental discharge from water closets.
   3. Intercepting trap: These are special types of trap provided at the junction of house drain with the public sewer or septic tank. It is thus provided in the last manhole of the house drainage system. It has a deep water seal of 100 mm so as to effectively prevent the entry of sewer gases from public sewer line into the house drain.
   4. Grease traps: Such traps are used only in large hotels, restaurants, or industries where large quantities of oily wastes are expected to enter the water flow. If the oily or greasy matter is not separated, it will stick to the building drainage system resulting in the formation of ugly scum and consequent obstruction to re-aeration.

19.9.3 SANITARY FITTINGS

The following fittings are commonly used in buildings for efficient collection and removal of wastewater to the house drain.

a. **Wash basin**: Wash basins are usually made of pottery or porcelain ware. An ordinary wash basin is mounted on brackets fixed on wall while a pedestal type basin is mounted on pedestal rising from wall. The waste pipe with a metallic strainer is provided at the bottom of the bowl.

b. **Sink**: While a wash basin is used for washing hands, face, etc., a sink is used in kitchen or laboratory. These may be made of glazed fire clay, stainless steel, and metal porcelain. It may have a dashboard attached to it.

c. **Bath tub**: Bath tubs are usually made of iron or steel coated with enamel, enameled porcelain, or of plastic material. It has a length varying from 1.7 to 1.85 m, width between 0.7 and 0.75 m, and depth near waste pipe varies between 0.43 and 0.45 m. It is provided with an outlet and overflow pipes. A trap with proper water seal is used at the outlet.

d. **Water closets**: Water closets are designated to receive and discharge human excreta directly from the person using it. The appliance is connected to the soil pipe by means of a suitable trap. It is usually connected to a flushing cistern to flush the closet and discharge human excreta to the soil pipe.

e. **Urinals**: Urinals are usually of two types. They are bowl type and slab type. The former type is used in residential buildings while the latter type is used in public buildings. The best types of urinals are made of enameled fireclay and others of salt-glazed stoneware, marble, and slate. The contents of urinals are collected and discharged into the soil pipe through floor trap.

f. **Flushing cisterns**: Flushing cisterns are used for flushing out water closets and urinals. These are made of either cast iron (CI) or porcelain. For Indian-type water closet, it is normally used, fixed at about 2 m above the floor level. For European-type and Anglo-Indian-type closets, they are fixed at about 60 cm above floor levels.

19.10 QUALITY OF SEWAGE

Characterization of wastes is essential for an effective and economical waste management program. It helps in the choice of treatment methods deciding the extent of treatment, assessing the beneficial uses of wastes, and utilizing the waste purification capacity of natural bodies of water in a planned and controlled manner. The factors which contribute to variations in characteristics of the domestic sewage are daily per capita use of water; quality of water supply and the type, condition, and extent of sewerage system; and habits of the people.

19.10.1 CHARACTERISTICS OF SEWAGE

The quality of sewage can be checked and analyzed by studying its physical, chemical, and bacteriological (biological) characteristics as explained below.

19.10.1.1 Physical Characteristics of Sewage

The physical characteristics of sewage involve turbidity, color, odor, and temperature.

a. **Color**: The color of sewage can normally be detected by the naked eye, and it indicates the freshness of sewage. If the color is yellowish, grey, or light brown, the sewage is fresh. However, if the color is black or dark brown, it is stale and septic sewage. Other colors may also be formed due to the presence of some specific industrial wastes.

b. **Odor**: Fresh sewage is practically odorless. But, however, in 3–4 hours, it becomes stale with all oxygen present in sewage being practically exhausted. If then starts omitting offensive odors especially that of hydrogen sulfide gas, which is formed due to decomposition of sewage, the odor of wastewater can be measured by a term called the threshold odour number (TON), which represent the extent of dilution required to just make the sample free of odor.

$$TON = \frac{V_S + V_D}{V_S} \qquad (19.15)$$

where

TON = threshold odor number

V_S = volume of sewage

V_D = volume of distilled or odorless water added to just make the sewage sample lose its odor

c. **Temperature**: The average temperature of sewage in India is 20°C, which is near about ideal temperature for the biological activities. The temperature has an effect on the biological activity of bacteria present in sewage, and it also affects the biological activity of bacteria present in sewage, and it also affects the solubility of gases in sewage. In addition, temperature also affects the viscosity of sewage, which, in turn, affects the sedimentation process in its treatment.

The normal temperature of sewage is generally slightly higher than the temperature of water, because of additional heat added during the utilization of water.

d. **Turbidity**: The degree of turbidity is measured and tested by turbidity rods or by turbidimeters. Sewage is normally turbid, resembling dirty dish water or wastewater from baths having floating matter like fecal matter, pieces of paper, cigarette ends, match sticks, greases, vegetable debris, etc., the turbidity increases as sewage becomes stronger.

19.10.1.2 Chemical Characteristics of Sewage

The chemical analysis is carried out on sewage to determine its chemical characteristics. It includes tests for determining the following:

a. pH value

The pH of the fresh sewage is slightly more than the water supplied to the community and is measured using potentiometer. However, decomposition of organic matter may lower the pH, while the presence of industrial wastewater may produce extreme fluctuations. pH value of sewage affects the efficiency of certain treatment method. Generally the pH of raw sewage is in the range of 5.5–8.0.

b. Total solids

Sewage normally contains very small amount of solids in relation to the huge quantity of water (99.9%). It only contains about 0.05%–0.1% (i.e., 500–1000 mg/l) of total solids. Solids present in sewage may be in any of the four forms: suspended solids, dissolved solids, colloidal solids, and settleable solids.

i. Total solids: The sample of sewage is taken and most of the liquid is evaporated from it. The residue is dried and its weight represents the amount of total solids in sewage. Now the solids may either be volatile or fixed. In order to find out the proportion of volatile and fixed solids in total solids, the dry residue as obtained above is heated or ignited. The loss of weight due to heating represents volatile matter and the rest is fixed matter.

ii. Suspended and dissolved solids: The sample of sewage is passed through an asbestos filter. The quantity of solids in the effluent is then found out as above. The difference with the amount of solids obtained in (i) indicates the quantity of suspended solids. The remaining quantity of total solids as obtained in (i) indicates the dissolved solids. Thus,

Total solids = Suspended solids + Dissolved solids

Suspended and dissolved organic solids are responsible for creating nuisance if left untreated.

iii. Settleable solids: It is the portion of solid matter that settles out from sewage when left undisturbed for a period of 2 hours. To find out the quantity of settleable solids in a sample of sewage, the Imhoff cone is used. The capacity of cone is 1 liter and is graduated up to about 50 ml as shown in Figure 19.17.

It is in the form of a specially designed conical glass vessel. The sewage is allowed to stand in the Imhoff cone for a period of 2 hours, and the quantity of settled solids in the cone is then read to find out the exact proportion of the settleable solids, the liquid should be removed or decanted off and the settleable solids, collected at the bottom of cone, should be weighed.

c. Chloride content

The chloride concentration in excess than the water supplied can be used as an index of the strength of the sewage. The daily contribution of chloride averages to about 8 gm/person. Normal chloride content

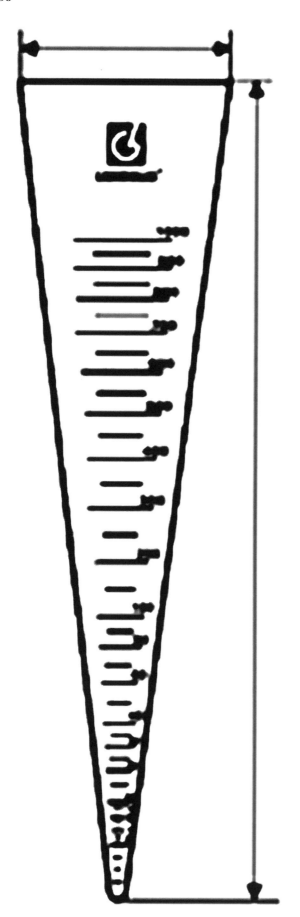

FIGURE 19.17   Imhoff cone.

of domestic sewage is 120 mg/l which is less than the permissible chloride content for water supplies. Any abnormal increase should indicate discharge of chloride-bearing wastes or sea water infiltration, the latter adding to the sulfates as well, which may lead to excessive generation of hydrogen sulfide.

The chloride content can be measured by titrating the wastewater with standard silver nitrate solution, using potassium chromate as an indicator.

d. Nitrogen content

The presence of organic matter in sewage is usually reflected as the presence of nitrogen in sewage. The nitrogen is present in sewage either as *free ammonia* or *albuminoid nitrogen* or *nitrates* or *nitrites*. The presence of considerable amount of *free ammonia* indicates stale or old sewage. It is also known as *Ammonia Nitrogen*. It can be measured by boiling the sample and measuring the ammonia gas liberated.

The *albuminoid nitrogen* indicates the amount of undecomposed nitrogenous material in sewage. It is also known as *Organic nitrogen*. Its presence is measured by adding strong alkaline solution of $KMnO_4$ to already boiled sewage. Free ammonia and albuminoid nitrogen are together knows as *Kjeldahl Nitrogen*.

The *nitrites* and *nitrates* are found in very small amount in fresh sewage. The nitrates are found in very small amount in fresh sewage. The nitrites indicate the intermediate stage of conversion of organic matter of sewage into stable forms. The nitrates indicate the most stable form of nitrogenous matter contained in sewage. Thus, the nitrate will predominate in stale sewage and nitrates will predominate in well-oxidized sewage. The increase in proportion of nitrates during the process of sewage treated serves as a guide for measuring the progress achieved in sewage treatment. The amount of nitrites and nitrates is found out by color-matching methods.

e. Presence of fats, oils, and greases

Greases, fats, and oils are derived in sewage from the discharges of animals and vegetable matter, or from the industries like garages, kitchens, and restaurants, etc. Such matter form scum on the top of the sedimentation tanks and clog the voids of the filtering media.

In order to estimate the amount of fats and greases, sample sewage is, first of all, evaporated. The residual solids left are then mixed with ether (hexane). The solution is then poured off and evaporated, leaving behind the fats and greases as a residue, which can be easily weighed.

f. Sulfides, sulfates, and hydrogen gas

In aerobic digestion of sewage, the aerobic and facultative bacteria oxidize the sulfur and its compounds present in the sewage to initially form sulfides, which ultimately break down to form sulfates ions, which is stable and unobjectionable end products. In anaerobic digestion of sewage the anaerobic and facultative bacteria reduce the sulfur and its compounds into sulfides, with the evolution of H_2S gas along with methane and carbon dioxide, thus causing very obnoxious odors.

g. Dissolved oxygen

Dissolved oxygen is the amount of oxygen in the dissolved state in the wastewater. Though the wastewater generally does not have dissolved oxygen (DO), its presence in untreated wastewater indicates that the wastewater is fresh. Similarly, its presence in the treated wastewater effluent indicates that the considerable oxidation has been accomplished during the treatment stages. While discharging the treated wastewater into receiving waters, it is essential to ensure that at least 4 mg/l of DO is present in it. If DO is less, the aquatic animals like fish, etc., are likely to be killed near the vicinity of disposal. DO determination also helps to find the efficiency of biological treatment. The amount of DO present will be less if the temperature of the sewage is more.

The DO content of sewage is generally determined by the Winkler's method which is an oxidation-reduction process carried out chemically to liberate iodine in amount equivalent to the quantity of DO originally present.

h. Biochemical oxygen demand

There are two types of organic matter: (1) biodegradable or biologically active and (2) nonbiodegradable or biologically inactive. Organic matter is often assessed in terms of oxygen required to complete oxidizing the organic matter to CO_2, H_2O, and other end products of oxidation. Biochemical oxygen demand (BOD) is defined as the amount of oxygen required by the microorganisms (mostly bacteria) to carry out decomposition of biodegradable organic matter under aerobic conditions.

The BOD test is widely used to determine the pollution strength of domestic and industrial wastes in terms of the oxygen that they will require if discharged into natural watercourses. This test is of prime importance in regulatory work and in studies designed to evaluate the purification capacity of receiving bodies of water. It is also useful in the design of wastewater treatment plant and also to measure the efficiency of some treatment processes.

The organic matter present in the wastewater may belong to two groups:
a. Carbonaceous matter
b. Nitrogenous matter

The ultimate carbonaceous BOD of a waste is the amount of oxygen necessary for microorganisms in the sample to decompose the biodegradable carbonaceous material. This is the first stage of oxidation and the corresponding BOD is called as first-stage BOD.

In the second stage the nitrogenous matter is oxidized by autotrophic bacteria, and the corresponding BOD or nitrification demand. In fact, polluted water will continue to absorb oxygen for many months, and it is not practically feasible to determine this ultimate oxygen demand. Hence the 5-day period is generally chosen for the standard BOD test, during which oxidation is about 60%–70% complete, while within 20 days period oxidation is about 95%–99% complete. A constant temperature of 20°C is maintained during incubation. The BOD value of 5-day incubation period is commonly written as BOD_5 or simply as BOD.

Another reason for selecting 5 days as standard duration is to avoid interference of nitrification bacteria. Nitrification starts after 6th or 7th day. Sanitary engineers are generally interested in carbonaceous BOD only, so by selecting 5 days, we generally get only the carbonaceous BOD. Interference of nitrification can be eliminated by pretreatment of sample or by using inhibitory agents like methylene blue. Nowadays BOD test is also done at 27°C and a duration of 3 days (BOD_3^{27}); results can be obtained faster and it is more nearer to the actual field conditions in India.

BOD test: The sample is first diluted with a known volume of specially prepared dilution water. Dilution water contains salts and nutrients necessary for biological activity and phosphate buffer to maintain pH around 7–7.5. The diluted water is fully aerated and initial DO of the diluted sample is measured. The diluted sample is then incubated for 5 days at 20°C. The DO of the diluted sample after the incubation period is found out. The difference between the initial DO of the diluted sample after the incubation period is found out. The difference between the initial DO value and the final DO value will indicate the oxygen consumed by the sewage sample in aerobic decomposition in 5 days. The BOD in mg/l or ppm is then calculated by using the equation:

BOD = DO consumed in the test by the diluted

$$\text{sample} \times \frac{\text{Volume of diluted sample}}{\text{Volume of undiluted sample}} \quad (19.16)$$

The ratio of volume of diluted sample to undiluted sample is known as the dilution factor.

Theory of BOD reaction: The first stage of BOD can be expressed for practical purposes as per first-order reaction kinetics in the form of the following mathematical equation:

Notations used:
L_t = amount of first stage BOD remaining in the water at time t
R = constant and is taken as negative because BOD remaining will be decreasing

L = BOD remaining when $t=0$, i.e., BOD present in the sewage sample

The amount of BOD remaining at any time can be obtained from the equation:

$$L_t = L\left(10^{-Rt}\right) \qquad (19.17)$$

The amount of BOD that has been exerted at any time t, X can be found out from the following equation:

$$X = L\left(1 - 10^{-Rt}\right) \qquad (19.18)$$

For instance, the 5-day BOD equals to

$$X_5 = L - L_5 = L\left(1 - 10^{-5t}\right) \qquad (19.19)$$

For polluted water and sewage, a typical value of 0.1 per day for R is taken at 20°C and the value of R at any other temperature T per day can be obtained by applying the following approximate equation:

$$R_t = R_{20}\left(1.047^{T-20}\right) \qquad (19.20)$$

It may be noted that greater the temperature, greater the rate constant up to 45°C.

Significance and applications of BOD test:

- BOD is the principle test, which gives an idea of the biodegradability of any sample and strength of the waste.
- BOD is considered an important parameter in the design of treatment plant to determine the size of certain units particularly trickling filters and activated sludge units.
- BOD is used to estimate the population equivalent of any industrial waste, which is useful to collect cess from industrialist for purification of industrial wastes in municipal sewage treatment plants.
- From BOD of the influent and effluent discharged, the efficiency of the treatment plant can be judged.

Limitation of BOD test:
- It can measure only the biodegradable organic matter.
- It consumes more time, i.e., 5 days, so if quick results are needed it is not useful.
- Pretreatment is needed when the sample contains toxic waste.
- Nitrifying bacteria causes interferences and provides higher results. Proper care must be taken to avoid them.
- It is essential to have high concentration of active bacteria present in the sample.

Chemical oxygen demand (COD): The amount of oxygen required to chemically oxidize organic matter using a strong oxidizing agent like potassium dichromate under acidic condition. The COD test can be conducted to measure organic matter present in industrial wastes having toxic compounds likely to interfere with the biological life. It is used to measure both biodegradable and nonbiodegradable organic matter. COD test takes 3 hours in comparison to 5 days for BOD test. In COD test, a strong chemical oxidizing agent like potassium dichromate is used in acidic medium to oxidize the organic matter present in the waste. Almost all types of organic matter, with a few exceptions, can be oxidized by the action of strong oxidizing agents under acidic conditions.

In the case of several types of wastes, a relation between BOD and COD can be established. Once this is done, it becomes convenient and simple to rely on the COD test because it can be determined in 3 hours as against five days taken for the BOD. For typical domestic wastes, the ratio of COD to BOD varies between 1.2 and 1.5. If the ratio exceeds 3, the sewage is considered difficult to biodegrade; nonbiodegradable sewage typical has a ratio that exceeds 10. The limiting value of COD of wastewater is taken as 250 mg/l.

Significance and applications of the COD test are as follows:

- The COD test is widely used instead of the BOD test in the operation of treatment facilities because results can be obtained at a high speed.
- The COD test is useful in assessing the strength of wastes that contains toxic and biologically resistant organic matter.
- The ratio of BOD and COD is a useful means to evaluate the amenability of waste for biological treatment.
- The analysis of industrial waste widely uses the COD test.
- The COD test is used to identify the performance at various stages of the treatment plant.
- It determines the strength of wastewater which cannot be determined by the BOD test.

Advantages of COD over the BOD test:
- When toxic matters are present and conditions are not favorable for the growth of microbes, BOD cannot be determined accurately.
- The COD test gives speedy result as it takes about 3 hours as against 5 days for the BOD test.
- The COD test determines the strength of certain wastes which cannot be determined by BOD test.
- The COD test is very easy as compared with the BOD test.
- The COD test is unaffected by interferences as in case of the BOD test.

19.10.1.3 Bacteriological Characteristics

Several microorganisms such as bacteria, algae, fungi, protozoa, etc., are found in sewage. The bacteria in sewage are largely nonpathogenic or harmless, and aid the oxidation and decomposition of sewage. However, a few of them are pathogens which produce diseases and these are the real danger to public health. Routine bacteriological tests that are performed for water samples are generally not performed for sewage samples owing to the high concentration of bacteria present in it. But at the time of epidemic outbreak, specific tests may be done to find the type of pathogens.

19.10.2 Relative Stability

Relative stability of sewage effluent is the ratio of oxygen available in the effluent (in the form of DO, nitrite, or nitrate) to the total oxygen needed to satisfy its first-stage BOD demand. It is expressed as a percentage of the total oxygen required and is expressed by the equation:

$$S = 100\left(1 - 0.794^{t_{20}}\right) \qquad (19.21)$$

or

$$S = 100\left(1 - 0.630^{t_{37}}\right) \qquad (19.22)$$

where

S = the relative stability,

t_{20} and t_{37} denote the time (in days) it takes for a sewage sample to decolorize a standard methylene blue solution when incubated at 20°C and 37°C, respectively.

In fact, the decoloration caused by enzymes, which is produced by anaerobic bacteria, is an indication of the available oxygen in oxidizing the unstable organic matter. The sooner the decolorization occurs the earlier the anaerobic condition develops, which means the availability of oxygen is lesser. Hence, if the decolorization occurs sooner (within 4 days), the effluent sample may be taken as relatively unstable. Samples which do not decolorize in 4 days can be taken as relatively stable and can be discharged into streams without any trouble or concerns.

19.10.3 Population Equivalent

Industrial wastewaters are generally compared with per capita normal domestic wastewaters in order to rationally charge the industries for the pollution caused by them. The strength of the industrial sewage is worked out as follows:

Standard BOD$_5$ of industrial sewage = [Standard BOD$_5$ of

domestic sewage per person per day] × [Population equivalent]
$$(19.23)$$

The standard average BOD$_5$ of domestic sewage is worked out to be about 0.08 kg/day/person. Now, if the BOD$_5$ of the sewage coming from an industry is worked out to be 350 kg/day, then

$$\text{Population equivalent} = \frac{\text{Total BOD}_5 \text{ of the industry in kg/day}}{0.08 \text{ kg/day/person}}$$

$$= \frac{350}{0.08}$$

$$= 4375 \qquad (19.24)$$

Thus, the population equivalent indicates the strength of the industrial wastewaters for estimating the treatment that is required at the municipal sewage treatment plant. This helps in arriving at a realistic assessment of what industries must be charged, instead of charging them simply by volume of sewage.

19.11 NATURAL METHODS OF SEWAGE DISPOSAL

As it becomes stale, sewage begins to cause nuisance to the public. If it is possible to dispose off sewage within 4 or 5 hours after its production, the treatment required is less in magnitude. The methods of sewage disposal can be broadly be categorized as artificial methods and natural methods.

In artificial methods, the sewage is given such treatment that it becomes harmless and can therefore be safely discharged into natural waters. These methods are broadly divided into two categories:

1. Disposal by dilution
2. Disposal by land treatment

19.11.1 Disposal by Dilution

1. **Meaning**: In this process, the raw sewage or partially treated sewage is thrown into natural waters having large volume. Over time, sewage is purified by what is known as the self-purification capacity of natural waters.
2. **Conditions favorable for dilution**:
   • When sewage is relatively fresh (4–5 hours old) and free from floating and settleable solids.
   • When the diluting water (i.e., the source of disposal) is high in DO content.
   • Where diluting waters are not used for navigation on the downstream from the point of sewage disposal.
   • When the city's outfall sewer or its treatment plant is located near some natural waters having large volumes
3. **Standards of dilution**: Dilution factor is the ratio of the quantity of diluting water to the quantity of sewage, and depending upon the dilution factor, the Royal Commission on Sewage Disposal framed the standards of dilution in 1912. These standards are given in Table 19.4.
4. **Preliminary investigations**: When the method of disposal by dilution is to be adopted, a study of the

TABLE 19.4
Standards of Dilution

Dilution Factor	Standards of Purification Required
Above 500	No treatment is required. Raw sewage can be directly discharged into the volume of dilution water
Between 300 and 500	Primary treatment such as plain sedimentation should be given to the sewage, and the effluents should not contain suspended solids more than 150 ppm
Between 150 and 300	Treatments such a sedimentation, screening, and essentially chemical precipitation are required. The sewage effluent should not contain suspended solids more than 60 ppm
Less than 150	Complete thorough treatment should be given to sewage. The sewage effluent should not contain suspended solids more than 30 ppm, and its 5-day BOD at 18.3°C should not exceed 20 ppm

effluent to be discharged and the nature of the receiving body of water should be made. Such preliminary investigations should include the following aspects:

- analysis of diluting water with special reference to the available dissolve oxygen;
- effect of sewage disposal on water used for aquatic life, bathing beaches, etc.;
- hydrographic survey to determine the intensity and direction of winds, currents, and tides;
- minimum quantity of diluting water available at all the time during the year;
- the degree of treatment given to the sewage, etc.;

5. **Types of natural waters**: Following are the natural waters into which the sewage can be discharged for dilution:
    i. Creeks
    ii. Estuaries
    iii. Groundwaters
    iv. Lakes
    v. Ocean or sea
    vi. Perennial rivers and streams
    i. Creeks: A creek is an inlet on sea coast and it is likely to not have dry weather flow for some periods in the year.
    ii. Estuaries: The river's wide lower tidal part is known as an estuary and the dilution of sewage in estuaries is affected by ocean water in addition to the river water. Therefore, the process of dilution of sewage in estuaries is generally satisfactory.
    iii. Groundwaters: The sewage, when applied on land, ultimately filters out through different layers of solids and it meets groundwaters at great depths. If groundwater flows through favorable strata of soil, the dilution of sewage is satisfactory.
    iv. Lakes: A lake is an enclosed water space and it may be used for the purpose of dilution of sewage. The various characteristics of lake such as its size, flowing in it, etc., must be studied carefully before its self-purifying capacity is decided.
    v. Ocean or sea: The ocean or sea has water in abundance, and hence, its capacity to dilute sewage is practically unlimited. The sewage of any quality can be diluted into ocean. It can be

observed that the sewage reacts with sea water and forms precipitates giving milky color to the sea water. This is known as sludge banks. These sludge banks are undesirable because its interaction with the sulfate-rich sea water produces hydrogen sulfide gas, resulting in bad odor.

vi. Perennial rivers and streams: The perennial rivers and streams possess some flow throughout the year. But it may have maximum and minimum limits. The minimum limit generally occurs in summer. The dilution of sewage in summer becomes difficult due to the fact that in summer the solubility of oxygen is reduced because of the high temperature of water. The sewage, under such circumstances, should be properly treated before allowing dilution with perennial rivers and streams.

19.11.2 Disposal by Land Treatment

It is also known as effluent irrigation or broad irrigation or sewage farming. In this process, the raw or partly treated sewage is applied on land, where a part of sewage evaporates and the portion that remains percolates through the ground. It is then caught by the underground drains for disposal into the natural waters. The sewage adds to the fertilizing value of land and the crops can be profitably raised on such land.

When the sewage is applied continuously to a land, the voids may get filled up with sewage solids leading to the clogging of the soil. This phenomenon of soil getting clogged is known as sewage sickness.

1. **Conditions favorable for land treatment**:
   - The area for land treatment is composed of sandy, loamy, or alluvial soils over soft murum. Such soils are easily aerated and it is easy to maintain aerobic conditions in them.
   - The depth of water table is low even in rainy season so that there are no chances of pollution of underground water sources by land treatment.
   - The rainfall in the area is low as it will assist in maintaining good absorption capacity of soil.
   - There is absence of river or other natural water sources in the vicinity for disposal of sewage.

- There is demand for cash crops that can be grown on sewage farms.
- There is availability of large open areas in the surrounding locality for practicing broad irrigation by sewage.

2. **Methods of applying sewage**:
   i. Sprinkler or spray irrigation
   ii. Subsurface irrigation
   iii. Surface irrigation

   i. Sprinkler or spray irrigation: In this method, nozzles fitted on the tip of pipes are used to spread sewage over the soil. The sewage is sprinkled under pressure and such a process is used for watering gardens and lawns. This method of sewage application is useful for sandy soils and also for hilly land having steep slopes.

   ii. Subsurface irrigation: In this method, a system of underground pipes (with open joints) is used to supply sewage directly to the root zone of plants. As it flows through these pipes, the sewage percolates through the open joints. Capillary action ensures that it is distributed in the areas surrounding the pipes. This method is useful for places where rainfall is poor and demand for irrigation is high and subsoil water level is low.

   iii. Surface irrigation: In this method, the sewage is applied on the surface of land and it is also known as the broad irrigation. This method is widely adopted in practice and the following are its different modes of application.

   - Basin method: In this method, the basins are constructed around the plants and they are filled by sewage. The sewage slowly percolates to the root zone of plants and maintains the root zone in moist or damp conditions. This method is useful for orchards or gardens of fruit trees.

   - Flooding method: In this method, the land to be treated with sewage is divided into rectangular plots of convenient dimensions. The sewage is distributed over these plots to a depth of about 300–600 mm. The subsoil drain pipes are provided to achieve two purposes, namely, to supply air to the soil and to remove the percolated effluent through the soil.

   - Furrow method: In this method, the furrows and ridges are formed. The furrows are very small ditches having depth of about 300–500 mm and width of about 1.20–1.50 m. The ridges have length of about 15–30 m and width of about 1.20–2.50 m. The subsoil drain pipes collect percolated effluent and lead it to nearby natural waters for disposal.

   - Managed turf method: In this method, the turf is prepared before application of sewage to the land. The sewage is applied from pipe laid underground. Thus, the suspended solids are deposited at the roots of grass plants. The microbes attack the suspended solids and convert them into simpler forms of CO_2 and H_2O.

   - Zigzag method: In this method, the sewage is conveyed in the form of corrugations. The ridges are made of small earthen banks and the sewage flows in a zigzag path. The subsoil drains are provided to collect the percolated effluent. This method is very useful for lands which are practically level.

19.11.3 Self-Purification of Natural Waters

The self-purification of natural water systems is a complex process. Its complexity lies in the fact that it involves the simultaneous working of physical, biological, and chemical processes. One of the commonly used indicators of river health is the amount of DO in water. At DO levels below a threshold of 4 or 5 mg/l, the forms of life that can survive begin to be reduced. Higher life forms require a minimum of about 2.0 mg/l of DO to survive.

1. **Factors affecting self-purification**
   a. Dilution: When there is sufficient dilution water in the receiving water body where the wastewater is discharged, the DO level in the receiving stream may not reach zero or critical DO. This owes to the availability of sufficient DO initially in the river water before receiving discharge of wastewater.

   b. Current: When the water current is strong, the discharged wastewater will mix thoroughly with stream water preventing deposition of solids. In the case of smaller currents, the solid matter from the wastewater gets deposited at the bed following decomposition and reduction in DO.

   c. Temperature: In cold temperature, the quantity of DO available in stream water is more than that available in hot temperature. Also, at higher temperature levels, the activity of microorganisms is more; therefore, the self-purification will take lesser time in summer than in winter.

   d. Sunlight: Due to photosynthesis, algae produce oxygen in the presence of sunlight. Thus, sunlight helps purify the stream by adding oxygen through photosynthesis.

   e. Rate of oxidation: DO depletion occurs because of the oxidation of organic matter discharged in the river. The rate of depletion increases with temperature, i.e., it is faster at higher levels of temperature. The rate of oxidation of organic matter thus depends on the chemical composition of the organic matter.

2. **Zone of pollution in a stream**:

The self-purification process of a polluted stream can be divided into four zones: Zone of degradation, zone of active decomposition, zone of recovery, and zone of cleaner water.

Zone of degradation: It is characterized by the water turning dark and turbid with sludge deposition at the bottom. DO is reduced to about 40% of the saturation value and increase in carbon dioxide content.

Zone of active decomposition: It is marked by heavy pollution. Water turns grayish and darker than in the previous zone. DO concentration becomes zero. Anaerobic organic decomposition takes place. This zone will not have fish life but bacteria flora will be present. Near the end of the zone DO concentration rises to 40%.

Zone of recovery: The water becomes clearer and algae reappear. BOD falls down and DO rises above 40% of saturation value.

Zone of cleaner water: In this zone river attains original conditions and DO rises to saturation level. Water becomes clearer and attractive. It will have some pathogens so treatment is required before its use.

3. **Oxygen sag analysis**: The oxygen sag or oxygen deficit in the stream at any given point of time during the process of self-purification is the difference between the saturation DO content and the actual DO content at that point of time.

Oxygen deficit, D = Saturation DO – Actual DO

The saturation DO value of fresh water depends upon the total dissolved salts present in it and the temperature; its value varies from 14.62 mg/l at 0°C to 7.63 mg/l at 30°C, and lower DO at higher temperatures. The DO in the stream may not be at saturation level and there could be an initial oxygen deficit "D_o." When the effluent with initial BOD load L_o is discharged in to stream, there is depletion in the DO content of the stream resulting in an increase in the oxygen deficit (D). The "Oxygen Sag Curve" (Figure 19.18) illustrates the variation of oxygen deficit (D) with the distance along the stream and hence with the time of flow from the point of pollution. The point of minimum DO, i.e., maximum deficit, is the major point in sag analysis. The inflexion points of the oxygen sag curve are when maximum or critical deficit (D_c) occurs.

4. **Deoxygenation and reoxygenation curves**: The DO level in the stream depletes when wastewater is discharged into the stream. This is known as deoxygenation. The amount of organic matter remaining (L_t) to be oxidized at any time t, as well as temperature (T) at which reaction occurs, determines the rate of deoxygenation. The deoxygenation curve in the absence of aeration depicts the variation in the depletion of the DO content of the stream with time. The ordinates below the deoxygenation curve (Figure 19.18) indicate the oxygen that remains in the natural stream after the BOD of oxidizable matter is satisfied.

When the DO content of the stream is gradually consumed owing to the BOD load, the atmosphere continuously supplies the water with oxygen through the process of re-aeration or re-oxygenation. Thus along with deoxygenation, re-aeration is a continuous process.

The rate of re-oxygenation depends upon

- depth of water in the stream: decreases with depth;
- velocity of flow in the stream: less for water that is stagnant;
- oxygen deficit (D) below saturation DO: because the solubility rate depends on the difference

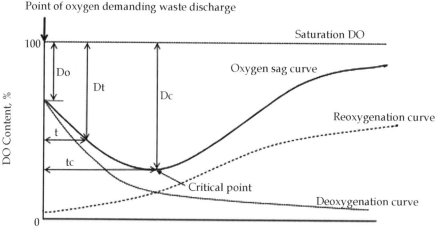

FIGURE 19.18 Deoxygenation, reoxygenation, and oxygen sag curve.

between saturation concentration and the existing concentration of DO;

- temperature of water: at higher temperatures, the solubility of oxygen and saturation concentration are lower.

5. **Mathematical analysis of oxygen sag curve: Streeter–Phelps equation**

The Streeter–Phelps analysis suggests that the analysis of the oxygen sag curve can be done by superimposing the rates of deoxygenation and reoxygenation. The rate of change in DO deficit is the sum of the two reactions as shown below:

$$dD_t/dt = f \text{ (deoxygenation and reoxygenation)}$$

or

$$dD_t/dt = K'L_t - R'D_t \tag{19.25}$$

where

$D_t=$ DO deficit at any time t,
$L_t=$ amount of first-stage BOD remaining at any given time t,
$K'=$ BOD reaction rate constant or deoxygenation constant (to the base e),
$R'=$ re-oxygenation constant (to the base e),
$t=$ time (in days),
$dD_t/dt=$ rate of change of DO deficit.

Now,

$$L_t = L_0 e^{-K't} \tag{19.26}$$

where $L_0=$ BOD remaining at time $t=0$, i.e., ultimate first-stage BOD.

Hence, DO deficit at time "t"

$$D_t = \frac{KL_0}{R-K}\left[10^{-Kt} - 10^{-Rt}\right] + D_0 10^{-Rt} \tag{19.27}$$

where

$K=$ BOD reaction rate constant, to the base 10;
$R=$ re-oxygenation constant to the base 10;
$D_0=$ initial oxygen deficit at the point of waste discharge at time $t=0$;
$t=$ time of travel in the stream from the point of discharge $=x/u$.

This is Streeter–Phelps oxygen sag equation.

19.12 PRIMARY TREATMENT OF SEWAGE

The sewage before being disposed in rivers streams or on land has to be treated in order to make it safe. However, the degree of treatment that is required depends on the characteristics of the source of disposal. Sewage can be treated in different ways. Treatment processes are generally classified as

1. Preliminary treatment
2. Primary treatment

3. Secondary or biological treatment
4. Complete final treatment

19.12.1 PRELIMINARY TREATMENT

The equipment and facilities used to remove items such as rags, grit, sticks, other debris, and foreign objects are an important part of any wastewater treatment plant. Pretreatment or preliminary treatment includes methods of removing such materials prior to primary and subsequent treatment. Grit chamber and Detritus tanks are for removing grit and sand and skimming tanks are for removal of oil and grease.

1. **Screening and comminution**: Screening and comminution are preliminary treatment processes that are used to protect mechanical equipment in treatment works, to aid downstream treatment processes by intercepting unacceptable solids, and to alter the solids to render them acceptable for treatment.
   a. Screening: Screening devices (Figure 19.19) are used to remove materials which could damage equipment or could interfere with a process or piece of equipment. Screens are classified as fine or medium or coarse; these can be further classified as manually or mechanically cleaned. Preliminary treatment uses coarse screens having openings of size 50 mm, while fine screens having perforations of 1.5–3 mm are used as a step in advanced wastewater treatment or in lieu of sedimentation preceding secondary treatment. Fine screens as preliminary or primary treatment are more applicable to process or industrial wastes.
   b. Comminution: A comminutor acts as cutter and as a screen. Its aim is shred (comminute) the solids, not to remove it. Comminutors, like most screens, are mounted in a channel through which wastewater flows. The cutting teeth shred the rags and other debris it can pass through the openings. Some units require specially shaped channels for proper hydraulic conditions, resulting in more expensive construction. All comminutors require a bypass channel to allow for the maintenance of equipment.
2. **Grit removal**: Grit represents the heavier inert matter in wastewater which does not decompose in treatment processes. Grit is identified as matter that has a specific gravity of about 2.65. The grit chambers are designed to remove all particles that are 0.011 in. or larger (65 mesh). Some sludge handling processes require the removal of grit of 0.007 in. or larger (100 mesh), as a minimum. Compared to other unit treatment processes, grit removal is economical. It is used to achieve the following results:
   - Prevent excessive abrasive wear of equipment such as sludge scrapers and pumps.

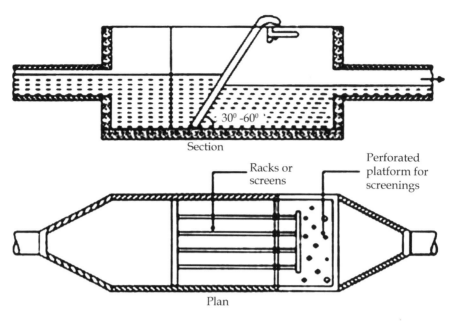

Section

Racks or screens

Perforated platform for screenings

Plan

Bar screen with chamber

FIGURE 19.19 Screening device.

- Prevent deposition and subsequent operating problems in pipes, channels, and basins.
- Prevent reduction of the capacity in sludge handling facilities.

Grit removal facilities should be used for combined sewer systems or for separate sanitary systems that may have excessive quantities of inert material. Equipment to remove grit should be placed after bar screens and comminutors, and ahead of raw sewage pumps. In some cases, owing to the depth of the influent line, it is not practical to locate the grit removal system ahead of the raw sewage pumps. Thus wastewater containing grit is required to be pumped; in these modes, pumps that are capable of handling grit should be utilized.

a. Horizontal flow grit chambers: This type of grit chamber (Figure 19.20) is designed in order to allow wastewater to pass through channels or tanks at a horizontal velocity of around one foot per second. This horizontal velocity facilitates the settling of grit in the channel or tank bottom, while the lighter organic solids are kept in suspension.

b. Detritus tanks: A grit chamber can be designed with a lower velocity. This is done in order to allow organic matter to settle with the grit. This mixture of grit and organic is known as detritus and the devices that remove this mixture are removal devices that are called as detritus tanks. In detritus tanks, grit is separated from organic matter either by gentle aeration or by washing the removal detritus in order to resuspend the organic matter. The configuration of the tank is simple and grit is continually removed in this system.

c. Aerated grit chambers (Figure 19.21): As the name implies, in these chambers, diffused air is used to separate grit from other matter. That this freshens the wastewater prior to further treatment is a secondary benefit of the aeration method. This is commonly used in conjunction with a preaeration facility.

3. **Skimming tanks**: Skimming tanks facilitate the removal of floating substances which include grease, soap, wood pieces, fruit skins, etc., which are prominent in industrial sewage or in wastewaters emanating from restaurant kitchens, oil refineries, motor garages, soap factories, candle factories, etc. It has a detention period of about 3–5 minutes.

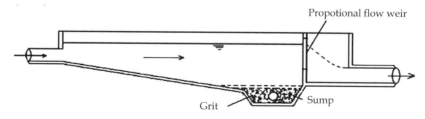

FIGURE 19.20 Horizontal flow grit chamber.

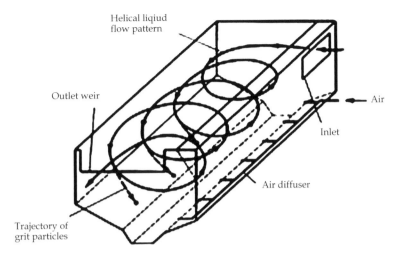

FIGURE 19.21   Aerated grid chamber.

Its efficiency can be improved by passing chlorine gas along with compressed air. A skimming tank is shown in Figure 19.22.

19.12.2 PRIMARY TREATMENT

Primary treatment pertains to the removal of large suspended organic solids. This is usually done by sedimentation in settling basin. The preliminary as well as primary treatments could be sometimes classified together as primary treatment.

1. **Plain sedimentation**: After preliminary treatment, wastewater undergoes sedimentation by gravity in a basin or tank that is sized to produce near quiescent conditions. In this facility, the settleable solids and most of the suspended solids settle to the bottom of the basin. Mechanical collectors are provided, which continuously sweep the sludge to a sump where it is removed for further treatment and subsequent disposal. Skimming equipment should be provided to remove floatable substances such as scum, greases, and oils which accumulate at the liquid surface. These skimmings are then combined with the sludge for disposal. Removals from domestic wastewaters that are put through plain sedimentation range from about 30% to 40% for BOD and about 40% to 70%

for suspended solids. The overflow rate is one of the most significant design parameters, expressed in liters/day/m². This is obtained by dividing the flow in liters/day by the settling surface area of the basin (in m²). Typically, average daily flow rates are used for sizing facilities. Figure 19.23 shows a typical Rectangular Settling Tank.

2. **Sedimentation with chemical coagulation**: Sedimentation using chemical coagulation is mainly used for the pretreatment of industrial or process wastewaters and for the removal of phosphorus from domestic wastewaters. Chemical coagulating agents have not been used extensively on domestic wastewaters (to enhance the removal of BOD and suspended solids) as it is not economical or operationally desirable. However, some installations may have special applications. The advantages of increased solids separation in primary sedimentation facilities are as follows:
   - A decrease in organic loading to the secondary treatment process units.
   - A decrease in quantity of secondary sludge that is produced.
   - An increase in the quantity of the primary sludge produced, which can be thickened and dewatered more readily than the secondary sludge.

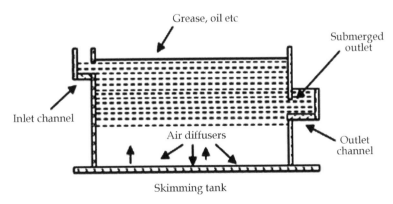

FIGURE 19.22   Skimming tank.

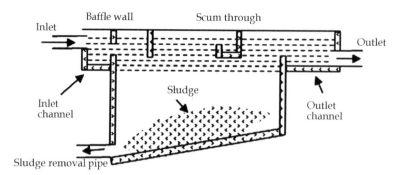

FIGURE 19.23   Rectangular settling tank.

Chemicals commonly used, either singularly or in combination, are lime, the salts of iron and aluminum, synthetic organic poly electrolytes. It is recommended that jar studies be run in order to ascertain optimal chemicals and dosage levels.

19.13   SECONDARY TREATMENT (FILTRATION) OF SEWAGE

In secondary treatment, effluents from primary treatment are further treated in order to remove dissolved and colloidal organic matter. Generally, this is accomplished through biochemical decomposition of organic matter, carried out either under aerobic or anaerobic conditions. The main end products of aerobic decomposition are carbon dioxide and bacterial cells. The end products of anaerobic process are CH_4, CO_2, and bacterial cells. The organic matter is decomposed (oxidized) by aerobic bacteria in a biological reaction that comprises

1. filters,
2. activated Sludge Process (ASP),
3. oxidation ponds, etc.

The bacterial cells separated out in secondary setting tanks are disposed after they are stabilized under aerobic or anaerobic process in a sludge digestion tank along with the solids settled in primary sedimentation tanks.

19.13.1   FILTERS

There are four types of filters that are commonly employed in the secondary treatment of sewage:

a. Contact beds
b. Intermittent sand filters
c. Trickling filters
d. Miscellaneous filters

a. **Contact beds**: Contact beds or contact filters are the point at which the sewage comes in contact with the filtering media for some period of time. As the sewage effluents pass through, an organic film forms around the particles of the filtering media. This film comprises a large number of aerobic bacteria, the presence of which results in the oxidation of organic matter. In the second contact period, i.e., when the bed stands empty, the filter comes in contact with oxygen from the atmosphere and the organic matter caught in the voids of the filtering media gets oxidized and will be washed away by the fresh sewage in the next cycle.

Constructional features: A contact bed is a watertight tank. It is filled with the filtering media which may be of gravel, ballast, or broken stone. The size of particles of filtering media varies from 15 to 40 mm. The tanks are generally dug below ground level and are provided with concrete lining. A syphonic dosing tank serves two or three contact beds. The sewage moves from the settling tank to the dosing tank. The sewage after passing through the filtering media is collected and conveyed through the under drainage system to the effluent pipe. The depth of bed is about 1–1.8 m, the common being 1.2 m. The area of one bed generally does not exceed 0.20 hectare. The beds are usually designed for a dosing rate of 300–500 liters per loading per m^3 of the filtering media. The contact beds are generally given two loadings in a day.

Working: The complete cycle of operating a contact bed is usually carried out in the following four stages:

- The tank is filled with sewage effluent. This could take around 2 hours.
- The sewage effluent is made to stand on the filtering media for around 2 hours.
- The tank is emptied and the sewage effluent flows through the effluent pipe in a manner that it does not disturb the organic film of the bed. This may require around 2 hours or so.
- The contact bed is then allowed to stand empty for a period of 6 hours or so.

The actual period of contact will depend on the quality and the type of effluent required. The cycle of operation is generally completed in 12 hours.

If the effluent of better quality is required, the contact beds may be arranged in series and the effluent from one contact bed is taken to the next one for further treatment.

b. **Intermittent sand filter**: Earlier, these were used as the biological units of sewage treatment. These are similar to contact beds but differ when it comes to its contact media. Here, the contact media does not have a concrete lining around it and is finer when compared to that used in contact beds. The sand used should have an effective size of about 0.2–0.5 mm and uniformity coefficient between 2 and 5.

The rate of filtration of an intermittent sand filter depends mainly on the following three factors:
- Depth and size of filtering media
- Nature of the influent
- Quantity of effluent required

The effluent obtained is excellent in quality. The suspended solids in the effluent are less than 10 ppm and the BOD is less than 5 ppm. These filters, however, need large quantities of sand and land for its installation due to the low rate of loading. Therefore, the use of these filters at larger plants is not economical.

c. **Trickling filter**: These filters (Figure 19.24) can be used for the complete treatment of domestic waste. It can also be used as roughing filter for strong industrial waste in the stage preceding the activated sludge process. The primary sedimentation tank is utilized before the trickling filter is used so that the filter is not clogged by the settleable solids in the sewage. This is followed by the secondary settling tank so that the settleable biosolids produced in filtration process can be removed.

When the wastewater passes through or comes in contact with the filter media (which consists of rocks sized between 40 and 100 mm, or plastic media), its surface witnesses a building up of a biological slime that consists of aerobic bacteria and other biota. This biological slime absorbs organic material in the sewage, where the biota partially degrade the material, thereby resulting in a thickening of the bio film. Eventually, the biofilm scours and this is replaced by fresh biofilm that again grows on the media; the phenomenon involving the detachment of the biofilm is called *sloughing* of the filter.

Depending on the organic and hydraulic loadings, trickling filters can be classified as low rate and high rate. Low-rate filters are designed for hydraulic loading of 1–4 m^3/m^2 day and organic loadings as 80–320 g BOD/m^3/day, while filters classified as high rate are designed for hydraulic loading of 10–30 m^3/m^2 day (including recirculation) and organic loading of 500–1000 g BOD/m^3 day (excluding recirculation). In general, low-rate filters do not adopt recirculation. In high-rate trickling filters, recirculation ratio of 0.5–3.0 or higher is used. The depth of media varies from 1.0 to 1.8 m for high-rate filters; for low-rate filters, this is between 2.0 and 3.0 m. The slope of the bed of the trickling filter is 1 in 100 to 1 in 50. The underdrainage system consists of "V" shaped or half round channels, which are cast in concrete floor during construction. Revolving distributors are provided at the top. Two or four of the horizontal arms of the pipe have perforations or holes. These rotating arms remain 15–25 cm above the top surface of the media. The electric motor rotates the distribution arms that are rotated at about 2 rpm. They may also be rotated by the back-reaction on the arms by the wastewater. The head of 30–80 cm of wastewater is needed to rotate the arms.

d. **Miscellaneous filters**: In addition to the filters discussed above, many other filters are also sometimes used for the treatment of sewage.

i. Dunbar filters: These are similar in construction and operation to the intermittent sand filters.

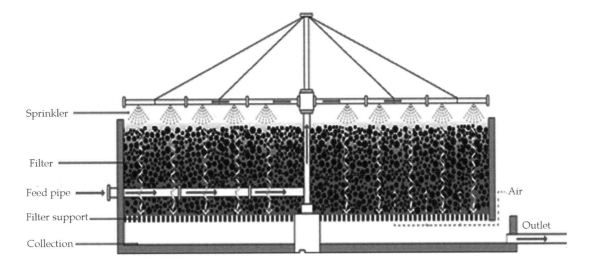

FIGURE 19.24 Trickling filter.

The filter media is graded from top to bottom from 1 to 150 mm; at each dose of sewage, the filter is flooded and the process is repeated until the filter is clogged. The surface of filter is then cleaned. The rate of filter loading may be about 25 million liters per hectare of surface area per day. The BOD removal is to the extent of about 85% or so.

ii. Magnetic filters: In this type of filters, a layer of crushed magnetite iron ore is provided. The depth of this layer is about 80 mm and it is supported on a nonmagnetic metal wire screen. The sewage is filtered through the magnetic layer. The filter is cleaned when the loss of head becomes excessive. These filters are not in common use.

iii. Rapid sand filters: These filters are adopted in the treatment of sewage for reclaiming the used water and the water thus reclaimed can be consumed for purposes other than drinking. However, such filters get clogged in very short time.

19.14   ACTIVATED SLUDGE PROCESS

The activated sludge process is an aerobic biological system of treatment. The settled wastewater is aerated in a tank for a few hours, during which the organic matter in the aeration tank is stabilized by microorganisms. During this process, a part of the organic matter is synthesized into new cells and a part of it is oxidized in order to derive energy. The main mechanism of BOD removal in the activated sludge process comprises this synthesis reaction followed by the separation of the resulting biological mass and the oxidation reaction.

Generally, the biomass produced in the aeration tank is flocculent. A secondary settling tank is then used to separate this from the aerated wastewater and is partially recycled to the aeration tank. The mixture of wastewater in the aeration tank and recycled sludge is known as mixed liquor. The recycling of sludge aids the initial buildup of a high concentration of active microorganism in the mixed liquor, which then accelerates the removal of BOD. BOD removal is up to 80%–95% and bacteria removal is up to 90%–95%. Once the mixed liquor has the required concentration of microorganism, its further increase is prevented by the regulating the quantity of sludge recycled and wasting the excess sludge.

Aeration units are main units of the activated sludge process. The main aim of this process is to supply oxygen to the wastewater in order to keep the reactor content aerobic and to facilitate a thorough mixing of the return sludge with wastewater. The detention period is usually kept between 6 and 8 hours for treatment of sewage or similar industrial wastewater. The volume of aeration tank may be determined after taking into consideration the return sludge, which is typically about 25%–50% of the wastewater volume.

Liquid depth provided is normally between 3 and 4.5 m. Along with this, a free board (0.3 to 0.6 m) is provided. Air supply in the aeration tank may be provided in three modes: by supplementing compressed air from tank bottom, which is known as diffused aeration, or by mechanical aerators provided at the surface, or using a combination of the two, i.e., diffused aeration and mechanical aerators. The activated sludge process may be classified according to flow regimes: *conventional (plug flow)* and *completely mixed activated sludge process.* Wastewater treatment popularly uses a modification of activated sludge process such as extended aeration, which is designed for higher hydraulic retention time (18 hours).

19.14.1   ADVANTAGES OF THE ACTIVATED SLUDGE PROCESS

- Initial cost is lower than that of trickling filter.
- Land area requirements are smaller because the design is compact.
- The effluent is clearer and is free from odor.
- It gets rid of nuisance from flies.
- Result provides very high efficiency with over 90% removal of total solids and BOD.
- Compared to the trickling filter, the head required for operation is small.

19.14.2   DISADVANTAGES OF ACTIVATED SLUDGE PROCESS

- The operational cost is higher compared to trickling filter.
- The process is very sensitive. Setting it right is troublesome if the system has gone out of order.
- Produces large quantity of sludge with high moisture content, disposing which is a problem.
- It needs skilled attendance.

19.15   SLUDGE TREATMENT AND DISPOSAL

The suspended solids which accumulate at the bottom of clarifiers or settling tanks form what is known as the sludge. There are different types of sludge produced from different treatment units such as trickling filter sludge, chemically precipitated sludge, digested sludge, sludge from primary setting tank or from activated sludge process, etc.

It is thus observed that the sewage after treatment is separated into two distinct parts: effluent and sludge. The effluent is clear sparkling liquid and the sludge is a combination of sewage solids with different proportions of water. The disposal of effluent does not create any problem, but disposal of sludge has to be carefully planned.

19.15.1   METHODS OF SLUDGE DISPOSAL

The following methods may be used to dispose sludge:

1. Disposal on land
2. Distribution by pipe line
3. Drying on drying beds
4. Dumping into the sea

5. Heat drying
6. Incineration
7. Lagooning or ponding
8. Press filters and vacuum filters
9. Digestion followed by drying

1. **Disposal on land**: The sludge can be disposed off on land in two ways (1) ploughing and (2) trenching.

In the *ploughing method*, the sludge is mixed either with milk of lime or with powdered lime. It is then spread on the land which is then ploughed. The crops can be raised on such land, if convenient.

In *trenching method*, the trenches about 900 mm wide and 600 mm deep are dug in parallel rows at a distance of about 1.5 m. The trenches are filled with sludge and a thin layer of excavated earth is placed over it. The process is repeated by digging new trenches between the old ones and then digging trenches that are at right angles to the direction of previous ones.

The disposal of sludge on land requires considerable amount of good land. For the normal conditions, one hectare of good land will be required to dispose sludge obtained from about 10,000 to 12,000 persons.

2. **Distribution by pipe line**: This is a simple method of sludge disposal. In this method, the sludge is conveyed through pipe line to nearby farm and it is utilized as fertilizer. The success of this method depends on the presence of one of the following conditions:

- There should be sufficient quantity of land in possession of sewage treatment plant for this purpose (or)

- The owners of the nearby land give co-operation in receiving the sludge as and when it is discharged on their land.

3. **Drying on drying beds**: In small wastewater treatment plants, the sludge is dewatered before it is disposed on beds that are known as sludge drying beds. This involves two mechanisms: filtration of water through sand and evaporation of water from the sludge surface. The water filtered thus is sent back to the plant for treatment. Sludge from the conventional activated sludge, contacted stabilization, trickling filter, and rotating biological contactor processes usually contain a significant amount of volatile solid, which tends to emanate an unpleasant odor. Therefore, this method is generally not suitable for handling this sludge without prior stabilization, making the digestion of sludge prior to the application of sludge on sludge-drying beds is essential.

Typically, a sludge-drying bed (Figure 19.25) consists of 15–30 cm of coarse sand layer, which is placed over approximately 20–45 cm of grade gravel that ranges between 0.6 and 4 cm in size. The gravel has open jointed tubes (10–15 cm diameter and spaced at 2.5–6 cm) to provide for drainage of the liquid that passes through the bed. Depending upon the climatic conditions locally, the sludge (a layer of 20–30 cm spread over the drying bed) is permitted to dry for a period of two to four weeks. The dewatering process may be improved by enclosing the drying beds with glass: this is deployed in cold or wet climates and doing so can reduce the area required for the open beds by a third.

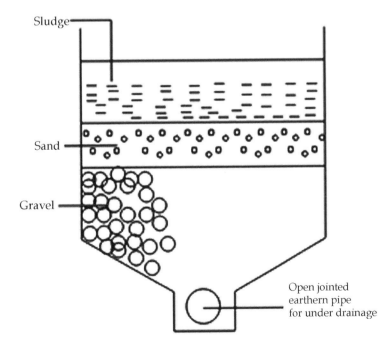

FIGURE 19.25   Sludge drying bed.

4. **Dumping into the sea**: In this method, the sludge is conveyed and discharged into the sea. The sludge must be taken sufficiently deep into the sea from the shore so as to avoid any chances of possible nuisance by the sludge being washed back on the shore. This method of sludge disposal is adopted for many large cities, and even though the method requires a fleet of special sludge tankers, it proves to be the most economic one when adopted on large scale.

5. **Heat-drying**: In this method, the sludge is actually heated so that it may become dry. This method is commonly adopted in America to convert the sludge obtained from the activated sludge process into fertilizer. Hence, this method may be seen as a method of producing fertilizer rather than as a method for disposal of sludge. This is due to the fact that if this method is extremely costly, and hence, it becomes impracticable.

6. **Incineration**: The process by which biosolids can be destroyed by heat is known as incineration. It is used as a volume reduction to ash, rather than a means of disposal. Incineration of biosolids produces a useful reuse product: biosolids ash.

   The advantages of incineration are
   • considerably reduces area required for sludge management, compared to land burial or lagoons;
   • reduces the volume and weight of biosolids;
   • provides immediate reduction in volume of sludge;
   • on-site incineration avoids transportation costs.

   The solids content of sludge is a key factor to be considered in incineration. The sludge should be dewatered (by centrifugation) before incineration because despite the relatively high heat value of sludge, excessive water content will lead to consumption of auxiliary fuel in order to maintain incinerator combustion.

7. **Lagooning or ponding**: This is an economical mode of disposal. It requires a remote area where an earthen basin on a lagoon can be used for depositing untreated or digested sludge. The stabilized sludge settles at the bottom of the lagoon, where it accumulates. Any excess liquid from the lagoon may be returned to the wastewater treatment plant for primary sludge treatment. Sludge can either be stored indefinitely or can be periodically removed after it is drained and dried.

8. **Press filters and vacuum filters**: A *filter press* consists of series of cast-iron plates which can be pressed tightly. The sludge is filled in jute or cotton bags and these bags are then placed between the plates. Then the plates are pressed under a pressure of about 0.4–0.5 N/mm^2. The pressing of plates removes the water from sludge and consequently the sludge cakes are formed. The sludge cake obtained by this process has a moisture content of about 75% or so.

   A *vacuum filter* comprises a rotating drum covered with filter cloth. A pump is worked to evacuate the air and the water from the inside of drum. It results into vacuum and the suction of sludge takes place through the filter cloth. The drum rotates partly submerging into the sludge. During its rotation, it picks up a thin film which dries upon the surface of cloth and it is removed by a scraper.

   The liquid obtained during the working of filters is collected and returned to the inlet of the sewage treatment plant.

9. **Digestion followed by drying**: In this method, the sludge is first digested in specially designed sludge digestion tanks and it is then dried on sludge drying beds. The term *sludge digestion* refers to the decomposition of complex organic substances that are present in sludge into simpler stable compounds; these biochemical reactions are carried out by anaerobic bacteria. At the end of decomposition, digested sludge, gas, and supernatant liquids are obtained.

   The biological action involved in the process of sludge digestion occurs in the following three distinct stages:

   i. Acid production stage: In this stage, the simple compounds like cellulose, starch, sugar, soluble nitrogenous compounds, etc., are attacked by bacteria. Such action of bacteria starts fermentation and the products of decomposition are organic acids and gases. This stage lasts for about 15 day and BOD increases to some extent.

   ii. Acid regression stage: Organic acids and nitrogenous compounds are attacked by bacteria and converted into acid carbonates and ammonia compounds in this stage. The decomposed sludge has very offensive odor. It entraps the gases of decomposition, and hence, it is foamy in character and tends to rise to form scum at the surface. This is the intermediate stage. This stage continues for 3 months or so and BOD remains high even during this stage.

   iii. Alkaline fermentation stage: In this stage, the more resistant substances like proteins and some organic acids like amino acids are attacked by bacteria and they are broken down into ammonia, organic acids, and gases. This is the final stage of sludge digestion. Here, the liquid separates out from the solids and the digested sludge is formed, which is alkaline in nature. This stage extends for a period of about 1 month or so. BOD rapidly falls during this stage.

19.16 MISCELLANEOUS METHODS OF SEWAGE TREATMENT

The following miscellaneous methods of sewage treatment will be discussed:

1. Cesspools
2. Chlorination of sewage

3. Imhoff tanks
4. Oxidation ponds
5. Septic tanks
6. Treatment of industrial wastes
7. Wastes from fertilizer factories

1. **Cesspools**: Cesspools are pits excavated in soil with water tight lining and loose lining by stone or brick. This is done in order to provide for leaching of wastewater. The pit is covered. The capacity of the pit should not be less than what is estimated as one day's flow into the pit. The soil considered most suitable for cesspools in one in which all the water in a test pit of one meter diameter and two meter depth disappears in 24 hours. It is required that the bottom of the cesspool be well above groundwater level. At the level of overflow, an outlet is provided to take-off unleached liquid from the cesspool into a seepage pit. At regular intervals, the settled matter is removed. Water-tight cesspools are cleaned every 6 months and their capacity is at minimum 70 liters/person/month.

2. **Chlorination of sewage**: The disinfection of sewage may be carried out by various methods as in case of the water treatment process. Of these, chlorination is the most common method used to disinfect sewage. The objectives of the chlorination of sewage are
   • to assist the process of treatment of some type of industrial waste,
   • to control foaming in the sludge digestion tanks,
   • to control the possible fly nuisance due to the sewage,
   • to increase the efficiency of sewage treatment units,
   • to prevent corrosion of sewers,
   • to prevent spread of epidemic,
   • to reduce the BOD,
   • to remove the oil, grease, etc.

3. **Imhoff tanks**: Named after a German engineer Karl Imhoff (1876–1965), the Imhoff tank is a chamber designed for receiving and processing sewage. It is used for the clarification of sewage by settling and sedimentation, along with anaerobic digestion of the extracted sludge. The sedimentation takes place in an upper chamber. The collected solids then slide down the incline of bottom slopes that lead to the entrance of a lower chamber where sludge is both collected and digested. There is no flow of sewage in the lower digestion chamber and the two chambers are disconnected barring for the inclined slop that connects the two chambers. The lower chamber comprises biogas vents and pipes that are used to remove the digested sludge; the removal is typically carried out after 6–9 months of digestion of the sludge. In effect, the Imhoff tank is a two-storey septic tank. It retains the simplicity of operation and structure of the septic tank even as it eliminates its shortcomings, mainly that of problems that arise due to the mixing of sewage and septic sludge in the same chamber.

Plain sedimentation tanks supersede Imhoff tanks in sewage treatment. These tanks use mechanical methods for continuous collection of the sludge and for its movement to separate digestion tanks; this arrangement allows for improved sedimentation results and better temperature control in the digestion process, leading to a more rapid and complete digestion of the sludge. Figure 19.26 illustrates a typical Imhoff tank.

4. **Oxidation ponds**: Oxidation ponds are stabilization ponds that receive sewage that is partially treated. It is a pond of shallow depth that is dug into the ground. The earthen pond maintains a minimum depth of 1.0 m to deter aquatic weed growth, with a maximum depth of 1.8 m. The partially treated sewage may be detained in the pond typically between one to four weeks; this can vary according to availability of sunlight or conditions of temperature. Treatment can be more efficient or optimal if multiple ponds are placed in a series such that sewage can flow from one unit to the other until it is finally discharged.

   The surface area of the pond may be ascertained by assuming a suitable value of organic loading (in hot tropical countries such as India this may range from 150 to 300 kg/ha/day), with each unit occupying an area ranging between 0.5 and 1.0 ha.

   Figure 19.27 illustrates the plan of an oxidation pond. The length is half the width of the tank. A free board (about 1 m) has to be provided above a capacity that corresponds to 20–30 days of detention period. The pond has BOD removal efficiency up to 90% and Coliform removal efficiency up to 99%.

5. **Septic tank**: Septic tank is a type of sedimentation tank. This is ordinarily designed so that liquid can be retained for 24 hours at average daily flow. Accounting for sludge and scum accumulation, the septic tank is typically designed so it can retain wastewater for 1–2 days. Table 19.5 details the flow and characteristics of the wastewater that must be taken into account while designing a skeptical tank. These tanks may be made out of masonry, concrete, or fiberglass. In order to retain the floating matter and grease in the tank, the inlet and outlet of the tank are baffled. At the bottom of the tank, the organic fraction of the heavy solids that settle decompose following anaerobic pathway. As the production of biogas could interfere with solid sedimentation, septic tanks are provided with ventilation pipes that are covered with mosquito-proof wire mesh. It is required that the top portion of the pop extend at least 2 m above the highest building height present in the vicinity of 20 m from the location of the septic tank.

   For small septic tanks, the ratio of peak to average flow could be very high, leading to flow surges that can disturb the functioning of the tank, sometimes leading to the washing out of settled solids. As Figure 19.28 shows, the liquid depth of the tank

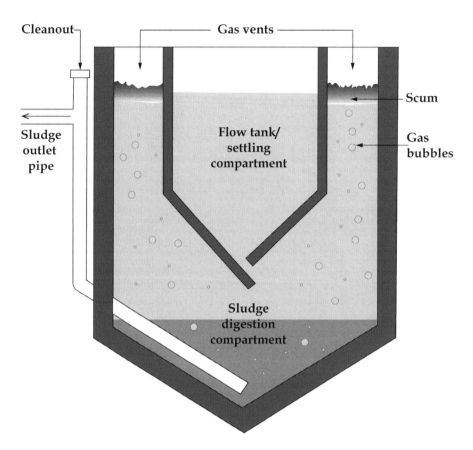

FIGURE 19.26   Imhoff tank.

is 1–2 m and the ration of the tank's length to width can range between 2:1 and 4:1 (Figure 19.28). At least 300 mm of free board is provided in the tank. The tank is cleaned out once in 2–3 years. Given that the treatment of the sewage is known to be inadequate, it is recommended that septic tanks only be used for individual houses or for a cluster of houses

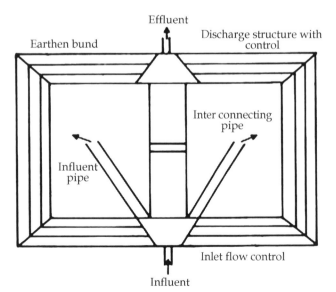

FIGURE 19.27   Plan of oxidation pond.

or institutes that cater to a maximum population of 300 people.

Posttreatment can be accomplished by aerobic treatment or by subsurface disposal. Septic tank effluents may be treated using methods such as diffused air aeration with solids recycling (known as extended aeration) or using sand filter or synthetic media filter (known as the attached growth process). Hypochlorite addition and the use of filter bags are also deemed as suitable for treatment.

6. **Treatment of industrial wastes**: Methods used for the treatment of industrial wastes depend upon their characteristics and various other factors. For the purpose of treatment, the industrial wastes may broadly classify into three categories as shown in Table 19.6.

TABLE 19.5

Characteristics of Household Wastewater to be Considered for Septic Tank Design

Average Flow per capita	100–160 litre/day
Peak flow per capita	170–270 litre/day
BOD per capita	0.045 kg/day
Suspended solids per capita	0.070–0.090 kg/day
Soluble solids per capita	0.035 kg/day
Sludge accumulation per capita	0.073 m³/year

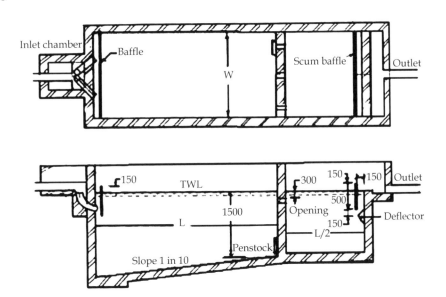

FIGURE 19.28 Construction details of the septic tank.

TABLE 19.6
Classification of Industrial Waste

Category	Industries
Wastes of animal origin	Milk dairies, fertilizer industry, glue industry, leather industry, soap industry, woolen textile mills, etc.
Wastes of mineral origin	Chemical industry, dye works, gas plants, mines, oil refineries, metal industry, water treatment plants, etc.
Wastes of vegetable origin	Beet-sugar industry, cotton textile mills, canning industry, paper industry, rubber industry, saw mills, etc.

The following methods may be adopted for the treatment of industrial wastes:

- Aerobic biological oxidation
- Anaerobic digestion
- Chemical coagulation
- Chemical oxidation
- Chlorination
- Deionization
- Dewatering and drying
- Filtration
- Incineration
- Lagooning or ponding
- Screening
- Sedimentation
- Spray irrigation
- Vacuum filtration, etc.

7. **Wastes from fertilizer factories**: The effluent from a fertilizer industry is found to be of complex character and it contains various toxic and undesirable constituents. The fertilizer employs processes for the manufacture of nitrogenous and phosphatic fertilizers and substances in the form of chemicals, intermediates, and final products formed during such processes are carried in the fertilizer waste. The usual complex constituents of fertilizer wastes can be listed as follows:

- Acids and alkalies
- Ammonia and ammonium salts
- Arsenic
- Carbon slurry
- Fluorides
- Monoethanolamine
- Phosphates, etc.

The most ideal situation for a fertilizer factory would be the nearness of sea because it will then become possible to easily discharge the effluent to the sea. Similarly the presence of perennial rivers or availability of land to receive effluent for irrigation purposes would also be considered as favorable conditions for the proper sitting of a fertilizer factory.

19.17 MISCELLANEOUS TOPICS

The following seven miscellaneous topics will be discussed:

1. Biogas
2. Elutriation
3. Garbage collection and removal
4. Garbage disposal
5. Micro-organisms
6. Night soil disposal without water carriage
7. Rural sanitation

19.17.1   Biogas

The biogas is essentially a mixture of gases containing two-thirds of utilizable gas methane and remainder is mainly carbon dioxide along with traces of nitrogen, hydrogen sulfide, hydrogen, oxygen, and ammonia. The biogas is generated from locally available wide range of materials like animal dung, human excreta, vegetable wastes, water hyacinth, agricultural wastes like banana stems, de-oiled seed cakes, willow dust, etc., as feed material.

It is claimed that 10 m^3 of biogas has an energy equivalent to 6 m^3 of natural gas, 7.0 liters of gasoline, 3.6 liters of butane, or 6.1 liters of diesel oil. An ordinary normal family of four will require about 4.25 m^3 of biogas per day for cooking and lighting, and this much quantity of biogas can be generated easily from the night soil of family and the dung of three cows.

It is estimated that the efficiency of the direct burning of dung cakes is only about 11% and the efficiency of biogas is as high as 60%. Thus, the generation of biogas grants about five times more energy than that direct burning of the same quantity of the dung cakes. Thus, the biogas installations prove to be economically sound, and in addition, the housewives will be saved from the irritating smoke nuisance resulting from the combustion of firewood, cattle-dung, detritus of raw vegetables, etc.

The ferro-cement technique can be used for fabricating the biogas digesters to improve the efficiency of gas production. Following are the four commonly adopted designs of the biogas digesters:

1. Fixed dome digester
2. Flexible bag digester
3. Floating gas holder digester
4. Prototype digester

1. **Fixed dome digester**: This is the most popular type of biogas digester mainly due to low construction cost, local availability of required building materials, its hidden underground structure, and reduced maintenance costs. However, it is likely for the cracks to be developed in plastered masonry dome within a few years, leading to the gas leakage and repair of such cracks may prove to be a difficult job. The inside of dome may be sealed with appropriate lining material to avoid this difficulty.
2. **Flexible bag digester**: This type of biogas digester is rare in India. It appears to be promising in future due to few operational problems, easy and cheap installation, high durability, and stable gas pressure.
3. **Floating gas holder digester**: This type of biogas digester has lost its widespread popularity due to high construction cost, problems of corrosion, and gas leakage.
4. **Prototype digester**: These digesters are smaller and cheaper. They are more appropriate where requirements to be satisfied are not very important.

19.17.2   Elutriation

The washing of sludge with clean water or sewage effluent is known as the elutriation and it is adopted to remove organic and fatty acids from the sludge. The process of elutriation is applied just before the coagulation of sludge and during washing. The solids must be subjected to continuous mechanical agitation or air diffusion in order to be kept in suspension. The elutriation of sludge results in the reduction of alkalinity of sludge, and consequently, less quantity of lime is required in sludge coagulation.

The amount of water or sewage effluent required in the process of elutriation is about 2–3 times the volume of sludge. The process may be adopted for raw sludge or digested sludge. When elutriation is adopted, there is a considerable reduction in the quantity of chemicals required, especially when ferric coagulants are used. Thus, it results in the saving of chemicals and makes coagulation of sludge easy and economical. The elutriation is usually recommended for treatment plants designed for 1000 population or so.

19.17.3   Garbage Collection and Removal

The dry refuse or garbage should be properly collected by the concerned local authorities. For this purpose, the usual arrangement made is that dust bins of suitable sizes are placed at convenient places in the locality. The garbage of nearby houses is discharged into such dust bins by the individual house owners. The dry refuse fallen on the public streets and roads is collected, usually twice a day, by the laborers employed by the local authority.

The garbage thus collected requires careful removal. Following points should be observed:

- The dust bins should be of proper size and of approved pattern and they should be properly maintained.
- The garbage should be removed with least nuisance to the nearby inhabitants.
- The period of removal of garbage should be previously decided and made known to the public. By such arrangement, the garbage will be detained for the least possible time in dust bins. The public mind should be trained to contribute the garbage just before its period of removal.
- The removal of garbage should be speedy.

The dry refuse is removed by employing vehicles of various types and models. The governing factors in the choice of a particular vehicle are its initial and maintenance costs, strength of vehicle required. and condition of road. Auto-rickshaws, trailors, and trucks are some of the vehicles which are commonly employed for the removal of garbage.

19.17.4   Garbage Disposal

The unsatisfactory disposal of garbage results in unhygienic conditions and poses public health hazard/risk. It may be noted that the nature of dry refuse varies from place to place and it depends on climatic conditions, extent of industrialization, living standards, etc. Even changes in social, economic, and technical progress are also reflected in the nature of dry refuse.

Following are the various methods of garbage disposal:

1. **Controlled tipping**: This method is adopted where site for redevelopment is available. In this method, the garbage is tipped in hollows to a depth of about 1–2 m. The coarse material is tipped at the bottom and the fine material is laid at the top. The tips are covered with earth or ash in about 24 hours. The fermentation by anaerobic bacteria occurs under the seal and it is completed within 12 months or so. The tip usually settles down by about 300 mm due to fermentation, and at the end of fermentation period, the site is free from germs and is available for further development.

2. **Disposal into sea**: This method of garbage disposal is available to the coastal town only. It is cheap. The garbage should be discharged at such points and the periods that it does not return to the shore to create the unhygienic conditions.

3. **Incineration**: In this method, the dry refuse is burnt in the incinerator plant. This method is useful where suitable site is not available for the safe disposal of garbage. It is also adopted in case of disposal of hazardous garbage from hospitals and industrial plants. Such garbage contains dangerous materials which are to be burned to render them harmless.

4. **Mechanical compost plant**: The compost plant converts the garbage into manure which is rich in nitrogen contents. This is the most hygienic method of dry refuse disposal and the apparent looking rubbish is transformed into useful chemical fertilizer. The main aim of setting up of compost plant in various cities of our country is to supplement the supply of chemical fertilizers, the import of which is becoming prohibitive in cost. It has been observed that the mechanical compost plants have immense capacity to increase the food production, conserve soil properties, and improve ecological objectives.

5. **Pulverization**: In this method, the dry refuse is simply pulverized into powder. The powder thus formed may be used as manure. This method may prove to be uneconomical, if there is no steady demand for such type of fertilizer.

6. **Trenching**: Dry refuse is dumped into trenches and converted into a type of compost in this method. The trenches are 3–12 m long, 2–3 m wide, and 1–2 m deep. The trenches are excavated and then they are filled with garbage. Finally, a layer of non-combustible material is spread over the garbage. Within about 6 month's interval, the garbage is converted into a type of compost by fermentation,

carried out by the anaerobic bacteria. The contents are taken out from the trenches and then sold as manure. The compost obtained from trenching has small agricultural value and it contains substances which may be harmful to the soil. The compost is however cheap and can therefore create good demand.

19.17.5 Micro-Organisms

The micro-organisms are of considerable importance in public health engineering because of the fact that they are responsible for the stabilization of organic wastes. According to the type of biological metabolism, micro-organisms may be broadly divided into two categories:

  i. Autotrophs
 ii. Heterotrophs

  i. **Autotrophs**: These organisms can grow independently of an external source of organic matter. They do so because they are capable of synthesizing their organic requirements from inorganic sources. Their cell carbon is derived from carbon dioxide. The metabolic methods adopted may be chemosynthesis or photosynthesis, and accordingly the autotrophs are known as the *chemotrophs* or *phototrophs*, respectively. In chemosynthesis, the energy is supplied by an inorganic oxidation-reduction reaction and in photosynthesis; the energy is supplied by the light of the sun.

 ii. **Heterotrophs**: These are organisms that require an external source of organic matter for their growth, and thus, they obtain their energy from the breakdown of complex organic substances. They derive their cell carbon from organic carbon. They are of three types:

 - Paratrophs: They obtain soluble organic matter from the tissues of other living organisms. Therefore, they are likely to be pathogenic or harmful to human life.
 - Phagotrophs or holozoic forms: They utilize solid organic matter.
 - Saprophobs: They obtain soluble organic matter from the environment directly or by extracellular digestion of insoluble materials.

19.17.6 Night Soil Disposal Without Water Carriage

The different types of privies or latrines that are constructed to dispose off human excreta, without the help of water carriage systems, are as follows:

  i. Aqua or wet privies
 ii. Bore-hole privies
iii. Pit privies
 iv. Trench privies

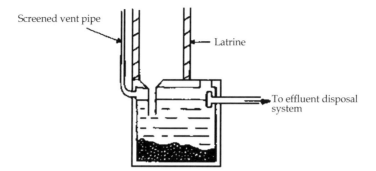

FIGURE 19.29 Aqua privy.

i. **Aqua or wet privies**: Set above on adjacent to a septic tank, an aqua-privy is a latrine that is beneficial in situations in where water supply is limited (Figure 19.29). When the latrine is placed above the tank, a chute drop-pipe that is 100–150 mm in diameter is made to hang below the squat hole or latrine seat in such a manner that the excreta is dropped directly into the tank below water level. The bottom of the pipe is placed 75 mm below the liquid level in the tank; this is done to prevent gases from escaping into the latrine superstructure and to limit the access of mosquitoes and flies into the tank. Another method is to fit the toilet with a pan with a water seal, where the latrine is adjacent to the tank and the pan with water seal is connected with a short pipe. Here, the effluent from the tank will go into a soak pit, sewer, or drainage trench. The flow of effluent then is small and is therefore also concentrated.

ii. **Bore-hole privies**: Borehole latrines have an augered hole in lieu of a dug pit and may be sunk to a depth of 10 m or more, although a depth of 4–6 m is usual. Augered holes, 300–500 mm in diameter, could be dug quickly by hand or by machine in areas where the soil is firm, stable, and free from rocks or large stones. The drawbacks include increased likelihood of blockage owing to its small diameter and increased danger of groundwater contamination because of the depth of the augered hole. Even when the hole is not blocked, the sides of the hole tend to get soiled near the top, increasing the probability of fly infestation. Borehole latrines are however convenient for emergency or short-term use because they can be made using light portable slabs and can be made rapidly in large numbers.

iii. **Pit privies**: One of the most widely used sanitation technology is the single pit. Excreta, along with anal cleansing materials (water or solids), are deposited into a pit. To avoid people falling into the pit, increase convenience and reduce odor, a slab with a hole is used to cover the pit. Lining the pit prevents it from collapsing and provides support to the superstructure, but the underground of the latrine should be water permeable. When the pit is filled up, it needs to be emptied or closed and relocated.

By constructing twin pits (the double pit latrine), it should be possible to dig out a filled pit, after it has stood for a year, without any objectionable smell, while the other pit is in use. Consequently, the pits can be used more than once (no need to construct a new one every year) and there is less risk of groundwater pollution in densely populated areas. The content of the pit however still contains nutrients and it could also contain some pathogens after 1 year. Therefore, it needs to be reused or disposed in an appropriate manner. However, it should be noted that due to soil infiltration there always remains the danger of groundwater contamination, especially in densely populated areas. Further, odor and fly nuisance are also very common.

iv. **Trench privies**: In this arrangement, a long trench of about 60 mm width is excavated and its depth is kept as about 400–600 mm. The trench is covered with wooden planks. An enclosed space of about 1 m width is made with the help of jute rags supported by wooden ballies, etc., and a hole is made in the wooden plank to receive the human excreta. The trench privy is useful as a temporary measure for occasions which are to last for few days only such as festivals, fairs, etc.

19.17.7 RURAL SANITATION

India is a country of villages and more than 70% of population of our country resides in villages and small towns. The term rural sanitation is used to denote the development or maintenance of sanitary conditions in rural areas. It mainly refers to situations in which the following two conditions are absent:

- Piped water supply
- Sewerage system of waste disposal

The absence of the above two conditions in the rural areas is mainly due to the following factors:

- The installations of piped water supply and sewerage system prove to be very costly and it is not possible to raise the necessary funds for the same.

- The population to be served in the rural areas is considerably small and it is not worthwhile to go for large projects of water supply and sanitary engineering.
- The maintenance aspect of projects should also be considered as skilled supervisory staff will not be easily available for rural areas.

However, the open space is available in abundance in rural areas and it can be used in the best profitable manner to grant the satisfactory environmental conditions in rural areas. The aspects of rural sanitation may be categorized in four ways:

   i. Collection and disposal of dry refuse
  ii. Collection and disposal of sullage
 iii. Disposal of night soil
 iv. Supply of potable or wholesome water for domestic use.

20 Environmental Engineering

20.1 THE ENVIRONMENT

Environment (Figure 20.1) of an organism indicates surrounding and everything that affects that organism during its lifetime. It can further be designated as the sum total of water, air, and land interrelationships among themselves and also with the human being, other living organisms and property. It comprises all the physical and biological surroundings and their interactions. Environmental studies give a path to realizing the environment of our planet and the impact of our human life on the environment.

Natural Environment

As per ecology and biology, the environment is the sum of natural materials and living things including sunlight. This is also called the natural environment. Natural resources are the things in the natural environment that have some value. For example, sunlight, fish and forests. As these are reappeared naturally after our use, these can be listed as renewable resources. In the environment, nonrenewable resources such as fossil fuels and ores are important things because of their limited availability.

Historical Environment

The historical environment is the culture and circumstances that a person lived in. A person's belief and actions depend on his environment. For example, Julius Caesar and Thomas Jefferson have owned slaves. Modern people mostly feel that it is not good to own slaves. But slavery was normal in Caesar's and Jefferson's environments. So, their actions did not look as wrong in their society.

20.1.1 COMPONENTS OF ENVIRONMENT

Our environment has been classified into four major components (Figure 20.2):

a. **Hydrosphere**: All water bodies such as rivers, lakes, streams, ponds, ocean, etc., are included in the hydrosphere. The ocean water or the saltwater accounts for 97.5% of the total while the freshwater resources account only for 2.5%. But most of the freshwater resources are in the form of ice glaciers and fresh ground water. So only 0.3% of the freshwater on Earth is available as lakes, reservoirs, and river systems. The hydrosphere functions in a cyclic manner, from precipitation, condensation, infiltration, and evaporation which takes place in a cycle, thus getting the name hydrologic cycle or water cycle.

b. **Lithosphere**: Earth has a solid inner core, a liquid outer core, a highly viscous (but less viscous than outer core) mantle, and outer silicate solid crust. The lithosphere is mainly divided into tectonic plates. These tectonic plates are responsible for the formation of continents and ocean floors. Based on this there are mainly two different types of lithosphere-oceanic lithosphere and continental lithosphere. Lithosphere mainly contains earth rocks, soil, mountain etc.

c. **Atmosphere**: Atmosphere is a narrow layer which protects the solid earth and human beings from the harmful radiations of the sun and which encloses gases like oxygen, carbon dioxide, etc., on the basis of temperature one can differentiate atmosphere into

FIGURE 20.1 Environment.

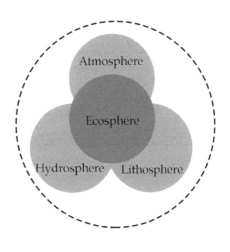

Biosphere

FIGURE 20.2 Component of environment.

five concentric layers. These include the troposphere, the stratosphere, the mesosphere, the thermosphere, and the exosphere.

The chrematistics of each layer is different.

d. **Biosphere**: Biosphere is otherwise known as the life layer, specifies to all organisms on the surface of the earth and their interaction with air and water. It is composed of animals, plants, and micro-organisms, varying from the teeny microscopic organism to the largest whales in the sea. The abundance of biosphere depends on several factors like temperature, rainfall, geographical reference, etc.

20.2 ECOLOGY AND ECOSYSTEM

20.2.1 Ecosystem

An ecosystem can be visualized as a functional unit of nature, where living organisms interact among themselves and also with the surrounding physical environment. Ecosystem varies incredibly in size from a small lake to a large woodland or ocean. Many ecologists consider the entire biosphere as a composite of all local ecosystems on Earth that is as a global ecosystem. Since this system is too much massive and complicated to be studied at one time, it is comfortable to divide it into two basic categories, namely, the terrestrial and the aquatic. Desert, Forest, and grassland are some examples of terrestrial ecosystems; lake, pond, river, wetland, and tidewater are some examples of aquatic ecosystems. Man-made ecosystems consist of an aquarium and crop fields.

- **Balance of ecosystem**: All ecosystems, even the ultimate biosphere, are open systems. There must be at least an inflow and outflow of energy. The balance of ecosystem means the balance of autotrophs and heterotrophs in an ecosystem, to maintain a uniform distribution of sustainable energy through food chain without any external interference. Any disturbance in autotrophic-heterotrophic balance leads to ecosystem imbalance, more the imbalance more external is required to balance. Bigger the ecosystem lesser is the imbalance. Balance, in fact, indicates to lesser consumers than primary producers.
- **Stability of ecosystem**: Very stable ecosystem can be defined as an ecosystem which has attained maturity. It is reserved by feedback mechanism which may be either positive or negative.

20.2.1.1 Structure and Function of an Ecosystem

An ecosystem has two types of components.

1. **Abiotic components**:
   a. Physical components: It includes soil (provide nutrients and base), water (essential for living beings), sunlight (for photosynthesis), and temperature (necessary to get survive), altitude, turbidity, wind etc.

   b. Chemical components: It includes carbohydrates, proteins, minerals, fats, etc.
2. **Biotic components**: It includes autotrophs (producers), heterotrophs (consumers), and decomposers (microorganisms). Autotrophs are called producers since they produce food through the process of photosynthesis. Heterotrophs are organisms which depend on other organisms for food. They are mainly divided into the following:

Primary consumers: Herbivores that rely on the producers for food.

Secondary consumers: Carnivores or omnivores which rely on the primary consumers for energy.

Tertiary consumers: These are organisms which depend on the secondary organisms for energy.

Quaternary consumers: They prey on the tertiary consumers for food.

Decomposers are mainly the microorganisms mainly bacteria that thrive on the dead and decaying matter. They complete the biocycle. The decomposers also include fungi. They are essential in a biotic system since they supply the necessary nutrients needed for the plant nutrition (autotroph nutrition).

Species composition of an ecosystem can be found out by identification and enumeration of plant and animal species in that ecosystem. Vertical distribution of different species occupying different levels is called stratification. For example, trees take up top vertical strata or layer of wood land, shrubs the second, and grasses and herbs take up the bottom layer.

The consideration of following aspects is necessary for the function of components of the ecosystem as a unit. These aspects include energy flow, productivity, decomposition, and nutrient cycle. Figure 20.3 shows the representation of cycle in a terrestrial ecosystem.

a. **Productivity**: The basic requirement for the proper functioning and sustainability of any ecosystem is a steady input of solar energy. The amount of organic matter or biomass produced per unit area over a time period during photosynthesis by plants is defined as primary production. It is expressed in terms of weight (g^{-2}) or energy ($kcal/m^2$). The rate of biomass production is called productivity. It is expressed in terms of g^{-2}/year or ($kcal/m^2$)/year to compare the productivity of different ecosystems. It can be classified into gross primary productivity (GPP) and net primary productivity (NPP). GPP of an ecosystem is the rate of production of organic matter during photosynthesis. Respiration of plants utilizes a noticeable amount of GPP. Subtraction of respiration losses (R) from GPP gives NPP.

$$GPP - R = NPP \qquad (20.1)$$

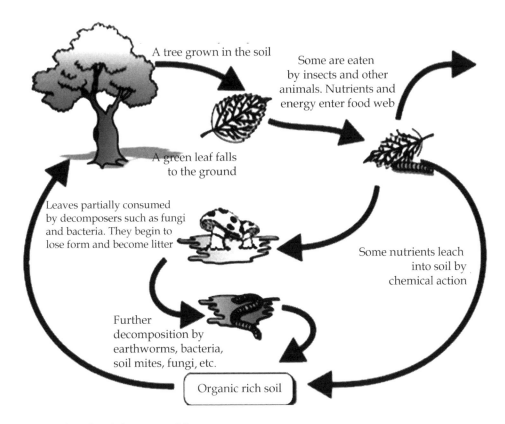

FIGURE 20.3   Representation of cycle in a terrestrial ecosystem.

The accessible biomass for utilization heterotrophs (herbivores and decomposers) is NPP. The rate of formation of new organic matter by consumers is known as secondary productivity.

b. **Decomposition**: Decomposition is the main function of decomposers. Decomposition is the process in which complex organic matter is broken down into an inorganic substance like water, nutrients, and carbon dioxide. The raw materials for decomposition are dead plant remains such as bark, flowers, leaves, and dead remain of animals, including fecal matter which constitutes detritus. The important steps in the process of decomposition are fragmentation, leaching, catabolism, humification, and mineralization.

Fragmentation is the process in which detritus are broken down into smaller particles by detritivores (example earthworm). During the process of leaching, water-soluble inorganic nutrients penetrate into soil horizon and get precipitated as unavailable salts. In catabolism, detritus is degraded into simpler inorganic substances by bacterial and fungal enzymes.

Humification leads to the formation of humus which is a dark-colored amorphous substance. The humus undergoes decomposition at an extremely slow rate due to its high resistivity to microbial action. The colloidal nature of it serves as nutrients reservoir. The humus is further degraded by some microbes and release of inorganic nutrients occurs by the process known as mineralization.

c. **Energy flow**: The unique source of energy for all ecosystems on Earth is the Sun. Less than 50% of the incident solar radiation only converted into photosynthetically active radiation (PAR). Autotrophs (plants and photosynthetic bacteria) capture the sun's radiant energy to make food, for example, inorganic materials. This captured energy by plants is only 2%–10% of the PAR and this minute quantity of energy keeps up the whole living world. So, it is very essential to learn how the solar energy captured by plants flows through different organisms of an ecosystem. All organisms are dependent on producers (either directly or indirectly) for their food.

As per ecosystem terminology, the producers in the ecosystem are green plants. Woody plants and herbaceous are the major producers in a terrestrial ecosystem. Similarly, in an aquatic ecosystem, primary producers are different species like algae, phytoplankton, and higher plants. All animals are dependent on their food on plants either directly or indirectly. Therefore, they are known as consumers and also heterotrophs. If they consume plants (producers) directly, they are called primary consumers, and if the animals feed other animals which turn eat the plants, they are called secondary consumers. The primary consumers will be herbivores. Figure 20.4 represents the energy flow through different tropic levels.

d. **Nutrients cycle**: The producers prepare food by using nutrients, the consumers consume it, and the

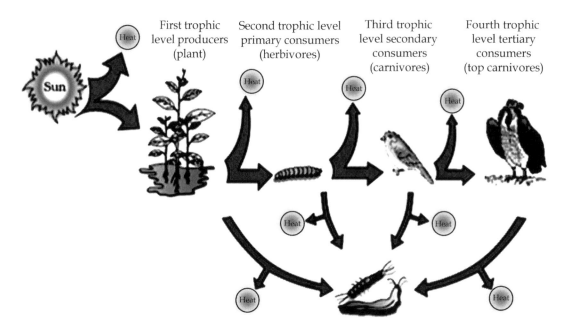

FIGURE 20.4 Energy flow through different tropic levels.

decomposers restore the nutrients keep running between biotic and abiotic units, forming nutrients cycle termed as a biological cycle. The movement of nutrient elements through the different components of an ecosystem is called nutrient cycle or biochemical cycle. Nutrient cycles are of two types—gaseous and sedimentary. The source of the gaseous type of nutrient cycle (e.g., carbon, nitrogen cycle) occurs in the atmosphere, and for the sedimentary cycle (e.g., sulfur and phosphorus cycle), the source is located within the Earth's crust. The rate of release of nutrients into the atmosphere can be regulated by environmental factors such as soil, moisture, temperature, pH, etc. The function of the reservoir is to prevent shortage that occurs due to the imbalance in the rate of influx and efflux.

i. Ecosystem—carbon cycle

The carbon cycle is mainly divided upon five pathways of exchange. They are the atmosphere, the terrestrial biosphere, the ocean floor, the sediments, and inside the earth. It mainly occurs due to various physical, geological, and biological processes.

Carbon cycle (Figure 20.5) takes place through living and dead organisms and through atmosphere and ocean. According to one evaluation, annually, 4×10^{13} kg of carbon is fixed in the biosphere through photosynthesis. Respiratory activities of the producers and consumers cause the return of considerable amount of carbon to the atmosphere as CO_2. To a large extent, decomposers also contribute to CO_2 pool through their processing of dead organic matter of land or oceans and waste materials. Some quantity of the fixed carbon is lost to sediments

and removed from circulation. Forest fire, burning of wood, combustion of fossil fuel, organic matter, and volcanic activity are further sources for the discharge of CO_2 to the atmosphere.

ii. Ecosystem—phosphorus cycle

Phosphorus is a major component of nucleic acids, cellular energy transfer systems, and biological membranes. Phosphorous cycle is a biogeochemical cycle that explains the movement of phosphorous through hydrosphere, lithosphere, and the biosphere. Figure 20.6 represents a phosphorus cycle. Many animals also demand a large volume of this element to make teeth, bones, and shells. The rock which contains phosphorus in the form of phosphates is the natural source of phosphorus. When rocks are weathered, very small quantities of these phosphates are dissolved in soil solution and which are taken by the root of the plants. Hence, herbivores and other animals obtain this element from plants. The phosphate-solubilizing bacteria decompose dead organisms and waste products and phosphorous again reaches the soil.

iii. Ecosystem—nitrogen cycle

The nitrogen cycle is a biogeochemical cycle in which the nitrogen is converted to other chemical forms that pass through the atmosphere, terrestrial, and marine ecosystems. Nitrogen takes up 78% of the earth's atmosphere making it the highest. But this atmospheric nitrogen is difficult for the ecosystems to use, so they should be converted to usable form. The processes required to convert the atmospheric nitrogen to usable form (nitrates and nitrites) is known as nitrogen fixation so that the plants utilize them

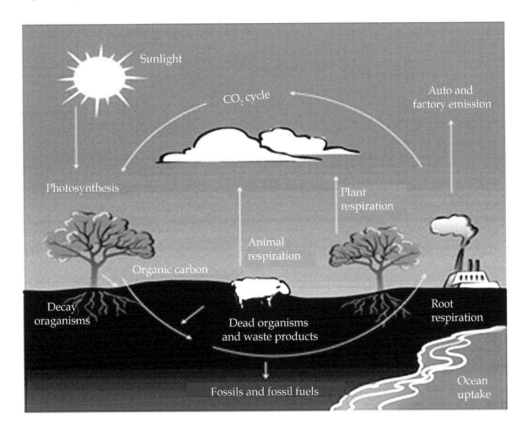

FIGURE 20.5   Carbon cycle.

The Phosphorus cycle

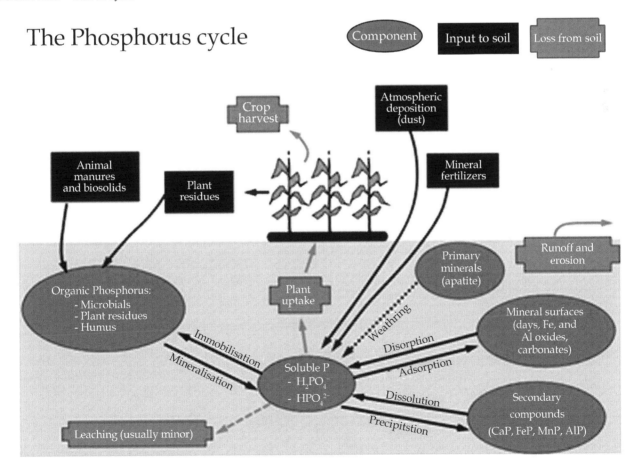

FIGURE 20.6   Phosphorus cycle.

efficiently. Other important processes in the nitrogen cycle are ammonification, nitrification, and denitrification.

iv. Ecosystem—oxygen cycle

Oxygen cycle is a biogeochemical process between the molecule oxygen, its oxides, and its ions. Oxygen is the most abundant element on the earth since it can react with almost all elements like carbon, nitrogen, magnesium, etc. In the atmosphere oxygen takes up 20.9% of the total air surface. Oxygen production is done in two ways, by biotic production where plants produce the atmospheric free oxygen by the process of photosynthesis and by abiotic production in which oxygen is produced by photolysis (ultraviolet radiation breaks the atmospheric water and nitrous oxide components into individual atoms).

20.2.1.2   Functions of Ecosystem

1. It balances the nutrients cycle
2. It balances the rate of biological energy flow

Sun → Producer → Consumer → Decomposers

20.2.1.3   Food Chain

Food chain (Figure 20.7) is the process of eating and being eaten by the successive creatures. In brief, it is the flow of energy from producer to tertiary consumer.

20.2.1.4   Types of Food Chain

1. **Grazing food chain**: It begins with plants. Herbivores eat plants. Then it is eaten by carnivores and through carnivores, and it reaches to the decomposers. Finally, carnivores breakdown the complex organic matter into a simple one. Example for grazing food chain is pond ecosystem, grassland ecosystem, etc.
2. **Parasitic food chain**: It occurs when plants and animals get affected by parasites. These smaller organisms consume them without killing them; for example, bug, nematode, etc.
3. **Detritus food chain**: This type of food chain begins from dead and decayed organisms. Then these are consumed by the micro-organisms (detritivorous or saprovorous).

20.2.2   ECOLOGY

Ecology means the scientific study of the abundance and distribution of organisms and interactions that decide abundance and distribution. Ecology deals with the following three levels: the individual organism, the population (includes individuals of the same species), and the community (consisting of a greater or lesser number of species populations). At the level of the organism, ecology deals with how individuals are affected by (and how they affect) their environment. At the level of population, ecology is concerned with the presence or absence of particular species, their abundance or rarity, and with the trends and fluctuations in their number. Community ecology then deals with the composition and organization of ecological communities.

Ecosystems consisted of actively cooperating parts including organisms, the communities they created, and the nonliving things of their environment. Ecosystem processes, such as

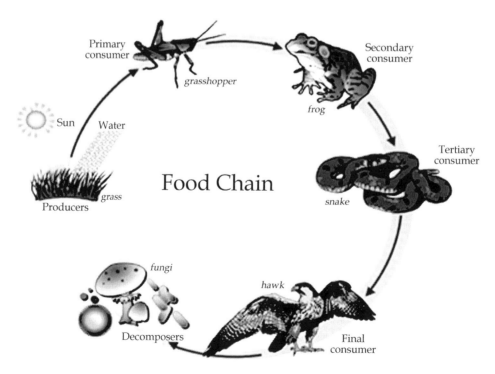

FIGURE 20.7   Food chain.

primary production, pedogenesis, nutrient cycling, and various niche construction activities, regulate the flux of energy and matter through an environment. These processes are maintained by organisms with specific life history features, and the variety of organisms is called biodiversity. In short, biodiversity deals with varieties of genes, species, and ecosystems, which improve certain ecosystem aids.

20.2.2.1 Hierarchical Ecology

The scale of ecological actions may be a closed system or an open system. Blackflies migrating on a single tree is a closed system; while at the same time, it remains open with regard to wider scale influences, like climate or atmosphere. Hence, ecologists classify ecosystems hierarchically by analyzing data accumulated from finer scale units, such as soil types, climate, and vegetation associations, and consolidate these data to identify resulting patterns of consistent organization and processes that operate on local to regional, landscape, and chronological scales.

- Biodiversity
  Biodiversity refers to a collection of life and its activities. It consists of a wide range of living organisms, the communities, and ecosystems in which they occur, the genetic differences among them, and the evolutionary and ecological processes that preserve them active.
- Habitat
  The environment in which a species occurs and the kind of community that is formed as a result is described as a habitat. More precisely, habitats deal with the zones in environmental territory that includes multiple dimensions, each representing a biotic or abiotic environmental variable; that is, any component or characteristic of the environment related directly (e.g., forage biomass and quality) or indirectly (e.g., elevation) to the use of a location by the animal. For example, habitat must be a terrestrial or aquatic environment that can be again classified as an alpine or montane ecosystem. Habitat shifts contribute major evidence of competition in nature where one population changes relative to the habitats that consist of other individuals of the species.

  For example, one population of a species of tropical lizards (*Tropidurus hispidus*) has a flattened body relative to the main populations that live in open flat grassland. The population that lives in an isolated rock outcrop hides in crevasses where its flattened body offers a selective advantage. Habitat shifts also happen in the advancing life history of amphibians and in insects that transit from aquatic to terrestrial habitats.
- Niche
  A niche is a group of the biotic and abiotic condition in which a species is able to remain and keeps up stable population sizes. The ecological niche can be subdivided into the fundamental and the realized niche. The fundamental niche is the set of

environmental conditions under which a species is able to prevail. The realized niche is the set of environmental and ecological conditions under which a species prevails. Niche has a key role in the ecology of organisms.

20.2.2.2 Individual Ecology

One can predict patterns and processes of an organization including populations, ecosystems, and communities by understanding the characteristics of individual organisms. Such characteristics also include features of an organism life cycle such as lifespan, age to maturity, and metabolic costs of reproduction. Various fields of the ecology of evolution that focus on such characteristics are life history theory, metabolic theory of ecophysiology, ethology, and ecology. Other characteristics including emergent properties which have at least a part of interactions with the surrounding environment such as resource uptake rate, growth rate, winter, and deciduous vs. drought-deciduous trees and shrubs may be connected to structure, like dorsal spines of a bluegill sunfish or the spines of a cactus or behaviors such as pair bonding or courtship displays.

20.2.2.3 Population Ecology

Population ecology studies the activities of species populations and interaction of these populations with the extensive environment. A population comprises individuals of the same species that live, interact, and migrate through the same habitat and niche. A fundamental law of population ecology is the Malthusian growth model which states, a population will grow (or decline) exponentially as long as the environment experienced by all individuals in the population remains constant. Simplified population models commonly begin with four variables: death, birth, immigration, and emigration.

20.3 AIR POLLUTION

According to the United States Environmental Protection Agency (USEPA), air pollution is defined as the presence of contaminants or pollutant substances in the air that interfere with human health or welfare, or produce other harmful environmental effects.

Air pollution (Figure 20.8) indicates the atmospheric condition which can produce undesirable effects on man and environment due to the presence of certain substances in certain concentrations. Gases (SO_x, NO_x, CO, HCs, etc.), particulate matter (fumes, dust, smoke, aerosols), radioactive materials, and many others are belonging to these substances. We can consider a particular substance as an air pollutant only if its concentration is relatively high compared to the background value and causes adverse effects.

20.3.1 AIR POLLUTANTS

1. **Sulfur dioxide**: Sulfur dioxide was a severe air pollutant in the earliest days of industrialization. It has been a major problem in reducing or acidifying air pollution during the period of rapid economic

FIGURE 20.8 Air pollution.

growth in many countries. It is a colorless gas with extreme pungent odor which is suffocating in nature. It is mainly generated by the ignition of sulfur-containing fuels such as coal and petroleum which is in the form of vehicles, power production and heating. The main disadvantage of sulfur dioxide is that it travels long distances and affects the formation of ozone layer.

Sulfate, the sulfur-containing ion present in water, remains a major constituent of air pollution capable of forming acid. Sulfate is a major source of ultrafine particulates and it is capable of triggering bronchoconstriction in persons with airways reactivity. They can cause irritation in the lining of the nose, throat, and lungs. For asthma patients the intake of excess sulfur oxide can lead to chest infections. Acid ingredients like nitric acid also exist but the knowledge about them is limited. Due to its emission into the air by motor vehicles and industries, these acids cause a phenomenon known as acid rain.

Acid aerosols like nitrogen dioxide, sulfur dioxide, and sulfates deposit deeply in the airspace and distal lung as a result of their small size and tendency to ride along on particulates. They arise to aggravate airway responses in an additive or synergistic manner with ozone. Primary chemical species that cause acid rain are sulfur dioxide and sulfates. They may be transported away from their source to a long distance through the atmosphere and result in acidification of water and soils.

2. **Nitrogen dioxide**: Nitric oxide (NO) is formed by ignition. The oxidation of NO under conditions of sunlight or higher temperature combustion of NO in power plants or indoors from gas stoves causes the formation of a secondary pollutant namely nitrogen dioxide (NO_2), which originates greater health effects. Levels of exposure to nitrogen dioxide that should not be exceeded are, respectively, 400 $\mu g/m^3$

(0.21 parts per million (ppm) for 1 hour and 150 $\mu g/m^3$ (0.08 ppm) for 24 hours. The direct effects of nitrogen oxide include increased infectious lower respiratory disease in children (including long-term exposure as in houses with gas stoves) and increased asthmatic problems. Extensive studies of the oxides of nitrogen have shown that they increase the incidence and severity of bacterial infections after exposure and damage host defenses in the respiratory tract. They have a marked effect in reducing the capacity of the lung to clear particles and bacteria.

3. **Particulates matter**: It is a mixture of solids, liquids which includes carbon, complex organic chemicals, sulfates, mineral dust as well as water present in the air. Particle matter present in the atmosphere (aerosols) is correlated with an exalted risk of morbidity and mortality (including cough and bronchitis), particularly among populations such as aged and asthmatics. As pointed out, they are released from wood and coal stoves, diesel, fireplaces, tobacco smoke, automotive exhaust, and other sources of ignition. The USEPA sets a standard of 265 $\mu g/m^3$ in circulating air but does not have a standard for indoor air levels. Usual concentrations range from about 50 $\mu g/m^3$ in homes to 500 $\mu g/m^3$ in waiting rooms and bars. Tobacco smoke is also a primary donor to respirable particles indoors in developed countries. Particulate matter varies in size. Some of these are seen by naked eye while others are not large enough to be seen by naked eye. The latter ones are the most dangerous ones, called PM_{10} and $PM_{2.5}$. They are discussed below.

4. **Particulate matter (PM_{10})**: PM10 (particulates 10 μm and smaller) are larger particulates. These comprise predominantly of carbon-containing substances which are formed by combustion; wind blowing soil also contributes it in the air. These larger particulates have less effect on human health as compared to the smaller particulates.

5. **Particulate matter ($PM_{2.5}$)**: $PM_{2.5}$ (2.5 μm and below) are smaller particulates. These particulates in urban air pollution are different in their chemical composition than larger particles due to its smaller diameter. They are also known as fine particles. Particulates in the fraction $PM_{2.5}$ contain a relatively larger amount of water and acid forming chemicals such as nitrate and sulfate, also trace metals. These particulates nearly evenly circulate throughout urban regions where they are produced as it is small in size and it can penetrate completely and easily into buildings. One cannot easily separate $PM_{2.5}$ sulfate and ozone cannot as they are likely to occur together in urban air pollution. Recent research strongly conveys that at least $PM_{2.5}$ and sulfate, and probably ozone as well, cause an increase in deaths in affected cities. Of these, ozone is linked with episodes of asthma. All others are associated with higher rates of deaths

from and complaints about lung diseases and heart diseases.

6. **Hydrocarbons (HCs)**: Most HCs such as salicylic and aliphatic HCs generally cause little hazards because of its biochemical inert nature at ambient levels. On the other hand, aromatic HCs are more irritating to mucous membranes and compounds like benzopyrene are potent carcinogens because of are biochemical and biologically active nature. As HCs promoting the formation of photochemical smog, these are included among the criteria air pollutants.

7. **Lead**: Lead is noted as a highly harmful substance that particularly causes nerve damage. This can result in neurobehavioral problems and learning disabilities in children. About 80%–90% of lead in ambient air is thought to be evolved from the ignition of leaded petrol. Attempts for removing lead from gasoline are going on in all around the world because of its impacts on the learning abilities and the behavior of children even at low levels of exposure.

8. **Asbestos**: Asbestos is a mineral fiber that has been used in the early days as a fire retardant and as insulation in buildings. Many asbestos products are now banned and presently its use is limited. But in older buildings asbestos is still found in floor tiles, asbestos shingles, pipe, and furnace insulation, textured paints, and other construction materials. If these materials are disturbed by sanding, cutting or other activities, enormous airborne asbestos levels can be generated. Inappropriate attempts to clear away these materials cause health hazards such as lung cancer, mesothelioma (cancer of the lung and the abdominal lining), and asbestosis (lung scarring) due to release of asbestos fibers into the indoor air.

9. **Mercury**: It is present in the atmosphere in gaseous form due to its relatively high vapor pressure. During monsoon seasons the gaseous mercury is washed from the atmosphere by rain and a portion of it arrives in the water sources and the remaining is bound to the soil over the land. By the action of bacteria, the inorganic mercury is generally concurred into its methyl or diethyl cpds.

10. **Beryllium**: Most beryllium emissions are in the form of a metallic powder of beryllium oxide particulate. Beryllium concentration as low as 0.01–0.1 $\mu g/m^3$ causes berylliosis or chronic beryllium disease which is a systematic poisoning which starts with progressive suffocation, weight loss, cough and leads to cardiac failure.

11. **Fluorides**: Fluoride from copper smelters, superphosphate, enamel, and glass factories and aluminum plants (fluoride manufactory plant) causes damages to vegetation. Accumulation of fluorides on bones of animals could result in loss of weight, lameness, and dental fluorosis.

12. **Ozone**: Ozone is a highly toxic compound that makes changes in the reflex breathing mechanism which in turn cause unusual changes in breathing patterns. It also irritates airways in the lungs and disrupts host defense mechanisms in the body.

Ozone is created by the reaction of oxygen with nitrogen compounds and volatile HCs in the lower atmosphere in the presence of light from the sun as a source of energy. This occurs especially in stagnant weather conditions and inversions under conditions of sunshine, where there is ample time for the photochemical reactions to take place. Ozone will react with a set of substances as it is chemically unstable.

Ozone affects humans in a complicated manner which is dependent on pollutant concentration and activity level among other factors. Ozone appears to attack the epithelial cells in the bronchial tree, which in turn may cause airway inflammation and hyperresponsiveness in the first place, although this has been hard to prove.

20.3.2 PHYSICAL FORMS OF POLLUTANTS

The elements of air pollution may happen in any of the three phases of matter; they may be solid, liquid, or gas. Often all three are present at once, especially in very small particulates.

a. **Aerosols**: Aerosols are the small solid or liquid substances (fine drops or droplets) that are suspended in air. They may be a combination of solid phase particles, mixed solid- and liquid-phase particles, or sometimes liquid droplets. Even aerosols that are generally solid may contain absorbed water.

Composition and size are the most important trait that can anticipate the behavior of aerosols. Composition predicts what will happen when it lands or settles on something and size determines how the particle will travel in air. The individual particles in aerosols may be almost uniform in size (monodispersed) or highly changeable in size (polydispersed). Aerosols in air pollution are all polydispersed. Many little particulates may form one big particle by the aggregation of fumes of polydispersed fine aerosols containing solid particles.

b. **Liquid phase**: Liquid components of air pollution exist as aerosols, either in the form of liquid-phase particles, which are droplets or in association with solid-phase particles. As the evaporation of droplets of more volatile organic compounds (VOCs) to the gaseous phase is very quick, liquids that are constituents of air pollution are always, water-based, or aqueous. A dense collection or cloud of droplets is termed as a mist.

Less quantity solid-phase particles also contain a less amount of absorbed water. Both liquid and gasphase components of air pollution usually are fixed to and drive on the surface of solid particles; this is called adsorption. The major determinant of the water content of particles is humidity in the atmosphere. Lower humidity promotes the drying out of the water; as a result of it, the particle is reduced to a solid phase. When dry particles are released into a humid atmosphere, they may take on water. Generally, this water absorption is more in smaller particles, so they are known to be hygroscopic. This boosts up the mass of the particle which in turn increases its capacity to carry other dissolved constituents. Therefore, in different climates (humid climates and dry climates) rate of air pollution may be different even it is from the same types of sources.

There are processes in the atmosphere that convert liquid to gas and vice versa or again convert liquid to solid. Evaporation of volatile liquids converts it into gases. The evaporated compound in the gas phase is termed vapor and behaves like gas in air pollution. Condensation of vapor in a saturated atmosphere causes the formation of droplets. One of the well-known examples of an aerosol of water droplets is fog which is formed by condensation in an atmosphere saturated with liquid around a small solid particle.

c. **Gas phase**: Gaseous components of air pollution are dissolved in air. Chemical reactivity and solubility in water are important features while considering gaseous components of air pollution.

Solubility is a major deciding factor of the health effects of gases. Comparably soluble elements of air pollution contain sulfur dioxide, sulfates, and nitrates. They may also blend to form ultrafine particles. (Further, there are so many gases more common as occupational exposures which are water-soluble, including ammonia and hydrochloric acid vapor.) Comparably insoluble elements contain the oxides of nitrogen and ozone.

Solubility for gases is influenced by particle size; it is a property that evaluates the efficiency with which they infiltrate deeply into the respiratory tract. A gas that is soluble in water will be dissolved in the water coating the mucous membrane of the lungs and upper respiratory tract and will be removed from air passing more deeply. A gas that is insoluble in water will not be so removed and will penetrate to the alveoli, the deepest structures of the lung, more precisely.

d. **Inhalation**: The most rapid path of entry of toxicants into the body is inhalation. It is because of the close bond of air passages in the lungs with the circulatory system. On inhalation, soluble gases tend to dissolve into the water surface of the pulmonary tract while the insoluble gases generally penetrate to the alveolar level. Because the alveolus brings the blood into a very near and immediate response to air, gases may pass directly across the alveolar membrane and into the bloodstream very efficiently. Particles once deposited in the alveoli may dissolve and deliver

their constituent compounds. The degree to which they enter the blood, circulate, and then release to the body's tissues depends on the duration of exposure, the reactivity of the compound concentration inhaled, the reactivity of the compound and tissue, respiratory rate, and the reactivity of the compound.

Chemical asphyxiants are the compounds that inhibit the transfer of oxygen to the tissues or the adsorption of oxygen once it reaches the tissues. The two most common examples of such inhibitors of oxygen transfer or adsorption are carbon monoxide (CO), which obstructs the location on hemoglobin that absorbs and transports oxygen and hydrogen cyanide, which (in the form of cyanide) obstructs the way by which the tissues utilize oxygen.

e. **Volatile organic compounds**: VOCs consist of methanol, benzene, formaldehyde, carbon tetrachloride, chloroform, and hundreds of other compounds. Gasoline is a mixture of many such compounds. Some of these VOCs are highly reactive. They cause direct human physiological effects as well as indirect effects (such as helping to the formation of ozone). They may develop from household activities such as refineries, dry cleaning establishments, painting supplies, gasoline stations, and many other sources. They lead to headaches as well as irritation to the respiratory tract (starts as a runny nose or increased rhinitis and developed to asthma) and other nonspecific problems. The effects of VOCs are varying depending on compound and concentrations (high concentration causes considerably toxic effects), but in all cases, these lead to neurological problems. Indoor air pollution problem and occupational hazards are the primary reasons for direct toxicity from VOCs. Pollution rate from indoors and in the workplace is much greater than that from outdoor levels.

f. **Trace metals**: The trace metals or elements are elements which are present in the body or environment or only in small quantities. Lead, mercury, copper, cadmium, zinc, and many others are examples of trace metals. Trace metals cause direct health hazards on the nervous and respiratory systems. The release of such metals into the environment is increasing by human activities.

20.3.3 Sources of Air Pollution

There are several locations, factors, or activities which cause discharge of pollutants into the atmosphere. These sources can be classified into two major categories.

1. **Man-made (Anthropogenic) sources**: These are often linked to the burning of numerous types of fuel.
   a. Stationary sources: Manufacturing facilities (factories), waste incinerators, and smokestacks of

power plants as well as furnaces and other types of fuel-burning heating devices are included in this type. Traditional biomass (includes crop waste, wood, and dung) burning is the major source of air pollutants in poor and developing countries.

b. Mobile sources: Aircraft, marine vessels, and motor vehicles.

c. Controlled burn: Controlled or prescribed burning is an approach sometimes used in greenhouse gas reduction, agriculture, forest management, or prairie restoration. Fire is a natural part of both forest and grassland ecology and controlled fire can be a tool for foresters. Controlled burning encourages the growth of some desirable trees in a forest, thus restoring the forest.

d. Fumes: Sources are hair spray paint, aerosol sprays, varnish, and other solvents.

e. Waste deposition: Methane generated in a landfill is highly combustible and may form explosive mixture with air. Methane is also an asphyxiant and may displace oxygen in a confined space. Reduction of oxygen concentration below 19.5% may cause suffocation or asphyxia.

2. Natural sources

a. Dust from large areas of land with very less or no vegetation.

b. Methane: emanated by the digestion of food by cattle-like animals.

c. Radon gas: Radon is a naturally occurring, colorless, odorless, radioactive noble gas that is formed from the decay of radium within the Earth's crust. Radon gas from natural sources can accumulate in buildings, especially in enclosed areas such as the basement and it is the second most common cause of lung cancer, after cigarette smoking.

d. CO and smoke from forest fires.

e. VOCs of environmentally significant amounts are released by vegetation in some regions (such as poplar, oak, willow, black gum, etc.) during warmer days. The VOC production from these plants leads to an increase in ozone levels up to eight times greater than the low-impact tree species. These VOCs react with primary anthropogenic pollutants, particularly, SO_2, NO_x, and organic carbon compounds to produce a seasonal cloud of secondary pollutants.

f. Volcanic activity, which generates chlorine, sulfur, and ash particulates.

20.3.4 GLOBAL ENVIRONMENTAL PROBLEMS DUE TO AIR POLLUTION

Burning of chemical and fossil fuel and large-scale deforestation have created an increase of CO_2 concentration which is now 25% greater than what it was at the beginning of the century. Climatologists expect that even a small change in

atmospheric CO_2 can have crucial effects on climate. CO_2 is a major cause of greenhouse effect. It occurs due to its transparency to incoming visible sun energy and absorption of infrared heat re-radiated from the Earth's surface like glass.

20.3.4.1 Global Warming

Air pollutants consist of greenhouse gases. Carbon dioxide is one among such gases which is a component of exhaust from trucks and cars. Greenhouse gases lead to global warming by trapping heat from the Sun in the Earth's atmosphere. Greenhouse gases naturally occur in Earth's atmosphere is essential for the survival of the ecosystem, but the increase in greenhouses gases is the cause of most of the global warming. This increase occurs from car exhaust and pollutants released from smokestacks at factories and power plants. The trucks, smokestacks, and cars also release tiny particles, called aerosols, into the atmosphere. Dust lifted into the atmosphere from evaporating droplets, from deserts, from the ocean, released by the smoke from forest fires, and erupting volcanoes also causes the formation of aerosols. Air pollutants released by humans by burning of fossil fuels again add them to the atmosphere. Generally, in the last 150 years or so, the amount of CO_2 in atmosphere has increased. While different types of aerosols act differently in the atmosphere, the overall effect of aerosols is cooling.

Greenhouse gases remain in the atmosphere for years and cause warming around the world. Computer models deduce that, worldwide, the tiny aerosols cause about half as much cooling as greenhouse gases cause warming.

20.3.4.2 Ozone Depletion

Ozone is a kind of super-charged oxygen (O_3). This gas forms a layer below the stratosphere. This layer functions as a cover to the Earth against ultraviolet radiation from the Sun. The canopy of the ozone layer is with the variable range less dense near the surface of the Earth compared to the height of 30 km.

Several pollutants in the atmosphere such as chlorofluorocarbons (CFCs) cause the depletion of the ozone layer. When hit near the stratosphere these CFCs and other similar gases are split by the ultraviolet radiation, and as a result of it, they deliver free atoms of bromine or chlorine. These atoms are highly sensitive with ozone and disturb the stratospheric chemistry. Further reactions cause depletion ozone layer which leads to direct contact of the earth to ultraviolet radiation. These rays cause harmful effects to living beings on the Earth such as a rise in the temperature, decrease of immunity, various skin diseases, etc., and it also affects the process of photosynthesis in plants.

20.3.4.3 Acid Rain

Acid deposition occurs in earth, particularly by acid rain. Mist, snow, dew, and fog also catch and discharge atmospheric contaminants. In addition, fallouts of dry sulfate, chloride, and nitrate particles can contribute to about half of the acidic deposition in some areas.

Regional and continental scale of acid rain is the factor which increases the complications of the acid rain problems and which can further make difficulty in finding a solution. Much oxides of nitrogen and sulfur are diffusing from tall stacks at power plants in order to increase the dilution and dispersion of the stack gases. Gases from tall chimneys lead the pollutants to be travelled for a long distance through the atmosphere. This will protect adjacent communities from the instantaneous effect of air pollution. The pollution in its effect is air-transited to other regions and even to other continents.

Formation of acid rain:

$$SO \rightarrow H_2SO_4 \quad 62\%$$

$$NO_2 \rightarrow HNO_3 \quad 32\%$$

$$Cl \rightarrow HCl \quad 6\%$$

In urban areas, where transportation is the major sources of pollution, nitric acid is equal to or slightly greater than sulfuric acid in the air.

20.3.4.4 Indoor Air Pollution

In many traditional and developing societies, disturbing resources of severe air pollution are incomplete combustion and smoky fuels burned for cooking and heating. The utilization of such fuels induces air pollution issues both indoors and outdoors. In developed countries as many buildings were built to be energy efficient and airtight, the quality of air indoors is a major problem. Indoor air pollution is also a serious problem in many developing societies. In homes where open fires burn, particularly in cold climate, the pollution from the fires concentrates and exposes the occupant, especially women, to the risks associated with smoke inhalation. This will cause serious lung disease and an increased risk of cancer, as happened in some parts of China among women who tend fires in homes heated with coal.

In urban households the exposure to indoor air pollution is severe due to tightly sealed buildings, no ventilation within the building, use of synthetic materials as well as many household products. These lead to the release of many harmful gases in the air.

The remarkable indoor air pollutants in developed countries are combustion products (like CO, carbon dioxide, NO_x, SO_x, and polycyclic aromatic HCs), tobacco smoke, formaldehyde, radon decay products, asbestos fibers, and other chemicals used in the household. VOCs present in perfumes, hairsprays, furniture polish, and air fresheners are mainly found in households which lead to air pollution. Formaldehyde is a gas that is produced from the carpet rugs, particle boards, and insulation boards. The usage of tobacco is another factor leading to indoor air pollution. Asbestos products are a major concern since it may lead to cancer. Many microbiological air pollutants or biological pollutants are also prominent including fungi and moulds, bacteria, viruses, pollen, algae, spores, and their derivatives. Indoor air pollutants can accumulate in airtight buildings especially (e.g., buildings which are energy efficient, but with poor ventilation) which causes tight building syndrome.

20.3.5 Air Pollution Prevention and Control

The activities for controlling and preventing of air pollution are given below:

a. Increasing public awareness about sources and effects of air pollution and how to control them.

b. **Substitution measures**: The current industrial or combusting practices which produce pollutants are substituted by the nonhazardous or less hazardous process. Replace electric power in place of fossil fuels and coal replaced by the biogas plants. The use of hydraulic operations and solar energy for industrial functions should be encouraged. The use of smokeless churlish in place of the age-old churlish effectively enhances indoor air quality.

c. **Containment action**: Escape of pollutants in the air from the industrial operations can be controlled by operating local exhaust ventilation, trapping, and then the disposal of pollutants.

d. **Dilution**: Conservation of green belt among and around the industries, in between industries and civilian habitat, cleans out a lot of air pollutants and keeps the air quality.

e. **Legislative action**: The Smoke Nuisance Act and Factories Act prescribe measures for the height of chimney stacks and use of arrestors, for industrial areas. The governmental regulatory approach can minimize the lead levels in petrol or increase the availability of leadless petrol.

f. **Active community involvement**: Cleanliness of streets and open areas, maintenance public places, use of smokeless churlish, forbidding tobacco smoking, maintenance of green belts, community action for collection and final disposal of refuse and solid wastes, proper effective ventilation in home and at workplace, use of solar and biogas energy, use of LPG gas; screening of windows, complete combustion of coke, coal and wood, regular maintenance of automobile vehicles, use of arrestors to the exhaust from automobiles, and a lot many actions can be undertaken for the control and prevention of air pollution.

20.4 NOISE POLLUTION

Unwanted sound that is not pleasing to the ear is called as noise. When there is lot of noise in the environment, it is called noise pollution. Noise pollution disrupts the daily activities of human life like working, sleeping, and during conversations. The normal level of tolerance for a human being is up to 80 dbA. If it exceeds this decibel, it is considered to be noise pollution. Side-by-side residential and industrial buildings can result in noise pollution in the residential areas. Thus, poor urban planning may lead to noise pollution.

20.4.1 SOURCES OF NOISE

Public address systems, inter alia, vehicular traffic, neighborhood, electrical appliances, TV and music system, railway and air traffic, and generating sets are the major sources of noise. Most of the people occupying in big towns or metropolitan cities and those working in factories are vulnerable to the negative effects of noise. Typically, it affects the poor and the rich alike. Aimless use of horn by the vehicles and widespread use of loudspeakers in Indian religious and social ceremonies cause several health hazards to the urban inhabitants. It may cause headaches, mental disorder, deafness, heart troubles, nervous breakdown, dizziness, inefficiency, insomnia, and high blood pressure. The noise level and exposure area depend on its source and its strength.

The noise from various sources may either be steady for a long period or changes over a particularized period remarkably. Road noise, especially at some distance from the road, can be described as a steady state noise that does not oscillate much. But rail and aircraft noise are acoustically characterized by high noise levels of relatively short duration. Noise from construction sites, industrial installations, and fixed recreation facilities radiate from a point source and the shape of the vulnerable area is generally a circle.

In big cities, a fundamental source of noise is road traffic. The speed and exhaust system determines the noise released by road traffic. The closeness between tires and the road surface is the superior source of noise at speeds above 60 km/hour for light vehicles. In urban areas, fast acceleration and re-starting the engine in traffic could result in noise up to 15 dB more than the normal levels of noise resulting from smooth driving. Another important source of noise is the public address system used by temples, mosques, etc.

20.4.2 EFFECT OF NOISE POLLUTION

a. **Humans**: Noise pollution affects both behavior and health. Unwanted noise (sound) can damage psychological health. Noise pollution can cause sleep disturbances, hearing loss, hypertension, tinnitus, high-stress levels, and other detrimental effects. Sound becomes unwanted when it either interferes with normal activities such as conversation, sleeping, or disturbs or diminishes one's quality of life.

Prolonged disclosure to noise may cause noise-induced hearing loss. There may be a feeling of fullness in the ear due to this. Older males exposed to significant occupational noise demonstrate more significantly reduced hearing sensitivity than their nonexposed peers; however, differences in hearing sensitivity decrease with time and the two groups are identical by age 79. This can even lead to occupational hearing loss.

Increased sound levels can result in cardiovascular effects, and disclosure to reasonably high levels during a single 8-hour period causes a statistical rise in blood pressure of five to ten points and an increase in stress, and vasoconstriction leading to the increased blood pressure noted above, as well as to the increased occurrence of coronary artery disease. It can also lead to severe depression, loss of memory, and panic attacks.

b. **Wildlife**: Noise can have a harmful effect on wild animals, increasing the risk of death by changing the delicate balance in predator or prey detection and avoidance, and interfering the use of the sounds in communication, especially in relation to reproduction and in navigation. Acoustic overexposure can lead to temporary or permanent loss of hearing.

An impact of noise on wild animal life is the reduction of usable habitat that noisy areas may cause, which in the case of endangered species may be part of the path to extinction. Noise pollution may have caused the death of certain species of whales that beached themselves after being exposed to the loud sound of military sonar.

Noise also makes species communicate more loudly, which is called Lombard vocal response. Scientists and researchers have conducted experiments that show whales' song length is longer when submarine-detectors are on. If creatures do not speak loudly enough, their voice will be masked by anthropogenic sounds. These unheard voices might be warnings, finding of prey, or preparations of net-bubbling. When one species begins speaking more loudly, it will mask another species' voice, causing the whole ecosystem eventually to speak more loudly.

20.4.3 NOISE CONTROL

The main areas of noise removal or reduction are occupational noise control, architectural design, transportation noise control, and urban planning through zoning codes. Several approaches have been developed to find out interior sound levels, many of them are promoted by local building codes; in the best case of project designs, planners are encouraged to work with design engineers to examine trade-offs of roadway design and architectural design. These approaches consist of the design of exterior walls, party walls, and ceiling and floor assemblies; moreover, there are an array of specially designed modes for damping vibrations from special-purpose rooms such as an auditorium, entertainment, and social venues, concert halls, meeting rooms, dining areas, and audio recording rooms.

Many of these techniques trust upon materials science applications of establishing sound baffles or using sound-absorbing liners for interior spaces. Industrial noise control is really a subset of interior architectural control of noise, which is prominent upon certain methods of sound isolation from industrial machinery and to safeguard workers at their workstations.

20.4.3.1 Basic Technologies

1. **Sound insulation**: It prevents the transmission of noise by the introduction of a mass barrier. Common materials such as thick glass, metal, brick, concrete, etc., must have high-density properties to use it as a barrier.
2. **Sound absorption**: A porous material which acts as a noise sponge by converting the sound energy into heat within the material. Common sound absorption materials include fiberglass, open cell foams, and decoupled lead-based tiles.
3. **Vibration damping**: It is suitable for large vibrating surfaces. The damping mechanism works by extracting the vibration energy from the thin sheet and disperses it as heat. A common material is sound-deadened steel.
4. **Vibration isolation**: It restricts the transmission of vibration energy from a source to a receiver by introducing a physical break or a flexible element. Familiar vibration isolators are springs, cork, rubber mounts, etc.
   a. Road ways:

      Source control in roadway noise has provided a little reduction in vehicle noise, except for the development of the hybrid vehicle; nevertheless, hybrid use will need to attain a market share of roughly fifty percent to have a major impact on noise source reduction of city streets. Highway noise is today less affected by motor type since the effects in higher speed are aerodynamic and tire noise related.

      The most fertile areas for roadway noise mitigation are in urban planning decisions, roadway design, noise barrier design, speed control, surface pavement selection, and truck restrictions. Speed control is effective since the lowest sound emissions arise from vehicles moving smoothly at 30–60 km/hour. Above that range, sound emissions double with every 5 miles/ hour of speed. At the lowest speeds, braking and (engine) acceleration noise dominate.

      Selection of road surface pavement can make a difference of a factor of two in sound levels, for the speed regime above 30 km/hour. Quieter pavements are porous with a negative surface texture and use a medium to small aggregates; the loudest pavements have a transversely grooved surface and/or positive surface texture, and use larger aggregates. Surface friction and roadway safety are important considerations as well for pavement decisions.

      Noise barriers can be applicable to existing or planned surface transportation projects. They are probably the single most effective weapon in retrofitting an existing roadway and commonly can reduce adjacent land use sound levels by up to 10 dB.

A computer model is required to design the barrier since contour, micrometeorology, and other local specialized factors make the effort a very complicated operation.
   b. Aircraft:

      As in the case of roadway noise, little change has been made to mitigate aircraft noise at the source, other than the rejection of heavy engine designs. Because of its velocity and volume, jet turbine engine exhaust noise reduction by any simple means.

      The most suitable forms of aircraft noise reduction are residential soundproofing, flight operations restrictions, and through land planning. Flight restrictions can be in the form of time-of-day restrictions, departure flight path and slope, and preferred runway use. The propagation of aircraft noise and its penetration into buildings can be stimulated by using the underlying technology which is a computer model. Variations in aircraft types, local meteorology, and flight patterns can be interpreted along with benefits of alternative building modifying programs such as window glazing improvement, roof upgrading, caulking construction seams, fireplace baffling, and other measures. The computer model allows cost-effectiveness evaluations of an array of alternative approaches.
   c. Architectural solutions:

      Architectural acoustics control practices include room noise transfer mitigation, interior sound reverberation reduction, interior and exterior building skin augmentation.

      With regard to exterior noise in the building, the codes frequently require analysis of the exterior acoustic environment for determining the performance standard required for exterior building skin design. To reach the most cost-effective means of developing a silent interior, the architect has to work with the acoustical scientist. The most important elements of design of the building cover are usually glazing (glass thickness, double pane design, etc.), perforated metal (used internally or externally), exterior door design, chimney baffles, roof material, caulking standards, attic ventilation ports, mail slots, and mounting of through-the-wall air conditioners.

      There are two fundamental types of transmission in concerning with noise produced from inside of the building. The first type is an airborne sound that travels through walls or floor and ceiling assemblies and can arise from either human activity in neighboring living spaces or from mechanical sounds within the building systems. The second type of interior sound is called Impact Insulation Class (IIC) transmission. This effect arises not from airborne transmission but significantly from the transfer of sound through the building itself. The most common concept of IIC noise is from footfall of occupants in living spaces above. Low-frequency noise is transmitted easily through the ground and buildings. This type of noise is more difficult to reduce, but consideration must be given to isolating the floor assembly above or hanging the lower ceiling on resilient channel.

Both the transmission effects noted above may arise either from building occupants or from building mechanical systems such as ventilating and air conditioning units, elevators, plumbing systems, or heating. In some cases, it is barely necessary to point out the best feasible damping technology in selecting such building hardware. In other cases, shock mounting of systems to control vibration may be in order. There are specific protocols in the case of plumbing systems especially for water supply lines, to create isolation clamping of pipes within building walls. It is important to baffle any ducts that could transmit sound between different building areas, in the case of central air systems.

d. Industrial:

This situation is simply a thought to cover primarily manufacturing framework where industrial machinery produces severe sound levels, not unusual in the 75–85 dB range. While this situation is the most dramatic, there are many other office type environments where sound levels may lie in the range of 70–75 dB, entirely composed of office equipment, public address systems, music, and also exterior noise interruption. The other environments can also generate noise health effects provided that exposures are long term.

In the case of industrial machinery, the usual approach for noise protection of workers consists of the provision of ear protection equipment, generation of acrylic glass or other solid barriers, and shock mounting source machinery. In particular cases, the machinery itself can be redesigned to operate in a manner less liable to produce frictional, grinding, grating, or other motions that induce sound propagation. Buying quiet programs and promoting the purchase of quieter tools and equipment can inspire manufacturers to design quieter equipment.

e. Commercial:

Noise control technologies are now in use not only in recording studios and performance facilities but also in noise-sensitive small businesses such as restaurants due to reductions in the cost of technology. Noise absorbing materials such as wood fiber panels, fiberglass duct liner, and recycled denim jeans serve as artwork-bearing canvasses in environments in which aesthetics are important.

Using a sequence of sound absorption materials, set of microphones and speakers, and a digital processor, a restaurant operator can use a tablet computer to selectively control noise levels at different places in the restaurant: the microphone sets pick up sound and transmit it to the digital processor, which controls the speakers to output sound signals on command.

f. Urban planning:

Society may use zoning codes to isolate noisy urban activities from areas that should be protected from such unhealthy exposures and to set up noise standards in areas that may not be conducive to such isolation plans. Because low-income neighborhoods are often at greater risk of noise pollution, the establishment of such zoning codes is often an environmental justice issue. Mixed-use areas present especially difficult conflicts that require special care to the need to protect people from the harmful effects of noise pollution.

20.5 NATURAL RESOURCES AND POPULATION

Human lifestyle illustrates that a strong will to survive, to reproduce, and to secure some level of prosperity and quality of life. However, individuals as well as societies differ in their views of what they feel like a satisfactory lifestyle. In addition, most of these fundamental resources (water, land, energy, and biota) are not unlimited in their supplies and many are finite. As human populations continue to increasing, prosperity and the quality of life can be proposed to decrease because resources must be subdivided among more people. As the dependency of human population on natural resources increases, the quantity as well as quality of these natural resources diminishes.

The quantity and quality of cultivable land, water, energy, and biodiversity balanced against human population numbers determine the current and future status of the environmental resources that hold up human life.

Our environment contributes us with a collection of goods and services necessary for our day-to-day lives. These natural resources include abiotic and biotic parts. The abiotic or nonliving parts of nature consist of air, water, soil, minerals, along with climate and solar energy. The biotic or living parts of nature comprise plants and animals, including microbes. Plants and animals can only survive as a colony of different organisms, all closely linked to each in their own habitat, requiring specific abiotic conditions. Thus, grasslands, forests, deserts, lakes, mountains, rivers, and the marine environment all form habitats for specialized communities of plants and animals to live in. Interactions between the abiotic aspects of nature and specific living organisms together form ecosystems of different types. Many of these living organisms are used as our food resources. Others are linked to our food less directly, such as dispersers and pollinators of plants, soil animals like worms, which recycle nutrients for plant growth, and fungi and termites that break up dead plant material so that micro-organisms can act on the detritus to rectify soil nutrients.

20.5.1　Changes in Land and Resource Use

During the last 100 years, the increase in human population led to great demands on the earth's natural resources. A better healthcare delivery system and an improved nutritional status have led to rapid population growth, chiefly in the developing countries. Large extent of land such as forests, grasslands, and wetlands has been converted into intensive agriculture and has been taken for industry and the urban sectors. These transformations have brought about dramatic changes in land-use patterns and rapid loss of relevant natural ecosystems. The increase in the need for water, food, energy, consumer goods, etc., is not only the result of an increased population but also the result of overusage of resources by people from the wealthier societies, and the society sections of our own.

Industrial development is aimed at meeting growing demands for all consumer items. However, these consumer goods also develop waste in very larger quantities. The growth of industrial complexes relocates people from their traditional, sustainable, rural way of life to urban centers. During the last few decades, several small urban centers have become large cities, and some have even become giant mega cities. This has increased the gap between what the surrounding land can produce and what a large number of increasingly consumer-oriented people in these areas of high population density consume. Urban centers cannot exist without resources such as water from rivers and lakes, food from agricultural areas, domestic animals from pasture lands and timber, fuelwood, construction material, and other resources from forests. Rural agricultural systems are dependent on forests, wetlands, grasslands, rivers, and lakes. The result is a shift of natural resources from the agricultural sector and forest ecosystems to the urban user. The degree of the shift of resources has been increasing in parallel with the growth of industry and urbanization and has changed natural landscapes all over the world. In many cases, this has led to the rapid development of the urban economy, but to a far slower economic development for rural people and serious poverty of the lives of wilderness dwellers. The result is a serious inequality in the distribution of resources among human beings, which is both improper and unsustainable.

20.5.2　Renewable and Nonrenewable Resources

Ecosystems act as resource producers and processors. The main driving force of ecological systems is solar energy which provides energy for the growth of plants in aquatic ecosystems forests, and grasslands. All the aquatic ecosystems are solar energy dependent and have cycles of growth when plant life spreads and aquatic animals breed. A forest recycles

its plant material slowly but continuously returning its dead material, branches, leaves, etc., to the soil. Grasslands recycle material much faster than forests as the grass dries up after the rains are over every year.

20.5.2.1　Renewable Resources

Renewable energy sources are those resources which are constantly replenished. Water and biological living resources are treated as renewable. They are really renewable only within particular limits as they are connected with natural cycles such as the water cycle. The other examples include sunlight and wind.

i. Solar energy:

　Human beings have been using sun as the source of energy for thousands of years. There is more sunlight falling on the earth's surface than it is being utilized. Solar energy can be properly utilized by converting them into direct electricity. This can be done by using materials made up of silicon. Thus it can generate electricity for local homes as well as for commercial projects. One major merit of solar energy power source is that no air pollutants and greenhouse gases are produced during its production.

ii. Wind energy:

　The high speed winds can be utilized to produce electricity by rotating the turbine's blades and can be used as inputs to the generator. One disadvantage of the wind energy source is that the turbines should be very tall and should have large width as well. Another demerit is that it can only be placed in areas having high speed winds such as hilltops or open plains.

iii. Water energy:

　These resources are constantly renewed and replenished by water cycle. So water resources are a type of renewable energy source. But they can also be treated as a nonrenewable resource if its consumption is huge. The use and quality of the freshwater resources should be monitored to ensure future use.

iv. Biomass fuels:

　They are organic matter (wood, plants, animal, residues) which contain the stored solar energy. It is used to supply 15% of the world's energy supply.

v. Geothermal energy:

　The word geo means earth and thermal means heat, so geothermal energy is the heat produced from inner depth of the earth. It is usually fueled by the decay of radioactive elements. The main purpose of geothermal energy is to heat water.

Some of the points to be remembered for renewable energy resources are given below:

a. Freshwater (even after being used) is evaporated by the sun's energy, forms water vapor, and is transformed into clouds and falls to earth as rain.

Nevertheless, if water sources are wasted or over-used to such an extent that they locally run dry and if water sources are heavily polluted by sewage and toxic substances then it becomes impossible to use the water.

b. If overused, the forests are said to act like nonrenewable resources. It is because it will take thousands of years to regrow into fully developed natural ecosystems with their full set of species if destroyed at once.

c. Fish are today being overharvested until the catch has become a fraction of the original resource and the fish are incapable of breeding successfully to restore the population.

d. The output of agricultural land if mishandled drops terribly.

e. If a population of species of plant or animal is reduced by human activities, the species becomes extinct until it cannot reproduce fast enough to maintain a reasonable number.

f. Many species are possibly becoming extinct without us even knowing that other linked species are affected by their loss.

20.5.2.2 Nonrenewable Resources

The nonrenewable resources are materials which will become waste material after one use. After its use, it will remain on earth in a different form and one can recycle it to make it as usable. These are minerals that have been formed in the lithosphere over millions of years and constitute a closed system. Nonrenewable resources comprise fossil fuels such as coal and oil which will be rapidly over if they are extracted at the present rate. The end products of fossil fuels are in the form of chemical compounds, heat, and mechanical energy which cannot be reformed as a resource.

i. Fossil fuels:

The natural resources like petroleum, natural gas, and coal take millions of years for generation. Coal is mainly the remains of wetland plants that have compressed for millions of years. Some types of coal are peat, lignite, bituminous, and anthracite. Petroleum and natural gas are remains of marine organisms. These extracted from the depths of the earth are mainly used for the consumption of energy for humans in automobiles or internal combustion engines, large industries, power plants, etc. Ultimately this leads to its depletion and the shift focuses on the nonrenewable technologies.

ii. Earth minerals and ores:

They is another type of nonrenewable resource. The earth minerals become valuable only if they are concentrated in nature. The concentration of the earth minerals is by natural geological processes which take up tens or thousands of year. Ores are mineral deposits from which metals and nonmetals can be extracted. The metallic ores include gold, silver, platinum, etc. The nonmetallic ores include salt, sand, gravel, clay, etc.

20.5.3 Use and Overexploitation of Resources

a. Forest resources

Scientists evaluate that India should ideally have 33% of its land under forests. Today we have only about 12%. Thus, we need not only to protect existent forests but also to increase our forest cover.

The communities which live in or near forests know the value of forest resources first hand because their lives and occupation depend directly on these resources. However, the rest of us also acquire considerable benefits from the forests which we are hardly aware of. The water we use depends on the existence of forests on the watersheds around river valleys. Our homes, furniture, and paper are made from wood from the forest. We use many medicines that are based on forest produce. And we depend on the oxygen that plants give out and the removal of carbon dioxide we breathe out from the air.

The tribal people hounded animals and gathered plants and lived entirely on forest resources. As agriculture spread, the forests were left in patches which were controlled mostly by these people. In British times, a large amount of timber was cleared for construction of their ships; hence, deforestation became a major problem. This led the British to develop scientific forestry in India. After all they isolate local people by creating protected and reserved forests which curtailed access to the resources. Due to this loss of pole in the conservation of the forests occurs which further led to a progressive degradation and fragmentation of forests across the length and breadth of the country.

Deforestation: Today cutting and mining are crucial causes of loss of forests in our country and all over the world. One of India's major environmental problems is forest degradation due to timber extraction and our dependence on fuelwood. A large number of poor rural people are still highly dependent on wood to cook their meals and heat their homes. We have not been able to plant enough trees to support the need for timber and fuelwood.

Timber cutting, mining, and dams are strong parts of the needs of a developing country. If timber is overcollected the ecological features of the forest are lost. Unfortunately, forests are located in areas where there are rich mineral resources. Forests also cover the steep embankments of river valleys, which are typically suited to develop hydro and irrigation projects. Thus, there is a uniform conflict of interests between the conservation interests of environmental scientists and the Mining and Irrigation Departments. What needs to be realized is that long-term ecological gains cannot be given up for short-term economic gains that sadly led to deforestation. The planning and execution of development projects in forests can cause migration of thousands of tribal people.

b. Water resources

The water cycle, through precipitation and evaporation, maintains hydrological systems which form rivers and lakes and support in a variety of aquatic ecosystems. Wetlands are transitional forms between terrestrial and aquatic ecosystems and contain species of plants and animals that are more moisture-dependent. All aquatic ecosystems are used by a large number of people for their daily needs such as drinking water, cooking, washing, irrigating fields, and watering animals. Water covers 70% of the earth's surface but only 3% is freshwater. Of this, 2% is in polar ice covers and only 1% is usable water in rivers, lakes, and subsoil aquifers. Only a fraction of this can be actually used. At a global level, 70% of water is used for agriculture, about 25% for industry, and only 5% for domestic use. India uses 90% for agriculture, 7% for industry, and 3% for domestic use.

One of the biggest challenges that face the world in this century is the need to re-establish the global management of water resources. Based on the proportion of young people in developing countries, this will keep on increasing significantly during the next few decades. This places excessive demands on the world's limited freshwater supply. World Commission on Dams, 2000, evaluated that the total annual freshwater withdrawals today are at $3800 \, km^3$, which is twice of it at just 50 years ago. Studies represent that a person needs a minimum of 20–40 liters of water per day for drinking and sanitation. More than one billion people worldwide have no access to clean water, and to many more, supplies are not trustable.

1. Overuse and pollution of surface and groundwater: Along the growth of the human population there is an increasing need for larger amounts of water to fulfill a variety of basic needs. Most people use more water than they really need. Most of us waste water during a bath by using a shower or during washing of clothes. Many agriculturists use more water than necessary to grow crops. There are many ways in which farmers can use less water without reducing yields such as the use of drip irrigation systems.

   Agriculture also pollutes surface water and underground water sources by the enormous use of chemical pesticides and fertilizers. Methods such as the use of nontoxic pesticides such as neem products, biomass as fertilizer, and using integrated pest management systems reduce the agricultural pollution of surface and groundwater. As people begin to learn about the serious health hazards caused by pesticides in their food,

public awareness can begin putting pressures on farmers to reduce the use of chemicals that are harmful to health.

Industry turns to maximize short-term economic gains by not considering the release of liquid waste into rivers, streams, and the sea. In the longer term, as people become more aware of using green products made by eco-sensitive industries, the polluter's products may not be used. The polluting industry that does not care for the environment and illegally skips from the cost needed to use effluent treatment plants may eventually be caught, punished, and even closed down. Public awareness may increasingly put pressures on industry to produce only eco-friendly products which are already gaining popularity.

2. Global climate change: Increasing air pollution now started to affect our climate. In some regions, global warming and the El Nino winds have created unprecedented storms. In other areas, they cause long droughts. Everywhere the greenhouse effect due to atmospheric pollution is leading to increasingly abnormal and unpredictable climatic effects. This has negatively affected the regional hydrological conditions.

3. Floods: Floods have been a serious environmental hazard for centuries. However, the calamity produced by rivers overflowing their banks has become gradually more hurtful, as people have deforested catchments and enhance the use of river floodplains that once acted as safety valves. Wetlands in floodplains are nature's flood control systems into which overfilled rivers could spill and act like a temporary sponge holding the water, and restricting fast flowing water from damaging surrounding land.

   As the forests are destroyed, rainwater no longer infiltrates slowly into the subsoil but runs off down the mountainside carrying large amounts of topsoil. This obstructs rivers temporarily but gives way as the pressure rise allowing large quantities of water to wash suddenly down into the plains below. Thereby, rivers swell, burst their banks, and floodwaters spread to consume peoples' homes and farms.

4. Drought: In most desert parts of the world the precipitation rate is very less which leads to periods when there is a serious water scarcity to drink, use in farms, or provide for urban and industrial uses. It alternates in frequency in various parts of our country.

   It is practically impossible to prevent the failure of the monsoon but its ill effects can be reduced by good environmental management. The shortage of water during drought years affects homes, agriculture, and industry. It also leads to food shortages and malnutrition which chiefly affects children. Various control measures can be taken to minimize the heavy impacts of drought. However, this must be done as a preventive measure so that if the monsoon is inadequate its impact on local people's lives can be minimized. In years when the monsoon is adequate, we use up the good supply of water without trying to conserve it and use the water carefully. Thus, during a year when the rains are poor, there is no water even for drinking in the drought area. Deforestation is the major among several factors that increase the effect of drought. Forest cover allows water to be retained in the area allowing it to seep into the ground. The loss forest cover on the hill slope leads to rushing of rainwater into the rivers. This affects the underground stores of water in natural aquifers which can be used in drought years if the stores have been filled during a good monsoon. If water from the underground stores is overused, the water table drops and vegetation suffers. This soil and water management and afforestation are long-term measures that reduce the impact of droughts.

20.5.3.1 Sustainable Water Management

Save water campaigns are necessary to make people everywhere aware of the dangers of water scarcity. A number of measures need to be taken for the better management of the world's water resources. These include measures such as:

a. Constructing several small reservoirs instead of a few mega projects.
b. Builds up small catchment dams to protect wetlands.
c. Afforestation, soil management, and micro-catchment development allow recharging of underground aquifers thus minimizing the need for large dams.
d. Treating and recycling municipal wastewater for agricultural use.
e. Preventing a loss in municipal pipes.
f. Preventing leakages from dams and canals.
g. Effective rainwater harvesting in urban environments.
h. Water management in agriculture such as using drip irrigation.
i. Pricing water at its real value makes people use it more responsibly and efficiently and reduces water wasting.

In deforested areas where land has been degraded, soil management by bunding along the hill slopes and making nala plugs can help hold moisture and make it feasible to re-vegetate degraded areas. Leaving the course of the river as undisturbed as possible is the best river management technique. Canals and dams cause major floods in the monsoon and the drainage of wetlands seriously affects areas that get flooded when there is high rainfall.

20.5.4 Natural Resources and Associated Problems

1. **The uneven consumption of natural resources**: The consumption of resources per capita (per individual) of the developed countries is up to 50 times greater than in most developing countries. Advanced countries produce over 75% of global industrial waste and greenhouse gases.

   Energy from fossil fuels is consumed in relatively much greater quantities in developed countries. Their per capita consumption of food too is much greater as well as their waste of enormous quantities of food and other products, such as packaging material, used in the food industry. Producing animal food for human consumption requires more land than growing crops. Thus, countries that are highly dependent on nonvegetarian diets need much larger areas for pastureland than those where the people are mainly vegetarian.

2. **Planning land use**: Land itself is a dominant resource, needed for food production, animal husbandry, industry, and for our growing human settlements. The land is now under serious pressure due to an increasing land hunger—to produce sufficient quantities of food for an exploding human population. It is also affected by degradation due to misuse. Land and water resources are polluted by industrial waste and rural and urban sewage. They are increasingly being diverted for short-term economic gains to agriculture and industry. Natural wetlands of good value are being drained for agriculture and other purposes. Semi-arid land is being irrigated and overused.

3. **The need for sustainable lifestyles**: The indicators of the sustainable use of resources are the quality of human life and the quality of ecosystems on earth. There are clear indicators of sustainable lifestyles in human life.
   • Increased longevity
   • An increase in knowledge
   • Enhancement of income.

These three together are known as the Human development index. The quality of the ecosystems has indicators that are tougher to assess.

   • A stabilized population.
   • The long-term conservation of biodiversity.
   • The careful long-term use of natural resources.
   • The prevention of degradation and pollution of the environment.

20.6 MISCELLANEOUS TOPICS OF ENVIRONMENTAL ENGINEERING

20.6.1 Pipelines

The cleaning and disinfecting of new or repaired pipelines can be accelerated and greatly simplified if special care is applied in the handling and laying of the pipe during installation. Trenches should be kept dry and a tight-fitting plug provided at the end of the line to keep out foreign matter. Lengths of pipe that have soiled interiors should be cleaned and disinfected before being connected. Each continuous length of main should be disinfected separately with a heavy chlorine dose or other effective disinfecting agents. This can be done by using a portable hypo chlorinator, a hand-operated pump, or an inexpensive mechanical electric- or gasoline-driven pump throttled down to inject the chlorine solution at the beginning of the section to be disinfected through a hydrant, corporation cock, or other temporary valve connection. Hypochlorite tablets can also be used to disinfect small systems, but water must be introduced very slowly to prevent the tablets being carried to the end of the line.

The first step in disinfecting a main is to close all service connections, then flush out the line thoroughly by opening a hydrant or drain valve below the section to be treated until the water runs clear. A velocity of at least 3 fps should be obtained. After the flushing is completed, the valve is partly closed so as to waste water at some known rate. The rate of flow can be evaluated with a flow gauge or by running the water into a barrel, can, or other containers of known capacity and measuring the time to fill it. It is a simple matter to approximate the time, in minutes, it would take for the chlorine to reach the open hydrant or valve at the end of the line being treated by dividing the capacity of the main in gallons by the rate of flow in gallons per minute. In any case, injection of the strong chlorine solution should be continued at the rate indicated until samples of the water at the end of the main show at least 50 mg/l residual chlorine. The hydrant should then be closed, chlorination treatment stopped, and the water system let stand at least 24 hours. At the end of this time, the treated water should show the presence of 25 mg/l residual chlorine. If no residual chlorine is found, the operation should be repeated. Following disinfection, the water main should be thoroughly flushed out, to where it will do no harm, with the water to be used and samples collected for bacterial examination for a period of several days. The disinfection should be repeated if the laboratory reports the presence of Coliform bacteria until two successive acceptable results are received. The water should not be used until all evidence of contamination has been removed as evaluated by the test for Coliform bacteria.

If the pipeline being disinfected is known to have been used to carry polluted water, flush the line thoroughly and double the strength of the chlorine solution injected into the mains. Let the thickly chlorinated water stand in the mains at least 48 hours before flushing it out to waste and proceed as explained in the preceding paragraph. Cleansing of heavily contaminated pipe by the use of a biodegradable, nontoxic, nonfoaming detergent and a pig, followed by flushing and then disinfection, may verify to be the quickest method. Tubercles found in a cast-iron pipe in water distribution systems protect microorganisms against the action of residual chlorine.

Where pipe breaks are repaired, flush out the isolated section of pipe completely and dose the section with 200 mg/l chlorine solution and try to keep the line out of service at least 2–4 hours before flushing out the section and returning it to service.

Potassium permanganate can also be used as a main disinfectant. The presence and then the absence of the purple color can determine when the disinfectant is applied and then when it has been flushed out.

20.6.2 Septic Tank

A septic tank is a waterproof tank planned to slow down the movement of raw sewage passing through it so that solids can settle out and be broken down by liquefaction and anaerobic bacterial action. Septic tanks do not purify the sewage, eliminate odors, or destroy all solid matter, but rather, simply condition sewage so that it can be disposed of using a subsurface absorption system. Suspended solids removal in such tanks is 50%–70%; 5-day BOD removal is about 60%.

The septic tank should have a liquid volume of not less than 750 gallons. A plastic sludge and gas deflector on the outlet is highly recommended. The detention time for large septic tanks should not be less than 24–72 hours. Schools, theatres, camps, factories, and parks are examples of places where the total or a very large proportion of the daily flow takes place within a few hours. Septic tanks should be constructed of good-quality reinforced concrete. Precast-reinforced concrete and commercial tanks of metal, polyethylene, fiberglass, and other composition materials are also available. Because some metal tanks have a limited life, it is recommended that their purchase is predicated on their meeting certain minimum specifications (e.g., 12- or 14-gauge metal thickness, guaranteed 20-year minimum life expectancy, 12- or 14-gauge metal thickness, and acid-resistant coating). Figure 20.9 shows the cross-section of septic tank disposal system.

If the septic tank is to afford ground garbage, its capacity should be raised by at least 50%. Some expert advice a 30% increase. Others stand against garbage disposal to a septic tank.

Open-tee inlets and outlets are generally used in small tanks, and high quality reinforced concrete baffled inlets and outlets are suggested for the larger tanks. Precast open concrete baffles or tees have, in some instances, disintegrated or fallen off; cast-iron, vitrified clay, polyvinyl chloride (PVC), acrylonitrile butadiene styrene (ABS), or polyethylene (PE) tees should be used. Cement mortar joints are inadequate. Compartmented tanks are somewhat more efficient. The first compartment should have 60%–75% of the total volume. A better distribution of flow and detention is obtained in the

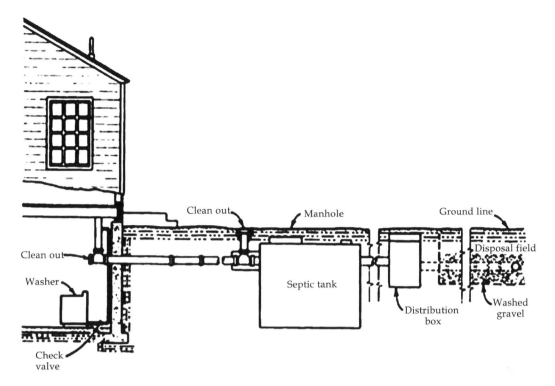

FIGURE 20.9 Cross-section of septic tank disposal system.

larger tank with a baffle arrangement of preferably rigid acid-resistant plastic. The preferred size of manhole over the inlet of a small tank is a minimum of 16-in. and that over both the inlet and outlet of a larger tank is of 20- to 24-in.

An effective septic tank design should provide for a detention period longer than 24 hours; an outlet configuration with a gas baffle to minimize suspended solids carryover; maximized surface area to depth ratio for all chambers (ratio more than 2); and a multichamber tank with interconnections similar to the outlet design (open-tee inlet and outlet).

20.6.2.1 Construction of Small Septic Tanks (Figure 20.10)

1. **Top**: Reinforced concrete poured 3–4 in. thickness with two 3/8 in. steel rods/ft, or equivalent, and a 20 in.×20 in. manhole over inlet, or precast reinforced concrete 1 ft slabs with sealed joints.
2. **Bottom**: Reinforced concrete 4 in. thick with reinforcing as in top or plain poured concrete 6 in. thick.
3. **Walls**: Reinforced concrete poured 4 in. thickness with 3/8 inc. steel rods on 6 in. centers both ways, or equivalent; plain poured concrete 6 in. thick; 8 in. brick masonry with 1 in. cement plaster inside finish and block cells filled with mortar.
4. **Concrete mix**: One bag of cement to 2.25 in. of sand to 3 ft of gravel with 5 gal of water for moist sand. Use 1:3 cement mortar for masonry and 1:2 mortar for plaster finish. Gravel or crushed stone and sand shall be clean, hard material. Gravel shall be 0.25–0.5 in. in size; sand from fine to 0.25 in.
5. **Bedding**: At least 3 in. of sand or pea gravel, levelled.

20.6.2.2 Care of Septic Tank and Subsurface Absorption Systems

A septic tank for a private home will generally require cleaning every 3–5 years, depending on occupancy, but in any case, it should be inspected once a year. If a garbage disposal unit is used, more frequent cleaning is needed. Septic tanks serving commercial operations should be inspected at least every 6 months. When the depth of settled sludge or floating scum approaches the depth given in the table below, the tank needs cleaning. Sludge accumulation in a normal home septic tank has been approximated at 69–80 liters (18–21 gallons) per person per year. The appearance of scum or particles in the effluent from a septic tank going through a distribution box is also an indication of the need for cleaning. Regular inspection and cleaning will prevent solids from being carried over and clogging leaching systems.

Septic tanks are generally cleaned by septic tank–cleaning firms. Sludge sticking to the inside of a tank that has just been cleaned would have a seeding effect and assist in renewing the bacterial activity in the septic tank. The use of septic tank cleaning solvents or additives containing halogenated HC, aromatic HC, or hazardous chemicals can cause carryover of solids, and clogging of absorption field, as well as contamination of groundwater can be prevented. An individual should never enter a septic tank that has been cleared, nevertheless, whether it is open or covered. Cases of asphyxiation and death have been reported due to the lack of adequate oxygen or the presence of toxic gases in the empty tank. If it should become necessary to inspect or make repairs, the tank should first be checked with a gas detector for oxygen and toxic gases and thoroughly ventilated using a blower, which is kept operating.

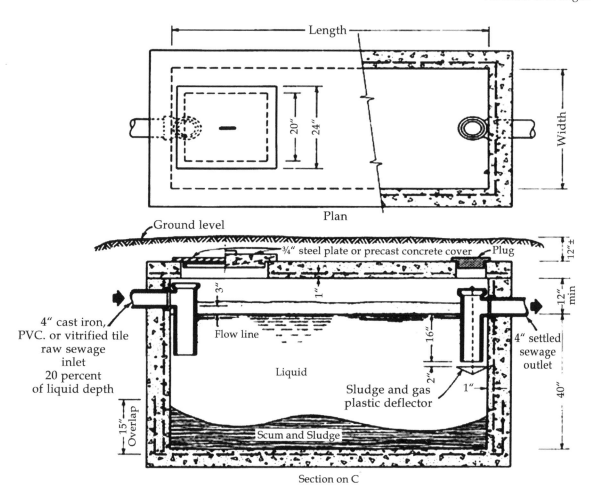

FIGURE 20.10   Construction of septic tank.

Although soap, disinfectants, drain solvents, and related materials used individually for household purposes are not harmful to septic tank operation unless used in large quantities, organic solvents and cleaners, pesticides, and compounds containing heavy metals could contaminate the groundwater and well-water supplies and should not be a dump in a septic tank system. Also, sanitary napkins, absorbent pads, and tampons should not be disposed of in septic systems.

High grasses, brush, shrubbery, and trees should not be allowed to grow over an absorption system or sand filter system. It is better to cover the bed and seed the area with grass. If trees are near the sewage disposal system, difficulty with roots entering poorly joined sewer lines can be foreseen. About 2–3 pounds of copper sulfate crystals flushed down the toilet bowl once a year will destroy roots that the solution comes into contact with but will not prevent new roots from entering. The application of the chemical should be done at a time, such as late in the evening when the maximum contact time can be obtained before dilution. Copper sulfate will corrode chrome, iron, and brass; hence it should not be allowed to come into contact with these metals. Copper sulfate in the selected dosage will not interfere with the operation of the septic tank.

Typical reasons for the failure of septic tank system are seasonal high groundwater; lack of routine cleaning of the septic tank, or outlet baffle disintegration or loss; excessive water use or hydraulic overloading; carryover of solids into the absorption field due to use of septic tank cleaning compounds; settlement of the septic tank, distribution box, or connecting pipe; and improper design and construction of the absorption system, including compaction and smearing of absorption trench bottom and sidewalls.

Remedial measures, once the cause is identified, include water conservation measures such as reduced water usage, reduced water pressure, and low-flow toilets and showerheads. Other measures to examine are cleaning of septic tank and flushing out distribution lines, and installation of additional leaching lines; lowering the water table with curtain drains; installation of a separate absorption system and division box or gate for alternate use with the annual resting of existing system; discontinuation of use of septic tank cleaning compounds; replacement of corroded or disintegrated baffles; replacement or releveling of distribution box; cleaning of septic tank at least every 3 years; and disconnection of roof, footing, and area drains.

21 Quantity Surveying and Valuation

21.1 INTRODUCTION

One of the most important duties of an engineer or an architect is to prepare the probable cost of the proposed building. This is also called as an estimate. An estimate is the expected cost of a work and is generally prepared prior to construction. The estimate is prepared by calculating the required quantities and calculating the costs at the suitable rates, in order to obtain the probable expenditure incurred during execution of the work or structure.

The actual costs are only known after the work is completed from the completed work account. If the estimate is properly and thoroughly prepared, there will be little difference between the estimated cost and the actual cost. The estimate can be prepared by various methods to know the approximate or rough cost.

21.2 TYPE OF ESTIMATES

The various methods of estimates are as follows:

1. Preliminary estimate or approximate or abstract or rough cost estimate
2. Plinth area estimate
3. Cube rate estimate or cubical content estimate
4. Approximate quantity method estimate
5. Detailed estimate or item rate estimate
6. Revised estimate
7. Supplementary Estimate
8. Supplementary and revised estimate
9. Annual repair or maintenance estimate (AR or AM estimate)

21.2.1 PRELIMINARY ESTIMATE OR APPROXIMATE OR ABSTRACT OR ROUGH COST ESTIMATE

For preliminary studies of various aspects of a work or project, a preliminary estimate is required to determine the financial position and the administrative sanction policy of the competent administrative authority. However, in the case of large-scale projects such as irrigation, residential construction projects, and similar projects earning revenue, the probable revenue can be calculated and the approximate cost can be determined from the preliminary estimate and it can be seen whether the funding on the project is reasonable or not. For noncommercial projects or projects that do not return directly, their necessity, utility, availability of money, etc., may be considered before final decision is taken.

The preliminary estimate for various structures and works can be prepared in various ways.

a. **Buildings**:
   i. Per unit basis: Per school and hostel student, per school classroom, per hospital bed, per cinema and theater seat, per factory bay, per apartment for residential buildings, and so on.
   ii. Plinth area basis
   iii. Cubic content basis
   iv. Approximate quantity method
b. **Roads and Highways**: Depending on the nature of the road, the width and thickness of the metalwork, per kilometer (per mile).
c. **Irrigation channels**:
   i. Based per kilometer
   ii. Commanded land area. i.e., per hectare basis
d. **Bridges and culverts**: It is based on per running meter of span depending on the roadway, nature and depth of foundation, type of structure, etc., approximate cost may also be for small culverts of different widths per number of culverts.
e. **Sewerage and water supply**:
   i. Based on the population served per head
   ii. Based on the area covered, i.e., per hectare basis
f. **Overhead water tank**: Actually based on the capacity of the tank per liter depending on the type of tank, height of tank, material with which the tank is constructed, etc.

21.2.2 PLINTH AREA ESTIMATE

This is done on the basis of the plinth area of the building. The rate is deduced from the cost of similar nearby buildings with similar specifications, height, and construction. The estimation using plinth area is done by finding the plinth area of the building and multiplying it by the plinth area rate. Courtyard and other open spaces are usually eliminated. For the covered area, the plinth area should be calculated by the external dimension of the building at the floor level.

21.2.3 CUBE RATE ESTIMATE

Cube rate estimate is a preliminary estimate or an approximate estimate. It is prepared on the basis of the cubic content of the construction. The cube rate is obtained from the cost of a similar nearby building with similar specifications and construction. The estimate is calculated by finding the cubic content (length × width × height) of the building and multiplying it by the cubic rate. The length and width should be taken as the external dimensions of the buildings on the floor level. The height has to be measured from the floor level to the top level of the roof (or half the sloping roof).

21.2.4 Approximate Quantity Method Estimate

Approximate total wall length is found in this method in the running meter. This total length is then multiplied by the rate per running meter of the wall to get a fairly accurate cost. For this method, the structure can be divided into two parts: (1) the base with plinth and (2) the superstructure. The plan or line plan of the structure should be available for using this method. The running meter cost for the foundation and superstructure should first be calculated. The total length of the walls should be multiplied by this running meter rate.

21.2.5 Detailed Estimate or Item Rate Estimate

The detailed estimate is an accurate estimate and consists of assessing the volume and cost of each work item. From the drawing, the dimensions—length, breadth and height—of each item are correctly taken. The quantity of each item is then calculated after which the abstracting and the billing steps are carried out.

In two steps, the detailed estimate is prepared:

i. Details of measurement and quantity calculation
ii. Estimated cost abstract

21.2.6 Revised Estimate

Revised estimate is a detailed estimate and should be carried out on any one of the following circumstances:

i. If the original approved estimate is exceeded or is expected to exceed 5%.
ii. If the expenditure on a work exceeds or is likely to exceed the administrative sanction by more than 10%.
iii. If there are material deviations from the original proposal, even though the cost may be met from the sanctioned amount.

21.2.7 Supplementary Estimate

The supplementary estimate is a detailed estimate. It is prepared when additional work is required to complete the original works or when further development is necessary during the work. In addition to the original estimate, this is a fresh detailed estimate of additional works. The abstract should show the amount of the original estimate and the total amount, including the additional amount required for the sanction.

21.2.8 Supplementary and Revised Estimate

If a work is partially abandoned and the estimated cost of the remaining work is less than 95% of the original work, or if there are material deviations and design changes that can lead to substantial savings in the estimate, the amount of the original estimate can be revised by the competent authority. A supplementary and revised estimate is then prepared and the competent authority receives fresh technical consent.

21.2.9 Annual Repair or Maintenance Estimate (AR or AM Estimate)

The AR or AM estimate is a detailed estimate that is prepared to maintain an orderly and safe structure or work. White washing, color washing, painting, minor repairs, etc., are part of the building. The AR estimate provides for patch repair, renewal, bridge, and culvert repairs, etc., for road works.

21.3 ESTIMATING PROCESS

Estimation: a four-stage process.

Initiation: Analogous estimates are used at this phase. They are big picture estimates based on similar projects that have been documented in the corporation's project archives. These estimates are stated as order of magnitude estimates.

Early planning: Project-level and major deliverable-level estimates are often analogous or 3-point estimates. During this phase, you may also use parametric estimating techniques.

Final project plan: You use information from the team members and include them in bottom-up estimating of their deliverables.

Weekly status: You use rolling estimates every week until the project is complete.

21.4 TAKING OUT QUANTITIES

The procedure by which the quantities of the different items in a particular structure are determined is called the quantities taken out. The quantities are obtained by studying the structure drawings in detail.

21.4.1 Methods of Taking Out Quantities

Every country has developed its own method for taking out quantities by estimating suitable conventions. We use our own method, known as the Method of Department of Public Works, to take quantities. Work items such as earthwork in foundation excavation, concrete foundation stone masonry in foundation and basement, stone or brick masonry in super structure can be estimated using either of the following methods:

1. Long-wall and short-wall method (or) General method
2. Center line method
   a. **Long-wall and short-wall method:**
   In this method, the wall is considered to be a long wall along the length of the room, while the wall perpendicular to the long wall is said to be a short wall. To calculate the length of the long wall or short wall, first calculate the length of the center line of the individual walls. The length of the long wall (out to out) can then be calculated by adding half a breadth to its center line length at each end. The length of the short wall is measured in and can be found by deducting half a width from the length of the center line at each end. The length of the long wall usually decreases in

super-structure from earthwork to brickwork while the short wall increases. These lengths are increased in width and depth to obtain quantities.

Long out-to-out wall length = centre to centre length

+ half breadth on one side + half breadth on the other side

= centre to centre length + one breadth

$$(21.1)$$

Short wall in-to-in length

= Centre to centre length − one breadth.   (21.2)

b. **Center line method**

This method is suitable for cross-sectional walls. In this case, the total length of the center line is multiplied by the width and depth of the respective item to obtain the total amount at once. When cross walls or partitions or veranda walls are mainly connected, the length of the center line is reduced by half the width for each junction. Such joints are carefully studied in the calculation of the total center line length. The estimates prepared are most accurate and fast.

c. **Partly center line and partly cross-wall method**

This method is used when the external wall (i.e., around the building) is of one thickness and the internal walls of various thicknesses. In such cases, the center line method is used on external walls and long wall–short walls are used on internal walls. This method is suitable for different thicknesses and foundations. For this reason, all Departments of Engineering practice this method.

21.5 MODES AND UNIT OF MEASUREMENT

In any civil engineering work, the measurement of the item plays an important role and slight ambiguity may lead to serious financial implications absolutely necessary to indicate the mode of measurement in the specification of the item. The calibration modes and unit for different types of trades are mentioned below, and they are used to prepare the rate analysis of various items and for working out the detailed estimates of the typical civil engineering works:

1. **Preliminaries**: These are the items such as removing trees, demolishing existing structure, clearing the site, etc., normally such items are paid on a lump sum basis. But if the demolishing work consists of reinforced cement concrete (RCC) or brickwork, it is possible to work out the cubical contents, and in such cases, the actual measurements of the work to be demolished are taken. The unit of measurement is cubic meter.

2. **Excavation**: The item of excavation is paid per cubic meter, and hence, the length and depth are measured. The length and the width are measured as per exact length and width according to the drawings or as instructed by an engineer.

3. **Concrete**: The foundation concrete is measured in cubic meters, the length and width being the same as in excavation. The depth is measured as per actual concrete laid.
   - The plain cement concrete work is paid in cubic meters.
   - RCC slab up to 15 cm depth, RCC parade, RCC partition wall, and RCC box are measured in square meters.
   - RCC lintels, beams, and columns are measured in cubic meters.
   - In case of RCC beams, clear rib section projecting below or above the roof slab is taken for beam measurements.
   - RCC stair is measured in terms of number of steps. The half-landing and quarter-landing are considered as equivalent to four steps and two steps, respectively.
   - RCC weather shed is measured in square meters and clear projection is taken for the measurement.
   - The damp-proof course is measured in square meters and the depth of the course is specified.
   - The precast cement concrete jali work or louvers are measured in square meters and the thickness is specified. The precast concrete, posts, struts, etc., are measured in cubic meters.

4. **Brickwork**: The brickwork is measured in cubic meters for one brick wall to three brick walls. For modular bricks of size 190 mm × 90 mm × 90 mm, the thickness of walls for one to three brick walls shall be, respectively, 200, 300, 400 mm, 500, and 600 mm.
   - If the thickness of the wall is more than three bricks, the actual thickness is measured to the nearest 1 cm.
   - The half-brick partition wall and the walls having thickness less than half brick are measured in square meters.
   - The brickwork for arch is measured separately in cubic meters.

5. **Stonework**: The stonework is measured in cubic meters. The full specification gives description for the cut-stonework in steps, cornices, etc. The wall thickness shall be measured as close as 1 cm.
   - The stonework for the sills and copings of parapets is measured in running meters.
   - The stonework for shelves, weather sheds, and slabs is measured in square meters.
   - The deductions in stonework are made as per item of the brickwork.

6. **Woodwork**: In case of wooden work, the type of the material used and the quality of finish required should be clearly mentioned in the specifications. The rate for any type of woodwork includes fabrication, fixation, fastenings, fixtures, and three oil paint coats. The measurements are taken from network carried out and no consideration is given to the material wastage.

- The rectangular wooden beams and vertical wooden posts are measured in cubic meters.
- The door, window, fan, and cupboard frames are measured in cubic meters.
- The wooden trusses are measured in cubic meters.
- The wooden door, window, ventilator, and cupboard shutters are measured in square meters.
- The wooden trellis work is measured in square meters.

7. The wooden stair is measured by the number of steps and the stair type and width are defined by the specification.

8. **Plastering**: The thickness of the plastering and the type of mortar are mentioned in the specification. The measurement of the plastering is recorded in square meters.

9. **Pointing**: The type of pointing and quality of the mortar are mentioned in the specifications. The measurements are taken of the whole flat surface in square meters. The deductions are made as per item of plastering.

10. **Steel work**: The grillwork in windows is measured in square meters. The rolling steel shutters are paid per square meter of the opening covered or actual area of the shutters. Sometimes, they are paid per number and the size is mentioned.

   In square meters, the steel doors, windows, and ventilators are measured and the measurements are taken overall. The plain rolled sections, built-up work, steel bars, etc., are paid by weight in quintals.

11. **Roof coverings**: In general, the roof coverings are measured flat without any laps and are paid per square meter and no extra amount is paid for any wastage, cutting, etc. The ridges, hips, and valleys are measured in running meters.

12. **Floor finishes**: The measurements of floor finishes are taken in square meters for the net area covered. No extra amount is paid for wastage of the materials.

13. **White-washing, color-washing, and distempering**: The items of whitewashing, color-washing, and distempering are measured in square meters. The corrugated surfaces are measured flat, and then, a suitable coefficient is applied to compensate for the corrugations as follows:

Corrugated iron sheets   1.14
Corrugated asbestos cement sheets with large corrugations   1.20
Semi-corrugated asbestos cement sheets   1.10

14. **Painting**: The item of painting is paid in square meters. The measurements are taken flat and then multiplied by a suitable coefficient to compensate for mouldings, rebates, corrugations, etc.

15. **Water supply connections**: The item of galvanized iron or cast-iron pipeline is measured in running meters for the length laid and no extra amount is paid for the specials such as bends, tees, etc. The measurement is taken along the center line of the pipeline.

16. **Drainage connections**: The item of pipeline is measured in running meters for the actual length laid and no extra amount is paid for the specials such as bends, tees, etc. The measurement is taken along the center line of the pipeline.

17. **Road-work**: Unless otherwise mentioned, the road work is measured in square meters and the specification should state the quality of the materials to be used and building method. The expansion and construction joints are measured in running meters and the specification should state the thickness and the depth of the joint and the quality of filler. The side berms are measured in kilometers and the average depth of filling and width are mentioned in the specification.

21.6 SPECIFICATIONS

The specifications deal with the methods of execution, qualities of material, nature of labor to be employed, proportions in which materials are to be mixed, measurement of work, etc.

The structure drawing shows the proportions and relative positions of the different parts. Data on the quality of materials and workmanship on drawings due to lack of space cannot be provided. Therefore, this information on material quality and workmanship is transmitted in a separate contract document, known as the specifications for the work. The term is derived from old French word *specifier* or Latin *specificare* meaning to *describe*. Thus, it is intended that the combination of the drawings and specifications will completely define the structure, physically as well as technically.

The main purposes of giving specification are as follows:

- The cost of a unit work quantity is subject to its specifications.
- Work specification is required to describe the quality and quantity of the various materials required for a construction work and is one of the most important contract documents.
- This also specifies the workmanship and work method. The specification of a work serves as a guide for a contractor's supervisory staff and the owner to carry out the work to their satisfaction.
- Work shall be done in accordance with its specifications and the contractor shall be paid for it. Any modification in the specification changes the offered rate.
- Because the work rate is based on the specification, the contractor can calculate the rates of various items of works in the tender with its material and labor procurement rates. The tender rate is therefore baseless, incomplete, and invalid without the specification of the works.

- Specification is required to specify the equipment tools and plants to be used in the work and thus enable them to be obtained in advance.
- The need for specifications is to verify the strength of the materials for a project work.

21.6.1 TYPES OF SPECIFICATIONS

The specifications are broadly divided into the following two categories:

1. Brief specifications
2. Detailed specifications

21.6.1.1 Brief Specification

The nature and class of the works and the names of the materials to be used are described in short specifications. Only a brief description of each item is provided. It is useful for the project estimation. The general requirements are not a part of the contract document.

21.6.1.2 Detailed Specifications

The detailed description is included in the contract document and is divided into three groups.

i. General provisions
ii. Technical provisions
iii. Standard specifications

i. **General provisions**: These are also known as contractual terms and conditions and apply to work as a whole. The conditions governing the contract are written in this document.
ii. **Technical provisions**: These specifications describe the technical necessities of each construction type and contain elaborated directions concerning the specified quality of the ultimate product. There are three types of technical provisions.
   a. Specifications for materials and workmanship:
      The subsequent properties should be enclosed with in the specifications.
      – Physical properties like size, shape, grade, strength, hardness, etc.
      – Material chemical composition
      – Electrical, thermal, and acoustic properties
      – Material appearance
      – Clear statement of inspection and test procedure
      The following important features should be included in the specifications for workmanship.
      – The desired result
      – Tools and plants to be used
      – Detailed description of the construction method for each item
      – Protection instructions for the finished work as well as the adjacent property.

   b. Specifications for performance:
      These specifications are written for the overall performance of the finished product and are therefore written for the supply of equipment and machinery such as pumps, engines, etc. The general description, design, installation, and guarantee of the equipment, etc., are specified in these specifications.
   c. Specifications for proprietary commodities:
      Commercial products standardized or patented are referred to as proprietary goods. The specifications for these materials should contain the name of a specific brand or company (e.g., Brand Sun, Brand Everest, etc.). It is not desirable, however, that certain trade names or brands should be specified for public works. In order to avoid monopoly and favoritism, it is a common practice to specify the brand chosen and then follow the phrase "or equal."
iii. **Standard Specifications**: Specifications are rarely written entirely for all work items. It is possible to standardize specifications for most of the items occurring in works of similar nature. Hence every engineering department prepares the detailed specifications of the various items of works and gets them printed in book form under the name "Detailed Standard Specifications." When the work or structure or project is taken up, the printed standard specifications are referred to in the contract instead of writing detailed specifications each time.

21.6.2 POINTS TO BE INCLUDED IN THE SPECIFICATIONS

The specifications contain the following points for civil engineering works:

a. Quality of materials to be used in accordance with strength/size requirements.
b. Quantity of materials to be used and the measurement methods to be used.
c. Method of mixing various materials.
d. Methods of construction to be followed, including the equipment and machinery to use.
e. Works dimensions such as width, thickness, etc.
f. Measurement methods of works for payments.

The points to be included in the specifications of some of the civil engineering construction work are given below.

1. **Lime mortar concrete, surkhi mortar concrete, and cement mortar concrete specification**
   Specifications for lime, cement, fine aggregate, and coarse aggregate—water quality and quantity—lime mortar and surkhi mortar specification—batching volume or weight batching and proportion of ingredients—mixing method—hand mixing ding or machine mixing—mixing platform—transportation of

concrete—placement—ramming and compaction—curing—measurement and payment method.

2. **Cement Concrete specification for RCC Work**

   Specifications for cement, fine aggregate, and coarse aggregate—proportion of ingredients—quality of water—consistency—volume batching or weigh batching—mixing—hand mixing or machine mixing—mixing time –transporting and placing of concrete—thickness of each layer—compaction—use of vibrator—construction joints—specification of reinforcement—fabrication of reinforcement—centering and forming—curing the removal of forming—finishing—measurement and payment method.

3. **Specification for Brick masonry in lime mortar, surkhi mortar or cement mortar**

   Specification for brick masonry in lime mortar, surkhi mortar, or cement mortar. Quality and size of bricks—proportion and size of bricks—soaking of bricks in water—setting of bricks in mortar—thickness of joints—bond—raking joints for plaster—uniform elevation—maximum height for a day's work—scaffolding—corbelling and cornices—rounding off corners—plinth offsets—brick on edge coping—curing unit of measurement and payment.

4. **Specification for Stone masonry in lime mortar, surkhi mortar or cement mortar**

   Building stone requirements—size and dressing of stones—hammer dressing of chisel dressing—bond stones—methods of laying stones—wetting of stones before placing—mortar specification—thickness of mortar bed and thickness of joints filling the voids—uniform raising—scaffolding—curing—measuring and payment methods.

5. **Specification for Plastering stone masonry or brick masonry with lime mortar, surkhi mortar or cement mortar**

   Surface preparation—surface cleaning—surface wetting and washing—mortar specification—mix ratio—plaster thickness—number of coats—surface application of mortar—finishing—curing measurement and payment.

6. **Specification for pointing stone masonry, brick masonry with lime mortar, surkhi mortar, cement mortar**

   Raking out of joint—brushing and cleaning—washing with water –mortar specification—application of mortar in the joints—finishing thickness of joints—curing measurements and payment.

21.6.3   Typical Specifications

21.6.3.1   Examples of General Specifications

1. **General specifications of some works concerned within the construction of a residential building**

   i. Concrete foundation: cement concrete 1:4:8 with a size of 40 mm of broken stone granite.

   ii. Foundation and basement: cement mortar brickwork 1:5 with bricks of 7.5 grades.

   iii. Super structure: cement mortar brickwork 1:6 using 7.5 grade bricks.

   iv. Flooring: mosaic flooring over 100 mm thick concrete base 1:5:10 with 40 mm bat size.

   v. Roofing: RCC thick at 120 mm 1:2:4 concrete roof with 20 mm blur granite broken stone.

   vi. Finishing: plastering 1:3, 12 mm thick cement mortar walls and ceilings, and finishing the same with three coats of white washing.

   vii. Doors and windows: doors and windows of country wood painted two coats of ready-mixed paint over the first coat.

2. **General details of some works involved in the construction of a village road**

   i. Subgrade: Leveling and compacting the surface with a camber of 1:48 for 8 m width, uniform on the total length, with watering.

   ii. Soling: 150 mm granite boulders, fully gravel packed and compacted with hand rollers, dry and wet rollers.

   iii. Spreading gravel: Red gravel 20 mm thick, watered, and rolled over the base.

   iv. Finish: lined with a thin layer of sand.

21.6.3.2   Examples of Detailed Material Specifications

1. **Detailed Mortar Sand Specification**

   The sand used in the mortar is clean, sharp, heavy, and grimy. Free of clay, salt, mica, and organic impurities. It shall in no way contain harmful chemicals. In mortars, medium and fine sand is used. Coarse sand is sieved through 600 μm and used in plastering works in mortars.

2. **Detailed first-class brick specification**

   The earth used to mold bricks is free of salts and chemicals from organic matter. The size, weight, and color of the bricks burned should be consistent. The adjoining faces of the bricks must be at right angles. The bricks are free from cracks, imperfections, and lumps. They should not break down from 1 m in height on the ground. They should not absorb more than 15% of their weight if they are immersed in water for an hour. Not less than 7.5 N/mm^2 shall be the average compressive strength of the bricks. The dry weight of a single brick does not exceed 3 kg.

3. **Detailed specification for cement**

   Ordinary Portland cement or rapid hardening Portland cement confirming to IS: 269-1989 and IS: 8041-1990 shall be used. The cement fineness is not less than 30 minutes and the final set time is not more than 10 hours. The average compressive strength of 1:3 cement mortar cubes after 7 days of healing shall not be less than 33 N/mm^2 (33°).

21.6.3.3 Examples of Detailed Works Specifications

1. **Detailed earth work specification for foundation:**
   a. Leveling the surface: The entire construction area should be cleared of tees, grass, trees, roots, etc., completed, and horizontally leveled to allow easy marking of the center of the building.
   b. Dimensions: The excavation shall be carried out in accordance with the trench dimensions shown in the working drawings.
   c. Shoring: The trenches should be vertical on the sides and flat on the bottom of the trenches. The sides of the trenches should be shored with sheets of steel in case of loose soils.
   d. Fencing: Appropriate temporary clamping must be provided around the excavation site to prevent any accidental fall into the trenches.
   e. Dumping the soil (soil dumping): The excavated soil must be dumped and heaped at a distance of at least 1.5 m from the trenches so that it does not slide back into the trenches.
   f. Bottom treatment: The bottom of the trench is watered and compacted by ramming before the base concrete is laid. The filling with loose excavated soils should not adjust excessive excavations. Sand or plain concrete can also be used with proper compaction to adjust levels.

2. **Detailed specification of random rubble masonry in foundation and basement:**
   a. Stone: The stone is obtained through the approved queries. It is sound, free of cracks, and decline and has a specific gravity not less than 2.5.
   b. Mortar: using cement mortar 1:6.
   c. Method of laying: The stones must be placed on the widest face, giving the mortar a better chance of filling the spaces between stones. The stones are laid layer by layer with enough mortar to bind better. The basement's exterior should be vertical and the joints should be staggered. There is no gap, between the stones, unfulfilled with mortar.
   d. Curing: Masonry should be kept wet by sprinkling water three times a day for a minimum of seven days after construction.

3. **Detailed specification of for brickwork in cement mortar 1: X using first class bricks in super structure:**

   Bricks are moulded, well-burned in approved kiln, colored in copper, free of cracks, and with sharp and square edges. Bricks are uniform in shape and are standard in size and give a clear ringing sound when struck.

   Bricks must be soaked in water at least 12 hours prior to use, preferably in a tank provided at the work site.

   The mortar proportion shall be one part of cement to five parts of sand by volume and shall be prepared for cement mortar according to the standard specification. Cement and sand must comply with the standard specification.

   Broken bricks are only used as closers. All corners must really plumb. Mortar joints break for bonding and do not exceed the thickness of 10 mm. Only skilled masons are used for the work, and the work is kept well-watered for a minimum of 15 days. All brickwork shall be done in such a way that no portion is excessively higher than another.

   Length and height are measured on the premises. The thickness of the walls is paid as one, one and a half bricks, two bricks, etc. The brickwork rate includes necessary scaffolding.

21.7 MARKET SURVEY

It is the net amount that a seller, having technically verified its characteristics and assuming an appropriate marketing, could reasonably receive for the sale of a property on the date of valuation, if at least a potential buyer and a seller are correctly informed, both acting freely for an economic interest and without a particular condition during the transaction. The intersection of their demand and supply curves always determines the market value of a property. The demand curve results from the costs of replacing potential buyers. The market value can be achieved by comparing the physical and technical characteristics with those of similar property, the value of which is already known, taking into account the impact that the differences detected on their market value can have. The net income that the market value could generate and the rate of return obtained with similar property in that market.

21.8 RATE ANALYSIS

To determine the rate of a specific item, the factors affecting the rate of that item are carefully studied and a rate for that item is finally decided. This process for determining an item's rates is referred to as a rate analysis.

The rate of particular item of work depends on the following:

1. Works and material specifications concerning their quality, proportion, and method of construction operation.
2. Quantity of materials and their costs.
3. Labor costs and wages.
4. Work site location and distances from source and transportation charges.
5. Overhead and charges for the establishment.
6. Profit.

21.8.1 Purpose of Analysis of Rates

Following are the main purposes of carrying out the rate analysis of an item:

1. To determine the actual cost of the items per unit.
2. To establish the cost-effective use of materials and processes in the completion of the item.
3. To determine the costs of additional items not provided in the contract bond, but to be carried out in accordance with the department's directions.
4. To revise the tariff schedule due to increased material and labor costs or technical changes.

21.8.2 Factors Affecting the Rate Analysis

The factors which affect the rate analysis of an item can broadly be divided into the following two categories:

1. Major factors
2. Minor factors

1. **Major factors**: There are mainly two factors on which the rate of an item depends are as follows:
   i. Materials: The different amount of materials required in order for the construction of an item can be easily worked out by knowing the specification of that item. The prices of various materials will depend on the market conditions. Thus, the quantities of the various materials required are fixed. But their prices are variable from place to place and from time to time as they depend on the prevailing market conditions. Hence, before starting the rate analysis of an item, it is essential to collect the prices of such materials from the market at that instant.
   ii. Labor: The labor force will be necessary to arrange the materials in a proper way so that the item can be completed. For instance, heaps of bricks, cement, and sand cannot be used in the construction of brick wall, unless some masons, coolies, and bhisties are available. The efficiency of the laborer and the wage of the laborer should be properly studied before starting the rate analysis for a particular item. By knowing the amount of labor force and wage of laborer, the cost of labor of a particular item is calculated.
2. **Minor factors**: These are the factors which come into force under special circumstances and should, therefore, be given proper weight, when such conditions are to exist. Following are some of such factors:
   i. Special equipment: If the execution of an item requires the use of some special equipment or plant, the cost of using special equipment on the rental basis should be included in the item's rate analysis.
   ii. Place of work: The site of work will also have some effect on the rate of an item under certain conditions. If it is too far, more amounts will have been spent on carting. Similarly, if it is situated in a highly congested area, it will not be possible to take the materials directly to the site.

iii. Nature of work: If the work consists of large quantities of the items, the rates may be less and vice versa.
iv. Contract conditions: If the contract conditions are very stringent, the rate of the different items is high and vice versa.
v. Profit of the contractor: The usual percentage of the profit of the contractor is ten. But if it is more or less, the rate of the item will be correspondingly affected.
vi. Specifications: If the specifications of work provide for rigid type tolerance and superior quality turn out, the rates will be on the higher side.
vii. Site conditions: If the site conditions are such that difficulties will be experienced during execution of work, such as foundations involving water troubles, the rates will be on the higher side. On the other hand, if site conditions are ideally suited for the construction activities, the contractor may quote slightly lower rates.
viii. Miscellaneous: The other remaining miscellaneous factors affecting rates of items include time of completion of the project, climatic conditions, reputation of the contracting firm, discipline of the organization, etc.

21.8.3 Task or Out-Turn Work

This is the amount of work that an artisan can do for 8 hours of work. Although the task is different from person to person depending on their physical and mental abilities, the average task or work per unit is taken into account.

21.9 ESTIMATES OF VARIOUS TYPES OF BUILDINGS

In all the estimates, it will be assumed that the rate for woodworks includes three coats of oil paint, and in items of brickwork, modular type bricks are considered unless otherwise specified. While preparing the detailed estimates for buildings and other structures, suitable provision will have to be made for electrification also. However, as a civil engineer, one should have an elementary idea of electrical installation and wherever necessary, a provisional sum to meet with the likely expenses on electrical installation may be provided in the estimate, and subsequently, the detailed estimate for electric installation including wiring, plug points, etc., can be got prepared from a qualified electric consultant. It should be noted that no such special provision is made for electrification in all the estimates.

21.10 ESTIMATES OF DIFFERENT RCC STRUCTURES AND THEIR FORM WORK

The work of estimates of various reinforced cement concrete structures is estimated under three items:

TABLE 21.1
Weights of Steel Reinforcement of Different Diameters

Diameter of Bar in mm	Weight in kg per meter	Diameter of Bar in mm	Weight in kg per meter
6	0.22	16	1.58
8	0.39	18	2.00
9	0.50	20	2.46
10	0.62	22	2.98
12	0.89	25	3.85

i. The concrete work excluding centering and shuttering is measured in cubic meter and paid separately.

ii. The centering and shuttering or form work is measured in square meter and paid separately.

iii. Reinforcement of steel including cutting, bending, binding, and positioning is quintal measured and paid separately.

Steel reinforcement is calculated according to the actual requirements of the position, including overlaps, hooks, etc. No deduction for steel is made from the volume of concrete. The cost of binding wire and wastage of steel is considered in the item of steel reinforcement,

The steel reinforcements used in RCC construction are of two types: (1) mild steel and (2) tor steel, i.e., high yield strength deformed (HYSD) steel. In conventional RCC works, the mild steel bars have been replaced by HYSD bars except 6 mm diameter bars. The properties like diameter, area, weight per meter, etc., are same for both types of bars. The weights of bars of different diameters are as shown in Table 21.1.

21.11 ESTIMATES OF VARIOUS ROOF TYPES AND STEEL STRUCTURES

Roof may be of RCC slab, reinforced brick (RB) slab, Jack arches, flat terraced, sloping or pitched roof, etc.

1. **Jack arch roof**

   Arches of Jack are segmental arches. The span of the jack arches varies from 90 cm to 1.8 m and rises from 1/6th to 1/10th of the span, usually consisting of a single 10 cm ring of brickwork in lime mortar or cement mortar. The jack arches are stretched onto RS, RCC, or Joist, spaced beams at regular intervals.

   For calculation, the length of the arch is considered to be equal to the width of the room plus twice the thickness of the arch as the wall cover. The arc length of the jack arch, i.e., the breadth is taken as the curved length of the intrados. The mean length of arc is same as that of the intrados. For the end bays the mean breadth is slightly greater than the central bays, but for practical purposes, this may be taken as same as that of central bays. If the rise of jack arch is 1/8 of the span, then the length of arc may be taken as 1.07

times the clear span. Clear span may be taken as center to center of beams. For haunch filling or spandrel filling the average thickness of concrete is taken as 1/8 of the rise of jack arch. The haunch filling is done with weak cement concrete or lime concrete.

The jack arches are usually supported on the lower flange of the joists, but it is better to support the jack arches over the upper flange of the joists so that the joists are not liable to rust and corrosion. Steel joists may also be encased with cement concrete to protect against rusting. Instead of RS Joist, RCC beam may also be provided.

2. **Roof truss shed**

   Roof truss shed consist of timber or steel roof trusses supported over walls or purlins or columns, spaced at regular intervals. Over the trusses run the purlins and over the purlins roof covering is provided. Common rafters may also be provided over the purlins over which battens are fixed to support the roof covering, which are required especially in the case of tile or flat roof. Timber trusses are usually spaced 2.50–3.00 m apart, and steel trusses are spaced 3.00–4.50 m apart.

   For estimating timber truss, the length of each member is taken from the drawing by measuring with scale if not mentioned, and the quantity is calculated in m³ by multiplying the length by the dimensions of the section.

   For steel truss, the length of each member is similarly taken from the drawing and the weight calculated in kg by multiplying the length by the weight per meter which is obtained from steel table, and the weight is converted into quintal. For rivets, bolts, and nuts usually 5% of the total steel work is provided. Gusset plates are considered as rectangular, as cut pieces are wasted, and the weight is calculated by the help of steel table. For gusset plate 18% of the weight of truss may also be taken. The number of rivet and their length may also be determined from the drawing, and their weight may be calculated by consulting steel table.

   The latest method is to join the members of truss by welding instead of riveting with gusset plates. This leads to rigid joint substantial economy in steel. In case of welded joint, the estimate for joints is made at suitable rate per joint.

The roof covering is calculated in m² for finished work including over laps multiplying the length of roof by the sloping breadth. Hook bolts, J-bolts, coach screws, washers, nails, etc., which are required for fixing the roof covering, battens, etc., are not recounted separately but included in the rate of the respective work item.

21.12 ESTIMATES OF WATER SUPPLY AND SANITARY WORKS

21.12.1 SANITARY WORKS

Sanitary works usually involve the provision of flush latrines and the connection to septic tank sewer lines. For estimation, the number of different fittings is determined and the rates are taken for each supply and fixation number. The latrine seat with a flushing cistern, a flushing pipe, etc., is usually taken as a single set and the rate per set is estimated for the complete work. The flushing cistern can also be taken separately with a flushing pipe, wash basins, reservoirs, urinals, etc. For complete work, they are also estimated per number. The fittings, such as mica valve, cowl, gully trap, master trap, etc., are also calculated by number. The pipelines of various materials of various diameters are estimated on the basis of the running meter as specified. Fittings like bends, interfaces, etc., not separately measured. Masonry manholes and inspection chambers for various sizes equipped with GI are estimated per number. Detailed estimates for manholes and inspection chambers may also be prepared for the complete work if required.

a. **Septic tank**

The septic tank is designed to keep the sewage in the tank for 24 hours during which a certain biological decomposition occurs by the action of anaerobic bacteria that breaks and liquefies the night soil leaving a small amount of solid that settles in the form of sludge at the bottom of the tank and clears water flows from the septic tank. The septic tank's capacity depends on the sludge removal or cleaning interval. Normally sludge is removed once in every 2 years and the liquid capacity of the septic tank may be taken as 0.13 m³ to 0.07 m³ per head. For small number of users 0.13 m³ per head and for large number of users 0.07 m³ per head may be taken as the capacity of the tank. The septic tank shall be at least 60 cm wide and a minimum liquid depth of one meter below water level with a minimum free board of 30 cm above water level.

The septic tank is usually made of cement mortar brick wall not less than 20 cm thick and the base floor is concrete 1:3:6 or 1:2:4. Both the wall and floor faces inside and outside are plastered with a minimum thickness of 12 mm thick cement mortar1:3 and all the corners inside are rounded. For convenience of collection and removal of sludge,

the floor should be given a slope of about 1 in 20. The septic tank cover is from RCC slab with suitable circular openings for cleaning and inspection with cast iron manhole cover.

Connecting pipe should be 100 mm minimum diameter and may be stone ware (SW) pipe, RCC, or hume pipe or cast iron pipe. Inlet and outlet can be made by means of a T-junction pipe or a prefabricated RCC wall. The septic tank may be provided at a distance of 1/5 of the length so that the inlet sewage cannot disturb the operation of the tank. Ventilation pipe of 50 mm minimum diameter is provided up to a height of 1.80 m. If the septic tank is within 15 m of a habitable building, the ventilating pipe should be carried to a height of 1.80 m above the roof of the building.

21.12.2 WATER SUPPLY WORKS

Water supply works consist mainly of pipelines. The lines are estimated on the running meter basis for different diameters for the complete work with fittings, including digging, laying, refilling, joining, etc. Fittings as stop cock, bib cock, ferrule, etc., are estimated number wise. Galvanized iron (GI) overhead tanks of various capacities, fitted with ball cock, are estimated numberwise supplying and fixing in position for complete work.

Water main pipes may be galvanized iron, cast iron, steel, hume steel, RCC, etc. For small diameter usually GI pipes are used; for diameters up to 60 cm cast iron (CI) pipes are used. Unplasticized PVC pipes are also in use now a day. It is available in various sizes ranging from 16 to 315 mm. The main pipeline is also estimated on running meter basis. Since valves, stop valves, air valves, etc., are provided as required which are estimated numberwise specifying their sizes. The service connections and internal connections are usually made with GI pipe.

21.13 ESTIMATES OF CULVERTS, BRIDGES, AND PIERS

21.13.1 CULVERTS

The different parts of the culvert should be separately considered for estimation. It is necessary to estimate first the two abutments with foundations up to the spring level and then the portions of haunch or spandrel above the spring level. The four walls with the foundation up to the haunch level should then be mounted and the parapet walls estimated. Masonry Arch should be computed separately. The finishing work of the surfaces is finally taken up.

Earthwork only for the excavation of foundation is generally taken up with the estimating of culvert. The filling up to the road level after the construction of the culvert is done later on and is usually taken up together with the earthwork of the road work. If required the earthwork may also be estimated together with culvert.

The portion from the spring level up to the crown of arch is usually taken as solid rectangular block and then to get the masonry in the spandrel or haunch filling deduction is made for

i. arch opening
ii. arch masonry work
iii. triangular portion above spandrel

Spandrel or haunch filling may also be lime concrete or weak cement concrete and may be taken separately. The wing walls four in number calculated are step by step from foundations upward. For high bridges or culverts weep holes are provided in the abutment and wing walls for the seepage of water in the soil, but no deduction needs to be made for these holes.

21.13.2 CULVERT WITH SERIES OF PIERS

When there are numbers of span and arches, the two abutments and four wing walls can be estimated in the same manner as for a single span culvert. The piers shall have to be calculated separately and multiplied by the number of piers if they are similar, to get the total quantity. Similarly, one arch can be calculated and multiplied by the number of arches. The spandrel and the portion above the springing point should be calculated as solid rectangular block for one span and necessary deduction made and then multiplied by the number of span.

21.13 3 STEEL BRIDGES

Steel bridges may be various types, may be of I- beams, plate girders, latticed, or trussed girders, etc., supported over abutments and piers of masonry. The steel work may be estimated on the same principles of steel roof trusses and steel stanchions. The length of individual members can be found from the drawings and the weight with the help of steel table.

21.14 ESTIMATES OF IRRIGATION WORKS

In irrigation works generally there are three types of canal sections:

1. Canal fully in excavation
2. Canal partly in excavation and partly in embankment
3. Canal fully in embankment

The volume of earthwork for irrigational canals is calculated by the trapezoidal formula which is known as end-area formula or mid-section formula. Generally irrigation canals are given certain longitudinal slope to develop certain velocities of water depending on the nature of soil and silt content in water. Steep slope develops higher velocities which causes scouring in the canal bed. If the general ground level has steep slope and the canal is given a flat slope, it may meet the ground level and further may move the ground level requiring high bank. To overcome this difficulty, falls or drops are provided in canal at suitable points.

21.15 ESTIMATES ON ROAD WORK

Universally applicable common method for the computation of volumes in road work is cross-section. In this method, the total volume to be computed is split up into a series of solids by the planes of cross-sections. The area of cross-sections taken along the line is worked out from the standard formulas and then the volume between successive cross-sections is computed by the application either of trapezoid or prismoidal formula.

The main three types of cross-sections are level section, two-level section, and side hill two level sections. The quantity of earthwork worked out by applying any one of the following method:

1. Mid-sectional area method
2. Mean-sectional area method
3. Trapezoidal formula
4. Prismoidal formula
5. Spot levels

21.16 CONTRACTS AND TENDERS

21.16.1 CONTRACT

When a contracting party's tender is accepted, an agreement is concluded between the contracting party and the owner and documents defining the rights and obligations of the owner and the contracting party are attached to the contract bond, which is called a contract document. Each page of the contract document shall be signed by the contractor and the accepting authority and any correction shall be initialled therein.

The document on the contract must contain:

1. Title page—work name, owner's name, contractor's name, contract no., content, etc.
2. Index page—agreement contents with reference pages.
3. Tender notice—brief description of work, estimated cost of work, tender date and time, deposit of EMD and security, completion time, etc.
4. Tender form—quantity bill, contractor rate, total work cost, completion time, deposit amount, etc.
5. Schedule of issuance of materials—list of materials to be issued to the contractor by the owner/department at rates and location of issue.
6. Drawings—full set of drawings with plan, elevation, sections, detailed drawings, etc. Everything fully dimensioned.
7. Specifications:
   a. General specifications: specifying the working class and type, material quality, etc.
   b. Detailed specifications: detailed description of each work item, including the material and method to be used, together with the required quality.

8. Conditions of contract:
   - Tariffs for each item of work including materials, labor, transport, plant/equipment and other arrangements required to complete the work.
   - Amount and form of earnest deposit of money and security.
   - Payment method to the contractor, including running payment, final payment and security money reimbursement, etc.
   - Work completion time.
   - Extension of work completion time.
   - Engagement of subcontractors and other agencies at the expense and risk of the contractor
   - Low-quality punishment and unsatisfactory work progress.
   - Contract termination.
   - Arbitration for dispute settlement.
9. Special conditions: depending on the nature of the taxes and royalties included in the rates, labor camps, labor facilities, work compensation in the event of accidents, etc.
10. Deed of pledge.

21.16.1.1 Types of Contract

There are different types of contracts which can be employed in any of the delivery methods. Owner can pay the money to the contractor, in lump-sum, based on measured work with unit price, based on percentage plus quantity involved.

1. **Lump-sum contract**: This is a fixed price contract in which the contractor agrees to perform a fixed amount job. The owner provides the contractor with exact work requirements. In this contract, the parties attempt to fix the working conditions as precisely as possible.

   Advantages:
   - The owner is aware of the project costs before construction begins.
   - A lot of details and accounting are avoided by both the owner and the contractor.
   - Contractor shall be given free hand to perform the work.
   - If this contract is used with the delivery design—build method, the contractor will be given the opportunity to use value engineering.

   Disadvantages:
   - Every charge in design and specification is very difficult to accommodate.
   - This contract is as good as the document's accuracy. If there are errors in the contract document, the contract must be renegotiated and there is therefore more risk from the owner's ride.
   - The contractor can be placed in an adverse situation in the event of unforeseen hazards during construction.

   This type of contract is suited for small job, precisely specified job, and low risk with construction job. Lump-sum contract should be avoided for underground work. Lump-sum contract with design construct method of delivery is often called *turnkey contract*.

2. **Unit price or item-rate contracts**: In this contract type, the price is paid per unit of work performed. There are different variations of this type of contract. Some of them are mentioned below.

   i. Bill of quantities contract: In this type of contract, owner provides the drawing, quantities of work to be done, and specification. The contractor bids based on the unit cost of the items of construction. The contractor's overhead, profit, and other expenses can be included in the unit cost of the item of work. Sometimes contractor quotes the unit price of the work and lump-sum amount separately as profit overhead. The estimated quantities of the work to be done called bill of the quantities are fixed. Minor variation in the quantities is admissible in this type of contract. The drawing of the work is not supposed to change. Although change and deviations from original drawing could be accepted during construction but even then unit price does not change.

   This type of construction is usually followed in government sector for large infrastructure construction. This type of contract provides owner a competitive bid. Disadvantages of the methods are as follows:
   - Owner needs to measure the quantity of work done in the field, hence requires owner presence at the site.
   - Final price of the construction is not known precisely until last price of work is completed. If the estimated quantities differ significantly from the reality of the situation, the owner is placed in a negative situation. Mistaken quantities are called *unbalanced bid*. Significant unbalanced bid is now considered as unethical.

   ii. Schedule of rate contract: Many a time, the quantity of work to be executed is not known before. Contract is signed based on the unit cost of the item of work. Generally more items are inserted in the contract than to be executed because it becomes sometimes difficult to exactly specify all the items. There is no guarantee that all the items mentioned will be used in the construction. This type of contract is widely used in underground work, flood control, and road constructions.

3. **Cost-plus or percentage contracts**: In this contract, the payment is made based on the work carried out plus the fee which includes overhead, profit, etc. Sometimes a cap is put on the type of contract by providing maximum and minimum cost limit such as guaranteed maximum cost contract. If project cost exceeds this limit, the contractor is responsible

for that. Sometimes incentive clause is also included if the contractor brings the project before a certain specified limit.

The advantage of this type of contract is that considerable overlap is providing between design and construction. Hence the project can be executed in the fast-tract basis. This contract is suitable for the work where it is difficult to define the task to be done before awarding the contract.

21.16.1.2 Termination of Contract

A contract may be terminated or discharged in any one of the following five ways:

- **Impossibility of performance**: Typically, a contract requires one or more parties to do something called performance. A company can, for example, hire and sign a contract to talk to a public speaker at a business event. Once the public speaker fulfils his contractual duties, the possibility of performance is called. If it is impossible for the public speaker to carry out his duties for any reason, it is called impossible to perform. In the event of impossibility of performance, the company has the right to terminate the contract.
- **Breach of contract**: If a contract is deliberately not honored by one party, it is called a breach of contract and is the reason for terminating the contract. There may be a breach of contract because one party has failed to fulfil its obligations or has not fully fulfilled its obligations. A material breach of contract enables the employer to seek monetary damage and an immaterial breach of contract does not allow the employee to seek monetary damage.
- **Prior agreement**: You can terminate a contract if you and the other party have a prior written agreement calling for a termination of the contract for a particular reason. The agreement must provide details of what qualifies as a reason to terminate the contract. It should also state what measures must be taken to terminate the contract by one of the parties. In most cases, one party must provide the other party with a written notice to terminate the contract.
- **Rescission of the contract**: A withdrawal from a contract is the termination of a contract because an individual misrepresented himself, acted illegally, or made a mistake. A withdrawal from the contract may occur if one party is not old enough to enter into a contract or if an elderly person is unable to make any legal decisions due to disability.
- **Completion of the contract**: A contract is terminated essentially once the obligations specified in the contract have been fulfilled. The parties should keep records showing that they have fulfilled their contractual obligations. Documentation is helpful if the other party attempts to dispute your contractual obligations later on. A court of law shall require proof of fulfilment of the contract in the event of a dispute.

21.16.1.3 Earnest Money Deposit

The amount of money has to be deposited with the tender document by the contractors who quote a tender to the department. This money is a guarantee against any contractor's refusal to take up the work after the tender has been accepted. If rejected, this amount is forfeited. The Earnest Money Deposit shall be refunded to the contractors whose tenders are not accepted.

21.16.1.4 Mobilization Funding

Financing for mobilization is a financial tool commonly used by construction contractors to raise capital to cover costs before starting the work on a project or before invoicing. Funding depends on the preapproval of contractors by a qualified subcontractor evaluation process such as subguard. Mobilization funding typically advances up to 10% of the contract amount and up to 85% of the invoices issued and a total of 40% of the contract amount.

21.16.2 TENDER

A "tender" is required to perform certain work or to supply certain materials; subject to certain terms and conditions, such as rates, time limits, etc. It is a written offer: in legal terms, it is an offer to receive an offer for the work within the limits specified.

Before inviting tenders, an engineer must observe the following:

- Quantity bill accuracy as far as possible.
- All plans, specifications, and details with no ambiguity are correct.
- No item and special condition should be overlooked.

Each form must include the following:

1. Special notice given on behalf of either the owner or the architect.
2. Letter of acceptance from the contractor.
3. Articles of understanding with special contractual conditions.
4. Specification general.
5. Bill of quantities in case of tender of item rate.
6. In the case of a lump sum tender, special conditions, and detailed work specifications.

21.16.2.1 Classification of Tenders

The majority of builders still receive much of their work through the tendering system. There are three methods of selecting a contractor. They are as follows:

1. Open Tendering
2. Selective Tendering
3. Negotiated Tendering
4. E-Tender

1. **Open Tendering**: This is usually in the form of advertisements in national or local newspapers inviting contractors to apply to tender for work. The deposit is returnable when a tender is submitted. This method enables any contractor to submit an offer to a project advertized.

    Process:

    i. Customer publicly advertizes in the press or in commercial publications inviting contractors to apply for the project.

    ii. The contractor who can undertake the project would ask for a tendering document.

    iii. The contractor may be required to give a deposit and ensure a bonafide tender after receiving the tender from the architect. This is done to filter out contractors that do not want to submit an offer.

    Advantages:

    - Keep a large number of applicants competing.
    - Possibility of pricing competitive.
    - No favorism opportunity.
    - Provide new entrants, unknown contractors, with the opportunity to enter the market.
    - Disadvantages:
    - Review time and cost of the large number of applicants.
    - The industry-accumulated tendering costs and the general industry costs are high.
    - Hard to evaluate and choose the best of a wide range of applicants.

2. **Selective Tendering:** The contractor shall be included in the approved list, often maintaining an architectural practice, a local authority, or a statutory body by an employing body. These organizations have extensive knowledge and experience of individual contractors and periodically review the performance of contractors and the list of approvals.

    Advantages:

    - Tendering costs are reduced.
    - Time to evaluate the tender is reduced.
    - Allow tendering only from capable and approved companies.
    - Limit competition and easy screening and selection.

    Disadvantages:

    - Lets high tender sum into competition.
    - Opportunities of collusion and misleading time between competitors.
    - Opportunities missing (new comers to come).
    - Industry complains of not short-listed people.

3. **Negotiated Tendering**: Only well-known contractors were negotiated in this process by inviting them to submit their tenders. Customers can determine the negotiating perimeters before the start of the project. Importantly, maintaining positive relationships during the negotiation process.

    Advantages:

    - More can be achieved with expected quality and functions.
    - Accelerate the entire project.

    - The aim and definition of the project may be managed. (Participation of builders in the design stage.)

    Disadvantages:

    - Noncompetition forces for high tender sum.
    - Less formality makes course more contract administration inquiries.
    - Lack of accountability.

4. **E-Tender**: The traditional systems of procurement in government departments through manual modes suffered from various problems such as inordinate delays (approximately 4–6 months) in tender/order processing, heavy work, multilevel scrutiny that consumes a lot of time, physical threats to bidders, cartel formation by the contractors to suppress competition, human interface at every stage, inadequate transparency, discretionary treatment in the entire tender process, etc. Though it is known to the departments that their traditional processes are inefficient, hardly any effort was taken to improve the system for obvious reasons.

    E-Tendering is a process of carrying out entire the Tendering Cycle Online, including submission of Price Bid to harness Efficiency, Economy, and Speed of Internet. Despite the apparent benefits of information technology (IT), many organizations have been slow to adopt e-construction and in particular e-tendering. Many companies are approaching the use of e-tendering with caution in order to test its practical advantages and encourage age confidence among staff before implementation to a general concern over security and legality issues. Tendering is a practice involving a complex web of legal issues, which must be known before tendering. The unguarded use of electronic technology in electronic tendering and posttendering project management has created contradictory effects, such as the tradeoff between efficiency and security. The people are unsure of the legal impact of using the existing e-tendering project management system. For this reason, industry is reluctant to conduct contracting activities. Business has traditionally incorporated many elaborate procedures into their regular business processes to seek legal protections. E-Tendering systems should also include appropriate security mechanisms for increasing the system's reliability.

21.16.2.2   Invitation Procedure for Tender

1. Preparation of bidding papers.
2. Announcement of inviting tender or call notice.
3. Offering and opening and scrutiny of tenders.
4. Tender acceptance and contract award.

Information to be given in a tender notice:

1. Name of the department inviting tender.
2. Name of work and location
3. Designation of officer inviting tender
4. Last date and time of receipt of tender

5. Period of availability of tender document
6. Cost of tender document
7. Time of completion and type of contract
8. Earnest Money Deposit to be paid
9. Date, time, and place of tender opening
10. Designation of the tendering officer

21.16.2.3 Opening of Tenders

While opening the tenders, the following points should be observed:

1. The tender should be opened in the presence of the owner or committee members.
2. For the list of tenderers and the money they deposit, a record must be kept. Instead, earnest money should be accepted in the form of a draft.
3. If the owner is not present, the architect should open the tender together with an assistant. The architect shall examine the same, prepare comparative statements, and forward them to his client, who should be awarded the work and why.
4. After a close investigation of the reputation and status of the contractor, the lowest tender should be accepted.
5. Public tenders are always open in the presence of tenderers who choose to stay.
6. It is important to make sure that they are thoroughly checked and without errors.

21.16.2.4 Scrutiny of Tenders

After the tenders are received, they should be properly scrutinized with respect to the following items:

1. A list of tenders which are received should be made with the details of the deposit cheques.
2. It should be checked and verified that each tender is duly signed and contains the address of the contractor.
3. It should be seen that the contractor has not altered the terms and conditions of the tender, and at the same time, he has not mentioned some other terms and conditions.
4. The rates for each item should be entered in figures and words. The calculations showing the amount of each item should be carefully checked.
5. The totaling of amounts from page to page also should be done carefully in order to arrive at the final cost of construction in each case.
6. After all the tenders are scrutinized, a comparative list should be made. The tender showing the lowest amount should be placed first and the tender showing the highest amount should be placed last.

21.16.2.5 Acceptance of Tender

Generally tender with the lowest amount is accepted. But in some cases, the rates of the lowest tender are so low that it becomes evident that the work will be spoiled, if contract is given to such tenderer. In such cases, the owner decides

for other negotiates with the second lowest contractor. In any case, the acceptance of tender should be done in specified time limit or within reasonable time and the contractors who have filled up the tenders should be informed about the successful tenderer. The earnest money of unsuccessful contractors should be returned.

Following are considered to be the modes of acceptance of tender in the eyes of law:

1. Conduct of the parties
2. Telegram acceptance
3. Written acceptance
4. Total acceptance

21.16.2.6 Revocation of Tender

A tender is converted to a legally binding contract, the moment it is accepted by the owner. The owner calls for a tender which is merely an offer and he is entitled to reject it without any liability. In similar way, the law has granted protection to the contractor also. If after filling up the tender and before its acceptance by the owner, the contractor comes to know his mistake or intends to withdraw from the work, the contractor is entitled to revoke his tender.

21.16.2.7 Unbalanced Tender

In case of unit-price contracts, the contractor has to quote his rate for each item. If these rates cited by the contractor are reasonable, the offer is referred to as a balanced offer. But sometimes, the contractor puts up higher rates for certain items and lower rates for other items so that the total amount of the tender remains practically unaffected. Such a tender is known as an unbalanced tender and some of the purposes of filling an unbalanced are as follows:

1. Equipment utilization and organization
2. Capital working
3. Speculation on estimate errors

21.16.2.8 Liquidated Damages and Unliquidated Damages

Liquidated damage is an amount of compensation due to delayed construction by the contractor to the owner. This does not relieve the contractor of its contractual obligations. If a part of the project or premise is often used by the owner before its completion, the amount to be paid shall be reduced in proportion to the value of the part used after the certificate of occupancy has been issued.

An unliquidated damage is a compensation amount due when a contract has been broken. The party suffering such a violation has the right to receive this amount from the party that has broken the contract.

21.16.2.9 Indirect vs. Direct Costs

Direct costs can be attributed directly to the object. In construction, the costs of materials, work, equipment, etc., and all efforts or expenses directly related to the object are direct costs. In manufacturing or other nonbuilding industries, the portion of operating costs that can be directly attributed to a particular product or process is direct. Direct costs are those for activities or services that benefit specific projects, e.g., project staff salaries and materials required for a specific project.

Indirect cost cannot be attributed directly to a cost object. Indirect costs are usually assigned to a cost object on a certain basis. In construction, all costs necessary to complete the installation but not directly attributable to the cost object, such as overhead, are indirect. These can be, for example, management, insurance, taxes, or maintenance costs. Indirect costs are those associated with activities or services benefiting more than one project. Their precise advantages for a particular project are often difficult or unacceptable. For example, it can be difficult to determine exactly how the activities of an organization's director benefit a particular project. Indirect costs do not vary significantly in certain production volumes or other activity indicators and can therefore sometimes be regarded as fixed costs.

21.17 CONDITIONS OF CONTRACT

The clauses which relate to the work as a whole are written in a separate contract document, known as the general provisions or conditions of contract. The main object of framing conditions of contract is to avoid dispute between the parties concerned and thus, to keep them out of the court of law. In case of most of the civil engineering contracts, the following groups of contract conditions are generally accepted:

Conditions relating to documents:

  i. Quantity bill and price schedule
 ii. Drawings
iii. Indian standard specifications
 iv. Notices
  v. Provisional and prime cost sums, etc.

  1. **Conditions relating to the general obligations of the contractor**:
       i. Access to works
      ii. Acts, bye-laws, and regulations
     iii. Fencing, watching, and lightning
      iv. Instructions of engineers
       v. Insurance
      vi. Setting out
     vii. Site, etc.
  2. **Conditions relating to labor and personnel**:
       i. Accidents to workmen
      ii. Contractor's representative
     iii. Engineer's representative
      iv. First aid
       v. Rates of wages
      vi. Removal of the employees of the contractor, etc.
  3. **Conditions relating to assignments and subletting**:
       i. Assignments
      ii. Subletting
     iii. Specialist contractors, etc.

  4. **Conditions relating to the execution of work**:
       i. Alterations, additions, and omissions during progress of work
      ii. Amount of extra items
     iii. Damages
      iv. Defensive work
       v. Defects
      vi. Materials
     vii. Protections of trees and shrubs
    viii. Public travel
      ix. Safety by shoring and during blasting
       x. Water for construction
      xi. Work at night and on holidays
     xii. Workmanship, etc.
  5. **Conditions relating to measurements and payments**:
       i. Method of measurement of completed works
      ii. Payment methods
     iii. Payment to subcontractors, etc.
  6. **Conditions relating to default and noncompletion**:
       i. Abandonment of the work by the contractor
      ii. Bankruptcy of contractor
     iii. Engineer during construction
      iv. Failure to complete the work in time
       v. Right to suspend the work by the owner
      vi. Time of completion, etc.
  7. **Conditions relating to settlement of disputes**:
       i. Arbitration
      ii. When engineer's decision is to be final, etc.
  8. **Special conditions**:
       i. Equipment
      ii. Names of firms supplying materials
     iii. Pollution of streams
      iv. Use of intoxicants, etc.

21.17.1 TYPICAL CLAUSES OF THE CONDITIONS OF CONTRACT

Some of the clauses of the conditions of contract which are normally found in all civil engineering contracts are given below.

  1. **Definitions**: This clause contains the definitions of the words which are frequently used in the specifications. Thus, its purpose is to avoid confusion and to fix the meaning of a particular word.
  2. **Establishment and preservation of the points**: Normally, the lines and the levels of the works are given by the engineer. But all the materials required in connection with the setting of them are to be supplied by the contractor, free of charge. Also, it is the duty of the contractor to preserve all such points.
  3. **Storage of tools and materials**: The contractor is supposed to make suitable arrangements at the site of work for the storage of his tools, plants, materials, etc., and his own temporary office also. It is to be clarified that the rates quoted by the contractor in his tender also include the expenses of all such arrangements.

4. **Contractor to study the particulars of the work in detail**: The engineer prepares plans and specifications for the works with the hope that the structure will function satisfactorily when completed. It is, however, difficult for the engineer to give a guarantee for the results. For this purpose, this clause is included in the conditions of contract. It should be noted that this provision in the conditions of contract will be waived by the court of law, if it is proved that it was not possible to collect the unknown details by a reasonably careful and diligent contractor. The main points to be included would be as follows:
   i. Contractor to investigate personally into the details of the works.
   ii. Contractor not to rely solely on the information supplied by the engineer or owner.

5. **Engineer during construction**: This clause is mainly included to give some powers to the engineer. During construction, the engineer has to act in various capacities, and in order that his decisions are honored, this clause is inserted in the conditions of contract. It also intends that besides completing the works as per specifications and plans, the contractor has to satisfy the engineer also. However, if the contractor or owner wants to over-rule the decision of the engineer by the court of law, it becomes essential to prove that the decision was arrived at due to dishonesty or unfairness or prejudice or incompetency. The points to be included in this clause would be as follows:
   i. Engineer to act as a referee in all disputable questions.
   ii. Meaning of the specifications to be explained by the engineer.
   iii. The decision of the engineer to be final and binding to both the parties.

6. **Receiving instructions from the engineer**: It is expected that the contractor will remain present at all the times on the works. Hence, the contractor should put up a qualified person on the works to receive the engineer's directions and instructions. The main point to be included in this clause would be to stress that the person thus appointed shall be familiar with the works to be executed.

7. **Removal of employees of the contractor**: Sometimes, the persons employed by the contractor are not competent for the works or are found to behave rudely with the engineer. In such circumstances, the engineer can discharge such employees and this clause is intended to give such power or authority to the engineer.

8. **Alterations, additions, and omissions**: Practically in every contract, some alterations, additions, or omissions will be required at some stage, and unless provision is made for this fact, the contractor can refuse to carry out such alterations, additions, or omissions. The main points to be included in this clause would be as follows:

   i. The engineer's authority to do reasonable alterations, additions, and omissions.
   ii. The changes not to affect the original contract.
   iii. No extension of the time limit.
   iv. The adjustment in prices to be done by the engineer.

9. **Arbitration**: The main purpose of including this clause is to avoid the tedious procedure of litigation and thus to avoid legal formalities, delays, and expenses. It is a procedure in which the parties concerned submit their disputes to an impartial person or a committee of experts. The important points to be included in this clause would be as follows:
   i. Appointment of the arbitrator
   ii. Powers of the arbitrator
   iii. Period of the notice
   iv. Delay of the work
   v. Remuneration of the arbitrator
   vi. Award of the arbitrator

10. **Inferior materials and workmanship**: This clause is provided to reject the materials and workmanship which are not confirming with the specifications of the works.

    The main points to be included in this clause would be as follows:
    i. Marketing and removal of the rejected materials.
    ii. Replacing or repairing the condemned workmanship to the satisfaction of the engineer.

11. **Extension of time for the completion of the works**: This clause is intended to indicate the circumstances under which an extension of the time limit may be granted to the contractor by the owner. Some specifications are very rigid in not granting a single day's extension. Even the death of the contractor is not considered as the legitimate cause for delay of the works. The points to be included in this clause would be as follows:
    i. Probable causes of delay.
    ii. Original contract not to be affected by the extension.
    iii. Contractor to give notice for the demand of the extension.
    iv. Contractor not to claim for any compensation.

12. **Failure to complete the works in time**: This clause is inserted just to give stress that the works are to be completed in time. A certain amount is fixed which the contractor has to pay, not as penalty but as liquidated damages for the delay of the works. In case the matter goes to the court of law, it becomes essential to justify the amount of liquidated damages.

13. **Right to suspend the work by the owner**: Due to some unavoidable circumstances, the owner may request the contractor to stop the work for a certain period. A clause, therefore, insets to give such power to the owner and it should contain two points:
    i. Extension of the time to be granted for the actual period of suspension.

ii. Owner to pay amount to the contractor for damages due to such delay.

14. **Compliance of laws, etc.**: Generally, the responsibility of respecting all national, state and local laws, ordinances, etc., is placed on the contractor and a clause for this purpose is therefore included in the conditions of contract.

15. **Patent rights and royalties**: This clause is included in the conditions of contract to warn the contractor regarding the use of patented articles, processes, etc., the main points to be included would be as follows:
    i. Royalties to be included in the tender amount.
    ii. Owner to be kept harmless.
    iii. This clause to operate even after the contract period.

16. **Labor laws**: This clause is included to force the contractor to observe prevailing labor laws. Sometimes, extracts from the labor laws are given in this clause to stress some of them. Such extracts are normally for the wages of labor, working hours, compensation in case of accidents, etc.

17. **Public travel**: This clause is provided to prohibit the contractor from interfering with the traffic. The points to be included would be as follows:
    i. Written permission of the engineer required to stop the traffic.
    ii. Provisions of signs and lamps at night.
    iii. Owner to be kept free from all claims, etc., due to such actions.

18. **Abandonment**: This clause is provided to meet with the situation when the contractor deliberately stops the works. Normally, provision is made in this clause, stating that the remaining portion of the works is to be carried out by the owner by employing other person or persons and the difference in the cost is to be deducted from the security deposit of the contractor.

19. **Subletting**: In order to guard against the behavior of subcontractors appointed by the main contractor, this clause is included in the conditions of contract. The clause should provide for two important points:
    i. Written permission to appoint subcontractors to be obtained from the owner.
    ii. Main contractor to be considered responsible for the works as a whole.

20. **Interim payment to the contractor**: This clause is provided to make partial payments to the contractor. The main points to be included would be as follows:
    i. Interim payment to be made at the end of each calendar month.
    ii. 10% to be retained as deposit out of which 5% to be returned at the time of final payment and the other 5% at the end of the maintenance period.
    iii. No amount to be given for the materials brought on site (in case, if it is desired for the reverse, it should be stated that the materials shall become the property of the owner and the contractor cannot remove them).

iv. Interim payment not to be considered as final acceptance of the works up to that level.
    v. Engineer to be given power to withhold the interim payments under special circumstances.

21. **Final payment**: After the works are completed in all respects, the engineer accurately prepares the final amount of the works and then, after deducting all previous payments, the owner pays the final amount to the contractor. The usual provisions to be made under this clause would be as follows:
    i. Period to be given to the engineer for the preparation of the final bill, usually fortnight to 1 month.
    ii. Period to be given to the owner for paying the final amount, usually fortnight to 1 month.
    iii. Mention of certain percentage of amount during maintenance period, if any.

22. **Possession prior to completion**: Sometimes, the project is quite big and the owner desires to have the possession of the portions which are completed during the progress of work. It is to be mentioned that the acceptance of part of the project does not necessary mean the final acceptance of the portion whose possession is taken by the owner.

23. **Rates of contractor to remain firm during contract period**: This clause is provided to make sure that the rates quoted by the contractor in his tender are to remain firm during the whole period of contract, i.e., up to the completion of the work and they are not to be raised even due to various circumstances such as increase in prices of materials and wages of labor, increase in railway freight, etc.

24. **Payments of the extra work**: Theoretically, a well-prepared contract should not allow any extra work or items. But practically, in all contract, some items which are not covered in the original contract are to be executed. The clause mainly includes the following points:
    i. Written order of the engineer or owner necessary to carry out extra items.
    ii. Rate to be decided by the engineer either with comparison with contract prices or from schedule of rates or by actual rate analysis.
    iii. Contractor to supply the necessary information as required by the engineer to arrive at a reasonable rate of the extra item.
    iv. No extension of time limit due to extra items.

25. **Fencing, watching, and lighting**: This clause is provided to make the contractor responsible for protecting the works by suitable arrangements. The main points to be included in this clause would be as follows:
    i. All required materials to be provided by the contractor free of cost.
    ii. Arrangements to be made as per instructions of the engineer or requirements of the local authority.
    iii. Contractor to be responsible for any accidents.

26. **Method of measurement of completed works**: This clause is inserted to inform the contractor regarding the method of measurement of the completed works; it includes the following points:
   i. Engineer to take the net dimensions of the completed works.
   ii. Contractor to remain present and assist the engineer.

27. **Bankruptcy of contractor**: This clause is provided to meet with the situation bankruptcy or death of the contractor. Following important points are to be included this clause:
   i. Engineer to be satisfied regarding the bankruptcy of the contractor.
   ii. Owner to take over the charge of the works and get it completed by any means.
   iii. Provision regarding materials, equipment, etc., of the contractor at the site of works.
   iv. Valuation on the date of the bankruptcy.
   v. Payment to the contractor after bankruptcy.

28. **Maintenance**: This clause is included in the conditions of contract to ensure that the works carried out under the contract shall remain in good condition for a certain period after its completion. Usually the period of maintenance varies from 6 to 12 months but can be of any length, depending upon the nature of works. The main points to be included in this clause would be as follows:
   i. Period of maintenance to be stated.
   ii. Amount of deposit to be kept with the owner.
   iii. Period of notice to be given to the contractor to carry out repairs.
   iv. Provision for urgent repairs.

29. **Site clearance on completion of the works**: This clause is provided to put stress on the fact that the contractor is responsible for clearing the site of works after the works under this contract are completed.

30. **Clerk of works**: When the construction project is complicated and to sizeable amount, it becomes necessary to make provision of clerk of works who is a qualified person approved by the architect or engineer and who is appointed and paid by the owner. The clerk of works inspects and supervises the works on behalf of the architect and the contractor is supposed to give him co-operation and assistance to perform his duty well.

31. **Protection of trees and shrubs**: The engineer or architect sometimes desires to retain the existing trees and shrubs for getting some architectural effect. It includes the following points:
   i. Engineer should designate the trees and shrubs to be retained.
   ii. Minimum distance of one meter should be kept undisturbed around such trees and shrubs.
   iii. Contractor should provide fencing, if found necessary.

32. **Water for construction**: This clause is inserted to inform the contractor that he has to make necessary arrangements at his cost to get sufficient quantity of water for the construction work. It included the following points:
   i. Application to the concerned authority is to be made by the contractor for getting temporary water connection.
   ii. Temporary pipes are to be laid and taps are to be fixed by the contractor.
   iii. For supply of water other than authorized one, sample of water is to be analyzed and to be approved by the engineer.

21.18 ARBITRATION

Arbitration is a process in which a dispute is submitted to an impartial outsider who usually takes a binding decision on both parties. The unbiased person or persons are referred to as arbitrators. The arbitrator reinforces his own views on the contentious parties and the views of the participants do not prevail. Arbitral proceedings are judicial. The award of the arbitrator is binding and is based on fairness and justice, i.e., no compromise is possible.

21.18.1 ARBITRATOR AND REFEREE

It sometimes happens that the parties to a suit refer the matter to a referee and they agree to honor the statement made by the referee. In this case, the referee makes only the statement or pronounces his opinion from the facts and data available to him. The court of law decides the case and gives judgement on the basis of the contents made by the referee.

21.18.2 MATTERS FOR REFERENCE TO ARBITRATION

It is generally understood that all matters may be referred to the arbitration except those which are prohibited by a statute or which are against the public policy. The matters which can be referred to the arbitration under the provisions of the act can be summarized as follows:

1. All disputes except those of criminal nature.
2. Matrimonial disputes under certain circumstances.
3. Pure questions of law and fact including questions of territorial jurisdiction.
4. The matter pertaining to the protection of private rights under civil litigation, etc.

Following are, however, the matters which cannot be referred to the arbitration under the provisions of the act:

1. Disputes which are purely criminal.
2. Insolvency proceedings.
3. Lunacy proceedings.
4. References for illegal transactions.
5. Suit for divorce.
6. Testamentary matters such as genuineness of a will, etc.

21.18.3 Types of Arbitration

Following are the two main types of arbitration:

1. Voluntary arbitration
2. Compulsory Arbitration

1. **Voluntary arbitration**: It implies that the two contentious parties, unable to make their own differences, agree to submit the conflict/disagreement to an impartial authority whose decision they are ready to accept.

   The essential elements of voluntary arbitration are as follows:
   - Voluntary dispute submission to the arbitrator.
   - Participation of witnesses and investigations afterward.
   - It may not be necessary and binding to enforce an award.
   - For disputes arising under agreements/contracts, voluntary arbitration may be especially necessary.

2. **Compulsory arbitration**: The obligatory arbitration is one in which the parties must accept arbitration without their willingness. If one of the parties to an industrial dispute is grieved by an act of the other, the appropriate government may request that the dispute be referred to the adjudication machinery. The essential requirements for mandatory arbitration are as follows:
   - The country is experiencing serious economic crisis.
   - Strategically important industries are involved unbalanced parties.
   - Mandatory arbitration leaves no scope for strikes and lockouts; both parties are deprived of their very important and fundamental rights.

The other types of arbitration are as follows:

1. Ad-hoc arbitration
2. Institutional arbitration
3. Statutory arbitration
4. Domestic or international arbitration
5. Foreign arbitration

1. **Ad-hoc arbitration**: When there is a dispute or difference between the parties during trade transactions. This arbitration is agreed to bring justice to only the unresolved part of the dispute.
2. **Institutional arbitration**: There is prior agreement between the parties that in the event of future differences or disputes between the parties during their commercial transactions, such differences or disputes shall be resolved by arbitration in accordance with the provisions of the agreement.
3. **Statutory arbitration**: It is mandatory arbitration which is imposed by law on the parties. In such a case, the parties have no option but to comply with land law.

4. **Domestic or international arbitration**: Arbitration taking place in India and involving all parties in India is referred to as domestic arbitration. An arbitration in which any party belongs to another party than India and the dispute in India is to be resolved is called international arbitration.
5. **Foreign arbitration**: When arbitration proceedings are held outside India and the award is to be implemented in India, it is referred to as foreign arbitration.

21.18.4 Sole Arbitrator, Joint Arbitrator, and Umpires

Unless otherwise clearly mentioned in the arbitration agreement, the reference is only to a *sole arbitrator.*

When each party to the arbitration agreement appoints one arbitrator, the proceedings of arbitration are conducted by the *joint arbitrators.* If the number of joint arbitrators is odd, the award given by the majority of joint arbitrators prevails.

The term *umpire* is used to denote arbitrator appointed by the two appointed arbitrators, one by each party to the agreement. Thus, when the reference made to an even number of arbitrators, it will be necessary for the appointed arbitrators to appoint an umpire. If the arbitrators are equally divided in their opinions, the award by the umpire shall prevail, unless otherwise provided in the arbitration agreement.

Following are the four different ways of appointing arbitrators by the parties:

1. A person may be designated as an arbitrator either by name or as the holder of a particular office.
2. Each party may choose one arbitrator, and if there are two parties to the agreement, the two appointed arbitrators may select an umpire agreeable to both.
3. The agreement may provide for the name or holder of any post who in turn will appoint arbitrator or arbitrators.
4. A panel of persons may be mentioned in the agreement and the parties have to select sole arbitrator from this list of arbitrators.

21.18.5 Powers of an Arbitrator

- To give oath
- Refer to the court for opinion
- Declaring the award
- Correct clerical mistakes
- Managing interrogatories

21.18.6 Arbitration Agreement

An arbitration agreement is a written agreement between two or more parties to resolve a dispute outside the court. The agreement on arbitration is usually a clause in a bigger contract. The dispute may be about the execution of a specific contract, an unfair or unlawful treatment in the workplace, a defective product, among other issues. People are free to agree to arbitrate in respect of anything else that they could resolve through legal proceedings. An arbitration agreement can be as simple as a provision in a contract stating that you agree to arbitrate in case of future disputes by signing that contract.

21.18.7 Power of Court to Appoint Arbitrator or Umpire

The cases in which the court has discretion to appoint an arbitrator or umpire are as follows.

- where an arbitration agreement stipulates that a reference shall be made to one or more arbitrators to be appointed by the consent of the parties and all parties do not agree to the appointment or appointments after differences have arisen; or
- if any appointed arbitrator or umpire neglects or refuses to act or is unable to act or dies and the arbitration agreement does not show that the vacancy was not intended to be provided and the parties or arbitrators, as the case may be, do not provide the vacancy; or
- where parties or arbitrators are required to nominate an umpire and not nominate an umpire;

Any party may, as the case may be, serve the other parties or the arbitrators with a written notice to agree to an appointment or to provide a vacancy. If the appointment is not made within fifteen clear days after the notice is served, the court may, at the request of the party giving the notice and after giving the other parties the opportunity to be heard, appoint an arbitrator or arbitrators or umpire, as the case may be, who has the power to act in the reference and make an award as if he or she were appointed.

21.18.8 Arbitration Award

An arbitration award (or arbitral award) is a ruling on the merits of an arbitration tribunal and is analogous to a court of law ruling. It is referred to as an " award" even if the claims of the whole claimant fail (and therefore no money has to be paid by either party) or the award is nonmonetary in nature.

In most jurisdictions, several subcategories of awards exist:

1. A **provisional award** is an award on a provisional basis subject to the final determination of the merits.
2. A **partial award** is an award of only part of the claims or cross-claims, or a determination between the parties of only certain issues. This makes it possible for the parties to solve or continue to arbitrate (or dispute) the remaining issues.
3. An **agreed award** is usually in the form of settlement (equivalent to a judgment by consent) between the parties to their dispute. However, by incorporating the settlement into an award, it can have a number of advantages.

4. A **reasoned award** is not a subcategory of award, but is used to describe an award in which the court explains its reasons for its decision.
5. An **additional award** is an award made by the tribunal, on its own initiative or at the request of a party, in respect of any claim submitted to the tribunal but not resolved by the main award.
6. A **draft award** is not an award as such and does not bind the parties until the tribunal has confirmed it.

Most countries in the world allow arbitration awards to be "challenged" in court, although they usually restrict the circumstances under which such challenges may arise. The two most commonly accepted challenges are as follows:

1. the tribunal was not competent to award; or
2. serious irregularity on the part of the tribunal.

Awards for arbitration are not justifiable. Distinguish from an "expert determination" where the expert determines a factual matter (which is usually not subject to any form of appeal, except in cases of obvious prejudice or manifest error or mistrust). Furthermore, although not as a challenge, many countries allow appeals on a legal basis (although practically no countries allow appeals to be made in relation to factual findings). This right is normally restricted so as not to undermine the commercial effectiveness of arbitration.

21.18.9 Advantages of Arbitration

- It is private; although not necessarily confidential, there is no public record of any proceedings.
- Speed, though this depends heavily on the way the arbitrator conducts the arbitrage.
- The parties may agree with the relevant expertise on the matter on the arbitrator. The judgment of the arbitrators can be enforced as judgment of the court.

21.18.10 Disadvantages of Arbitration

- The parties must bear both the arbitrator and the venue's costs.
- Sometimes arbitration simply mimics court proceedings and you do not benefit from informality and speed.
- Limited powers of compulsion or penalty if one party fails to comply with the directions of the arbitrator, which can significantly slow the process down.
- The arbitrator has no power to take interim measures, such as property preservation.
- Limited right to appeal.

21.19 ACCOUNTS

In every business, the accounts play an important role. In construction works also, it is extremely necessary to maintain a proper record of the materials purchased and used, and also, the payments made to the labor. In case of a public body, a special officer is appointed to make purchases for the department and then to distribute these materials as and when required. The officers entrusted with such duties should be extremely careful in maintaining proper record of all the purchases made and of all the materials supplied. Following points are to be noted:

1. **Receipts**: The receipts of goods purchased or supplied should be kept in order of occurrence and should be strictly in accordance with the prevailing rules. The received materials should be examined, counted, weighted, or measured as the case may be when delivery is made and an acknowledgement receipt should be approved by the person responsible.

2. **Fictitious adjustments**: The term fictitious adjustment is used to mean the showing of transaction although there may not be actual necessity of initiating the transaction.

3. **Accounting procedure of stores**: The cost of each item is ultimately debited the final head of the account concerned or the particular work for which it is required is the basic principle of the accounts of stores. In case of the materials, tools and equipment, road metals, machines, and such other items which are required for a specific work, such booking is possible immediately at the time of acquisition. But in case of purchase of cement, steel, sand, etc., required for general use, it is not possible to debit the cost to the proper head of account. At the time cost is temporarily debited to suspense head of account "Stock."

4. **Suspense heads**: Suspense heads are reserved for the temporary booking of the transactions.

5. **Suspense subheads**: The suspense head is divided into the following five subheads:
   i. Purchases
   ii. Stock
   iii. Miscellaneous public works (PW) advances
   iv. London stores
   v. Workshop suspense

6. **Classes of stores**: The stores are divided into the following four classes:
   i. Stock of general store
   ii. Materials charged direct to works
   iii. Road metal
   iv. Tools and plants

7. **Reserve limit of stock**: The financial limit up to which the materials in a particular store of the division can be stored is known as the reserve limit of stock and as soon as this limit is crossed, the officer concerned should make arrangements to get approval of the proper authority to temporarily increase the financial limit or he has to stop making new purchases.

8. **Issue rate**: The rate fixed for issue of each article from the store is known as its issue rate. This issue rate is fixed on the basis of no profit, no loss, and should include all the expenditure up to the point of issue of the article. Thus, the issue rate is derived by keeping in view the actual cost of the article, transport charges, overhead charges, losses due to depreciation and wastage, storage charges, etc.

9. **Subheads of stock**: For the purpose of keeping suspense account of stock, the subheads of stock are given below with their appropriate items:
   • Building materials: cement, sand, bricks, lime, asbestos cement (AC) sheets, etc.
   • Fuel: steam, coal, soft coke, charcoal, fire wood, etc.
   • House fittings: hinges, handles, tower bolts, steel windows, glass panes, etc.
   • Land kilns: hiring charges of land and other expenditure connecting departmental kilns for bricks and lime, etc.
   • Manufacture: the items which are manufactured in the departmental workshop.
   • Metals: MS bars, tor steel, rolled steel (RS) joist channel sections, expanded metal.
   • Miscellaneous: bitumen, bleaching powder, empty bags, empty drums, hume pipe, S.W. pipe, etc.
   • Painters stores: red oxide paint, varnish, lead paint, enamel paint, turpentine, etc.
   • Small stores: nails, screws, bolts and nuts, battery cells, etc.
   • Storage: rent and maintenance of storage godown, salary of watchman, etc.
   • Timber: teak wood, sal wood, plywood, deodar, sesame, ballies, wooden posts, etc.

10. **Storage rate**: The expenses to be incurred for preserving the materials include rent, electricity charges, staff for handling the goods, cleaning the godown, etc., to reimburse the storage expenditure, the total amount to be spent per year, and total issues of the year are worked out and then a percentage is decided to increase the issue rate in such a way that the end of the year, there is neither profit nor loss on storage expenditure. Such percentage is known as the storage rate and it is applied to all items of stock.

11. **Supervision charges**: These are the charges over and above storage rates in respect of stock materials sold or transferred and are intended to cover such items of expenditure and are neither entered in their book value nor included in the storage rates.

12. **Market rate**: The term market rate means the rate of an item at the store godown from the public market at a given time. The market rate shall include carting, incidental charges, depreciation, and reasonable provision of wastage.

13. **Indent**: The materials from the store are procured by what is known as the process of indenting. The indents are in the form of a book and they are serially numbered. Usually, the form consists of three parts- counterfoil, indent, and invoice.

14. **Stock taking**: The stock of materials in a store should be checked at least once in a year by counting, weighing, or measuring. The surplus or shortage of any materials should be properly adjusted by the sanction of the competent authority.

15. **Quantity accounts**: Usually, at the end of every month, an abstract is prepared indicating the materials received and the materials issued, and thus, this monthly stock account will be of very much use to the indenting officer who will then indent for the materials available in the store.

16. **Value accounts**: The value accounts show the money value of the materials. The amount spent to purchase the materials is entered in the payment column. The amount recovered by issuing the materials at rates determined by competent authorities is entered on the receipt column. At the end of each half year, the value of the materials on hand is determined by assuming reasonable accurate prices and the amount is put up in the receipt column. Then, the difference between the columns indicates profit or loss and it serves as a sufficient guide in re-adjusting the issue prices of the materials.

17. **Stock account**: The term stock account is used to mean quantity accounts and value account of materials in the stores. The former is maintained by the subdivisional office and the latter is kept by the divisional office.

18. **Bin card**: A bin card keeps a record of receipts, issues, and running balance of certain items of stock, especially of fitting items. Immediately after transactions, the entries of receipts or entries of issues are posted by bin card from the register of "Stock Receipts and Issues" and balances are worked out and also entered in bin card.

19. **Road metal**: For new roads or for maintaining existing roads, the road metal in the form of rubbles, metal, kankar, murum, etc., is required and the indenting officer should make this point clear in his indent. This will help the issuing officer to charge the costs of the road metal to the proper head.

20. **Tools and plants**: In the store, proper tools and plants are also kept and issued as they are required by the various indenting officers or contractors as the case may be. A proper record should be maintained for the receipts and issues of the various tools and a monthly balance should be prepared. At the end of each year, the stock of tools and plants should be checked and in case of any deficiency, proper adjustment should be made by getting approval of the competent authority.

21. **Rate contract**: The superintendent of store purchase department of Government invites the tenders of the manufactured items or materials which are commonly required by the various departments, fixes the rates, and makes contract with the manufacturer or supplier, which is known as "rate contract." The final rate contract with the rate of an item, its specification, description, etc., is notified in the Gazette of information to all departments.

22. **Survey reports**: When any item of store is in deficit or has to be written off on account of becoming unserviceable, survey report is prepared, mentioning the cause, and is submitted to the competent authority for sanction of the write off. After the write-off is sanctioned, the unserviceable items are disposed by the public auction.

23. **Sale account**: When surplus materials or items, unserviceable machinery or items, etc., are sold by public auction, the sale account is prepared. Sale account contains the description of the items, quantity, book value, amount of sale, etc.

21.19.1 ISSUE NOTES

The issue notes are provided in the form of books or pads and they are of convenient size so that they can be easily carried in pockets. Each note or book or pad should be numbered consequently and it should be issued to authorized responsible persons only. The receiver of the issue note should sign for every note given to him. All unused issue notes should be kept under lock and key by a responsible officer.

21.19.2 VOUCHERS

A voucher is a document used by the accounting department of a company to collect and file all the supporting documents necessary to approve the payment of a liability. The voucher is an internal accounting check that ensures that all payments are properly authorized and that the purchased goods or services are actually received. The voucher should include a brief description of the work or item to be paid. If the voucher is for recoverable payment, the name of the contractor or other person from whom the expenditure can be recovered should be included.

21.19.3 ADMINISTRATIVE APPROVAL AND TECHNICAL SANCTION

For every work, it is necessary to obtain, in the first instances, the concurrence of the competent authority of the administrative department requiring the work. The formal acceptance of the proposals by the authority is termed as *Administrative Approval* of the work. It is the duty of the engineering department requiring the work by the administration to obtain the requisite approval to it. An approximate estimate and such preliminary plans are necessary to explain the proposals are submitted by an engineering department to the administration to obtain administrative approval to take up the work within the sanctioned amount. After receiving the administrative approval, detailed drawings, design, and the estimated cost, etc., are prepared by the engineering department and submitted to the administrative department for sanction.

After receipt of administrative approval and expenditure sanction, a detailed estimate is further sanctioned by a competent technical authority of the engineering department empowered by the Government, which ensures that the proposals are structurally sound and the estimate is accurately calculated based on adequate data. Such sanction is known as *Technical Sanction* and should be taken before inviting tenders to execute the work.

21.19.4 MEASUREMENT BOOK

It is a book in which the measurements of all sort of construction work undertaken by the department and material supplied are recorded. The measurements pertaining to the normal periodical services are entered in this book by a main person like junior engineer/overseer (JE/OS) and offer due checking accuracy of the measurements which is certified by the assistant executive engineer/assistant Garison engineer/assistant engineer/senior divisional officer (AEE/AGE/AE/SDO) concerned.

The important features of the Measurement Books are as follows:

- The calculations of periodical services are scrutinized and certified.
- The quantity of work done is recorded in the bill of the contractor.
- Measurements are taken accurately. Preferably steel tape is used for taking measurements.
- All measurements are recorded in ink.
- No entry is allowed to be erased.
- If correction is required, it is to be properly authenticated.
- Measurements are taken in the presence of the contractor and his signature is taken in the measurement book, in token of having agreed to the measurements.
- Measurements are continuously recorded without interruption.
- Measurement books form a basis of payments of all the items of work which can be measured.
- It forms the original record of all works/supplies which are measurable.
- It is an auditable document.
- The measurement book contains:
  1. Work's name
  2. Contractor name
  3. Measurement date
  4. Location of location
  5. Work order date
  6. Measurement number

21.20 CONSTRUCTION MANAGEMENT AND PLANNING

Construction Project Management (CM) is a professional service that uses specialized project management techniques from the beginning to the end to supervise the planning, design, and construction of a project. The purpose of CM is to control the time, cost, and quality of a project. CM supports all project delivery systems, including design-bid-build, design-build, CM At-Risk and Public Private Partnerships. Professional construction managers can be reserved for large, large-scale, high-budget undertakings (commercial real estate, transport infrastructure, industrial facilities, and military infrastructure).

The functions of construction management typically include the following:

- Specifying project objectives and plans, including scope delineation, budgeting, planning, performance requirements, and selection of participants in the project.
- Maximize resource efficiency through labor, material, and equipment procurement.
- Implementation of different operations through proper planning, design, estimation, contracting, and construction coordination and control over the whole process.
- Developing effective communication and conflict resolution mechanisms.

21.20.1 NEED FOR CONSTRUCTION MANAGEMENT

The construction management is required for the following important factors during construction:

1. **Co-ordination between different agencies**: An effective construction management is required to co-ordinate between various agencies engaged during construction and organize the inputs in an efficient way to maximize the outputs at minimum cost.
2. **Development of manpower and machinery**: The construction management aims to co-ordinate and organize sufficient and efficient manpower and machinery. Also, new tools of management such as Bar charts, Critical Path Method (CPM), and Program Evaluation Review Technique (PERT) are available for planning and monitoring construction activities.
3. **Economy in construction:** The construction management also works out the best alternative to get the best products or result at minimum cost.
4. **Speed of construction**: For the speed of construction also, we need the construction management as it gives the best alternative to complete the project in optimum time and cost using the latest management techniques.
5. **Quality control of materials and workmanship**: To control the quality of material and workmanship, construction management is required.

21.20.2 FACTORS AFFECTING CONSTRUCTION MANAGEMENT AND PLANNING

1. Time
2. Skilled and unskilled labor
3. Machinery

4. Materials

5. Money

The above factors can be remembered in short as T+4M.

21.20.3 METHODS FOR PLANNING CONSTRUCTION ACTIVITY

Whenever a construction project is initiated, it is desirable to fix up a target date of completion and to meet the target; it becomes essential to plan all the activities that comprise the project in a systematic manner.

Following are the different methods adopted for planning construction activities:

1. Gantt bar charts
2. Network technique
   i. CPM
   ii. PERT
3. Time-grid diagram

21.20.3.1 Gantt Bar Charts

The Gantt chart is a horizontal bar chart that Henry L developed as a control tool for production in 1917. A Gantt chart is often used in project management to illustrate a schedule that helps plan, coordinate, and track specific tasks in a project.

The Gantt chart is a type of bar chart used to illustrate a project schedule, including activity start and end dates and a summary of project activities. This is a common technique used to plan construction activities by project managers or quantity surveyors. Bar chart shows how a contractor can plan construction activities, including ordering supplies, planning subcontractors, and administrative tasks related to work.

A Gantt chart offers the following advantages:

- Bar charts can be used to represent project phases and activities so that the general public can understand them.
- It can be helpful to show the critical points on the chart with bold or colored bar outlines.
- An updated Gantt chart helps to manage the project and to solve problems on schedule.
- The computer software can make it easier to build and update a Gantt chart.
- You really do not need specialized software to create a Gantt chart, because it can easily be prepared using Excel or similar software.

How to prepare a Gantt chart for Construction Project

The Gantt chart lists the building activities or tasks that must be carried out in one column. A second column indicates how long each activity will be completed. The dates for the construction projects in the horizontal row are shown at the top of the chart. A line or bar is drawn on the right of each activity from the start date to the completion date of the activity. All activities in the first column are carried out in this way across the chart rows, with the bar for each activity starting as soon as possible. Many activities must be carried out in sequence and one activity must be completed before the next activity can be started. Some tasks are autonomously completed. The Gantt chart appears to be a horizontal bar chart in most cases. Each task associated with the project is taken into account in the chart, because the Gantt chart is concerned with the most efficient execution of the project. This also helps to define the amount of time needed to perform each task. The chart also helps to define tasks which can be completed simultaneously and which cannot be completed until other tasks are completed. This function helps to project the sequence and the length of each project task. Figure 21.1 shows the example of a Gantt chart for a Residential Construction Work

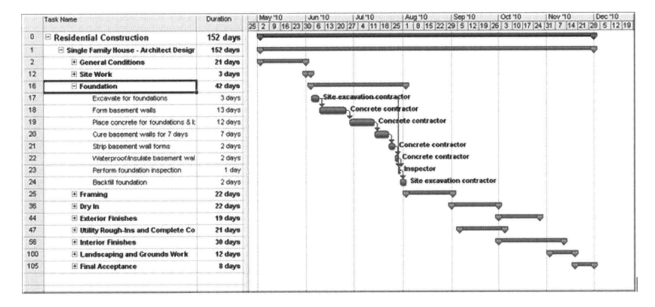

FIGURE 21.1 Example of a Gantt chart for a residential building.

21.20.3.2 Critical Path Method

The CPM is a step-by-step process planning project management technique that identifies critical and noncritical tasks, prevents time frame problems, and processes bottlenecks. Complex projects require a range of activities, some of which must be carried out sequentially and others can be carried out in conjunction with other activities. This series and parallel tasks collection can be modelled on a network. CPM models project activities and events as a network. Activities are shown as network nodes and events which mean the start or end of activities are shown as arcs or lines between nodes. Figure 21.2 shows an example of a CPM network diagram:

21.10.3.2.1   Steps in CPM Project Planning

- Specify the different activities
- Determine the activities' sequence
- Draw a diagram of the network
- Estimate time to complete each activity
- Identify the critical path (longest network path)
- When the project progresses, update the CPM diagram

Critical path: The critical path is the longest path via the network. The importance of the critical path is that the activities which lie there cannot be delayed without the project being delayed. The critical path can be determined by four parameters for each activity:

- ES—earliest starting time: the earliest starting time for the activity, given that its precedent activities must first be completed.
- EF—the earliest time to complete the activity, equal to the earliest start time plus the time required to complete the activity.
- LF—the latest time to finish, the latest time to complete the activity without delaying the project.
- LS—latest start time, equal to the latest finish time minus the time needed to finish the activity.

The slack time for an activity is the time between its earliest and latest start time, or between its earliest and latest finish time. Slack is the amount of time that an activity can be delayed past its earliest start or earliest finish without delaying the project. The critical path is the path through the project network in which none of the activities have slack, that is, the path for which ES=LS and EF=LF for all activities in the path. A delay in the critical path delays the project. Similarly, to accelerate the project, it is necessary to reduce the total time required for the activities in the critical path.

21.10.3.2.2   CPM Benefits

- Provides the project's graphical view.
- The time required to complete the project is predicted.
- Shows which activities are critical to schedule maintenance and not.

21.10.3.2.3   CPM Limitations

Although CPM is easy to understand and use, it does not take into account the time changes that can have a significant impact on the completion time of a complex project. CPM has been developed for complex but rather routine projects with minimum uncertainty in the completion times of the project. There is more uncertainty in the completion times for less routine projects, and this uncertainty limits its usefulness.

21.20.3.3   Program Evaluation and Review Technique

The PERT is a network model that allows random activity times to be completed. PERT was developed for the USA in the late 1950s. Polaris Navy Project with thousands of contractors. It has the potential to reduce both the time and the cost of a project.

An activity in a project is a task to be carried out and an event is a milestone in the completion of one or more activities. All its predecessor activities must be completed before an activity can begin. Project network models represent arc and node activities and milestones. PERT is typically represented as an activity on the arc network, where the activities on the

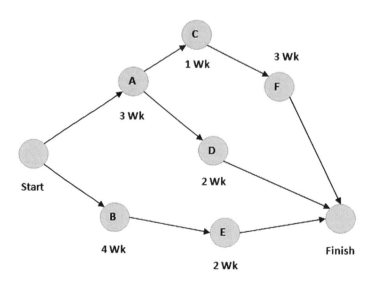

FIGURE 21.2   CPM diagram.

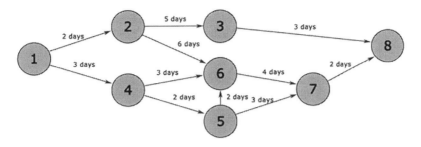

FIGURE 21.3 PERT chart.

lines and milestones on the nodes are represented. Figure 21.3 shows a simple PERT diagram example.

The milestones are generally numbered so that the end node of an activity is greater than the start node. Increasing the numbers by 10 allows the insertion of new numbers without changing the numbering of the whole diagram. The activities in the diagram above are marked with letters and the time expected to complete the activity.

21.20.3.3.1 Steps in the PERT Planning Process
The following steps are covered by PERT planning:

1. Identify the specific milestones and activities
2. Determine the right activity sequence
3. Build a network diagram
4. Estimate the duration of each activity
5. Identify the critical path
6. Update the PERT chart with progress of the project

21.20.3.3.2 Activity Completion Times
The model usually includes three time estimates for each activity:

- **Optimistic time (OT)**: usually the shortest time to complete the activity. This is believed by an inexperienced manager.
- **Most likely time (MT)**: maximum probability of completion time. This is different from the time planned. Experienced managers have an incredible way to estimate the actual data from previous estimation errors very closely.
- **Pessimistic time (PT)**: the longest time it could require an activity. The expected time for each activity can be approximated using the following weighted average:

$$\text{Expected time} = (OT + 4 \times MT + PT) / 6 \qquad (21.3)$$

This expected time might be displayed on the network diagram.

Variance for each activity is given by:

$$[(PT - OT) / 6]2 \qquad (21.4)$$

Critical path determination: The critical path is determined by adding the time for each sequence of activities and determining the longest path of the project. The critical path determines the total project time required. If the critical path is not immediately clear, it can be helpful for each activity to determine the following four quantities:

- ES—earliest time to start
- EF—earliest time to finish
- LS—latest time to start
- LF—latest time to finish

The variance in the completion time of the project can be calculated by summing up the variances in the completion time of the critical activity. In view of this variation, the probability that the project will be completed by a certain date can be calculated. Since the critical path determines the project's completion date, the project can be accelerated by adding the necessary resources to reduce the time spent on the critical path. Such a project shortening is sometimes referred to as a project crash.

21.20.3.3.3 Benefits of PERT
PERT is useful because the following information is provided:

- Expected completion time of the project
- Completion probability before a specified date
- Critical path activities impacting the completion time directly
- Slow time activities that can lend resources to critical activities
- Start and end dates of activities

21.20.3.3.4 Limitations of PERT
The following are some of the limitations of PERT:

- The time estimates for the activity are rather subjective and depend on judgment. If there is little experience in carrying out an activity, numbers can only be a guess. In other cases, if the person or group conducting the activity estimates the time, the estimate may be prejudiced.
- The underestimation of the completion time of the project due to the criticality of alternative paths is perhaps the most serious.

21.20.3.4 Time-Grid Diagram

In the CPM, the length of an arrow does not indicate the duration of the activity. In the time-grid diagram this aspect is considered.

The essential points of the time-grid diagram are as follows:

1. The length of the activity indicates the duration of that activity.
2. All the arrows indicating activities and float are drawn horizontally. The tail of each arrow starts at the head of the arrow for the immediately preceding activity and then continues to the right side of the following activities.
3. When numbers of activities are starting from one point in the time-grid diagram, they are represented horizontally, but by keeping vertical displacement. This vertical displacement is indicated by vertical dotted line in the diagram. The vertical lines do not indicate any elapsed time but indicate precedence of the activity.
4. When drawing the diagram, it is desirable to draw the critical path at or near the middle of the diagram.

21.20.4 JOB LAYOUT OF CONSTRUCTION SITE

The layout of the job/site can be defined as the allocation of space for material storage, work areas, accommodation units, plant positions, general circulation areas, and access and exit for deliveries and emergency services. This is a scaled drawing of the proposed construction site showing all the relevant features such as entrance and exit points to the site, storage areas for materials, contractors' offices, equipment storage areas such as mixer, bar bending area, labor, toilets, etc. It is also necessity that first aid kits are kept in the site engineer's office.

The following factors should be considered in preparing site layout:

- **Nature, scope, and type of work**: The nature of the layout of the site depends mainly on the type of work to be performed. For example, the site layout for a dam would be much more complicated and extensive as compared to that of a residential building.
- **Topography of the site**: The topography of the site, i.e., location, size, nature of terrain, etc., of the site also affects the site layout. For example, the site layout in case of Rocky Mountains terrain would be vastly different from that in case of a plain terrain.
- **Methods adopted for execution of work**: The method selected for the execution of work deeply affects the nature of site layout; for example, if the work is carried out mostly mechanically, i.e., using

more equipment and less labor, more emphasis has to be placed on providing facilities for the machinery than for the workers.

The following are the advantages of a good layout of work:

- Smooth and cost-effective project work
- Reduces project completion time
- Makes the work of the project more secure
- Waste and deterioration of materials are reduced
- Material transactions are easy, fast, and cost-efficient
- Increases labor and machinery output

21.20.5 STORAGE OF MATERIALS

A building site has many materials at one point. Most of these are usually in their rough condition, which means that they will go through some process before they can be entered into the building to form part of the building. They come in various shapes and can be categorized as follows:

- **Factory goods**: These are mostly off-shelf items; they are unique because they can be easily resold and are therefore easy pilfering targets. The fact that they have unique storage requirements is also delicate.
- **Cement**: The most important attribute to be taken into account in the storage of cement is its chemical reaction when it is in contact with moisture. It should therefore be kept in the shade and on a platform, away from excessive humidity.
- **Ceramics**: These include cupboards, wash basins, tiles, and so on. They are extremely delicate and they break easily. This is also shared by glasses. They should therefore be properly packed in padded cartons and away from areas with a great deal of activity, usually under lock.
- **Ironmongery**: These include locks, hooks, handles, and similar things. They are prone to pilfering due to their small sizes. They should also be kept well locked and only accounted for strictly.
- **Raw materials**: This category is one of the most important items, such as stone, ballast, and sand. These are not prone to the preceding weather and pilfering problems. They have, however, one attribute which is bulky. They take up a lot of space on site and require a generous storage space allocation. They are best stored in bays and contained in stones, for example, in sand.
- **Workshop finished items**: This category also contains semi-finished items, e.g., wood. Items here are generally ready to be installed in works and are mainly intended. Some may have been imported from and measured exactly in foreign countries. This means that damage or loss of such damage leads to very costly replacement work. Examples of this include fittings, wood, roofing materials, etc.

21.20.6 STOCK CONTROL

The scientific methods of stock control have developed and the two such methods which are commonly used for materials management are as follows:

1. ABC analysis
2. VED analysis

21.20.6.1 ABC Analysis

ABC analysis is a categorization technique for inventory items based on their substantial impact on an organization's total expenditure. It provides a solution to faulty inventory management in the purchased items or services.

It is based on the Pareto principle that "80% of the total value of consumption is based on only 20% of the total items." The breakdown suggests that inventories have different values, which means that different tactics and controls are necessary. The categories arrangement is based on their expected value.

ABC analysis is an "inventory categorization method" that divides items into three categories: A, B, and C: "A" contains the "most valuable items" and "C" is the "least valuable items," while "B" contains items ranging from "A" to "C." It aims to concentrate on the critical few (A-pieces) and not the trivial ones (C-pieces). Different items are listed according to their total use in this analysis; unit costs and then total item costs are calculated.

This approach states that items should be rated between A and C by the company when reviewing the inventory, establishing its ratings on the following rules:

i. **A-items**: have the "highest annual value for consumption" of goods, i.e., 70%–80% of the company's annual consumption value. Ironically, it only accounts for 10%–20% of the total inventory items. They require strict inventory control, more protected storage areas, and improved sales forecasts, frequent reordering, weekly or even daily reordering; avoiding storage on A items is a priority.
ii. **B-items**: are the interclass items, having medium consumption value, i.e., 15%–25% of annual consumption value. It consumes around 30% of the total inventory items.
iii. **C-items**: have an annual "lowest consumption value" of goods, i.e., 10%–15% of the annual consumption value. In contrast, it represents 50% of the total inventory.

21.20.6.2 VED Analysis

VED analysis tries to classify the items in three broad categories: vital, essential, and desirable. The analysis classifies items based on their industrial or company criticality.

i. **Vital**: Vital category items are those items without which the company's production or other activities would be stopped or at least dramatically affected.

ii. **Essential**: Essential items are items that have a very high stock—outgoing cost for the company.
iii. **Desirable**: Desirable items are those items whose storage or shortage causes only a minor disruption in the production schedule for a short period of time. The cost is very nominal.

21.20.7 STAGES OF MATERIAL MANAGEMENT

Following are the stages of material management:

1. Register of reputed suppliers
2. Quantities of materials
3. Preparing the supply schedule
4. Attaching priorities
5. Initiating purchase formalities
6. Placing of orders
7. Inspection
8. Acceptance and issue
9. Use of network schedule

21.20.8 DISPOSAL OF SURPLUS MATERIALS

The circumstances which lead to the surplus of building materials at site of project are as follows:

1. Due to reduction in the scope of construction
2. Due to subsequent changes in the design and specifications of the work
3. Excess ordering of materials than required

Following are the two ways of disposal of surplus materials:

1. They are transferred to the main store of the organization or they are diverted to other site where they can be fruitfully used. The cartage of the surplus materials is suitably adjusted in the books of account.
2. When the surplus material is unserviceable or it cannot be disposed off as mentioned above, it is auctioned out in public or is sold to private parties after getting necessary permission from the concerned responsible authority of the organization.

21.21 COST, PRICE, AND VALUE

The term cost, price, and value are used for the exchange of goods or products. It is imperative that a real estate valuer knows clearly the meaning of these three terms.

21.21.1 COST

The term cost indicates the actual amount incurred in the production of a commodity that has a certain value. Such value loss is technically referred to as the depreciation.

The expenditure or charges incurred in terms of quantitative as well as qualitative labor and capital for production of a commodity are termed as the costs of the commodity.

The expenditures or charges represented directly in the commodity produced are called the *prime costs* and other expenditures or charges like rent, management, salaries and services, depreciation, etc., represented indirectly in the production of the commodity are called the *supplementary costs*.

21.21.2 PRICE

The term price is used to indicate the cost of the commodity plus profit of the manufacturer. As labor and capital are required to produce a commodity, the manufacturer is entitled to have some reward over and above the cost of commodity. The price of a commodity is determined by the conditions of supply and demand on the market for that particular commodity, and a commodity can therefore be sold above or below its production cost.

21.21.3 VALUE

In economics, the term value is defined as the corresponding exchange of one commodity into any other commodity. Value in the most basic sense can be referred to as "Real Value" or "Actual Value." This is the measure of value based solely on the utility derived from a product or service consumption.

21.21.3.1 Purpose of Valuation

Valuation is the estimation technique or the determination of the fair price or value of property such as building, factory, other engineering structures of different types, land, etc. The present value (PV) of the property can be determined by its sales price or its income or rent. The main purposes of evaluation are as follows:

1. **Buying or selling property**: Its valuation is required when it is required to buy or sell a property.
2. **Taxation**: The valuation of the property tax is required. Taxes may be municipal taxes, wealth taxes, property taxes, etc.. and all taxes are subject to property valuation.
3. **Rent fixation**: Valuation is required to determine the rent of a property. Rent is usually determined by a certain percentage of the valuation (6%–10%).
4. **Loan security or mortgage security**: Its valuation is required when the loans are taken against the security of the property.
5. **Compulsory acquisition**: Wherever the property is acquired, the owner shall be compensated by law. To determine the amount of property compensation valuation is necessary.

21.21.3.2 Different Forms of Value

Following are the different terms or names which are used in connection with the value of a property:

1. Accommodation value
2. Annual value
3. Book value
4. Distress value
5. Market value
6. Monopoly value
7. Potential value
8. Replacement value
9. Salvage value
10. Scrap value
11. Sentimental value
12. Speculative value

1. **Accommodation value**: As the city or town develops, the surrounding agricultural area is to be converted into the accommodation land. For this purpose, the owner of the land has to obtain the permission from the competent authority to convert his agricultural land into non-agricultural or accommodation land. Such accommodation land possesses building potential value, and hence, its value lies between the values of nearby building land and agricultural land. Thus, the accommodation value is greater than that of agricultural land and less than that of building land.
2. **Annual value**: The local authority has to decide the annual value of the property so that taxes can be calculated on that basis. Such annual value of the property has to be fixed by observing the principles of rating valuation.
3. **Book value**: The book value is the amount shown in the account book after the required depreciation has been allowed. The book value of the property in a given year shall be the original cost minus the depreciation year amount. The book value is only scrap value at the end of the utility period of the property.
4. **Distress value**: A property is said to have distress value when it can fetch lower value than the market value. The distress value is developed due to various reasons such as:
   i. Fear of war, riots, earthquakes, etc.
   ii. Financial difficulties of seller
   iii. Intention to favor purchaser
5. **Market value**: It is the net amount that a seller, having technically verified its characteristics and assuming an appropriate marketing, could reasonably receive for the sale of a property on the date of valuation, if at least a potential buyer and a seller are correctly informed, both acting freely for an economic interest and without a particular condition during the transaction.
6. **Monopoly value**: In some cases, the property possesses certain advantages with respect to the adjoining properties due to its size, shape, frontage, location, etc. The owner of such property may demand a fancy price for that property. Such a price is known as the monopoly value of the property and it does not represent the market value of the property.
7. **Potential value**: The potential value is used to indicate the potential possibilities of the property when developed in its most advantageous manner.

8. **Replacement value**: The cost to be incurred to replace the property, either fully or in part, at the prevailing market rates for labor and materials, is known as the replacement value.

9. **Salvage value**: It is the value of the period without being decommissioned. We can sell it as a second handle.

10. **Scrape value**: Scrape value is the dismantled material value. This means that after dismantling we will get steel, brick, wood, etc. The scrape value is metal or dismantle parts in the case of machines. The scrape value is generally about 10% of total construction costs. Scrape value=sale of material for use – cost of dismantling and removing the rubbish.

11. **Sentimental value**: Sentimental value is the value of an object derived from its personal or emotional value rather than its material value. It is the excessive value of opinion based on what the sellers want. The fair market value differs from the sentimental value, as both parties must agree to its value.

12. **Speculative value**: The speculative value is the value that an investor believes an investment will achieve.

21.22 MORTGAGE, FREEHOLD, AND LEASEHOLD PROPERTY

21.22.1 Types of Interests

Following are the two principal interests or rights in land or land and buildings with which the valuer is concerned:

1. Freehold interests
2. Leasehold interests

21.22.2 Freehold Interests

Freehold property may be defined as any property that is "free of holding" of an entity other than the owner. Therefore, the owner of such a property has free ownership for life and can use the property for any purpose in compliance with local regulations. The sale of a freehold property does not require state consent and therefore requires less paperwork, which makes it more expensive than the property of the leasehold. A legal guardian can inherit a freehold property. A freehold property may be transferred by registering a sales certificate.

21.22.3 Leasehold Interests

The freeholder may give permission to any other person to use his property for a certain number of years. This is known as the *granting property on lease* and it is subject to some annual payment and other conditions mentioned in the lease document. The person who takes the lease is called the *lessee* or the *leaseholder* and the owner who grants such lease is known as the *lessor*.

1. **Reasons for creating leasehold interest**: The above arrangement is adopted by the lessor under the following circumstances:

   • He is in possession of a large piece of valuable land and it is not possible to find out a suitable purchaser for the same.

   • The owner is in possession of the property for a petty long time and he is not sentimentally prepared to transfer the ownership to somebody else.

   • The sale of property has been made virtually legally impossible due to various reasons such as restrictions imposed by the will of the deceased.

   • The tax planning may encourage the owner to enter into a lease contract. The owner has either to pay capital gain tax (to be worked out on the basis of sale price) or the income tax on lease rent received as per the terms of lease document. In some cases, the capital gain tax proves to be substantial or prohibitive in quantum.

2. **Nature of leasehold interest**: It is quite evident that the leaseholder has got restricted powers to use the property. In certain cases, the leaseholder is required to maintain the property in good condition and redecorate it at regular intervals. Also, the leaseholder is required to obtain the consent of the lessor before carrying out any alterations or additions or modifications in the property. Hence, owing to such restrictions on the part of the lessee, the leaseholds are found to be less appealing or less attractive than freeholds from the investment point of view, and therefore, the investors in leasehold properties require higher yields on their investments.

3. **Renewal and extension of leases**: A renewal lease is a new lease agreement. Under the renewal of the lease, there is a legal moment between the expiry of the original term and the beginning of the renewal period. In contrast to a lease renewal, an extension of the original lease is a continuation without interruption of the original lease. Some of the main differences between the consequences of the renewal and the extension of the lease are as follows:

   • A term of renewal may extinguish or limit personal covenants and rights that benefit the landlord or the tenant, unless specified in the wording, whereas they are likely to continue during an extension of the term (personal rights include a period of free rent, a tenant improvement allowance, or a right of first refusal held by the tenant in additional space).

   • In the absence of a language to the contrary, a renewal term may result in the improvement of the tenant being taken into account in determining the fair market rent for the renewal period, where an extension of the term would probably not

- Where a lease has been assigned, there is a risk that the liability of the original tenant will cease when the original term ends, but with an extension of term, the previous tenant may remain liable unless the lease provides for its release.
- To the extent that a breach by the tenant gives the landlord the right to terminate the lease, the landlord may terminate the lease for a breach that occurred during the original term during the extension term.

4. **Theory of sinking fund**: The fund which the leaseholder will set aside annually from his annual income from the property so that when the lease terminates, he will have sufficient savings which will permit him to purchase another property having a similar income flow. This fund is known as the *sinking fund* as it grows in size as the leasehold interest sinks in value. Thus, the main aim of sinking fund is to recoup the capital which is invested in a wasting asset.

   The sinking fund theory for terminable incomes is often criticized on the following two grounds:

   - The theory is unrealistic in practice as no one would be ready to invest money to earn very low rate of interest as granted on sinking fund. This argument is not valid in the sense that amounts of sinking fund are very small and an investor has to make choice between higher yields and absolute security of his capital.
   - The allowance for recoupment of capital should be sufficient to come over the impact of inflation. To avoid the provision of a large sinking fund, it will be reasonable for the leaseholder to accept a lower yield on his investment and that will partly eliminate the risk of inflationary trends.

5. **Lessor and lessee**:
   i. Rights and liabilities of the lessor:
      - The lessor shall reveal to the lessee all material defects in the property. He must also reveal to the lessee all known latent defects in the property. However, he does not need to reveal the defects that can be discovered with ordinary care. This is based on the "Caveat Emptor" principle.
      - The lessor is bound to place the leasee in the possession of the property leased. However, the lessee must first ask the lessor to advise him in possession. If the tenant does not do this and does not take possession, he must not pay the rent for the rental.
      - The lessor shall indemnify the lessee for all losses incurred by the latter because of the interruption of the property's enjoyment. In other words, it is an implicit agreement that leasee can enjoy the property freely and quietly. The title covenant is with the land.
      - If the lessor transfers the property leased, the lessor shall be entitled to all the rights and obligations of the lessor. However, in order to make the transferee responsible, the lessee must choose to treat the transferee as the person responsible to him. In addition, the transferor is not entitled to any rental delays due prior to the transfer.

   ii. Rights and Liabilities of the Lessee:

      - If any accession to the property is made during the continuation of the lease, such accessed property must also be returned to the lessor with the main property. In other words, "accessories follow the principle" in the countries accessed.
      - If a part of the property is destroyed by fire, the lease shall be void at the leasee's option. Such destruction of the property must not, however, be caused by the misdeed of the lessee.
      - If the lessor fails to make adequate repairs by the lessee even after notice. The tenant can make these repairs himself and can deduct the costs of such repairs from the rental.
      - If the lessor fails to pay the income, taxes, etc., the lessee may make such payment and with interest deduct such payments from the rent or recover the interest from the lessor.
      - After the termination of the lease, the lessee can remove all the things, which he had attached to the earth but they must leave the property in the same state as he received it.
      - If the lease is terminated by certain uncertain events, the lessee or his legal representatives may obtain and carry all the crops harvested by them.
      - In the absence of any express provision in the lease deed, the lessee may sublease or mortgage the property.
      - The lessor shall reveal to the lessor any material increase in the value of the property known only to him.
      - The lessor is bound to pay the premium or the rent to the lessor or his agent at the proper time or place.
      - The landlord is bound to maintain the property in good condition. The property must be returned in the same condition as it was received. He must allow the lessor and his agent to enter the country and at all reasonable times inspect the condition of the land. He must give all the defense notice.
      - If the lessee becomes aware of any encroachment on the property, he must give immediate notice to the lessor for taking steps against such encroachment.

- The leasee shall not build a permanent structure on the property leased without the consent of the leasor. But for agricultural purposes, even without the consent of the leasor, it can build such a permanent structure.
- After the leasee has been terminated, the leasee shall return the leased property to the leasor's possession.

Forms of lease: Following are the five principal forms of lease:

i. Building lease
ii. Occupation lease
iii. Sublease
iv. Life lease
v. Perpetual lease

i. **Building lease**: If the owner of a freehold plot of land commanding a good site for the construction of a building is unable to develop his land, he gives the plot of land on lease to somebody who is eager to construct a suitable building on that particular piece of land. In such cases, the leaseholder pays yearly ground rent to the freeholder. The leaseholder then proceeds to erect the building, and after the construction work is completed, he maintains it in good condition and he also bears all the expenses in connection with the building.

The *ground rent* is the amount of rental value of the open plot of land at the time of granting the lease. After the construction of the building, the leaseholder will, if rented, get some return from the property. The difference between the available rent from the property and the ground rent paid by the leaseholder represents the *leaseholder's net income or profit*.

It is clear that the leaseholder will have to invest a good amount in the construction of the building over the plot of land and hence, such leases will not become popular, if sufficient time is not given to the leaseholder to enjoy the fruits of such investment. The building lease, therefore, is normally granted for 99 or 999 years. However, it is interesting to note that, unless contrary is provided, the lessor at the end of lease period has not only right on his land, but also on the building constructed and maintained by the leaseholder during the tenure of the lease.

ii. **Occupation lease**: In case of an occupation lease, the owner constructs a suitable structure on the land and together with land; he grants lease for certain occupation such as residential, factory, shop, office, etc.

Following points are to be noted in case of the occupation lease:

- Period of lease: The period of the occupation lease will depend on the use of the structure. For instance, if considerable amount is to be spent on the decoration at the start of the lease, as in case of show room, the lease period will be quite long, say 20–30 years. In case of residential buildings, the lease period is about 2–5 years or so.
- Rack rent and head rent: If the rent agreed in an occupation lease is very nearly equal to the prevailing rent for such similar land and structure, it is known as the rack rent. But if the agreed rent is less than the prevailing rent for such similar land and structure, it is known as the head rent. Consequently, the difference between the rack rent and head rent will represent the leaseholder's interest.

iii. **Sublease**: Depending on the terms and conditions of the original lease, the tenant may grant a lease to other persons. In all such cases, the original leaseholder may grant a sublease to other persons. In all such cases, the original leaseholder becomes the head lease for the sublease holder. It should be noted that the sublease can be granted only for a period which is less than the original lease period. Usually a gap of a few days is kept between the periods of the original lease and the sublease so as to facilitate the reversion of the property.

iv. **Life lease**: In case of a life lease, the lease is granted for the duration of the life or lives of one or more persons. The lease comes to an end on the death of such person or persons.

v. **Perpetual lease**: In case of a perpetual lease, the lease is granted for a year or a number of years. But it is renewable from time to time at the discretion or will or desire of the leaseholder in perpetuity. The lessee is however required to respect the provisions of the lease document and as long as there is no violation of lease conditions, the lease document and as long as there is no violation of lease conditions, the lessor cannot terminate the lease.

6. **Lease or licence**: A lease or license is a contractual agreement between the council (the lessor or the licensee) and another party (the lessee or the licensee) which binds both parties to the agreement. The individual circumstances surrounding the land and buildings and the needs of the users will help guide the appropriateness of leasing or licensing. The key differences between a lease and a license are explained below.

Lease:

- A lease is a transfer by the lessee of the right to enjoyment (exclusive possession) of that property for a certain period of time subject to the terms of the lease agreement.
- For a fixed period (term) a lease grants exclusive possession.

- A lease creates a land interest that can be transferred to the leasee during the lease period.
- A lease may be transferred (assigned) to another party and binding on a new land owner if registered on the title.
- A lease cannot be revoked (other than under any conditions laid down in the lease (e.g. A clause for redevelopment)).

License:
- A license shall be the granting of permission to use the land for a fee, subject to the conditions laid down in the license.
- There is no license to grant exclusive possession.
- A license shall not create or transfer land interests.
- A license cannot be transferred.
- A license can be revoked

21.22.4 MORTGAGE

When the owner of freehold or leasehold property grants an interest in his property to other person against the security of a loan advanced by that person, he is said to perform a *mortgage deed*. The person who grants such interest is known as the *mortgagor* and the person who advances loan for such interest is known as the *mortgagee*.

In the mortgage deed, the mortgagor agrees to repay the loan, and when the loan is fully repaid, he has got a power to recover his property from the mortgagee. This is termed as the equity of redemption.

Following points should be noted in this connection:

1. **Amount of loan**: The amount of the loan which the mortgagee will advance to the mortgagor will depend on the capitalized value of the property. Usually, about 70 to 80 per cent of such value is advanced while 50 to 60 per cent will be a safe limit for such purpose.
2. **Insurance**: It is desirable to have an insurance of the property in the name of both the mortgagor and the mortgagee. The insurance policy is kept with the mortgagee and the mortgagor is required to pay the premium regularly and to show the receipts of such payments to the mortgagee.
3. **Leasehold property**: It is possible to have a mortgage deed for a leasehold property. But in such cases, care should be taken to revalue the property periodically so as to see that the amount of the remaining loan does not exceed the value of the property.
4. **Period of loan**: The period of repayment of loan is generally more, and hence, the property must stand as security for the loan over a longer period of time.
5. **Remedies to recover loan**: If the mortgagor regularly pays the instalments of the loan and interest on the loan, this question does not arise. However, when the mortgagor fails to do so, the mortgagee can take the possession of the property and sell it so as to get the amount of the loan and the interest.

6. **Subsequent mortgages**: A property can be mortgaged more than once. In such cases, the first registered mortgage deed will have first claim than the subsequent mortgage deeds. Hence, the mortgagee of the subsequent mortgages should be careful in deciding the amount of the loan against the property.
7. **Third party guarantee**: Normally, the mortgagee is more interested in the recovery of his loan rather than the possession of the mortgagor's property. Hence sometimes, a personal guarantee from a reputable party is included in the mortgage deed so that in case of emergency. The mortgagee may request such party to repay the loan advanced to the mortgagor.
8. **Types of mortgage lenders**: The institutional lenders include banks, financial organizations, LIC, Government agencies, etc. The noninstitutional lenders include individuals, estate developers, real estate brokers, semi-public institutions, etc.
9. **Valuation**: The net income should be carefully worked out so as to see that it at least covers the interest of the loan that is granted. Also, the value of the materials which can be easily removed by the mortgagor without the notice of the mortgagee should not be included in the valuation.

21.23 OUTGOINGS AND NET INCOME

21.23.1 OUTGOINGS

The term *outgoing* is used to indicate the expenses which are to be incurred in connection with the property so as to maintain the revenue from it. As a matter of fact, all types of the properties are subject to the various types of outgoings and it is very important for a valuer to estimate the outgoings as correctly as possible because on it depends the net income of the property.

Following are the usual types of outgoings connected with immovable or real properties:

   i. Municipal taxes
  ii. Government taxes
 iii. Annual repairs and maintenance
  iv. Management and collection
   v. Insurance
  vi. Vacancies and bad debts
 vii. Sinking fund
viii. Miscellaneous

1. **Municipal taxes**: The taxes which are paid to the local authority for various services such as water supply, sanitation, etc., are deducted from the gross rent. The actual amount for taxes on the material date of valuation should be worked out and the same should be allowed in the outgoing. These charges are usually determined by a certain percentage of the property's rateable value. The rateable value is derived from the gross income by deducting the amount of

annual repairs. It should, however, be noted that if the tenant pays the municipal taxes, no deduction for such amount is allowed in the outgoing.

2. **Government taxes**: The amount of tax imposed on the property by the government and paid by the owner should be deducted from the gross rent. Such taxes include land taxes, cessation of education, etc.

3. **Annual repairs and maintenance**: The amount to be spent on the repairs and maintenance is variable and it depends on various factors such as age of the building, situation of the building with respect to climatic conditions of the locality, specifications of work, category and number of tenants occupying the building, condition of the building with respect to degree of maintenance, use of the building, etc.

4. **Management and collection**: The term *management* is used in the sense of attending to and disposing off the complaints regarding the essential services and amenities of the property and it usually does not include the costs of services and amenities. The term *collection* denotes the labor involved in contacting the tenants for realizing rents from them.

   The amount to be spent for the management and collection of rents from the property should be suitably estimated and the same should be included as one of the item of outgoing. Such an amount depends on the class of property, number of tenants and gross rent.

5. **Insurance**: The premium paid by the owner for fire insurance of the property is to be considered as one of the item of outgoings. The fire may occur due to accidents or due to any other event such as riot. Following points should be noted in connection with the insurance of real estates:

   i. Provision: It is the general practice to include the reasonable amount for premium of fire insurance policy of the property, even though the owner might not have insured the property at all. This is due to the fact that every hypothetical prudent purchaser of a property is supposed to insure it against fire contingency. Also, if the property is under-insured, the valuer has to work out fair insurance premium for the full value of the property and allow that amount in the outgoings of the property.

   ii. Rate: The rate of premium depends on various factors such as the type of construction of the building, proximity of structures which are easily liable to catch fire, facilities for fire protection and fire-fighting, the nature of occupancy, etc. The insurance companies have their own schedules for rates of premiums for different types of properties and depending upon the circumstances of each case, a suitable amount for premium is decided.

   iii. Value: It is obvious that the land, plinth, foundations, and pavements in open spaces are not destroyed due to fire. Hence, for working out the value of property for the purpose of fire insurance, only the reinstatement cost of superstructure of existing building is to be considered

6. **Vacancies and bad debts**: It is not possible to keep some types of property fully or continuously occupied e.g. hostels, hotels, etc. Thus, some properties will be unlet and nonrevenue producing during certain interval of time. These periods are referred to as the *vacancies* or *voids*. In such case, a suitable reduction for loss of rents should be made from the gross income of the property for these periods.

7. **Sinking fund**: This is an arrangement in which uniform or equal deposits must be contributed to a specified compound interest rate within fixed periods in order to have a specified amount available in a given future period. For example, a sinking fund could be created as a reserve to cover the future cost of some capital expenditure, such as a roof refurbishment, replacing the windows it has planned for its rental property.

8. **Miscellaneous**: The expenses, which are not included under any of the above heads, are to be treated as miscellaneous. Such charges include lighting charges for common amenities, ground rent for leasehold properties, service charges for lift and pump, salaries to lift-man, sweepers, etc. These miscellaneous expenses will have to be estimated in each case by the valuer with the facts and data supplied to him.

21.23.2 Gross Income and Net Income

The total amount of revenue received from a property during a year is known as the **gross income**.

The net income from a property is an amount left at the end of the year after all ordinary outgoings have been deducted.

Thus, we get the equation

$$\text{Net income} = \text{Gross income} - \text{Outgoings} \qquad (21.5)$$

21.24 EASEMENTS

A facilitation is a legal right to occupy or use the land of another person for certain purposes. The land use is limited and the original owner retains the land's legal title. A legally binding facility must be made in writing, the exact location specified in the deed of the property. Services most often grant access to utility companies for the installation and maintenance of power, telephone, and cable lines as well as for water drainage purposes.

Services are created if a property owner expresses a legal document in the language. Although an oral agreement can be concluded to create an easement, it does not always hold up in court. This is referred to as an "implicit facilitation" and often happens when two neighbors simply agree that one property with no legal documentation needs a driveway or other access. This can be a complicated situation if the crossed property is sold to someone who does not want to grant servicing rights under an implicit servicing system and is not always held in court. In order to create a facility that is legally binding, the document must be submitted to the county registrar. The exact location, nature, and purpose of servitude in an act or other legal document create an "express servitude."

21.24.1 Types of Easement

An easement can be classified as either a gross easement or as a gross easement.

1. **Easement appurtenant**: An easement appurtenant is an easement that benefits one parcel of land, known as the dominant tenement, to the detriment of another parcel of land, known as the *servient tenement*. Easements appurtenant are attached to the land and are transferred automatically when the servient or dominant tenement is sold to a new owner.
2. **Easement in gross**: Gross servicing benefits a person or entity, not a parcel of land. If the property is sold to a new owner, the property is usually transferred to the facility. However, the owner of the service has a personal right to the service and is prohibited from transferring the service.

21.24.2 Creation of Easements

An easement may be expressed or may be implied or prescribed.

1. **Express easements**: Express servicing is created through a written agreement between the landowners who grant or reserve a servicing. Express easements must be signed by both parties and are usually recorded on each property with the deeds.
2. **Implied easements**: An implicit relief can only be created if two parcels of land were treated at one time as a single tract or belonged to a common owner. Existing use implies an easement if the easement is necessary for the use and enjoyment of one parcel of land and the parties involved in the division of the tract into two parcels of land intended to continue the use after the division. When a parcel of land is sold, the other parcel is deprived of access to a public road.
3. **Prescriptive easements**: Prescriptive easements, also known by prescription as easements, occur if a person has used an easement for a number of years in a certain way. In most states, a prescription relief is created if the use of the property by the individual complies with the following requirements:

- The use is open and notorious, i.e., apparent and not secret.
- Actually the individual uses the property.
- Use for the statutory period is continuous—usually between 5 and 30 years.
- Usage is contrary to the real owner, i.e., without permission of the owner.

21.24.3 Easement Rights

A servicing holder has the right to do everything reasonably necessary to make full use of the property for the purpose for which the servicing was created. However, his rights under an easement do not allow an unreasonable burden to be imposed on the owner of the property. The owner of the property may also use the property until such use interferes with the purpose of servicing. Easements are recognized as following rights, even if there are no official documents or agreements:

- **Aviation easement**: The right to use airspace over a property flying above a certain altitude to spray property or other agricultural purposes where necessary.
- **Storm drain easement**: The right to install a storm drain for the transport of rainwater to a river, wetland, or other body.
- **Sidewalk easement**: The public right to use sidewalks outside the public area.
- **Beach access easement**: The right to access the public beach to neighboring residents, even if access is crossed by private property.
- **Dead end easement**: A landowner's requirement to grant public access to the next public route, even if such access to its property crosses.
- **Conservation easement**: A landowner's requirement to grant public access to the next public route, even if such access to its property crosses.

21.24.4 Terminating an Easement

The termination of an installation requires the approval of the court. To terminate a property owner, at least one of the following facts must be proved in court:

- Both parties have agreed to the termination
- The easement has reached its expiration date
- The easement holder discontinued use the of property
- The need for the easement no longer exists
- The easement holder is creating a hostile use of the property
- The owner of an easement has died

If the holder of the property seeking to terminate a servicing is unable to prove one or more of the required facts, the court may order him to continue to allow the servicing holder to use the property until such facts can be proven.

21.25 VALUE OF LICENSED PREMISES

21.25.1 VALUATION TABLES

In order to save time and reduce the chances of error in elaborate and laborious mathematical calculations, the valuation tables are prepared so that by referring to them, suitable coefficients can be found out.

The valuer should know the mathematics of valuation tables mainly for the following three reasons:

1. He should be able to derive any coefficient he desires from first principles even though the same is available from ready-made published valuation tables.
2. He should be able to work out the coefficients for unusual terms of years or unusual rates of interest which may not be available in the published valuation tables.
3. The clear conception of construction of the valuation tables will enable him to use the tables efficiently.

It should be noted that in valuations, the investments are on the principles of compound interest and as such, the valuation tables are constructed on the basis of compound interest principles. The basic concept is that when money is invested, the original capital sum is interest-bearing and this interest is added to the original capital in subsequent years. This process is repeated every year and therefore the capital is regularly increased and the interest earned increases with each period or compounds as a result of this process.

Following valuation tables are considered:

1. **To find the amount accumulated on Re. 1 at the rate of interest i after n years**

$$P = (1+i)^n \qquad (21.6)$$

where
P = amount at the end of the term,
i = rate of interest per annum compound,
n = number of years.

2. **To find the PV of Re. 1/- receivable at the end of a given term, n with rate of interest, i.**

$$\text{P.V.} = \frac{1}{(1+i)^n} \qquad (21.7)$$

where
i = rate of interest p.a. compound,
n = number of years.

3. **To find the amount to which Re. 1/- p.a. invested at the end of each year will accumulate in a given time at a rate of interest i p.a.**

$$A = \frac{(1+i)^n - 1}{i} \qquad (21.8)$$

where
A = amount to which Re. 1/- p.a. invested at the end of each year will accumulate in a given time,
i = rate of interest, compound, and
n = Number of years.

The following three points are to be noted:

- Re. 1/- becomes due to be invested at the end of the year,
- Next Re. 1/- gets added at the end of each year,
- Last Re. 1/- will be added at the end of the period.

1. **To find the amount of the annual sinking funds for the redemption of Re. 1/- capital.**

   Annual sinking fund may be defined as the fund set aside annually (at the end of a year) out of the annual income from the capital invested which at the compound rate of interest will grow by the end of the term to replace the capital invested or for a major repair or replacing an important part after expiry of its life.

   In case of a building, a sinking fund may be created to replace the building after expiry of its calculated life fully or for major rehabilitation and renovation work after a lapse of a definite period.

   The formula for sinking fund is the reverse of the formula of the amount of Re. 1/- per annum as in (3). The amount set aside for sinking fund is deposited in bank to carry an interest

$$S = \frac{s}{(1+s)^n - 1} \qquad (21.9)$$

where
S = amount of sinking fund,
s = compound rate of interest p.a. on the amount deposited,
n = number of years.

2. **Year's purchase**

   Year purchase is the figure or multiplier that gives the capitalized value of a property on the material date of valuation when multiplied by net income, year Capitalized value = net income × YR.

3. **To find the annuity of Re. 1/- will purchase on single rate principle.**

   Annuity = Return on capital + Return of capital,

$$A = i + S \qquad (21.10)$$

$$A = i + \frac{i}{(1+i)^n - 1} = \frac{i(1+i)^n}{(1+i)^n - 1} \qquad (21.11)$$

where
i = rate of interest,

S = sinking fund required to replace Re. 1/- at the end of 7 years on single rate basis,

n = number of years.

This formula is useful in calculation of annual equal instalments with capital and interest and amortization cost with interest. Again, annuity is the inverse of year's purchase.

4. **To find year purchase (YP) of a reversion to perpetuity after a given number of years**

Single rate of YR is used for incomes produced by freehold properties and the role of sinking fund in the case would be to make arithmetical adjustment to the YR merely to permit valuations to be made for terminal incomes. As a matter of fact, the annual sinking fund is a reducing agent in the formula and it devalues YP to such an extent that no value is attached to the period during which there is no income flow.

Finding YP of reversion to perpetuity after a given number of years: In this case, the YR for incomes continues up to the perpetuity, but starts at some future date. Such incomes are known as deferred incomes, and as YP for nonincome period is not available, YR for deferred income can be worked out from the following equation:

YP for deferred income of a reversion to a perpetuity = YP in perpetuity − YP for the nonincome period on single rate basis

$$= \frac{1}{i} - \frac{1}{i + \dfrac{i}{(1+i)^n - 1}} \qquad (21.12)$$

where

i = rate of interest

n = number of years for the nonincome period.

5. **Finding the depreciation percentage based on sinking fund method**

$$\text{Depreciation percentage} = (S \times A)100 \qquad (21.13)$$

As per formula (4) sinking fund at given rate of interest for the total life of a building

$$S = \frac{s}{(1+s)^n - 1} \qquad (21.14)$$

where

n = age of building,

s = rate of interest of investment in sinking fund (compound).

21.26 DEPRECIATION

The term depreciation is derived from the Latin word *depretiatum* meaning the fall in value or becoming less in worth. At present the word depreciation is used in the following two distinct meanings:

1. Depreciation as cost in operation
2. Depreciation as decrease in worth

21.26.1 DEPRECIATION AS COST IN OPERATION

In a business enterprise, the physical capital in the form of machines, buildings, and the like is used in carrying out production activities. In business and trade activities, the depreciation is cost of production because an asset is actually consumed in the production operation. If the cost of capital consumption is neglected, the profit will appear to be higher than it is an amount equal to the depreciation that has taken place during production period.

21.26.2 DEPRECIATION AS DECREASE IN WORTH

The term depreciation is used to decrease or decrease or decrease in value. It is quite obvious that cost is not worthwhile. The value of an asset depends on its ability to provide services to the satisfaction of its estimated or expected future life. The current value of these expected services in a tangible or intangible way represents their value at a certain time.

The causes of depreciation can be divided up between physical deterioration, economic factors, the time factor, and depletion. These are briefly explained below:

1. **Physical Deterioration**:
   - Wear and Tear: When a motor vehicle or machinery or equipment and fittings are used, wear them out. Some years have passed, others only last a few years. This also applies to buildings, some of which may last a long time.
   - Erosion, Rust, Rot, and Decay: Land can be eroded or destroyed by wind, rain, sun, and other natural elements. Likewise, metals rust away in motor vehicles or machinery. Wood is not going to rust eventually. Decay is a process that is also present because of the elements of nature and the lack of care.
2. **Economic factors**: These are the reasons why an asset is not used even if it is in good physical condition. The two most important factors are usually obsolescence and insufficiency.
   - Obsolescence: This is the out-of-date process. For example, the development of synthesizers and electronic devices used by leading commercial musicians has made considerable progress over the years. Therefore, the old equipment has become obsolete and much of it has been taken out of use by such musicians.
   - Inadequacy: This occurs when assets are no longer used due to growth and changes in business size.
3. **The time factor**: There is obviously a need for time and obsolescence and inadequacy. There are, however, fixed assets to which the time factor is linked in another way. These endorsements have a legal life

of years. For example, you may decide to rent certain buildings for ten years. Usually this is called a lease. When the years have passed, the lease does not matter to you, it is over. Anything you paid for the lease is now worthless. A similar assertion is that you buy a patent so that only you can produce something when the time of the patent is over, it has no value. Instead of using the term depreciation, these approvals are often used by the term amortization.

4. **Depletion**: Other assets are of a wasteful nature, possibly because raw materials are extracted from them. These materials are then either used by the company to produce something else or sold to other companies in their original state. This heading is covered by natural resources, such as mines, quarries, and oil wells. The provision for depletion is called to provide for this consumption for an asset of a wasteful character.

21.26.3 Methods for Estimating Cost Depreciation

Following are the six methods adopted to calculate the cost depreciation of a property:

1. Straight-line method
2. Constant percentage method
3. Quantity survey method
4. Sinking fund method
5. Sum of the digits method
6. Unit cost method

1. **Straight-line method**: It is a very simple and common method of calculating the cost. In the case of straight-line depreciation, the amount of expenditure is the same each year for the life of the asset.

   Depreciation Formula for the Straight Line Method:

   Depreciation Expense = (Cost – Salvage value)/Useful life
   (21.15)

   The straight-line method is usually the slowest system of depreciation. All other systems of depreciation are treated as accelerated depreciation methods and they are to be used under certain conditions and even then with a great deal of risk.

2. **Declining balance method**: It is an accelerated depreciation method. In this method the annual depreciation is expressed as a fixed percentage of the book value at the beginning of the year and is calculated by multiplying the book value at the beginning of each year with a fixed percentage. Thus this method is also sometimes known as fixed *percentage method of depreciation*. The ratio of depreciation amount in a given year to the book value at the beginning of that year is constant for all the years of useful life of the asset. When this ratio is twice the

straight-line depreciation rate, i.e., 2/*n*, the method is known as *double declining balance (DDB) method*. In other words the depreciation rate is 200% of the straight-line depreciation rate. DDB method is the most commonly used declining balance method. Another declining balance method that uses depreciation rate equal to 150% of the straight-line depreciation rate, i.e., 1.5/*n* is also used for calculation of depreciation. In declining balance methods the depreciation during the early years is more as compared to that in later years of the asset's useful life. In case of declining balance method, for calculating annual depreciation amount, the salvage value is not subtracted from the initial cost. It is important to ensure that the asset is not depreciated below the estimated salvage value. In declining-balance method the calculated book value of the asset at the end of useful life does not match with the salvage value. If the book value of the asset reaches its estimated salvage value before the end of useful life, then the asset is not depreciated further.

3. **Quantity survey method**: This method has been widely used in practice. The property is examined in detail and value losses due to the physical deterioration are determined. The amount to be spent on upgrading, etc. So that obsolescence is offset. Each step is based on a logical basis without reference to a certain fixed percentage of the cost of the property. Therefore, observation is the basis of this method and not assumption.

4. **Sinking fund method**: In this method it is assumed that money is deposited in a sinking fund over the useful life that will enable to replace the asset at the end of its useful life. For this purpose, a fixed amount is set aside every year from the revenue generated and this fixed sum is considered to earn interest at an interest rate compounded annually over the useful life of the asset, so that the total amount accumulated at the end of useful life is equal to the total depreciation amount, i.e., initial cost less salvage value of the asset. This means that each year's annual depreciation has two components. The first is the fixed sum deposited in the sinking fund and the second is the interest earned on the amount accumulated in the sinking fund until the beginning of that year.

   For this purpose, the uniform depreciation amount (i.e., fixed amount deposited in the sinking fund) at the end of each year is calculated by multiplying the total depreciation amount (i.e., initial cost less salvage value) by the sinking fund factor over the useful lifetime. After that the interest earned on the accumulated amount is calculated.

5. **Sum of the digits method**: It is also an accelerated depreciation method. In this method the annual depreciation rate for any year is calculated by dividing the number of years left (from the beginning of that year for which the depreciation is calculated) in

the useful life of the asset by the sum of years over the useful life.

6. **Unit cost method**: This method of calculating depreciation is applicable to machines and other equipment where it is possible to work out the cost per unit output. In this method, it is assumed that a machine after n years of use will give the same unit cost of output during its remaining years of life.

21.27 STANDARD RENT

Sometimes, the rental of a property must be fixed. The capitalized value of the property is known and the net income from the property can be calculated by dividing this capitalized value by a proper figure of the year of purchase. All possible outputs are added to this net income, which then results in the gross rent or gross income of the property per year. The rent per month can be calculated by dividing this gross income by twelve. The rent worked out by this procedure is known as the standard rent while the actual rent of a property may be higher or lower than this rent depending upon various uncertain elements such as situation of the property, type of construction, demand and supply of such properties, etc.

The actual rent is known as the *contractual rent*, and in legal parlance, the rent which can be lawfully charged on a tenant is known as the *standard rent*. If a tenant seeks redress at law against contractual rent, the law will reduce the contractual rent, if it is lower than the standard rent. Every state has formed and adopted its own act for the fixation of standard rent.

21.28 METHODS OF VALUATION

The properties can generally be classified in the following two categories for evaluation purposes:

1. Methods of valuation for open lands
2. Assessment methods for lands with buildings

21.28.1 METHODS OF VALUATION FOR OPEN LANDS

For the purposes of income tax, wealth tax, banks and embassies, valuation of an open property or any other property is required. In order to submit annual returns to the income tax department, valuation is necessary for income tax and wealth tax purposes. Even in order to obtain loans from banks and to show assets to the embassies (to obtain visas), the valuation is necessary. There are different valuation methods:

1. Abstractive method
2. Belting method
3. Developmental method
4. Comparative method.

1. **Abstractive method**: The abstract method becomes useful when there is no information about land transactions in the neighboring area, or in other words,

the value of land where sales do not occur frequently can be determined by applying this method. There are three separate steps involved in this method:

- A nearby properly hired rent shall be considered and its capitalized value shall be calculated by multiplying its net income by purchase year (Say S).
- The estimated cost of replacing the above-mentioned building is calculated and a figure representing at present the cost of the building alone is obtained after making the due allowance for the depreciation (Say T).

The difference $D = (S - T)$ gives the land value

and the cost per unit area $= D/A$ if A is the land area

2. **Belting method**: When a large plot is to be evaluated or when a plot with fewer fronts and more depth is to be evaluated, it is logical to use the belting method. It is because of the principle that the value of the land generally decreases as the depth of the plot increases, or in other words, the road to the front of the land is more valuable than the road to the back.

The main problem facing the valuer during the adoption of this method is to determine the depth to which the maximum land value extends and then decreases or decreases. The next step would then be to fix the value of the land back to the front. In this method, the land area under consideration is divided into different sections or zones and for each section or zone different land rates are estimated. The plot of land is usually divided into three belts. The depth of the first belt near the road is adjusted appropriately. The depth of the second belt is 50% higher than the depth of the first belt, and the depth of the third belt is 50% higher than the depth of the second.

Consider the size, shape, location and various other factors affecting the land rate, an appropriate land rate for the first belt is estimated. Two thirds of the rate of the first belt is taken for the second belt and one-half of the rate of the first belt is taken for the third belt. There is no hard and fast rule regarding the ratio of front and back land values. Each case must be investigated independently, depending on the merit of each case.

3. **Developmental method**: This method is used for land that is not developed but has the potential to appreciate a significant value if it is converted to a residential/commercial or industrial layout depending on the location, size, shape, frontage, and depth, etc. When a NA country is being developed, the net plottable area is only 50%. Working back, the current value of NA residential area land can be reached.

4. **Comparative method**: In this method, the different transactions in the neighboring countries are properly studied and a fair land rate is then decided. The comparative method is therefore only useful in

the case of an active market where there are a large number of comparable statistics. After a thorough inspection of all the underlying factors in the market, the valuer must be satisfied that there have been no changes in conditions since the transaction took place. The time element plays a decisive role in this method. In the case of volatile markets, it is found that the evidence for the sale chosen for comparison is unreliable within a very short period of time. During the analysis using the comparative method, the following factors should be taken into account:

- Location: The value of land in a busy town or city center or shopping district will certainly be more than the value of land far from the town.
- Size: The size of the plot has an important role to play in determining its value. The rate of a large country cannot be compared to a small country. There is usually a strong demand for plots of certain sizes in a particular location. The land rate with such sizes should be considered as the trend to reach the value of open land under consideration in this location.
- Shape: People prefer to have regular land. Plots with regular shapes are observed to be sold at a higher price than those with irregular shapes.
- Frontage and depth: It is quite clear that the most important part of a parcel of land is its street front and the value of the back part of the parcel decreases as the distance from the street increases.
- Return frontage: A corner plot receives an additional return frontage and the value of such plots will increase accordingly, depending on the importance of crossing streets or roads. The corner plot has better access and exit and provides more light and ventilation. A corner plot in the residential area offers a wide range for better layout of shops or offices with more space for showrooms and advertising.
- Level: If the natural land level is lower than the road level, significant amounts will have to be spent on filling and the cost of the foundations will also increase substantially. On the contrary, if the natural level of the land is significantly higher than the level of the road, it will be difficult to lay water/drainage lines, and therefore, the extra earth will have to be excavated in order to make the land reasonably level.
- Nature of soil: The laying capacity of the soil should also be considered for lands with building potential. The land with a good carrying capacity will have a higher rate than the land with a poor carrying capacity.
- Land-locked land: Sometimes it happens that a parcel of land has no well-defined legal access and is surrounded by parcels belonging to other owners of less value on all sides.

- Restriction on development: For the plot to be evaluated and examined, the allowable floor space index (FSI) should be studied. The land plot with more permissible FSI will be sold at higher prices, of course, compared to the less permissible FSI.
- Encumbrances: Land plots subject to air, light, or passage rights will be less attractive to prospective buyers, and depending on the inconvenience caused, the values of such land will be reduced.
- Miscellaneous advantages: in addition to the above consideration, if the property has certain special advantages due to its location, or for any other reason, the same should be taken into account when arriving at a reasonable open land rate.

21.28.2 METHODS OF VALUATION FOR LANDS WITH BUILDINGS

The assessment of a building depends on the type, structure, and durability of the building, the situation, size, shape, frontage, the width of the roads, the quality of the materials used in the construction and the current prices of the materials. The assessment also depends on the building's height, plinth height, wall thickness, floor nature, roof, doors, windows, etc. The valuation of a building is determined by calculating its construction costs at the present rate and allowing an appropriate depreciation.

When lands are valued with buildings, the following six methods of valuation are adopted:

1. Rental method of valuation
2. Direct comparisons of the capital value
3. Valuation based on the profit
4. Valuation based on the cost
5. Development method of valuation
6. Depreciation method of valuation

1. **Rental method of valuation**: In this method, the net rental income is determined by deducting all outgoings from the gross rent. An appropriate interest rate is assumed as prevailing in the market and the year of purchase is calculated. This net income multiplied by the year in which the property is purchased gives the capitalized value or valuation. This method only applies if the rent is known or the likely rent is determined by inquiries.
2. **Direct comparison with the capital value**: This method can be used if the rental value is not available from the property in question, but evidence of the sale price of the property as a whole is provided. The capitalized value of the property in such cases is determined by direct comparison with the capitalized value of similar property in the locality.
3. **Valuation based on profit**: This evaluation method is appropriate for buildings such as hotels, cinemas, theaters, etc. For which profit depends the capitalized value. In such cases, net income is calculated

following the deduction of gross income; any possible work expenses, outgoings, interest on the invested capital, etc. The net profit is multiplied by the purchase of the year in order to achieve the capitalized value. In such cases, the valuation may be high in relation to the construction costs.

4. **Valuation based on cost**: In this method, the actual cost of building or owning the property is used to determine the value of the property. In such cases, the necessary depreciation and the obsolescence points should also be considered..

5. **Development method of valuation**: This evaluation method is used for properties that are either underdeveloped or partially developed and partially underdeveloped. If a large area of land is to be divided into plots after roads, parks, etc., are provided, this method of valuation must be adopted. In such cases, the likely selling price of the parcels, the area for roads, parks, etc., and other development expenditures should be known.

6. **Depreciation method of valuation**: According to this method of valuation, the building should be divided into four parts:
   • Walls
   • Roofs
   • Floors
   • Doors and windows

   And the cost of each part should first be determined by detailed measurements on the current rates; the current value of the supply of land and water, electrical and sanitary equipment, etc. The valuation of the building should be added to arrive at the total valuation of the property.

21.29 MISCELLANEOUS TOPICS

The important topics connected with the subject of valuation will be briefly described:

1. Accommodation land and accommodation works
2. Amortization
3. Annuity
4. Capitalized value
5. Cost inflation index (CII)
6. Deferred or reversionary land value
7. Dilapidations
8. Discounted cash flow (DCF)
9. Encumbrance factor
10. Floating FSI
11. Life of structure
12. Mesne profit
13. Mobilization fund
14. Rate of interest
15. Rating
16. Records of rights
17. Rent fixation
18. Year's purchase

21.29.1 ACCOMMODATION LAND AND ACCOMMODATION WORKS

As the city or town develops, the surrounding agricultural area is to be converted into the accommodation land. For this purpose, the owner of the land has to obtain the permission from the competent authority to convert his agricultural land into nonagricultural or accommodation land. The government will lose revenue due to such conversion known as the nonagricultural assessment.

When the agricultural or accommodation land is required for public use such as railways, highways, etc., the acquiring authority is required to provide what are known as the *accommodation works*. These works include provisions of bridges, retaining walls, fencing, etc.

21.29.2 AMORTIZATION

When the payment of a debt is made by a series of equal periodic payments, it is known as the amortization. Each periodic payment includes two provisions:

1. A certain portion of the principal
2. An account of interest on the outstanding debt

Thus, it is clear that as the number of payments increases, the share of the principal also increases and the amount of interest decreases correspondingly. The last payment will thus be for the remaining debt and the interest on this debt. Due to amortization procedure, a higher rate of interest is realized on the payments than in case of the sinking fund procedure.

21.29.3 ANNUITY

The term annuity is used for a series of periodic payments. Generally, the word annuity indicates annual payments. But at present, it is also used for monthly, quarterly, or semi-annual payments. The total amount of payment in 1 year constitutes what is known as the annual rent. The annuities are of four types:

1. Annuity certain
2. Annuity due
3. Deferred annuity
4. Perpetuity

• A certain annuity is an annuity whose payments shall be made at the end of each period and which shall be continued for a certain number of periods.
• An annuity due is an annuity whose payments are made at the beginning of each period and for a certain number of periods such payments are continued.
• A *deferred annuity* is an annuity whose payments begin at some date in future.
• A *perpetuity* is an annuity which continue indefinitely.

21.29.4 Capitalized Value

The term capitalized value is defined as the amount of money whose annual interest at the highest prevailing interest rate, corresponding to the type of property referred to, is equal to the net income derived from the property concerned. In other words, the amount of capital to be paid for a periodic payment in the form of a net return in perpetuity or for a specified period is indicated.

In order to arrive at the capitalized value of the property, the data to be collected would therefore be as follows:

1. The net income from the property after deducting all the outgoings from the gross income.
2. The highest prevailing rate of interest for investment in such type of property by giving due considerations for nature of property, type of income, and other allied factors.

The rate of capitalization depends upon the following factors:

1. Divisibility of holding
2. Ease of liquidity of the asset
3. Ease of management and transfer
4. Likelihood of capital gains
5. Nature of use of property, i.e., for business, residence, etc.
6. Presence of legal hazards
7. Presence of rent review clause
8. Security of capital
9. Security in real terms
10. Security and regularity of income

21.29.5 Cost Inflation Index

CII is an index used to factor in the effect of inflation in the prices of capital assets. CII is used while calculating long-term capital gains.

For example, a property purchased in 2014 would normally cost more than it would have in 2008. However, the increased value also includes inflation. So, while calculating real profit (capital gains) from the sale of a capital asset, we need to adjust the cost of acquisition of property and cost of improvement to factor in the inflation. For this purpose, every year Central Government notifies CII to adjust for inflation the value of assets which are held for more than 36 months (long term capital assets).

21.29.5.1 Calculation of Capital Gains Using CII

If the asset is sold within 36 months of acquisition, capital gains are calculated on actual cost of acquisition and selling price. These are called *short-term capital gains*.

However, if the asset is sold after 36 months from the date of acquisition, capital gains are calculated using the following formula: indexed acquisition cost, indexed improvement cost, and actual selling price

$$\text{Capital Gain} = \text{Sale Consideration} - \text{Indexed Cost of Acquisition} - \text{Indexed Cost of Improvement} \quad (21.16)$$

where

$$\text{Indexed Cost of Acquisition} = \text{Actual Purchase Price} * (\text{CII of Year of Sale} / \text{CII of Year of purchase}) \quad (21.17)$$

$$\text{Indexed Cost of Improvement} = \text{Actual Cost of Improvement} * (\text{CII of Year of Sale} / \text{CII of year of improvement}) \quad (21.18)$$

21.29.6 Deferred or Reversionary Land Value

The term *reversionary value* is very popular with English real estate tenures and it means that part of the value which depends on the future realization of an expected improvement in the asset which may be due to various reasons like:

1. Future scarcity
2. Increased anticipated demand
3. Increased rent under an existing lease
4. Maturing of a crop as with timber
5. Return to a freeholder of vacant possession of land or buildings on the expiry of a long lease at a rent, etc.

The term is more properly expressed as the PV of reversion of the monetary equivalent of whatever the future asset is worth or that percentage of what it would be if it were in hand at the present time.

If the structure is standing on the land, the full value of land can be realized only when the structure is demolished. For valuation purposes, the PV of the land is determined and referred to an appropriate valuation table; the deferred or reversive value of the land up to the estimated structural life is obtained. If the income is estimated to continue in perpetuity, the problem of deferred land value does not arise and it is automatically seen reflected in the YP for such property.

It is thus seen that the concept of reversionary land value is based on the following two assumptions:

1. The land value of the material date of valuation is taken to work out the reversionary land.
2. The land is open to the owner at the end of the estimated structural life and can be developed as he subsequently likes.

21.29.7 Dilapidations

Dilapidation is the term used to describe decay/damage or waste, the condition of a building, or premises. It means decay/waste disrepair caused willfully or otherwise by continuous neglect of maintenance and repair.

Dilapidating a building means that a building's physical life tends to expire, i.e., the building approaches a condition that would make it unfit for use. Such a building condition can occur due to various reasons. It can be due to natural deterioration. Whatever materials are used, the universal ageing phenomenon, which applies to every living organism, also applies to a building structure over time. Apart from natural decay, other causes may be created. A building may be used by the owner himself, he may not be able to carry out proper repairs in time due to financial difficulties and allow the building to be dilapidated against his will.

Again, it can happen that the owner of a building is more than one, and because of their internal dispute, no one is willing to spend money on the building's repairs and its decline begins and dilapidates.

The building may be leased or abandoned and either the tenant or the lessor may be responsible for the necessary repairs in accordance with the contract; however, due to a dispute, the party responsible neglects the repairs and causes the building's waste and gradually dilapidates its condition.

Waste is an act of damage or destruction caused by a breach of the contractual provisions. It may be voluntary or allowed.

- *Voluntary waste* involves an act that tends to destroy the premises/buildings, as if it were pulling the house down. The destruction must be deliberate or inadvertent. If the building is damaged because of its reasonable use, it is not waste. It is waste however to change the nature of the building.
- *Allowable waste* is where the tenant suffers from damage to a building resulting from the omission to do something that the owner should have done, for example. Failure to repair.

21.29.7.1   Causes of Building Dilapidation

The physical causes that reduce a building to a delayed condition are as follows:

1. **Natural decay and ageing**: The materials used in the building have certain lives to remain functional, after which the parts disappear and the building gradually drift into a dilapidated state.
2. **Inadequate or no maintenance**: Proper maintenance and prompt repair or replacement of declined or damaged structural parts could delay early decline and dilapidation.
3. **Bad use of building**: Bad use of a building or abuse may aggravate damage to a building and accelerate dilapidation; e.g., use of a residential building for convenience purposes, internal alteration, etc.
4. **Use of underspecified materials**: It is clear that underspecified or bad materials will have less life and less strength than necessary and that the building will be dilapidated earlier in the natural course of time.
5. **Bad workmanship**: Bad workmanship will reduce the life of the structure and may also cause an imbalance in the load bearing system, causing unforeseen stresses resulting in the structure's early decay.

6. **Physical influence**: Any kind of physical influence would cause injury and exacerbate damage and decline and accelerate dilapidation.
7. **Effect of aggressive environment**: Aggressive environment such as strong salty winds, polluted with smoke and aggressive chemicals, gasses, etc., will cause structural damage in various ways. Aggressive chemicals entering a structure through the cracks would speed up corrosion and dilapidate more quickly.
8. **Force majeure**: Unforeseen events such as cyclones, floods, earthquakes, fires, warfare, etc. The buildings can cause serious damage, rendering them destructive or dilapidated. Besides these, there may also be other reasons for the dilapidation of a building.

21.29.8   Discounted Cash Flow

The term "discounted cash flow" refers to a technique for evaluating a company's likely future cash returns. The company's analytical values are based on the value of all future cash available to investors. It is described as "discounted," as the value of cash in the future is less than cash today (due to its reduced return capacity, such as interest, and inflation).

DCF is a similar concept to Internal Rate of Return and Net Present Value.

There are three elements in calculating DCF:

- The time period used for the evaluation.
- A precise estimate of the annual cash flows that occur during this period.
- The amount of money that could be gained by investing in something other than equivalent risk.

Very broadly, the assessment of the annual cash flow to be discounted involves:

- Add depreciation for the year on net income after tax.
- Subtract from previous year's change in working capital.
- Subtract expenditure on capital.

While DCF is a valuable tool, it has disadvantages. Small changes in input values in particular can lead to major changes in the company's value. The owner's ability to project future cash flow accurately depends on the accurate valuation using DCF, which can be difficult. It is also more suitable for long-term investments than for short-term investments, and investors should consider other valuations and DCF prior to decision-making.

21.29.9   Encumbrance Factor

When there are two plots at the same place, their values are nearly equal with a slight variation according to the individual characteristics. When structures are placed on such lands, the land takes on the value imposed by the type of structure because it is the building's earning capacity that determines the land's value. Hence, for getting full market value of the land, the type of structure suitable for fully exploiting the land has to be constructed over it so that full economic benefit can be derived from the total unit of land and structure. If any other type of structure is constructed, then the economic yield will be less, and hence, part of the value of the land will be wasted. This wastage is due to the wrong type of structure, what is called, encumbers the land.

The encumbrance factor involves economic considerations and the phenomena of pulling down of unsuitable old existing structures, though otherwise sound, profitable and in some cases even modern, can be seen in the developing pockets of any city or town.

The value of the available usable FSI is obtained by multiplying the encumbrance factor to the prevailing rate of open land in the locality. Hence, in such cases, there can be only either of the two values of the property, namely, the highest and the best value of land plus scrap value of the structure or the value as a composite entity of land and existing structure combined. It cannot have the highest land value plus what is erroneously called the depreciated value of the existing structure.

21.29.10   Floating FSI

The ratio of the total built-up area, including all floors, to the area on which the building stands is known as the FSI. In some towns it is also known as the floor area ratio (FAR). The value of FAR is determined by the local authority and it may be different for different areas and for different buildings of the town.

The novel idea of using the available FSI of a particular plot of land on some other land has been advanced in recent times and this right is known as the floating FSI or a system of transferable development rights. The essential condition would be of course that both the lands must fall within the jurisdiction of the local authority which permits the use of floating FSI.

The term floating FSI should not be confused with the balance or available FSI the difference between permissible FSI and the FSI utilized indicates the balance or available FSI on the plot itself. However, if the circumstances permit, the balance FSI can be made to float and settle on the adjoining property.

21.29.11   Life of Structures

The life cycle of a building consists of its construction, the serving necessary to maintain it in a usual condition over a period of years and its demolition, and the clearance of the site. The servicing mainly consists of maintenance, adaption to changing needs, ventilation, lighting, cleaning, insurance, drainage, etc. Thus, a building presents a flow of costs which begin with the construction, include various amounts arising each year for servicing and end with demolition and site clearance.

While valuing the property with structure, a valuer is required to estimate the future life of structure, i.e., its utility period during which income from it can be treated as assured. The estimation of future structural life requires the valuer's personal skills, which is why it is essential that the valuer visits the site and examines the structure personally.

Depending upon the materials used in the construction, specification of work, degree of maintenance, climatic conditions, use of the structure, approximate year of construction and various other factors which affect a life of a structure, the valuer estimates the suitable future life of property under consideration. The estimated future life of building should be realistic and normally be the period over which the building is expected to earn an income because it is during this period, rather than the period for which it is required by the first owner, than the cost will be recovered.

Normally, the buildings are considered to have a substantial life of something of the order of 40–80 years. It can be observed that the physical life of buildings is often much bigger. However, they are often demolished before the end of this period to make the site more profitable or because it is cheaper to clear and rebuild than to adapt the building to a change in requirements. Thus, the economic life of a building can be defined as the period during which a capital value of the building exceeds the capital value of the site.

21.29.12   Mesne Profits

The term mesne profit is used for damages for trespass, a wrongful act relating to immovable and the said wrongful act forms one of the torts affecting reality that is immovable property when the tenant or occupier of a property is disposed legally and possession has been passed in favor of the owner of the property, still the tenant or occupier holds over the property for a specified period before handling over the possession to the rightful owner.

21.29.12.1   Ingredients of Mesne Profit

1. Claim for mesne profit may remain floating till the competent court declares the possession of the person as wrongful, but it is not absolute proposition.
2. The measure of mesne profit is not what the owner has lost, but it is the value of the user to the person in incorrect possession with ordinary diligence.
3. The measure of mesne's profits can be much more than the rent paid by the person in wrongful possession before the date of the expulsion decree. This actual rent prior to the date of decree will lose its efficacies and will be replaced by the estimated rent or profit if the owner could prove that higher rent or profit could have been obtained with due diligence.
4. Value of user of the property to the person in wrongful occupation will be at what rent that wrongful person could find equivalent accommodation on the date of decree of possession; the market rent will form measure of Mesne profit.

5. Mesne profit begins from the date on which an eviction decree is passed until the date of possession and not from the date on which contract tenancy is terminated, the contract tenant becomes a statutory tenant. During the period of termination of tenancy until the date of the passing of an eviction decree, the current rent shall be a fair measure of Mesne.

21.29.13 MOBILIZATION FUND

When mobilization financing is a financial tool commonly used by building contractors to raise capital to cover costs before starting work on a project or before invoicing. Funding depends on the preapproval of contractors by a qualified subcontractor evaluation process such as subguard. Mobilization funding typically advances up to 10% of the contract amount and up to 85% of the invoices issued and a total of 40% of the contract amount.

21.29.14 RATE OF INTEREST

When the rental method of valuation is adopted, an appropriate interest rate is to be assumed to achieve the market value of the udder consideration. A real estate is type of investment and its value is related to the return it gives to the owner.

It is quite evident that the safest and the most certain securities are those which are guaranteed by the respective governments. Hence, the government securities are treated as first class securities or gilt-edged securities.

In olden days, the share certificates which were issued for government securities were provided with a border of golden wire and hence, they come to be known as the securities with golden touch or gilt-edged securities.

1. **Money and banking**: A commodity used as a means of exchange is termed the money and it acts as a mode of payment. The money consists of *commodity money*, *paper money*, and *bank money*. These three types indicate the different stages of development of money.

   Bank is the institution which transacts the business of banking which means accepting for the purpose of lending or investment of deposits of money from the public, repayable on demand or otherwise and withdrawal by cheque, draft, and order or otherwise.

   A bank is thus a financial institution. It is an intermediary between a depositor and an entrepreneur or businessman.

2. **Bank rate**: In money market, a rate is charged for purchase of bills of exchange and treasury bills. This rate is referred to as the *discount rate*. The discount, as the word implies, is the amount of money deducted from advance money at the time of making a loan. Compared to this, the *interest* is the amount of money added to the amount advanced as loan, either at certain periods or at the time of repayment of loan. However, there should be very little or no difference between the rates of discount and interest as otherwise the flow of loanable capital will be for one operation only

3. **Shops**: The modern shops located in popular business centers are likely to be more attractive and the investors would be satisfied with low return on their investments. In addition to the location, the types and number of tenants, the types of business run by the tenants, and such other factors are to be carefully weighed to ascertain the degree of security of rents from the tenants. The well-located shops, though of old design, can even be considered as the secured investments, if improvement and modernization can be effected without much trouble.

4. **Offices**: The same remarks as mentioned in case of shops apply in the case of office blocks also. The modern office units placed in office blocks located in commercial zone of the town are more attractive to the investors from the point of security, and hence, the low return can be expected from such units.

   The net income from the office property varies to a wide extent as it depends on a number of factors. It is therefore necessary to have a careful examination of the rent documents, understanding of covenants of each party, thorough survey of the condition of building, etc.

5. **Factories and warehouses**: The rents received from the industrial units and warehouses will depend on factors such as the access to main roads, sources of raw materials, availability of labor, nearness to railways, location of market for the finished goods, etc. The rents from factories and warehouses will be affected seriously by the economic fluctuations in the market, and in this respect, they are considered to be less secured than those from shops and offices. Further, such premises are subjected to heavy wear and tear and they are likely to become rapidly obsolete because of changes in technology.

   For all these respond, the industrial properties have not been so popular to the investors as shops or offices with the result that investors in industrial properties expect high return over their investments.

6. **Residential properties**: The residential units cover a wide range of properties from ownership flats to the village cottages. These properties have been attacked by a stream of legislative measures by government of all colors and day-by-day; the restrictions on the ownership of residential properties are increasing.

7. **Agricultural properties**: The agricultural properties are usually changing hands with low yields. There may be several factors for this tendency of investors and one of them being that the food is essential for human habitation and that there will always be demand for farms to produce the agricultural products.

21.29.15   RATING

A rate is a type of tax which is assessed on the property by a local authority and the collection from rates is used for carrying out activities of public use. The term rateable value is used to indicate the amount arrived at after deducting statutory provisions for repairs and maintenance from the annual value of a property. The usual provision for repairs and maintenance is 10% of the annual value.

22 Sustainable Technology and Green Building

22.1 ENVIRONMENTAL ISSUES

The general meaning of sustainable development is "the creation and proper administration of a sound environment dependent on principles of resource efficacy and ecology." Sustainable structures focus on diminishing their effect on our condition through productive utilization of resources and energy. "Sustainable building" by and large characterized as building practices, which comprehensively searches forward for fundamental quality (counting monetary, social, and environmental performance). In along these lines, the balanced utilization of natural assets and suitable administration of the building stock will add to saving rare assets, decreasing energy utilization (energy protection), and enhancing environmental quality.

Contemporarily, the idea and thought of Sustainable Architecture/Development are particularly relevant in two ecological viewpoints. These include the following:

- Environmental and Ecological emergency
- Disasters which are unavoidable and their prompt administration

The significant causes contributing to these two angles are as follows:

a. **Rapid industrialization and urbanization**: This will prompt topographical stores of sewage and rubbish, population explosion, unsustainable examples of living and development, degradation of natural resources (contamination of air, water, soil, and so forth, and sustenance web interruption). In this way maintainable urban advancement is significant to enhance the lives of urban population and the rest of earth. Both individuals and environment will be affected upon by their actions.

b. **Natural disasters**: Natural disaster like earthquakes, flood, volcanic emissions, starvation, and so forth which are further exasperated by humanity add to the rundown of other sick impacts like atomic explosion, green house impact, ozone depletion, and so forth. Suitable planned endeavors have to be made so as to get an idea of the natural processes and the effect of design in the environment. Making regular cycles and procedures obvious breathes life into the structured condition back.

c. **Exhaustion of nonrenewable sources**: Rapid utilization of nonresources sources prompts significant issues related to energy and water preservation and

so on. Therefore utilization of natural assets in an appropriate way and proper management of building stock can contribute to reduction in the energy utilization, saving rare assets, and radically enhancing the quality of environment.

22.2 CARBON TRADING

Carbon dioxide is a harmful ozone harming substance; it is in charge of about portion of the climatic warmth held by follow gases causing a worldwide temperature alteration. It is created by consuming fossils fuels and deforestation pursued by biodegradation and burning of biomass.

Different firms scour the world searching for environmental administrations that can counter balance the discharge of its customer. These administrations include afforestation and are referred to in the business as carbon sinks or carbon resources, since trees expel carbon from the climate and sequester it in their wood. This movement of the sinks is regularly referred to as carbon sequestration.

In the economy point of view, carbon trading is a type of emission exchanging that enables a nation to meet its carbon dioxide emanations decrease duties, frequently to meet Kyoto Treaty prerequisites, in as minimal effort as conceivable by making utilization of the free market. By this, people in general expense or societal expense of contamination via carbon dioxide is privatized. Carbon trading means exchanging of authentications that represent different manners by which carbon-related emanations decrease targets may be met. Members in carbon exchanging purchase and move legally binding duties or endorsements that speak to indicated measures of carbon-related emanations that either are permitted to be radiated; contain decreases in discharges (new innovation, vitality effectiveness, sustainable power source); or include counterbalances against outflows, for example, carbon sequestration (catch of carbon in biomass).

Individuals purchase and move such items since it being the best cost-effective method through which a general decrease in the amount of emissions can be accomplished, with the assumption that transaction costs engaged with market interest are kept at reasonable price. The way that it is savvy is on the grounds that the substances that have accomplished their very own outflow decrease target effectively will have the capacity to make discharge decrease declarations "excess" to their own prerequisites. These elements can sell those surpluses to different elements that would cause high expenses by looking to accomplish their requirement of emission reduction inside their very own business. So also, venders of carbon sequestration give elements another option, to be specific counterbalancing their outflows against carbon sequestered in biomass.

There are two sorts of carbon exchanging. The first is emissions trading. The second is trading in project-based credits. These categories are put together in hybrid trading systems.

22.2.1 EMISSIONS TRADING

Emanations' exchanging is otherwise called "cap-and-trade." In this system, all emissions are constrained or "capped." By the Kyoto Protocol, which is a cap and trade framework, the emissions from Annex B nations are capped and overabundance permits might be exchanged. However, ordinary cap and trade systems never incorporate instruments like CDM, which enables more permits to enter the system, beyond the cap (Point Carbon).

State for instance, there are two organizations, A and B. Every one of them emanates 100,000 tons of carbon dioxide yearly. The administration anticipates cut their emanations by 5%. The legislature along these lines gives each organization rights, or "stipends," to discharge 95,000 tons this year. Each organization should either decrease its emanations by 5,000 tons or purchase remittances worth 5,000 tons from another person. The market cost for these stipends is $ 10 for each tone. Organization A can lessen its discharges for a large portion of this expense per ton. So it is sensible for it to cut its outflows by 10,000 tons: on the off chance that it moves additional 5,000 tons (for $ 50,000) and it will have the capacity to recuperate its whole consumption. So the organization will have a reserve fund of $ 25,000. For organization B, making decreases is much costly. Cutting each ton of emanations will cost it $ 15. So it chooses as opposed to lessening its outflows, it will purchase the 5,000 tons of surplus stipends that organization A is putting forth. On the off chance that organization B diminished its own emanations, it would cost $ 75,000. Be that as it may, if it chooses to purchase organization A's surplus remittances, the cost may be $ 50,000. So organization B additionally spares $ 25,000 on the arrangement. The two firms, in the end, spare $ 25,000 over what they would have needed to spend without exchanging. In the event that they are the main two organizations in the nation, this implies the nation's business area ends up cutting emanations the same amount of as it would have under common control. In any case, by conveying the decreases over the nation's whole private segment, it costs the division in general $ 50,000 less to do as such. A few discharges exchanging plans enable organizations to spare any surplus stipends they have for their own utilization in future years, instead of moving them.

22.2.2 TRADING IN PROJECT-BASED CREDITS

Assume there are two organizations, A and B, each transmitting 100,000 tons of carbon dioxide a year. What's more, the administration needs to cut their outflows by 5%, so it gives each organization recompenses to transmit just 95,000 tons. Be that as it may, presently the administration tells each organization that on the off chance that it would not like to cut its emanations by every 5000 tons, it has another choice. It can put abroad in undertakings that "decrease" outflows of carbon dioxide 5000 tons beneath what might have happened something else. Such undertakings may incorporate developing products to create biopowers that can be utilized rather than oil; introducing apparatus at a substantial industrial facility to demolish ozone harming substances; consuming methane leaking out of a coal mine or waste dump with the goal that it does not disappear to the air; or building a breeze control generator. The cost of credits from such activities is just $4 for each ton, because of low work costs, a plenty of "messy" industrial facilities, and government and World Bank endowments taking care of part of the expenses of building the tasks and figuring how much carbon dioxide equal they spare. In this circumstance, it bodes well for both organization A and organization B to purchase credits from abroad as opposed to make decreases themselves. Organization A recovers $5000 by purchasing credits from activities abroad instead of cutting its very own discharges. Organization B then spares $55,000. The complete putting something aside for the local private part is $60,000. Other names for undertaking-based credit exchanging incorporate "standard and credit" exchanging and "offset" exchanging.

22.2.3 HYBRID TRADING FRAMEWORKS

Half-and-half frameworks make utilization of the two emanations exchanging and "offset" exchanging, and endeavor to present a trade between "stipends" for task-based "credits." The US sulfur dioxide advertisement utilizes emanations exchanging only. Whereas both the Kyoto Protocol and the EU Emissions Trading System blend "top and-exchange" recompenses and task-based attributes, and attempt to make them commonly interchangeable. Such frameworks are extremely perplexing. It is hard to make believable "attributes" and make them identical to "remittances." Blending the two additionally changes the financial aspects.

For instance, envision that organization A and organization B above are given three alternatives in any mix: cutting their very own emanations, exchanging remittances with one another, or purchasing credits from abroad. For organization B, the best alternative would be, by and by, to purchase $20,000 worth credits from abroad as opposed to burning through $75,000 to diminish its very own emanations. For organization A, the best alternative is to chop down its very own emanations by 10,000 tons—however just on the off chance that it can discover a purchaser who will pay $10 for each ton for the 5,000 stipends it would need to save. As opposed to paying $20,000 for carbon credits from abroad, it would not need to spend anything. Tragically for organization A, it cannot locate any such purchaser. In the event that organization B can spare $ 5000 by traveling to another country for credits, it will not purchase organization A's extra recompenses. Be that as it may, organization B is the main other firm in the discharges exchanging plan. So without organization B as a purchaser, it is not beneficial for organization A to make any cuts whatsoever, and it also will end up purchasing credits abroad.

22.3 LIFE CYCLE ASSESSMENT

Life cycle assessment (LCA, otherwise called life-cycle analysis, eco-balance, and cradle-to-grave analysis) is a technique which is generally used to quantify the different natural effects that are related with every phase of an item's life beginning from raw material extraction through materials preparing, produce, distribution, use, fix and upkeep, and transfer or reusing. Figure 22.1 shows the life cycle evaluation graph.

Designers utilize this procedure to help scrutinize their items. LCAs can help maintain a strategic distance from a limited attitude toward environmental problems by:

- compiling a stock of important energy and material sources and environmental releases;
- evaluating the potential effects related with distinguished sources and releases;
- interpreting the outcomes to help settle on a well-informed choice.

22.3.1 GOALS AND PURPOSE

The objective of LCA is to think about the full scope of natural impacts assignable to items and administrations by measuring all data sources and outputs of material flows and evaluating how nature is affected by these material flows. These data are utilized to enhance forms, bolster arrangement, and give a sound premise to educated choices.

The term life cycle alludes to the thought that a reasonable, holistic evaluation requires the evaluation of raw material generation, fabricate, conveyance, use, and transfer including all interceding transportation steps important or caused by the item's presence.

22.3.2 FOUR FUNDAMENTAL STAGES

LCA is in a general sense done in four particular stages as appeared in Figure 22.2 shown underneath. The stages are frequently related so that the aftereffects of one stage will portray how other stages are finished.

a. Goal and scope

An LCA begins with an express articulation of the objective and scope of the study, which sets out the context and discloses how and to whom the results are to be conveyed. The objective and goal incorporate specialized details that guide consequent work:

- The functional unit, which precisely characterizes what is being studied and measures the service by the product system, gives a reference to the data inputs and outputs that can be connected. Further, the functional unit is an essential basis that let alternates goods and services to be compared.
- The framework limits; these incorporate delimitations of which processes that ought to be incorporated in the examination of the product system.
- Any assumptions and constraints.
- The allocation strategies used to divide the environmental processes when a few items or functions share similar process; allotment is regularly managed in any of the following three different ways: framework extension, substitution, and segment.

b. Life cycle inventory

Life cycle inventory (LCI) examination is a dynamic type of investigation that includes making a stock of streams from and to nature for an item framework. These inventory flows will incorporate contributions of water, energy, and crude materials, and discharges to air, land, and water. To build

FIGURE 22.1 Life cycle appraisal.

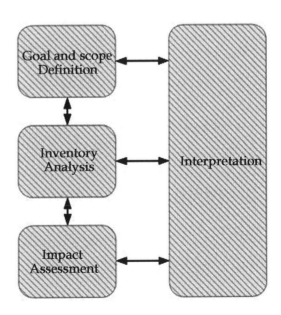

FIGURE 22.2 Illustration of LCA stages.

up the above stock, a flow model of the specialized framework is for the most part built dependent on the input and yield information. The flow model is normally represented with a flow chart that incorporates the activities which will be examined in the applicable store network and gives an unmistakable image of the specialized framework limits. A case of a life cycle stock or life cycle inventory (LCI) graph is shown in Figure 22.3.

The information must be identified with the practical unit characterized in the objective and scope definition. Information can be introduced in tables and a few translations can be made as of now at this stage. The consequences of the inventory is an LCI which gives data pretty much all information sources and yields as rudimentary stream to and from the earth from all the unit forms engaged with the investigation.

c. Life cycle impact assessment:

This period of LCA is essentially focusing on assessing the pertinence of potential natural effects as per the LCI stream results. Established life cycle impact assessment (LCIA) incorporates for the most part the accompanying compulsory components:

- selection of classification pointers, impact classifications, and portrayal models;
- the grouping stage, wherein the stock parameters are arranged and appointed to explicit effect classifications; and
- impact measurement, wherein the categorized LCI flows are characterized, using one among the many possible LCIA methodologies, into common equivalence units which are then summed to provide an overall impact category total.

d. Uses of LCA:

LCA is commonly used in supporting business strategies (18%) and R&D (18%), as input to product or designing process (15%), in education (13%) and for labeling or declarations of product (11%).

LCA will be continuously integrated into the built environment as tools such as the European ENSLIC Building project guidelines for buildings or developed and implemented, which provide practitioners guidance on methods to implement LCI data into the planning and design process.

22.4 ENERGY CONSERVATION

Energy conservation means the reduction in energy utilization through utilizing less of energy service. Energy conservation varies from effective energy use, which alludes to utilizing less energy for a steady administration. Energy protection is a piece of the idea of adequacy.

As such, preservation (sparing) can be found in two different ways: in the positive feeling of advancing the more effective utilization of energy (utilizing less to accomplish more) and in the negative feeling of purposely confining use by appeal ("Switch off Something"), by valuing, including tax collection (making it too costly to even consider wasting), or by apportioning (impulse). It is not just an issue of applying innovation; it likewise includes inducing individuals not exclusively to utilize the available technology yet in addition to change their behavior when they prize the accommodation and comfort brought by higher energy utilization.

It is made progressively complex since energy conservation is observed to be a way to different closures:

1. Reducing emanations of ozone-harming substance to counter a dangerous atmospheric deviation.
2. Reducing the petroleum product utilization, in this manner contributing both to decrease of carbon discharge and local vitality security.
3. Helping those living in the alleged fuel neediness by diminishing utilization of vitality for the most part by methods for better protection, controls, draft sealing, and so on.
4. Eliminating the requirement for power station substitution.

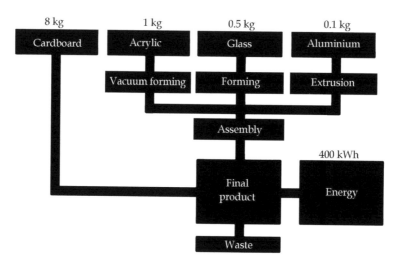

FIGURE 22.3 A case of a life cycle stock (LCI) chart.

22.4.1 ENERGY EFFICIENCY

Researchers, architects, and technologists are continuously trying to enhance the effectiveness with which machines, apparatuses, structures, and vehicles use energy and have been breathtakingly fruitful; however, crushing progressively out of existing innovation gets more diligently with each enhancement.

Like other Western countries, Britain is a substantially more vitality proficient country than 40 years back. That is a truly dynamite advance by any principle; however, some of it will have come about because of the decay of substantial industry. In any case, power request has kept on ascending by a normal of 1%–1.5% a year; however, the yearly figure fluctuates with the economic activity. At the end of the day, enhanced energy effectiveness does not really diminish energy utilization.

22.4.2 CUTTING OUT WASTE

There are numerous ways to lessen utilization by removing waste. It is surely conceivable to cut its utilization in crises, for example, the multiday week in 1974 through limitations, yet it is hard to continue the investment funds. In a democratic society, voters get extremely anxious if their convenience and comfort are meddled with. It pursues that reserve funds accomplished amid an emergency are entirely present moment and never kept up.

As a device for diminishing interest, vitality preservation must be something in excess of a concise reaction to national troubles. It requires a difference in demeanor and conduct toward the utilization of vitality.

It is conceivable to assist purchasers with economizing through such things as space and cavity divider protection, two-fold coating, draft sealing, and lighting and focal warming controls. They make for the more effective utilization of vitality, gave, obviously, individuals do not utilize windows or entryways as a type of cooling and enable their warming frameworks to warm the flying creatures. Regularly, be that as it may, expanded productivity urges individuals to turn up their indoor regulators to appreciate more solace at a similar expense. It is likewise conceivable to decrease utilization by increasingly proficient innovation, direction, and expense weight similarly as with mixture vehicles, operational models, building controls, and special duty rates for progressively practical frameworks.

22.4.3 ELECTRIC AND HEAT ENERGY

Heat energy implies the vitality that can be acquired through consuming of resources, for example, oil, coal, and charcoal. Heat energy turns the motive energy in the motor of a vehicle or vessel, or steam train through air or steam. Electric energy implies the "power" that is delivered in a power plant, transmitted by transmission lines, and can be gotten by paying for utilities. By and large, power can be acquired through an outlet or battery and can be utilized as a thought process control for electrical items, for example, TVs, and iceboxes.

Heat energy and electric energy are utilized in different forms, depending upon equipment or facilities in use. Typically, they cannot be utilized at 100% of full effectiveness, and a few misfortunes happen. For example, at workplaces, if PCs are worked just while clients take a gander at their screens, the best proficiency will occur. In any case, indeed, while clients are on the phone, or serving clients, screens are still shown. Electric energy is wasted amid such time. Albeit more PCs have the capacity of energy protection naturally getting to be backup mode when immaculate after a specific timeframe, it is difficult to turn control on just while clients take a gander at screens. A similar kind of energy misfortunes jumps out at numerous parts of energy utilization in bigger size, for example, influence plants in industrial facilities all through society. Energy preservation is relied upon to decrease these sorts of misfortunes of vitality in the whole society however much as could be expected and mean to raise the proficiency of vitality use as near 100% of the full rate.

22.5 RENEWABLE ENERGY ASSETS

Sustainable power source is energy that is gathered from inexhaustible assets, which are normally recharged on a human timescale, for example, daylight, wind, rain, tides, waves, and geothermal warmth. Sustainable power source regularly gives energy in four vital zones: power age, air and water warming/cooling, transportation, and provincial (off-network) vitality administrations.

Sustainable power source assets and noteworthy open doors for vitality productivity exist over wide land zones, rather than other vitality sources, which are amassed in a set number of nations. Fast organization of sustainable power source and energy effectiveness, and mechanical enhancement of vitality sources, would result in noteworthy vitality security and financial advantages. It would likewise diminish ecological contamination, for example, air contamination caused by consuming nonrenewable energy sources and enhance general well-being.

Coming up next are the instances of inexhaustible assets.

22.5.1 WIND POWER

Wind currents can be utilized to run wind turbines. The present day utility-scale wind turbines extend from around 600 kW to 5 MW of evaluated control, in spite of the fact that turbines with appraised yield of 1.5–3 MW have turned into the most well-known for business use. The biggest generator limit of a solitary introduced inland wind turbine achieved 7.5 MW in 2015. The power accessible from the breeze is an element of the 3D square of the breeze speed, so as wind speed builds and controls yield increments up to the most extreme yield for the specific turbine. Zones where winds are more grounded and progressively consistent, for example, seaward and high-elevation locales are favored areas for wind ranches.

Internationally, the long-term specialized capability of wind energy is accepted to be five times all out flow worldwide total energy generation, or multiple times flow power request, expecting every single useful boundary required were survived. This would require wind turbines to be introduced over vast zones, especially in territories of higher breeze assets, for example, seaward. As seaward wind speeds normal 90% more prominent than that of land, so seaward assets can contribute considerably more energy than land-positioned turbines.

22.5.2 Hydropower

In 2015 hydropower produced 16.6% of the world's all out power and 70% of all inexhaustible power. Since water is around multiple times denser than air, even a moderate streaming stream of water, or moderate ocean swell, can yield extensive measures of vitality.

Hydropower is delivered in 150 nations, with the Asia-Pacific district producing 32% of worldwide hydropower in 2010. For regions having the biggest level of power from renewables, the best 50 are principally hydroelectric. China is the biggest hydroelectricity maker, with 721 terawatt-long stretches of generation in 2010, speaking to around 17% of household power use.

Wave control (catches the vitality of sea surface waves) and tidal power (changing over the vitality of tides) are the two types of hydropower with future potential; notwithstanding, they are not yet broadly utilized monetarily. A show venture worked by the Ocean Renewable Power Company on the shoreline of Maine, and associated with the framework, outfits tidal power from the Bay of Fundy, area of world's most astounding tidal stream. Ocean thermal energy conservation, which utilizes the temperature distinction between cooler profound and hotter surface waters, has as of now no financial attainability.

22.5.3 Solar Energy

Sunlight-based energy, light, and heat from the sun are harnessed by utilizing a scope of consistently advancing technologies, for example, solar heating, photovoltaic (PV), concentrated solar power (CSP), concentrator PV, sun-based engineering and artificial photosynthesis. Solar technologies are comprehensively characterized as either passive or active sunlight-based relying upon the manner in which they collect, convert, and distribute sun-based energy. Passive solar systems incorporate orientation of building to sun, choosing materials with ideal warm mass or light scattering properties, and structuring spaces that normally flow air. Active solar technology incorporates sun-powered heat energy, and sun-based power, changing over daylight into power either specifically utilizing PV, or indirectly utilizing CSP.

A PV system changes light into electrical direct current by exploiting the photoelectric impact. Concentrated sun-based power (CSP) system uses lenses or mirrors and following frameworks to center a huge territory of daylight into a little shaft. CSP-Stirling has by a wide margin the most noteworthy proficiency among all sunlight-based energy technology.

22.5.4 Geothermal Energy

High-temperature geothermal energy is from thermal energy created and contained within the Earth. Heat energy is the energy that decides the temperature of matter. Earth's geothermal energy starts from the first development of the planet and from radioactive decay of minerals. The geothermal gradient, which is the distinction in temperature between the center of the planet and its surface, drives a constant conduction of warm energy from the center to the surface. The word geothermal starts from the Greek roots geo, which means earth, which means heat.

Low-temperature geothermal refers to the utilization of the outer crust of the earth as a thermal battery to facilitate renewable warm vitality for warming and cooling structures, and other refrigeration and mechanical employments. In this form of geothermal, a Geothermal Heat Pump and Ground-coupled warmth exchanger are utilized together to move warm energy into the earth (for cooling) and out of the earth (for warming) on a changing occasional premise. Green house potential (referred to as "GHP") is an important sustainable innovation since it both decreases yearly energy consumption related with heating and cooling, and it likewise levels the electric interest bend dispensing with the extraordinary summer and winter crest electric supply necessities. Along these lines low temperature geothermal is turning into an expanding national need with numerous tax credit support and center as a part of the continuous development toward Net Zero Energy.

22.5.5 Bioenergy

Biomass is a natural material obtained from living, or as of late living life forms. It refers to plants or plant-inferred materials which are explicitly called ligno-cellulosic biomass. As an energy source, biomass can either be utilized straightforwardly by means of burning to create heat, or converting it into different types of biofuel. Transformation of biomass to biofuel can be accomplished by various strategies which are extensively arranged into thermal, chemical, and biochemical methods. Wood remains the biggest biomass energy source today, models incorporate backwoods deposits, for example, dead trees, branches, and tree stumps, yard clippings, wood chips, and even metropolitan strong waste. In the second sense, biomass incorporates plant or creature matter that can be changed over into strands or other mechanical synthetic concoctions, including biofuels. Mechanical biomass can be developed from various sorts of plants, including miscanthus, switch grass, hemp, corn, poplar, willow, sorghum, sugarcane, bamboo, and an assortment of tree animal groups, extending from eucalyptus to oil palm.

Plant energy is created fundamentally by products that are especially developed for utilization as fuel that provides high biomass yield per hectare with low amount of input energy. The grain can be utilized for transportation of fluid powers while the straw can be scorched so as to create warmth or power. Plant biomass can likewise be corrupted from cellulose to glucose through a progression of concoction medications, and the sugar along these lines acquired can be utilized as an original biofuel.

Biomass can be converted to other usable types of energy like methane gas or transportation powers like ethanol and biodiesel. Decaying refuse, and farming and human waste, all discharge methane gas—also known as landfill gas or bio-gas. Harvests like corn and sugarcane can be matured which will create the transportation fuel, ethanol. Biodiesel, which is another transportation fuel, can be created from left-over items like vegetable oils and animal fats. Likewise, biomass to liquids and cellulosic ethanol are ongoing research works. There is a lot of research including algal fuel or green growth inferred biomass as a result of the way that it is a nonfood resource and can be delivered five to ten times than those of different kinds of land-based agriculture, for example, corn and soy. Once collected, it very well can be fermented to deliver biofuels, for example, ethanol, butanol, and meth-ane, just as biodiesel and hydrogen. The biomass utilized for power age shifts relying on the district. Backwoods side-effects, for example, wood buildups, are basic in the United States. Agrarian waste is regular in Mauritius (sugar stick buildup) and Southeast Asia (rice husks). Creature farming buildups, for example, poultry litter, are basic in the United Kingdom.

22.6 INTRODUCTION TO GREEN BUILDINGS

Worldwide sustainability objectives have at last brought about the improvement of the green building development. Sustainable developments and green structures are frequently utilized reciprocally.

There are many green building rating systems. United States Green Building Council managed "Leadership in Energy and Environment Design (LEED)" is the worldwide market pioneer in the rating systems. LEED is an honorable and great effort in moving toward sustainable environment by changing built environment green. In any case, it has certain entanglements and difficulties. A portion of these difficulties are regarding strategies on material determina-tion and monitoring of performance. The materials utilized in a project are considered at a typical beginning stage and no thought is given to the existence cycle execution of the material. Explanations concerning manageability require approval, and life cycle analysis (LCA) is an apparatus that can give such legitimacy.

In the field of building, sustainable design is a design ide-ology, which harbors the thought of sustainable human and societal advancement. One of the remarkable viewpoints in supportability is economic development. Reasonable devel-opment rehearses are done so that they depend on natural standards, with no ecological effects, have a shut material circle, and have full coordination into the scene after the administration life of the structure is finished. The idea of green structures is the effort of our efforts in accomplish-ing the idealistic sustainable construction practices. As indi-cated by Environmental Protection Agency in the United States, Green Building is the "practice of making structures and utilizing processes that are environmentally dependable and asset proficient all through a building life-cycle from sitting to plan, development, activity, upkeep, remodel, and deconstruction."

22.6.1 GOALS OF GREEN BUILDINGS

Green building unites a wide range of practices, methods, and abilities to reduce and eliminate the effect of structures on the earth and human well-being. It underlines on exploiting sustainable assets, e.g., utilizing daylight through aloof sun powered, and PV gear, and utilizing plants and trees through green rooftops, rain patio nurseries, and decrease of water runoff. Numerous different strategies like low-impact build-ing materials or packed gravel or porous concrete rather than ordinary concrete or black-top are utilized to improve ground water renewal.

While the practices or technologies utilized in green build-ing are always developing and may vary from locations to locations, key standards endure from which the technique is inferred: siting and structure plan effectiveness, vitality pro-ficiency, water productivity, materials productivity, indoor natural quality upgrade, activities and support enhancement, and squander and toxics reduction.

On the aesthetic side of green engineering or practical structure is the rationality of planning a building that is in agreement with the regular highlights and assets encompass-ing the site. There are a few key strides in structuring rea-sonable structures: indicate "green" building materials from nearby sources, diminish loads, enhance systems and create nearby sustainable power source.

22.7 GREEN BUILDING FOUNDATIONS

The establishment is maybe a building's most challeng-ing part from energy transport and hygrothermal points of view. The foundation can represent as much as 40% of the building envelope conduction misfortune and, all the more significantly, in vacant conditions, practically the majority of the inactive load (or vitality required for dehumidifica-tion). From a materials point of view, the foundation needs to give a durable interface between the ground and interior of the building inside within the sight of mass water, soil gasses (for example, radon, water vapor, and even, every so often, hydrogen sulfide), ice, biotic action, and pest invasion while at the same time being the building's structural prem-ise. From a material point of view, the foundation must pro-tect the inside from the encompassing soil and surrounding condition, and give an inside surface temperature creating agreeable conditions for the tenants while maintaining both reasonable and latent thermal loads to the best conceivable degree.

A well-planned foundation can make a significant con-tribution to controlling heat and cooling costs while dispos-ing problems of potential dampness and shape issues. There are a wide range of foundation to choose, depending upon atmosphere, soils condition, water table, and different elements. Heat is lost through foundation walls, soil conditions, etc. Foundations should be protected, ideally outside to lessen the danger of condensation and make the mass of cement or block walls part of the conditioned space. An insulated foundation turns out to be a piece of the house's warm envelope, adding to energy preservation and making an increasingly agreeable condition. Although exterior protection offers more favorable circumstances, basements can be protected from inside, as well. The key is controlling the development of dampness and mold-inducing condensation as warm air relocates to a cooler surface.

22.7.1 Green Foundation Transition

a. Build frames from plywood or aluminum:
  1. Initial expenses are higher, yet they spare the inconvenience and cost of supplanting shapes produced using dimensional wood.
  2. Wood structures produced using 2×12's must be utilized a couple of times and are normally too ruined to even consider being reused.
b. Reuse frames however much as could be expected:
  1. Modular structures are made for most establishment applications.
  2. Keep however much material out of the landfill as could be expected can diminish landfill costs.
  3. Clean and stack utilized structures after use and reuse them on another venture.
c. Use protected solid structures:
  1. Combine protection and cement to shape one divider.
  2. Polystyrene squares are stacked to make an establishment divider and cement has filled the void for unbending.
  3. Save on both cement and work costs.
  4. There are hinders that are produced using reused woodchips and mineral fleece. Most are produced using polystyrene
d. Install an edge deplete at the base of an establishment divider:
  1. A edge deplete assembles water that may some way or another harm the establishment divider and channels it far from the establishment.
  2. A punctured pipe enveloped by arranging felt and put in a bed of smashed shake situated around the balance of the establishment.
e. Insulate establishment dividers and pieces with unbending froth:
  1. Fly ash is not constantly appropriate for level work as it takes more time to fix.
  2. Fly ash makes concrete more grounded, increasingly strong and water safe.
  3. Fly ash is a result of coal let go vitality plants. When requesting concrete, request it to be added to your blend. Advise the auxiliary designer that you expect to utilize fly ash.

22.8 ECOLOGICAL DESIGN

Biological plan is characterized by Sim Van der Ryn and Stuart Cowan as any type of structure that limits earth ruinous effects by incorporating itself with living procedures.

An eco-plan item will have a support-to-support life cycle guaranteeing that zero waste is made in the entire procedure. By taunting life cycles in nature, eco-structure is a principal idea in accomplishing a genuinely round economy.

Ecological angles which should be broke down in each phase of the existence cycle are as follows:

1. Resource utilization (vitality, materials, water, or land zone).
2. Emissions to air, water, and ground (our Earth) as being significant for the earth and human well-being.
3. Miscellaneous (for example, clamor and vibration).

Squander (unsafe waste and other waste characterized in natural enactment) is just a middle of the road step and the last emanations to the earth (for example, methane and draining from landfills) are stocked. All consumables, materials, and parts utilized in the existence cycle stages are considered responsible for, and all roundabout natural viewpoints are connected to their creation.

The ecological parts of the periods of the existence cycle are assessed dependent on their natural effect based on various parameters, for example, degree of natural effect, potential for development, or capability of progress.

22.8.1 Applications in Design

Eco-design ideas as of now affect numerous parts of structure; the effect of an unnatural weather change and an increase in CO_2 discharges have driven organizations to consider more naturally cognizant methodology in their planning and process. In building plan and development, originators are utilizing the idea of Eco-design all through the structure procedure, from the material choice to the sort of vitality that is being expended and the transfer of waste.

As for these ideas, online stages that bargain with Eco-design items are approaching, with the extra economical reason for disposing of conveyance steps which are a bit much between the originator and the last client.

Eco-Materials, for example, the utilization of nearby crude materials, are less expensive and diminish the ecological expenses of delivery, utilization of fuel, and CO_2 outflows created from transportation. Ensured green building materials, for example, wood from economically overseen woodland estates, with accreditations from organizations, for example, the Forest Stewardship Council, or the Pan-European Forest Certification Council are prudent.

Recyclable and reused materials are predominantly utilized in development; however, guarantee that they do not create any harmful waste items amid make or after their life cycle closes. Recovered materials, for example, timber at a building site or junkyard can be given a second life by utilizing them for platforms, bolster shafts in another building, or as furniture. Stones from an unearthing can be utilized in a holding divider. The reuse of these things implies that less energy consumption will be involved in making new items.

Water reusing frameworks, for example, water tanks that reap water for various purposes. Reusing dim water produced by family units is a helpful method for not squandering drinking water.

22.9 ASSESSING HIGH-PERFORMANCE GREEN BUILDINGS

A green building utilizes extensively less vitality and water than a customary building has less site impacts and by and large larger amounts of indoor air quality (IAQ). It additionally represents some proportion of the life cycle effect of building materials, furniture, and decorations. These advantages result from better site advancement rehearses; plan and development decisions; and the aggregate impacts of activity, support, evacuation, and conceivable reuse of building materials and framework.

Superior green structures can

1. substantially lower the vitality utilization of the building and cost;
2. improve understudy and laborer well-being and efficiency through better quality of indoor condition;
3. generate on location inexhaustible power which is likewise less inclined to catastrophes and national security dangers;
4. reduce the natural effects of the "manufactured" condition; and
5. provide occupation opportunity in the sustainable power source and bio-based item enterprises (lessening the nation's dependence on imported oil).

22.10 ASSESSMENT OF GREEN BUILDINGS

Manageability appraisal incorporates the ecological, the social, and the practical effects of our structures on present and future ages just as the collaborations between these three circles. Structures are not maintainable. An "economical building" is an interesting expression. There is no such thing as a building that can continue itself in time. The expression "maintainable structures" is utilized not to allude to self-supporting structures but rather to allude to structures that contribute decidedly to practical advancement.

Structures cooperate with their prompt surroundings in a direct and once in a while unintervened way. The greater part of their effect is nearby. Building configuration decides open space in urban areas, structures impact the urban frameworks adding to the weight on foundation and metabolic frameworks; they divert wind, cast shadows, create waste, and so forth. All effects specifically identified with urban personal satisfaction. Despite the fact that we cannot overlook that every single neighborhood activity has an effect on worldwide frameworks, it very well may be said that most quantifiable effects of structures are on the nearby scale. Other than the natural effects, the financial and social execution of structures must be surveyed on the off chance that we relate them to a city that is the framework to which those structures have a place. As expressed in the Agenda 21 on Sustainable Construction (CIB 1999) "obviously the entire development industry has a critical effect, both straightforwardly and in a roundabout way, on accomplishing Sustainable Development in the urban condition."

22.11 GREEN BUILDING RATING SYSTEMS

A green building rating framework is a method that assesses the execution of a building and its effect on nature. It contains a predefined set of criteria identifying with the structure, development, and tasks of green structures.

We can characterize Green Buildings as structures that guarantee proficient utilization of common assets like building materials, water, vitality, and different assets with insignificant age of nondegradable waste. Advancements like proficient cooling frameworks have sensors that can detect the warmth produced from the human body and naturally alter the room temperature, sparing vitality. It applies to lighting frameworks as well. Utilization of vitality productive light emitting diode (LED) lamps and compact fluorescent lamps (CFLs) rather than ordinary glowing light, new age apparatuses that devour less vitality, and numerous different choices help in making the structures green and make them not quite the same as regular ones.

In India, there are transcendently three rating frameworks Indian Green Building Council (IGBC), Green Rating for Integrated Habitat Assessment (GRIHA), and Bureau of Energy Efficiency (BEE).

22.11.1 INDIAN GREEN BUILDING COUNCIL

The LEED is the rating framework created for ensuring Green Buildings. LEED is created by the U.S. Green Building Council (USGBC), the association advancing maintainability through Green Buildings. LEED is a structure for surveying building execution against set criteria and standard purposes of references. The benchmarks for the LEED Green Building Rating System were created in year 2000 and are presently accessible for new and existing developments.

Confederation of Indian Industry (CII) founded the IGBC in year 2001. IGBC is the charitable research foundation having its workplaces in CII-Sohrabji Godrej Green Business Center, which is itself an LEED-guaranteed Green building. IGBC has authorized the LEED Green Building Standard from the USGBC. IGBC encourages Indian green structures to end up one of the green buildings. Figure 22.4 speaks to the Indian green structure.

IGBC has built up the accompanying green building rating frameworks for various kinds of working in line and similarity with US Green Building Council. Till date, the following Green Building rating frameworks are accessible under IGBC:

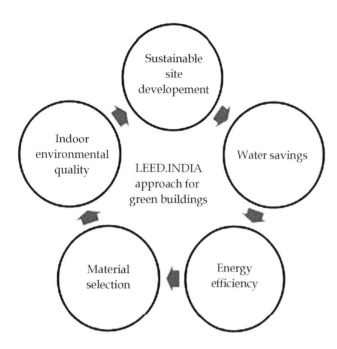

FIGURE 22.4 Indian green structure.

1. LEED India for New Construction
2. LEED India for Core and Shell
3. IGBC Green Homes
4. IGBC Green Factory Building
5. IGBC Green SEZ
6. IGBC Green Townships

22.11.2 GREEN RATING FOR INTEGRATED HABITAT ASSESSMENT

GRIHA is India's very own rating framework together created by The Energy and Resources Institute (TERI) and the Ministry of New and Renewable Energy, Government of India. It is a green building structure assessment framework where structures are evaluated in a three-level process. The procedure starts with the online accommodation of records according to the endorsed criteria pursued by on-location visit and assessment of the working by a group of experts and specialists from GRIHA Secretariat. GRIHA rating framework comprises 34 criteria classified in four distinct segments. Some of them are site choice and site arranging, preservation and proficient usage of resources, building task and upkeep and advancement.

22.11.3 BUREAU OF ENERGY EFFICIENCY

BEE built up its very own rating framework for the structures dependent on a 1–5 star scale. More stars mean more vitality proficiency. It has built up the Energy Performance Index. The unit of kilo watt hours per square meter every year is considered for rating the building and particularly targets cooled and noncooled places of business. The Reserve Bank of India's structures in Delhi and Bhubaneshwar, the CII Sohrabji

Godrej Green Business Center, and numerous different structures have gotten BEE 5-star appraisals.

22.12 THE GREEN BUILDING DESIGN PROCESS

Integrative plan is an extremely dynamic process that can be joined to elevate an undertaking's general achievement. It includes different regions of a venture co-operating from the beginning toward one noteworthy objective. With respect to green building, this methodology is normally taken to enable a building structure to accomplish greatest effectiveness, bring down expenses, and increment generally execution. It can likewise help accomplish LEED focuses if a building is looking for that accreditation.

For green buildings, integrative structure has two definitions. The regular definition depicts colleagues from all territories cooperating through a venture's advancement and movement. The improved definition, be that as it may, incorporates the cooperation of what these different colleagues are working near: atmosphere, building configuration, use, and frameworks. Around 70% of the choices related with natural effects are made inside the principal 10% of the plan procedure. In this manner, it is through the blend of these two definitions that a venture can be arranged and executed to its greatest potential.

One expression stands consistent with the accomplishment of integrative plan: the more, the merrier. All gatherings required with a specific undertaking ought to work together, including customers, modelers, venture proprietors, engineers, general contractual workers, and that is only the tip of the iceberg.

The structure procedure itself accentuates the accompanying expansive succession.

a. Establish execution focuses for a wide scope of parameters and create primary strategies to accomplish these objectives. This sounds self-evident; however, with regard to a coordinated plan group approach, it will bring building aptitudes accordingly helping the proprietor and draftsman to abstain from surrendering to a problematic structure arrangement.
b. Minimize warming and cooling loads and boost day lighting potential through introduction, building arrangement, an effective building envelope, and cautious thought of the sum, type, and area of fenestration.
c. Meet warming and cooling loads through the greatest utilization of sun-based and other inexhaustible advancements and the utilization of effective HVAC frameworks, while keeping up execution focus for IAQ, warm solace, brightening levels and quality, and commotion control.
d. Iterate the procedure to deliver no less than two, and ideally three, idea structure options, utilizing vitality reproductions as a trial of advancement, and after that select the most encouraging of these for further improvement.

22.13 THE SUSTAINABLE SITE AND LANDSCAPE

Manageable site configuration is a standout among the most affordable and basic instruments for consolidating green frameworks into advancement ventures. Green framework serves to ensure water quality in streams, lakes, and wetlands and in this way can address network issues and administrative necessities. In particular, manageable site configuration joins stream supports, bioswales, rain gardens, diminished hard surfaces, and comparable low-effect advancement rehearses into a system of green spaces can help enhance or sensibly emulate preconstruction spillover conditions. Endeavors to sensibly imitate preconstruction spillover in new advancement can avert streak surges, store, and treat storm water overflow as nature would and accommodate network esteems, for example, amusement and stylish green space. Green framework can be consolidated into new, redevelopment or retrofit ventures. Sadly, numerous networks have discovered that their very own advancement codes and measures can really neutralize this objective. For instance, nearby codes and benchmarks regularly make unnecessary impenetrable cover as wide avenues, broad parking garages, and extensive part subdivisions and they frequently require intemperate clearing and evaluating.

Coordinating green foundation into the site configuration requires incorporating its standards into site ace arranging. This will upgrade arrive use, walker and vehicular dissemination and access, common asset safeguarding and security, stopping, utilities, overflow the executives, diversion and finishing. Ace site arranging must consider green framework structure objectives toward the start of the procedure where the best open door exists for the following:

1. To preserve the natural biosphere.
2. Engineer administration frameworks to improve common frameworks and sensibly impersonate normal, predevelopment works through practices that
   a. enhance evapotranspiration
   b. enhance penetration
   c. minimize increments in surface spillover rates and volume

A manageable scene is more than the cognizant game plan of open air space for human happiness and fulfillment. It is a scene that utilizes negligible water, manures, pesticides, work, and building materials. Making a feasible scene implies moving in the direction of a keen harmony between assets utilized both in development and upkeep and results picked up.

The most widely recognized strides in building up a landscape include

1. drawing a scale delineate (plan) of the property
2. completing a site investigation and surveying family needs
3. determining use zones
4. brainstorming elective formats and structure thoughts
5. creating a scaled illustration of the structure (scene plan)
6. selecting plants

An end-all strategy is fundamental to guarantee that all work done on the property will mix into the ideal ultimate result. Remember that scene advancement can be a long haul process. There is no compelling reason to build up your whole part on the double; finishing the scene over a 5-year time frame may be increasingly practical. This time period enables you to assess plants as they develop and by and large is all the more monetarily reasonable.

22.14 ENERGY AND CARBON FOOTPRINT REDUCTION

A carbon impression is truly characterized as "the all-out arrangement of ozone depleting substance outflows caused by an individual, occasion, association, item communicated as carbon dioxide equivalent." A proportion of the aggregate sum of carbon dioxide (CO_2) and methane (CH_4) emanations of a characterized populace, framework or movement, thinking about every important source, sinks and capacity inside the spatial and worldly limit of the populace, framework, or action of intrigue. It is determined as carbon dioxide identical utilizing the pertinent 100-year an Earth-wide temperature boost potential (GWP100).

Ozone-depleting substances (green house gases (GHGs)) are radiated through transport, arrive freedom, and the generation and utilization of sustenance, energizes, made products, materials, wood, streets, structures, and administrations. For straightforwardness of revealing, usually communicated as far as the measure of carbon dioxide, or GHGs, produced.

An individual's, nation's, or association's carbon impression can be apportioned via conveying a GHG outflows evaluation or other calculative exercises indicated as carbon accounting. When the span of a carbon impression is known, an appropriate system can be conceived to lessen it, for example, by mechanical advancements, better process and item the board, changed Green Public or Private Procurement, carbon catch, utilization techniques, carbon balancing, and others.

Through the advancement of elective activities, for example, sun oriented and wind vitality, which are condition agreeable, inexhaustible assets, or reforestation, the restocking of existing backwoods or forests that have recently been depleted, carbon footprints can be decreased. These models are known as Carbon Offsetting, which is the checking of carbon dioxide outflows with a proportionate decrease of carbon dioxide in the climate.

The primary effects on carbon impressions incorporate populace, monetary yield, and vitality and carbon force of the economy. These variables are the principle focuses of people and organizations so as to diminish carbon impressions. Generation makes a substantial carbon impression, and researchers recommend that diminishing the measure of vitality required for creation would be a standout among the best approaches to diminish a carbon impression. This is because of the way that electricity is in charge of generally 37% of carbon dioxide discharges.

22.14.1 Ways to Lessen Carbon Impression

The most widely recognized approach to diminish the carbon impression of people is to Reduce, Reuse, Recycle, and Refuse. In assembling this should be possible by reusing the pressing materials, by pitching the out-of-date stock of one industry to the business that is hoping to purchase unused things at lesser cost to wind up aggressive.

This should likewise be possible by utilizing reusable things, for example, canteens for day-by-day espresso, or plastic holders for water and other cold drinks as opposed to expendable ones. In the event that that alternative is not accessible, it is best to legitimately reuse the expendable things after use.

Another simple choice is to drive less. By strolling or biking to the goal instead of driving, not exclusively is an individual going to get a good deal on gas; however, they will consume less fuel and discharging less emanations into the climate.

One more alternative for lessening the carbon impression of people is to utilize less cooling and warming in the home. By adding protection to the dividers and storage room of one's home, and introducing climate stripping or caulking around entryways and windows, one can bring down their warming costs in excess of 25%. So also, one can modestly overhaul the "protection" (garments) worn by occupants of the home.

The carbon impression development underscores singular types of carbon balancing, such as utilizing increasingly open transportation or planting trees in deforested areas, to diminish one's carbon impression and increment their "imprint."

Besides, the carbon impression in the nourishment business can be diminished by advancing the production network. A real existence cycle or production network carbon impression study can give valuable information which will assist the business with identifying basic zones for development and gives a core interest.

22.15 BUILT ENVIRONMENT HYDROLOGIC CYCLE

The development of water on the world's surface and through the climate is known as the hydrologic cycle. The hydrologic cycle (Figure 22.5) is used to demonstrate the capacity and development of water between the biosphere, climate, lithosphere, and hydrosphere. Water is put away in the accompanying repositories: climate, seas, lakes, streams, ice sheets, soils, snowfields, and groundwater. It moves starting with one repository then onto the next by procedures like vanishing, buildup, precipitation, testimony, overflow, invasion, sublimation, transpiration, and groundwater stream.

Water is circulated inside the hydrological cycle in three forms. Water vapor in the climate is regularly alluded to as humidity. In the event that fluid and strong types of water can beat environmental updrafts they can tumble to the Earth's surface as precipitation. The development of ice gems and water beads happens when the air is cooled to a temperature that causes buildup or testimony.

- Precipitation can be characterized as any watery store, in fluid or strong shape, that creates in an immersed air condition and for the most part tumbles from mists.
- Evaporation is development of free water to the climate as a gas. It requires a lot of vitality.

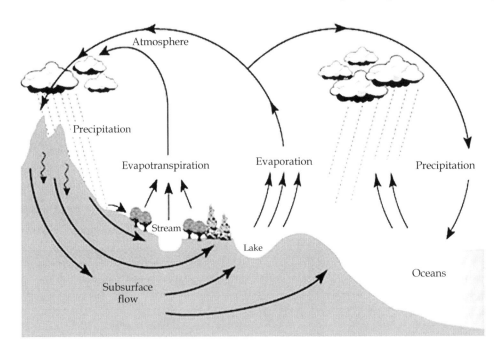

FIGURE 22.5 Hydrologic cycle.

- Transpiration is the development of water through a plant to the climate. Researchers utilize the term evapotranspiration to portray the two procedures.
- Infiltration is the development of water from precipitation into the dirt layer. Invasion differs both spatially and transiently because of various natural components.
- Runoff is the surface stream of water to territories of lower rise or the measure of water that penetrates the dirt differs with the level of land incline, the sum and kind of vegetation, soil type and shake type, and whether the dirt is now soaked by water. The more openings in the surface (splits, pores, joints), the more invasion happens. Water that does not penetrate the dirt streams superficially as spillover.

Surface stream fuses development of water inside the earth, either inside the energize zone, or aquifers. Subsequent to invading, subsurface water may come back to the surface or in the end saturate the sea.

22.16 LCA OF BUILDING MATERIALS AND PRODUCTS

The building business utilizes extraordinary amounts of crude materials that likewise include high vitality utilization. Picking materials with high substance in typified vitality involves an underlying abnormal state of vitality utilization in the building creation organize yet in addition decides future vitality utilization so as to satisfy warming, ventilation, and cooling requests.

The existence cycle center must help basic leadership while choosing the best innovation accessible and limiting the ecological effect of the structures through their plan or repairing. Regularly, items that are introduced as shoddy in the medium term can have high upkeep or waste administration costs and profoundly mechanical items can have high generation costs that are never recovered. Conflictingly, it might be that when we consider the entire life cycle, materials with huge CO_2 outflows, for example, concrete, can see their discharges decreased by allowing them a second life as a filler material in foundation, with a twofold impact: the decrease of emanations contrasted and acquiring filler materials from quarries and the retention of CO_2 because of the recarbonation forms. Accordingly, it is central to apply the existing cycle vision and consider both the financial and natural costs while distinguishing the most eco-effective innovation.

22.17 INDOOR ENVIRONMENTAL QUALITY

Indoor Environmental Quality (IEQ) incorporates the conditions inside a building air quality, lighting, warm conditions, ergonomic, and their consequences for tenants or occupants. Systems for tending to IEQ incorporate those that ensure human well-being, enhance personal satisfaction, and decrease pressure and potential wounds. Better indoor natural quality can upgrade the lives of building inhabitants,

increment the resale estimation of the building, and diminish risk for building proprietors.

Indoor environmental quality (IEQ) is once in a while considered as a need in most advancement arranging and the executives. IEQ components cover 21% of green building assessment criteria for nonprivate building, for example, scholastic structures in advanced education instinct. An unevenness of IEQ can likewise give negative effect to offices, building, and inhabitants. This adds to the well-being nature of tenant and the instance of scholarly building will influence the educating and learning process. Other than that, it will contribute for Sick Building Syndrome (SBS), which implies an expansive definition that incorporates an assortment of side effects viewed as experienced when the tenants invested energy in a specific building. A building will be sorted as a sick building when more than 30% inhabitants have grumbled of related manifestations. SBS is firmly much identified with the building administrations that incorporate mechanical ventilation relative moistness and natural air ventilation rates. Ordinary individual invest 90% of the energy in the building either in work environment and home, both indoor ecological (IEQ) of home or working environment, or equivalent significance on human well-being. IEQ essentially centers around warm temperature (°C), relative mugginess (%), acoustic solace estimated in decibel (dB), lighting estimated in lux level (lux), and IAQ (CO_2 focus level and velocity development).

22.18 GREEN BUILDING ECONOMICS

Manageability has turned into an undeniably critical trait of financial exercises depicting techniques for generation, yet additionally showing characteristics of utilization and qualities of capital venture. To some extent, this reflects prevalent worry with ecological protection, yet it might reflect changes in tastes among buyers and financial specialists. Supportability may likewise be a promoting gadget that substantial partnerships and independent companies alike can utilize effectively.

The assembled condition and supportability are nearly interwoven, and prominent consideration regarding green building has expanded over the previous decade. This mirrors the potential significance of genuine property in issues of ecological preservation. Powerful investigations of atmosphere alleviation strategies have underscored that the manufactured condition offers an extraordinary potential for ozone depleting substance reduction. In this manner, little increments in the maintainability of structures, or all the more explicitly in the vitality effectiveness of their task, can affect their present utilization of vitality and their life cycle vitality utilization. Anticipated patterns in urban development in created nations and in the urbanization of creating economies propose that the significance of vitality proficiency in building will keep on expanding.

Be that as it may, the effect of vitality costs straightforwardly influences tenants and speculators too. Vitality costs speak to about 30% of working costs in the run of the mill place of business in the United States. This is the single biggest and most sensible cost in the arrangement of office space.

Rising vitality costs will just expand the remarkable quality of this issue for the private benefit of interest in genuine capital.

Instead of stipulating explicit measures or advancements, green building accreditation programs consider the appropriation of tweaked answers for individual structures. This adaptability is a key component of green building programs. Accreditation programs, for example, LEED, offer a menu of building advances and development works on including Energy and Atmosphere, Water Efficiency, Materials and Resources, just as different classifications. Developers procure credits by satisfying certain criteria over a scope of classes to accomplish a layered confirmation of the existence cycle effect of a building (Certified, Silver, Gold, and Platinum). Be that as it may, green building confirmation programs regularly do not unveil explicit building enhancements or individual credits. This adaptability additionally implies that the evaluative criteria for green building traverses numerous classifications of natural and human wellbeing impacts, leaving the meaning of what establishes a green building vague.

22.18.1 Economic Benefits of Green Buildings

Green structures can be conveyed at a value tantamount to traditional structures, with ventures recovered through operational cost reserve funds and, with the correct plan highlights, make an increasingly gainful working environment, says the World Green Building Council (World GBC).

- **Design and construction costs**: Research demonstrates that building green does not really need to cost more, especially when cost techniques, program management, and natural systems are incorporated into the improvement procedure from the beginning.
- **Asset value**: As speculators and occupiers turn out to be increasingly well informed of the ecological and social effects of the environment, structures with better maintainability qualifications will have expanded attractiveness. Truth be told, greener structures have the capacity to effectively pull in occupants and to order higher rents and deal costs;
- **Operating costs**: Green structures have been appeared to set aside some cash through diminished vitality and water use and lower long haul activities and upkeep costs. The vitality investment funds alone commonly surpass any expense premiums related with their plan and development inside a sensible compensation period.
- **Workplace productivity and health**: Research demonstrates that the green structure qualities of structures and indoor situations can enhance specialist efficiency and tenant wellbeing and prosperity, bringing about primary concern benefits for organizations.
- **Risk mitigation**: Sustainability hazard variables can essentially influence the rental salary and the future estimation of land resources, thus influencing their arrival on speculation. Administrative dangers have

turned out to be progressively evident in nations and urban communities around the globe, including required revelation, construction standards, and laws prohibiting wasteful structures.

22.19 SUSTAINABLE CONSTRUCTION

Practical Construction (otherwise called green development or manageable building) alludes to a structure, the development procedure, and inhabitance forms that are ecologically dependable and asset proficient all through a building's life cycle from area to plan, development, task, support, redesign, and devastation.

Assets are the foundation of each economy and give two essential capacities crude materials for the generation of products and ventures, and ecological administrations. A typical characterization of assets is as appeared.

- Nonsustainable and nonrecyclable assets, for example, petroleum products.
- Nonsustainable yet recyclable assets, for example, minerals.
- Quickly sustainable assets, for example, angles.
- Slowly sustainable assets, for example, woods.
- Environmental assets, for example, air, water, and soil.
- Flow assets, for example, sunlight-based and wind vitality.

The issue of consumption assumes a critical job in the utilization of sustainable and noninexhaustible assets. In the inexhaustible assets exhaustion happens when extraction surpasses the restoration rate.

Supportable development ought to likewise

1. enhance living, working, and relaxation conditions for people and networks,
2. consume least vitality over its life cycle,
3. generate least waste over its life cycle,
4. integrate with the common habitat,
5. use inexhaustible assets where conceivable.

Manageable development ought NOT

1. cause harm to the common habitat or devour a lot of assets amid development, use, or destruction;
2. cause superfluous misuse of vitality, water or materials because of short life, poor plan, wastefulness or low standard development systems;
3. create reliance on high-effect transport frameworks with their related contamination;
4. use materials from compromised species/situations.

22.19.1 Sustainable Materials in Construction

A valuable marker of the natural effect of development materials is Embodied Energy. Typified vitality is the whole of all the vitality required all through the lifecycle of a material.

Vitality utilized is vitality in the following:

1. Acquisition of the crude material
2. Manufacture of the completed item
3. Transportation of the item to site
4. Construction of a building
5. Maintenance through the term of a building
6. Demolition of a building

To lessen the Embodied Energy it is critical to utilize a decent plan in the structure as well as in the administration and usage of materials.

Timber outline development can profit this circumstance on the grounds that:

1. Timber is an inexhaustible asset
2. Timber has low encapsulated vitality

3. Timber has great warm properties
4. Timber has high solidarity to weight proportion

Also we can plan for standard lengths/sizes, pick rescue materials or pick dependable low upkeep materials.

Saving assets is a foundation of green building methods. There are numerous approaches to monitor assets amid the building procedure. For instance, choosing materials that have probably some reused substance can preserve characteristic assets and virgin materials. Limiting development squanders can facilitate the effect on landfills and assets. Introducing water- and vitality-effective items can monitor assets, while lessening working expenses. Picking a green (plant-shrouded) rooftop can lessen vitality use, cool urban warmth islands, and avoid storm water overflow. It also improves the quality of wildlife habitat and air quality.

Bibliography

Allen, R.T. and Edwards, S.C. (1987). *Repair of Concrete Structures*, Blakie and Sons, UK.

Anil, G.K. and Nair, S. S. (2011). *Environmental Knowledge for Disaster Risk Management*, NIDM, New Delhi.

Anu, K. (2010). *Vulnerable India: A Geographical Study of Disasters*, IIAS and Sage Publishers, New Delhi.

Arora, K.R. (2017). *Soil Mechanics and Foundation Engineering*, Standard Publishers and Distributors, New Delhi, 7th Edition.

Asawa, G.L. (2000). *Irrigation Engineering*, NewAge International Publishers, New Delhi.

Atkinson, J. (2007). *Mechanics of Soils and Foundations*, Taylor & Francis.

Atkinson, J.H. and Bransby, P.L. (1978). *The Mechanics of Soil: An Introduction to Critical State Soil Mechanics*, McGraw-Hill, London.

Ayyar, T.S.R. (2000). *Soil Engineering in Relation to environment*, LBS Centre for Science and Technology, Trivandrum.

Babu, G.L.S. (2006). *An Introduction to Soil Reinforcement and Geosynthetics*. United Press (India) Pvt Ltd.

Bagchi, A. (2004). *Design of Landfills and Integrated Solid Waste Management*, John Wiley & Sons, USA.

Baker, K.H. and Herson, D.S., (1994). *Bioremediation*, McGraw-Hill, Inc., New York.

Bandyopadhya, J.N. (2008). *Design of Concrete Structures*, Prentice Hall of India Pvt Ltd, New Delhi.

Bansal, R.K. (2010). *Strength of Materials*, Laxmi Publications Pvt Ltd, New Delhi.

Bansal, R.K. (2013). *Fluid Mechanics and Hydraulic Machines*, Laxmi Publications Pvt Ltd, New Delhi.

Barry, P. A.D. (2004). *Watershed: Processes, Assessment and Management*, John Wiley & Sons, New York.

Basak, N.N. (1999). *Irrigation Engineering*, Tata McGraw-Hill Publishing Co., New Delhi.

Basavarajaiah, B.S. and Mahadevappa, P. (2010). *Strength of Materials*, Universities Press, Hyderabad.

Beer, F.P and Johnston, Jr. E.R. (2004). *Vector Mechanics for Engineers (In SI Units): Statics and Dynamics*, Tata McGraw-Hill Publishing Company, New Delhi, 8th Edition.

Bharucha, E. (2015). *Textbook of Environmental Studies*, Universities Press(I) Pvt Ltd, Hyderabad.

Bhave, P. R. (1991). *Analysis of Flow in water Distribution Networks*, Technomic Publishing Co., Lancaster, PA.

Bhavikatti, S.S. (2014). *Matrix Method of Structural Analysis*, I. K. International Publishing House Pvt Ltd, New Delhi.

Bhavikatti, S.S. (2014). *Structural Analysis, Vol. 1, & 2*, Vikas Publishing House Pvt Ltd, New Delhi.

Bhavikatti, S.S. (2015). *Concrete Technology*, I.K. International Publishing House Pvt Ltd, New Delhi.

Biswas, A. K. (1976). *Systems Approach to Water Management*, McGraw-Hill Publication, New York.

Blyth, F.G.H. and de Freitas, M.H. (2010). *Geology for Engineers*, Edward Arnold, London.

Bowles, J.E. (1997). *Foundation Analysis and Design*, McGraw-Hill International Edition, New York.

Bruggeling, A.S.G and Huyghe, G.F. (1991). *Prefabrication with Concrete*, A.A. Balkema Publishers, USA.

Budhu, M. (2000). *Soil Mechanics and Foundations*, John Wiley & Sons, New York.

Budhu, M. (2010). *Soil Mechanics and Foundations*, John Wiley & Sons.

Campbell, D., Allen and Roper, H. (1991). Concrete Structures, Materials, Maintenance and Repair, Longman Scientific and Technical, UK.

Canter, L.R. (1996). *Environmental Impact Assessment*, McGraw-Hill, New Delhi.

Cedergren, H.R. (1980). *Seepage, Drainage, and Flow Nets*, John Wiley & Sons, New York, 3rd Edition.

Cernica, J. N. (1995). *Geotechnical Engineering: Foundation Design*, John Wiley & Sons, New York.

Chang, H. H. (1988). *Fluvial Processes in River Engineering*, John Wiley & Sons, New York.

Chapra, S. C. and Canale, R. P. (2007). *Numerical Methods for Engineers*, Tata McGraw-Hill, 5th Edition.

Chaturvedi, M.C. (1997). *Water Resources Systems Planning and Management*, Tata McGraw-Hill Inc., New Delhi.

Chitkara, K.K. (2009). *Construction Project Management Planning, Scheduling and Control*, Tata McGraw-Hill Publishing Co., New Delhi.

Chow, V.T. (1959). *Open Channel Hydraulics*, McGraw-Hill, New York.

Chow, V.T. (1964). *Handbook of Applied Hydrology: A Compendium of Water Resources Technology*, McGraw-Hill, New York.

Chow, V.T., David, R. M. and Larry, W. (1985). *Applied Hydrology*, McGraw-Hill Company, New York.

Clayton, C. R. I., Woods, R. I., Bond, A. J. and Milititsky, J. (2013). *Earth Pressure and Earth-Retaining Structures*, CRC Press, Boca Raton, FL.

Cobb, F. (2014). *Structural Engineer's Pocket Book: Eurocodes*, CRC Press, New York, 3rd Edition.

Coduto, D.P. (2001). *Foundation Design Principles and Practices*, Prentice Hall, Upper Saddle River, NJ.

Coduto, D.P. (2010). *Geotechnical Engineering – Principles and Practices*, Prentice Hall of India Pvt Ltd, New Delhi.

Cook, R. D., Malkus, D. S., Plesha, M. E. and Witt, R. J. (2001). *Concepts and Applications of Finite Element Analysis*, John Wiley & Sons, New York.

Cox, H.L. (1963). *The Buckling of Plates and Shells*, Macmillam, New York.

Craig, R.F. (2012). *Soil Mechanics*, E & FN Spon, London and New York.

Daniel, B.E. (1993). *Geotechnical Practice for Waste Disposal*, Chapman & Hall, London.

Das, B.M. (1983). *Fundamentals of Soil Dynamics*, Elsevier Publishers, New York.

Das, B.M. (2010). *Principles of Geotechnical Engineering*, Cengage Learning Inc, Stamford, 7th Edition.

Das, B.M. (2011). *Principles of Foundation Engineering*, PWS Publishing, Pacific Grove, CA.

Das, B.M. (2013). *Advanced Soil Mechanics*, CRC Press, New York, 4th Edition.

David, E. D. (1993). *Geotechnical Practice for Waste Disposal*, Chapman & Hall, London.

Davis, M.L. and Cornell, D.A (1998). *Introduction to Environmental Engineering*.

Davis, R.O. and Selvadurai, A.P.S. (1996). *Elasticity and Geomechanics*, Cambridge University Press, New York.

Day, R. W. (2005). *Foundation Engineering Handbook*, McGraw-Hill, New York.

Dean, R.G. and Dalrymple, R.A. (1994). *Water Wave Mechanics for Engineers and Scientists*, Prentice-Hall, Inc., Englewood Cliffs, NJ.

Deodhar, S.V. (2012). *Construction Equipment and Job Planning*, Khanna Publishers, New Delhi.

Desai, C.S. and Christian, J.T. (1977). *Numerical Methods in Geotechnical Engineering*, McGrew-Hill, New York.

Desai, C.S. and Abel, J. F. (1987). *Introduction to FEM*, CBS Publishers and Distributors, New Delhi.

Duggal, K.N. and Soni, J.P. (2005). *Elements of Water Resources Engineering*, New Age International Publishers.

Duncan, M.J. and Wright, S. J. (2005). *Soil Strength and Slope Stability*, John Wiley & Sons, Inc., New Jersey.

Fanella, D. A. (2016). *Reinforced Concrete Structures: Analysis and Design*, McGraw-Hill Education, New York, 2nd Edition.

Fredlund, D.G. and Rahardjo, R. (1993). *Soil Mechanics for Unsaturated Soils*, John Wiley & Sons, New York.

Gambhir, M.L. (2006). *Fundamentals of Reinforced Concrete Design*, Prentice Hall of India Pvt Ltd, New Delhi.

Gambhir, M.L. (2009). *Fundamentals of Solid Mechanics*, PHI Learning Pvt Ltd, New Delhi.

Gambhir, M.L. and Jamwal, N. (2012). *Building Materials, Products, Properties and Systems*, Tata McGraw Hill Educations Pvt Ltd, New Delhi.

Garber and Hoel. (2010). *Principles of Traffic and Highway Engineering*, CENGAGE Learning, New Delhi.

Garg, S. K. (2009). *Irrigation Engineering and Hydraulic Structures*, Khanna Publishers, New Delhi, 23rd Revised Edition.

Garg, S.K. (2010). *Environmental Engineering, Vol. I*, Khanna Publishers, New Delhi.

Geiger, W. F., Marsalek, J. Z., and Rawls, G. J. (1987). *Manual on Drainage in Urban Areas, 2 Volumes*, UNESCO, Paris.

Gokhale, K.V.G.K. (2011). *Principles of Engineering Geology*, B.S. Publications, Hyderabad.

Goodman, R.E. (1989). *Introduction to Rock Mechanics*, John Wiley & Sons.

Grady, Jr. C.P.L. and Lin, H.C. (1980). *Biological Wastewater Treatment: Theory and Applications*, Marcel Dekker, Inc., New York.

Graf, W.H. (1971). *Hydraulics of Sediment Transport*, McGraw-Hill, New York.

Granger, R.A. (1995). *Fluid Mechanics*, Courier Dover Publications, New York.

Grewal, B.S. (2008). *Higher Engineering Mathematics*, Khanna Publishers.

Gupta, R. S. (1987). *Hydrology and Hydraulic Systems*, Prentice Hall Publishers, New Jersey.

Guyon, V. (1995). *Limit State Design of Prestressed Concrete, Vols 1 & 2*, Applied Science Publishers, London.

Haith, D.A. (1982). *Environmental Systems Optimization*, John Wiley & Sons, New York.

Hall, M.J. (1984). *Urban Hydrology*, Elsevier Applied Science Publishers, New York.

Hall, W.A. and Dracup, J. A. (1970). *Water Resources Systems Engineering*, McGraw-Hill Publication, New York.

Halpin, D.W. (1985). *Financial and Cost Concepts for Construction Management*, John Wiley & Sons, New York.

Hayward, A., Weare, F. and Oakhill, A. C. (2011). *Steel Detailers' Manual*, Wiley-Blackwell, USA, 3rd Edition.

Head, K. H. (1998). *Manual of Soil Laboratory Testing*, John Wiley & Sons, England.

Heathcote, I. W. (1998). *Integrated Watershed Management: Principles and Practices*, John Wiley & Sons, New York.

Henderson, F.M. (1966). *Open Channel Flow*, MacMillan Publishing Co., New York.

Hendrickson, C. and Au, T. (2000). *Project Management for Construction - Fundamental Concepts for Owners, Engineers, Architects and Builders*, Second Edition, prepared for world-wide-web publication in 2000. Version 2.2 prepared Summer, 2008. [This textbook is available free at: http://pmbook. ce.cmu.edu/]. (First Edition originally printed by Prentice Hall Inc. NJ, USA)

Henn, W. (1995). *Buildings for Industry, Vol. I and II*, Hill Books, London.

Herschy, R. W. (1998). *Hydrometry: Principles and Practice*, John Wiley & Sons, New York, 2nd Edition.

Hewlett, P. C. (1972). *Concrete Admixtures use and applications, ed M R Rixom*, The Concrete Press, London.

Hicks, T. G. (2010). *Civil Engineering Formulas*, McGraw-Hill Inc., New York, 2nd Edition.

Hicks, T. G. (2016). *Handbook of Civil Engineering Calculations*, McGraw-Hill Education, New York, 3rd Edition.

Holtz, R. D., Kovacs, W. D. and Sheahan, T. C. (2010).*An Introduction to Geotechnical Engineering*, Prentice-Hall, Englewood Cliffs, NJ, 2nd Edition.

Hough, B. K. (1957). *Basic Soil Engineering*, The Ronald Press Co.

Hunt, R.E. (2005). *Geotechnical Engineering Investigation Manual*, McGraw-Hill, New York, 2nd Edition.

Ingold, T. (1982). *Reinforced Earth*, Thomas Telford, London.

Ippen, A.T. (1978). *Estuary and Coastline Hydrodynamics*, McGraw-Hill, Inc., New York.

Jagadeesh, T.R. and Jayaram, M.A. (2013). *Design of Bridge Structures*, Prentice Hall of India Pvt Ltd/Learning Pvt Ltd.

Jain, A.K. (2016). *Fluid Mechanics (Including Hydraulic Machines)*, Khanna Publishers, 12th Edition.

Jain, S. K and Singh, V. P. (2003). *Water Resources Systems Planning and Management*, Elsevier, Netherlands.

James, D. and Lee, R. (1971). *Economics of Water Resources Planning*, McGraw-Hill Publication, New York.

Jewell, R.A. (1996). *Soil Reinforcement with Geotextiles, Special Publication No. 123*, CIRIA, Thomas Telford, London, UK.

John, N.W.M. (1987). *Geotextiles*, Blackie & Son Ltd, London, UK.

Johnson, R.P. (2012). *Composite Structures of Steel and Concrete, Vol.1 Beams, Slabs, Columns and Frames in Buildings*, Oxford Blackwell Scientific Publications, London.

Jones, C.J.F.P. (2010). *Earth Reinforcement and Soil Structures*, Thomas Telford, London, UK.

Joseph, B. (2006). *Environmental Science and Engineering*, Tata McGraw-Hill, New Delhi.

Junnarkar, S.B. and Shah, H.J. (2016). *Mechanics of Structures, Vol I*, Charotar Publishing House, New Delhi.

Justin, J.W., Creagar, W.P and Hinds, J. (1945). *Engineering for Dams, Volume I, II & III*, John Wiley & Sons, New York.

Kadiyali, L.R. (2013). *Principles and Practice of Highway Engineering*, Khanna Technical Publications, New Delhi, 8th Edition.

Kanetkar, T.P and Kulkarni, S.V. (2008). *Surveying and Levelling, Parts 1 & 2*, Pune Vidyarthi Griha Prakashan, Pune.

Kazimi, S.M.A. (2003). *Solid Mechanics*, Tata McGraw-Hill Publishing Co., New Delhi.

Kennedy, M. (1996). *The Global Positioning System and GIS: An Introduction*, Ann Arbor Press, Chelsea.

Kenneth, W. and Warner, C.F. (1981). *Air Pollution Its Origin and Control*. Harper and Row Publishers, New York.

Kesavulu, C.N. (2009). *Textbook of Engineering Geology*, Macmillan India Ltd.

Kessler, J. (1979). Drainage Principles and Applications, *Volumes I to IV*, International Institute for Land Reclamation and Improvement (ILRI), Netherlands.

Khan, K. (2015). *Fluid Mechanics and Machinery*, Oxford University Press, New Delhi.

Koerner, R.M. (1999). *Designing with Geosynthetics*, Prentice Hall, New Jersey, 4th Edition.

Kramer, S. L. (1996). *Geotechnical Earthquake Engineering*, Prentice-Hall, Englewood Cliffs, NJ, First Edition.

Kreyzig, E. (1994). *Advanced Engineering Mathematics*, John Wiley & Sons.

Krishnamoorthy, C. S. (1994). *Finite Element Analysis-Theory and Programming*, Tata McGraw-Hill Publishing Company, New Delhi.

Krishnaraju, N. (2009). *Structural Design and Drawing*, Universities Press.

Krishnaraju, N. (2012). *Prestressed Concrete*, Tata McGraw-Hill Company, New Delhi, 5th Edition.

Krishnaraju, N. *Design of Reinforced Concrete Structures*, CBS Publishers & Distributors Pvt Ltd, New Delhi.

Kumar, K. (2009). *Basic Geotechnical Earthquake Engineering*, New Age International Publishers, New Delhi.

Kurien, N.P. (1992). *Design of foundation Systems: Principles & Practices*, Narosa, New Delhi.

Lambe, T.W. (1951). *Soil Testing for Engineers*, John Wiley and Sons, New York.

LeLiavsky, S. (1996). *An Introduction to Fluvial Hydraulics*, Courier Dover Publications, New York.

Lewitt, M. (1982). *Precast Concrete- Materials, Manufacture, Properties And Usage*, Applied Science Publishers, London and New Jersey.

Look, B.G. (2014). *Handbook of Geotechnical Investigation and Design Tables*, CRC Press.

Loucks, D.P., Stedinger, J. R. and Haith, D.A. (1981). *Water Resources Systems Planning and Analysis*, Prentice Hall, New Jersey.

Majumdar, D. K. (2008). *Irrigation Water Management*, Prentice-Hall of India, New Delhi.

Mani, J.S. (2012). *Coastal Hydrodynamics*. PHI Pvt Ltd, New Delhi.

Mannering, F.L., Washburn, S. S. and Kilareski, W. P. (2011). *Principles of Highway Engineering and Traffic Analysis*, Wiley India Pvt Ltd, New Delhi.

Martin, A.M. (1991). *Biological Degradation of Wastes*, Elsevier Applied Science, London.

Mays, L. W. and Tung, Y.K. (1992). *Hydrosystems Engineering and Management*, McGraw-Hill, New York.

McCormac, J. C. and Brown, R. H. (2013). *Design of Reinforced Concrete*, John Wiley & Sons, NJ, 9th Edition.

Meenakshi, P. (2006). *Elements of Environmental Science and Engineering*, Prentice Hall of India, New Delhi.

Mehta, K.P. and Monterio, P.J.M. (2016). *Concrete - Microstructure, Properties and Materials*, McGraw-Hill Education (India) Pvt Ltd, New Delhi.

Menon, D. and Pillai, S. U. (2009). *Reinforced Concrete Design*, Tata McGraw-Hill Education Pvt Ltd, New Delhi, Third Edition.

Metcalf and Eddy. (2010). *Wastewater Engineering, Treatment and Reuse*, Tata McGraw-Hill, New Delhi.

Metha, P.K. and Monteiro, P.J.M. (2006). *Concrete, Microstructure, Properties and Materials*, Tata McGraw-Hill Publishing Company Ltd, New Delhi, 4th Edition.

Miller, G.T. and Spoolman, S. E. (2014). *Environmental Science*, Cengage Learning India Pvt, Ltd, New Delhi.

Mitchel, J.K. and Saga, K. (2005). *Fundamentals of Soil Behaviour*, John Wiley & Sons.

Modi, P.N. (2010). *Water Supply Engineering, Vol. I*, Standard Book House, New Delhi.

Modi, P.N. and Seth, S.M. (2009). *Hydraulics and Fluid Mechanics including Hydraulic Machines*, Standard Book House, New Delhi.

Moseley, M.P. and Kirsch, K. (2004). *Ground Improvement*, Spon Press, Taylor and Francis Group, London, New York, 2nd Edition.

Mundrey, J. S. (2013). *Railway Track Engineering*, McGraw Hill Education (India) Pvt Ltd, New Delhi.

Murray, E. J. and Sivakumar, V. (2010). *Unsaturated Soils: A Fundamental Interpretation of Soil Behaviour*, John Wiley & Sons, New York.

Murthy, J.V.S. (1995). *Watershed Management in India*, John Wiley & Sons, New York.

Murthy, V.N.S. (1996). *Principles and Practices of Soil Mechanics and Foundation Engineering*, UBS Publishers and Distributors, New Delhi.

Murthy, V.N.S. (2014). *Text book of Soil Mechanics and Foundation Engineering*, CBS Publishers Distribution Ltd, New Delhi.

Nayak, N.V. (2012). *Foundation Design Manual*, Dhanpat Rai Publications, New Delhi.

Neville, A.M. (1995). *Properties of Concrete*, Pitman Publishing Ltd, London.

Neville, A.M. and Brooks, J.J. (2008). *Concrete Technology*, Pearson Education India.

Nilson, A. H., Darwin, D. and Dolan, C. W. (2004). *Design of Concrete Structures*, McGraw-Hill Inc., New York, 13th Edition.

Palanikumar, M. (2013). *Soil Mechanics*, Prentice Hall of India Pvt Ltd, Learning Pvt Ltd, New Delhi.

Paquette, R.J., et al. (1982). *Transportation Engineering Planning and Design*, John Wiley & Sons, New York.

Park, R. and Paulay, T. (1975). *Reinforced Concrete Design*, John Wiley & Sons, New York.

Peary, H.S., Rowe, D.R., Tchobanoglous, G. (1995). *Environmental Engineering*, McGraw-Hill Book Co., New Delhi.

Peck, R. B., Hanson, W. E. and Thornburn, T. H. (1974). *Foundation Engineering*, John Wiley & Sons, New York.

Petts, J. (1998). *Handbook of Environmental Impact Assessment Vol. I and II*, Blackwell Science, New York.

Peurifoy, R., Schexnayder, C. J., Shapira, A. and Schmitt, R. (2010). *Construction Planning, Equipment, and Methods*, McGraw-Hill Inc., New York, 8th Edition.

Pitchel, J. (2014). *Waste Management Practices-Municipal, Hazardous and Industrial*, CRC Press, Taylor and Francis, New York.

Ponnuswamy, S. (1996). *Bridge Engineering*, Tata McGraw-Hill, New Delhi.

Popov, E. P. (2012). *Engineering Mechanics of Solids*, PHI Learning Pvt Ltd, New Delhi, 2nd Edition.

Potts, D. M. and Zdravkovic, L. (1999). *Finite Element Analysis in Geotechnical Engineering: Theory and Application*, Thomas Telford, London.

Poulos, H.G. and Davis E.H. (1990). *Pile Foundation Analysis and Design*, John Wiley & Sons, New York.

Prakash, S. and Puri. (1988). *Foundations for Machines: Analysis and Design*, John Wiley & Sons, New York.

Prasad, B.B. (2010). *Advanced Soil Dynamics and Earthquake Engineering*, PH Publishers, New Delhi.

Puzrin, A. M. (2012). *Constitutive Modelling in Geomechanics*, Springer, New York.

Qasim, S. R. (2010). *Wastewater Treatment Plants*, CRC Press, Washington DC, New York.

Qasim, S.R., Motley, E.M. and Zhu, G. (2009). *Water works Engineering – Planning, Design and Operation*, Prentice Hall, New Delhi.

Quinn, A. D. (1961). *Design and Construction of Ports and Marine Structure*, McGraw-Hill, New York.

Raj, P.P. (2008). *Soil Mechanics & Foundation Engineering*, Pearson Education, India.

Rajasekharan, S. (1987). *Numerical Methods for Initial and Boundary Value Problems*, Wheeler and Co., Pvt Ltd.

Ranjan, G. and Rao, A. S. R. (2016). *Basic and Applied Soil Mechanics*, New Age International Publication, 3rd Edition.

Rao, S. S. (2004). *Mechanical Vibrations*, Pearson Education Inc., New Delhi.

Rattan, S. S. (2012). *Strength of Materials*, Tata McGraw Hill Education Pvt Ltd, New Delhi.

Reimold, R.J. (1998). *Watershed management: Practice, Policies and Co-ordination*, McGraw-Hill, New York.

Remson I., Hornberger, G.M. and Moltz, F.J. (1971). *Numerical Methods in Subsurface Hydrology*, John Wiley & Sons, New York.

Reynolds, C. E., Steedman, J. C., and Threlfall, A. J. (2007). *Reinforced Concrete Designer's Handbook*, CRC Press, Taylor & Francis, London, UK, 11th Edition.

Richard, F. H. (1985). *Open channel hydraulics*, MacMillan Publishing Co., New York.

Ricketts, J. T., Loftin, M. K. and Merritt, F. S. (2004). *Standard Handbook for Civil Engineers*, McGraw-Hill Inc., New York, 5th Edition.

Ritzema, H.P. (1994). *Drainage Principles and Applications, Publication No. 16*, International Institute of Land Reclamation and Improvement, Netherlands.

Roy, S.K. (2004). *Fundamentals of Surveying*, Prentice Hall of India, 2nd Edition.

Rushton, K. R. and Redshaw, S. C. (1979). *Seepage and Groundwater Flow, Numerical Analysis by Analog and Digital Methods*, John Wiley & Sons, New York.

Sabatini, P.J., Bachus, R.C., Mayne, P.W., Schneider, J. A. and Zettler, T.E., Evaluation of soil and rock properties, Geotechnical Engineering circular No. 5, US dept. of transportation, Federal highway administration.

Sai Ram, K.S. (2015). *Design of Steel Structures*, Dorling Kindersley (India) Pvt Ltd, New Delhi, 2nd Edition, www.pearsoned.co.in/kssairam.

Santhakumar, A.R. (2006). *Concrete Technology*, Oxford University Press, India, 1st Edition.

Santhi, S. (2011). *Integral Equations*, Krishna Prakasan Media.

Sastri, S.S. (1977). *Introductory Methods of Numerical Analysis*, Prentice Hall of India, New Delhi.

Schlitching, H. (1979). *Boundary Layer Theory*, McGraw Hill, New York.

Schnaid, F. (2009). *In-Situ Testing in Geomechanics*, Taylor and Francis.

Scott, F. (1963). *Principles of Mechanics*, Addison – Wesley, London.

Scott, J. S. (1991). *Dictionary of Civil Engineering*, Penguin Books, UK, 4th Edition.

Sengar, D. S. (2007). *Environmental Law*, Prentice Hall of India Pvt Ltd, New Delhi.

Shah, V. L. and Karve, S. R. (2013). *Limit State Theory and Design of Reinforced Concrete*, Structures Publications, Pune.

Sharma, H. D. and Reddy, K. R. (2004). *Geo-Environmental Engineering*, John Wiley & Sons, USA.

Sharma, R.K. (2002). *Irrigation Engineering and Hydraulic Structures*, Oxford and IBH Publishing Co., New Delhi.

Sharma, S.C. (2002). Construction Equipment and Management, Khanna Publishers, New Delhi.

Shetty, M. (2003). *Concrete Technology*, S. Chand and Company Ltd, New Delhi.

Shukla, S.K. (2012). *Handbook of Geosynthetic Engineering*, ICE Publishing, London, UK, 2nd Edition.

Sickle, J.V. (1996). *GPS for Land Surveyors*, Ann Arbor Press, Chelsea.

Simons, N., Menzies B. and Matthews M. (2002). *A Short Course in Geotechnical Site Investigation*, Thomas Telford.

Singh, A. and Ward, O.P. (2004). *Biodegradation and Bioremediation*, Springer-Verlag, Berlin Heidelberg, New York.

Singh, V.P. (1996). *Kinematic Wave Modelling in Water resources-Surface Water Hydrology*, John Wiley & Sons, New York.

Sinha, S.N. (2002). *Reinforced Concrete Design*, Tata McGraw-Hill Publishing Company Ltd, New Delhi.

Slater, R.A.C. (1977). *Engineering Plasticity*, John Wiley & Sons, New York.

Sneddon, I.N. (1957). *Partial Differential Equations*, McGraw-Hill.

Sokol, N. (1951). *Tensor Analysis*, John Wiley & Sons, New York.

Sorenson, R.M. (1978). *Basic Coastal Engineering*, A Wiley-Interscience Pub, New York.

Srivastava, R. (2008). *Flow through Open Channels*, Oxford University Press, New Delhi.

Stahre, P. and Urbonas, B. (1983). *Stormwater Detention for Drainage, Water Quality and CSO Management*, Prentice Hall Publishers, New Jersey.

Streeter, V. L. and Wylie, E. B. (1985). *Fluid Mechanics*, McGraw-Hill, New York.

Subramanya, K. (2000). *Flow in Open Channels*, Tata McGraw-Hill, New Delhi.

Subramanya, K. (2010). *Fluid Mechanics and Hydraulic Machines*, Tata McGraw-Hill Education Pvt Ltd, New Delhi.

Swami Saran, S. (2006). *Reinforced Soil and Its Engineering Applications*, I.K. International Pvt Ltd, New Delhi.

Tattersall, G. H. (1991). *Workability and Quality Control of Concrete*, E&FN Spon, London.

Taylor, F.W., Thomson, S.E. and Smulski, E. (1955). *Reinforced Concrete Bridges*, John Wiley & Sons, New York.

Tchobanoglous, G. and Frank, K. (2002). *Handbook of Solid Waste Management*, McGraw-Hill, New York.

Tchobanoglous, G., Theisen, H. and Samuel, A. (1993). *Vigil, Integrated Solid Waste Management*, McGraw-Hill, New York.

Terzaghi, K. (1943). *Soil Mechanics*, John Wiley & Sons, New York.

Terzaghi, K., Peck, R. B. and Mesri, G. (1996). *Soil Mechanics in Engineering Practice*, John Wiley & Sons, New York.

Thomann, R. V. (1974). *Systems Analysis and Water Quality Management*, McGraw-Hill, New York.

Thomas, C.V. (2003). *Principles of Water Resources: History, Development, Management and Policy*, John Wiley & Sons, New York.

Thomas, J. (2015). *Concrete Technology*, Cengage Learning India Pvt Ltd, New Delhi.

Thomas, T. C. H. (2010). *Unified Theory of Reinforced Concrete*, CRC Press, London.

Tideman, E.M. (1996). *Watershed Management*, Omega Scientific Publishers, New Delhi.

Timoshenko, S.B. and Gere, J.M. (1999). *Mechanics of Materials*, Van Nos Reinbhold, New Delhi.

Todd, D.K. (2000). *Ground Water Hydrology*, John Wiley & Sons, New York.

Tomlinson, M.J. (1996). *Foundation design and Construction*, John Wiley & Sons, New York.

Varghese, P.C. (1998). *Design of Reinforced Concrete Foundations*, PHI-Ltd, New Delhi.

Varghese, P.C. (2002). *Limit State Design of Reinforced Concrete*, Prentice Hall of India, Pvt Ltd, New Delhi.

Varghese, P.C. (2007). *Building Construction*, Prentice Hall of India, Pvt Ltd, New Delhi.

Varghese, P.C. (2012). *Engineering Geology for Civil Engineering*, Prentice Hall of India, Pvt Ltd, New Delhi.

Varghese, P.C. (2015). *Building Materials*, Prentice Hall of India, Pvt Ltd, New Delhi.

Vazirani, V.N and Ratwani, M.M. (1995). *Analysis of Structures, Vol I*, Khanna Publishers, New Delhi.

Venkataraman, M.K. (1992). *Higher Engineering Mathematics*, National Publishers.

Venkatramaiah, C. (2014). *Text Book of Surveying*, University Press, New Delhi.

Venkatramaiah, C. (2015). *Transportation Engineering-Vol.2 Railways, Airports, Docks and Harbours, Bridges and Tunnels*, Universities Press (India) Pvt Ltd, Hyderabad.

Venkatramaiah, C. (2017). *Geotechnical Engineering*, New Age International Pvt Ltd, New Delhi.

Ventsel, E. and Krauthammer, Th. (2001). *Thin Plates and Shells: Theory, Analysis and Applications*, Marcel Dekker, Inc., New York.

Vesilind, Worrell, Reihhart. (2006). Solid Waste Engineering, RCRA Orientation Manual, US EPA.

Victor, J. D. (2009). *Essentials of Bridge Engineering*, Oxford and IBH Publishing Co., New Delhi.

Vinson, J. R. and Chou, P.C. (1975). *Composite Materials and Their Use in Structures*, Applied Science Publishers, Ltd, London.

Walsh, I. D. (2011). *ICE Manual of Highway Design and Management*, ICE Publishers, USA, 1st Edition.

Walski, T.M. (1984). *Analysis of Water Distribution System*, Van Nostrand Reinhold Company, New York.

Wanielista, M. P. and Yousef, Y. A. (1993). *Stormwater Management*, John Wiley & Sons, New York.

Warren, H.A. and Dracup, J. A. (1998). *Water Resources System Engineering*, Tata McGraw-Hill Publishing Company Ltd, New Delhi.

Weaver, W. and Gere, J. J. M. (1995). *Matrix Analysis of Framed Structures*, CBS Publishers & Distributors, New Delhi.

Wellenhof, H. B. (1997). *GPS theory and Practice*, Springer, Wien, New York.

Wells, D.E. (1988). *Guide to GPS Positioning*, Canadian GPS Association, New Brunswick, Canada.

Wesley, L.D. (2010). *Geotechnical Engineering in Residual Soils*, John Wiley & Sons.

Westlake, K. (1995). *Landfill Waste Pollution and Control*, Albion Publishing Ltd, England.

White, F.M. (2017). *Fluid Mechanics*, Tata McGraw-Hill, New Delhi, 5th Edition.

White, P. (2008). *Public Transport: Its Planning, Management and Operation*, Taylor & Francis, London.

Wilbur, J. B. and Norris, C. H. (2012). Elementary Structural Analysis, Literary Licensing, LLC, MT 59937 USA.

Wood, D. M. (2012). *Civil Engineering: A Very Short Introduction*, Oxford University Press, UK.

Zienkiewicz, O.C. (1979). *The Finite Element Method*, Tata McGraw-Hill Publishing Company, New Delhi.

Index

Note: **Bold** page numbers refer to tables and *Italic* page numbers refer to figures.

Printed and bound by CPI Group (UK) Ltd, Croydon, CR0 4YY

25/10/2024

01779171-0001